water

W9-AAL-594

LIST of ELEMENTS with their SYMBOLS and ATOMIC MASSES

Name	Symbol	Atomic Number	Atomic Mass	Name	Symbol	Atomic Number	Atomic Mass
actinium	Ac	89	(227)	neon	Ne	10	20.1797
aluminum	Al	13	26.981539	neptunium	Np	93	(237)
americium	Am	95	(243)	nickel	Ni	28	58.69
antimony	Sb	51	121.75	niobium	Nb	41	92.90638
argon	Ar	18	39.948	nitrogen	N	7	14.00674
arsenic	As	33	74.92159	nobelium	No	102	(259)
astatine	At	85	(210)	osmium	Os	76	190.2
barium	Ba	56	137.327	oxygen	O	8	15.9994
berkelium	Bk	97	(247)	palladium	Pd	46	106.42
beryllium	Be	4	9.012182	phosphorus	P	15	30.973762
bismuth	Bi	83	208.98037	platinum	Pt	78	195.08
boron	B	5	10.811	plutonium	Pu	94	(244)
bromine	Br	35	79.904	polonium	Po	84	(209)
cadmium	Cd	48	112.411	potassium	K	19	39.0983
calcium	Ca	20	40.078	praseodymium	Pr	59	140.90765
californium	Cf	98	(251)	promethium	Pm	61	(147)
carbon	C	6	12.011	protactinium	Pa	91	(231)
cerium	Ce	58	140.115	radium	Ra	88	(226)
cesium	Cs	55	132.90543	radon	Rn	86	(222)
chlorine	Cl	17	35.4527	rhenium	Re	75	186.207
chromium	Cr	24	51.9961	rhodium	Rh	45	102.90550
cobalt	Co	27	58.93320	rubidium	Rb	37	85.4678
copper	Cu	29	63.546	ruthenium	Ru	44	101.07
curium	Cm	96	(247)	samarium	Sm	62	150.36
dysprosium	Dy	66	162.50	scandium	Sc	21	44.955910
einsteinium	Es	99	(252)	selenium	Se	34	78.96
erbium	Er	68	167.26	silicon	Si	14	28.0855
europium	Eu	63	151.965	silver	Ag	47	107.8682
fermium	Fm	100	(257)	sodium	Na	11	22.989768
fluorine	F	9	18.9984032	strontium	Sr	38	87.62
francium	Fr	87	(223)	sulfur	S	16	32.066
gadolinium	Gd	64	157.25	tantalum	Ta	73	180.9479
gallium	Ga	31	69.723	technetium	Tc	43	(99)
germanium	Ge	32	72.61	tellurium	Te	52	127.60
gold	Au	79	196.96654	terbium	Tb	65	158.92534
hafnium	Hf	72	178.49	thallium	Tl	81	204.3833
helium	He	2	4.002602	thorium	Th	90	232.0381
holmium	Ho	67	164.93032	thulium	Tm	69	168.93421
hydrogen	H	1	1.00794	tin	Sn	50	118.710
indium	In	49	114.82	titanium	Ti	22	47.88
iodine	I	53	126.90447	tungsten	W	74	183.85
iridium	Ir	77	192.22	unnilennium	Une	109	(266)
iron	Fe	26	55.847	unnilhexium	Unh	106	(263)
krypton	Kr	36	83.80	unniloctium	Uno	108	(265)
lanthanum	La	57	138.9055	unnilpentium	Unp	105	(262)
lawrencium	Lr	103	(260)	unnilquadium	Unq	104	(261)
lead	Pb	82	207.2	unnilseptium	Uns	107	(262)
lithium	Li	3	6.941	uranium	U	92	238.0289
lutetium	Lu	71	174.967	vanadium	V	23	50.9415
magnesium	Mg	12	24.3050	xenon	Xe	54	131.29
manganese	Mn	25	54.93805	ytterbium	Yb	70	173.04
mendelevium	Md	101	(258)	yttrium	Y	39	88.90585
mercury	Hg	80	200.59	zinc	Zn	30	65.39
molybdenum	Mo	42	95.94	zirconium	Zr	40	91.224
neodymium	Nd	60	144.24				

Chemistry

AN ENVIRONMENTAL PERSPECTIVE

Chemistry

AN ENVIRONMENTAL PERSPECTIVE

Phyllis Buell

James Girard

American University

Prentice Hall
Englewood Cliffs, New Jersey 07632

Library of Congress Cataloging-in-Publication Data

Buell, Phyllis E.
    Chemistry, an environmental perspective/Phyllis E. Buell, James
  E. Girard.
  p.   cm.
    Includes index.
    ISBN 0-13-644659-0
    1. Chemistry.   I. Girard, James E. II. Title.
  QD33.B8885 1994
  540—dc20   93-33454
                                         CIP

Acquisitions editor: Paul Banks
Editor in Chief: Tim Bozik
Development editors: Ray Mullaney, Anne-Marie Shaw
Editorial/production supervision: Robert C. Walters
Design director: Florence Dara Silverman
Interior design: Judith A. Matz-Coniglio
Cover design: Jeannette Jacobs Design
Production coordinator: Trudy Pisciotti
Page formatting: Michael Bertrand, John Jordan, John Nestor
Photo editor: Lorinda Morris-Nantz
Photo researcher: Tobi Zausner
Cover photo: The Image Bank/Alex Stewart
Art Program: Hans & Cassady, Inc., Westerville, Ohio

©1994 by Prentice-Hall, Inc.
A Paramount Communications Company
Englewood Cliffs, New Jersey 07632

Printed in the United States of America

10 9 8 7 6 5 4 3 2 1

ISBN 0-13-644659-0

Prentice-Hall International (UK) Limited, *London*
Prentice-Hall of Australia Pty. Limited, *Sydney*
Prentice-Hall Canada Inc., *Toronto*
Prentice-Hall Hispanoamericana, S.A., *Mexico*
Prentice-Hall of India Private Limited, *New Delhi*
Prentice-Hall of Japan, Inc., *Tokyo*
Simon & Schuster Asia Pte. Ltd., *Singapore*
Editora Prentice-Hall do Brasil, Ltda., *Rio de Janeiro*

Dedicated to the memory of
William C. Buell IV
and Mary Catherine Girard

BRIEF CONTENTS

CONTENTS

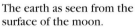

The earth as seen from the surface of the moon.

3

ECOSYSTEMS: The Flow of Energy and Materials through the Natural World *61*

Ecosystem: an interaction of plants, animals, and their surroundings.

4

ATOMS AND ATOMIC STRUCTURE *89*

5 ELECTRON CONFIGURATION AND CHEMICAL BONDING *117*

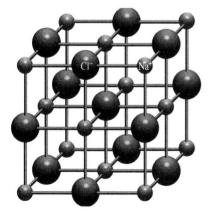

The structure of a crystal of sodium chloride, common table salt.

Using an isotope of the element carbon, scientists have shown that this Persian bowl was made 5000 years ago.

8 REACTIONS IN SOLUTION: Acids and Bases, and Oxidation-Reduction Reactions *211*

Acids are present in many consumer products.

In diamonds, carbon atoms are joined to form a strong three-dimensional network.

Polyurethanes are used to make diaphragms in artificial hearts.

13   THE AIR WE BREATHE *395*

Levels of carbon monoxide
build quicly in traffic jams.

14 FOSSIL FUELS: Our Major Source of Energy *429*

A California oil field.

15 ENERGY SOURCES FOR THE FUTURE *457*

Arrays of wind turbines are an efficient means of producing energy.

16   TOXICOLOGY *487*

17   FOOD AND NUTRITION *517*

18 AGRICULTURAL CHEMICALS: Feeding the Earth's People *551*

Swarms of locusts, if not controlled, can completely destroy crops.

Dangerous chemicals leak from drums at "Valley of the Drums," in Kentucky.

LIST OF CONSUMER BRIEFS

LIST OF EXPLORATIONS

PREFACE

At present, there is worldwide concern that many of our human activities are endangering—perhaps permanently—the quality of the environment and that the time for action to address these problems is running out. The public is becoming increasingly aware of the environmental damage caused by pesticides, toxic wastes, chlorofluorocarbons, nuclear radiation, oil spills, the greenhouse effect, and other human inputs. Grassroots organizations like Greenpeace and the Natural Resources Defense Fund are booming in popularity—especially on college campuses—and are becoming a major influence in the political arena. Articles on environmental issues appear daily in the newspapers, and Congress is introducing legislation to combat threats to the environment.

Paralleling this widespread concern for the environment is the realization that the majority of United States citizens, including those with college degrees, are virtually science-illiterate and are not, therefore, well-equipped to make informed decisions on environmental issues. As a response to this crisis in science education, many universities and colleges, the American University included, are altering their general curriculums to require science courses for all entering students—for nonscience majors as well as for science majors. To meet the new science requirement, more and more chemistry departments are offering courses with an environmental perspective, hoping to capture the interest of students who otherwise would not choose a chemistry course.

OBJECTIVES

We designed *Chemistry: An Environmental Perspective* for nonscience majors who have little or no previous knowledge of chemistry. Our primary objective in this text is to enable students to make informed judgments on crucial issues that are of current concern worldwide while providing a basic understanding of chemical principles and practices. We aim to guide students to the knowledge that humans live in a chemical environment and that chemistry affects every aspect of our lives. The text emphasizes that all the living and non-living parts of our environment are made up of chemicals, and that the natural processes continuously occurring in the environment all involve chemical reactions. With a grasp of this notion of interdependence, students begin to see that without some understanding of chemistry, it is impossible to fully understand environmental issues such as ozone depletion, global warming, air and water pollution, and the hazards of radioactivity.

ORGANIZATION

The organization and approach of this text differ in several ways from other chemistry books intended for nonscience majors. It places an environmental perspective within the proven framework of basic chemistry, assuming little or no scientific background.

The opening chapter introduces basic chemical principles and familiarizes students with the language of chemistry. The two chapters that follow provide an early orientation to the earth and its ecosystems. This early coverage of the dynamic nature of the earth and its natural cycles not only establishes the importance of maintaining a sustainable natural world but also gives students a firm foundation for the further study of chemical principles.

The middle portion of the text focuses primarily on chemical processes and principles, with applications to the environment and other areas interwoven throughout. Chapters 4-8 offer a solid introduction to core chemical principles normally included in chemistry texts for nonscience majors. Topics include the atom, electron configuration, bonding, radiation, the mole, nuclear chemistry, acids and bases, and oxidation-reduction reactions. Chapters 9, 10, and 11 present the more advanced topics of organic chemistry, biochemistry, and synthetic polymers. Wherever possible within this section, the chemistry is discussed in the context of natural processes, building a bridge to understanding the chemistry-related environmental issues examined in the chapters that follow.

The final eight chapters (Chapters 12-19) show students the relevance of the basic chemistry just covered by focusing on specific applications that pertain to the environment. Chapter 12 covers water—its properties, its importance to life on earth, and the dangers of polluting and misusing it. The focus in Chapter 13 turns to the earth's atmosphere, with discussions of pollution, ozone depletion, global warming, and the greenhouse effect. Chapters 14 and 15 are devoted to energy, addressing the chemistry of fossil fuels and their use as our major energy source. Emphasized are the dangers of depleting these nonrenewable resources and the need to explore other energy sources such as nuclear and wind power, and geothermal and solar energy. Chapter 16 examines toxic chemicals and their effect on our health. Food and nutrition are the focus of Chapters 17 and 18, with insightful discussions of the risks and benefits of food additives, synthetic fertilizers, and pesticides as they relate both to human health and environmental well-being. The final chapter considers hazardous and radioactive chemicals and the laws governing their proper disposal. For example, the Clean Air Act, the Clean Water Act, EPA regulations, and Superfund are all considered.

CHAPTER ELEMENTS

• **Introduction**   Each chapter begins with an introduction that explains the importance of the subject matter to our understanding of the environment and outlines the material that will be covered.

• **Learning Goals**   The introduction is followed by a list of the concepts and subject matter that students should understand after reading the chapter. Stating learning objectives provides goals that students should strive for and serves as a useful guide for reviewing each chapter.

• **Examples and Exercises**   Illustrative worked examples, each one accompanied by a challenging practice exercise, are included throughout the text, particularly in the chapters covering basic chemical principles.

• **Keywords and Concepts**   Lists of keywords and concepts introduced in the chapter are included at chapter's end to help reinforce the most important information.

• **Questions and Problems**   Each chapter includes a wide selection (30-40) of problems and questions. These include quantitative, review, and discussion types of questions, with answers to all the even-numbered ones given in an appendix.

• **Consumer Briefs**   Each chapter includes one or two short sections that investigate the chemistry at work in common consumer products such as smoke detectors, antacids, shampoos and conditioners, sunscreens, bottled mineral water, and the breathalyzer test for alcohol.

• **Explorations**   The two-page Explorations essays at the end of each chapter explore ways in which chemistry affects our dynamic world. They profile people such as Marie Curie and Linus Pauling, advances such as the pacemaker and the electric car, and events such as the eruption of Krakatoa and environmental terrorism during the Gulf War.

COURSE USE

Chemistry: An Environmental Perspective offers the flexibility to tailor a course to suit both instructors' preferences and the needs of particular audiences. The full text may be used for a comprehensive two-semester course, or the book may be broken down in several ways for a one-semester course. One option for a one-semester course is to use the first eight chapters followed by choices from the remaining chapters on more advanced chemistry and environmental applications according to the teacher's preferences. For a more traditional course, Chapters 2 and 3 may be omitted, and following Chapters 4-8, the course can be rounded out with organic chemistry (Chapter 9) and biochemistry (Chapter 10), and choices from Chapters 13-19. If students already have a strong chemistry background, Chapters 4-8 may be omitted and the course devoted to the environmental issues discussed in Chapters 2, 3, and later chapters.

SUPPLEMENTS

• **Instructor's Manual**   Contains solutions to in-text exercises, plus chapter overviews and outlines, and suggested video supplements.

• **Study Guide**   A tool for students, containing learning goals, chapter reviews, solutions to practice problems, and self-tests. Prepared by Dr. Dale Marko, Siena College.

• **Laboratory Manual**   Includes a number of interesting experiments that parallel the text. Prepared by the authors.

• **Instructor's Manual to the Laboratory Manual**   Contains directions, and a list of equipment and chemicals needed for each experiment, plus answers to questions and exercises.

- **Transparencies**   One hundred four-color acetates of illustrations from the text.
- **Test Item File**   Over 2000 questions—multiple choice and short answer—prepared by Dr. Thomas VanKoevering, University of Wisconsin at Green Bay.
- **Prentice Hall TestManager, Version 2.0 (Macintosh and IBM)**   Computerized version of the Test Item File.
- *The New York Times* **Themes of the Times Program**   Compilation of current articles from *The New York Times* illustrating issues pertaining to chemistry in the world around us.
- *ABC News*/**Prentice Hall Video Library**   Video segments selected from ABC's documentary and news coverage to accompany 20 case studies related to environmental issues. The case studies, which are presented in an appendix to the text, relate to material in the text. Each has a chapter reference and includes questions for discussion.

ACKNOWLEDGMENTS

We would like to express our gratitude to our colleagues at the American University who supported us through the development of the university's Environmental Chemistry course and this textbook which grew out of this effort. We would particularly like to thank Dr. Ann Ferren, Director of General Education, for development support for the course; Dr. Nina Roscher, Chemistry Department Chair, who provided support and encouragement during the various stages of the project; and Dr. Constance Diamant, who assisted in the design of the pedagogic features of the text.

We also wish to thank Dr. Janet Morrison of the National Institute of Standards and Technology, who checked the manuscript very carefully for errors, and Dr. Suzan Potuznik of the University of San Diego, who checked the answers to the end-of-chapter questions.

The following reviewers provided many helpful suggestions and much constructive criticism throughout the development of the text:

STEVE H. ALBRECHT, *Ball State University*

ANIMA BOSE, *Kent State University*

RUDOLPH S. BOTTEI, *University of Notre Dame*

FRED W. BREITBEIL, III, *University of Colorado at Boulder*

VIRGINIA R. BRYAN, *Southern Illinois University at Edwardsville*

EDWARD COOK, *Tunxis Community College*

MALCOLM J. D'SOUZA, *Ball State University*

MARVIN PORTER DIXON, *William Jewel College*

GERY J. ESSENMACHER, *University of Wisconsin—Madison*

GREG FOSTER, *George Mason University*

JAMES A. GOLEN, *University of Massachusetts*

KEITH J. HARPER, *University of North Texas*

DAVID KATZ, *Rockhurst College*

LESLIE N. KINSLAND, *University of Southwestern Louisiana*

ROBLEY J. LIGHT, *Florida State University*

DEAN F. MARTIN, *University of South Florida*

LILY NG, *Cleveland State University*

THOMAS J. OUELLETTE, *University of New Hampshire*

KAREN I. PETERSON, *University of Rhode Island*

SUZAN POTUZNIK, *University of San Diego*

ANN RATCLIFFE, *Oklahoma State University*

PHILIP REEDY, *Metropolitan Community College*

SALVATORE F. RUSSO, *Western Washington University*

KIMBERLY A. SCHUGART, *California State University—Long Beach*

PAUL G. SEYBOLD, *Wright University*

DAN M. SULLIVAN, *University of Nebraska at Omaha*

CONRAD N. TRUMBORE, *University of Delaware*

H. DAVID WOHLERS, *Northeast Missouri State University*

For all their help, we are indeed grateful.

We would like to thank the many people at Prentice Hall who worked with us to produce this book, particularly editor in chief of College Development Ray Mullaney, who masterfully organized our original manuscript into its final form; development editor, Anne-Marie Shaw, who gave us many valuable ideas for improving the text; production editor Robert C. Walters, who managed to remain calm and unruffled during the difficult task of bringing all the elements of the book together on schedule; photo researcher Tobi Zausner, who did a wonderful job of locating the photos; designer Judy Matz-Coniglio, who gave the book its attractive design; and page formatters Michael Bertrand, John Jordan, and John Nestor—and their proofreader Christina Mahon—who gave each page of the book its final form.

We and our team at Prentice Hall have labored long and hard to make this book as error-free as possible. If, despite all our efforts, a reader discovers an error or has suggestions for ways to improve this book, we would be delighted to hear from her or him.

Phyllis Buell
James Girard

WHAT IS CHEMISTRY?
An Introduction to the Central Science

A magnified image of a leaf surface made with an electron microscope. Although we cannot see them, all materials, including this leaf fragment, are composed of atoms.

Chemistry is the science concerned with the composition of matter and with the changes that matter undergoes. We practice chemistry all the time in our daily lives: when we cook, do the laundry, take medicine, fertilize the lawn, paint the house, or strike a match. In all these activities, substances interact, and chemical changes occur. In our bodies, as we breathe, walk, and digest food, chemical reactions constantly occur. The environmental problems we must deal with today, such as disposing of toxic wastes, controlling smog, and removing asbestos, are all essentially chemical problems.

In this chapter, we will look at some of the basic properties of matter and see how matter is classified, and we will identify the basic units of all matter: compounds, elements, atoms, and ions. We will introduce the language of chemistry: the symbols and names that describe and identify chemical substances and the equations that explain and quantify the chemical reactions that substances undergo. We will examine the two major types of changes—physical and chemical—that matter can undergo. We will also learn about the radiant energy that reaches us from the sun and sustains all life on earth.

Chemistry is an experimental science. At the end of this chapter, we will discuss the scientific method, the method of gaining knowledge that is fundamental to the advancement of chemistry.

Learning Goals:

In this chapter, you should gain an understanding of:

1. How technology is applied to basic scientific discoveries.
2. The differences between physical and chemical changes.
3. The different states of matter.
4. How all matter can be divided into pure substances and mixtures.
5. The symbols used to represent different elements.
6. The names chemists use to describe elements, compounds, and subatomic particles.
7. The application of the scientific method.

SCIENCE, TECHNOLOGY, AND CHEMISTRY

What Is Science?

Science is the study of all aspects of the world around us. It seeks to understand the order in the natural world and to explain in a logical way how and why things happen as they do. Because science covers such an enormous body of knowledge and because it can be studied in so many different ways, it is subdivided into different disciplines. Some of the major ones are astronomy, botany, chemistry, geology, physics, and zoology. There are no definite boundaries between the different scientific disciplines. Figure 1-1 shows how chemistry overlaps with several other disciplines.

What Is Technology?

Technology is often defined as *applied* science. It is the application of knowledge to the development of new materials and new processes, primarily those that will have a practical use. Technology involves the modification of natural

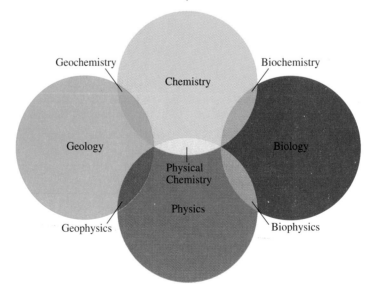

Figure 1-1   Chemistry overlaps several other scientific disciplines.

Figure 1-2 Glass was being made from a mixture of sand, limestone, and soda ash long before the scientific basis for the process was understood.

materials to generate products that improve the quality of life or satisfy society's wants.

Today, nearly all technological advances are based on scientific knowledge, but most of the early developments were achieved accidentally or by trial and error. For example, people knew how to make alcohol from grain, and glass (Fig. 1-2) from a mixture of limestone, sand, and soda ash (sodium carbonate), long before there was any understanding of the scientific basis for these processes.

Technological progress has brought both benefits and problems. Recent advances in chemistry include penicillin, artificial fertilizers, plastics, the microchip, the pesticide DDT, Freon (used as a refrigerant), and nuclear weapons, not all of which have been of lasting benefit to humans. When DDT and Freon were first introduced, only their many advantages were apparent. It was not until later that their harmful effects on the environment were recognized. With each new technological development, the costs must be carefully balanced against the benefits.

What Is Chemistry?

Chemistry is the study of the composition and structure of matter and the changes that matter undergoes. **Matter** is anything that occupies space and has mass. **Mass** is a measure of the *quantity* of matter. (The *weight* of an object is related to the pull of gravity on the object and is not the same as its mass. Mass and weight are explained in Appendix B.)

Because everything in the universe is composed of matter, chemistry is the study of our material world. Chemistry touches our lives and influences our activities in so many ways that it is often called the **central science**.

Our environment is the planet Earth, a small sphere in the vastness of space. Everything in this environment is made of chemicals.

▶ PROPERTIES OF MATTER

The distinguishing characteristics that we use to identify different samples of matter are called **properties**. We can recognize our own car among hundreds of other cars in a parking lot by characteristics such as make, model, color, dents in the body, and the belongings left on the back seat. In much the same way, we can recognize different substances by their characteristics or properties. Properties of substances can be grouped into two main categories: physical and chemical.

Physical Properties

Each chemical substance has a unique set of properties that distinguishes it from every other substance.

The **physical properties** of a substance are those properties that can be observed without changing the substance into another substance. Color, odor, taste, hardness, density, solubility, melting point (the temperature at

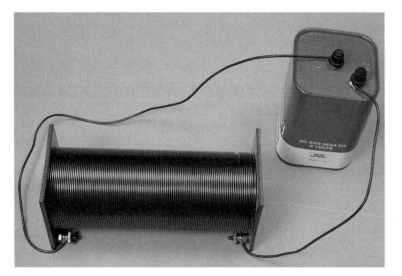

Figure 1-3 Some physical properties of copper. Copper is ductile and can be drawn into thin wire that can be used to conduct an electrical current. It is malleable and can be beaten into thin sheets.

which a substance melts), and boiling point (the temperature at which a substance boils or vaporizes) are all physical properties. For example, some physical properties of pure copper are as follows. Copper is a bright, shiny metal. It is malleable (can be beaten into thin sheets) and ductile (it can be drawn into fine wire), and it is a good conductor of electricity (Fig. 1-3). It melts at l083°C (1981°F) and boils at 2567°C (4653°F); its density is 8.92 grams per milliliter (g/mL) (units of measurement are explained in Appendix B). No matter what its source, copper, if pure, always has these properties.

To determine the melting point of a substance, we must change it from a solid to a liquid, but when we do so, we are not changing the composition of the substance. For example, to determine the melting point of ice, we must change the ice into water. In this process, the appearance changes, but there is no change in composition.

Chemical Properties

The **chemical properties** of a substance are those properties that can be observed when the substance undergoes a change in its chemical composition. It is a chemical property of water that it undergoes a process called electrolysis: If an electrical current is passed through a container of water (H_2O), a change occurs, and the water is broken down to yield the gases hydrogen (H_2) and oxygen (O_2) (Fig. 1-4).

A chemical property is often observed when a substance reacts with another substance. For example, copper turns green when it is exposed to the atmosphere for a long time; the surface reacts with moisture, oxygen, and carbon dioxide in the atmosphere to form a new substance (copper carbonate). The fact that a substance does *not* react with another substance is also a chemical property. For example, it is a chemical property of gold that, unlike copper, it does not tarnish or change when it is exposed to the atmosphere.

Some physical properties of water, iron, and gold are shown in Table 1-1.

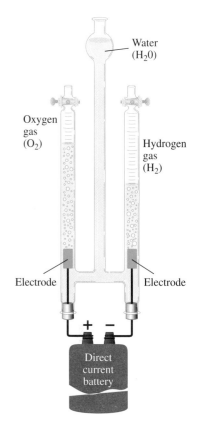

Figure 1-4 A chemical change. The breakdown of water (H_2O) into the gases hydrogen (H_2) and oxygen (O_2) by the process of electrolysis. (Hydrogen and oxygen are produced in a volume ratio of 2:1.)

TABLE 1-1   Some Physical and Chemical Properties of Water, Iron, and Gold

	Physical Properties			Chemical Properties
Substance	Color	Melting Point (°C)	Boiling Point (°C)	
Water (liquid)	Colorless	0	100	Breaks down to hydrogen and oxygen in the process of electrolysis
Iron (solid)	Gray	1535	3000	Reacts with moisture and oxygen in the air to form rust, an iron oxide
Gold (solid)	Yellow	1065	2808	Dissolves in a mixture of hydrochloric and nitric acids to form a yellow solution

▶ CHANGES IN MATTER

Changes in matter occur all the time in the natural world: Snow melts, leaves on trees change color in the fall, dead creatures decompose, and iron rusts. And men and women have been purposely changing natural substances into new products since well before the beginning of recorded history. Early changes brought about by humans for their own advantage included the conversion of natural clays into pottery and limestone rock into building materials.

Most of the manufactured goods produced today involve chemical changes. For example, chemical changes convert iron ore into the steel that is used to make car bodies, and they convert simple chemicals into Dacron, nylon, and other synthetic materials used in the manufacture of clothing and carpeting. Controlled chemical change is thus a major factor in the attainment of our present standard of living. Changes in matter can be divided into physical changes, chemical changes, and nuclear changes.

Physical Changes

A **physical change** in a substance is one that does not alter the composition of the substance. The commonest type of physical change is a change in physical state, for example, a change from solid to liquid. The freezing of water, the melting of ice, the evaporation of water, and the condensation of steam are all examples of physical changes (Fig. 1-5).

Figure 1-5   The melting of ice is an example of a physical change.

Figure 1-6 In the three forms of gold shown, the properties of gold remain the same.

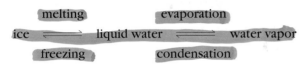

In none of these changes is there a change in composition. The proportions of hydrogen and oxygen in ice, liquid water, and water vapor are identical.

Certain changes in the physical appearance of a substance are also classified as physical changes (Fig. 1-6). The physical weathering of rock as it is broken down into smaller and smaller fragments by the action of wind or by the freezing and melting of water in cracks in the rock is a physical change—assuming that no change in the composition of the rock occurs. Chopping up cabbage in a food processor brings about a physical change, as does dissolving salt in water to form a salt solution. The salt does not change into another substance when it dissolves in water; if we evaporate the water, we can always recover the salt unchanged.

No new substance is ever formed in a physical change.

Chemical Changes

A **chemical change**, or **chemical reaction**, occurs when the composition of a substance is changed. In the process, the substance is changed into one or more other substances that have different physical and chemical properties from the original substance. For example, when iron rusts, a red-brown compound, quite different from the original iron, is formed. The decomposition of water into hydrogen and oxygen by electrolysis is another example of a chemical change. Examples of chemical reactions in the natural world are the combustion of wood in a forest fire, the digestion of food in an animal's stomach, and the dissolution of limestone rocks by rain.

Most of the materials we use in our daily lives, such as medicines, plastics, and synthetic fibers, are produced by controlled chemical changes.

> **EXAMPLE 1-1** Identify the following as physical or chemical changes: **a.** boiling an egg; **b.** silver becoming tarnished; **c.** boiling water.
>
> *Solution:* **a.** Chemical change. New substances are formed as the egg hardens. **b.** Chemical change. Silver and the tarnish have different compositions. **c.** Physical change. Water changes to a vapor, a different physical state, but its composition is not changed.
>
> **PRACTICE EXERCISE:** Which of the following is a physical change and which a chemical change?
> **a.** fruit ripening; **b.** dissolving sugar in tea; **c.** chopping up an onion.
>
> *Answer:* **a.** Chemical change; **b.** physical change; **c.** physical change.

Nuclear Changes

The splitting of uranium atoms in an atomic bomb and the generation of elements in stars are examples of nuclear reactions.

Any pure substance is composed of basic units called elements. In any chemical change, the pure substances involved are changed, but the elements that make up the substances are not changed; they are rearranged to form new substances. In a **nuclear change**, or **nuclear reaction**, however, the nucleus (or core) of the atoms that make up an element is changed, and a new element may be formed. Nuclear reactions, which are sometimes associated with the production of enormous amounts of energy, will be discussed in Chapter 6.

Energy and Changes in Matter

Chemical energy is stored in a substance; it can be released during a chemical reaction.

Energy is the capacity to do work, and for any change in matter to occur, energy is required. Energy exists in many forms, including light, heat, and electricity. Energy is not made of matter, but energy and matter are closely related. Energy is associated with all changes in matter; in nuclear reactions, some matter is actually converted into energy. Heat energy is required to change water into steam, and green plants need light energy from the sun to convert water and carbon dioxide into sugars in the process of photosynthesis.

Radiant energy from the sun will be discussed later in the chapter. The flow of energy through the natural environment will be studied in Chapter 3, and the energy needs of our society will be discussed in Chapters 14 and 15.

STATES OF MATTER

Matter can be classified according to the three familiar states of matter: solid, liquid, and gas (Fig. 1-7).

Solid objects such as ice cubes, pebbles, books, and most household objects we can think of occupy a fixed volume and retain their shape no matter where they are located. The structural particles that solids are made of are held tightly together in fixed positions, thus limiting motion.

Liquids have a fixed volume but not a fixed shape. They take the shape of the container in which they are placed. The structural particles of liquids are held together less tightly than those of solids, and liquids have considerable freedom of motion. If milk is poured into a glass, the milk takes the shape of the glass up to the point at which no more milk is added.

Gases are easily compressed, and are often compressed into steel cylinders for storage and transportation.

Gases maintain neither volume nor shape. The forces holding their structural particles together are relatively weak, and the particles move freely and completely fill whatever vessel contains them.

Whether a substance exists as a solid, a liquid, or a gas, at any particular time, depends on its temperature, the surrounding pressure (the force per unit area exerted on it), and the strength of the forces holding its structural particles together. Water, the most common liquid in our environment, is a very unusual substance in that it exists naturally on Earth in all three states: ice, liquid water, and water vapor. Many substances, however, can be changed from one state to

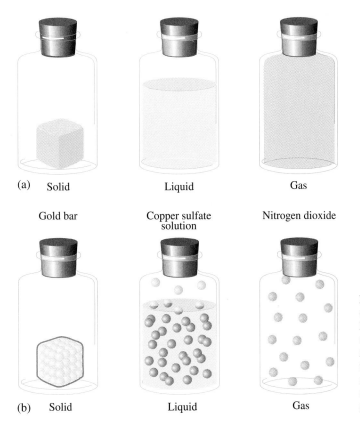

(a)   Solid   Liquid   Gas

Gold bar   Copper sulfate   Nitrogen dioxide
                   solution

(b)   Solid   Liquid   Gas

Figure 1-7   The three physical states of matter. (a) Shows how the shape and volume of a solid, a liquid, and a gas are influenced by the vessels containing them. (b) Shows the arrangement of structural particles in the three states (diagrammatic and not to scale).

another only by the use of extremes of temperature or pressure or both. Iron, for example, is a solid at normal temperatures and must be heated to 1535°C (2795°F) to change it to a liquid and to 3000°C (5432°F) to change it to a vapor (Table 1-1).

Plasmas are sometimes called the fourth state of matter. Plasmas do not exist naturally in our own environment, but in the universe as a whole, the plasma state is probably the most common state of matter. A plasma is similar to a gas except that its structural particles are charged and extremely energetic. The atmosphere of the stars and much of the matter in outer space exist in the plasma state, and our own solar system was formed out of a swirling cloud of plasma.

PURE SUBSTANCES AND MIXTURES

All matter can be divided into two classes: pure substances and mixtures. A **pure substance** always has a definite and constant composition and, regardless of source, always has the same properties under a certain set of conditions. Water is a pure substance; no matter what the source, water is always 11 percent by mass hydrogen, 89 percent by mass oxygen, and, at sea level, it always

Figure 1-8   In a heterogeneous mixture, the components are clearly visible, as in chocolate-chip cookies, oil and vinegar, and a piece of labradorite.

Substance is a general term used to describe any type of matter. *Pure substance* is used to denote a single substance with a specific set of properties.

melts at 0°C (32°F) and boils at 100°C (212°F). Gold is a pure substance; pure gold is always 100 percent gold. Other common examples of pure substances are salt, sugar, aluminum, and nitrogen gas.

Mixtures are combinations of two or more pure substances in which each substance retains its own identity. The components of a mixture are physically mixed but are not chemically combined. If we make a mixture of salt and ordinary sand, we can still distinguish the colorless salt crystals from the larger light brown sand particles. Mixing changes neither the salt nor the sand in any way. Mixtures have variable composition. In our example, we obviously can alter the proportions of salt and sand and make any number of different mixtures.

Another characteristic of mixtures is that the components can be separated by physical means. In the salt-sand mixture, it would be possible, although tedious, to separate the components by picking out all the sand particles with a pair of tweezers. Alternatively, we could add water to the mixture to dissolve the salt, then remove the sand from the salt solution using a fine strainer, and finally evaporate off the water to recover the salt.

Heterogeneous and Homogeneous Mixtures

Examples of mixtures in the natural world include rocks, soil, milk, blood, ocean water, and air. In many mixtures (rocks, soil, and salt-sand mixtures, for example), the different components are visibly distinguishable from one another (Fig. 1-8). Such mixtures are called **heterogeneous** mixtures. Other mixtures (clear ocean water, samples of air, and water containing dissolved sugar) are uniform in appearance; the separate ingredients in the mixtures cannot be seen. Mixtures such as these that have uniform composition and uniform properties throughout the sample are called **homogeneous mixtures**, or **solutions**. The composition of a solution can vary from one sample to another. A given sugar solution, for example, is uniformly sweet throughout the solution, but the sweetness of another sugar solution may be different.

ELEMENTS AND COMPOUNDS

Pure substances can be subdivided into elements and compounds. **Elements** are the building blocks from which all matter is constructed. Elements cannot be decomposed into simpler substances by ordinary chemical means. **Com-**

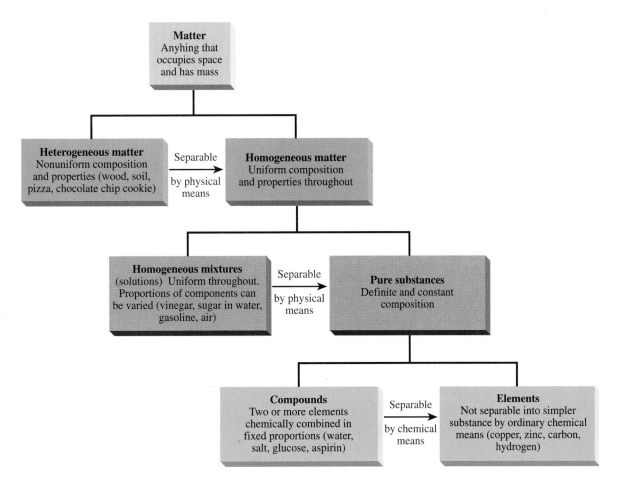

Figure 1-9   A classification of matter.

pounds are pure substances that are composed of two or more elements combined chemically in fixed proportions. A classification of matter based on composition and properties is outlined in Figure 1-9.

Compounds can be decomposed by chemical means to yield the elements from which they are constructed. The properties of a particular compound are always the same under a given set of conditions but are different from the properties of the elements from which they are constructed (Fig. 1-10).

Examples of simple compounds (together with the elements from which they are composed) are water (hydrogen and oxygen), carbon dioxide (carbon and oxygen), salt (sodium and chlorine), and calcium carbonate (calcium, carbon, and oxygen). In all these compounds, two or more elements are chemically combined in fixed proportions.

More than 6 billion different compounds are known.

There are 92 naturally occurring elements and 19 others that have been made by scientists in nuclear reactions (Chapter 6). Every compound that exists is made up of some combination of these 109 elements. Figure 1-11 shows the distribution in the universe, the earth, and the human body of the most abundant of the elements. The universe is composed almost entirely of hydrogen, while on the earth, atoms of oxygen predominate. Just 5 of the 92 naturally occurring elements make up over 90 percent of all the matter in the earth's crust. The three elements oxygen, carbon, and hydrogen make up over 90 percent of our bodies. A complete list of the elements can be found inside the back cover of this book.

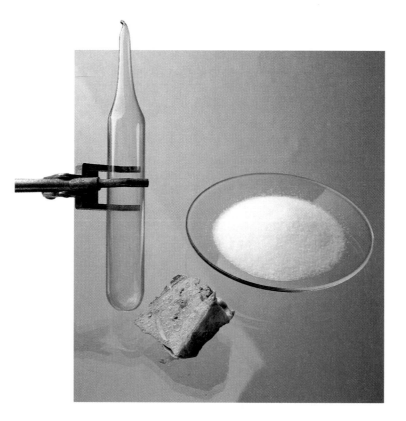

Figure 1-10 The properties of a compound are very different from the properties of the elements from which it is composed. For example, chlorine gas (in the glass container) combines with sodium metal to form sodium chloride (table salt).

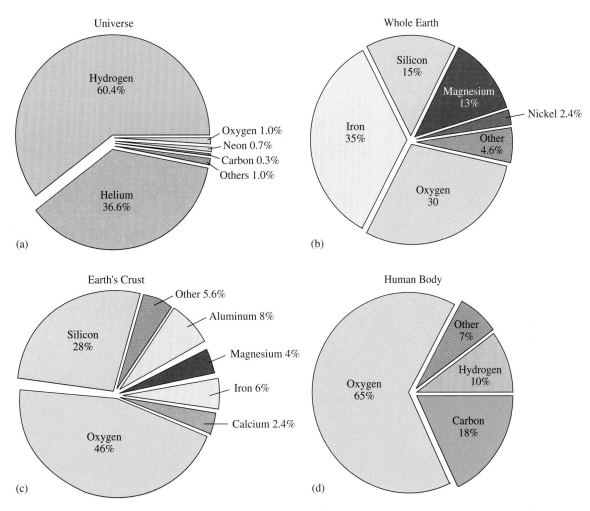

Figure 1-11 The relative abundance, in percent by mass, of the major elements in (a) the universe, (b) the earth as a whole, (c) the earth's crust (including the oceans and the atmosphere), and (d) the human body. Notice the significant differences between the composition of the earth's crust (the part of the planet with which we are in contact), the earth as a whole, and the universe.

Symbols

Chemists use one- and two-letter **symbols** to represent the different elements. The first letter of the symbol is always capitalized, and the second letter is always lower case. Co is the symbol for the element cobalt, whereas CO denotes the compound carbon monoxide. Table 1-2 lists some common elements with their symbols.

The first (or, in some cases, only) letter of a symbol is usually the first letter of the element's name (e.g., O, oxygen; Mg, magnesium); the second letter is often the second letter in the element's name (e.g., Br, bromine; Si, silicon). Where elements share the same second letter, another letter from the name of the element is chosen (e.g., Ca, calcium; Cd, cadmium). Some of the elements shown in Table 1-2—mainly those that have been known for hundreds of years—have symbols that are derived from their Latin names (Table 1-3).

Some of the elements created by scientists in nuclear reactions have three-letter symbols. These elements are of little general interest and will not be considered.

EXAMPLE 1-2   Which of the following is an element and which a compound?

Be   BN   CO   Cr

Solution: Be and Cr, which represent beryllium and chromium, are elements. BN is a compound of B (boron) and N (nitrogen). CO is a compound of C (carbon) and O (oxygen).

PRACTICE EXERCISE: Identify the element(s) and compound(s)

Si   KI   He   HI

Answer: Si and He are elements. KI and HI are compounds.

TABLE 1-2   Common Elements and Their Symbols

Element	Symbol	Element	Symbol
Aluminum	Al	Iron	Fe
Antimony	Sb	Lead	Pb
Argon	Ar	Lithium	Li
Arsenic	As	Magnesium	Mg
Barium	Ba	Mercury	Hg
Beryllium	Be	Neon	Ne
Bismuth	Bi	Nickel	Ni
Boron	B	Nitrogen	N
Bromine	Br	Oxygen	O
Cadmium	Cd	Phosphorus	P
Calcium	Ca	Platinum	Pt
Carbon	C	Potassium	K
Chlorine	Cl	Radium	Ra
Chromium	Cr	Selenium	Se
Cobalt	Co	Silicon	Si
Copper	Cu	Silver	Ag
Fluorine	F	Sodium	Na
Gold	Au	Sulfur	S
Helium	He	Tin	Sn
Hydrogen	H	Uranium	U
Iodine	I	Zinc	Zn

TABLE 1-3   Elements That Have Symbols Derived from Their Latin Name

Element	Latin name	Symbol
Antimony	*Stibium*	Sb
Copper	*Cuprum*	Cu
Gold	*Aurum*	Au
Iron	*Ferrum*	Fe
Lead	*Plumbum*	Pb
Mercury	*Hydrargyrum*	Hg
Potassium	*Kalium*	K
Silver	*Argentum*	Ag
Sodium	*Natrium*	Na
Tin	*Stannum*	Sn

ATOMS AND MOLECULES

Atoms

Elements are composed of atoms. Copper is an element, and even a very small piece of pure copper wire is made up of billions and billions of copper atoms. An **atom** can be defined as the smallest unit of an element that can take part in a chemical change. An atom of any one element differs from the atom of any other element (Chapter 4).

Molecules

Very few elements exist in nature as isolated atoms. Most matter is composed of molecules. A **molecule** is a combination of two or more atoms. It can be defined as the smallest unit of a pure substance (element or compound) that can exist and still retain its physical and chemical properties. Molecules are the particles that undergo chemical changes in a reaction. They are composed of atoms held together by forces known as chemical bonds (Chapter 5).

Molecules may be composed of two or more *identical* atoms or two or more *different* kinds of atoms. For example, two atoms of oxygen bond together to form a molecule of ordinary oxygen gas, O_2. A molecule of water consists of two atoms of hydrogen bonded to a single atom of oxygen, H_2O. A molecule of glucose comprises 6 atoms of carbon, 12 atoms of hydrogen, and 6 atoms of oxygen, $C_6H_{12}O_6$. Proteins and many other biological molecules are made up of several thousands of atoms bonded together in unique ways.

SUBATOMIC PARTICLES AND IONS

Atoms are composed of three major subatomic particles: **protons**, **electrons**, and **neutrons**. A proton carries a one-unit positive charge, an electron carries a one-unit negative charge, and a neutron is uncharged. In an atom of a neutral (uncharged) element, the number of protons is balanced by an equal number of electrons. An element is defined by the number of protons it contains. We will discuss the structure of the atom in detail in Chapter 4.

In certain circumstances, atoms can lose or gain one or more electrons. When electrons (negatively charged) are removed from or added to neutral atoms (or molecules), charged particles called **ions** are formed. A positively charged particle is called a **cation**; a negatively charged particle is called an **anion**. For example, a sodium atom (Na) can lose an electron to become the cation Na^+, and a chlorine atom (Cl) can gain an electron to become the anion Cl^-. Sodium chloride (NaCl), common table salt, is a compound formed from Na^+ and Cl^- ions.

Ionic solids such as NaCl do not exist as individual molecules. As we shall see in Chapter 5, they are more accurately described in terms of formula units.

CHEMICAL FORMULAS

The composition of a compound can be represented by a **chemical formula**. The formula indicates the number of atoms of each element that are present in a molecule of the compound. The chemical formulas of some familiar compounds are given in Table 1-4.

A chemical formula is the chemist's shorthand for representing the composition of a pure substance.

TABLE 1-4   Some Substances and Their Formulas

Substance	Formula
Aluminum sulfate	$Al_2(SO_4)_3$
Ammonia	NH_3
Ammonium phosphate	$(NH_4)_3PO_4$
Calcium carbonate	$CaCO_3$
Carbon dioxide	CO_2
Carbon monoxide	CO
Glucose	$C_6H_{12}O_6$
Helium	He
Kaolinite	$Al_2Si_2O_5(OH)_4$
Methane	CH_4
Nitrogen gas	N_2
Silica	SiO_2
Water	H_2O

The numerical subscripts to the right of the symbols for the elements indicate the number of atoms of each element present in a molecule of the compound. If there is no subscript following the element (as for the O in H_2O), it means that only one atom is present.

It might seem that the formula for glucose ($C_6H_{12}O_6$) could be simplified to CH_2O. In this simpler formula, the C, H, and O atoms are in the same ratios to each other (1:2:1) as in $C_6H_{12}O_6$. However, for glucose to have its characteristic properties, there must be 6 atoms of carbon, 6 atoms of oxygen, and 12 atoms of hydrogen bonded together in a unique way.

The manner in which atoms are joined in a molecule can be depicted in a number of ways (Fig. 1-12). A molecular formula shows the number of atoms of each element that is present. A structural formula shows which atoms are joined to which other atoms in the formula. Ball-and-stick and space-filling models give more information: They show the relative sizes of the atoms and their geometric arrangement. The space-filling models show the relative distances between the atoms.

Some of the chemical formulas shown in Table 1-4 include parentheses. These formulas indicate that the grouping of atoms within the parentheses is repeated more than once in the molecule. For example, aluminum sulfate, which is used in water purification, includes three sulfate, SO_4, units. Ammonium phosphate, a fertilizer, includes three ammonium, NH_4, units; and kaolinite, a clay, includes four hydroxyl, OH, units. In chemical reactions, multiatom units—NO_3, SO_4, PO_4, OH, and NH_4—remain intact. They will be considered in more detail in Chapter 5.

CHEMICAL EQUATIONS

The changes that occur during chemical reactions can be described by means of **chemical equations**. For example, when coal—which is composed mainly of carbon—is burned, it combines with oxygen gas in the atmosphere to produce the gas carbon dioxide. This information can be summarized concisely in the following chemical equation:

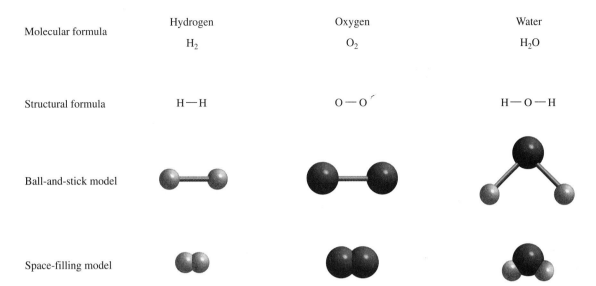

	Hydrogen	Oxygen	Water
Molecular formula	H_2	O_2	H_2O
Structural formula	H—H	O—O	H—O—H
Ball-and-stick model			
Space-filling model			

Figure 1-12 Structural formulas, ball-and-stick models, and space-filling models showing how H and O atoms join together to form H_2, O_2, and H_2O. Structural formulas show which atoms are joined together; ball-and-stick and space-filling models show the relative sizes of the atoms, and their geometric arrangement; space-filling models show the relative distances between atoms and give the most accurate representation of the actual molecule.

$$C \quad + \quad O_2 \quad \longrightarrow \quad CO_2$$
carbon   oxygen   carbon dioxide

The symbols for the elements are not only abbreviations of the names of the elements; each symbol represents one atom of the element. The above equation tells us that one atom of carbon reacts with one molecule of oxygen (two O atoms bonded together) to produce one molecule of carbon dioxide (one carbon and two O atoms bonded together).

Coal often contains some sulfur, and during burning, the sulfur will react with atmospheric oxygen to produce sulfur trioxide (SO_3), a toxic gas that is a major source of air pollution.

$$2\,S + 3\,O_2 \quad \longrightarrow \quad 2\,SO_3$$

The numbers in front of the formulas in this equation ($\underline{2}$ S, $\underline{3}$ O_2, $\underline{2}$ SO_3) are called **coefficients**. They indicate the minimum numbers of atoms and molecules that take part in the reaction. In the above example, the coefficients tell us that two atoms of sulfur react with three molecules of oxygen to produce two molecules of sulfur trioxide. If there is no number in front of a formula—as in the previous equation for the burning of coal—it is understood that one atom or one molecule is involved.

Sulfur trioxide released into the atmosphere as a result of the burning of sulfur-containing coal reacts with moisture in the atmosphere to form sulfuric acid (H_2SO_4). The acid falls to earth as acid rain.

$$SO_3 + H_2O \quad \longrightarrow \quad H_2SO_4$$

ACIDS AND BASES

Acid in rain is harmful in many ways. It slowly dissolves monuments and statues made of limestone, and it may make lake water too acidic for fish to survive. Sulfuric acid and two other common acids, hydrochloric acid (HCl) and nitric acid (HNO_3), are strong acids. However, most acids we encounter are weak acids that are not harmful to the environment. Common weak acids include citric acid found in citrus fruits (e.g., oranges and grapefruits), vinegar, and ascorbic acid (vitamin C).

Bases are compounds that can neutralize the effects of acids. For example, the base lime (calcium oxide, CaO) is used to neutralize acid soil. Other common bases are the strong base sodium hydroxide (NaOH), which is present in drain cleaners, and ammonia (NH_3), which is a common ingredient in many household cleaners. Acids and bases are discussed in Chapter 8.

ELECTROMAGNETIC RADIATION

The energy source that sustains all life on Earth is the sun. It provides the light energy that plants need for photosynthesis and the heat that warms the earth. It controls our climate, drives our weather systems, and regulates the life cycles of all plant and animal species. The radiant energy, or **electromagnetic radiation**, (EMR), that the sun continuously transmits through space, reaches the earth in many forms. The most familiar forms are visible light and radiant heat; other forms include cosmic rays, X-rays, ultraviolet (UV) radiation, microwaves, and radio waves (Fig. 1-13).

EMR travels in waves that can be compared to the waves that ripple outward on the surface of a pond when a stone is dropped into the water. Electromagnetic waves differ from water ripples, however, in that they travel outward in all directions. All electromagnetic waves, whether they are light waves, microwaves, radio waves, or any other type of electromagnetic wave, travel at the same rate. This rate is at a maximum (300,000 kilometers per second, or 186,000 miles per second) when the waves travel in a vacuum; waves are slowed very slightly when they travel through air or any medium where they encounter atoms or molecules.

Interactions with gases and particles prevent about half the EMR transmitted by the sun from reaching the earth's surface (Chapter 13).

For convenience, the electromagnetic spectrum is divided into different types of radiation that change from one type to another at specified intervals. In reality, the spectrum is continuous, with the **wavelength** of the radiation gradually increasing from cosmic rays to radio waves, as shown in Figure 1-13. The wavelength (λ) is the distance between adjacent wave crests (Fig. 1-14), and it is usually measured in nanometers (nm). The number of crests that pass a fixed point in 1 second is termed the **frequency** (v). A frequency of 1 cycle (passage of one complete wave) per second is equal to 1 **hertz** (Hz). Wavelength and frequency are related as shown in the following equation:

$$v = \frac{c}{\lambda} \text{ (where } c = \text{ the speed of light in a vacuum)}$$

The various forms of radiation seem very different, but they are all manifestations of the same phenomenon; they differ from each other only in

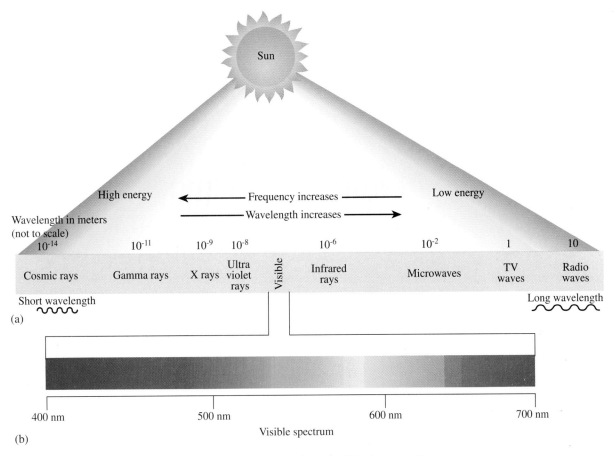

Figure 1-13   The electromagnetic spectrum. As the wavelength of electromagnetic radiation increases from cosmic rays to radio waves, the energy and frequency of the waves decrease. The visible wavelengths make up a very small part of the electromagnetic spectrum. [One nanometer (nm) is equal to 10^{-9} meters (m). Exponential notation, as in 10^{-9}, is explained in Appendix A.]

energy, wavelength, and frequency. The shorter the wavelength (and higher the frequency) of the radiation, the greater the energy it transmits. High-energy radiation is very damaging to living tissues. The effects of gamma rays (which are emitted by radioactive elements) and X-rays will be discussed in Chapter 6. We are protected from the harmful effects of UV radiation from the sun primarily by the ozone layer in the upper atmosphere. The potentially dangerous results of depletion of the ozone layer will be covered in Chapter 13.

Visible light, the kind we see, makes up only a small part of the entire electromagnetic spectrum. Its wavelength ranges from approximately 400 nm (violet) to approximately 700 nm (red). Infrared radiation (which has a wavelength greater than 700 nm) is invisible but can be felt as heat when, for example, it is emitted from a stove or an electric heater. Microwave radiation is used for radio communication, microwave cooking, and weather tracking (radar). Long-wavelength radio waves are emitted by large antennae at broadcasting stations.

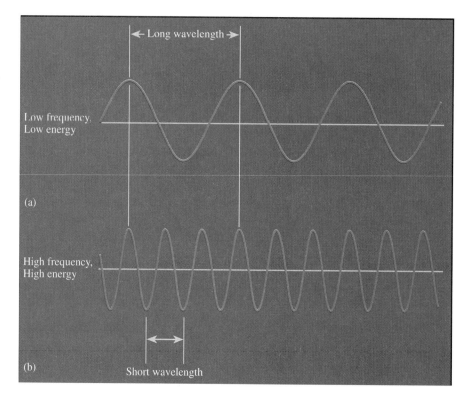

Figure 1-14 Wavelength and frequency of electromagnetic radiation. The frequency of the high-frequency (high-energy) radiation shown in (b) is three times that of the low-frequency (low-energy) radiation shown in (a).

THE SCIENTIFIC METHOD

If a scientist wishes to find the answer to a specific problem, he or she will apply the **scientific method**, an orderly step-by-step procedure involving observation, measurement, speculation, experimentation, and deduction. To explain how this method of inquiry is carried out, we will consider a simple environmental problem.

The first step is to identify the problem to be solved. Let us suppose that people living near a lake have reported that, recently, trout in the lake have been dying at an alarming rate, but that less desirable fish, such as catfish, appear to be unaffected. The problem in this case is to find the cause of the decline in the trout population.

The next step is to obtain as much information about the problem as possible. This would include verifying the validity of the observations made by the people living near the lake, checking the present populations of the different species of fish in the lake and comparing these figures with data that might have been gathered earlier, looking for evidence of dead fish, examining fish for evidence of disease, and studying the possible sources of pollutants that might be entering the lake. It would be relevant to know if there is a chemical plant, mining operation, extensive logging, or new construction in the area. Let us assume that (1) the decline in the population is validated; (2) there is no evidence of disease caused by fungi or bacteria; and (3) there are two chemical plants, A and B, quite nearby on either side of a small river that feeds into the lake. Plant

A has end products that include toxic polychlorinated biphenyls, PCBs. Plant B is not in operation. On the basis of the information that has been gathered, it is suggested that the fish are being killed because plant A is discharging PCBs into the river that feeds the lake. This suggestion, which is still speculation and has not been proved, is termed a **hypothesis**.

The next step is to devise and carry out experiments that will prove or disprove the hypothesis. PCB levels in the water in the river near the plant and in the lake must be monitored at regular intervals to determine if the levels increase at times when the factory discharges wastes into the river. PCB levels in the sediment of the lake, in bottom-dwelling creatures, and in different species of fish must also be determined. Let us suppose that PCBs cannot be detected in any of the samples tested. It will have to be concluded that the original hypothesis is incorrect (and that the operators of plant A deserve an apology).

The next step is to revise the original hypothesis. We will assume that the scientists conducting the inquiry now discover that drums leaking acid have been found buried on plant B's property near the river. Armed with this new information, another hypothesis is formulated: Namely, acid leaking from the drums is killing the fish. A new set of experiments reveals that samples of water from the river near plant B and from the lake are unusually acidic. It is determined that the lake is not in a region where acid rain is a problem and, further, that water in two nearby lakes that are not fed by the river is of normal acidity. It is therefore tentatively concluded that the second hypothesis is correct. To prove conclusively that this hypothesis is correct, further experiments must be performed. For example, trout and other species of fish would have to be kept in tanks of water of different acidity to determine if water as acid as that found in the lake (1) killed the trout and (2) had no apparent effect on less-desirable fish. If this were found to be the case, the second hypothesis, on the basis of the available data, would have been proved to be correct.

It is possible that the owners of plant B might question the experimental findings and conduct their own investigation. Let us assume that their findings agree with those of the initial study and thus strengthen the conclusion that increased acidity is causing the decrease in the trout population. The hypothesis, now that repeated experimentation has shown it to be valid, can be termed a **theory**.

If our environmental problem were a real one, the final step, which is not a part of the scientific method, would probably be the most difficult to accomplish. This would be to remove the leaking drums and get someone to pay for the job to be done.

The scientific method can be summarized as follows:

1. **Observation** As a result of observation, a question is raised. In our simple environmental example, people noticed (observed) that fish were dying and wanted to know why. In science, many of the often complex questions to be answered originate in the laboratory. Observations are not confined to what can be seen, heard, smelled, touched, and tasted with the unaided senses; they also include information gained from analytical instruments.

2. **Hypothesis**   Based on the relevant information that is available, an educated guess is made to explain the initial observations. This guess is termed a hypothesis.

3. **Experimentation**   A series of tests (experiments) is devised to prove or disprove the validity of the hypothesis. The experiments must be well designed and carefully controlled, and the exact conditions under which they are conducted must be recorded.

4. **Revision of Original Hypothesis**   If the results of the experiments are inconclusive or prove that predictions based on the original hypothesis are incorrect, a new hypothesis must be formulated, and another set of experiments must be devised and conducted.

5. **Repetition**   Once it has been established that the experimental data uphold the hypothesis, it must be shown that repeated experimentation, under the same conditions, always yields the same results.

6. **Theory**   A theory is a hypothesis that has been tested by repeated experimentation and proved to be correct. The term *theory*, in science, is usually reserved for statements that cover a wide body of knowledge rather than answers to small regional problems such as the environmental one we cited.

▶ KEY WORDS AND CONCEPTS

anion	energy	nuclear reaction
atom	frequency	physical change
cation	gas	physical properties
chemical change	hertz (Hz)	plasma
chemical equation	heterogeneous	proton
chemical formula	homogeneous	pure substance
chemical properties	hypothesis	science
chemical reaction	ion	scientific method
chemistry	liquid	solid
coefficient	matter	solution
compound	mixture	symbol
electromagnetic radiation	molecule	technology
electron	neutron	theory
element	nuclear change	wavelength

▶ QUESTIONS AND PROBLEMS

1. Classify the following as physical or chemical changes:
   a. dissolving sugar in water   **b.** setting gasoline on fire   **c.** pulverizing (smashing) a rock   **d.** a bronze statue turning green on exposure to wind and rain   **e.** discharging carbon dioxide into the air from a smoke stack   **f.** milk souring

2. Dry ice (solid CO_2) evaporates to gaseous CO_2 by a process called sublimation. Is sublimation a physical or chemical change?

3. Define and give an example of each of the following:
   a. pure substance   **b.** mixture   **c.** solution

4. Classify the following as either homogeneous or heterogeneous mixtures:
   a. sand and water   **b.** sand and salt   **c.** salt dissolved in water   **d.** filtered apple juice   **e.** orange juice   **f.** coffee with sugar

5. How would you separate the components of the following mixtures?

a. sand and water **b.** sugar and water **c.** oil and water **d.** salt and sand

6. Classify the following as compounds or elements.
   a. sodium, Na **b.** carbon dioxide, CO_2 **c.** arsenic, As **d.** helium, He **e.** sulfur dioxide, SO_2 **f.** aluminum, Al **g.** water, H_2O **h.** octane, C_8H_{18} **i.** zinc, Zn **j.** nitrogen dioxide, NO_2 **k.** sulfuric acid, H_2SO_4 **l.** mercury, Hg **m.** calcium carbonate, $CaCO_3$ **n.** iron, Fe

7. Write the chemical symbol for the following elements:
   a. uranium **b.** sulfur **c.** oxygen **d.** carbon **e.** calcium **f.** potassium **g.** radon **h.** nitrogen **i.** nickel **j.** lead **k.** gold **l.** chromium **m.** cadmium **n.** silver

8. Write the names of the following elements:
   a. Mg **b.** Cu **c.** Co **d.** C **e.** B **f.** Ni **g.** P **h.** Si **i.** S **j.** Al **k.** Br **l.** I **m.** Hg **n.** O

9. Classify the following as anions, cations, or neutral atoms:
   a. Cl^- **b.** Na^+ **c.** He **d.** Ca^{2+} **e.** F **f.** C **g.** O^{2-} **h.** Fe^{3+} **i.** Si **j.** K^+ **k.** I^- **l.** Ne

10. Classify the following as anions, cations or neutral molecules:
    a. CO_2 **b.** PO_4^{3-} **c.** Fe^{2+} **d.** CO_3^{2-} **e.** NO **f.** Co^{2+} **g.** SO_2 **h.** NH_4^+ **i.** $C_6H_{12}O_6$ **j.** NO_3^-

11. Describe in words the chemical reactions that are summarized below:
    a. $C + O_2 \rightarrow CO_2$ **b.** $4\,Fe + 3\,O_2 \rightarrow 2\,Fe_2O_3$ **c.** $2\,Na + Cl_2 \rightarrow 2\,NaCl$

12. How do the gas, liquid and solid forms of the same material differ in physical properties?

13. List two chemical reactions you carried out today.

14. Define the following. Give the appropriate units for each. How are a, b, and c related to each other?
    a. wavelength **b.** frequency **c.** speed of light

15. The visible spectrum that the human eye responds to goes from red through orange, yellow, green, blue, and indigo to violet. Is an orange beam of light of higher or lower energy than a blue beam?

16. Define the following:
    a. electromagnetic radiation **b.** energy **c.** nanometer (nm) **d.** wavelength **e.** spectrum **f.** vacuum

17. Sunscreen creams absorb high energy ultraviolet (UV) radiation and thus protect the skin from damage.
    a. What is the wavelength of UV radiation? **b.** Is UV radiation of shorter or longer wavelength than visible light? **c.** Is UV radiation of higher or lower energy than infrared (IR) radiation?

18. X-rays are useful for diagnostic medical evaluation. The Surgeon General has warned that excessive X-ray exposure may pose a risk to health. Referring to Figure 1-13:
    a. Explain why X-rays are a risk. **b.** Are X-rays more dangerous than microwaves? **c.** TV and radio waves permeate our environment. Are they dangerous? **d.** Cosmic rays bombard our planet from the sun. Should we be concerned about cosmic rays? **e.** Tanning salons use UV lights to darken (tan) the skin. Why is UV light used? Is there a risk?

19. The earth is warmed by sunlight during the day. On a clear winter night, without clouds, the temperature drops lower than if clouds were present. Explain.

20. Can you think of anything you have done today that did not involve a chemical reaction?

THE ORIGIN OF THE EARTH AND THE EARTH'S MINERAL RESOURCES

2

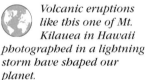
Volcanic eruptions like this one of Mt. Kilauea in Hawaii photographed in a lightning storm have shaped our planet.

To understand how our environment works, we must first look back billions of years and try to find out how the earth was born and how it evolved into the life-supporting planet we inhabit today.

In this chapter, we will consider the most generally accepted theory of the formation of the universe, including the origin of the galaxies, the stars, and our own planet earth. We will learn how the oceans and the atmosphere evolved, and how the rocky surface on which we live was formed. We will find out how the continents were shaped, and how eventually it became possible for life to develop on the earth. We will also examine the earth's mineral resources and discuss the ways we use them for the needs of society.

Learning Goals:

In this chapter, you should gain an understanding of:

1. How the earth was formed and developed into three layers (or zones).
2. How the oceans and atmosphere were formed.
3. The theory of plate tectonics.
4. The origin of life on earth.
5. The uniqueness of planet earth.

6. The three main classes of rock that make up the earth's surface.
7. Silicate minerals and how they are used in our society.
8. Important metal elements that are extracted from ores.

THE FORMATION OF THE UNIVERSE

If we gaze up at the sky on a clear night, away from the lights of the city, we see a myriad of stars. All the stars we see are just a part of our galaxy, the **Milky Way**. This pinwheel-shaped body, which is made up of clouds of gas and cosmic dust and billions and billions of stars, includes our own solar system—the sun and its nine orbiting planets. What we see is only a minute fraction of the whole **universe**. Beyond the Milky Way are countless other galaxies extending out into space for distances beyond our comprehension. It was probably only when we humans first ventured out into space in the 1960s that we began to appreciate the smallness and insignificance of our planet in relation to the universe as a whole. The first photographs of the earth taken from the moon showed us the earth suspended, small and round, in the black vastness of space (Fig. 2-1).

It is generally believed that the universe began some 15 to 20 billion years ago. Although differences of opinion still exist among scientists, many believe that all the matter in the universe today was once compressed into an infinitesimally small and infinitely dense mass that exploded with tremendous force. This explosion of unimaginable proportions, appropriately called the **Big Bang**, generated enormous amounts of energy and released the cosmic matter

Figure 2-1   The earth as seen from the surface of the moon.

from which the galaxies and stars were eventually formed. The universe expanded in all directions, and according to most astronomers has been expanding ever since.

Galaxies and Stars

As the universe expanded, it cooled very, very slowly, and cosmic matter gradually condensed to form the first **galaxies**. Atoms of hydrogen—the simplest and lightest of all the elements—formed in the swirling clouds of condensing matter. Over billions of years, cosmic matter in the galaxies gave birth to the early **stars**, and these stars generated sufficient heat to cause hydrogen atoms to fuse (join) together to form atoms of helium, the second lightest of the elements. The energy released during fusion initiated further fusion reactions in which all 90 of the remaining naturally occurring elements found on earth were formed. In the universe as a whole, 90 percent of all atoms are hydrogen, 9 percent are helium, and all the other elements together make up the remaining 1 percent. Scientists believe that subsequent explosions of the early stars scattered the elements, and it is from the debris of one of these explosions that our sun was born. The sun, which to us appears so very bright, is an average-sized star located toward the edge of the Milky Way.

The Planets in Our Solar System

Scientists still do not know with any certainty how the planets developed, but it is generally believed that they began to form about 5 billion years ago from hot, mainly gaseous, matter rotating about the sun. With time, the matter slowly cooled, and solid particles condensed out from the gases. The particles gradually coalesced into clumps of matter. Larger clumps, with their stronger gravity, gradually drew in and retained additional particles and grew to form the planets in our **solar system** (a solar system is a group of planets that revolve around a star). Our solar system consists of nine unique planets—Mercury, Venus, Earth, Mars, Jupiter, Saturn, Uranus, Neptune, and Pluto—that revolve around the sun in different elliptical, but nearly circular, orbits that lie roughly in the same plane (Fig. 2-2). Each planet is approximately twice as far from the sun as the next inner one. The orbits of the planets can be defined so precisely that it is possible to predict the exact position of any planet with respect to the earth at any time for years to come.

The four planets closest to the sun—Mercury, Venus, Earth, and Mars—are called **terrestrial planets** because, like the earth, they are small, dense, and rocky and are composed mainly of metals. The more distant, **giant planets**—Jupiter, Saturn, Uranus, and Neptune—are much larger and are gaseous bodies of low density.

The earth and the other terrestrial planets that formed close to the sun were so hot that the lighter, easily evaporated materials could not condense and were swept away. Only substances with extremely high boiling points, such as metals and minerals, condensed on these planets. Mercury, the planet closest to the sun and therefore the hottest, is composed mainly of the metal iron. On the earth, which formed at a somewhat lower temperature, silicates and other metals besides iron were able to condense. (Silicates are minerals formed from the elements silicon, oxygen, and a variety of metals.) Substances such as water, ammonia, and methane, which evaporated from the hot terrestrial planets, condensed into solids on the cold, distant giant planets. The larger planets, with

In 1992, the Hubble Space Telescope detected new evidence that the universe will not collapse back on itself billions of years from now but will probably go on expanding forever.

The clumps of matter formed by the coalescing particles are called planetesimals.

The planets revolve around the sun in a counterclockwise direction.

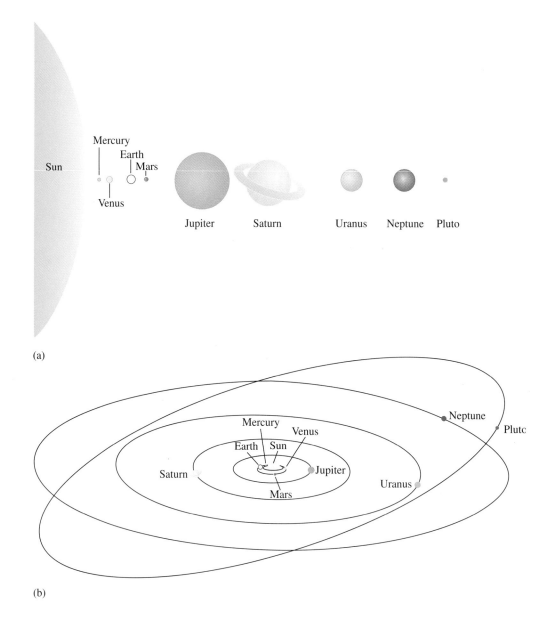

(a)

(b)

Figure 2-2   The solar system. (a) The relative sizes of the planets. (b) The planets in their orbits around the sun.

their greater mass and thus greater gravitational pull, retained gases—mostly hydrogen and helium—in the atmospheres surrounding them. Some important features of the planets, as they exist today, are listed in Table 2-1.

The Sun

The **sun** is the ultimate source of all the energy on Earth. It makes up 99.9 percent of the mass of the solar system, and its diameter is approximately 110 times as great as that of the earth. Scientists estimate that temperatures near the center of this immense rotating sphere of white-hot gases reach almost 15 million degrees Celsius (27 million degrees Fahrenheit). Fusion reactions occur at

TABLE 2-1 Vital Statistics of Planets

Planet	Diameter (km)	Diameter (mi)	Mass (earth=1)	Density (water=1)	Gravity (earth=1)	Time for One Rotation on Axis (earth days or hours)	Time for One Revolution Around Sun (earth years)	Distance from Sun (million km)	Distance from Sun (million mi)	Composition of Atmosphere
Terrestrial										
Mercury	4,835	3,004	0.055	5.69	0.38	59 days	0.24	57.7	36.8	None
Venus	12,194	7,577	0.815	5.16	0.89	243 days	0.62	107	66.9	CO_2
Earth	12,756	7,926	1.00	5.52	1.00	1.00 days	1.0	149	92.6	N_2, O_2
Mars	6,760	4,200	0.108	3.89	0.38	1.03 days	1.9	226	141	CO_2, N_2, Ar
Giant										
Jupiter	141,600	87,986	318	1.25	2.64	9.83 hours	12	775	482	H_2, He
Saturn	120,800	75,061	95.1	0.62	1.17	10.23 hours	29	1421	883	H_2, He
Uranus	47,100	29,266	14.5	1.60	1.03	23.00 hours	84	2861	1777	H_2, He, CH_4
Neptune	44,600	27,713	17.0	2.21	1.50	22.00 hours	165	4485	2787	H_2, He, CH_4

these incredibly high temperatures, and the tremendous amounts of energy continually being released in the form of heat and light have allowed the sun to shine brightly for billions of years.

 ## ➡ DIFFERENTIATION OF THE EARTH INTO ZONES

Exactly how the earth evolved to its present state is not known, but earth scientists believe that, when first formed some 4.7 billion years ago, the earth was homogeneous in composition: a dense rocky sphere with no water on its surface and no atmosphere. Then, over time, the earth gradually grew hotter and became differentiated into layers, each layer having a different chemical composition. This crucial period in the development of the earth led to the formation of the earth's magnetic field, the atmosphere surrounding the earth, the oceans, and the continents—and, ultimately, to life.

Why the Earth Heated Up

Three factors are believed to have caused the earth to heat up. The cosmic particles that collided and clumped together to form the earth were drawn toward the earth's center by the pull of gravity. As more particles collided with the developing planet, heat was released. Some of the heat was retained within the earth; this heat gradually built up as more and more material accumulated.

As the earth grew, material in the center was compressed by the weight of new material that struck the surface and was retained. Some of the energy expended in compression was converted to heat and caused a further rise in temperature within the earth.

The third, and very significant, factor in the warming of the earth's interior was the decay of radioactive elements within the earth that released energy in the form of heat. The atoms in radioactive elements are unstable and disintegrate spontaneously, emitting atomic particles or energy or both. In this process, which continues today, these radioactive elements are converted into atoms of other elements (this topic will be discussed in Chapter 6). Only a very small percentage of naturally occurring elements have atoms that disintegrate in this way, and the heat generated with each disintegration is extremely small. Nonetheless, geophysicists have calculated that the retention of this heat within the earth over billions of years (together with the heat released as new material accumulated and was compressed) would have been sufficient to raise the temperature of the material at the center of the earth to the point where it became molten.

It seems probable that this critical temperature was reached about 1 billion years after the earth was born. Metallic iron, which melts at 1535°C (2795°F) and makes up over 30 percent of the earth's mass, began to melt, and this heavy molten iron, together with some molten nickel, sank to the center of the earth. As the molten iron sank, less-dense material was displaced and rose toward the surface. As a result, the earth ceased to be homogeneous and eventually became differentiated into three distinct layers, or **zones**: the **core**, the **mantle**, and the **crust** (Fig. 2-3).

The Core

The core, which extends 3500 km (2200 mi) out from the center of the earth, is believed to be composed of iron and small amounts of nickel. The metals are thought to be in solid form in the inner core and molten in the surrounding outer core (Fig. 2-3). Because the core is inaccessible to us, we have no direct proof that it consists primarily of iron, but there is considerable indirect evidence to support this view. For example, analysis of light emitted by the sun and stars has revealed that iron is the most abundant metal in the universe; analysis of waves generated by earthquakes has shown that the core is very dense, and iron is the densest metal found in any quantity on the earth; and most of the meteorites that have landed on the earth from outer space are composed of iron.

The Mantle

The mantle, which lies between the core and the crust, is approximately 2900 km (1800 mi) thick. The relatively thin upper part of the mantle is solid and rigid, but the layer below it—called the **asthenosphere**—although essentially solid, is able to flow extremely slowly like a very thick viscous liquid. In the deep mantle below the asthenosphere, the rock is believed to be rigid again.

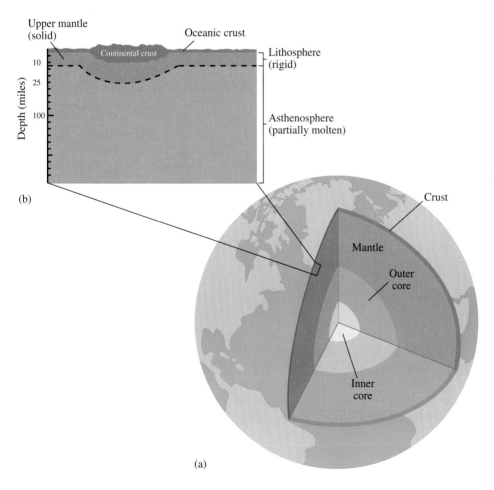

Figure 2-3   Diagrammatic representation of the structure of the earth. (a) The earth is differentiated into distinct layers: the core, the mantle, and the crust. (b) The lithosphere, which comprises the continental and oceanic crust together with the solid upper part of the mantle, rests on the partially molten asthenosphere.

The Crust

Above the mantle is the crust, which forms the thin outer skin of the earth. The crust is thicker beneath the continents than beneath the oceans. Its thickness ranges from 6 km (4 mi) under the oceans to 70 km (45 mi) under mountainous regions. Although the crust makes up a very small part of the earth as a whole, we gather from it practically all the resources that sustain our way of life.

The crust and the solid upper part of the mantle make up the relatively cool and rigid **lithosphere**, which floats on the hotter partially molten asthenosphere. The boundary between the lithosphere and the asthenosphere is not caused by a difference in the chemical composition of the rocks in the two layers but by a change in the properties of the rocks that occurs as temperature and pressure increase with depth.

Lithosphere is derived from *lithos*, the Greek word for "stone."

Relative Abundance of the Elements on the Earth

By mass, the four most abundant elements in the earth are iron, oxygen, silicon, and magnesium, which together account for about 93 percent of the earth's mass (Fig. 2-4a). Nickel, sulfur, calcium, and aluminum make up another 6.5 percent. The remaining 0.5 percent or so of the earth's mass is made up of the other 84 naturally occurring elements.

Primarily because most iron sank to the earth's center during the period of differentiation, the types and abundance of the elements in the crust (Fig. 2-4b) are very different from what is found in the earth as a whole. Seventy-four percent of the crust consists of oxygen and silicon, while aluminum, iron, magnesium, calcium, potassium, and sodium together account for 25 percent.

It might have been expected that as the earth became differentiated into zones, the elements would have been distributed strictly according to mass, with the heavier elements falling to the earth's center and the lighter ones rising to the surface. This distribution did not occur, however, because some ele-

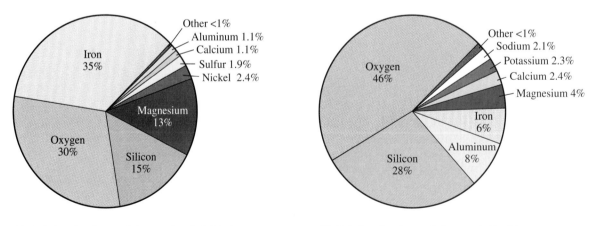

(a) Relative abundance of elements in whole Earth   (b) Relative abundance of elements in Earth's crust

Figure 2-4   The relative abundance (by mass) of elements in the whole earth and in the earth's crust. Because of the differentiation that occurred early in the earth's history, the percentage of iron in the crust (b) is lower than that in the whole earth (a), and the percentages of aluminum, silicon, and oxygen (the elements that combine to form silicates) are higher.

TABLE 2-2   The Relative Abundance of Economically Important Elements in the Earth's Crust

Name	Chemical Symbol	Crustal Abundance (percent by Mass)
Aluminum	Al	8.00
Iron	Fe	5.80
Magnesium	Mg	2.77
Potassium	K	1.68
Titanium	Ti	0.86
Hydrogen	H	0.14
Phosphorus	P	0.101
Fluorine	F	0.0460
Sulfur	S	0.030
Chlorine	Cl	0.019
Chromium	Cr	0.0096
Zinc	Zn	0.0082
Nickel	Ni	0.0072
Copper	Cu	0.0058
Cobalt	Co	0.0028
Lead	Pb	0.00010
Arsenic	As	0.00020
Tin	Sn	0.00015
Uranium	U	0.00016
Tungsten	W	0.00010
Silver	Ag	0.000008
Mercury	Hg	0.000002
Platinum	Pt	0.0000005
Gold	Au	0.0000002

Source: From F. Press and R. Siever, *Earth,* 3d ed. (New York: W. H. Freeman, 1982), p. 553.

ments combined with other elements to form compounds, and the melting points and densities of these compounds (rather than those of the elements from which they were formed) primarily determined how the elements were distributed in the earth. For example, silicon, oxygen, and various metals combined to form silicates, which are relatively light and melt at relatively low temperatures. When the earth's interior was hot, these silicates rose to the surface. Today, they are the most abundant minerals in the earth's crust.

As a result of the chemical changes that occurred during the period of differentiation, the distribution of the elements on the earth is very uneven. The relative abundance in the earth's crust of elements that are economically valuable is shown in Table 2-2. Of these, only four—aluminum, iron, magnesium, and potassium—are present at levels greater than 1 percent of the total mass of the crust. It is fortunate for us that, as the result of geologic processes that have been occurring for millions of years, the less abundant (but valuable) elements such as gold and silver are concentrated in specific regions of the world. If the elements had been distributed evenly throughout the earth's crust, their concentrations would be too low to make extraction technically or economically feasible.

FORMATION OF THE OCEANS AND THE ATMOSPHERE

It is generally accepted that for billions of years after the earth was formed, there was no water on its surface. Then, as the interior of the earth heated up, minerals below the earth's surface became molten. The molten material rose to the surface, and oxygen (O) and hydrogen (H) atoms that were chemically bound to certain minerals escaped explosively into the atmosphere as clouds of water (H_2O) vapor. In these tremendous volcanic eruptions, which were widespread and numerous, carbon dioxide and other gases were also released from the earth's interior (Fig. 2-5). The lighter gases escaped into space, but the heavier ones, including water vapor and carbon dioxide (CO_2), were held by gravity as a thick blanket of clouds surrounding the earth. In time, as the earth's surface cooled, the water vapor condensed, and the clouds released their moisture. For the first time, rain fell on the earth. During the next several million years, volcanoes continued to erupt, and, as more rain fell, the oceans filled with water.

It is interesting to note that although the early development of Venus was similar to that of the earth, Venus never formed oceans. Like the earth, Venus had a period when its interior warmed, and water vapor and carbon dioxide were released and collected near the surface. But, because Venus was closer to the sun than the earth, its surface never cooled sufficiently to allow the water vapor to condense and form rain. Carbon dioxide had no ocean in which to dis-

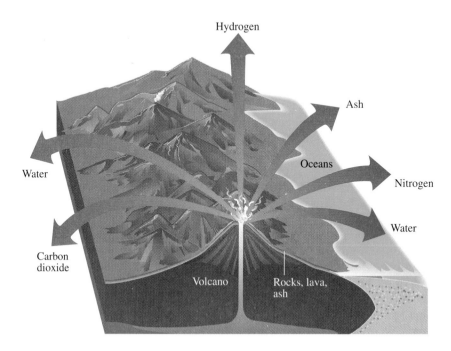

Figure 2-5 In volcanic eruptions, huge quantities of rocks, lava, ash, and gases (mainly water vapor, carbon dioxide, CO_2, and nitrogen, N_2) are ejected. Early in the earth's history, when volcanic eruptions were frequent and widespread, water vapor released to the atmosphere condensed and fell to earth as rain, and the oceans were formed; CO_2 and N_2 became the main constituents of the early atmosphere.

solve, and it remained in the atmosphere surrounding Venus. Today, the carbon dioxide concentration in Venus's atmosphere is 200 times as great as that in the earth's atmosphere. The carbon dioxide prevents heat from escaping into space, and the temperature on Venus's surface is known to be above 300°C (572°F). It is therefore not surprising that life never developed on Venus.

The earth's first atmosphere was quite different from the one that surrounds it today. Volcanic eruptions (Fig. 2-5) continued to occur long after the earth's surface had cooled to the point where water vapor condensed to form oceans. Evidence suggests that, in addition to water vapor and carbon dioxide, the enormous volumes of gases emitted would have included nitrogen, with smaller amounts of carbon monoxide, hydrogen, and hydrogen chloride—the same gases that are emitted by erupting volcanoes today. Hydrogen gas, being very light, would have been lost into space, but other gases would have been held near the surface by the earth's gravitational pull.

After millions of years of volcanic activity, the atmosphere would have been rich in nitrogen and carbon dioxide but completely devoid of oxygen. Today, our atmosphere is still rich in nitrogen (78 percent), but only 0.03 percent is carbon dioxide, and oxygen accounts for 21 percent. There were two main ways in which the excess carbon dioxide was removed. First, when rain began to fall on the earth, very large quantities of carbon dioxide dissolved in the oceans that were formed, and much of it combined with calcium in the water to form limestone (calcium carbonate). Then, some 3 billion years ago, the first primitive blue-green algae, or **cyanobacteria**, developed in shallow waters. Like the more advanced plants on earth today, these organisms used energy from the sun to convert carbon dioxide and water into simple carbohydrates (compounds of carbon, hydrogen, and oxygen that are the main food for plants) and oxygen by the process of **photosynthesis**. The oxygen escaped from the water, and for the first time, oxygen entered the atmosphere. Life had created the conditions for its own success.

As cyanobacteria multiplied, more and more carbon dioxide was removed from the atmosphere and replaced with oxygen. Eventually, when photosynthesis had been going on for millions of years, the earth's atmosphere attained its present composition. Although photosynthesis was important in reducing the carbon dioxide content of the atmosphere, dissolution in the oceans and the formation of limestone were the major factors in its removal.

In Shark Bay in Australia today, cyanobacteria are still at work producing oxygen and forming limestone. Round rocky structures called **stromatolites** that have been built up over thousands of years are present in the shallow waters (Fig. 2-6). Cyanobacteria on the surfaces of the stromatolites photosynthesize by day and at night secrete calcium carbonate that, together with sand stirred up by waves, collects on the sticky surfaces of the stromatolites, gradually increasing their size—the same process that began billions of years ago.

It is ironic that ever since the beginning of the Industrial Revolution in the latter part of the eighteenth century, we have been pouring unprecedented amounts of carbon dioxide back into the atmosphere by burning carbon-containing fuels such as coal and petroleum. There is now growing concern that the continuing increase in atmospheric carbon dioxide may lead to a rise in the earth's temperature, which could have serious consequences for our planet. This problem will be discussed in Chapter 13.

Small amounts of oxygen gas may have been formed by the breakdown of water vapor into its elements. If formed, the oxygen would have combined with other compounds and elements (for example, with iron to form iron oxides) and been removed from the atmosphere.

Combustion of coal provides the energy needed to convert water to steam. The steam is used to turn the turbines that generate electrical power.

Figure 2-6   Using the same process that began billions of years ago, cyanobacteria have built up rocky reefs called stromatolites in Shark Bay, Australia. The cyanobacteria produce oxygen in photosynthesis by day and calcium carbonate (limestone) by night. Fossilized stromatolites dating back more than 3 billion years have been discovered.

THE THEORY OF PLATE TECTONICS

In the previous sections we saw how, according to most earth scientists, the earth and its oceans and atmosphere were formed millions of years ago. We will now examine the origin of the geologic activity that continues to shape and change the earth's surface and is responsible for familiar features such as mountains and volcanoes.

Until the 1970s, geologists had no completely satisfactory explanation for many of the earth's internally generated geologic phenomena. Now, with the general acceptance of the theory of **plate tectonics**, geologists have a single unifying theory that can explain the relationships of the continents to each other, the development of mountains, the distribution of rocks, and the continued occurrence of volcanic eruptions and earthquakes in many parts of the world today.

Tectonics is derived from *tektonikos*, the Greek word for "builder."

According to this theory, very large segments of the rigid lithosphere, called plates, drift extremely slowly over the weak pliable asthenosphere. As they drift, the plates grind against each other or move apart, rather like ice floes moving on the ocean. These movements have been occurring for millions of years and continue today. The boundaries of the plates, and the directions in which they move, are shown in Figure 2-7. The boundaries are sites of great geologic activity: Mountain ranges are formed, volcanoes erupt, and earthquakes occur almost exclusively at plate boundaries.

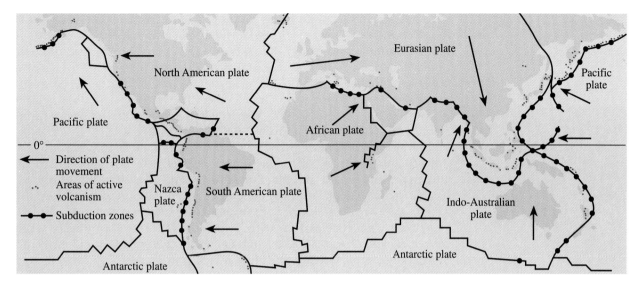

Figure 2-7 The earth's lithosphere is divided into rigid segments, called plates, which drift extremely slowly over the athenosphere. The boundaries of the plates are regions of great geologic activity.

Formation of the Continents

According to the theory of plate tectonics, the continents as they exist today were formed from a single supercontinent called **Pangea** (which means "all lands" in Greek). The name was given to the supercontinent by the German geophysicist Alfred Wegener (1880–1930) who was the first to put forward the idea of continental drift. Pangea began to split up about 200 million years ago, and the land masses that were formed drifted apart, as shown in Figure 2-8. North America broke away from Eurasia, South America and Africa separated, Australia separated from Antarctica, and India "bumped" into the Eurasian land mass and, in so doing, pushed up the Himalayan mountains. After about 135 million years of drifting, the continents reached the positions they occupy today. Strong evidence supporting the idea of continental drift is that the edges of today's continents fit together remarkably well, like the pieces of a jigsaw puzzle. Figure 2-9 shows a computer fit of the continents that border the Atlantic Ocean. The good fit between South America and Africa strongly suggests that these two continents were once joined together.

Other evidence for continental drift is the finding that rock cores taken from southwest Africa and southeast Brazil are very similar in composition.

Activity at Plate Boundaries

Today, the earth is still a very restless planet because plates continue to drift. At their boundaries, one of three geologic phenomena occurs: The plates either diverge, converge, or move in directions parallel to one another.

At **divergent boundaries**, adjacent plates pull apart, as indicated in Figure 2-10. This divergence occurs in the Atlantic Ocean between the North and South American plates on one side and the Eurasian and African plates on the other (Fig. 2-7). As a result of this spreading apart of the sea floor, molten rock, termed **magma**, rises up from below the lithosphere to fill the gap between the receding plates. New sea floor is continually, but extremely slowly, being formed and added to the Mid-Atlantic Ridge (Fig. 2-10).

At **convergent boundaries**, plates grind together, and one plate usually buckles and slides downward beneath the other plate (Fig. 2-11) at what is termed a **subduction zone**. Here lithosphere is consumed instead of being produced as it is at divergent boundaries. The overriding plate, which also buckles, is uplifted, leading to the formation of mountain ranges. The Andes mountains, for example, were formed where the Nazca and South American plates ground together (Fig. 2-7). Convergent boundaries are sites of frequent severe earthquakes and explosive volcanic eruptions (Fig. 2-7).

Plates may also move parallel to each other, as they do along the San Andreas fault in California. There, the North American plate moves southward relative to the Pacific plate, creating a **transform fault** (Fig. 2-12). It has been calculated that 25 million years ago the part of the Pacific plate that included the San Francisco area was located at the latitude where Los Angeles is now situated. If the plates slide past each other smoothly, little or no disturbance at the surface results. However, plates often become locked together and unable to move; pressure then builds up until the lock is eventually broken. When the lock is broken, the sudden slippage triggers earthquakes that may be very violent, as was the case in the 1906 quake in San Francisco.

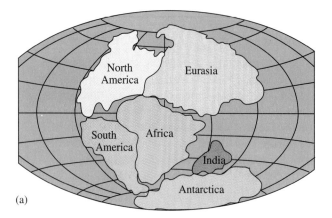

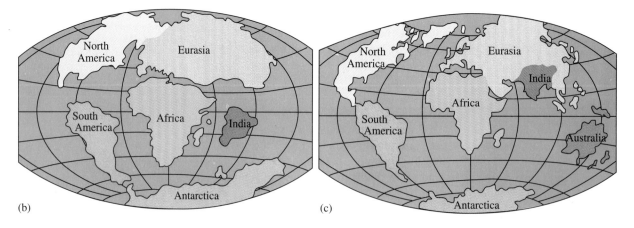

Figure 2-8 Formation of the continents as the supercontinent Pangea gradually broke up and the pieces drifted apart. (a) Approximately 200 million years ago. (b) Approximately 65 million years ago. (c) Today.

Figure 2-9 A computer fit of the continents bordering the Atlantic Ocean.

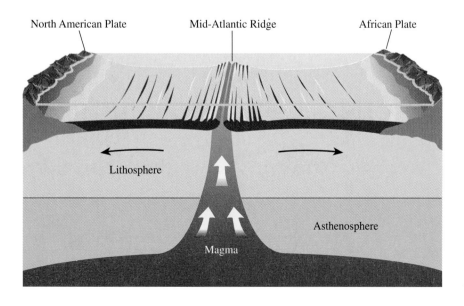

Figure 2-10 A divergent plate boundary. As adjacent plates move apart, magma rises up from the mantle to fill the space between the retreating plates. The magma solidifies to form new ocean crust. In this way, new sea floor is continually being added to the Mid-Atlantic Ridge.

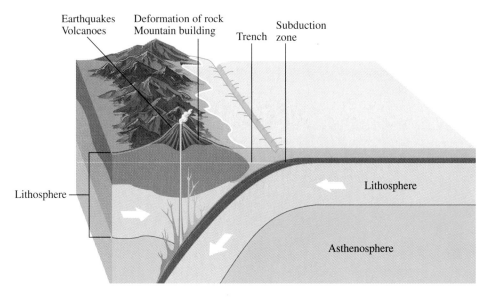

Figure 2-11   A convergent plate boundary. As plates collide, one plate slides under the other, and crust disappears into the mantle. The overriding plate is deformed and uplifted into mountain ranges. Earthquakes and volcanoes occur frequently at convergent boundaries.

The Earth's Internal Heat Engine

Convection currents generated in the earth's interior are believed to be the driving force that keeps the plates in motion (Fig. 2-13). Convection currents can be observed when soup is heated in a saucepan. Heat below the pan heats the bottom layer of soup, causing it to expand, become less dense (lighter) than the cooler soup around it, and rise to the surface, where it moves laterally. As the soup rises away from the heat source it cools slightly, becomes denser (heavier), and sinks. An essentially circular pattern of heat transfer is set in motion.

Within the earth, the continuing decay of radioactive elements provides the source of heat that sets up convection currents in the mantle. It is thought that the hot viscous material of the asthenosphere is conveyed upward to the

Figure 2-12   A transform fault. When plates move parallel to each other in opposite directions, a transform fault develops. If the slippage is sudden and jerky, earthquakes occur along the fault, and the opposing land masses are offset as indicated in the right-hand figure.

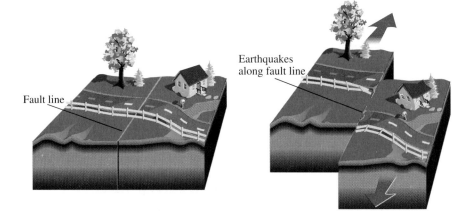

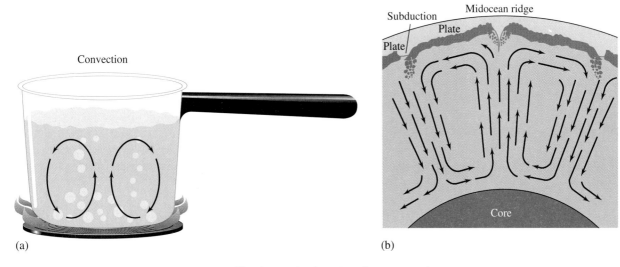

(a) (b)

Figure 2-13   Convection currents generated by decay of radioactive elements are the driving force in continental drift. (a) Convection currents are set up in soup heated in a pan. (b) In the earth, molten material carried upward by convection currents reaches the surface at midocean ridges and is then carried laterally to convergence zones, where it moves downward.

underside of the lithosphere, where it breaks through the surface at zones of divergence. Lateral movements carry the plates along, and downward flow occurs at zones of convergence.

THE ORIGIN OF LIFE ON EARTH

Exactly how life began on earth is probably something that we shall never fully understand. Today, most scientists believe that carbon dioxide (CO_2), nitrogen (N_2), and smaller amounts of other gases emitted with water vapor (H_2O) during the numerous volcanic eruptions that marked the earth's early development would have dissolved to some extent in the water vapor as it condensed and fell as rain. Intense ultraviolet radiation from the sun, together with lightning, which is always associated with volcanic eruptions, could then have provided the energy needed to cause the dissolved chemicals to combine to form simple amino acids (Chapter 10), compounds that contain carbon (C), oxygen (O), nitrogen (N), and hydrogen (H). Amino acids, which are basic building blocks of living tissues, might then have joined together to form simple proteins, and further reactions in the chemical "soup" of the early oceans could have produced other compounds essential for life.

In the 1950s, the American chemist Stanley Miller demonstrated that if a mixture of water, ammonia, hydrogen, and methane (CH_4) was subjected to electrical sparks, amino acids were formed.

It has been suggested that the essential compounds produced in the early oceans gradually clumped together to form units. Membranes formed around the units, separating them from the surrounding environment, and the primitive cells so formed gradually acquired the characteristics of living cells.

The earliest cells developed in an environment devoid of oxygen. Descendants of these primitive cells survive today in the oceans near regions where new sea floor is being formed. At such locations, the interaction of water with hot upwelling magma produces jets of dark water called **black smokers**. Numerous sulfide minerals, dissolved in the acidic superheated water, give the jets their dark color. In this oxygen-free, seemingly hostile environment, very primitive, nonnucleated bacteria flourish. Anaerobic (requiring no oxygen) bacteria also survive on land around hot springs and mud volcanoes at geothermally active places such as Yellowstone National Park, where conditions are similar to those on the early earth—hot, acidic, and lacking oxygen.

The development of the oxygen-producing cyanobacteria signaled the end for anaerobic bacteria. Oxygen was lethal to them, and except in a few specialized locations, they gradually died out. In Australia, fossils of primitive cells have been found in rock that is 3.5 billion years old. These cells are fairly complex, indicating that life on earth must have originated even earlier, possibly 4 billion years ago.

 ## THE UNIQUENESS OF THE EARTH

The earth is unique. It is the only planet in our solar system that developed an environment capable of supporting life. The position of the earth relative to the sun made possible the formation of the atmosphere and the oceans, which together maintain the heat on the earth's surface within a very narrow range—a range that extends approximately from the freezing point of water (0°C, 32°F) to the boiling point of water (100°C, 212°F). If the earth had formed a little closer to the sun, it would have been too hot to support life as we know it; if it had formed a little farther away, it would have been too cold.

The size of our planet is another important factor. If the earth were much smaller, the pull of gravity would be too weak to hold the atmosphere around the earth. Without an atmosphere, we would be exposed to life-destroying amounts of ultraviolet radiation from the sun. If the earth were much larger, the atmosphere would be thicker and would contain more gases, many of them poisonous.

Energy from the sun and the earth's internal heat engine were the two driving forces responsible for the development of our planet. These forces are still at work today and keep the earth in a state of dynamic equilibrium. Volcanoes and earthquakes continue to disturb the earth, and the flow of energy from the sun sustains all living organisms.

 ## ROCKS AND MINERALS

The rocks that make up the hard surface of the earth, the lithosphere, are composed of one or more substances called **minerals**. A mineral is a naturally occurring, usually crystalline, chemical substance that has a definite composition or restricted range of composition; it may be either an element or a compound.

Rocks and minerals form the **inorganic** part of the earth's crust; materials derived from the decayed remains of plants and animals make up the **organic** part. The terms "inorganic" and "organic" were introduced in the eighteenth century to distinguish between compounds derived from nonliving matter and compounds derived from plant and animal sources. At that time, chemists believed that the complex compounds that make up living matter—such as carbohydrates and proteins—could be produced only by living organisms. Once it was discovered that these compounds, which all contain carbon, could be synthesized in the laboratory, the definition of organic compounds was broadened to include all nonmineral compounds of carbon.

Some minerals have been known and used since ancient times. There is evidence that very early in human history, flint and obsidian (a volcanic glass) were shaped to make weapons and primitive knives, and clay was formed into pottery vessels and bricks. Gold, silver, copper, and brightly colored minerals such as jade and amethyst were fashioned into jewelry and other objects, and pigments were made from red and black iron oxides (oxides are compounds of oxygen and another element, in this case iron) in very ancient times.

Over 2500 distinct minerals have now been identified, but only a few of them are distributed widely over the earth's surface. Many of the more valuable minerals are found in only a few limited regions of the world, where they became concentrated as a result of tectonic activity (upheaval and subsidence of crust materials) and other rock-forming processes that have gone on for millions of years. The study of the composition of rocks has been an important factor in helping to explain how the earth was formed.

THE ROCK CYCLE

Although there are thousands of different types of rock, they can all be divided into one of three main classes—igneous, sedimentary, and metamorphic—according to the way they are formed. This system makes it possible to classify rocks of all ages found anywhere in the world.

Igneous Rocks

Igneous rocks are those that solidified, often in crystalline form, from hot molten magma originating deep below the earth's surface. In the earth's early development, when volcanic eruptions were numerous, a great deal of magma was extruded and spread as **lava** over much of the earth's surface (Fig. 2-14). As the lava cooled, a whole series of minerals crystallized out, forming **extrusive** rock.

Igneous is derived from *ignis,* the Latin word for "fire."

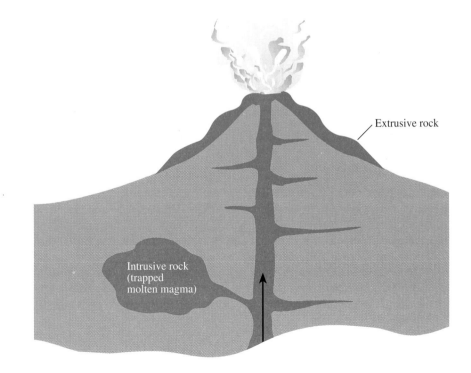

Extrusive rock

Intrusive rock (trapped molten magma)

Figure 2-14   Igneous rock, which is formed when magma cools, is of two types. Rock formed when lava (liquid magma) is ejected from volcanoes is called extrusive rock. Rock formed when magma below the earth's surface cools is called intrusive rock.

Figure 2-15 This granite rock was formed when molten magma beneath the earth's surface solidified. It became exposed as a result of erosion.

Often, molten magma became trapped in underground chambers (Fig. 2-14), where it cooled much more slowly than did magma extruded at the surface, and under these conditions, different minerals, including granite, were formed (Fig. 2-15). Rock formed in this way is called **intrusive** rock.

The majority of minerals that formed from magma were **silicates**, compounds that contain the two most abundant elements on the earth's crust: oxygen (O) and silicon (Si). In addition to oxygen and silicon, silicates also may contain other elements. Silicates formed at higher temperatures are rich in iron (Fe), magnesium (Mg), and calcium (Ca), while those formed at lower temperatures are rich in sodium (Na) and potassium (K). Nearly all of them contain aluminum (Al). We will consider silicates in more detail later in the chapter.

Igneous rock is still being formed. As we learned earlier in this chapter, magma comes to the surface where plate junctions diverge and where volcanoes erupt. Since its most recent eruption started in 1985, Kilauea volcano in Hawaii has added many tons of new igneous rock to the island's surface.

Sedimentary Rocks

Sedimentary rocks cover approximately 75 percent of the earth's surface. They form a thin layer over the more plentiful igneous and metamorphic rocks. Sedimentary rocks are of two types: detrital and chemical. **Detrital** sedimentary rocks are formed from particles eroded from other types of rocks. By the action of rain, freezing and thawing of ice, heat from the sun, ocean waves, and other natural means, igneous and metamorphic rocks (see below) are gradually worn down, broken into fragments, and often changed chemically. (This breakdown, known as weathering, will be discussed in more detail later in the chapter.) Sediments (small rock fragments in the form of sand and clay) are transported by rivers and wind to the oceans, where they are deposited in layers on the ocean floor. The layers gradually build up and are compressed and cemented into new rock. Common examples of detrital sedimentary rocks are **sandstone** (Fig. 2-16a) and **shale**.

Chemical sedimentary rocks are formed as a result of reactions between chemicals that precipitate out of ocean and lake waters and chemicals derived from the skeletal remains of coral and lime-secreting marine creatures. Important examples of chemical sedimentary rocks are **limestone** (calcium carbonate, $CaCO_3$) (Fig. 2-16b) and **dolomite** (calcium-magnesium carbonate, $CaMg(CO_3)_2$).

Metamorphic Rocks

Metamorphic rocks are formed by the action of temperature or pressure or both on either igneous or sedimentary rocks; the igneous and sedimentary rocks are thereby changed, or metamorphosed, into new minerals. Common metamorphic rocks are **marble** and **slate**: Marble is a metamorphosed form of limestone, and slate is a metamorphosed form of shale.

Igneous and metamorphic rocks form the foundations of the continents and ocean basins, but the sedimentary rocks that form the outer layer of much of the earth's crust are the ones that most directly influence our environment.

(a)   (b)

Figure 2-16 Sedimentary rock. (a) The sandstone cliff is an example of detrital sedimentary rock. It is formed by the erosion of igneous and metamorphic rock and is rich in silica. (b) Limestone, which is composed primarily of $CaCO_3$, is an example of chemical sedimentary rock. This cliff was formed from chemicals precipitated out of ocean water and the skeletons of marine creatures.

ROCKS AS NATURAL RESOURCES

Rocks and minerals are natural **resources**. A natural resource can be defined as anything taken from the physical environment to meet the needs of society. It may be a **renewable** or a **nonrenewable resource**. Resources such as soil, natural vegetation, fresh water, and wildlife are all renewable; if not depleted too rapidly, they are replaced in natural recycling processes. Rocks and minerals—and also oil, natural gas, and coal—are nonrenewable. They are present in the earth in fixed amounts and, once used up, are not replaced—at least not in a time frame that is relevant for the foreseeable future.

Rocks and minerals are quarried and widely used, often in modified form, in the construction and chemical industries and for making ceramics and many other products. In the next section, we will consider the chemical makeup, and some of the uses, of the most widespread of all minerals, the silicates.

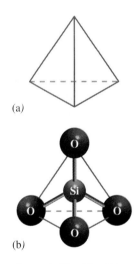

(a)

(b)

Figure 2-17 The SiO_4 tetrahedron present in silicate minerals. (a) A tetrahedron. (b) The tetrahedral arrangement of the SiO_4 unit in silicates. The silicon atom at the center is joined to four oxygen atoms located at the corners of the tetrahedron.

SILICATES

The basic unit of all silicates is the SiO_4 tetrahedron in which a silicon atom at the center is bonded (joined) to four oxygen atoms located at the corners (Fig. 2-17). The radius of the silicon atom is less than that of the oxygen atom and it fits comfortably between the four oxygen atoms. Other properties related to atomic structure, which will be discussed in Chapter 5, contribute to making this a very stable arrangement.

Bottled Mineral Water: Is it Just Water?

Bottled mineral water has become the beverage of the '90s. In the late 1970s Americans became increasingly aware of fitness and health. Baby-boomers were coming of age, and they wanted to hold onto their youth. Lighter fare in food and beverages was part of the solution. Nothing could be lighter than water. But tap water had become a hot political issue. The long-accepted belief that water that was piped into American homes was safe to drink was challenged by the Environmental Protection Agency, consumer groups, and millions of dissatisfied tap-water users. Tap water often tasted and smelled bad, and people were getting sick because of tap-water contamination.

Bottled-water sales continued to grow in the 1980s, sometimes at 15 percent per year. The sale of other beverages, especially alcoholic beverages, declined. Connoisseurs emerged who were able to distinguish the flavor and aroma of their favorite mineral water.

By the time water gushes from the ground in a mineral spring, it has traveled a considerable distance. The process begins when ordinary rainwater falls on porous, sandy soil. As it runs off the land and percolates through the soil, it dissolves soluble species, including minerals and organic substances from decaying plant and animal materials. Water with 500 parts per million (ppm) or more of dissolved solids are considered mineral waters; those with less are drinking water. Mineral water always contains more dissolved substances than rainwater. The dissolved ions in fresh water are derived from the weathering of rocks and soil. The concentration of these ions depends primarily on the type of bedrock with which the water has made contact. For example, if the bedrock is limestone (which is 31 percent calcium) it will yield more Ca^{2+} than sandstone (1–2 percent calcium) and more Mg^{2+} if the bedrock is dolomite, $CaMg(CO_3)_2$.

Mineral springs in regions rich in limestone or other carbonate rocks produce carbonated mineral water, which contains dissolved carbon dioxide. Water companies that bottle this water first remove the natural carbonation by running the water down a washboard-like chute and then collect the released carbon dioxide. As it is being bottled, the carbon dioxide is reinjected into the bottle and it dissolves in the water, producing a pressure in the bottle of three atmospheres. The removal and repressurization are done to assure that every bottle is charged with the same pressure. The natural pressure of the mineral spring changes with the season because the solubility of carbon dioxide in water varies as a function of temperature.

When dissolved in water, carbon dioxide forms carbonic acid, which is believed to increase the rate at which fluid is absorbed by the stomach. Carbonated beverages are known for relieving indigestion, and carbonation gives bottled water a pleasant sharp taste.

In 1960 the *Journal of the American Medical Society* reported that Dr. Henry Schroeder had found that death rates were lower in areas with hard water (water with high concentrations of Ca and Mg) than they were in areas with soft-water (low concentrations of Ca and Mg). It has since been shown that deficiencies in magnesium are capable of producing heart disturbances that range from premature beats to deadly ventricular fibrillation. Patients suffering from abnormal heart rhythms show marked improvement when given magnesium supplements. In 1975, Toronto researchers discovered a marked increase in the magnesium concentration of heart muscle of people living in hard-water areas.

Although all mineral water is considered to be hard water, the concentration of dissolved minerals varies widely from brand to brand. Shown on page 47 are the labels and composition of three popular brands of mineral water from three different parts of the world: Evian from France, Poland Spring from Maine, and San Pellegrino from Italy. As you compare the labels you will be able to tell which are springs that have limestone bedrock (high Ca^{2+}) and which have dolomite (more Mg^{2+}). The pH of the water tells if it is somewhat alkaline (pH>7) or acidic (pH<7). The carbon dioxide concentration in carbonated water is expressed as "bicarbonates," and you can also determine for yourself which are naturally carbonated.

When we purchase mineral water, we usually base our choice on flavor. Consider also the minerals dissolved in the water, its pH, and natural carbonation. Be sure to read the label.

Questions

1. You have indigestion. Would the three brands of mineral water shown here be good choices to provide relief?

2. The Nature Conservancy, a private nonprofit organization, raises funds to buy wetlands and land with natural springs to protect them from development. Would you support such an organization?

Reference: A. vonWiesenberger. *H₂O: The Guide to Quality Bottled Water* (Santa Barbara, CA: Woodbridge Press, 1988).

Evian Water Analysis

	Parts per Million
Calcium (Ca)	78
Magnesium (Mg)	23
Sodium (Na)	5.5
Potassium (K)	0.75
Chloride (Cl)	2.2
Bicarbonates (HCO_3)	357
Sulfates (SO_4)	10
Nitrates (NO_3)	3.8
pH (hydrogen ion concentration)	7.18

Poland Water Analysis

	Parts per Million
Calcium (Ca)	18.5
Magnesium (Mg)	7.2
Sodium (Na)	4.75
Potassium (K)	0.65
Chloride (Cl)	2.8
Iron (Fe)	<0.05
Bicarbonates (HCO_3)	78.0
Sulphates (SO_4)	7
Nitrates (NO_3)	13.6
Silica (SiO_2)	20.3
Total hardness, as $CaCO_3$	57.8
Total dissolved solids	124
pH (hydrogen ion concentration)	7.27

San Pellegrino Water Analysis

	Parts per Million
Calcium (Ca)	203.2
Magnesium (Mg)	59.4
Sodium (Na)	44.2
Potassium (K)	4.1
Chloride (Cl)	67.26
Fluoride (F)	0.58
Lithium (Li)	0.2
Strontium (Sr)	3.55
Bicarbonates (HCO_3)	225.65
Nitrates (NO_3)	.75
Sulphates (SO_4)	560
Free silicic residue (SiO_2)	11.9
pH (hydrogen ion concentration)	7.25

Because each oxygen atom can bond to two silicon atoms, adjacent SiO_4 units can join together by sharing oxygen atoms. In this way, long chains, sheets, or complex three-dimensional networks can be formed. Silicates are usually classified on the basis of the structural arrangement of their SiO_4 units and the major elements, in addition to silicon and oxygen, that are present.

Figure 2-18   Quartz crystals.

Three-dimensional Networks

The simplest formula for quartz is SiO_2. Although quartz is made from SiO_4 units, the ratio of Si to O atoms in its structure is 1:2 not 1:4 because all the O atoms are shared by adjacent tetrahedra.

The mineral **quartz** (Fig. 2-18) is composed entirely of SiO_4 units joined together to form a three-dimensional array (Fig. 2-19). **Feldspars**, which are the most common minerals in the earth's crust, are also based on a three-dimensional silicon-oxygen structure, but, unlike quartz, other elements besides silicon and oxygen are included in their chemical makeup. All feldspars contain aluminum (Al) associated with either sodium (Na), potassium (K), or calcium (Ca). Formulas for typical feldspars are shown below to give some idea of their variety and complexity. (It is not suggested that these formulas be memorized; they can always be looked up in tables.)

$$NaAlSi_3O_8 \qquad KAlSi_3O_8 \qquad CaAl_2Si_2O_8$$

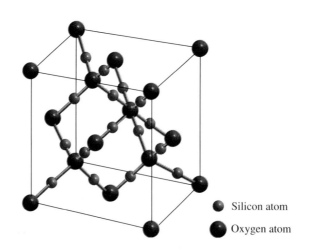

Silicon atom

Oxygen atom

Figure 2-19   In quartz, silicon and oxygen atoms join to form three-dimensional networks.

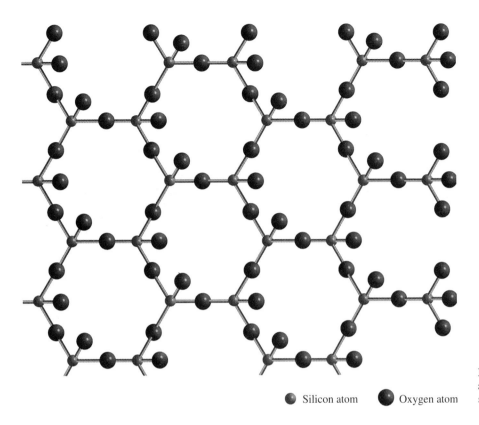

Silicon atom ● Oxygen atom

Figure 2-20 In micas, silicon and oxygen atoms join to form sheets.

Sheets

In **micas**, the SiO_4 units combine to form sheetlike arrays in which each tetrahedron is joined to three others (Fig. 2-20). Because of the planar arrangement of the SiO_4 tetrahedra, micas are easily cleaved into thin sheets as shown in Figure 2-21. Like the feldspars, micas are associated with a number of other elements besides silicon and oxygen, including potassium, aluminum, and magnesium (Mg). A common mica is muscovite, $KAl_3Si_3O_{10}(OH)_2$, a pearly white mineral that was used in medieval Europe to make window panes before glass became readily available. Other micas include talc, $Mg_3(Si_2O_5)_2(OH)_2$, a soft mineral used to make talcum powder, and kaolinite, $Al_2Si_2O_5(OH)_4$, a clay mineral.

Long Chains

Asbestos is a general term for a number of magnesium and calcium silicates formed from double chains of SiO_4 tetrahedra (Fig. 2-22). Chrysotile, the most abundant type, forms as curly fibers. Amphibole types of asbestos crystallize as sharp needles.

Asbestos is a very versatile material. It is strong, flexible, resistant to corrosion, and an excellent thermal insulator. Until the 1970s, when it became evident that it posed a serious health hazard, asbestos was widely used to insulate steam pipes and other heating units and to make protective clothing for firefighters, welders, and other people who are exposed to high temperatures. It was also used in brake linings, roofing materials, hair dryers, and other products.

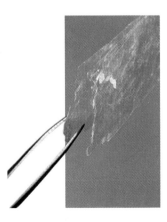

Figure 2-21 A specimen of mica showing how it can be separated into thin sheets.

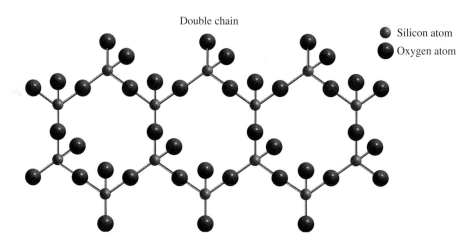

Double chain

● Silicon atom
● Oxygen atom

Figure 2-22   In asbestos, silicon and oxygen atoms join to form long chains. Two parallel rows of linked SiO$_4$ tetrahedra join through oxygen atoms to form double chains.

It is well established that workers exposed to large amounts of asbestos over a considerable period of time develop serious lung disorders, including lung cancer. Thousands of shipyard workers contracted crippling, often fatal, lung diseases as a result of exposure to asbestos in the 1940s. The risk of cancer for asbestos workers who smoke is very much higher than for asbestos workers who do not smoke. This relationship between asbestos inhalation and smoking is an example of **synergism**—the action of two factors that together have a greater effect than the sum of the two individual effects.

Because of the health risks, in 1978 the use of asbestos for insulation and fireproofing was prohibited in the United States, and many communities began removing asbestos from public buildings, particularly from schools. Now, as a result of new studies, some scientists claim that—except for people who work with asbestos—the risks have been exaggerated. The studies show that fiber concentration in most buildings, including those containing asbestos materials, is no higher than is the concentration in outdoor air. Further, their studies show that chrysotile, which accounts for 95 percent of the asbestos used in the United States, is not nearly as dangerous as the amphibole type of asbestos. They claim that straight amphibole fibers, and not the curly chrysotile fibers (which because of their shape are more easily filtered out in the body's air passages), are the cause of mesothelioma, a rare form of lung cancer. Because the process of asbestos removal is extremely costly and likely to release fibers into the air and expose removal workers to unusually high fiber levels, some authorities are now recommending that chrysotile asbestos be left in place, encapsulated in a plastic coating. Other authorities still maintain that any exposure to asbestos is a health risk, and the issue remains a matter of debate.

MODIFIED SILICATES AND THEIR USES

Since well before the dawn of recorded history, men and women have used natural silicates in one form or another in their daily lives. First, silicates were used to fashion primitive weapons and cutting tools; later, as new skills were learned, they were modified and transformed into pottery utensils. Today, silicates are starting materials for a broad variety of products including bricks, china, glass, cement, and other materials that have a wide range of specialized properties.

Pottery

Pottery is made of clays, which are complex mixtures of aluminum silicates produced by the weathering of feldspars. When wet, clays are pliable and readily shaped into objects that can be hardened by firing in an oven at high temperature after they have dried. The first kind of pottery, **earthenware**, was made about 9000 years ago and is still widely used today, for example, to make bricks and flower pots (Fig. 2-23a). Earthenware is porous unless it is glazed, a process in which the clay object is covered with various chemicals that, when fired, form a thin, nonporous, glasslike layer over the object's surface.

Finer quality pottery is made by using purified clays and adding various substances to the clay before firing. **Stoneware**, a glassy and very hard material, was first produced in China about 3000 years ago; **china**, a much finer material, was developed about 1000 years ago (Fig. 2-23b).

Early technological advances were largely the result of trial and error, but, more recently, a better understanding of the structure and properties of clays and glazes has led to the development of a great variety of new ceramic materi-

Fine oriental china, of a quality far superior to anything being produced in Europe, was first brought to the West in the Middle Ages.

Figure 2-23 (a) Native market pottery from Mexico (b) Chinese ceramic, Ming Dynasty, To Houa, teapot.

Figure 2-24 A skilled person can shape molten glass into many forms.

als, including very light materials and materials with special magnetic and optical properties. For example, ceramics able to withstand extremely high temperatures were developed to be used on rocket engine nozzles and as heat shields on reentry vehicles for the manned space programs.

Glass

Glass, like pottery, was developed in ancient times. The earliest glass objects found so far date from about 5000 years ago and were found in Egypt. The first glass was probably made by heating a mixture of sand (SiO_2), limestone ($CaCO_3$), and sodium carbonate (Na_2CO_3). As this mixture melts, it forms a homogeneous liquid that is transformed into a hard transparent material when cooled.

Quartz, the main component of sand, is unusual in that when it is heated it gradually melts over a wide temperature range to form a viscous liquid that, while soft, can be easily molded or blown into different shapes (Fig. 2-24). When most crystalline substances are heated, they melt suddenly to form liquids as soon as a fixed temperature (the melting point) is reached; when cooled, these liquids reform their crystalline structures. When quartz melts, bonds between oxygen and silicon break, and if the viscous liquid formed is cooled rapidly, the silicon-oxygen bonds are not able to recombine into the three-dimensional structure shown in Figure 2-19. The addition of limestone and sodium carbonate to sand lowers the melting point from 1710°C (3110°F) to about 700°C (1292°F), a temperature that is easy to attain.

A variety of properties can be added to glass by altering the proportions of its three main ingredients and mixing in certain metal oxides. For example, if boron oxide (B_2O_3) is added, glass that can withstand heat without shattering is obtained. One familiar borosilicate glass is Pyrex. Glass containing lead oxide (PbO) has a high refractive index—that is, the glass refracts or "bends" light, making it sparkle brilliantly when cut and polished (Fig. 2-25). Adding zinc oxide (ZnO) and selenium creates red glass; adding cobalt oxide (CoO) results in blue glass. Old glass is green because it contains small amounts of iron oxides as impurities.

Glass, which is a very durable material, makes up about 10 percent of municipal waste (Chapter 19), most of which ends up in landfills. Many communities concerned about the environmental impact of household trash have begun to recycle glass and other waste materials. Discarded glass can be crushed and made into new glass or building materials. Better still, glass containers can be returned for reuse.

Figure 2-25 Cut lead glass sparkles brilliantly because of the way it refracts light.

Cement

The first cements were made by the Greeks and Romans in classical times by mixing powdered limestone and volcanic ash. When mixed with water, the mixture reacted to form a hard material. Today, cement (often called Portland cement) is made by pulverizing limestone, clay (aluminum silicates), and iron oxide, and heating the mixture to about 1500°C (2732°F) in a rotary kiln. A small amount of gypsum (calcium sulfate) is added to the powdered product to control the time required for the cement to harden. When mixed with water, sand, and gravel, cement hardens within 24 hours into the synthetic rock **concrete**. Concrete is a very useful building material because, before it dries and hardens, it can be formed into any required shape.

Altering the composition and proportions of the starting materials produces cements with various properties, including color, waterproofing, rapid-hardening, and freeze-resistance. The chemistry of cement is complex and still not fully understood.

The manufacture and use of cement have deleterious impacts on the environment. Fine particles of the starting materials—which are readily dispersed by wind—and the gases and very fine particles released by burning fossil fuels to heat the kilns are all potential sources of air pollution. Ribbons of concrete highways that replace natural ground cover are one of the prices we pay for the convenience of automobile travel.

Portland cement got its name because the concrete made from it resembles a stone quarried on the Isle of Portland in England.

ORES AND METALS

Minerals in which a particular metallic element occurs in sufficiently high concentration to make mining and extracting it economically feasible are termed **ores**. Silicates, although very abundant, are seldom used as sources from which to extract metals because of the high costs and technological difficulties involved. Instead, most metals are extracted from sulfide, oxide, carbonate, chloride, and phosphate ores that are concentrated in a relatively small number of regions of the world. We will briefly consider the sources and uses of a few of the metals that are essential for our modern way of life.

Iron

Iron, the fourth most abundant element in the earth's crust (Fig. 2-4), is the metal used in greatest quantity by industrialized nations. The ores from which it is extracted usually contain a mixture of two iron oxides, hematite (Fe_2O_3) and magnetite (Fe_3O_4). Nearly all the iron extracted from the ores is used to manufacture **steel**, an alloy of iron with a small amount of carbon. The percentage of carbon determines the properties of the steel. Low-carbon steel (less than 0.25 percent carbon) is relatively soft and suitable for making cans and wire. High-carbon steel (up to 1.5 percent carbon) is very hard and strong and is used for making tools and surgical instruments. Steels with a variety of properties and uses are made by alloying iron with small amounts of other metals.

Although world reserves of iron are still large—and the United States has an abundant supply within its own borders—the demand for iron ore and steel continues to rise, and known sources of ore must eventually run out. Nodules on the sea floor, associated with black smokers, are a promising new source of iron that has not yet been tapped.

Large deposits of iron ores are located in Minnesota, Sweden, France, Venezuela, and the former Soviet Union.

An alloy is a mixture of a metal with another element.

Aluminum

Aluminum is the second most used metal in our society and the third most abundant element in the earth's crust (Fig. 2-4). Nearly all aluminum is a component of complex silicates, but, at present, there is no economically viable method for extracting it from them. Instead, the source of practically all aluminum is **bauxite**, an ore rich in aluminum oxide, that is found in quantity in only a few places in the world. Very large deposits of bauxite are found in Jamaica and Australia.

Aluminum is a light, strong metal used primarily for making beverage cans. In the building industry, it is used to make doors, windows, and siding; it is ideal for cooking utensils and many household appliances. Because it is a good conductor of electricity, aluminum is used extensively for high-voltage transmission lines, and, when alloyed with magnesium, it forms a light but strong material vital to the manufacture of airplane bodies.

Aluminum corrodes less easily than iron does; this feature is an important advantage for certain purposes, such as the construction of homes, but it also means that aluminum cans are very slow to degrade. Tin cans (which are made of steel and coated with tin) eventually rust away, but discarded aluminum cans remain unchanged in the environment for a very long time.

Aluminum oxides that include traces of certain metal impurities are valuable as gemstones. Rubies are crystalline aluminum oxide that is colored red with traces of chromium; the colors of sapphires—various shades of yellow, green, and blue—are due to traces of nickel, magnesium, cobalt, iron, or titanium.

Copper

Another valuable and extensively used metal is copper, which today is obtained from low-grade copper sulfide ores. Because the copper content of these ores is 1 percent or less, the cost of obtaining the pure metal is high. Valuable by-products of copper production are gold and silver, which are frequently present in very small quantities in the original ores. Increasingly, copper is being obtained by recycling copper-containing materials.

Copper is an excellent conductor of electricity and is used extensively for electrical wiring. It is also used for plumbing fixtures and as a constituent of alloys. **Brass** is an alloy of copper and zinc, and **bronze** is an alloy of copper and tin. Copper and copper alloys are also used for coinage.

Strategic Metals

Because metal ores are very unevenly distributed in the earth's crust, many countries must depend on imports for their supplies. Within the United States, for instance, there are no deposits of ores of many **strategic metals**, metals that are essential for industry and defense. Chromium (Cr), cobalt (Co), manganese (Mn), and the platinum (Pt) group of metals (platinum, palladium, rhodium, iridium, osmium, and ruthenium), which are needed for the manufacture of specialty steels, heat-resistant alloys, industrial catalysts, and parts for automobiles and aircraft, are found in significant quantities in only three countries: South Africa, Zaire, and the former Soviet Union.

MINERAL RESERVES

The world's human population continues to grow at an ever-increasing rate, and with it grows the demand not only for food but also for material goods. To meet these material needs, metals and minerals (nonrenewable resources) are being consumed at a tremendous rate, one that is bound to rise as the developing nations become increasingly industrialized.

At the present time, the industrialized nations consume a disproportionate amount of the earth's mineral reserves. The North American continent, for example, has less than 10 percent of the world's population, but it consumes almost 75 percent of the world's production of aluminum. The same disproportionate use is true for many other metals. At the present rate of consumption, supplies of many important metals are expected to be severely depleted by the first quarter of the twenty-first century.

It is possible that new mineral deposits will be discovered in the future, but since most of the earth's surface has already been thoroughly explored by geologists, it is unlikely that significant quantities of ores will be found. One source of minerals that has not yet been fully explored is the ocean floor. Areas where new sea floor is being formed have been shown to be rich sources of manganese and polymetallic sulfides. As deep-sea mining technology develops, sites on the sea floor may well become a much needed source of metals.

One of the most important ways to preserve our mineral supplies is to recover metals by recycling. More and more communities are collecting aluminum beverage cans (and also glass containers and paper) for recycling, which, in addition to conserving natural resources, saves energy. Approximately half as much energy is required to make new aluminum cans from old cans as is required to make them from bauxite.

WEATHERING OF ROCK AND THE FORMATION OF SOIL

We depend on rocks and minerals for many of our material needs, but for growing food we depend on the **soil**, the thin layer that covers the earth's rocky surface. Soil is a complex mixture of organic and inorganic materials. The organic part, which is derived from the decayed remains of plants and animals, is concentrated in the dark-colored uppermost layer of soil, the **topsoil**. The inorganic part, which is made up of rock fragments, was formed over thousands of years by the weathering of bedrock. **Weathering** is the result of a number of physical and chemical processes that occur simultaneously. (The composition of soil is covered in more detail in Chapter 18.)

Physical Weathering

Several physical processes contribute to the breakup of bedrock. When water gets trapped in cracks in rock and then expands on freezing, the rock is subjected to great pressure and it fractures. Similarly, roots of growing plants exert pressure on rocks and cause them to heave up and break apart. Temperature changes, particularly in areas such as deserts, where day and night temperatures vary widely, cause rock to expand and contract, and at lines of weakness, rocks crack and fragment. Rocks are also broken down by earthquakes, the action of wind and rain, the pounding of waves against shorelines, and the flow of water in rivers and streams. In these many different ways, rock is broken into boulders, rocks, pebbles, and finally sand and clay.

Chemical Weathering

Chemical weathering and physical weathering work together. Chemical weathering accelerates fragmentation, and physical fragmentation allows water and air to penetrate into rocks, where chemical reactions can then take place.

Chemical weathering changes the chemical composition of rock. It is brought about by the interaction of rock with air, water, and chemicals dissolved in water. For example, the mineral feldspar ($KAlSi_3O_8$) is very slowly converted to the clay kaolinite ($Al_2Si_2O_5(OH)_4$) by the action of water. In the process, potassium (K) is removed from the feldspar and dissolves in the water.

Pure rainwater is always slightly acidic because, as it falls, it dissolves some carbon dioxide (CO_2) from the atmosphere, and carbonic acid (H_2CO_3) is formed. Although carbonic acid is a very weak acid and is present in pure rainwater at very low concentration, it can react with certain types of rock to change their composition. For example, very dilute carbonic acid dissolves limestone and degrades clays into sand. In an unpolluted environment, these reactions occur extremely slowly, but in regions of the world where acid rain falls, they proceed at a greatly increased rate. Details of the chemical reactions involved, and the effects of acid rain on the environment, will be studied in Chapter 8.

Soil Erosion

In the natural weathering of crustal material, soil particles, primarily from topsoil, are dislodged and carried away by wind and flowing water. This process is termed **erosion**, and the extent to which it occurs in any region depends on wind strength and direction, the amount of precipitation (rain and snow), the type of rock and soil, the topography, and the amount of ground cover. Coarser particles such as sand settle out rapidly from water, but fine particles of soil remain suspended for much longer and are carried great distances in streams and rivers, often all the way to the sea.

Topsoil is a renewable resource if it is not removed more rapidly than it can be replaced. The roots of plants generally protect soil from undue erosion, but whenever ground cover is removed, erosion by both wind and rain occurs at an accelerated rate. Regeneration of topsoil is extremely slow—just a few inches every thousand years—and in many parts of the world, topsoil has become a nonrenewable resource. Soil scientists have estimated that already about one-third of the original topsoil on cropland in the United States has been blown away or washed into rivers and lakes. Human activities that lead to erosion and the loss of valuable and irreplaceable topsoil include construction, mining, clear-cutting of timber, agricultural development, and overgrazing. The impact of these activities on the environment will be discussed in more detail in Chapter 12.

EXPLORATIONS

The Eruption of Krakatoa

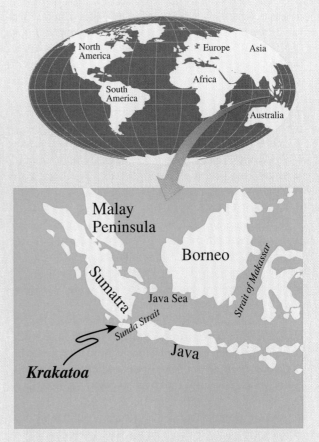

Figure 2-A The volcano Krakatoa is located between the islands of Java and Sumatra. When it erupted in 1883, volcanic ash fell over a large area in the Indian Ocean. Explosions were heard as far away as Australia and southern India.

On August 26, 1883, the small island of Krakatoa (Fig. 2-A), which lies in the Sunda Strait, 50 to 80 km (30 to 50 mi) from the coasts of Java and Sumatra, exploded in one of the most gigantic and catastrophic volcanic eruptions in recorded history. The explosion was heard over nearly 8 percent of the earth's surface, thousands of people died, and the aftereffects were apparent worldwide for many months.

The island of Krakatoa—8 km (5 mi) long by approximately 3 km (2 mi) wide—is situated in one of the most unstable regions of the world. Its volcano is just one of 500—over 100 of them still active—that lie in a 1600-km (1000-mi) arc across Java and Sumatra, the two main islands of modern Indonesia. The people on these densely populated islands have always lived with the menace of volcanoes. The dangers are great, but the volcanic ash makes the soil some of the richest in the world. Krakatoa was considered to be extinct, and the Dutch authorities who ruled Indonesia at the time of the eruption had no idea that the small island was a time bomb waiting to explode.

Why do some volcanoes erupt explosively while others, like the Hawaiian volcanoes, emit quietly flowing lava that does no harm except to objects in its path? It depends on the silica (SiO_2) and water content of the molten magma that is ejected. In an eruption, when molten magma rises toward the surface, the water in the magma turns to steam, expanding 1000 times as it does so. The force of the expanding gas propels the magma up through the volcano's chimney. Low-silica magma produces a liquid lava from which gases can escape easily. If its water content is low, the lava flows gently; if high, it surges up in lava fountains. High-silica magma forms a viscous lava from which gases escape with difficulty. The thick magma often solidifies and forms a plug in the chimney. As gases try to escape, pressure builds up until eventually the plug is blown out with explosive force, like the release of a

cork from a champagne bottle. The frothy, gas-saturated, high-silica magma is ejected as pumice, a light material quite unlike lava that is full of small gas-filled cavities and floats on water. If more had been understood about volcanoes in 1883, the emission of pumice in the early stages of the eruption would have been recognized as a clue that Krakatoa had the potential to be extremely dangerous.

The first signs that Krakatoa was awakening came on May 20 when rumblings and detonations were heard on shore and a dark column of ash appeared above the island. Pumice and ash fell on ships passing close by, and the forests of Krakatoa were seen to be on fire. Activity continued intermittently for 3 months, no one realizing that subterranean fires beneath Krakatoa were being stoked, and much more violent disturbances were to follow.

Then on Sunday, August 26, 1883, at one o'clock in the afternoon, Krakatoa erupted in one of the most awesome displays of nature's might ever witnessed. Ear-splitting blasts of increasing violence followed one another in quick succession, and an enormous black mushroom cloud billowed out above the Sunda Strait. In towns and villages along the coasts of Java and Sumatra, the inhabitants were stifled by sulfurous fumes and could barely see through the falling ash. To the accompaniment of strong winds and violent thunderstorms, tremendous roars continued through the night and into the next morning. The sea receded and then advanced. Boats at one moment were stranded on the beach, and minutes later were smashed into pieces by inrushing waves, which at times were 3 m (10 ft) high.

The captain of a British ship, the *Charles Bal,* that sailed within 16 km (10 mi) of Krakatoa during the night of August 26, told of blinding lightning and a hail of hot pumice stone. Holes were burned in the ship's sails and in the clothes of the sailors as they tried to clear pumice from the decks.

Shortly before 10 o'clock on the morning of August 27, after a terrifying night, it suddenly became ominously quiet, the explosions died down, and the frightened people on shore began to hope that the volcano was exhausted and that its eruption had ended. In fact, the worst was still to come. At exactly 10:02 a.m., almost the entire island of Krakatoa was obliterated in a titanic explosion greater than any that had gone before. An immense column of fiery rocks and ash rose into the atmosphere, and total darkness fell on Jakarta 160 km (100 mi) away. A gigantic sea wave was unleashed and sped outward toward the shores of Sumatra and Java. A great black wall of water, in places 40 m (130 ft) high, swept inland for distances up to 10 km (6 mi), destroying everything and everyone in its path. Not a house was left standing, and thousands of people were swept out to sea by the receding waters. For nearly two hours, a succession of mountainous waves followed the first one, and then, just before noon, Krakatoa roared for the last time. An area almost as large as the island of Manhattan had vanished, over 36,000 people were dead, and 165 towns and villages had been utterly destroyed. The aftereffects were felt around the world for months to come.

Soon after the eruption, the Dutch geologist R. D. M. Verbeek visited what remained of the island and established the sequence of events that had led to its final violent collapse. He was puzzled that, among the materials blasted from the volcano, he could find none of the rock that had once formed the cone—it was all either pumice or ash from the magma chamber. To explain this mystery, Verbeek concluded that 19 hours of continual eruption had emptied a huge shallow magma chamber that lay under the volcano. Once empty and with nothing left to support the rock walls, the volcanic cone and most of the surrounding island crashed with a deafening roar into the chasm below. The sea rushed in and then exploded violently outward in a tremendous column of steam and debris. The collapse set off the killer waves.

The sound from the most violent explosion was heard almost 5000 km (3000 mi) away—farther than any sound has ever been heard before or since. A pressure wave swept around the earth seven times, its passage recorded on instruments around the world. An increase in atmospheric pressure was first noted in New York 14 hours after the explosion. As immense sea waves sped outward, they raised sea levels many feet along coastlines worldwide. Six successive surges at 30-minute intervals were recorded as far away as the English Channel. A layer of pumice—which analysis showed to be very high in silica (72 percent)—spread into the Indian Ocean from the Sunda Strait, which for a time was impassable to shipping. Corpses and debris were found embedded in the pumice for many months. Ash from the cloud, which rose to a height estimated to have been 80 km (50 mi), fell on ships as far as 1600 km (1000 mi) away. Fine dust reached the stratosphere and, carried by the winds, circled the earth many times, creating spectacular sunsets and sunrises for several years.

A circular depression under the water, known as a caldera, was all that was left after most of Krakatoa vanished. But Krakatoa was not dead. Beginning with steam clouds in 1927, intermittent eruptions from within the caldera gradually built up a new island, named Anak Krakatoa (child of Krakatoa), which in 1952 emerged from the sea. A new cone has now appeared, and the island continues to grow. Perhaps several hundred years from now it will once again give a demonstration of the mighty forces that are at work beneath the surface of the earth.

Question

Discuss the probable results if one of the volcanos in the Cascade Range in the United States erupted with as much violence as Krakatoa in 1883.

References: R. Furneaux, *Krakatoa* (Englewood Cliffs, NJ: Prentice Hall, 1964); *Planet Earth: Volcano,* (Chicago, IL: Time/Life, 1982).

KEY WORDS AND CONCEPTS

asthenosphere	lava	renewable resource
Big Bang	lithosphere	sedimentary rock
black smokers	magma	silicates
convection currents	mantle	soil
convergent boundary	metamorphic rock	solar system
core	Milky Way	strategic metals
crust	minerals	stromatolites
cyanobacteria	nonrenewable resource	subduction zone
divergent boundary	ores	terrestrial planets
erosion	organic	topsoil
giant planets	Pangea	transform fault
igneous rock	photosynthesis	universe
inorganic	plate tectonics	weathering

QUESTIONS AND PROBLEMS

1. Draw a diagram of the solar system. Include the following:
   a. relative size of each planet   **b.** the distance between the planets   **c.** the size and position of the sun relative to the planets

2. List:
   a. the terrestrial planets   **b.** the giant planets
   c. the main differences between the two groups of planets

3. The sun makes up what percentage of the mass of the solar system?
   a. less than 1 percent   **b.** about 10 percent   **c.** nearly 30 percent   **d.** more than 99 percent

4. Name the three main layers (or zones) into which the earth can be divided.
   a. The center of the earth consists mostly of what element?   **b.** Make a diagram showing a cross section of the earth. Label the layers.

5. **a.** By mass, which are the four most abundant elements in the whole earth?   **b.** Explain why elements are not distributed uniformly throughout the earth.

6. How did water form on the earth's surface?

7. Why does not our sister planet, Venus, support life?

8. Describe the chemistry carried out by primitive cyanobacteria (blue-green algae). Name this chemical process.

9. The early earth is thought to have been devoid of oxygen. How did oxygen first appear in the earth's atmosphere?

10. Briefly describe the theory that accounts for the main geologic features of the earth.

11. Continental plates drift slowly. Define:
   a. divergent boundaries   **b.** convergent boundaries
   c. transform fault

12. What is magma? What is its source?

13. Describe briefly how the following geologic features were formed:
   a. Andes mountains   **b.** Mid-Atlantic Ridge   **c.** San Andreas fault

14. What provides the energy to move the huge continental plates?

15. What are anaerobic bacteria?

16. If you were an astronaut observing the many moons of Jupiter:
   a. What would you look for to determine if plate tectonics is an active process on their surfaces?
   b. What chemical compounds need to be present to support life as we know it?   **c.** What chemical compounds would suggest that some form of life once lived on the surface?

17. Is there any association between plate boundaries and volcanism?

18. Name the three main classes of rock. Describe briefly how each is formed.

19. Draw a diagram that explains the formation of sedimentary rock.

20. Give two examples of each of the following types of rocks:
   a. detrital sedimentary   **b.** chemical sedimentary
   c. metamorphic

21. What is the basic chemical unit that is present in all silicates? How are the atoms in this unit arranged? Illustrate your answer with a diagram.

22. The silicate units in the following minerals are arranged as either three-dimensional networks, two-dimensional sheets, or chains. For each mineral, indicate which arrangement is present:
   a. talc   **b.** chrysotile   **c.** quartz

23. There is great debate over the hazards posed by asbestos.
    a. What properties make asbestos a useful material? **b.** Give three examples of the industrial uses of asbestos.   **c.** "Friable" asbestos is defined as that which can be easily crumbled into a powder. Should we be more concerned about friable asbestos than asbestos that is encapsulated in a product (such as an asbestos floor tile)?

24. Describe the health hazards associated with asbestos for smokers and nonsmokers.

25. What is the starting material for making earthenware? How is the starting material converted to earthenware?

26. **a.** What are the three main ingredients that are heated together to make glass?   **b.** In what way is molten glass different from most other molten materials?   **c.** If you wanted to make blue glass, what would you add to your mixture?

27. Describe how you would make:
    a. cement   **b.** concrete

28. Name the mineral from which most of the world's aluminum is extracted. From what countries does the United States obtain its supply of aluminum?

29. If aluminum and steel cans are thrown into the city dump, which will degrade first? Briefly explain your answer.

30. Give three uses for the metal copper.

31. Give three uses for the metal aluminum.

32. What is the purpose of adding carbon to steel?

33. What is meant by the term strategic metal?

34. How do the following contribute to the physical weathering of rocks?
    a. water   **b.** temperature

35. Briefly describe the process of erosion. What factors are important in determining how rapidly erosion occurs?

36. Explain why you support or oppose the following:
    a. the passage of a national bottle and can deposit law
    b. recycling of metal, glass, and plastic containers
    c. tax incentives to mining companies for the extraction of minerals   **d.** curbside pickup of separated recyclable materials

37. The danger of asbestos in schools and office buildings is a matter of debate. Some argue that any exposure is to be carefully avoided, while others say that there is danger only from high concentrations of airborne asbestos fibers. What do you think should be done? Should asbestos already in school buildings be removed or coated over with a plastic sealant? Remember that removal will cause some of the fibers to become airborne during the removal process and that if the coating technique is used, the underlying asbestos may be disturbed during subsequent renovation.

38. Discuss each of the following statements.
    a. New discoveries of minerals will provide all we need in the future.   **b.** Our children will have sufficient mineral resources.   **c.** The ocean will supply all our mineral needs in the future.   **d.** Recycling scarce mineral resources will guarantee a continued ample supply of those minerals.

ECOSYSTEMS
The Flow of Energy and Materials through the Natural World

3

Energy and materials flow through all ecosystems. From the sun through plants, energy reaches these bull moose grazing in the wild in Wyoming.

In the previous chapter, we learned how the earth was formed and how it gradually evolved into the world we know today. We studied the rocks and minerals that make up the earth's solid, nonliving outer surface and examined the chemical structure of those materials and their importance in today's industrialized society. In this chapter, we will introduce the living creatures that inhabit our planet. We will see how all living organisms interact with their physical surroundings and with each other; how all these interactions are intertwined; and how the whole system is fueled by a continuing flow of energy throughout the environment.

Before starting this chapter, it is instructive to consider when in the development of life on earth the most advanced of living creatures—humans—appeared on its surface. If we present the history of life on earth, which began some 4 billion years ago, on a 1-year time scale as shown in Figure 3-1, we find that humans arrived on the last day of December. The first organic materials appeared on January 1, and complex cells, similar to the ones found in present-day plants and animals, had formed by September. During the latter part of December, the dinosaurs gave way to mammals, and at four o'clock in the afternoon of December 31, humans finally appeared. Since the start of the Industrial Revolution, which began 2 seconds before midnight on our time scale, we latecomers to the earth have already done more to change the environment than any other creatures since life began. We have a great responsibility to ensure a sustainable future for the earth.

Learning Goals:

In this chapter, you should gain an understanding of:

1. What is meant by an ecosystem.
2. The factors that determine the growth and survival of organisms in an ecosystem.
3. The producers and consumers of energy in the natural environment.
4. The flow of energy through ecosystems.
5. Food chains, food webs, and trophic levels.
6. The nutrient cycles that provide living organisms with the chemicals they need for life.

➡ THE ENVIRONMENT

Water, carbon dioxide, oxygen, and all minerals are abiotic factors.

When we speak of the **environment**, we are referring to all the factors, both living and nonliving, that in any way affect an organism during its lifetime. The living, or **biotic**, factors include plants, animals, fungi, and bacteria. The nonliving, or **abiotic**, factors include all the physical and chemical components such as temperature, rainfall, nutrient supplies, and sunlight.

➡ ECOSYSTEMS

For purposes of study and for the sake of simplicity, it is useful to subdivide the environment, which comprises the entire earth, into small functional units called **ecosystems**. An ecosystem consists of all the different organisms living within a finite geographic region, together with the nonliving surroundings. Interrelationships between the creatures and the surroundings are such that an ecosystem is usually self-contained and self-sustaining.

Natural geographic areas with more-or-less distinct boundaries, such as deserts, forests, lakes, and prairies, are examples of ecosystems; each has its own characteristic community of plants and animals. An ecosystem may be as small as a pond or a field (Fig. 3-2), or it may cover millions of acres, in which case it is termed a **biome** (Fig. 3-3). Major biomes in North America are shown in Figure 3-4. They include desert, grassland, deciduous forests, coniferous forests, and tundra.

A cornfield forms an ecosystem; in addition to corn, the field contains insects that eat part of the corn, birds and small animals that eat the insects, earthworms and bacteria that live in the soil, and weeds.

Areas that have been disturbed and modified by humans, such as agricultural land and cities, can also be considered to be ecosystems. Urban ecosystems, of course, unlike most natural ecosystems, are anything but self-sustaining as far as the human population is concerned. Humans depend on food being brought in and wastes being taken out.

Range of Tolerance and Limiting Factors

Each organism in an ecosystem, be it plant or animal, has a particular **range of tolerance** to variations in the abiotic factors in its environment. Plants, for example, tolerate a narrow range of air temperature (Fig. 3-5). If the upper or

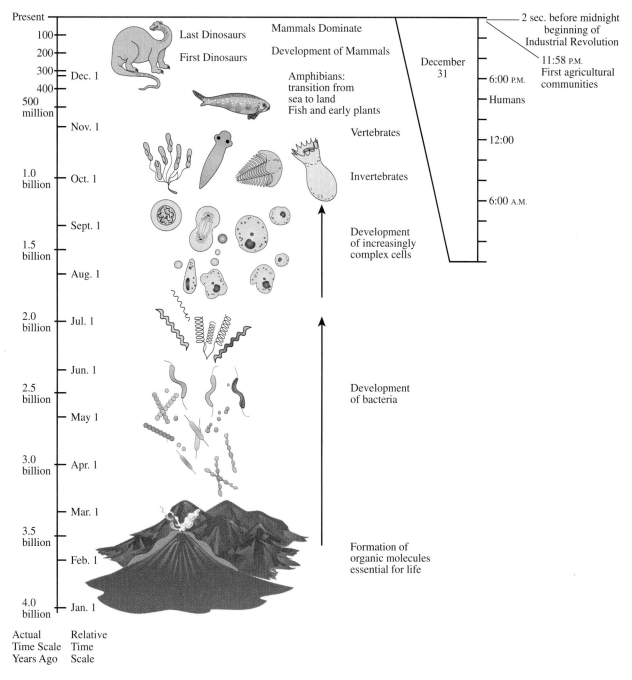

Figure 3-1   A representation of the 4-billion-year history of life on earth on a 1-year time scale. On this scale, almost 8 months is spent in the development of simple cells into cells similar to those present in existing plants and animals. The last 3 months sees the development of invertebrates, fish, amphibians, dinosaurs, and mammals. Human development occurs only during the last day of the year.

Figure 3-2   An ecosystem is a group of plants and animals interacting with one another and their surroundings. It may cover a small area, such as a pond.

For each physical factor in an environment, there exist minimum and maximum tolerance limits beyond which all members of a particular species will die.

lower limit of their range is exceeded, the plants will not survive. Between those limits, there is an optimum range where the plants thrive.

Although organisms are affected by all the abiotic factors in their environment, often one or two critical factors, called **limiting factors**, outweigh the others in determining growth and survival. For fish, oxygen is a limiting factor. Even when other factors such as temperature, availability of food, and the purity and flow of the water appear to be ideal, without an adequate supply of oxygen, fish will not survive.

Figure 3-3   An ecosystem that may cover millions of acres, like this forest in Ecuador, is called a biome.

1	Tundra
2	Boreal (Spruce-Fir) Forest
3	Mountain and Coniferous Forest
4	Coniferous Forest
5	Broadleaf Deciduous Forest
6	Prairie and Grasslands
7	Desert

Figure 3-4   The major biomes in Canada and the United States.

Some fish require more oxygen than others. Trout, for example, need a high level of dissolved oxygen. Because cold water dissolves more oxygen than warm water and because turbulent water is more readily reoxygenated with atmospheric oxygen than slow-moving water, trout flourish in high, fast-flowing, cold mountain streams but not in sunny, open, calm water at lower elevations. Carp and catfish, on the other hand, survive where dissolved oxygen levels are relatively low. These fish also have high tolerance limits for other factors, including increased water temperature, acidity, and toxic substances.

A limiting factor for plants is potassium. Despite having an adequate source of water and other nutrients, plants will not grow, or will not grow normally, if the supply of potassium is inadequate. An oversupply of a limiting factor can also be a problem. Plants can be killed by too much water and too much potassium as well as by too little.

One of the main aims of agriculture is to reduce the effects of limiting factors on crop yields.

Producers and Consumers

Ecosystems are sustained by the energy that flows through them. The biotic part of any ecosystem can be divided into producers of energy and consumers of energy. Green plants and blue-green algae are the **producers**; they are able

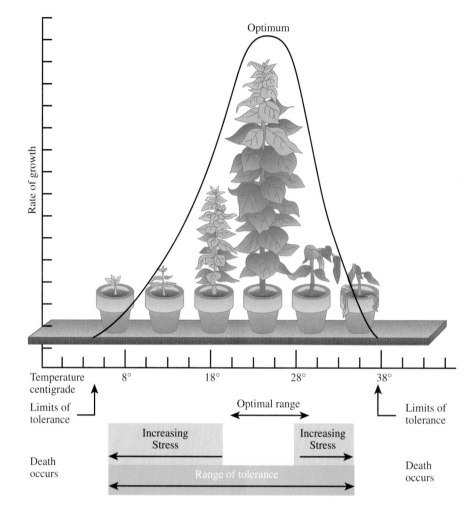

Figure 3-5 The range of tolerance of these plants to temperature lies between approximately 7°C (45°F) and 40°C (104°F). In the optimal range between 20°C (68°F) and 30°C (86°F), the plants thrive; growth is most abundant at about 25°C (77°F). At temperatures a few degrees below and above the optimum range, plants are under stress, and growth is limited.

to manufacture all their own food. By means of **photosynthesis**, they absorb light energy from the sun and use it to convert water and carbon dioxide (CO_2) from the air into the simple carbohydrate glucose ($C_6H_{12}O_6$). At the same time, oxygen gas is released to the atmosphere. By further reactions between glucose and chemicals obtained from water and soil, plants manufacture all the complex materials they need (Fig. 3-6). The plant world includes trees, flowers, grasses, mosses, and algae.

Consumers are unable to harness energy from the sun to manufacture their own food; they must consume plants or other creatures to obtain the nutrients and energy they need. Consumers can be divided into four main groups according to their food source: herbivores, carnivores, omnivores, and decomposers.

Herbivores feed directly on producers and are termed *primary consumers.* Deer, cows, mice, and grasshoppers are examples of herbivores. **Carnivores** eat other animals and are termed *secondary consumers.* Examples of carnivores are spiders, frogs, hawks, and all cats (including lions, tigers, and domestic cats). The animals that are eaten by carnivores may be herbivores, carni-

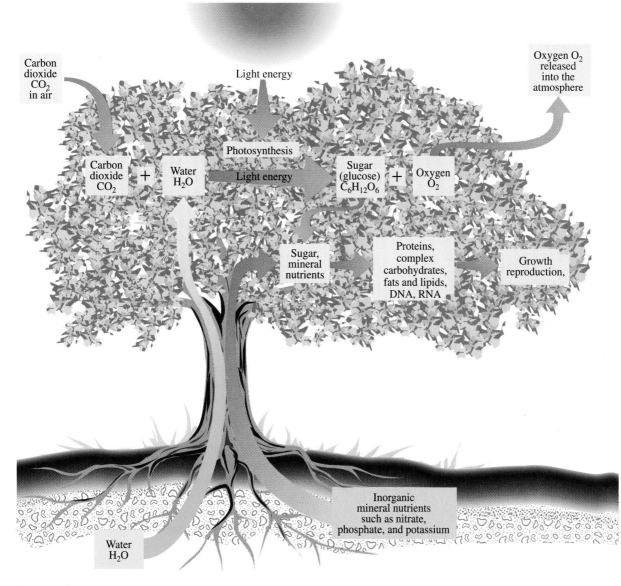

Figure 3-6 Green plants like this tree are producers. In the process of photosynthesis, they use light energy from the sun to convert carbon dioxide and water to glucose and oxygen. The oxygen is released to the atmosphere; the glucose, together with mineral nutrients from the soil, is used to produce the complex organic compounds that make up plant tissues.

vores, or omnivores. **Omnivores** are the creatures, including rats, raccoons, bears, and most humans, that feed on both plants and animals.

Decomposers feed on **detritus**, the freshly dead or partly decomposed remains of plants and animals, and include bacteria, fungi, earthworms, and many insects. Decomposers perform the very useful task of breaking down complex organic compounds in dead plants and animals into simpler chemicals and returning them to the soil for reuse by producers. In this way, many nutrients are endlessly recycled through an ecosystem.

Penicillin: The First Antibiotic

Antibiotics are substances that are produced by one microorganism and are toxic to other microorganisms. Penicillin was discovered by the bacteriologist Alexander Fleming in 1928. He was growing staphylococcus bacteria (which cause staph infections) in culture plates which accidentally became contaminated. Fleming noticed that the contaminant, a mold subsequently identified as *Penicillin notatum,* produced a substance that appeared to kill the staphylococcus. Fleming made crude extracts of the active substance, but Howard Florey and Ernst Boris Chain, working at Oxford University, purified and improved the method to isolate the valuable drug. Penicillin's use as a therapeutic treatment did not occur until fourteen years after its discovery, when the drug underwent its first important test on people.

On the night of November 28, 1942, a horrible fire occurred at the Coconut Grove, a famous Boston night club. The club was jammed with 800 to 1,000 partygoers, many of them college students. A young waiter with a lighted match accidentally ignited a fake palm tree, touching off one of the worst fires in American history. More than 450 people were burned to death or trampled as they attempted to escape through the club's only exit, a single revolving door.

Many of the injured were taken to Massachusetts General Hospital. Those who survived the first 24 hours received intravenous infusions of blood plasma, then a new procedure for combating dehydration. Several days after the fire, an 8 gallon sample of penicillin, a drug that until then had been reserved for military use, was rushed to the hospital. Dozens of surviving burn patients were given intravenous doses of penicillin which had a miraculous effect in combating or preventing staph infection. Usually when a protective layer of the skin is damaged, staph bacteria invade the body and multiply rapidly in the bloodstream, where they can cause systemic infection and death.

International publicity about the success of penicillin in treating the Coconut Grove victims spurred large-scale production of the drug. The wonders of penicillin were extolled in newsreels, newspapers and magazines. In 1945, Fleming, Florey, and Chain shared the Nobel prize in physiology and medicine for their work on penicillin. The publicity led to the widely held notion that penicillin's possibilities were limitless. Until the mid-1950s penicillin was available without a prescription and was used to treat colds and pneumonia, and to prevent infection.

Chemists working with penicillin soon recognized that it was not a single compound, but a group of compounds with related structures. The properties of the individual penicillins could be varied by altering the structure slightly. In this way, bacteria that became resistant to one penicillin could be killed by another.

Penicillin acts by preventing cell wall synthesis. The cell walls of bacteria are made up of mucoproteins, which are substances in which sugars are combined with protein molecules. Penicillin disrupts the cross-linking between these large molecules and prevents bacteria from forming cell walls. Human cells do not have mucoprotein walls; They have external membranes that are more sophisticated than the bacteria cell walls, and are not affected by penicillin. Penicillin can kill bacteria within a person without harming human cells.

Penicillin is truly a "wonder drug" that is arguably the most important therapeutic advance in the history of medicine. Prior to penicillin for instance, pneumonia was fatal in 85 percent of all cases. But there is evidence today that antibiotics, and especially penicillin, are being overused. For the past several years, Amoxicillin (a penicillin) has been one of the three most prescribed drugs in America. Ironically, it is the very success of these drugs that is proving to be their undoing.

It has been estimated that 50 percent of antibiotics are used unnecessarily or inappropriately.

Cattle are fed large doses of antibiotics to fatten them up for market. Some doctors prescribe antibiotics for short-lived sinus infections, colds, flu and other viral illnesses against which such drugs are useless. As a result many bacterial strains have developed a resistance to penicillin. Today, 95 percent of all staph infections worldwide are resistant to penicillin.

To stay ahead of resistant strains of bacteria, new antibiotics need to be constantly developed. But experts also agree that, for now, more judicious use of available antibiotics is also essential.

Questions
1. Do you think the Federal government should regulate the use of antibiotics?
2. What can be expected to happen to antibiotics that are administered to animals and enter the food chain?

Reference: Sandra G. Boodman, "Running Out of Wonder Drugs," Washington Post, March 16, 1993.

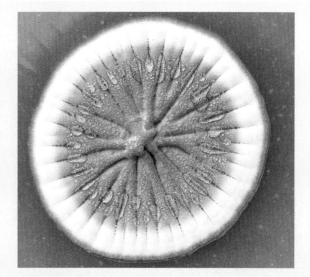

Penicillin mold.

THE FLOW OF ENERGY THROUGH ECOSYSTEMS

All the activities that go on in an ecosystem require **energy**. Without a constant flow of energy from producers to consumers, an ecosystem would not be able to maintain itself. Before we can understand these energy relationships we need to have some appreciation of what is meant by the term *energy*. We all know that it takes enormous amounts of energy to maintain our industrialized society. Energy is needed to run automobiles, heat and cool buildings, supply light, and grow food. Our bodies receive energy from the food we eat and use it up when we run or bicycle. Energy exists in many forms, including light and heat, and electrical, nuclear, and chemical energy. The ultimate source of energy for our planet is the sun. But what exactly is energy?

What Is Energy?

Energy is usually defined as the ability to do work or bring about change. **Work** is done whenever any form of matter is moved over a distance. Everything that occurs in the universe involves work in which one form of energy is transformed into one or more other forms of energy.

All forms of energy can be classified as either kinetic or potential. **Kinetic energy** is energy of motion. A moving car, wind, swiftly flowing water, and a falling rock all have kinetic energy and are all doing work. A flowing stream, for example, can turn a waterwheel (Fig. 3-7a), and wind can turn a windmill. **Potential energy** is stored energy, energy that when released is converted to kinetic energy. Water held behind a dam (Fig. 3-7b), or a rock at the edge of a cliff, has potential energy by virtue of its position. When the water is released or the rock falls off the cliff, the potential energy is converted to kinetic energy. Potential energy is also stored in chemical compounds, such as those present in food and gasoline. When food is digested, chemical bonds are broken, and energy that the body needs to function is released. Similarly, combustion of gasoline in a car engine releases energy to set the car in motion.

Energy Transformations

In the universe, energy transformations occur continuously. Stars convert nuclear energy into light and heat, plants convert light energy from the sun into chemical-bond energy in sugar molecules, and animals convert chemical-bond energy in sugars into energy of motion. None of these transformations is 100 percent efficient; in every case a part of the energy is converted to some useless form of energy, usually heat. But in any transformation, no new energy is created and no energy is ever destroyed. As stated in the **first law of thermodynamics** (also known as the law of conservation of energy): *Energy can neither be created nor destroyed; it can only be transformed from one form to another.*

The efficiency of any energy-transfer process is defined as the percentage of the total energy that is transformed into some useful form of energy. For example, the efficiency of an incandescent light bulb is very low. Only about 5

$$\% \text{ efficiency} = \frac{\text{usable energy}}{\text{total energy}} \times 100$$

Figure 3-7   (a) The kinetic energy of a flowing stream can be harnessed to do useful work. At this mill, falling water turns the waterwheel, which, through a series of gears, turns grindstones that mill the grain. (b) The potential energy of the water in this lake behind the Hoover Dam is converted to kinetic energy as it is released from the dam. The flow of water is used to drive turbines that produce electric power.

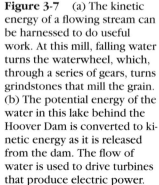

(a)

(b)

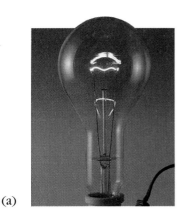

(a)

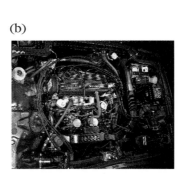

(b)

Figure 3-8 Most of the energy transformations we depend on are very inefficient and produce more waste heat than useful energy. (a) When electrical energy passes through the filament of a light bulb, only about 5 percent of it is converted to light; the rest is lost as heat. (b) An automobile engine converts about 10 percent of the chemical energy in gasoline to mechanical energy that can be used to drive the vehicle; 90 percent is lost as heat.

percent of the electrical input is converted into light energy; the remainder, as anyone who has touched a lighted bulb knows, is converted to heat (Fig. 3-8a). If the light bulb is in its normal surroundings in a home or office, there is no way for the heat energy to do useful work or be converted into some useful form of energy—it is essentially lost.

The loss of useful energy is summed up in the **second law of thermodynamics**: *In every energy transformation some energy is lost in the form of heat energy that thereafter is unavailable to do useful work*. This statement means that all systems tend to run "downhill," or, in other words, that high-quality energy is always converted into lower-quality energy. All the life-sustaining processes that go on in the human body and in all other living organisms follow this pattern of energy flow.

Although energy is never destroyed, the fact that energy is lost as heat in all transformations means that, unlike many material resources, energy cannot be recycled. This fact has important implications for our society, which is dependent on so many energy-inefficient machines. The gasoline motor, for instance, is only 10 percent efficient (Fig. 3-8b).

(The first and second laws of thermodynamics are considered more rigorously in Chapter 7, in which the related topic, entropy, is introduced.)

Food Chains and Trophic Levels

Green plants (producers) are the only organisms that can take energy from the sun in the process of photosynthesis and store some of that energy in chemical bonds in sugars, starches, and other large molecules. When an animal eats a plant, the sugars and other chemical substances in the plant are broken down in chemical reactions in the animal's body (Chapter 10). Bonds that had joined the atoms together in the plant's food molecules are broken, and energy is released. The energy is used to power the many activities that enable the animal to grow and survive.

In any ecosystem, there are innumerable feeding pathways, or **food chains**, through which energy flows. In one typical food chain, grasshoppers eat green leaves, frogs eat grasshoppers, fish eat frogs, and humans eat fish (Fig. 3-9). Each step in the chain is called a **trophic level**. In the example used, there are five trophic levels: green leaves (plants) at the first trophic level, grasshoppers (herbivores) at the second trophic level, frogs (carnivores eating herbivores) at the third trophic level, fish (carnivores eating carnivores) at the

Omnivores function at more than one trophic level. A bear eating berries is at a lower trophic level than a bear eating fish.

fourth trophic level, and humans, at the top of the food chain, at the fifth trophic level. A human eating a carrot would be at the second trophic level; a human eating beef that has been raised on corn would be at the third trophic level. Organisms at any one trophic level are dependent on the organisms at the level below them for their energy needs. Ultimately, all animals—including humans—are dependent on producers for their existence.

Energy and Biomass

At each trophic level in a food chain, energy is used by the organisms at that level to maintain their own life processes. Inevitably, because of the second law of thermodynamics, some energy is lost to the surroundings as heat. It is estimated that in going from one trophic level to the next, about 90 percent of the energy that was present at the lower level is lost (Figure 3-10). Thus, in any ecosystem, the energy at the second trophic level (herbivores) is only about 10 percent of the energy at the first trophic level (producers). The energy at the third trophic level (carnivores) is a mere 1 percent of that at the producer level. This progression has an important implication for humans. It means that it is much more efficient to eat grain than to eat beef that has been fed on grain (Fig. 3-11).

It is not practical to measure the amount of energy present in a trophic level, but the weight of the living material, or **biomass**, in the level can be used as a measure of the energy at that level. If, in a very simple ecosystem, all the producers, herbivores, and carnivores are weighed, it will be found that the weight relationships between the different trophic levels are the same as the energy relationships: Approximately 90 percent of the biomass is lost in passing from one trophic level to the next higher one. In other words, to support a given mass of herbivores, a mass of producers 10 times as large is required. To support a given mass of carnivores requires a mass of producers that is 100 times as large. Because of these mass requirements, food chains rarely go beyond four trophic levels.

Food Webs

In any natural ecosystem, many food chains are interlinked in complicated systems called **food webs**. A simple food web in which several food chains overlap is shown in Figure 3-12. Mice are a food source for many predators, including foxes, weasels, and hawks. These predators, though, depend on many different food sources. Hawks, for example, eat snakes, birds, and rabbits, as well as mice. Animals that have many sources of food have a far better chance of survival than those that depend on just one or two sources. In the wild, food webs are much more complex than the one shown in Figure 3-12.

Food webs can be subdivided into grazing food webs and detritus food webs. So far we have considered only **grazing food webs**, those in which living plants form the foundation. Eventually, of course, all plants and animals die, and their remains form the foundation of **detritus food webs**. A typical detritus food web starts at the bottom of a lake. Dead plants and animals accumulate in bottom sediments, where they are decomposed by a variety of microorganisms. The decomposers are consumed by scavengers such as worms and crayfish that, in turn, are consumed by fish. The fish may then become prey for fish-eating birds.

Fifth trophic level

Fourth trophic level

Third trophic level

Second trophic level

First trophic level

Figure 3-9   A typical food chain with five trophic levels: green leaves, grasshopper, frog, fish, and human.

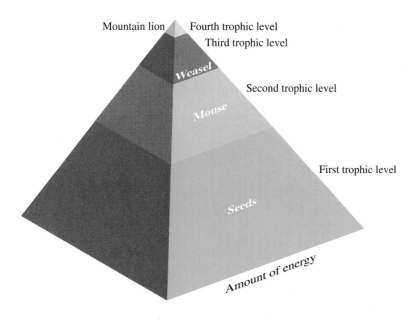

Figure 3-10   The flow of energy through a food chain. As energy passes to a higher trophic level in the food chain, approximately 90 percent of the useful energy is lost. High trophic levels contain less energy and fewer organisms than lower levels.

The leaves that accumulate on a forest floor are the basis for a land-based detritus food web. The leaves are decomposed by many organisms including bacteria, fungi, insect larvae, and earthworms, and these organisms, in turn, are consumed by many creatures including beetles, spiders, rodents, and birds. Interactions between detritus food webs and grazing food webs occur when, for example, a fox eats a rodent that has fed on a detritus feeder or a hawk eats a bird that has eaten a detritus-feeding earthworm.

Only about 10 percent of leaves in a forest is eaten by herbivores; 90 percent die and enter detritus pathways.

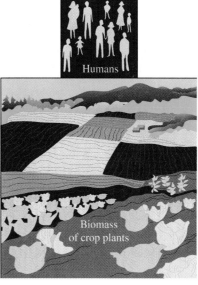

Figure 3-11   Energy is used much more efficiently if humans eat plants (first trophic level) instead of meat (second trophic level). A given area of farmland can support many more people if the crops are fed directly to people rather than to livestock that people then eat.

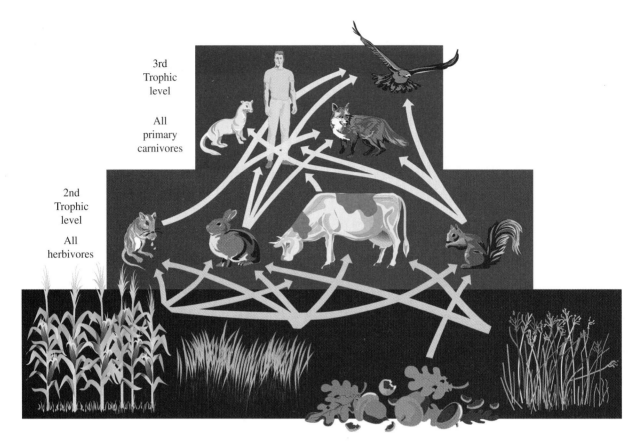

3rd
Trophic
level

All
primary
carnivores

2nd
Trophic
level

All
herbivores

Figure 3-12   A simplified food web showing how several food chains are interconnected.

    ## NUTRIENT CYCLES

To survive, a community of plants and animals in an ecosystem requires a constant supply of both energy and **nutrients**. The energy that sustains the system is not recycled. It flows endlessly from producers to consumers, entering as light from the sun and leaving as waste heat that cannot be reused. Nutrients, however, are continually recycled and reused. When living organisms die, their tissues are broken down, and vital chemicals are returned to the soil, water, and atmosphere.

If we analyze tissues from living organisms, we find that more than 95 percent of the mass of the tissues is made up of just six of the earth's 92 naturally occurring elements: carbon (C), hydrogen (H), oxygen (O), nitrogen (N), sulfur (S), and phosphorus (P). These six elements are the main building blocks for the manufacture of carbohydrates, proteins, and fats. These compounds, along with water, make up almost the entire mass of all living organisms. Plants are composed primarily of carbohydrates, while animals are composed primarily of proteins.

Small amounts of other elements are also required in order for plants and animals to survive and thrive. Iron (Fe), magnesium (Mg), and calcium (Ca) together make up most of the remaining 5 percent of the mass of living organisms. In many animals, iron is bound to hemoglobin, the protein in the blood that supplies oxygen to all parts of the body. In green plants, magnesium is bound to chlorophyll, the protein that absorbs light from the sun and is a vital part of the process of photosynthesis. Animals with skeletons need calcium as well as phosphorus to make bones and cartilage.

Trace amounts of approximately 16 other elements are also required. Copper (Cu) and zinc (Zn), for example, are components of certain enzymes, specialized proteins that are essential for the completion of many vital chemical reactions in the body.

Plants obtain the essential elements from the soil and the atmosphere. Animals obtain them from their food: plants and other animals. We will study the cycles by which supplies of four particularly important elements—carbon, nitrogen, phosphorus, and oxygen—are constantly renewed.

The Carbon Cycle

The **carbon cycle** is illustrated in Figure 3-13. The major sources of carbon for our planet are carbon dioxide gas in the atmosphere and carbon dioxide dissolved in oceans. Enormous quantities of carbon are also present in rocks, tied up in carbonates such as limestone, but this source recycles so slowly that it is not available to plants and animals for their daily needs.

Atmospheric carbon dioxide, although it makes up only 0.03 percent of the atmosphere by volume, is the starting material on which all living organisms depend. In the process of photosynthesis, carbon dioxide is taken into the leaves of green plants, where it combines with water to form sugar (glucose) and oxygen, as shown in the following equation.

Specialized structures in plant cells, called chloroplasts, capture radiant energy from the sun and convert it into chemical energy.

$$\text{solar energy} + \underset{\text{carbon dioxide}}{6\,CO_2} + \underset{\text{water}}{6\,H_2O} \overset{\text{chlorophyll}}{\longrightarrow} \underset{\text{glucose}}{C_6H_{12}O_6} + \underset{\text{oxygen}}{6\,O_2}$$

The needed water is obtained from the soil; it is absorbed through the plant's roots and then transported to the leaves. Part of the sugar that is formed is stored in the leaves, and part is converted into the complex carbohydrates and other large molecules that make up plant tissues. The oxygen is released to the atmosphere.

Photosynthesis is a very complex process that is still not fully understood. It involves many chemical reactions in which the green pigment chlorophyll plays an important role. We will consider only the overall result of photosynthesis as shown in the preceding equation.

When an animal such as a deer or a rabbit eats a green plant, the carbohydrates in the plant are digested and broken down into simple sugars, including glucose. Glucose is absorbed into the bloodstream and carried to the cells of the animal's body, where, in the process of **respiration**, it reacts with oxygen in the blood. Carbon dioxide and water are formed, and energy is released.

Photosynthesis consists of a series of *light* reactions, which occur only in the presence of light, and a series of *dark* reactions, which can occur in the dark.

$$C_6H_{12}O_6 + 6\,O_2 \longrightarrow 6\,CO_2 + 6\,H_2O + \text{energy}$$

Part of the energy released in respiration is used to power the many activities that go on in living cells and part is lost as heat.

Notice that the overall reaction for respiration is the same as the overall reaction for photosynthesis written backward. However, although cellular respiration is essentially the reverse of photosynthesis, the complex intermediate steps that are involved in the two processes are very different.

Plants engage in both photosynthesis and respiration. During the day, photosynthesis is the dominant process. At night, when there is no sunlight, respiration is dominant.

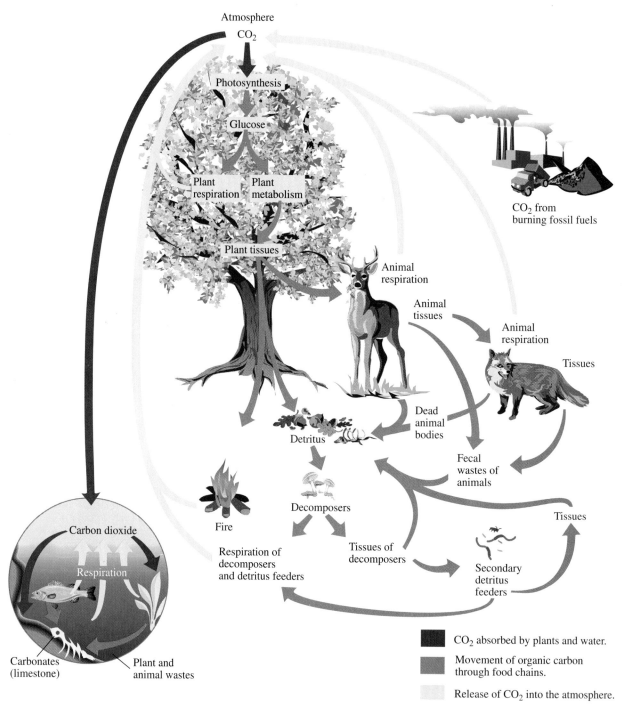

Figure 3-13 The carbon cycle. Atmospheric carbon dioxide is consumed by green plants in photosynthesis. Respiration in animals and plants and combustion of fossil fuels return carbon dioxide to the atmosphere. Carbon also cycles through water; dissolved carbon dioxide reacts with minerals and water to form carbonates, which are deposited in bottom sediments.

We have described the cycling of carbon through plants and animals in grazing food webs. Carbon is also cycled in much the same way through detritus food webs. Decomposers feed on the dead remains of plants and animals and by respiration release carbon dioxide and water to the atmosphere.

Sometimes, dead and decaying plant and animal remains become buried under sediments before they can be completely broken down by decomposers. Plant and animal remains were buried deeply and compressed on a very large scale about 280 to 345 million years ago. In time, chemical reactions transformed them into the fossil fuels: coal, oil, and natural gas. When these fuels are burned today to release the chemical-bond energy stored within them, they use oxygen from the atmosphere to convert their carbon into carbon dioxide.

$$C + O_2 \longrightarrow CO_2 + energy$$

The combustion of the fossil fuels that power our industrialized society therefore forms an integral part of the carbon cycle. Humans also intervene in the carbon cycle when they cut down more trees than they replace, thereby decreasing the amount of carbon dioxide that otherwise would be taken from the atmosphere and converted to nutrients. The climatic implications of these two activities, both of which tend to increase carbon dioxide concentration in the atmosphere, will be studied in Chapter 13.

Another important part of the carbon cycle is the continual exchange of carbon dioxide between the atmosphere and the oceans, a process that is important in maintaining the carbon dioxide concentration of the atmosphere at a constant level. The amount of carbon dioxide that dissolves in the oceans depends mainly on temperature and the relative concentrations of carbon dioxide in the atmosphere and in the oceans. When ocean temperature falls or when carbon dioxide concentration in the ocean becomes relatively low, more atmospheric carbon dioxide dissolves in the oceans. A very small portion of the dissolved carbon dioxide reacts with chemicals in the water, such as calcium, to form carbonates. The carbonates, primarily limestone ($CaCO_3$), are insoluble (do not dissolve) in water and settle on the ocean floor.

Rock formation and weathering are other aspects of the carbon cycle. As we saw in Chapter 2, sedimentary rocks such as limestone and dolomite were formed millions of years ago from the calcium carbonate–rich skeletal remains of coral and other marine creatures, and from the precipitation of calcium and magnesium carbonates from ancient seas. When chemically weathered by rain with a slight natural acidity, limestone rocks very gradually dissolve, releasing carbon dioxide into the atmosphere. This recycling of carbon through rock is a very slow process.

The Nitrogen Cycle

The **nitrogen cycle** is illustrated in Figure 3-14. Nitrogen is an essential component of proteins and of the genetic material that makes up DNA (deoxyribonucleic acid), and a constant supply is vital for all living organisms. Although 78 percent of the atmosphere is composed of nitrogen gas (N_2), plants and animals cannot use this source directly. Atmospheric nitrogen must be changed to other nitrogen compounds before it can be absorbed through the roots of plants. This change is achieved by **nitrogen fixation**, a process carried out by specialized bacteria that have the ability to transform atmospheric nitrogen into ammonia (NH_3). Some nitrogen-fixing bacteria live in soil. Others live in nodules on the roots of leguminous plants (peas, beans, clover, alfalfa; Fig. 3-14). The ammonia produced in root nodules is converted into a variety of nitrogen compounds that are then transported through the plant as needed.

Another means by which atmospheric nitrogen is converted to a usable form is lightning. The electric discharges in lightning cause nitrogen and oxygen in the atmosphere to combine and form oxides of nitrogen, which in turn

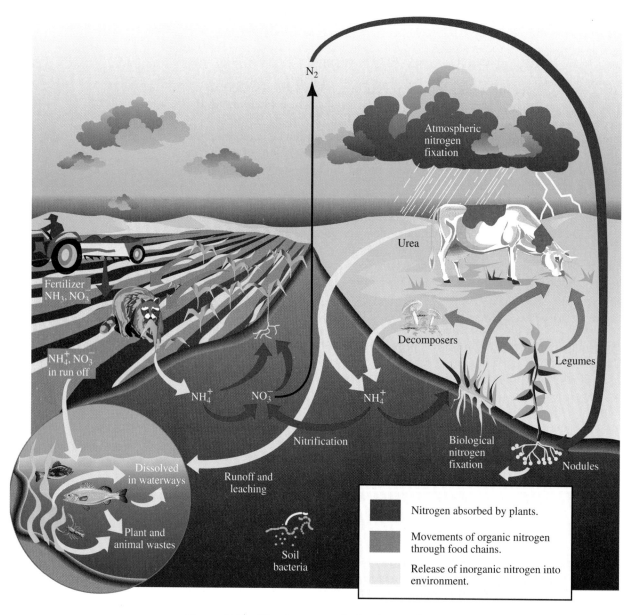

Figure 3-14   The nitrogen cycle. In nitrogen fixation, specialized bacteria convert atmospheric nitrogen to ammonia and nitrates, which plants absorb through their roots. Some nitrogen gas is fixed by lightning. Animals obtain the nitrogen they need to make tissues from plants. They return it to the soil in forms plants can use in their wastes and when they die. In denitrification, bacteria convert nitrates in soil back to nitrogen gas.

react with water in the atmosphere to form nitric acid (HNO_3). The nitric acid, which reaches the earth's surface dissolved in rainwater, reacts with salts in soil and water to form nitrates (NO_3^-) which are directly absorbed through plant roots. Compared with biological fixation, lightning accounts for only a small fraction of the usable nitrogen in soil.

Although some trees and grasses can absorb ammonia produced by nitrogen-fixing bacteria directly from the soil, most plants require their nitrogen to be in the form of nitrates. The transformation of ammonia into nitrates is carried out by specialized soil bacteria in the process known as **nitrification**.

Plants convert ammonia and nitrates that they get from soil or root nodules into proteins and other essential nitrogen-containing compounds. Animals get their essential nitrogen supplies by eating plants. When plants and animals die and decompose, the nitrogen-containing compounds in their tissues are broken down by detritus feeders until eventually ammonia is formed and is returned to the soil. Nitrogen is also returned to the soil in animal wastes. Both urine and feces have a high content of nitrogen-containing compounds. Urine, for example, contains urea (NH_2CONH_2), which by the action of specialized soil bacteria is converted to ammonia that can then be recycled by nitrification. In this way, nitrogen is continually cycled through food webs.

Not all the ammonia and nitrates that are formed in soil by the processes just described become available for plants. Both ammonia and nitrates are very soluble in water, and as water percolates down through the ground, these compounds are leached out of topsoil. They are often carried away in runoff into nearby streams, rivers, and lakes, where they are recycled through aquatic food webs. Another bacterial process that removes nitrates from the soil is **denitrification**, in which nitrates are converted through a series of reactions back to nitrogen gas.

In a few locations around the world, nitrates have accumulated in large mineral deposits. In Chile, for example, as mountain streams originating in the Andes mountains flowed across dry, hot desert toward the sea over thousands of years, much of the water evaporated, leaving behind huge deposits of sodium nitrate.

These sodium nitrate deposits are known as Chile saltpeter.

In the natural environment, a balance is maintained between the amount of nitrogen removed from the atmosphere and the amount returned. However, because nitrogen is often a limiting factor for plants grown on soils poor in nitrogen compounds, farmers frequently apply synthetic inorganic fertilizers containing ammonia and nitrates. As a result of runoff from fertilized land, the extra load of nitrogen compounds reaching rivers and lakes may upset the natural balance, sometimes with damaging consequences for the environment. We will discuss this problem in more detail in Chapter 12.

An alternative to the use of synthetic fertilizers for replenishing nitrogen in the soil is to plant a nitrogen-fixing crop, such as clover, and plow it back into the soil. Another method is to spread manure and allow the natural soil bacteria to degrade it and thus release nitrogen compounds that plants can absorb.

The Phosphorus Cycle

The **phosphorus cycle** is illustrated in Figure 3-15. Phosphorus is a component of many important biological compounds, including DNA and enzymes that play an essential role in the transfer of energy in living cells. It is a major constituent of cell membranes, and it is present in high concentration in bones, teeth, and shells.

Unlike the carbon and nitrogen cycles, the phosphorus cycle does not include an atmospheric phase, and the ultimate source of phosphorus for plants and animals is rock, nearly all of which contains small amounts of phosphorus, mostly in the form of phosphates (PO_4^{3-}). The phosphorus cycle is primarily a sedimentary cycle. As rocks weather, phosphates are very slowly dissolved and released to the soil.

Plants absorb dissolved phosphate directly from the soil through their roots and transport it to their leaves, where it is incorporated into large biological molecules. Animals get their phosphorus by eating plants. Not all the phosphate formed from rock by weathering becomes available to plants. Some of it becomes tightly bound to elements such as aluminum, iron, and calcium in compounds that are not very soluble in water.

Renewal of phosphorus through rock cycling is an extremely slow process; the main route for plants is through the recycling of the phosphorus in dead and decomposing organisms. When plants and animals die, microorganisms break down the organic phosphorus compounds in the dead tissues into inorganic phosphates, which immediately become available to plants.

Soil is often poor in the types of phosphates that plants can absorb readily, and, on agricultural land, synthetic fertilizers rich in available phosphate are used to replenish it. As was the case with nitrogen-containing fertilizers, runoff of large quantities of phosphate-containing fertilizers can have serious effects on aquatic ecosystems. We will also consider these effects in Chapter 12.

The droppings of sea birds are composed largely of phosphates because the food they obtain from the ocean has a high content of phosphates.

An important natural source of phosphates is **guano**. Droppings from thousands of sea birds that breed on islands off the west coast of South America have built up into huge deposits of phosphate-rich material called guano that, because of low rainfall, are not washed away. Large guano deposits have also been found in Arizona and New Mexico in dry caves where thousands of bats gather. These guano deposits are mined and are an important source of phosphorus for the manufacture of fertilizers. Also important for the manufacture of fertilizer are deposits of phosphate that formed in certain regions of the world millions of years ago from the skeletal remains of sea creatures.

Phosphorus is continually recycled through plants and animals but inevitably some is lost by leaching and erosion of soil into streams and rivers. This loss is often accelerated by human activities such as mining and farming. When phosphates reach the oceans they react with other chemicals in seawater, and most are converted to insoluble phosphates that sink to the ocean floor and, for all practical purposes, are permanently lost. It can take millions of years for phosphate lost to the oceans to be replenished by weathering of rock. Humans add to the problem of phosphorus loss by disposing of large amounts of their wastes—which contain significant quantities of phosphorus—into waterways instead of recycling them and returning them to the land.

Reserves of phosphorus in high-grade phosphate ores are estimated to be sufficient for several hundred years. We do not have an immediate problem, but we would be wise to conserve the reserves of phosphorus that we still have and reduce phosphorus loss as far as possible.

The Oxygen Cycle

Oxygen is all around us. As oxygen gas (O_2), it makes up 21 percent of the atmosphere and, like carbon, it is a component of all the important organic compounds in living organisms. Oxygen is a very reactive element and combines

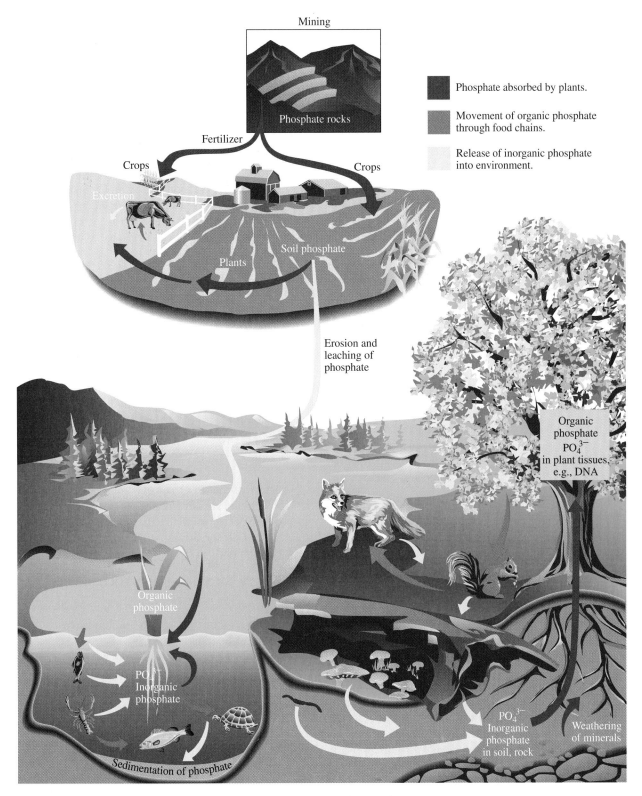

Figure 3-15 The phosphorus cycle. The main reservoir of phosphorus is phosphate in rock. Phosphate released to the soil during weathering is absorbed through plant roots. Animals obtain phosphorus from plants and return it to the soil in wastes and dead tissue. Phosphates leached from soil enter waterways and are carried to the oceans.

CONSUMER BRIEF — **Toothpaste: Scrub Away the Plaque**

Tooth decay has plagued humans for centuries, even though the enamel that makes up the outer layer of teeth is the hardest substance in the body. Most of the 2 mm thick enamel consists of hydroxyapatite, $Ca_{10}(PO_4)_6(OH)_2$. About 95 percent of the population is affected to some degree with tooth decay, a disease in which portions of the teeth become demineralized and cavities form. The demineralization process can be considered to be the dissolution of the solid enamel.

$$\text{demineralization}$$
$$Ca_{10}(PO_4)_6(OH)_2 \rightarrow 10\ Ca^{2+} + 6\ PO_4^{3-} + 2\ OH^-$$
hydroxyapatite   (dissolved material)
(solid)

The reverse process, which is called remineralization, in which calcium ion is incorporated into teeth, is the body's defense against tooth decay. In children, the growth of the enamel layer occurs faster than demineralization.

A thin, adhesive sugar-containing film called plaque is produced by bacteria that live in the mouth. Other bacteria convert the plaque into an acid that dissolves the calcium in the enamel. The bacteria live on energy obtained by decomposing sugars into acids. The more sugar that is present, the more the bacteria multiply, and the more acid they produce.

One important step a person can take to reduce cavities is to remove the accumulated plaque from the surface of the tooth. Tooth paste contains a surfactant (soap) such as sodium lauryl sulfate which makes suds as we brush, and an abrasive such as calcium phosphate $(CaHPO_4)$. The abrasive is gritty and grinds away the plaque. Together the grinding action of the abrasive and the detergent action of the surfactant act to remove the plaque.

Studies have shown that children develop markedly fewer cavities in regions where the water contains at least 0.1 ppm fluoride. The natural fluoride content of drinking water in the United States ranges from 0.05 ppm to 8 ppm. Fluoride ions in solution can replace some of the hydroxide ions in hydroxyapatite enamel to give fluorapatite:

$$Ca_{10}(PO_4)_6(OH)_2 + 2\ F^- \rightarrow Ca_{10}(PO_4)_6F_2 + 2\ OH^-$$
hydroxyapatite   fluorapatite

Fluorapatite is harder and more dense than hydroxyapatite and it is 100 times less soluble in acid. Fluoride ion also suppresses bacteria's ability to generate acid.

Fluoride treatment can be administered to young children by applying a gel containing fluoride to the teeth and keeping it in contact for about a half hour.

A typical toothpaste contains saccharin, and flavorings to make it taste good. They also contain food coloring to make them attractive looking.

COMPOSITION OF A TYPICAL TOOTHPASTE

Ingredients
Water
Calcium carbonate
Sorbitol
Sodium lauryl sulfate
Titanium dioxide
Cellulose gum
Saccharin
Sodium fluoride
Artificial colors

Some of the new "natural" toothpaste formulations contain abrasive materials, such as baking soda $(NaHCO_3)$ or chalk (powdered limestone, $CaCO_3$). No preservatives, saccharin or food colorings are included.

COMPOSITION OF A NATURAL TOOTHPASTE

Ingredients	*Source of Ingredient*
Sodium mono-fluorophosphate	The ore fluorospar
Calcium carbonate	Limestone
Glycerin	Vegetable oil by-product
Water	
Xylitol	Birch trees
Sodium lauryl sulfate	Lauryl alcohol from coconut
Carrageenan	Seaweed
Spearmint or Peppermint Oil	Spearmint or peppermint plant

readily with many other elements. It is a component of carbon dioxide (CO_2), nitrate (NO_3^-), and phosphate (PO_4^{3-}) and thus forms an integral part of the carbon, nitrogen, and phosphorus cycles. It is also a constituent of most rocks and minerals including silicates (e.g., $KAlSi_3O_8$, $CaAl_2Si_2O_8$), limestone ($CaCO_3$), and iron ores (Fe_2O_3, Fe_3O_4). The **oxygen cycle** is very complex and interconnected with many other cycles, and we will mention only some of the more important pathways in its cycle.

Photosynthesis and respiration are the basis of both the carbon cycle and the oxygen cycle. During photosynthesis, oxygen is released, and during respiration it is consumed. The oxygen and carbon cycles are also interlinked when coal, wood, or any other organic materials are burned. During burning, oxygen is consumed and carbon dioxide is released.

Another part of the oxygen cycle is the constant exchange of oxygen between the atmosphere and bodies of water, especially oceans. As we saw earlier, dissolved oxygen is essential for fish and other aquatic life. Oxygen dissolved at the surface of water is carried to deeper levels by currents.

Water, another vital part of our environment that is constantly being recycled, will be discussed in Chapter 12.

Nature's Cycles in Balance

Life on earth depends on a continual supply of energy from the sun and a continual recycling of materials. As we have learned in this chapter, the numerous and complex processes that sustain life are interlinked and interdependent. Because all aspects of an ecosystem are so interwoven, human intervention in any part of the system can easily lead to widespread disturbances of the natural balance in the environment.

EXPLORATIONS   **Dinosaur Extinction Controversy**

Through the study of fossils, scientists know that life on earth has been interrupted by periodic mass extinctions. We know that the dinosaurs disappeared from the face of the earth about 66 million years ago in a mass extinction. But exactly how that occurred is debated by scientists in fields as diverse as geophysics, astronomy, and paleontology. At one end of the spectrum are those who believe that mass extinctions were triggered by brief cataclysmic events such as the impact of an asteroid or periods of intense volcanic activity. Others argue that the extinction process is a gradual process brought on by environmental changes caused by tectonic, oceanic, or climatic fluctuations. Still others say the truth lies somewhere in between and is a combination of earthly and extraterrestrial causes.

The idea that large objects could strike our planet and cause mass extinctions was considered radical until 1978. That year a team of researchers at the University of California found a high level of iridium, which is a very rare element on earth but common in meteorites, in a thin layer of clay laid down on earth about the time dinosaurs became extinct. Analysis of this clay layer from more than 100 sites around the globe led them to the conclusion that an asteroid hit the earth 66 million years ago, wreaking environmental havoc on a global level. They suggested that the earth had been hit by an asteroid 6 mi (10 km) in diameter. The impact would be equivalent to 10,000 times the power of all the world's nuclear weapons hitting the earth simultaneously. So much debris would have been thrown up into the atmosphere that in the first days after the earth was hit, dust would have blanketed the entire world. It would have been pitch dark for a period of 1 to 3 months. If the asteroid had hit on land, it would probably have produced bitter cold. If it had hit at sea, the water vapor would have created a greenhouse effect, making it very hot. Soot particles, which are embedded in the same clay layer as the iridium, suggest that worldwide fires resulted from the impact. It has been estimated that as much as 90 percent of the world's forest burned as a result of this impact.

If the asteroid had hit in the ocean, the impact would have created a crater 185 mi (300 km) wide. A fireball would have had the radius of several thousand miles. Winds of hundreds of miles an hour would have swept the planet, drying trees like a giant hairdryer, and 2000°C rock vapor would have spread rapidly. This vapor would have condensed to white-hot grains that could have started additional fires.

One major question, however, is, "Where are the craters?" Unfortunately, craters are not always obvious. Many geophysicists suggest that this impact happened during a time when there was a large extrusion of lava. Some think that the meteorite might have struck in India, causing the great Deccan basaltic lava flows. Others suggest it was in a quiet area in Iowa where there is a 20-mile-wide (32-km) crater that glaciers filled with debris during the last ice age. Geologists have dated the Iowa crater at 66 million years, which perfectly coincides with the demise of the dinosaurs. Of course, the impact may have occurred in the ocean. If so, the scarred sea floor could have been buried by sediments or recycled into the planet's interior by plate tectonics.

Another theory holds that periods of intense volcanic activity could have disrupted the atmosphere in much the same way, blocking out sunlight and producing a global cooling effect as was described in the story about the volcano at Krakatoa in 1883 (Chapter 2). These same scientists have suggested that volcanism could have produced the iridium layer as well, by the eruption of a lava flow that was enriched by iridium (relative to normal ore bodies throughout the earth).

Still others contend that global sea levels have changed dramatically throughout time as a result of continental glaciation and plate tectonic activity. With rapid plate movement and glacial melting, sea levels rise. Periods of glaciation and tectonic quiet produce low sea levels. This variation can have disastrous effects on creatures living in some of our planet's prime habitats, estuaries and shallow seas. During the late Ordovician period, about 440 million years ago, warm seas dominated the world, and most land was sub-

Figure 3-A   "The real reason dinosaurs became extinct."

merged. Five million years later continental ice sheets had formed, and sea levels had dropped, exposing vast levels of land and cold water. As a result, it is estimated that more than 20 percent of the marine families perished. Such global climatic changes are found throughout the fossil record.

Some suggest that the extinction of dinosaurs was a slow process that took millennia to complete. The agent of extinction could have been one that would not be expected: flowers. About the time that dinosaurs disappeared, blooming and fruit-bearing plants known as angiosperms began to explode across the land. Attracting animals to spread their pollen and seeds, angiosperms colonized quickly and reproduced extremely rapidly. It is thought that over-grazing by dinosaurs threatened many low-growing plants—except for angiosperms—with extinction. The reproductive superiority of the angiosperms helped them not only to survive but to compete with the mulching jaws of the dinosaurs.

Changes in sea level might also have had an enormous impact on the dinosaurs' diet. At that time, it is thought, the receding sea level had created a land bridge between North America and the long-isolated Asian continent. Numerous little Asian mammals invaded North America and began eating the same flowering angiosperms that most dinosaurs were eating. Although the small mammals ate much less food per animal, they were so numerous that, it is thought, they ate the dinosaurs out of "house and home." Turtles, crocodiles, and lizards, along with mammals, survived, perhaps because they did not eat as much as dinosaurs and were small enough to find refuge. Scientists do not know why some groups survived several mass extinctions while others vanished—why fish and mammals have been successful and dinosaurs have not. But it is clear that mammals proliferated only after the mass extinction of dinosaurs.

From the fossil record, we can plot the episodes of mass extinction throughout geologic time. There appears to be an interval of roughly 26 million years between extinctions. This regularity suggests to some a celestial timetable for extinctions: a periodic comet shower, for example, triggered by the passing of a star or other celestial events. According to these calculations, with a 26-million-year period, Earth as we know it is safe for another 12 million years.

Extinction has claimed 99 percent of all species that have ever lived. During periods of mass extinctions, the rules of survival change. Characteristics that had been advantageous for a species may suddenly become liabilities. Tremendous size, for example, helped the dinosaurs dominate the world for some 140 million years. But they vanished during the great Cretaceous extinction, while many smaller animals, including mammals, survived.

Human beings, however, may turn out to be more deadly than any comet could ever be. Many scientists believe we are currently in a period of mass extinctions, generated by humans. Not only do humans have the power to kill other creatures, but they can also destroy their habitats, which is what we see happening in the rain forests of South America today. As the impact of technology on the biosphere worsens, humans become the environment's worst enemy.

It is easy to blame today's frightful extinctions on the developing countries in which such habitats are being destroyed. The crisis is acute in Brazil, Madagascar, and the Philippines, where rapidly expanding populations and the need for economic development can erase a forest in weeks. It is easy for wealthier countries to denounce and ignore the habitat losses. But we have only to ask who buys the timber or beef produced by these felled forests or who generates the acid rain that is wiping out New England's sugar maples and Germany's Black Forest to see that we share responsibility.

For destroyed habitats, we in the United States have only to look at Hawaii, which most of us regard as paradise. Biologists, however, consider it the endangered species capital of the world. Although occupying less than 0.2 percent of the nation's landmass, Hawaii contains 27 percent of its endangered birds and plants. Seventy-two percent of all the U.S. species that have become extinct in the past years did so on the Hawaiian islands.

Humans have become as destructive to the environment as an asteroid, a very big one.

Questions
1. Do you think that some species are more "worthy" of protection than others? Should humans be making such decisions?
2. Should NASA be doing more to monitor asteroids that may possibly hit the earth? What could we do if we found a large asteroid heading toward earth?

Reference: R. Gore, "Extinctions." *National Geographic* 175, (1989): pp. 662–699.

KEY WORDS AND CONCEPTS

abiotic
biomass
biome
biotic
carbon cycle
carnivore
consumer
decomposer
detritus
ecosystem
energy

environment
food chain
food web
fossil fuels
herbivore
kinetic energy
limiting factor
nitrogen cycle
nitrogen fixation
nitrification
omnivore

oxygen cycle
phosphorus cycle
potential energy
producer
range of tolerance
respiration
laws of thermodynamics
trophic level
work

QUESTIONS AND PROBLEMS

1. What is meant by biotic and abiotic? Give examples of each.
2. Define the word *ecosystem*. Give an example of a terrestrial and an aquatic ecosystem.
3. Living organisms can be divided into providers and consumers. Describe how:
   a. producers make their food.   **b.** consumers get the food they need.
4. Consumers can be divided into groups according to their diet. Name the main groups and give two examples of each.
5. Organisms that consume the remains of dead plants and animals are necessary for a balanced ecosystem. Explain their purpose.
6. What is meant by the term *limiting factor*? Give an example.
7. Is oxygen gas more or less soluble in cold water than in warm water?
8. Carp have a wider range of tolerance for oxygen and certain toxic substances than do trout or other sport fish. Discuss how this difference can affect the distribution of fish populations.
9. Define *energy*. Name four forms of energy.
10. What is the difference between potential energy and kinetic energy?
11. State the first law of thermodynamics. Give an example of the conversion of one form of energy to another.
12. State the second law of thermodynamics. Give an example that illustrates this law.
13. Make a diagram of a four-level food chain. In which direction does energy flow?
14. Suppose a bear has a choice between trout and blueberries for dinner. At which trophic level is the bear if it eats:
   a. the trout?   **b.** the blueberries?   Which food represents the more efficient use of energy?

15. Draw a diagram of a simple food web. Which type of food web does your diagram represent?
16. Each time energy flows from one trophic level to the one above, approximately what percentage of useful energy is lost?
17. What four elements are most abundant in the bodies of plants and animals? Name three other elements that are essential for life.
18. Draw a diagram to illustrate the carbon cycle.
19. Draw a diagram to illustrate the nitrogen cycle.
20. Draw a diagram to illustrate the phosphorus cycle.
21. What is the main source of carbon for living things? Describe the process by which plants use this source of carbon to make their food.
22. When a cow grazes, the grass is a fuel for the animal. Explain how the carbon cycle is involved.
23. What part do detritus feeders play in the carbon cycle?
24. What part do the following play in the carbon cycle?
   a. the oceans   **b.** limestone   **c.** crude oil
25. Name the process by which the nitrogen of the atmosphere is converted into a form that can be used by plants.
   a. Where are the bacteria that bring about this process found?   **b.** What chemical compound is produced by these bacteria?
26. Lightning can convert atmospheric nitrogen to a form that plants can use. Explain the process.
27. What is meant by *nitrification*? How is the nitrogen in animal wastes recycled by nature?
28. Describe two ways by which a farmer might replenish the nitrogen in her fields, without using inorganic fertilizers.
29. Where does the phosphorus needed for plant growth come from? How is phosphorus naturally recycled from decomposed organic material into living plants?

30. Most phosphorus on earth can be found in what form? Where are such deposits found?

31. Two parts of the oxygen cycle are intertwined with the carbon cycle. Name and briefly explain them.

32. As we Americans have become more aware of the delicate balance required to maintain our ecosystems, we have begun to bring environmental concerns into decision-making processes. The possible extinction of the snail darter, a small fish, almost stopped the development of a large dam in Tennessee. More recently, the Northern Spotted Owl's habitat has been threatened by the logging of mature timber in the Pacific Northwest.

a. Should the possible extinction of one species of animal be of such concern that development is halted?

b. Should new habitats (artificial, if necessary) be created for endangered species so that development may proceed?

ATOMS AND ATOMIC STRUCTURE

The idea that tiny, indivisible particles called atoms are the building blocks of all matter is an ancient one, but it was not until the early part of the eighteenth century that there was any convincing experimental evidence to support this theory. By the middle of the twentieth century, scientists realized that atoms, although fundamental particles, are not the ultimate, indivisible units of matter, as had been previously supposed. Atoms of all the elements that form our world are themselves built from three major fundamental subatomic particles: electrons, protons, and neutrons. The identity of each element is determined by the internal structure of its atoms. The atomic theory of matter is the foundation on which modern chemistry is built.

As more and more elements were discovered during the latter part of the nineteenth century, it became evident that certain groups of them had very similar properties. Scientists began to look for a systematic way to organize the elements on the basis of the observed similarities. This search led to the development of the periodic table of the elements, the table that is found in all chemistry textbooks today.

In this chapter, we will examine the experimental evidence that led to the development of the atomic theory and learn how the theory evolved to include our present knowledge of the structure of the atom. We will also study the periodic table, the invaluable reference guide that summarizes and correlates a great quantity of information about the elements from which every single thing—both living and nonliving—in our environment is made.

Learning Goals:

In this chapter, you should gain an understanding of:

1. The basic laws that govern the way matter reacts and elements combine.

2. The experimental evidence that led to the discovery of the composition of the atom.

3. The nature and properties of electrons, protons, and neutrons, the fundamental particles of which atoms are composed.

4. The relationships between electrons, protons, neutrons, and the mass of individual atoms.

5. The difference between isotopes of the same element.

6. The experimental basis for the arrangement of elements in the periodic table.

7. Similarities between elements in the same group in the periodic table.

8. The trends that occur in successive periods of the periodic table.

Figure 4–1 The Greek philosopher Democritus gave the name atoms to the ultimate indivisible particles of matter.

The Greeks believed that the natural world was constructed from four elements: earth, air, fire, and water.

Dalton was color blind and was the first to describe this condition.

THE ATOMIC NATURE OF MATTER

The proposal that matter is composed of ultimate particles that cannot be subdivided was first made in the fifth century B.C. by the Greek philosopher Leucippus and his pupil Democritus (Fig. 4-1). They argued that if one divided a piece of matter such as iron into ever smaller and smaller pieces, one would eventually reach an invisibly small particle of iron that could not be subdivided further and still preserve the properties of iron. Democritus gave these ultimate particles the name *atomos* (which literally means "uncuttable" in Greek). He believed that all matter was made up of various arrangements of **atoms** and that all atoms were composed of the same basic material. He asserted that the differences between materials—for example, between iron, gold, and lead—were due to differences in size and shape between the atoms that made up the materials. There was no way to verify or refute Democritus's hypothesis, and many influential philosophers of the time, including Plato (427–347 B.C.) and Aristotle (384–322 B.C.), rejected it and persisted in their belief that matter was continuous and could be endlessly subdivided.

Although the idea that matter was made up of atoms was discussed from time to time during the ensuing centuries, it was not until 1803 that the concept was seriously revived by John Dalton (1766-1844), an English schoolteacher and amateur meteorologist (Fig. 4-2). By that time, Dalton was able to present considerable experimental evidence to support the atomic theory.

DALTON'S ATOMIC THEORY

The main points in **Dalton's atomic theory** are as follows:

1. All matter is composed of tiny, indivisible particles called atoms.

2. All atoms of a particular element are identical, but the atoms of one element differ from the atoms of any other element.

3. Atoms of different elements combine with each other in simple numerical proportions to form compounds.

4. In a chemical reaction, atoms are rearranged to form new compounds; they are not created, destroyed, or changed into atoms of another element.

Not all these tenets could be verified in Dalton's time, but all were in agreement with the data then available to him.

EVIDENCE SUPPORTING THE ATOMIC THEORY

Evidence to support the atomic theory was provided by three fundamental laws of chemistry that were established by Dalton and other scientists of his time. The three laws are: (1) the law of conservation of mass, (2) the law of definite proportions, and (3) the law of multiple proportions.

The Law of Conservation of Mass

One of the first scientists to carry out systematic quantitative chemical experiments and draw logical conclusions from them was the Frenchman Antoine-Laurent Lavoisier (1743-1794) (Fig. 4-3). Lavoisier, often called "the father of modern chemistry," described his experiments in the first textbook of chemistry. He established chemistry as a *quantitative* science by showing the importance of accurately weighing the quantities of chemicals that were used up and produced in his experiments. It is one of the tragedies of history that at the age of 50 this brilliant man was guillotined during the excesses of the French Revolution.

Some of Lavoisier's most important experiments were concerned with combustion, or burning. He observed, for example, that when coal was burned in a closed container, the mass of the container plus its contents *before* combustion equaled the mass of the container plus its contents *after* combustion, even though the original substances in the container had been changed. Lavoisier realized that when the coal burned, it gained something from the air in the container, and that the amount gained by the coal was equal to the amount lost from the air. He named the substance lost from the air oxygen.

Lavoisier summarized the results of these and many other experiments in the form of a law known as the **law of conservation of mass**, which states that *in a chemical reaction, matter is neither created nor destroyed*. In modern terms, the law is more accurately written as follows: *There is no detectable change in mass in an ordinary chemical reaction*.

The law of conservation of mass is consistent with Dalton's atomic theory. If reacting substances are made up of atoms (statement 1), and if the atoms in the different elements in these substances are unique (statement 2) and can be neither created nor destroyed but only rearranged (statement 4), then it follows that the total mass of the products must equal the total mass of the reacting substances.

The Law of Definite Proportions

Another Frenchman, Joseph Proust (1754-1826), a contemporary of Lavoisier, is responsible for establishing the law of definite proportions. Proust showed that the substance copper carbonate, whether obtained from natural sources or synthesized in the laboratory, always had the same composition. It always con-

Figure 4–2 John Dalton (1766-1844) is regarded as the father of chemical theory in recognition of the importance of his atomic theory in the development of chemistry.

Lavoisier supported his scientific work with the profits from money invested in a private company authorized by Louis XVI to collect taxes. This activity, not his noble birth, lead to his death sentence during the French Revolution.

Figure 4–3 Antoine-Laurent Lavoisier (1743-1794) was one of the first scientists to appreciate the importance of accurate weight measurements in his experiments.

During the French Revolution, Proust escaped to Madrid where he continued his scientific work until his laboratory was destroyed during Napoleon's invasion of Spain.

tained the same three elements—copper, carbon, and oxygen—in the same proportions by mass, namely 5.3 parts of copper to 4 parts of oxygen to 1 part of carbon. These findings, and careful analysis of many other compounds, led Proust to formulate the **law of definite proportions** (sometimes called the **law of constant composition**), which can be stated as follows: *Different samples of any pure compound contain the same elements in the same proportions by mass.*

Proust's findings can also be explained in terms of Dalton's atomic theory. According to the theory, all atoms of copper are alike, all atoms of carbon are alike, and all atoms of oxygen are alike, but the three types of atoms differ from each other (statement 2). Therefore, atoms of the three elements will have different masses. Thus, if a fixed number of copper atoms combine with a fixed number of carbon atoms and a fixed number of oxygen atoms to form copper carbonate, it follows that in a pure sample of this substance, the three elements will always be in the same proportions by mass.

EXAMPLE 4–1   Three parts by mass of hydrogen combine with 14 parts by mass of nitrogen to form the gas ammonia. How many grams of hydrogen will combine with 70 g of nitrogen? (See Appendix C for an explanation of problem solving by dimensional analysis.)

Solution: Given: 3 g of hydrogen combine with 14 g of nitrogen.

$$70 \text{ g nitrogen} \times \frac{3 \text{ g hydrogen}}{14 \text{ g nitrogen}} = 15 \text{ g hydrogen.}$$

PRACTICE EXERCISE: When decomposed, the gas methane yields 3 parts by mass of carbon for every 1 part by mass of hydrogen. What mass of hydrogen will be obtained in the decomposition of 320 g of methane?

Answer: 80 g hydrogen.

Carbon monoxide
CO

Carbon dioxide
CO_2

Figure 4–4   The law of multiple proportions explained in terms of atomic theory. In carbon monoxide (CO), one atom of carbon is combined with one atom of oxygen; in carbon dioxide (CO_2), one atom of carbon is combined with two atoms of oxygen.

The Law of Multiple Proportions

Based on his own work and that of others, Dalton realized that two elements often combined to form more than one compound. The compounds were distinct from each other, and each obeyed the law of definite proportions. For example, two different compounds containing only carbon and oxygen were known. One was the poisonous gas carbon monoxide, and the other was carbon dioxide, a product of respiration and the burning of wood. Dalton analyzed both compounds and determined that 3 parts by mass of carbon combined with 4 parts by mass of oxygen to form carbon monoxide, and with 8 parts by mass of oxygen to form carbon dioxide. For equal quantities of carbon (3 parts), the ratio of oxygen in the two compounds was 4 to 8 (or 1 to 2). After analyzing other multiple compounds formed from other sets of elements, Dalton summarized his conclusions in the **law of multiple proportions**: *The masses of one element that can combine chemically with a fixed mass of another element are in a ratio of small whole numbers.*

The two compounds of carbon and oxygen can be explained in terms of atomic theory as follows. In carbon monoxide (CO), one atom of oxygen is combined with one atom of carbon, while in carbon dioxide (CO_2), two atoms of oxygen are combined with one atom of carbon (Fig. 4-4).

EXAMPLE 4-2   Nitrogen forms three compounds with oxygen: nitrous oxide (N_2O), nitric oxide (NO), and nitrogen dioxide (NO_2). Represent these oxides in terms of atomic theory, and show how the law of multiple proportions applies.

Solution: N_2O NO NO_2

    Atoms of N (●) and O(○)   ●●○   ●○   ●○○

    No. of atoms of O combined   ½   1   2
    with 1 atom of N

It is not possible to have *half* an atom of O. Therefore, the simplest ratio is given by the number of O atoms that combine with <u>2</u> atoms of N:

1   2   4

PRACTICE EXERCISE: Nitrogen reacts with hydrogen to form two compounds: ammonia (NH_3), a fertilizer, and hydrazine (N_2H_4), a rocket fuel. Represent these compounds in terms of atomic theory, and show how the law of multiple proportions applies (N = ●, H = ○).

Answer: H/N

    ammonia: ○○○●   3/1
    hydrazine: ○○○○●●   2/1

Dalton's atomic theory, although challenged, was widely accepted because its assumptions were supported by quantitative measurements and were consistent with a large body of experimental data. Although it later became evident that atoms are not the hard, indivisible particles that Dalton had envisioned, his basic conclusions are still valid. His atomic theory was a major breakthrough in the development of chemistry and remains a remarkable feat of scientific deduction.

THE STRUCTURE OF THE ATOM: THE EXPERIMENTAL EVIDENCE

By the end of the nineteenth century, it was evident that atoms were much more complex than Dalton had imagined. New work showed that an atom was not a hard uniform particle but was composed of a number of **subatomic particles**. In this section, we will review the experimental evidence that led to the discovery of the three fundamental subatomic particles that make up all atoms: electrons, protons, and neutrons.

The Electrical Nature of Matter

During the nineteenth century and the early part of the twentieth century, work by scientists in several different fields indicated that matter had an electrical component. As experimental observations were gradually pieced together, it became evident that atoms must be composed of electrically charged particles.

The English chemists Humphry Davy (1778–1829) and Michael Faraday (1791–1867) were among the first to demonstrate the electrical nature of matter. They showed that the passage of an electric current through certain molten

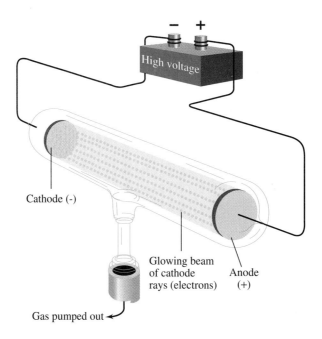

Figure 4–5   A gas-discharge tube, or cathode-ray tube. When the tube is partially evacuated and an electrical current is applied, the residual gas in the tube begins to glow. The glowing beam is composed of cathode rays (electrons).

Cathode (-)

Glowing beam of cathode rays (electrons)

Anode (+)

Gas pumped out

compounds and through water containing dissolved salts resulted in chemical changes in the compounds and the salts in solution. Faraday concluded that the electric current must be carried through the molten compounds and the solutions by charged atoms.

More information about the electrical nature of matter came in the latter part of the nineteenth century from experiments with **gas-discharge**, or **cathode-ray**, **tubes**, the forerunners of today's TV picture tubes, neon signs, and fluorescent lights. A simplified version of a cathode-ray tube is shown in Figure 4-5. It consists of a glass tube with a metal **electrode** sealed into each end and a sidearm that can be attached to a vacuum pump. When the tube is in use, the electrodes are connected to a source of electric power, and the vacuum pump removes gas from the tube. As gas in the tube is pumped out, pressure inside the tube decreases.

In experiments using a variety of gases in the tube, it was observed that when most of the gas had been pumped out, an electrical discharge was created between the electrodes, and the small amount of remaining gas began to glow. If a screen coated with a fluorescent material such as zinc sulfide was placed between the electrodes, pinpoint flashes of light were emitted randomly from the side of the screen facing the cathode, thus proving conclusively that the electrical discharge was coming from the **cathode** (negative electrode) and flowing to the **anode** (positive electrode). For this reason, the beam of current was said to be composed of cathode rays. This demonstration also indicated that the rays were not made up of waves, as many believed, but were composed of minute particles, each one producing a flash of light when it hit the screen. The particles later became known as electrons.

Fluorescent substances absorb electromagnetic radiation (EMR) of one wavelength (often invisible UV radiation) and then emit EMR of longer wavelength (often visible light of a particular color).

Electrons

The fundamental nature of cathode-ray particles was established by the English physicist J. J. Thomson (1856–1940) (Fig. 4-6). In 1897, in one of his most important experiments with gas-discharge tubes, Thomson showed that, in an electric field, a beam of cathode rays was deflected (or bent) toward the posi-

Seven of the men who worked under Thomson in the Cavendish Laboratory at Cambridge University (including Ernest Rutherford) went on to win Nobel Prizes.

tively charged plate (Fig. 4-7a). Because like charges repel and unlike charges attract, this deflection proved that cathode rays were negatively charged. The direction in which the rays were deflected when a magnetic field was applied further confirmed the negative charge (Fig. 4-7b).

Thomson also showed that, regardless of the nature of the residual gas in the discharge tube or the material composing the electrodes, cathode rays always had the same properties. His experimental findings led him to conclude that cathode rays must be composed of very energetic, negatively charged particles, or **electrons**, that are a fundamental part of all matter. For this work, Thomson was awarded the Nobel Prize in physics in 1906.

Thomson was not able to determine either the weight of an electron or its charge, but he was able to determine the charge-to-mass ratio. In 1909, the American physicist Robert A. Millikan (1868–1953) determined the charge on the electron, and from this value and Thomson's value of the charge-to-mass ratio, the mass of an electron was calculated to be 9.1×10^{-28} g, an extremely small number. Millikan received the Nobel Prize in physics in 1923.

Once electrons had been characterized, it was clear that, in the experiments of Davy and Faraday described at the beginning of this section, the current was carried through the molten compounds and the solutions by electrons.

Canal Rays

Another scientist who made important discoveries while experimenting with gas-discharge tubes was the German physicist Eugen Goldstein (1850–1930). In 1886, using a gas-discharge tube that had a perforated cathode (Fig. 4-8), Goldstein discovered that, in addition to cathode rays that flowed from cathode to anode, he could detect positively charged rays that flowed in the opposite direction through the holes (or canals) in the cathode. He called these rays **canal rays**.

Figure 4–6 Sir Joseph John Thomson (1856–1940) conducted experiments using gas-discharge tubes that led to the discovery of the electron.

When an electric charge is passed through a gas-discharge tube, electrons on the atoms of the gas are energized and escape, and move toward the anode (+). The atoms, stripped of electrons, acquire a positive charge and move toward the cathode (–) as cathode rays.

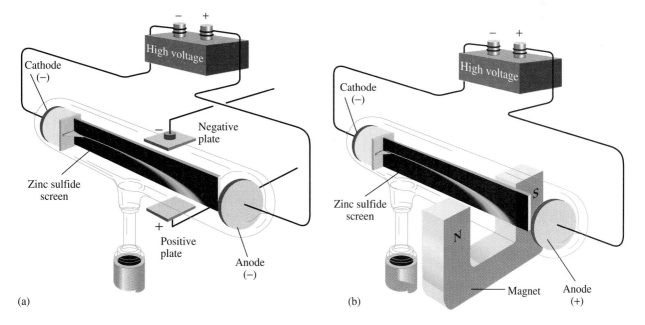

(a) (b)

Figure 4–7 The direction in which the cathode rays were deflected in (a) an electric field and (b) a magnetic field indicated that the rays were negatively charged.

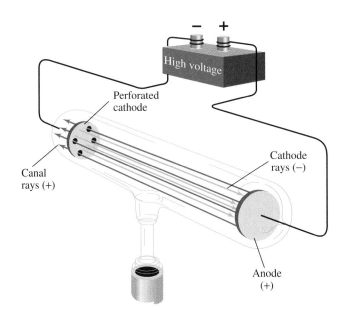

Figure 4–8   Using a perforated cathode in a gas-discharge tube, Goldstein showed that positively charged particles (canal rays) flowed from anode to cathode through the holes.

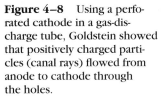

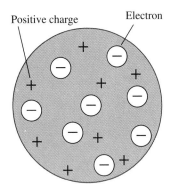

Figure 4–9   J. J. Thomson's plum pudding model (or the American blueberry muffin model) of the atom. Negatively charged electrons are embedded in a positively charged sphere of uniform density. The negative and positive charges are balanced to give an electrically neutral atom.

Gold was chosen because it can be beaten into extremely thin sheets.

Radioactivity is a spontaneous process in which certain unstable elements emit penetrating radiation.

Later research revealed that unlike electrons, which are all identical in mass, canal-ray particles are of varying mass. The lightest canal-ray particles were obtained when the residual gas in the gas-discharge tube was hydrogen. Even the lightest particles were very much heavier than an electron. Scientists had to wait until the discovery of the nucleus of the atom at the beginning of the twentieth century before the full significance of canal rays was understood.

THE DISCOVERY OF THE NUCLEUS OF THE ATOM

Based largely on the results of gas-discharge-tube experiments, Thomson concluded that an atom must be a positively charged sphere of almost uniform density in which negatively charged electrons are embedded much like "raisins in a plum pudding" (or, to use a more American analogy, like blueberries in a blueberry muffin) (Fig. 4-9). The positive and negative charges in the atom must be balanced so that the charge on the atom as a whole is zero. This view was generally accepted until 1911, when the New Zealand physicist Ernest Rutherford (1871–1937) (Fig. 4-10), who conducted most of his research in England and Canada, reported the results of an experiment that showed conclusively that Thomson's model of the atom was incorrect.

The classic experiment was carried out by Ernest Marsden, an undergraduate, and Johannes Geiger, a German physicist, who were working in Rutherford's laboratory. They bombarded a very thin sheet of gold foil with alpha particles, which are very energetic, positively charged particles emitted at high speed by certain radioactive elements (radioactivity and alpha particles will be discussed in Chapter 6). The alpha-particle-emitting radioactive source was placed in a lead-lined box that had a small hole in it. Alpha particles, which cannot penetrate lead, passed through the hole in the box in a narrow beam that was aimed at the extremely thin sheet of gold foil (Fig. 4-11). Almost completely surrounding the metal foil was a fluorescent zinc sulfide screen that recorded the impact of alpha particles on it as flashes of light. In this way, the paths followed by the alpha particles could be observed.

Assuming that the mass and positive charge of each gold atom were uniformly distributed throughout a sphere, as envisioned by Thomson (Fig. 4-9), Rutherford anticipated that this diffuse charge would have little effect on the fast-moving, energetic alpha particles. He expected the alpha particles to pass right through the thin gold foil without being deflected to any extent. He was, therefore, astonished to find that although most of the alpha particles (99.9 percent) passed straight through the foil as expected, a small but significant number were quite sharply deflected from their original paths. A very few were actually deflected directly back in the direction from which they had come. As Rutherford himself said, "It was about as credible as if you had fired a 15-inch shell at a piece of tissue paper and it came back and hit you."

Rutherford interpreted the results as follows: (1) Because almost all of the alpha particles passed through the foil without being deflected, the volume occupied by an atom must be mostly empty space, and (2) each of the alpha particles that was so dramatically deflected from its original path must have encountered a dense, similarly charged particle that repelled it and so caused it to change direction. The closer an alpha particle came to the dense positively charged particle, the more sharply it was deflected. Rutherford concluded that at the center of an atom, there must be a minute, very dense **nucleus** that accounts for almost all of the mass of the atom and contains all the positive charge. Electrons must be distributed in the space surrounding the nucleus; they would be too light to have any effect on the fast-moving, much heavier (approximately 8000 times heavier) alpha particles (Fig. 4-12).

The revolutionary model of the atom proposed by Rutherford has been verified many times by later experiments. Rutherford was awarded the Nobel Prize in chemistry in 1908 for his work on radioactivity, but his greatest contribution to science was undoubtedly his nuclear theory of the atom.

Figure 4–10 Ernest Rutherford (1871–1937) is best known for his discovery of the nucleus of the atom, but he also made important contributions to the understanding of radioactivity.

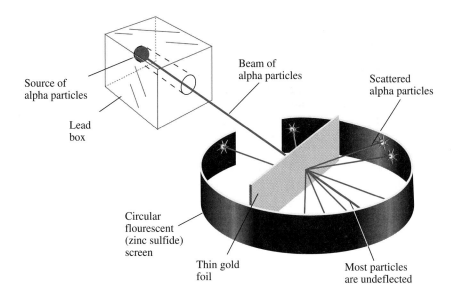

Source of alpha particles

Lead box

Beam of alpha particles

Scattered alpha particles

Circular flourescent (zinc sulfide) screen

Thin gold foil

Most particles are undeflected

Figure 4–11 Rutherford's gold foil experiment. Most of the alpha particles aimed at the gold foil passed straight through the foil, but a few were deflected at large angles.

The Structure of the Nucleus of the Atom

Once the existence of the nucleus of the atom had been established, the next step was to determine its exact composition. Rutherford suggested that the fundamental positively charged unit in the nucleus of all atoms was identical to the positively charged particles produced in Goldstein's canal-ray experiments when the gas in the discharge tube was hydrogen. He named these particles **protons**.

Rutherford deduced that atoms of hydrogen, the lightest of all the elements, contain one proton and one electron. In Goldstein's canal-ray experiments, the high-energy electrons streaming from cathode to anode knocked the single electron from some of the hydrogen atoms, leaving behind positively charged protons. Atoms of elements heavier than hydrogen contain increasingly larger numbers of protons. The heavier, positively charged particles observed in Goldstein's canal rays when a gas other than hydrogen was in the discharge tube were atoms containing more than one proton from which an electron had been removed. Rutherford later obtained confirmation of his theory that protons are fundamental particles present in all atoms by demonstrating that when nitrogen atoms are bombarded with alpha particles, protons are ejected (Chapter 6).

By 1914, it was possible to determine the number of protons in the nucleus of an atom, and Rutherford showed that, except for hydrogen, only about one-half the nuclear mass of the atoms of the lighter elements—elements such as boron, carbon, nitrogen, and oxygen—could be accounted for by the number of protons present. He therefore concluded that in addition to protons, the nuclei of atoms must contain electrically neutral (uncharged) particles with mass approximately equal to the mass of the protons. It was not until 1932 that the English scientist James Chadwick (1891–1974) confirmed the existence of this third basic unit of the atom, the **neutron**. Chadwick showed that when atoms of beryllium were bombarded with alpha particles, uncharged particles with mass almost identical to that of the proton were emitted (see Chapter 6).

H. G. J. Moseley, who was killed in World War I at the age of 27, discovered the relationship between the X-ray spectrum of an element and the number of charges in its nucleus. From this relationship, the number of protons in the atom of an element could be determined.

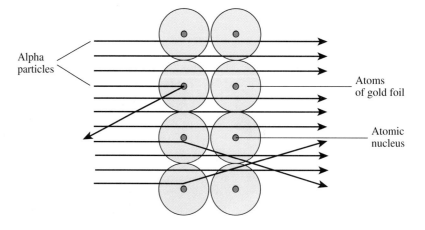

Figure 4–12   Rutherford's interpretation of the gold foil experiment. Most of an atom is empty space. The positive charge and mass of an atom are concentrated in a tiny nucleus at the center of the space. Most of the positively charged alpha particles passed undeflected through the empty space, but the very few that came close to a nucleus were repelled by the like charge and deflected.

THE SUBATOMIC PARTICLES: THEIR PROPERTIES AND ARRANGEMENT IN THE ATOM

In Chapter 2, we noted that it is difficult to believe that everything found on Earth, whether living, nonliving, natural, or synthesized by humans, is made up from fewer than 92 naturally occurring elements. Now we have to accept the even more startling fact that everything on earth is constructed from just three basic particles: electrons, protons, and neutrons. Let us summarize what we know about these three fundamental particles (Table 4-1).

Electrons

An electron, abbreviated e^-, is a negatively charged (-1) particle. It is the smallest of the subatomic particles, and compared with a proton or a neutron, its mass is negligible and is taken to be zero for most purposes.

Protons

A proton, abbreviated p, is a positively charged (+1) particle. The charge on a proton is opposite, but exactly equal to, the charge on an electron. A proton, although an extremely small particle, is approximately 1800 times heavier than an electron.

Neutrons

A neutron, abbreviated n, is a neutral particle; that is, it has no electrical charge. The mass of a neutron is very slightly greater than that of a proton, but for almost all purposes, it can be taken to be equal to that of a proton.

The Arrangement of Subatomic Particles in an Atom

An atom has two distinct regions: the center, or nucleus, and the area surrounding the nucleus. The very small, very dense nucleus is made up of protons and neutrons closely packed together. Because it contains protons, the nucleus is positively charged. The nucleus is so dense that 1 cubic centimeter (cm^3) of nu-

TABLE 4-1   Characteristics of the Three Basic Subatomic Particles

Name of Particle	Symbol	Location in Atom	Relative Electrical Charge	Mass (g)	Approximate Relative Mass (amu)[*]
Electron	e^-	Surrounding the nucleus	-1	9.1×10^{-28}	0
Proton	p	Nucleus	+1	1.7×10^{-24}	1
Neutron	n	Nucleus	0	1.7×10^{-24}	1

[*]Relative mass in amu is explained in the section "The Atomic Mass Unit and Atomic Masses (Atomic Weights)."

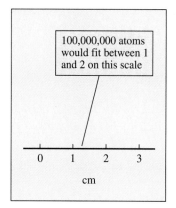

100,000,000 atoms would fit between 1 and 2 on this scale

| 0 | 1 | 2 | 3 |

cm

Figure 4–13   The size of an atom. If 100 million atoms (diameter 1×10^{-8}cm) are lined up in a row, they cover a distance of approximately one centimeter.

clear material would have a mass equal to 90 million metric tons. Almost all (99.9 percent) of the mass of an atom is concentrated in the nucleus.

Moving around the nucleus are the electrons. The region occupied by the electrons, often termed the **electron cloud**, is mostly empty space. Compared with the size of the nucleus, this region is very large. If we could expand the nucleus so that it was the size of the period at the end of this sentence, then the space occupied by the electrons would be approximately as large as a sphere with a diameter of 50 yards (46 m).

An atom, as a whole, has no charge because the number of positively charged protons in the nucleus of an atom is always balanced by an equal number of negatively charged electrons in its electron cloud. As we shall see in later chapters, this relationship between protons and electrons in an atom explains many of the physical and chemical properties of the elements.

The Size of the Atom

Although the dimensions of atoms cannot be measured directly because they are so small, it is possible, based on measurements made on large collections of atoms, to calculate the diameters and masses of individual atoms. It has been determined that the diameter of an atom of an element ranges from 1×10^{-8} to 5×10^{-8} cm. This means that if 100 million atoms, each with a diameter of 1×10^{-8} cm, were lined up side by side, they would occupy a distance of 1 cm (Fig. 4-13).

One of the reasons that Democritus's and Dalton's ideas about the atomic nature of matter were not immediately accepted was that, to many philosophers and scientists, it seemed irrelevant to speculate about particles that could

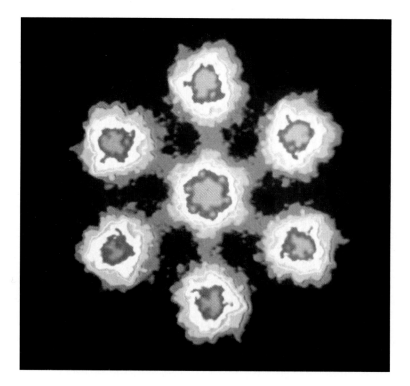

Figure 4–14   An electron microscope image of a uranyl acetate cluster on a thin film of carbon. The seven spots with orange-red centers are individual uranium atoms. The distance between the atoms is 3.4 nm.

not be seen. If atoms could not be seen, even with the most powerful optical microscope, how could one be certain that they existed? Recently, electron microscopes with magnification factors of several millions have made it possible to view images of atoms (Fig. 4-14), and the existence of atoms can no longer be disputed.

ATOMIC NUMBERS, MASS NUMBERS, AND ISOTOPES

We have seen that all atoms are made up of electrons, protons, and neutrons. Let us now learn how these three basic units are put together to form atoms of the different elements.

Atomic Number

All atoms of a particular element have the same number of protons in their nuclei. This number is the **atomic number** for that particular element. It is a basic property, and it determines the identity of an element.

Because atoms are electrically neutral, it follows that the number of protons in the nucleus of an atom will be balanced by an equal number of electrons in the space surrounding the nucleus:

number of protons = number of electrons = atomic number

Mass Number

The sum of the number of protons and the number of neutrons in the nucleus of an atom is termed the **mass number** of the element:

number of protons + number of neutrons = mass number

The mass of a proton is taken to be 1, and since the mass of a neutron is for all practical purposes identical to that of a proton, the mass of a neutron is also 1.

Table 4-2 shows how atomic numbers and mass numbers describe the structure of the atoms of four common elements—hydrogen (H), nitrogen (N), phosphorus (P), and radon (Rn).

TABLE 4-2   The Relationship between Atomic Number and Mass Number for Four Common Elements

Atom	Symbol	Number of Protons	Number of Electrons	Number of Neutrons	Atomic Number	Mass Number
Hydrogen	H	1	1	0	1	$(1 + 0) = 1$
Nitrogen	N	7	7	7	7	$(7 + 7) = 14$
Phosphorus	P	15	15	16	15	$(15 + 16) = 31$
Radon	Rn	86	86	136	86	$(86 + 136) = 222$

> **EXAMPLE 4-3**   An atom of iron (Fe) has 26 electrons. Its mass number is 56. What is the atomic number of Fe? How many protons and neutrons does the nucleus contain?
>
> ***Solution:***
>
> **1.** number of protons = number of electrons = atomic number
>
>   number of protons =   26   = atomic number
>
> Therefore, both the atomic number and the number of protons are 26.
>
> **2.** Rearranging the equation for mass number gives us
>
>   number of neutrons = mass number − number of protons
>
>   number of neutrons =   56   −   26   = 30
>
> Therefore, the number of neutrons is 30

> **PRACTICE EXERCISE:**   An atom of mercury, Hg, has a mass number of 200 and an atomic number of 80. How many electrons, protons, and neutrons are contained in this atom?
>
> ***Answer:*** e^- = 80; p = 80, n = 120.

Isotopes

The number of protons (and the number of electrons) in the atoms of a given element is always constant, but the number of neutrons may vary. Atoms of an element that have the same numbers of protons but different numbers of neutrons are called **isotopes**. It therefore follows that isotopes of a particular element all have the same atomic number but different mass numbers.

Research has shown that most naturally occurring elements are mixtures of isotopes in which one isotope usually predominates. For example, oxygen (atomic number 8) is a mixture of three isotopes, over 99.7 percent of which have eight neutrons. The remaining approximately 0.3 percent have either 9 or 10 neutrons.

The following notation is used to specify a particular isotope of an element.

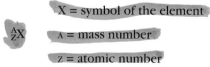

$$^A_Z X$$

X = symbol of the element

A = mass number

z = atomic number

The three isotopes of oxygen (O) are written

$$^{16}_{8}O, \qquad ^{17}_{8}O, \qquad ^{18}_{8}O$$

Numerous studies have established that, no matter what the source of a particular element, the percentage abundance of its isotopes will be very nearly constant. For example, samples of oxygen obtained from air, from the decomposition of water, and from calcium carbonate ($CaCO_3$) will contain practically the same percentages of the three isotopes of oxygen. Table 4-3 lists the isotopes of a number of naturally occurring elements together with their atomic numbers, numbers of protons, numbers of neutrons, and percent natural abundance (atomic mass, amu, listed in the last column will be explained in the next section).

EXAMPLE 4-4

a. Which of the following are isotopes of the same element?   **b.** Which have the same mass numbers?   **c.** Which have the same number of neutrons?

1. $^{12}_{6}X$ **2.** $^{14}_{6}X$ **3.** $^{14}_{7}X$ **4.** $^{15}_{7}X$ **5.** $^{11}_{5}X$

Identify the elements by referring to the atomic numbers listed in Table 4-3.

Solution:

a. 1 and 2 are isotopes of carbon (atomic number 6) with 6 (12 – 6) and 8 (14 – 6) neutrons, respectively. 3 and 4 are isotopes of nitrogen (atomic number 7) with 7 (14 – 7) and 8 (15 – 7) neutrons, respectively.
b. 2 and 3 have the same mass number (14).
c. 1 and 5 (an isotope of boron, B, atomic number 5) both have 6 neutrons (12 – 6 = 6; 11 – 5 = 6, respectively).

PRACTICE EXERCISE: Which of the following:
a. are isotopes? **b.** have the same mass number?   **c.** have the same number of neutrons?
In each case give the name(s) of the element(s).

1. $^{32}_{16}X$ **2.** $^{33}_{16}X$ **3.** $^{36}_{16}X$ **4.** $^{32}_{15}X$ **5.** $^{30}_{14}X$

Answer:

a. 1, 2, and 3 are isotopes of sulfur.   **b.** 1 (sulfur) and 4 (phosphorus).   **c.** 2 and 4 have the same number of neutrons (17); 1 and 5 (silicon) have the same number of neutrons (16)

In view of the existence of isotopes, Dalton's original atomic theory has to be modified to include the fact that the atoms of most elements are not always identical; they may differ in the number of neutrons in the nucleus. As we shall see later in the chapter, it is the number and arrangement of the electrons in an atom—not the number of neutrons—that determines the chemical properties of an element. Isotopes of a particular element share the same chemical properties; they differ from each other in mass only.

THE ATOMIC MASS UNIT AND ATOMIC MASSES (ATOMIC WEIGHTS)

The atomic mass of an element can be determined with great accuracy, but because atoms are so very small, it is not convenient to express the mass of an individual atom in grams. For example, the weight of one of the heaviest atoms known—the $^{238}_{92}U$ isotope of uranium—is only 3.95×10^{-22} g, an amount that is impossible to work with in the laboratory. To avoid this difficulty, scientists established an arbitrary unit of mass, the **atomic mass unit (amu)**. This unit expresses the mass of an atom *relative* to the mass of an atom of a different element. By international agreement, the mass of the most common isotope of carbon, $^{12}_{6}C$, is taken to be exactly 12 amu. The masses of the atoms of all other elements are determined relative to that value (see Table 4-3).

An atom of a specific element, if it exists in a number of isotopic forms, will have one of several different atomic masses. An atom of chlorine, for example, will have an atomic mass of either 34.97 or 36.97 because chlorine is made up of two isotopes, $^{35}_{17}$Cl and $^{37}_{17}$Cl (Table 4-3). When describing chemical reactions in quantitative terms, one rarely needs to specify the masses of individual isotopes because all the isotopes of an element behave in the same way in a chemical reaction. Instead, the *average* mass of the individual isotopes that make up an element is used. This average mass, which takes into account the relative abundances of the different isotopes, is termed the **atomic mass**, or **atomic weight** of the element. Chlorine, which is 75.53 percent $^{35}_{17}$Cl and 24.47 percent $^{37}_{17}$Cl, therefore has an atomic mass of 35.45 amu. The atomic masses of the 109 known elements are listed inside the back cover of this book.

TABLE 4-3   Isotopes of Selected Common Elements

Element	Isotope	Atomic Number	Number of Protons	Number of Neutrons	Natural Abundance (%)	Atomic Mass (amu)
Hydrogen	$^{1}_{1}$H	1	1	0	99.98	1.01
	$^{2}_{1}$H	1	1	1	0.02	2.01
Helium	$^{3}_{2}$He	2	2	1	0.0001	3.02
	$^{4}_{2}$He	2	2	2	100.00	4.00
Lithium	$^{6}_{3}$Li	3	3	3	7.42	6.01
	$^{7}_{3}$Li	3	3	4	92.58	7.02
Beryllium*	$^{9}_{4}$Be	4	4	5	100.00	9.01
Boron	$^{10}_{5}$B	5	5	5	19.6	10.01
	$^{11}_{5}$B	5	5	6	80.4	11.01
Carbon	$^{12}_{6}$C	6	6	6	98.89	12.00+
	$^{13}_{6}$C	6	6	7	1.11	13.00
	$^{14}_{6}$C	6	6	8	trace	14.00
Nitrogen	$^{14}_{7}$N	7	7	7	99.63	14.00
	$^{15}_{7}$N	7	7	8	0.37	15.00
Oxygen	$^{16}_{8}$O	8	8	8	99.76	15.99
	$^{17}_{8}$O	8	8	9	0.04	17.00
	$^{18}_{8}$O	8	8	10	0.20	18.00
Fluorine*	$^{19}_{9}$F	9	9	10	100.00	19.00
Sulfur	$^{32}_{16}$S	16	16	16	95.00	31.97
	$^{33}_{16}$S	16	16	17	0.76	32.97
	$^{34}_{16}$S	16	16	18	4.22	33.97
	$^{36}_{16}$S	16	16	20	0.01	35.97
Chlorine	$^{35}_{17}$Cl	17	17	18	75.53	34.97
	$^{37}_{17}$Cl	17	17	20	24.47	36.97
Copper	$^{63}_{29}$Cu	29	29	34	69.09	62.93
	$^{65}_{29}$Cu	29	29	36	30.91	64.93
Uranium	$^{234}_{92}$U	92	92	142	0.01	234.04
	$^{235}_{92}$U	92	92	143	0.72	235.04
	$^{238}_{92}$U	92	92	146	99.27	238.05

*Beryllium (Be) and fluorine (F) are 2 of the 20 elements that have only one naturally occurring isotope.
+By international agreement, the mass of $^{12}_{6}$C is exactly 12.0000 amu.

EXAMPLE 4-5   The element boron consists of two isotopes, $^{10}_{5}$ B with an atomic mass (amu) of 10.01 and $^{11}_{5}$ B with an atomic mass of 11.01. The natural abundance of $^{10}_{5}$ B is 20.0 percent. What is the atomic mass of naturally occurring boron?

Solution:

a. Calculate the abundance of $^{11}_{5}$ B. The sum of the abundances must add up to 100 percent. Therefore,

$$100 \text{ percent} = \text{abundance of } ^{10}_{5} \text{ B} + \text{abundance of } ^{11}_{5} \text{ B}$$

$$100 \text{ percent} = 20.0 \text{ percent} + \text{abundance of } ^{11}_{5} \text{ B}$$

$$80 \text{ percent} = \text{abundance of } ^{11}_{5} \text{ B}$$

b. Once the atomic mass and abundance of each isotope are known, the atomic mass of the element can be calculated as follows:

$$\text{atomic mass of boron} = (10.01 \times \frac{20.0}{100}) + (11.01 \times \frac{80.0}{100})$$

$$= 2.00 + 8.81 = 10.81$$

Note that 10.81 is the value listed for boron in the table inside the back cover of this book.

PRACTICE EXERCISE: The element carbon has two isotopes, $^{12}_{6}$ C and $^{13}_{6}$ C. The mass of $^{12}_{6}$ C is 12.00 and of $^{13}_{6}$ C is 13.00. The abundance of $^{13}_{6}$ C is 1.1 percent. What is the atomic mass of carbon?

Answer: Atomic mass of carbon (C) is 12.01.

THE PERIODIC TABLE: MENDELEEV'S CONTRIBUTION

In 1869, the first really useful classification of the elements, in the form of a **periodic table**, was published by the Russian chemist Dmitri Ivanovich Mendeleev (1834-1907) (Fig. 4-15). Mendeleev had noticed that if the 63 elements then known were arranged in order of increasing atomic mass, those sharing similar properties appeared at regular, or periodic, intervals. He constructed his table by first positioning hydrogen by itself and then arranging the remaining elements in order of increasing atomic mass in a series of horizontal rows so that elements with similar properties became grouped together one below the other. The arrangement of elements in the first two rows of Mendeleev's table was as follows:

Li	Be	B	C	N	O	F
Na	Mg	Al	Si	P	S	Cl

Each element in the first row is very similar to the element immediately below it: Lithium (Li) closely resembles sodium (Na), beryllium (Be) resembles magnesium (Mg), and so on, until we reach fluorine (F), which closely resembles chlorine (Cl).

Figure 4–15 Dmitri Mendeleev (1834-1907), Russia's most famous chemist, was born in Siberia, the youngest of 17 children. A popular teacher, he was for many years a professor of chemistry at the University of St. Petersburg.

TABLE 4-4   Comparison of Properties of Eka-silicon Predicted by Mendeleev in 1871 with Observed Properties of Germanium (discovered in 1886)

Property	Eka-silicon (X)	Germanium (Ge)
Color	Gray	Gray
Atomic mass	72	72.59
Density (g/cm^3)	5.5	5.35
Formula of oxide	XO_2	GeO_2
Density of oxide (g/cm^3)	4.7	4.70
Formula of chloride	XCl_4	$GeCl_4$
Melting point of chloride (°C)	below 100	84

A striking feature of Mendeleev's table is that, where necessary, he left spaces in it so that only elements with similar properties would be placed in the same column, and he correctly predicted that new elements would be discovered to fit into the spaces he had left.

Another important feature of Mendeleev's table was that, on the basis of the position of a missing element in his table, he was able to predict its properties with great accuracy, and this prediction helped to accelerate the discovery of the element. For example, knowing the properties of neighboring elements, Mendeleev correctly predicted the properties of germanium (Table 4-4), which he called eka-silicon. Mendeleev's table not only brought order to the study of chemistry, it stimulated the search for new elements.

Eka stands for "one" or "first" in Sanskrit. Eka-silicon means "one place from silicon."

THE MODERN PERIODIC TABLE

The modern periodic table shown on the inside cover of this textbook evolved directly from Mendeleev's original table. The main differences between the two tables are (1) the modern table includes 109 elements, and (2) the elements are arranged in order of atomic *number* instead of atomic mass as was done by Mendeleev. The atomic number, as we shall learn in the next chapter, explains the periodic repetition of similar properties more accurately than does atomic mass. It also gives each element its own identifying number.

Each element in the periodic table is represented by its symbol; included with the symbol are the element's atomic number and atomic mass. The first 92 elements—hydrogen through uranium—occur naturally; the remainder have been synthesized by laboratory-controlled nuclear reactions (Chapter 6).

Periods and Groups

The periodic table is divided into 7 horizontal rows called **periods** and 18 vertical columns called **groups**. Although in many ways it is simpler to use the new international system of numbering the groups 1 through 18, most textbooks still use the older system in which Roman numerals identify the groups, and the letters A and B distinguish families within a group.

Certain sets of elements are given special names. The eight A groups of elements (Groups IA through VIIIA) are called the **representative** elements, and the B groups of elements that lie between the two blocks of representative elements are called the **transition** elements. The table is kept compact by listing two groups of transition elements, the lanthanides (elements 58 to 71) and the actinides (elements 90 to 103), separately at the bottom of the table.

For historical reasons, some groups of representative elements are often referred to by common names. For example, Group IA elements are known as the **alkali metals**; Group IIA elements are the **alkaline earth metals**; Group VIIA elements are the **halogens**; and Group VIIIA elements are called the **noble gases**.

Elements within a group, particularly if they are representative elements, have similar physical and chemical properties (Fig. 4-16). Table 4-5 shows some of the similarities between the alkali metals—lithium (Li), sodium (Na), potassium (K), rubidium (Rb), and cesium (Cs)—and between the halogens—fluorine (F), chlorine (Cl), bromine (Br), and iodine (I). In addition to the similarities, the properties of the elements in each group vary in quite regular ways from top to bottom in the group. For instance, melting points and boiling points decrease from lithium to cesium in the alkali metals and increase from fluorine to iodine in the halogens. The increase in melting points and boiling points in the halogen group results in a change in state from gas (F and Cl) to liquid (Br) to solid (I) (Fig. 4-17). Chemical reactivity also shows a definite trend within the groups. As one proceeds down the alkali metal group, the elements become increasingly reactive from lithium to cesium; down the halogen group from fluorine to iodine, they become less reactive. Similar trends are found in other groups of elements in the periodic table.

For many years after their discovery, no compounds of the Group VIIIA elements were known, and these elements were called the inert gases. Following the discovery of krypton and xenon compounds, the group was renamed the noble gases.

Figure 4–16 Sodium, like the other metals in Group IA, is soft and easily cut. A freshly cut surface is shiny but tarnishes rapidly as the metal reacts with oxygen in the atmosphere. Because sodium is so reactive, it must be stored under oil.

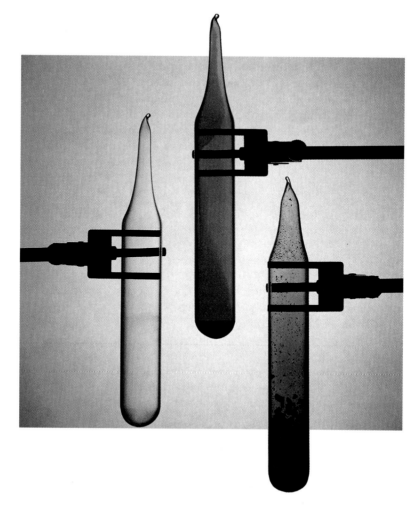

Figure 4–17 At room temperature, chlorine is a yellow-green gas, bromine is a red-brown liquid, and iodine is a purple-black solid.

TABLE 4-5 Some Properties of the Alkali Metals (Group IA or 1) and the Halogens (Group VIIA or 17)

Element	Symbol	Appearance	Boiling Point (°C)	Melting Point (°C)	Reactivity
Alkali Metals					
Lithium	Li	Shiny, silvery, soft metals	1336	181	Reactive
Sodium	Na		883	98	Increasing
Potassium	K		758	63	reactivity
Rubidium	Rb	Increasing softness	700	39	from Li to Cs
Cesium	Cs	from Li to Cs	670	29	Very reactive
Halogens					
Fluorine	F	Pale yellow gas	−188	−219	Very reactive
Chlorine	Cl	Yellow-green gas	−34	−101	Decreasing
Bromine	Br	Red-brown liquid	59	−7	reactivity
Iodine	I	Purple solid	184	114	from F to I Reactive

EXAMPLE 4–6   Which two elements in the following list would be most alike in their physical and chemical properties?

Sodium (Na)   Carbon (C)   Chlorine (Cl)   Sulfur (S)   Bromine (Br)

Solution: Chlorine (Cl) and bromine (Br) would be most alike because they are in the same group (VIIA) of the periodic table.

PRACTICE EXERCISE: Which two of the following elements would you expect to show similar physical and chemical properties?

Magnesium (Mg)   Potassium (K)   Aluminum (Al)   Fluorine (F)
Calcium (Ca)

Answer: Magnesium (Mg) and calcium (Ca) which are both in Group IIA.

CONSUMER BRIEF

Lithium: A Simple Salt That Affects a Person's Mood

Many people in our society suffer from the serious effects of manic depression. Typical symptoms of the manic phase include motor hyperactivity, reduced need for sleep, poor judgment, aggressiveness, and hostility. A significant number of these people have found that the lithium ion can help smooth out the high feelings of a manic episode and the low feelings of a depression. When given to a person experiencing a manic episode, lithium may produce normal behavior within 1–3 weeks.

It has been found that the easiest and least expensive way to get lithium ions into our nervous systems is to take the salt lithium carbonate, which must be prescribed by a doctor. The most common brand names for lithium carbonate are Eskalith, Lithobid, and Carbolithium. More than 3 million prescriptions for the 300-mg lithium carbonate tablets are filled by pharmacies in the United States each year. Lithium ions are released as the salt dissolves:

$$Li_2CO_3 \rightarrow 2\ Li^+ + CO_3^{2-}$$

Once dissolved, the salt is transported by the blood to the central nervous system.

Although its mechanism of action is not totally understood, it is thought that lithium ions interfere with complex reactions that relay and am-

plify messages carried to the cells of the brain. Lithium ions' interference with this cycle is thought to compensate for the overactivity in the brains of people afflicted with manic depression. It is believed that lithium treatment may reduce not only manic behavior, but also some forms of violent behavior.

One reason lithium carbonate takes more than a week to have its maximum effect is that the body has an elegant natural system for regulating salt concentration in body fluids, and it takes a relatively long period for the concentration of lithium ions in the brain to reach a constant level. When that point is reached, lithium excretion in the urine keeps pace with lithium intake. Because lithium and sodium are similar, the body cannot always distinguish between them, and to maintain the constant lithium level, the patient must maintain a relatively constant sodium intake. If sodium concentration in body tissues falls, more lithium is retained to compensate for the loss of sodium. Lithium may then reach toxic levels and cause kidney failure.

It is interesting that this simple ion has such a profound affect on the brain. So many of our modern-day pharmaceuticals are complex organic molecules with elegant structural features. For ex-

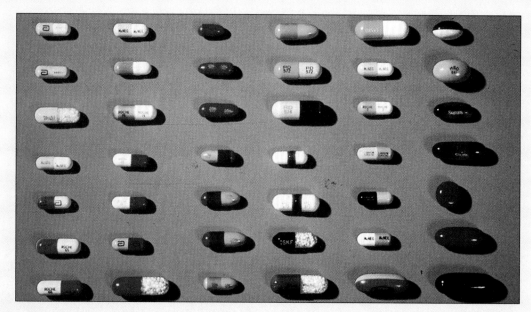

ample, the group of chemicals commonly called tranquilizers has been used to sedate manic people without inducing sleep. The most prescribed tranquilizers are diazepam, known by the trade name Valium, and chlordiazepoxide, known as Librium. Their structures are complex but similar. Librium is used to treat neuroses, behavior disturbances, and alcoholism. Valium is used to reduce the symptoms of anxiety. Unfortunately, both drugs can be addictive, and many patients find the cure worse than the sickness.

The discovery that lithium carbonate can affect mood has given mental health providers an effective, nonaddictive drug to treat the symptoms of anxiety. At present, lithium is the only inorganic ion known to have such a pronounced effect on the brain.

Questions
1. Why does the kidney have trouble distinguishing lithium ion from sodium ion? Refer to the periodic table.
2. Considering the effect the simple lithium ion has on the body, should the federal government fund more research focus on the importance of minerals in our diet?

Metals and Nonmetals

Elements are of two main types: metals and nonmetals. **Metals** have many distinctive properties. They are good conductors of heat and electricity, and most metals have a characteristic lustrous (shiny) appearance. Metals are ductile (i.e., they can be drawn out into a fine wire) and malleable (i.e., they can be rolled out into thin sheets). All metals, with the exception of mercury (Hg), which is a liquid, are solids at room temperature. Familiar metals are sodium (Na), aluminum (Al), calcium (Ca), chromium (Cr), iron (Fe), copper (Cu), silver (Ag), tin (Sn), platinum (Pt), and gold (Au).

Nonmetals, unlike metals, usually do not conduct heat or electricity to any significant extent. They have little or no luster and are neither ductile nor malleable. At room temperature, many—including hydrogen (H), nitrogen (N), oxygen (O), chlorine (Cl), and the noble gases—exist as gases. One nonmetal, bromine (Br), is a liquid. Solid nonmetals include carbon (C), phosphorus (P), sulfur (S), and iodine (I).

In the periodic table displayed inside the front cover of this book, metals and nonmetals are shown in different colors. The change from metal to nonmetal properties is not abrupt but gradual, and the elements shown in yellow in the table have properties that lie between those of metals and nonmetals. These elements, which are called **semimetals** or **metalloids,** include boron (B), which conducts electricity well only at a high temperature, and the **semiconductors**—silicon (Si), germanium (Ge), and arsenic (As)—which conduct electricity better than nonmetals but not as well as metals such as copper and silver. The special conducting ability of the semiconductors, particularly silicon, accounts for their use in computer chips and electronic calculators.

EXPLORATIONS

Joseph Priestley: Radical Thinker and Inspired Dabbler in Chemistry

Antoine-Laurent Lavoisier is rightly remembered and revered as the father of chemistry, but, as he himself acknowledged, he owed a great deal to his English contemporary, the clergyman Joseph Priestley (Fig. 4a). Lavoisier named oxygen and correctly explained its relationship to combustion and respiration, but it was Priestley's experiments that led him to his conclusions.

Priestley is scarcely remembered today outside the world of science, but in his own time he was a well-known philosopher, theologian, and politician who counted among his friends Adam Smith, Edmund Burke, Benjamin Franklin, John Adams, and Thomas Jefferson.

Priestley was born in England in 1733 in a small Yorkshire village, and from an early age was destined for the ministry. Because he was a Dissenter, a person who did not adhere to the established beliefs and practices of the Church of England, Priestley was denied admission to both Oxford and Cambridge universities. Instead, in 1752, he enrolled in one of the Dissenting Academies, which, in contrast to the conservative Oxford and Cambridge, were hotbeds of liberal thought and free inquiry. In this stimulating atmosphere, Priestley acquired many of the unorthodox religious beliefs and liberal ideas that later were to antagonize his parishioners and eventually force him to flee the country.

Priestley's life as a clergyman was difficult not only because his theology often offended his congregation but also because he stammered. His speech impediment had one good result. It took him to London in search of a cure, and although he never obtained more than temporary relief, his visits to the capital brought him in contact with persons whose views and intellects matched his own.

Priestley had always been interested in the study of nature, but it was not until he met Benjamin Franklin in London that he began to devote himself seriously to the study of science. Encouraged by Franklin, he published his first scientific work, a "History of Electricity" in 1767. His interest in science was further sparked by the fortunate circumstance that while a minister in Leeds, he lived next door to a brewery. He conducted many experiments with the "fixed air" (carbon dioxide) produced in the fermentation process and made a notable contribution to human enjoyment—he discovered a practical way of "impregnating water with fixed air" to produce "soda water" (carbonated water). This useful discovery led to his being asked to accompany Captain Cook on his second voyage to the South Seas. A supply of soda water was to be carried on the ship in the mistaken belief that it would prevent scurvy, the curse of all long sea voyages. However, when Priestley's unorthodox religious opinions became known, it was thought wiser not to tempt Providence by having a Dissenter on board, and so the invitation was withdrawn.

By 1773, Priestley's ministerial duties were such that he had more time to indulge his increasing interest

Figure 4–A Joseph Priestley

in chemistry. The fact that he had no prior knowledge of this subject did not deter him, and in the course of the next 7 years, he added five new gases to the list of only three—"air," carbon dioxide, and hydrogen—that were then recognized. Priestley discovered nitrous air (nitric oxide, NO), red nitrous vapor (nitrogen dioxide, NO_2), diminished nitrous air (nitrous oxide, N_2O), marine acid air (hydrogen chloride, HCl), and alkaline air (ammonia, NH_3). The main reasons for his success were his ability to design his own ingenious laboratory apparatus and the fact that he collected gases over mercury and thus could separate those that were water-soluble from those that were not.

In the eighteenth century, chemistry was not far removed from alchemy and was still dominated by the phlogiston theory, which held that any substance capable of being burned contained a fundamental component called phlogiston that escaped in the act of burning. When metals such as calcium or lead were strongly heated in air, for example, their changed appearance was said to be due to loss of phlogiston. The indisputable fact that the metal actually gained weight when heated was ingeniously explained by giving phlogiston a negative weight! A flame was said to move upward because it possessed "a quality of levity."

At this time, Priestley and other scientists realized that there was a connection between combustion and respiration. That a candle burning in a limited supply of air went out and that an animal placed in a closed container died were explained by assuming, in both cases, that phlogiston was transferred to the air. Both burning and life ceased when the air in the confined space became saturated with phlogiston. Priestley, in the course of many experiments, showed that during combustion or respiration in a confined space, the volume of air was reduced by about one-fifth; he also observed that a growing plant restored the air that had been used up by either a burning candle or a respiring animal (in this case, a mouse). His most important combustion experiment was performed in 1774. He heated red "calcinatus per se" (mercury(II) oxide, HgO) and obtained a colorless gas. He observed that a candle burned in this gas with a "remarkably bright flame." Priestley, a staunch believer in the phlogiston theory, called his gas "dephlogisticated air."

Priestley was better at devising experiments than at interpreting them, and he did not appreciate the significance of his discovery of this new "air." However, in that same year in Paris, he had his only meeting with Lavoisier—a meeting that was very significant for the future of chemistry. Priestley told Lavoisier about his new "air," and Lavoisier, who was a much more serious and careful scientist than Priestley, recognized the significance of the discovery. After repeating Priestley's experiment, and doing many other related experiments, he published his conclusions in

1779. Lavoisier realized that Priestley's dephlogisticated air was a component of the atmosphere. He gave it the name "oxygen," and he correctly deduced its role in respiration and combustion. Lavoisier's conclusions completely discredited the phlogiston theory, but despite all the opposing evidence, Priestley upheld the theory to the end of his life.

At its beginning in 1789, the French Revolution was generally hailed as the dawning of an era of peace and brotherhood. One of its first acts, the establishment of religious liberty, was warmly welcomed by the English Dissenters, who in their own country were still denied full rights of citizenship. Priestley, already unpopular because of his attacks on orthodox theology, made further enemies by openly and ardently supporting the French Revolution, which, as it progressed, appeared to many to threaten both the English crown and the Established Church. In 1791, in Priestley's hometown of Birmingham, a second anniversary celebration of the start of the revolution became the excuse for a mob to attack his home. Priestley and his family escaped unharmed, but his laboratory and library were completely destroyed. Continuing hostility eventually forced Priestley to flee to America, where he was warmly welcomed by his old friend Benjamin Franklin. He was offered the post of Professor of Chemistry at the University of Pennsylvania but declined it, preferring to devote himself to preaching, theology, and experimenting in his new laboratory. Priestley and his family settled in the Pennsylvania countryside, and, out of contact with scientific research and clinging to outdated theories, he made no further useful contributions to chemistry. He died in 1804.

Priestley was a true pioneer. A gentle, pious man, he devoted his life to the advancement of individual liberty, confident in the belief that political and religious freedom and the application of science would lead to human progress. He hoped to be remembered as a theologian and would doubtless be surprised to know that 200 years after his death he is revered instead for his contribution to chemistry.

Question
Explain Priestley's combustion and respiration experiments in modern terms. How did a growing plant restore the component of air used up in combustion and respiration?

References: A. Holt. *A Life of Joseph Priestley* (London: Oxford University Press, 1931); K. S. Davis. *The Cautionary Scientists (Priestley, Lavoisier, and the Founding of Modern Chemistry)* (New York: G. P. Putnam's Sons, 1966); W. R. Aykroyd. *The Three Philosophers: Lavoisier, Priestley and Cavendish* (Westport, CT: Greenwood Press, 1935).

KEY WORDS AND CONCEPTS

alkali metals	electrode	metals
alkaline earth metals	electron	neutron
anode	electron cloud	noble gases
atom	gas-discharge tube	nonmetals
atomic mass (weight)	group	nucleus
atomic mass unit (amu)	halogens	period
atomic number	isotope	periodic table
canal rays	law of conservation of mass	proton
cathode	law of definite proportions	representative elements
cathode-rays	law of multiple proportions	subatomic particle
Dalton's atomic theory	mass number	transition elements

QUESTIONS AND PROBLEMS

1. Many cities are having tremendous problems disposing of their garbage. Landfills are full, and there is no room for more trash. Many suggest burning (incinerating) the combustible portion. Others suggest recycling reusable materials or composting vegetable waste. How does the law of conservation of mass apply to this problem, and what does it suggest are possible solutions?

2. Medieval alchemists tried to change lead into gold by many different methods. Why didn't they succeed?

3. When a 5-kg piece of wood is burned, only 0.025 kg of solid ash is left. Does this result violate the law of conservation of mass?

4. How does Dalton's atomic theory explain:
   a. the law of conservation of mass? **b.** the law of definite proportions? **c.** the law of multiple proportions?

5. What experimental facts indicate that all atoms contain:
   a. at least one negatively charged electron? **b.** at least one positively charged proton?

6. Rutherford used the results of Marsden and Geiger's experiment to conclude that the nucleus contains most of the mass of the atom and that it has a positive charge. Describe the experimental results and show how they confirmed Rutherford's hypothesis (model).

7. How does Rutherford's model differ from Dalton's model?

8. Cholesterol is a compound suspected of causing hardening of the arteries. The formula of cholesterol is $C_{27}H_{46}O$.
   a. A sample of cholesterol is isolated from the arterial lining of a 60-year-old American male. What is the formula of this compound? **b.** Cholesterol is extracted from a chicken's egg. What is the formula of this compound?

9. Octane, a major component of gasoline, always contains 84 percent carbon and 16 percent hydrogen. What law does this fact illustrate?

10. State, in your own words, the law of multiple proportions.

11. Sulfur and oxygen form two compounds. By weight, one of the compounds is 40 percent sulfur and 60 percent oxygen; the other is 50 percent oxygen and 50 percent sulfur. What law is illustrated?

12. Which of the three laws described in this chapter is confirmed or violated by each of the following statements?
   a. Lavoisier found that when mercury oxide, HgO, decomposed, the mass of mercury and oxygen formed equaled the mass of HgO decomposed. **b.** The formula of sulfuric acid, an acid formed from some air pollutants, is $H_{1.2}SO_{4.9}$. **c.** The atom ratio of oxygen to nitrogen is twice as large in one compound as it is in another compound of the two elements. **d.** The formula of water found in China is H_4O, but water found commonly in the United States has the formula H_2O.

13. What are the distinguishing characteristics of:
   a. protons **b.** neutrons **c.** electrons

14. Indicate whether the following statements are true or false.
   a. An electron and a proton repel each other. **b.** A proton and a proton attract each other. **c.** An electron and an electron repel each other. **d.** A proton and a neutron repel each other.

15. a. Cathode rays are composed of _____.
   b. A neutral atom has an equal number of _____ and _____. **c.** The mass of a proton is _____ amu.

16. a. What is the difference between atomic number and mass number? **b.** What is the difference between mass number and atomic mass? **c.** Atomic mass and atomic number are expressed in what units?

17. Define the following terms:
   a. isotope of an element **b.** atomic number of an element **c.** mass number of an element

18. Sodium chloride containing the sodium isotope $^{24}_{11}Na$ is used to trace blood clots.
   a. How many protons are in its nucleus? **b.** How many neutrons are in its nucleus? **c.** How many electrons are in a sodium atom?

19. How many protons, neutrons, and electrons are present in each of the following atoms? (Obtain the atomic numbers from the table on the inside back cover of this book.)
   a. ^{208}Pb **b.** ^{204}Pb **c.** ^{37}Cl **d.** ^{27}Al **e.** ^{31}P

20. Complete the table below:

Atom	Atomic No.	Mass No.	Protons	Neutrons	Electrons
Zn	30	64	__	__	__
Eu	__	153	63	__	__
U	__	__	__	143	92
Pd	__	__	46	60	__

21. Complete the following table using the list of elements on the inside back cover of this book.

Atom	Protons	Neutrons	Electrons	Symbol
_____	10	11	__	__
Barium	__	82	__	__
_____	21	24	__	__
_____	15	16	__	__

22. Explain how the three isotopes of neon, ^{20}Ne, ^{21}Ne, and ^{22}Ne, differ.

23. Copper has two isotopes, ^{63}Cu and ^{65}Cu. The lighter isotope has a natural abundance of 70.5 percent and the heavier, 29.5 percent. What is the atomic mass of naturally occurring copper?

24. Arrange the following atoms in order of increasing mass (amu):
   a. N **b.** Zn **c.** Cl **d.** Xe **e.** Hg

25. Chlorine has two natural isotopes: ^{35}Cl and ^{37}Cl. Hydrogen reacts with chlorine to form hydrochloric acid, HCl.
   a. Would a given amount of hydrogen react with different masses of each chlorine isotope, if reacted separately? Explain. **b.** Does this outcome conflict with the law of definite proportions?

26. Describe what is meant by the terms *group* and *period* in relation to the periodic table.

27. Classify the following elements as metals or nonmetals.
   a. Mg **b.** Si **c.** Rn **d.** Ti **e.** Ge **f.** Eu **g.** Au **h.** B **i.** Am **j.** Bi **k.** At **l.** Br

28. Which of the following sets of elements are all in the same group in the periodic table?

a. Fe, Ru, Os **b.** Rh, Pd, Ag **c.** Sn, As, S **d.** Se, Te, Po **e.** N, P, O **f.** C, Si, Ge **g.** Rb, Sn **h.** Mg, Ca

29. List:
    a. the noble gases **b.** the alkali metals **c.** the halogens **d.** the alkaline earth metals

30. Using the periodic table, give the number of protons and neutrons in the nucleus of each of the following atoms:
    a. ^{15}N **b.** ^{3}H **c.** ^{207}Pb **d.** ^{151}Eu **e.** ^{107}Ag **f.** ^{109}Ag

31. Identify the following elements by referring to the periodic table:
    a. an element that has chemical properties similar to those of sulfur **b.** the halogen in the third period **c.** the alkaline earth metal in the second period **d.** the alkali metal in the fourth period **e.** the noble gas in the seventh period

32. The element with atomic number 22 forms crystals that melt at 1668°C, and the liquid boils at 3313°C. The crystals are hard, conduct heat and electricity, can be drawn into thin wires, and emit electrons when exposed to light. On the basis of these properties, classify the element as a metal or nonmetal. Which element is it?

33. If element 18 is an inert gas, in what groups would you expect to find elements 17 and 19?

ELECTRON CONFIGURATION AND CHEMICAL BONDING

Leaves on trees like these birches and maples change color at regular periodic intervals.

The naturally occurring elements are the building blocks from which everything on earth is constituted. Of these, only the noble gas elements exist as free separate atoms. Atoms of all other elements join (or bond) one to another to form the billions of substances that make up both the living and nonliving parts of our world. The forces that hold atoms together are called chemical bonds.

Why do atoms have a tendency to bond with other atoms to form distinctive units—units that may range in size from two atoms to thousands of atoms? The answer lies in the **electron configuration** of the individual atoms—that is, the way the electrons are arranged around the nucleus of the atom. By forming bonds, individual atoms achieve a stable structure.

In this chapter, we will first examine the way electrons are arranged in the atoms of different elements and how these arrangements determine the properties of the elements and explain their positions in the periodic table. We will then study the various types of chemical bonds that hold atoms together and learn how the electron configuration accounts for the different attractive forces responsible for the formation of the billions of substances that exist in the world. Our study will concentrate on bonds formed between representative elements (Chapter 4), those elements that form the majority of naturally occurring compounds.

Learning Goals:

In this chapter, you should gain an understanding of:

1. The arrangement of electrons in the atoms of the different elements.

2. The relationship between electron configuration and the periodic table.

3. The octet rule and how it is used to determine the formation of compounds.

4. How to draw Lewis electron-dot structures.

5. How ionic, covalent, and polar covalent bonds are formed.

6. Why some molecules are polar.

7. How forces of attraction operate between molecules.

THE ELECTRON CONFIGURATION OF ATOMS

Until the structure of the atom was elucidated, the theoretical basis for the orderly arrangement of the elements in the periodic table was not understood. Scientists could not explain why the properties of certain elements were repeated in a periodic manner, or why, for example, lithium (Li), sodium (Na), and potassium (K) in Group IA were so similar to each other but so different from the halogens and the unreactive noble gases.

By 1911, it was known that an atom consists of a nucleus of protons and neutrons surrounded by a cloud of rapidly moving electrons, but the way in which the electrons were arranged about the nucleus remained a mystery. Scientists knew that, reading from left to right across the periodic table, each successive element had one more electron in its atom than did the element before it. From this fact, they deduced that electron configuration must be the key to understanding the chemical behavior of the elements.

In 1913, the Danish physicist Niels Bohr (1885–1962) (Fig. 5-1) proposed the first useful model to explain the electron configuration of the atom. His concept was based on well established experimental evidence. Before we can understand Bohr's model, we need to consider the information available to him concerning the nature of light and the emission of light by different elements.

Continuous and Line Spectra

When white light from an incandescent light bulb is passed through a glass prism, the light is separated into a **continuous spectrum** of all the visible colors—violet, blue, green, yellow, orange, and red. The colors merge smoothly into one another in an unbroken band (Fig. 5-2a). The reason we see a rainbow in the sky when the sun reappears after a shower is that the falling raindrops act as prisms and disperse the sunlight. The different colors produced by a prism or by raindrops represent different amounts of radiant energy, with the shorter-wavelength violet end of the spectrum having more energy than the longer-wavelength red end (Chapter 1).

If instead of white light, the light from a gas discharge tube containing hydrogen, or some other element, is passed through a prism and focused onto a photographic film, a series of lines separated by black spaces is seen (Fig. 5-2b). This type of spectrum is called a discontinuous spectrum, or a **line spectrum**. The pattern of the lines produced is unique for each element and can be used to identify the element. Figure 5-3 shows the visible portions of the characteristic line spectra of hydrogen (H), sodium (Na), and neon (Ne). Other lines that

Figure 5-1 In the 1920s and 1930s Niels Bohr (1885–1962) headed the prestigious Institute of Theoretical Physics in Copenhagen, Denmark. To avoid imprisonment by the Nazis, who occupied Denmark in 1940, Bohr escaped to Sweden and then to the United States where he worked with other noted physicists on the development of the atomic bomb.

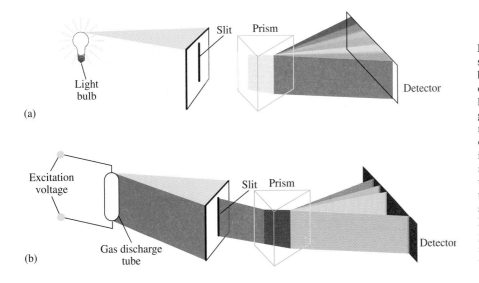

(a)

(b)

Figure 5-2   (a) Continuous spectrum. When a narrow beam of white light (sunlight or light from an incandescent light bulb) is passed through a glass prism, the light is separated into its component colors and a continuous spectrum is formed. Adjacent colors merge into one another in an unbroken band. (b) Line spectrum. When light emitted from a gas discharge tube containing hydrogen is passed through a prism, a discontinuous or line spectrum is formed. Lines in the visible region are colored.

cannot be seen occur in the ultraviolet and infrared parts of the spectra. By observing the line spectra of light coming from stars and planets, scientists have been able to identify elements that exist on these distant bodies.

Line spectra are also produced if the light from a flame in which an element is heated is passed through a prism. In such cases, the element imparts a characteristic color to the flame itself. For example, sodium compounds give a persistent bright yellow flame, while potassium gives lavender, copper gives blue-green, and strontium gives red. The presence of these and other compounds is responsible for the colors seen in fireworks displays.

The flame test can help identify an unknown compound. For example, if a compound emits a brilliant red color when heated in a flame, it is a strontium salt.

The Bohr Model of the Atom

As one requirement, a successful model of the atom had to be able to explain and predict the line spectra of different elements. Bohr's proposed model appeared to meet those requirements. He visualized the electrons in an atom as moving constantly around the nucleus in circular orbits in much the same way that the planets revolve around the sun. According to his model, each orbit was associated with a definite amount of energy.

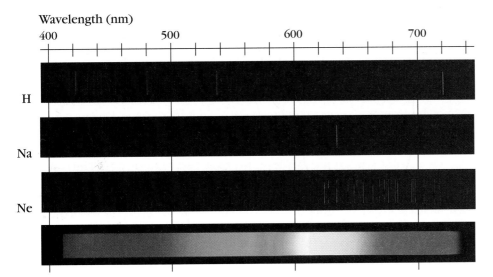

Figure 5-3   Line spectra of visible colors emitted by hydrogen, sodium, and neon. Each element has its own characteristic line spectrum that is different from that of any other element. A continuous spectrum is included for comparison.

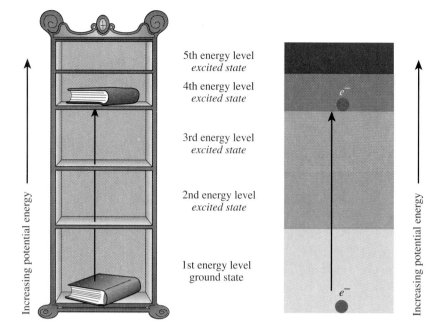

Figure 5-4 A book in a bookcase provides an approximate analogy for Bohr's theory of the arrangement of electrons in an atom. In the same way that a book can rest on any shelf in a bookcase but it cannot rest between shelves, electrons are restricted to certain permitted orbits or energy levels around the nucleus of the atom. If a book is moved to a higher shelf, it acquires potential energy. Similarly, if an electron is excited so that it moves to a higher energy level, its energy increases by a definite amount called a quantum of energy.

According to Bohr, electrons can exist only in certain specified energy states.

The analogy of a bookcase can be used to explain Bohr's concept (Fig. 5-4). A book can be placed on any shelf in a bookcase, but it cannot be placed between the shelves. In a similar way, he suggested, an electron can be in one orbit or another around the nucleus, but it cannot exist between orbits. Further, a book can be moved from a lower shelf to a higher one, but a certain amount of energy must be expended to do so. Similarly, if sufficient energy is supplied to an electron, the electron can move from a lower orbit to a higher one. Bohr suggested that a definite fixed amount of energy, called a **quantum** of energy, was needed to excite an electron so that it would jump from one orbit, or energy level, to a higher one. In going to a higher level, the electron—like the book—would acquire potential energy (energy of position). When the electron dropped back to a lower level, it would emit energy in the form of light—an experimental fact that could be recorded as a line spectrum.

Bohr explained the line spectrum of hydrogen as follows: When electricity was passed through a discharge tube containing hydrogen, the hydrogen atoms became energized and the single **excited-state electrons** in individual atoms jumped from their original orbit—the one closest to the nucleus—to a higher orbit. Depending on the degree of excitation, electrons jumped to different orbits, or **energy levels**. Some reached the second energy level; other, more excited ones reached fifth or sixth energy levels (Fig. 5-5). The characteristic line spectrum of hydrogen was formed when the excited electrons dropped back down to lower energy levels in a series of "jumps," eventually returning to the lowest level, the **ground state**. In each transition—from level 5 to level 1, from level 3 to level 2, from level 6 to level 3, and so on—a definite amount of light energy was released. Because light energy is related to the wavelength of the light, which in the visible range is related to the color of the light, those transitions that produced energy in the form of visible light were seen as lines in the spectrum (Fig. 5-3). On the basis of his model, Bohr calculated the expected positions of the lines for hydrogen. His calculated values agreed almost exactly with the observed values, and his model was generally accepted. Bohr was awarded the Nobel Prize in physics in 1922.

Short-wavelength violet light has more energy than does longer wavelength red light.

Building Atoms with the Bohr Model

Bohr visualized the single electron in the hydrogen atom as moving in a circular orbit at a fixed distance from the nucleus. Electrons in the atoms of heavier elements would circle around the nucleus in a series of wider orbits, according to certain rules. Bohr deduced that only a fixed number of electrons could be accommodated in any one orbit, and he calculated that this number was given by the formula $2n^2$, where n is equal to the number of the orbit, or energy level. For the first orbit—the lowest energy level, or ground state—n equals 1, and the maximum number of electrons allowed is $2(1)^2$, or 2. For the second energy level ($n = 2$), the maximum number of electrons is $2(2)^2$, or 8. Values for the first four orbits are given in Table 5-1.

Let us now see how, according to the Bohr model, electrons are arranged in the first few elements that follow hydrogen in the periodic table. Helium (He, atomic number 2) has two electrons, both of which can be accommodated in the first orbit. The first orbit is then filled. Lithium (Li, atomic number 3) has three electrons. The first two can go in the first orbit, but the third one must be placed in the second orbit. For the remaining seven elements in the second period— beryllium (Be, atomic number 4) through neon (Ne, atomic number 10)—two electrons are placed in the first orbit, and the remaining electrons are placed in the second orbit. With neon, the second orbit is filled, and for the next element, sodium (Na, atomic number 11), the additional electron must be placed in the third orbit. The arrangement of electrons in the first 20 elements in the periodic table, according to Bohr, is shown in Figure 5-6 (Bohr model column).

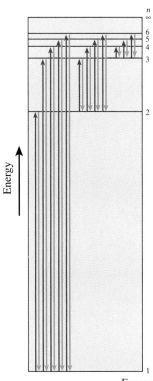

Energy
levels n

Figure 5-5   Red arrows show transitions of excited-state electrons from allowed energy levels to higher-energy levels. When the electrons return to their original energy levels (blue arrows), they produce line spectra. Some of the lines are in the visible part of the spectrum; others are in the ultraviolet and infrared regions.

EXAMPLE 5-1   How are electrons arranged in the element nitrogen according to the Bohr model?

Solution: The atomic number of nitrogen is 7. Therefore nitrogen has 7 electrons. The first orbit can accommodate 2 of the electrons. The remaining 5 are placed in the second orbit which can accommodate a total of 8 electrons. Representing this arrangement diagramatically gives:

PRACTICE EXERCISE: Draw a Bohr diagram to show the arrangement of electrons in sulfur.

Answer:

**TABLE 5-1   The Maximum Number of Electrons
That Can Occupy a Particular Energy Level
According to the Bohr Model**

Orbit or energy level (n)	1	2	3	4
Maximum number of electrons ($2n^2$)	2	8	18	32

Element	Atomic Number	Bohr Model	Wave Mechanical Model
Hydrogen (H)	1	1	$1s^1$
Helium (He)	2	2	$1s^2$
Lithium (Li)	3	2 1	$1s^2 2s^1$
Beryllium (Be)	4	2 2	$1s^2 2s^2$
Boron (B)	5	2 3	$1s^2 2s^2 2p^1$
Carbon (C)	6	2 4	$1s^2 2s^2 2p^2$
Nitrogen (N)	7	2 5	$1s^2 2s^2 2p^3$
Oxygen (O)	8	2 6	$1s^2 2s^2 2p^4$
Fluorine (F)	9	2 7	$1s^2 2s^2 2p^5$
Neon (Ne)	10	2 8	$1s^2 2s^2 2p^6$
Sodium (Na)	11	2 8 1	$1s^2 2s^2 2p^6 3s^1$
Magnesium (Mg)	12	2 8 2	$1s^2 2s^2 2p^6 3s^2$
Aluminum (Al)	13	2 8 3	$1s^2 2s^2 2p^6 3s^2 3p^1$
Silicon (Si)	14	2 8 4	$1s^2 2s^2 2p^6 3s^2 3p^2$
Phosphorus (P)	15	2 8 5	$1s^2 2s^2 2p^6 3s^2 3p^3$
Sulfur (S)	16	2 8 6	$1s^2 2s^2 2p^6 3s^2 3p^4$
Chlorine (Cl)	17	2 8 7	$1s^2 2s^2 2p^6 3s^2 3p^5$
Argon (Ar)	18	2 8 8	$1s^2 2s^2 2p^6 3s^2 3p^6$
Potassium (K)	19	2 8 8 1	$1s^2 2s^2 2p^6 3s^2 3p^6 4s^1$
Calcium (Ca)	20	2 8 8 2	$1s^2 2s^2 2p^6 3s^2 3p^6 4s^2$

Figure 5-6 Building atoms with the Bohr model and the wave-mechanical model. Electron arrangements of the first 20 elements in the periodic table are shown.

The Bohr model worked perfectly for hydrogen, and in many ways it worked well for the first 20 elements in the periodic table, but unfortunately it was a flawed model. It could not accurately predict the line spectrum of any element other than hydrogen, and it could not adequately explain the electron configuration of the transition elements, and therefore it had to be discarded.

The Wave-Mechanical Model of the Atom

Schrödinger used the very complex mathematical theory of quantum mechanics to develop mathematical equations that describe the motion of electrons.

In 1926, the Austrian physicist Erwin Schrödinger (1881–1961) (Fig. 5-7) developed the currently accepted **wave-mechanical model** of the atom. In this model, electrons are treated as both particles and waves, and complex mathematical equations are used to describe the arrangement and motion of electrons in atoms. We will consider a simplified form of Schrödinger's model.

The Schrödinger model retains Bohr's idea that electrons are limited to definite energy levels, but within each Bohr orbit, or principal energy level, it

TABLE 5-2   Orbitals in the First Four Energy Levels in the Wave-mechanical Model of the Atom

Principal Energy Level	Number of Orbitals	Orbitals
1	1	1s
2	2	2s 2p
3	3	3s 3p 3d
4	4	4s 4p 4d 4f

distinguishes sublevels called **orbitals**. The orbitals are identified by the letters s, p, d, and f. Each successive principal energy level going outward from the nucleus has one more orbital than the one before. Each orbital is described by the *number* of the energy level and the *letter* of the orbital as shown in Table 5-2.

The maximum number of electrons that can be accommodated in each principal energy level is the same as in the Bohr model (Table 5-1), but in the Schrödinger model the electrons are divided among the different orbitals in each energy level. Each orbital, like a principal energy level, can hold only a certain limited number of electrons. The maximum permitted number of electrons for each orbital is shown in Table 5-3. The sum of the electrons in each orbital at each energy level equals the maximum number of electrons permitted according to Bohr's model.

In the Schrödinger model, the term *orbit* is replaced by the term *principal energy level*. Orbitals are sublevels of the principal energy levels.

Building Atoms with the Wave-Mechanical Model

Building atoms using the wave-mechanical model is the same as building atoms with the Bohr model, except that now the electrons must be placed in the appropriate orbital in each principal energy level. Superscripts are used to indicate the number of electrons in each orbital. Hydrogen (H), with one electron, is designated $1s^1$; helium (He), with two electrons, is $1s^2$; lithium (Li), with three electrons, is $1s^2 2s^1$. The electron configurations of the first 20 elements according to the wave-mechanical model are shown in the last column (wave-mechanical model) in Figure 5-6. With argon (Ar, atomic number 18) the $3p$ orbital is filled. You will notice that for the next two elements—potassium (K, atomic number 19) and calcium (Ca, atomic number 20)—the additional electrons are located in the $4s$ orbital and not in the $3d$ orbital as might be expected. This location occurs because the $4s$ orbital is at a slightly lower energy

TABLE 5-3   The Maximum Numbers of Electrons Permitted in Each Orbital and in Each Principal Energy Level

Energy Level (n)	Maximum Number of Electrons in Each Orbital				Maximum Number of Electrons in Each Energy Level ($2n^2$)
	s	p	d	f	
1	2				2
2	2	6			2 + 6 = 8
3	2	6	10		2 + 6 + 10 = 18
4	2	6	10	14	2 + 6 + 10 + 14 = 32

Figure 5-7   In 1933, the Austrian physicist Erwin Schrödinger (1881–1961) was awarded the Nobel Prize in physics. Like many of his European colleagues, he was forced to flee his country when it was occupied by the Nazis in 1938. Schrödinger had many interests besides theoretical physics. He was a poet and sculptor with a wide knowledge and love of art and literature.

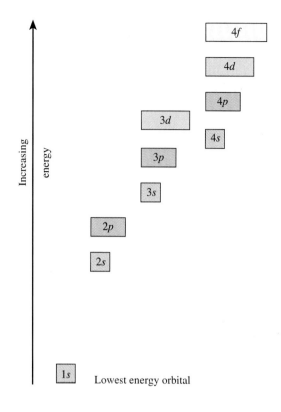

Figure 5-8 Energy levels of orbitals. Electrons are added to orbitals in order of increasing energy. Because the *4s* orbital is of lower energy than the *3d* orbital, it is filled first.

level than the *3d* orbital (Fig. 5-8), and orbitals are always filled in order of increasing energy. The *3d* orbital fills as electrons are added for the elements in the first row of transition elements—scandium (Sc, atomic number 21) through zinc (Zn, atomic number 30).

The order in which orbitals are filled does not have to be memorized; it can readily be determined by reference to an Aufbau diagram (Fig. 5-9). The

The German word *aufbau* means "building up."

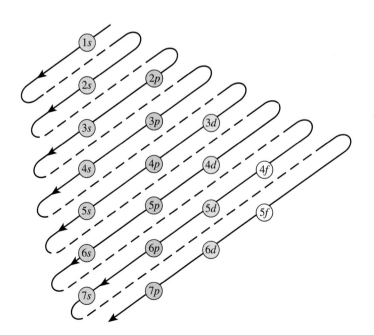

Figure 5-9 An Aufbau diagram can be used to determine the order in which orbitals are filled. Follow the arrows starting at 1*s*.

order in which orbitals are filled is found by following the diagonal arrows from top to bottom, starting with the top left arrow: $1s$, $2s$, $2p$, $3s$, $3p$, $4s$, $3d$, $4p$, $5s$, $4d$, $5p$, $6s$, and so on.

Writing electron configurations can often seem difficult and confusing. It need not be so, if the steps given in Example 5-2 are followed.

EXAMPLE 5-2   Write the electron configuration of sulfur, S, atomic number 16.

Solution:
1. Each S atom has 16 electrons.
2. Refer to Figure 5-9 to determine the *order* in which the orbitals are filled.
3. Refer to Table 5-3 to determine how many electrons can be placed in each orbital.

Order of filling orbitals	$1s$	$2s$	$2p$	$3s$	$3p$	$4s$	$3d$
Max. no. of electrons per orbital	2	2	6	2	6	2	10

Total number of electrons placed when orbitals $1s$ through $3s$ have been filled is: $2 + 2 + 6 + 2 = 12$. Four electrons remain to be placed. The next orbital to be filled, $3p$, can hold up to 6 electrons. The 4 remaining electrons are therefore placed in the $3p$ orbital. The electron configuration of sulfur (atomic number 16) is:

$$S = 1s^2 2s^2 2p^6 3s^2 3p^4$$

Note that the sum of the superscripts equals 16.

PRACTICE EXERCISE: Write the electron configurations for:
a. sodium (Na)   **b.** argon (Ar)   **c.** zinc (Zn)

Answer:
a. $(1s^2 2s^2 2p^6 3s^1)$ **b.** $(1s^2 2s^2 2p^6 3s^2 3p^6)$ **c.** $(1s^2 2s^2 2p^6 3s^2 3p^6 4s^2 3d^{10})$

The Position of Electrons in Orbitals

In the Bohr atom, electrons are visualized as circling the nucleus in well-defined orbits. If this model were correct, then in the same way that one can predict the position and speed of motion of a planet in space at any time, it should be possible to predict the exact location and speed of an electron in its orbit at a given time. In fact, this cannot be done. Schrödinger, using sophisticated mathematical explanations, concluded that, at any given time, it is possible to predict only the **probability** (or likelihood) of finding an electron in a certain volume of space around the nucleus. It is not possible to predict an electron's exact position.

We can use another analogy to explain the concept of probability. Let us assume that a bird feeder is set up in a yard and that there is one bird feeding from it. During the course of an hour the bird will eat at the feeder, fly off—often to nearby bushes and sometimes to more distant trees—and return to the feeder many times. If the bird is hungry, it will spend most of its time at, or close to, the feeder. By observing and recording the bird's position at intervals over the course of half an hour, for example, we can determine the area of the

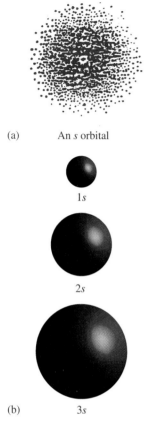

(a)   An *s* orbital

1*s*

2*s*

(b)   3*s*

Figure 5-10   *s* orbitals are spherical. (a) The area near the nucleus where the dots are most concentrated represents the region in the sphere where the probability of finding an electron is greatest. (b) Relative sizes of the 1*s*, 2*s*, and 3*s* orbitals.

yard in which the bird spends 90 percent of its time. In other words, we can determine the 90 percent probability of finding the bird in a definite area of the yard at a particular time. In a somewhat similar way, Schrödinger was able to calculate the volume of space in which an electron could be expected to be found 90 percent of the time.

For the hydrogen 1*s* orbital, the volume in which its electron can probably be found 90 percent of the time is in the shape of a sphere. The probability of finding the electron at a particular location within the sphere is greatest in the region close to the nucleus—that is, where the dots in Figure 5-10a are most concentrated. The sphere represents the 1*s* orbital. The 2*s*, 3*s*, and higher-numbered *s* orbitals are also spherical; they can be represented by increasingly larger spheres as shown in Figure 5-10b.

A *p* orbital is visualized as divided into three **suborbitals**, each shaped like a dumbbell. The dumbbells are aligned one in each of three spatial orientations designated *x*, *y*, and *z*, as shown in Figure 5-11a. Only two electrons can be present in any one of these *p* suborbitals, all of which are at the same energy level. As was the case for the *s* orbital, the probability of finding an electron at a particular location within the dumbbell is greatest in the regions closest to the nucleus (Fig. 5-11b).

The *p* electrons in any element are spread out as far as possible between the p_x, p_y, and p_z suborbitals. For example, carbon's two *p* electrons are located in different suborbitals—one in $2p_x$ and the other in $2p_y$ ($2p_x^1$, $2p_y^1$). Nitrogen's three *p* electrons are placed one in each of the $2p_x$, $2p_y$, and $2p_z$ suborbitals ($2p_x^1$, $2p_y^1$, $2p_z^1$). Only when we reach the next element, oxygen, which has four 2*p* electrons, are two electrons located in the same *p* suborbital ($2p_x^2$ $2p_y^1$, $2p_z^1$).

The *d* and *f* orbitals, like the *p* orbitals, are divided into suborbitals, each one containing two electrons. The shapes of the *d* and *f* suborbitals—the regions around the nucleus that define where the electrons are most likely to be located—are complex and need not be considered here.

ELECTRON CONFIGURATION AND THE PERIODIC TABLE

Once the electron configuration of the electrons in the atoms of the different elements was understood, it became evident that the electron configuration was the basis for the arrangement of the elements in the periodic table. When the elements are arranged in order of increasing atomic number, elements with similar properties recur at periodic intervals because elements with similar electron configurations recur at periodic intervals.

The correlation between the periodic table and the filling of electron orbitals is shown in Figure 5-12. The alkali and alkaline earth metals (Groups IA and IIA) are formed as one and two electrons are added to the *s* orbitals. The other representative elements (Groups IIIA–VIIIA) are formed as the *p* orbitals are filled. The main block of transition elements is formed as the *d* orbitals are filled, and the two rows of inner transition elements (the lanthanides and actinides) are formed as the *f* orbitals are filled.

To better understand this correlation between chemical properties and electron configuration, let us look at the first few members of three groups of elements (1) the alkali metals of Group IA, (2) the halogens of Group VIIA, and (3) the noble gases of Group VIIIA.

The alkali metals, which are very reactive, all have 1 electron in their outermost energy levels (shown in color):

Li $1s^2 2s^1$
Na $1s^2 2s^2 2p^6 3s^1$
K $1s^2 2s^2 2p^6 3s^2 3p^6 4s^1$

The halogens, which are also very reactive, all have 7 electrons in their outermost energy levels and are therefore one electron short of filled outermost orbitals:

F $1s^2 2s^2 2p^5$
Cl $1s^2 2s^2 2p^6 3s^2 3p^5$
Br $1s^2 2s^2 2p^6 3s^2 3p^6 4s^2 3d^{10} 4p^5$

The noble gases, which are all very inert and undergo only a few chemical reactions, have filled outermost orbitals. Except for helium, which has a total of only 2 electrons, they all have 8 electrons in their outermost energy levels.

He $1s^2$
Ne $1s^2 2s^2 2p^6$
Ar $1s^2 2s^2 2p^6 3s^2 3p^6$
Kr $1s^2 2s^2 2p^6 3s^2 3p^6 4s^2 3d^{10} 4p^6$

The above examples illustrate the repeating patterns in electron arrangements in the outermost energy levels. They also show us that elements with *filled* outermost energy levels (noble gases) are very inert and unreactive, whereas elements that have *one more* electron than a filled outermost energy level (alkali metals), or *one less* (halogens), are very reactive. As we shall see in the next section, the most important factor in determining how elements bond is the number of electrons in the outermost energy level.

ELECTRON CONFIGURATION AND BONDING

Valence Electrons

The electrons in an element's outermost energy level—that is, the electrons in the *highest-numbered* energy level—are known as **valence electrons**. As shown above, for potassium (K) the highest energy level is 4, and there is one electron in this level. Potassium, therefore, has one valence electron. The highest energy level for chlorine is 3 and includes both a $3s$ and a $3p$ orbital. The electrons in both these orbitals are valence electrons. Chlorine, therefore, has seven (2 + 5) valence electrons. For bromine, the highest energy level is 4. Only electrons in this level are valence electrons; the $3d$ electrons are not counted as valence electrons. Bromine, therefore—like chlorine—has seven valence electrons.

The concept of valence electrons is simplified when each valence electron is represented by a dot, as suggested by G. N. Lewis in 1916. Table 5-4 shows **electron-dot symbols** (or **Lewis symbols**) and electron configura-

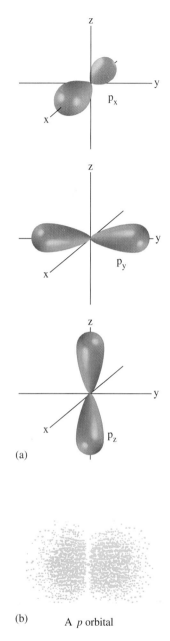

(a)

(b)   A *p* orbital

Figure 5-11   The *p* suborbitals are dumbbell-shaped. (a) The subscripts x, y, and z indicate the axes along which the suborbitals are oriented. (b) The areas near the nucleus where the dots are most concentrated represent the regions in the dumbbells where the probability of finding an electron is greatest.

The number of valence electrons is the same for each element in a particular group of the periodic table.

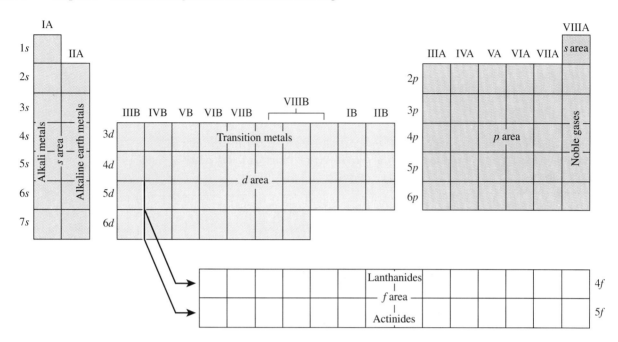

Figure 5-12 Correlation between electron configuration and the periodic table. There are 2, 6, 10, and 14 elements respectively in each period in the s, p, d, and f areas of the periodic table. These numbers are the maximum numbers of electrons that each of the orbitals can hold.

tions for the first 18 elements in the periodic table. The arrangement of the dots around the element symbol is arbitrary, except that dots are never paired unless more than four are present (except in the case of helium which has only 2 electrons). You will notice that the number of valence electrons (dots) is the same as the Roman numeral group number in the periodic table. Thus, the Group IA elements—lithium (Li), sodium (Na)—have one valence electron; the Group IIA elements—beryllium (Be), magnesium (Mg)—have two; the Group VIIA elements—fluorine (F), chlorine (Cl)—have seven; and so on.

EXAMPLE 5-3: Write electron dot symbols for aluminum (Al), nitrogen (N), and bromine (Br).

Solution: The number of dots in the symbol is equal to the Roman numeral group number.
Al (Group IIIA), 3 dots   N (Group VA), 5 dots   Br (Group VIIA), 7 dots

PRACTICE EXERCISE: Write electron dot symbols for Mg, C, S, Na, Ar, and Cl. Refer to Table 5-4 for the answers.

You will also notice in Table 5-4 that (1) the maximum number of valence electrons for any of the elements listed is eight, and (2) with the exception of helium, with a total of only two electrons, all the noble gas (Group VIIIA) elements have the maximum number of eight valence electrons. The noble gases

TABLE 5-4 Electron-dot Symbols and Electron Configurations for the First 18 Elements (Valence Electrons are Colored).

IA	IIA	IIIA	IVA	VA	VIA	VIIA	Noble Gases VIIIA
H•							He:
$1s^1$							$1s^2$
Li•	•Be•	•B•	•C•	•N•	•O•	•F•	•Ne•
$1s^22s^1$	$1s^22s^2$	$1s^22s^22p^1$	$1s^22s^22p^2$	$1s^22s^22p^3$	$1s^22s^22p^4$	$1s^22s^22p^5$	$1s^22s^22p^6$
Na•	•Mg•	•Al•	•Si•	•P•	•S•	•Cl•	•Ar•
$1s^22s^22p^63s^1$	$1s^22s^22p^63s^2$	$1s^22s^22p^63s^23p^1$	$1s^22s^22p^63s^23p^2$	$1s^22s^22p^63s^23p^3$	$1s^22s^22p^63s^23p^4$	$1s^22s^22p^63s^23p^5$	$1s^22s^22p^63s^23p^6$

are the most unreactive of all elements and rarely form compounds. In contrast to the noble gases, elements that have either one (Group IA) or seven valence electrons (Group VIIA) are extremely reactive and readily form compounds. The correlation between noble gas electron configuration and stability led Lewis to conclude that this structure, with eight valence electrons, in some way conferred great stability and prevented the noble gases from combining with other elements. He further concluded that other elements combined with one another in order to acquire a noble gas configuration. On the basis of these conclusions, Lewis formulated the octet rule—or rule of eight.

The Octet Rule

The **octet rule** can be stated as follows: *In forming compounds, atoms gain, lose, or share one or more valence electrons in such a way that they achieve the electron configuration of the nearest noble gas in the periodic table.* Although there are many exceptions to the octet rule, it provides a useful framework for explaining the many basic concepts of bonding that are described in the following sections.

IONIC BONDS: DONATING AND ACCEPTING ELECTRONS

When common salt (NaCl) is formed, sodium atoms transfer their single valence electrons to chlorine atoms. The sodium atom, in losing a negatively charged electron, becomes positively charged and achieves the stable electron configuration of neon (Ne) (see Table 5-4). The chlorine atom, in gaining an electron, becomes negatively charged and achieves the electron configuration of argon (Ar) (see Table 5-4). We can write the *transfer* of electrons as follows:

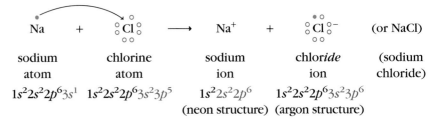

Na	+	$\ddot{\underset{..}{Cl}}$	→	Na$^+$	+	$\ddot{\underset{..}{Cl}}^-$	(or NaCl)
sodium atom		chlorine atom		sodium ion		chlor*ide* ion	(sodium chloride)
$1s^2 2s^2 2p^6 3s^1$		$1s^2 2s^2 2p^6 3s^2 3p^5$		$1s^2 2s^2 2p^6$		$1s^2 2s^2 2p^6 3s^2 3p^6$	
				(neon structure)		(argon structure)	

Although in the above equation, and in subsequent equations, electrons are given more than one symbol (in this case, ○ and •) in order to identify their source, it is very important to understand that all electrons are identical regardless of origin. For example, the electron added to a chlorine atom to form a chloride ion is indistinguishable from the other electrons already present in the chlorine atom.

The charged atoms of sodium and chlorine are called **ions**. Positively charged ions are termed **cations**, and negatively charged ions are termed **anions**.

It must be remembered that although the electron configurations of Na$^+$ and neon (Ne) and of Cl$^-$ and argon (Ar) are the same, the atoms are *not* identical. The number of protons does not change during electron transfer, and therefore the identity of the element does not change. Na$^+$ has 11 protons and is charged; Ne has 10 protons and is uncharged. Similarly, Cl^- has 17 protons and is charged, while Ar has 18 protons and is uncharged.

Because of their opposite charges, sodium ions (Na^+) and chloride ions (Cl^-) are attracted to each other. These electrostatic forces that hold the ions together are called **ionic bonds**, and the compounds that are formed, such as sodium chloride, are called ionic compounds.

Crystals of table salt (NaCl) are made up of billions and billions of Na^+ and Cl^- ions that, because of electrostatic attractions between oppositely charged ions, are arranged in an orderly fashion, with Na^+ ions and Cl^- ions alternating, as shown diagrammatically in Figure 5-13a. Sodium chloride crystals form as cubes. The ratio of Na^+ ions to Cl^- ions in the crystal is 1:1, and the simplest formula is NaCl. A more accurate representation of NaCl, showing the relative sizes of the Na^+ and Cl^- ions and the way the ions are packed together in a salt crystal, is shown in Figure 5-13b.

The octet rule predicts the number of atoms of one element that will react with the atom (or atoms) of another element. Let us see how magnesium (Mg), a Group IIA element, bonds with chlorine (Cl). To achieve the stable neon electron configuration, magnesium must lose two electrons (see Table 5-4). We have just seen that chlorine needs to gain one electron to achieve a stable structure. Therefore, to form a compound with Cl, Mg transfers one electron to each of two Cl atoms as shown below. In losing two electrons, Mg acquires a 2+ charge. The ionic compound formed ($MgCl_2$) has no net charge; the 2+ charge on Mg is balanced by the two 1– charges on Cl.

> A crystal can be defined as a solid whose internal arrangement of ions, atoms, or molecules is repeated regularly in any direction through the solid.

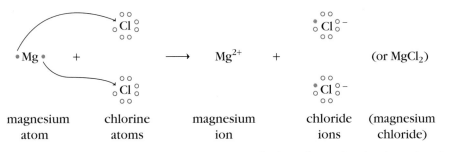

| magnesium atom | chlorine atoms | magnesium ion | chloride ions | (magnesium chloride) |

The numbers of valence electrons in Mg and Cl explain why the formula of magnesium chloride is $MgCl_2$ and not, for example, MgCl or Mg_2Cl.

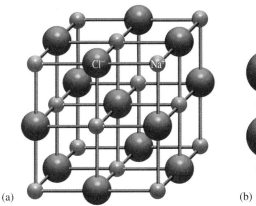

(a)

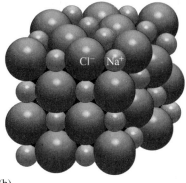

(b)

Figure 5-13   The structure of a sodium chloride (NaCl) crystal. In a crystal, each Na^+ ion is surrounded by six Cl^- ions and each Cl^- ion is surrounded by six Na^+ ions. (a) Diagram showing how the orderly arrangement of Na^+ and Cl^- ions leads to the formation of crystals in the shape of cubes. (b) Model showing the relative sizes of the Na^+ and Cl^- ions.

EXAMPLE 5-4:   What is the formula of the ionic compound formed between potassium (K) and oxygen (O)?

Solution:

1. Refer to Table 5-4 to find the electron-dot symbols of K and O. K needs to
lose one electron to achieve the Ar electron configuration; O needs to *gain*
two electrons to achieve the Ne electron configuration.

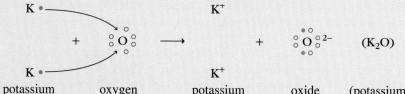

| potassium | oxygen | potassium | oxide | (potassium |
| atoms | atom | ions | ion | oxide) |

2. Since the oxygen atom has gained two electrons it has a 2– charge. Each
potassium atom has a 1+ charge because each has lost one electron.

3. The formula of potassium oxide is K_2O.

PRACTICE EXERCISE: What is the formula of the ionic compound formed
between:

a. lithium and sulfur?   **b.** calcium and fluorine?   **c.** magnesium and oxygen?

Answer: **a.** Li_2S **b.** CaF_2 **c.** MgO

Positive and negative ions always form together. It is a complementary
process. For ions to form, at least two atoms must be present: one to donate an
electron and another to accept the donated electron. For example, sodium will
not lose an electron to form Na^+ unless a chlorine (or some other) atom is present that can accept it. As we saw in the examples given above ($NaCl$, $MgCl_2$,
and K_2O), ionic compounds are neutral; in each case, the total positive charge
is balanced by the total negative charge.

Ionic compounds are formed primarily when metals on the left-hand side
of the periodic table donate electrons to nonmetals (excluding the noble gases)
on the right-hand side of the table.

Metals in Groups IA, IIA, and IIIA give up one, two, and three electrons to
form ions with 1+, 2+, and 3+ charges.

Examples: Li^+ (Group IA)

Ca^{2+} (Group IIA)

Al^{3+} (Group IIIA)

Nonmetals in Groups VA, VIA, and VIIA gain three, two, and one elec-
tron(s) to form ions with 3–, 2–, and 1– charges.

Examples: N^{3-} (Group VA)

S^{2-} (Group VIA)

F^- (Group VIIA)

Among the representative elements, the tendency to form ions is greatest
for elements in Group IA, at the far left of the table, and in Group VIIA, at the
far right. This tendency decreases as one approaches the center of the table and
is at a minimum in Group IVA.

Transition Element Ions

So far, we have discussed only ionic compounds formed by representative elements. The formation of these compounds can be predicted by the octet rule. Transition elements, however, and some representative elements in the lower part of the *p* area of the periodic table (Fig. 5-12), form ionic compounds in less predictable ways. For example, it is a characteristic of transition metals that they form more than one type of ion. Depending on conditions, iron forms Fe^{2+} or Fe^{3+} ions, and copper forms Cu^{+} or Cu^{2+} ions.

All transition elements are metals.

Examples of common ions and their positions in the periodic table are given in Figure 5-14.

Nomenclature of Two-Element Ionic Compounds

Ionic compounds containing just two different elements are named as follows: The name of the metallic element is given and then, as a separate word, the name stem of the nonmetallic element with the suffix *-ide*, as shown in Table 5-5. Thus, the compound CaO is named calcium oxide; AlF_3 is aluminum fluoride; Li_2S is lithium sulfide; and Na_3N is sodium nitride.

The compounds of metal elements that form more than one ion are named by writing within parentheses a Roman numeral corresponding to the magnitude of the positive charge, immediately following the name of the metal. Thus, the chloride of Fe^{2+} ($FeCl_2$) is named iron(II) chloride, and the chloride of Fe^{3+} ($FeCl_3$) is iron(III) chloride. Similarly, CuO and Cu_2O are named copper(II) oxide and copper(I) oxide. The appropriate charge on the metal ion in the compound can always be calculated from the known charge on the nonmetal ion. For example, the oxygen ion has a 2− charge. Copper in CuO must, therefore, have a 2+ charge to balance the negative charge. Similarly, each copper ion in Cu_2O must have a 1+ charge.

Period	1 IA	2 IIA	3 IIIB	4 IVB	5 VB	6 VIB	7 VIIB	8 VIIIB	9 VIIIB	10 VIIIB	11 IB	12 IIB	13 IIIA	14 IVA	15 VA	16 VIA	17 VIIA	18 VIIIA
1							H^+											2
2	3 Li^+	4											5	6	7 N^{3-}	8 O^{2-}	9 F^-	10
3	11 Na^+	12 Mg^{2+}											13 Al^{3+}	14	15 P^{3-}	16 S^{2-}	17 Cl^-	18
4	19 K^+	20 Ca^{2+}	21	22	23	24	25	26 Fe^{2+} Fe^{3+}	27	28	29 Cu^+ Cu^{2+}	30 Zn^{2+}	31	32	33	34	35 Br^-	36
5		38 Sr^{2+}	39	40	41	42	43	44	45	46	47 Ag^+	48	49	50	51	52	53 I^-	54
6		56 Ba^{2+}	57	72	73	74	75	76	77	78	79	80	81	82	83	84	85	86

Figure 5-14 Common ions and their positions in the periodic table.

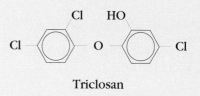

Triclosan

CONSUMER BRIEF

Antiperspirants and Deodorants

American adults are very concerned about body odor. The best way to control body odor is with good personal hygiene. Soaps and body creams are sold to scent and moisten our skin while bathing. Perfumes, which can be traced as far back as Cleopatra in ancient Egypt, can be considered sophisticated chemical body odor control systems. Modern chemistry provides us with deodorants and antiperspirants for preventing and covering up body odor.

Body odor is caused by the action of microscopic skin organisms on the secretions of the apocrine sweat glands, which occur in the armpits and groin. Apocrine sweat glands have a reservoir that reaches the surface of the skin via a duct. The duct usually empties into the upper part of the hair follicle. The milky apocrine secretion consists of water, sodium chloride, proteins, carbohydrates, and iron and ammonium salts. The ducts empty in response to emotional stimuli such as fear, pain, or anxiety. Because the apocrine gland does not develop until puberty, young children do not have a strong body odor.

Apocrine sweat is odorless when first secreted. The warm, moist environment on the surface of the skin has a reserve of organic and inorganic nutrients that promotes the proliferation of ubiquitous skin bacteria. The metabolites of the bacterial growth, short-chain carboxylic acids and steroidal compounds, produce the objectionable odor.

Deodorants and antiperspirants are two distinct types of products. Deodorants are cosmetic products that act either by controlling the growth of the bacteria that produce the odor or by masking the body odor as it forms. Most deodorants use both techniques. One of the first antimicrobials used in deodorants was hexachlorophene. But hexachlorophene was found to be toxic in animal testing, and it was banned by the FDA. Most deodorants now contain triclosan, or 2,4,4'-trichloro-2'-hydroxydiphenyl ether, a broad-spectrum bacteriostat that is effective at low concentrations against odor-causing bacteria. They also contain cosmetic oils, alcohol, and a distinctive fragrance.

The first antiperspirant, Mum, which was introduced in 1888, used zinc oxide (ZnO) as an active ingredient. Today such an ineffective product would not be allowed to be marketed as an antiperspirant. In 1902 Everdry, which was a solution of aluminum chloride ($AlCl_3$), was offered for sale. If it were available today it would probably be the most effective antiperspirant on the market. But not many consumers would want to use it. It was a messy liquid that irritated the underarm and damaged clothing. In the 1940's it was found that the aluminum ion (Al^{3+}) made the product effective and that if some of the chloride ions (Cl^-) in aluminum chloride were replaced with hydroxide ions (OH^-), the product was less irritating. Today, aluminum chlorohydrates such as aluminum dichlorohydrate, $Al_2(OH)_4Cl_2$, or aluminum chlorohydrate, $Al_2(OH)_5Cl$, are used as the drying agent in antiperspirants.

The aluminum chlorohydrates release aluminum ions (Al^{3+}). Aluminum ions (as alum or septic pencil) are known to coagulate proteins and are used to stop bleeding from small cuts caused by razors while shaving. The same process appears to happen in the sweat gland. The aluminum ions close the ducts that deliver the sweat. The aluminum chlorohydrates which are alkaline also kill the odor-causing bacteria on the surface of the skin.

Question

Given the composition of antiperspirants, do you think antiperspirant sprays (aerosol or pump sprays) pose a potential danger to the user? Should all antiperspirants be sold as roll-ons or solids?

TABLE 5-5   Names of Common Nonmetallic Ions

Element	Symbol	Name of Anion	Symbol
Fluorine	F	Fluoride	F^-
Chlorine	Cl	Chloride	Cl^-
Bromine	Br	Bromide	Br^-
Iodine	I	Iodide	I^-
Oxygen	O	Oxide	O^{2-}
Sulfur	S	Sulfide	S^{2-}
Nitrogen	N	Nitride	N^{3-}
Phosphorus	P	Phosphide	P^{3-}

COVALENT BONDS: SHARING ELECTRONS

In two-element ionic compounds, the two elements involved in forming the ionic bonds are a metal and a nonmetal. As we have seen, when metal atoms approach nonmetal atoms, there is a strong tendency for the metal atoms to lose electrons and for the nonmetal atoms to gain electrons in such a way that positive and negative ions are formed, and atoms of both elements attain a stable configuration with an octet of electrons. When two *identical* atoms come together, however, their tendencies, if any, to lose or gain electrons are the same. Neither atom is more likely than the other to transfer an electron. Identical atoms can achieve the stable noble gas electron configuration by *sharing* electrons.

Two chlorine atoms, for example, combine together to form a molecule of chlorine gas by sharing their single unpaired electrons as shown below.

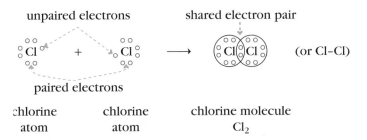

The chlorine molecule has no charge because there is no net gain or loss of electrons. The paired electrons in each chlorine atom that are not involved in bonding are called **unshared**, or **nonbonding**, electron pairs. The chlorine molecule can be represented as Cl–Cl. The dash represents the shared electrons, and the unshared electron pairs are not indicated.

Free chlorine atoms are very unstable, but by combining with other chlorine atoms they achieve the stable argon electron configuration. In accordance with the octet rule, each Cl atom in the chlorine molecule has eight electrons around it. Bonds formed in this way, by sharing one or more pairs of electrons, are called **covalent bonds**.

Hydrogen molecules, like chlorine molecules, are formed by sharing electrons.

A molecule can be defined as a unit of a pure substance in which the atoms are held together by covalent bonds.

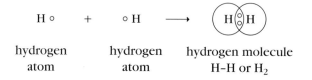

In sharing their single electrons, the hydrogen atoms achieve the stable two-electron helium configuration.

A more realistic picture of the covalent bond between two hydrogen atoms is obtained if we remember that the single electron in each hydrogen atom is located within a spherical 1s orbital. When two hydrogen atoms come close together, their 1s orbitals merge and overlap as shown in Figure 5-15.

Besides chlorine and hydrogen, all other elements with names ending in -*ine* or -*gen* form diatomic molecules by sharing electrons. The other halogens (fluorine, bromine, and iodine) and the common gases nitrogen and oxygen all form diatomic molecules: F_2, Br_2, I_2, N_2, and O_2.

Covalent bonds also form between atoms of *different* elements, usually between atoms of nonmetal elements. For example, the nonmetal carbon and the nonmetal chlorine combine by means of covalent bonds to form carbon tetrachloride, CCl_4. Carbon (Group IVA) needs four electrons to complete its octet; chlorine needs one. The number of covalent bonds that an atom forms equals the number of electrons it needs to attain the noble gas electron configuration. Carbon needs four electrons and will therefore form four bonds; each chlorine atom needs one electron and each will form one bond.

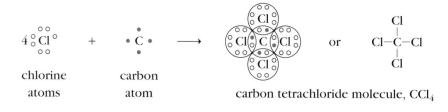

In the CCl_4 molecule, the C atom and each of the four Cl atoms have eight electrons surrounding them. Thus, each atom has achieved the stable noble gas configuration.

Figure 5-15 Overlap of 1s orbitals in the H_2 molecule. The two shared electrons are distributed within the overlapping orbitals. The probability of finding an electron at any particular moment is greatest in the space between the nuclei.

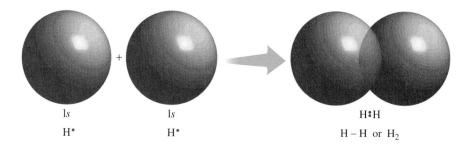

EXAMPLE 5-5:   Write an electron-dot structure to show the compound that is formed when nitrogen (N) and hydrogen (H) react to form covalent bonds.

Solution:
1. Determine the number of electrons that nitrogen and hydrogen need to attain the stable noble gas configuration.
2. Nitrogen (Group VA) has five valence electrons (see Table 5-4) and needs three more to complete its octet. A nitrogen atom will, therefore, form three covalent bonds.
3. Hydrogen has one valence electron and needs one more to attain the two-electron He structure. A hydrogen atom will, therefore, form one covalent bond.
4. Thus three H atoms will bond covalently with one N atom to form NH_3, as shown. In each bond, the nitrogen atom shares an electron with one of the three hydrogen atoms.

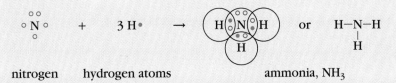

nitrogen   hydrogen atoms   ammonia, NH_3

PRACTICE EXERCISE: Write the electron-dot structure for the compound that is formed when the following elements combine:
a. silicon and hydrogen   **b.** two atoms of iodine   **c.** phosphorus and chlorine

Answer:

a. $H \overset{\overset{\displaystyle H}{\circ\ \bullet}}{\underset{\bullet\ \circ}{\circ\ Si\ \circ}} H$ SiH_4

       H

b. $\overset{\circ\circ\ \ \circ\circ}{\underset{\circ\circ\ \ \circ\circ}{\circ\ I\ \overset{\circ}{\underset{\circ}{}}\ I\ \circ}}$ I_2

c. $\overset{\circ\circ\ \ \bullet\bullet\ \ \circ\circ}{\underset{\circ\circ\ \ \bullet\circ\ \ \circ\circ}{\circ\ Cl\ \overset{\bullet}{\underset{\bullet}{}}\ P\ \overset{\bullet}{\underset{\circ}{}}\ Cl\ \circ}}$ PCl_3

       $\overset{\circ}{\underset{\circ\circ}{\circ\ Cl\ \circ}}$

Multiple Covalent Bonds

The sharing of electrons is not limited to one pair of electrons. Nitrogen, as we saw in Example 5-5, has five valence electrons and needs three more electrons to complete its octet. In forming N_2, the two N atoms in N_2 cannot attain the stable eight-electron structure by sharing a single pair of electrons.

$$\overset{\circ}{\underset{\circ}{\circ\ N\ \circ}} \quad + \quad \overset{\bullet}{\underset{\bullet}{\bullet\ N\ \bullet}} \quad \longrightarrow \quad \overset{\circ\quad\bullet}{\underset{\circ\quad\bullet}{\circ\ N\ \overset{\circ}{}\ N\ \bullet}} \quad \text{(incorrect)}$$

Each nitrogen in the above arrangement has only six electrons around it. However, the desired octet can be attained if *three* pairs of electrons are shared.

$$\overset{\circ}{\underset{\circ}{\circ\ N\ \circ}} \quad + \quad \overset{\bullet}{\underset{\bullet}{\bullet\ N\ \bullet}} \quad \longrightarrow \quad \overset{\bullet\bullet\bullet}{\underset{\circ\circ\circ}{\circ\ N\ \overset{}{}\ N\ \bullet}} \quad \text{or} \quad N{\equiv}N$$

The bond formed between the two N atoms is termed a **triple bond**. Each N atom is surrounded by an octet of electrons.

In a similar way, **double bonds** are formed when atoms share *two* pairs of electrons. Double bonds occur in carbon dioxide, CO_2. Carbon needs four electrons to complete its octet. It will, therefore, form four bonds. Each O needs two electrons, and each will form two bonds. Thus in forming CO_2, the C atom is joined to the two O atoms by double bonds.

$$:\!\overset{\circ\,\circ}{\underset{\circ\,\circ}{O}}\!:^{\circ} \quad + \quad ^{\circ}\!\overset{\circ}{\underset{\circ\,\circ}{O}}\!:^{\circ}_{\circ} \quad + \quad \cdot\overset{\bullet}{\underset{\bullet}{C}}\cdot \quad \longrightarrow \quad :\!\overset{\circ}{\underset{\circ}{O}}\!:^{\circ\,\circ}_{\circ\,\circ}\!C\!:^{\bullet\,\bullet}_{\bullet\,\bullet}\overset{\circ}{\underset{\circ}{O}}\!:^{\circ}_{\circ} \quad \text{or} \quad O{=}C{=}O$$

Although the octet rule cannot explain the formation of all covalent compounds, it accounts for a great many of them, and, as we will discuss in more detail in Chapter 9, it explains the enormous number and variety of compounds formed by carbon.

EXAMPLE 5-6:   Write the electron-dot structure to explain the formation of ethylene, C_2H_4, from carbon and hydrogen atoms.

Solution:

a. Draw the electron-dot symbols for carbon and hydrogen.

$$\cdot\overset{\bullet}{\underset{\bullet}{C}}\cdot \qquad H\,\circ$$

b. Each H needs one more electron to attain the stable two-electron helium structure. Each hydrogen will form one bond by sharing its electron with carbon.

c. Each C needs four more electrons to attain an octet. Each will form four bonds.

d. If the C atoms share one electron each to form a carbon–carbon single bond and the four hydrogen atoms share one electron with the C atoms as shown below, then the total number of electrons around each carbon will be only seven. This number does not satisfy the octet rule.

$$\begin{array}{c} H \quad H \\ \overset{\circ\,\bullet}{}\quad\overset{\bullet\,\circ}{} \\ H\!:\!C\!:\!C\!:\!H \\ \overset{\bullet}{}\quad\overset{\circ}{} \end{array} \qquad \text{(incorrect)}$$

e. The only way the C atoms can attain an octet is if each carbon shares two electrons with the other, forming a double bond (=).

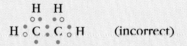

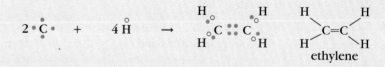

ethylene

PRACTICE EXERCISE   Draw electron-dot structures for each of the following:

a. Carbon monoxide, CO.

b. Nitrous oxide, N_2O (nitrogen–nitrogen–oxygen)

Answer:

a.

$$:\!C\!:^{\circ\,\circ}_{\circ\,\circ}\!\overset{\circ}{\underset{\circ}{O}}\!: \qquad C{\equiv}O$$

b.

$$:\!\overset{\bullet\,\bullet}{\underset{\bullet\,\bullet}{N}}\!:^{\bullet\,\bullet}\!\overset{\circ\,\circ}{\underset{}{N}}\!:^{\circ\,\circ}\!\overset{\circ\,\circ}{\underset{\circ\,\circ}{O}} \qquad N{=}N{=}O$$

POLAR COVALENT BONDS

In molecules such as H_2 and Cl_2, where both atoms are identical, the electron pair forming the bond is shared equally between the two atoms. The bond is a nonpolar covalent bond. Where two *different* atoms are covalently bonded, however, the bonding electrons are not shared equally but are shifted more toward one atom than the other. This shift occurs because one of the atoms attracts electrons more strongly to itself than the other.

Electronegativity

The ability to attract bonding electrons can be expressed in terms of electronegativity, a concept that was introduced by Linus Pauling (1901–) in 1934. Electronegativity can be defined as the relative tendency of an atom in a molecule to attract a shared pair of electrons in a bond to itself. Electronegativity values for representative elements are shown in Figure 5-16. These values are relative and are based on an arbitrary value of 4.0 for the most electronegative element, fluorine (F). The higher the electronegativity value for an element, the greater is the ability of that element to attract electrons to itself.

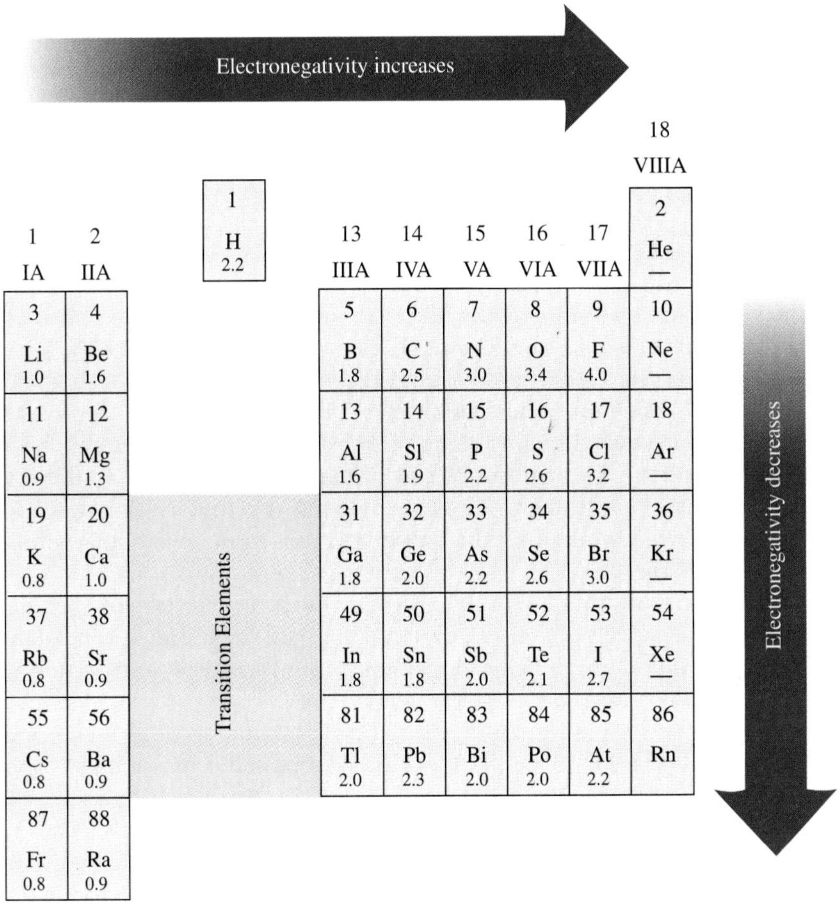

Figure 5-16   Electronegativity values for representative elements. In general, electronegativity decreases from top to bottom in a group and increases from left to right across a period. Fluorine (electronegativity = 4.0) is the most electronegative element.

With one or two exceptions, electronegativity increases from left to right across the periodic table and decreases from top to bottom down the table. The most electronegative element (fluorine) is located in the top right-hand corner of the periodic table. Other elements with high electronegativity values (3.0 or greater) are nitrogen and oxygen, located just to the left of fluorine, and chlorine and bromine, which are located just below it.

Let us consider the hydrogen chloride molecule, HCl. To achieve the stable noble gas electron structure, the H atom and the Cl atom both need one more electron; therefore, they share an electron as shown below. But Cl (electronegativity, 3.2) attracts electrons more strongly than does H (electronegativity, 2.2), and therefore the shared pair of electrons is drawn toward the chlorine atom. As a result, the chlorine end of the molecule acquires a *partial* negative charge, and the hydrogen end acquires a corresponding partial positive charge. The result is a **polar molecule**.

$$H \cdot \quad + \quad {}_{\circ}^{\circ\circ}Cl_{\circ}^{\circ} \quad \longrightarrow \quad H\,{}_{\circ}^{\circ\circ}Cl_{\circ}^{\circ} \quad \text{or} \quad H{-}Cl \quad \text{or} \quad \overset{\delta^+ \quad \delta^-}{H{-}Cl}$$

Partial charges are indicated either by the Greek letter symbols δ^+ and δ^- or by an arrow with a + on its tail. In magnitude, partial charges are always less than the 1– and 1+ charges of electrons and protons. A molecule that has an unequal distribution of charge has a **dipole**, and bonds formed by unequal sharing of bonding electrons are called **polar covalent bonds**.

The greater the difference in electronegativity between two atoms forming a bond, the more polar the bond is. In general, if the electronegativity difference between bonding atoms is greater than 1.7, the bond formed is ionic; if the difference lies between 0 and 1.7, the bond is polar covalent; and if the difference is zero, the bond is nonpolar covalent.

POLYATOMIC IONS

In each of the compounds having more than one bond that we have discussed so far, the bonds have been of the same kind. For example, in $MgCl_2$, the two bonds between the chlorine atoms and the magnesium atom are both ionic bonds. In CCl_4, the four bonds joining the chlorine atoms to the carbon atom are all covalent bonds. In many common substances, however, atoms are held together in part by covalent bonds and in part by ionic bonds. In these substances, atoms of more than one element bond together covalently to form **polyatomic** (many-atom) ions; the polyatomic ions form ionic bonds with ions of opposite charge.

Common polyatomic ions are listed in Table 5-6. All except the ammonium ion are negatively charged. Polyatomic ions are stable and in general maintain their identity during chemical reactions. They are always associated with an ion (or ions) of equal but opposite charge.

Polyatomic ions are found in many familiar commercial products and in the minerals that make up the earth's crust. Compounds containing phosphate ions (PO_4^{3-}), and nitrate ions (NO_3^-) are major constituents of fertilizers—and are also potential water pollutants (Chapter 12). Calcium carbonate ($CaCO_3$) is the main ingredient in limestone and marble. Sodium bicarbonate ($NaHCO_3$) is a common antacid and plays a vital role in maintaining acid-base balance in the

TABLE 5-6 Some Common Polyatomic Ions

Ammonium	NH_4^+
Hydroxide	OH^-
Nitrate	NO_3^-
Nitrite	NO_2^-
Bicarbonate	HCO_3^-
Acetate	$CH_3CO_2^-$
Cyanide	CN^-
Carbonate	CO_3^{2-}
Sulfate	SO_4^{2-}
Sulfite	SO_3^{2-}
Chromate	CrO_4^{2-}
Phosphate	PO_4^{3-}

body. Sodium hydroxide (NaOH), an important industrial chemical, is among the top 10 chemicals produced in greatest quantity in the United States. Silicates, which contain the SiO_4^{4-} ion, are the most abundant minerals on earth and are the main constituents of rock, sand, and clay.

The formation of covalent and ionic bonds in sodium hydroxide (NaOH) is readily understood if we write the electron-dot symbols for each of the atoms in the formula—Na, O, and H—and then determine how electrons must be donated, accepted, or shared in order for each atom to acquire a stable noble gas electron configuration.

$$Na \ + \ \overset{\circ\circ}{\underset{\circ}{\circ}O\circ} \ + \ H\bullet \ \longrightarrow \ \overset{\circ\circ}{\underset{\circ\times}{\circ}O{\bullet}H^-} \ + \ Na^+$$

hydroxide ion   sodium ion
(sodium hydroxide, NaOH)

The stable configuration is achieved if the H and O atoms form a covalent bond by sharing an electron pair and the Na atom transfers its single valence electron to the O atom. As a result of the electron transfer, the OH group acquires a negative charge and the Na atom acquires a positive charge. An ionic bond is formed between the OH^- ion and the Na^+ ion. Because both the O atom and the H atom in OH^- have the noble gas electron configuration, the hydroxide ion is very stable.

Calcium carbonate ($CaCO_3$) and sodium phosphate (Na_3PO_4) are formed in a similar way.

$$Ca \ + \ {\bullet}C{\bullet} \ + \ 3\,\overset{\circ\circ}{\underset{\circ}{\circ}O\circ} \ \longrightarrow \ Ca^{2+} \ + \ \left[\,\overset{\circ O\circ}{\underset{\circ\times}{\circ}O{\bullet}}C{\bullet}\overset{\circ}{\underset{\circ\times}{O}\circ}\,\right]^{2-}$$

calcium ion   carbonate ion
(calcium carbonate, $CaCO_3$)

$$3\,Na \ + \ {\bullet}P{\bullet} \ + \ 4\,\overset{\circ\circ}{\underset{\circ}{\circ}O\circ} \ \longrightarrow \ 3\,Na^+ \ + \ \left[\,O\ O\ P\ O\ O\,\right]^{3-}$$

sodium ions   phosphate ion
(sodium phosphate, Na_3PO_4)

The bonds forming the CO_3^{2-} and PO_4^{3-} ions are covalent bonds. The bonds between the Ca^{2+} ion and the CO_3^{2-} ion and between the Na^+ ions and the PO_4^{3-} ion are ionic bonds.

METALLIC BONDING

The bonds, or forces, that hold atoms together in metals are unlike any of the bonds we have described so far, and they are not completely understood. As we noted in Chapter 4, metals have many characteristic properties that distinguish them from other substances. They are, for example, good conductors of electricity and heat. Any model of bonding between metal atoms must account for these and other distinctive properties.

According to one model, a metallic element in its solid state consists of a regular lattice of positively charged metal ions surrounded by a "sea" of valence electrons. The electrons are not associated with particular positive ions but

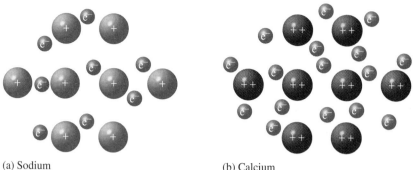

Figure 5-17   The electron sea model of bonding in metals. Electrons are uniformly distributed throughout the metal. They are free to wander and are not attached to particular positive ions.   (a) Sodium. (b) Calcium.

(a) Sodium

(b) Calcium

wander through the lattice of positive ions as shown in Figure 5-17. Because they can move freely, electrons are able to conduct electricity and heat through the metal.

POLAR MOLECULES

Hydrogen chloride (HCl) was used as an example of a polar molecule. As we saw, the H and Cl atoms in this molecule are joined by a single bond, and because of the unequal distribution of charges, the chlorine end of the molecule is slightly more negative than the hydrogen end. Many molecules, unlike HCl, are held together by more than one polar covalent bond. These molecules, as a whole, may be polar or nonpolar depending on the spatial arrangement of the bonds in the molecules.

Carbon dioxide is an example of a *nonpolar* molecule that is held together by two *polar* bonds. It is a linear molecule in which the two O atoms form polar covalent double bonds with the single C atom. Oxygen is more electronegative than carbon, and therefore the oxygen ends of the molecule have partial negative charges relative to the central carbon atom.

$$O=C=O$$
$$\longleftarrow\!\!+ \quad +\!\!\longrightarrow$$

Carbon dioxide: a linear nonpolar molecule

Despite the polarity of the bonds, the molecule as a whole is nonpolar because the partial charges on the two oxygen atoms cancel each other out. The pull on the bonding electrons toward one oxygen atom is exactly compensated by the pull, in the opposite direction, on the second pair of bonding electrons toward the other oxygen atom.

In water, the situation is different. Water is an angular, or bent, polar molecule. The partial charges associated with its two pairs of bonding electrons do not cancel each other out. The two bond polarities are of equal magnitude but are not directed in opposite directions. Thus, the molecule as a whole, is polar.

Water: an angular polar molecule

INTERMOLECULAR FORCES

So far we have discussed only *intra*molecular forces, the forces that hold atoms together *within* molecules. In addition, there are ***inter*molecular forces** that act *between* molecules and draw molecules toward one another. Although much weaker than ionic bonds or covalent bonds, intermolecular forces have a strong influence on the properties of a substance, particularly on its melting point, boiling point, and solubility in different solvents.

Intermolecular forces result from electrostatic attraction between dipoles, that is, attraction between the positive end of one molecule and the negative end of another. In order of increasing strength, the three main types of dipole-dipole interactions are (1) London forces, (2) dipole-dipole interactions between polar molecules, and (3) hydrogen bonds.

London forces and dipole-dipole interactions are called van der Waals forces, after the Dutch physicist who first explained how they affect the behavior of gases and liquids.

London Forces

London forces exist because of temporary spontaneous shifts in electron distribution that occur within atoms. For example, in nonpolar molecules such as H_2 and Cl_2, the electrons are, on average, as close to one nucleus in the molecule as the other, and the molecule is uniformly neutral. However, at any instant, the electrons, which are in constant motion, may be slightly more concentrated on one side of the molecule than on the other. Momentarily, one side of the molecule becomes very slightly negatively charged compared with the opposite side. An instantaneous dipole is formed that induces a similar dipole in an adjacent molecule (Fig. 5-18). A momentary weak electrostatic attractive force, called a **London force**, is established between the two dipoles.

London forces, which are approximately 1000 times weaker than single covalent bonds, exist between all chemical species: ions, atoms, and molecules. Although they are so weak, these forces, by drawing atoms and molecules toward each other, cause nonpolar substances such as the noble gases and the halogens to condense into liquids and freeze into solids at very low temperatures.

London forces are named after the German-American physicist who first explained them.

Dipole-Dipole Interactions between Polar Molecules

Dipole-dipole interactions exist between polar molecules, such as HCl, that have permanent dipoles. The negative end of one dipole (δ^-) is attracted toward the positive end (δ^+) of another dipole, and the molecules tend to orient themselves as shown in Figure 5-19. Dipole-dipole attractive forces between polar molecules are slightly stronger than London forces but are still much weaker than covalent bonds or the electrostatic forces that hold ions together in ionic compounds.

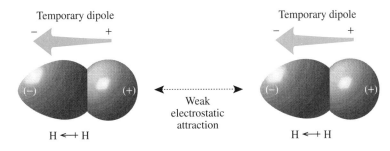

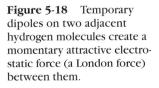

Figure 5-18 Temporary dipoles on two adjacent hydrogen molecules create a momentary attractive electrostatic force (a London force) between them.

Figure 5-19 Electrostatic forces called dipole-dipole interactions exist between polar molecules such as HCl. The positive end of one molecule tends to orient itself adjacent to the negative end of a second molecule.

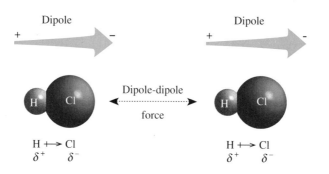

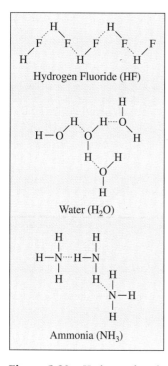

Hydrogen Fluoride (HF)

Water (H_2O)

Ammonia (NH_3)

Figure 5-20 Hydrogen bonding in hydrogen fluoride (HF), water (H_2O), and ammonia (NH_3). Hydrogen bonds are shown as dotted lines.

Hydrogen Bonds

Particularly strong dipole-dipole interactions exist between polar molecules that contain H atoms bonded to one of the following very electronegative elements: fluorine, oxygen, or nitrogen. These dipole-dipole interactions, which are approximately 1/10 as strong as covalent bonds, are called **hydrogen bonds**. They occur typically in hydrogen fluoride (HF), water (H_2O), and ammonia (NH_3). Because fluorine, oxygen, and nitrogen are very electronegative compared with hydrogen (Fig. 5-16), the F—H, O—H, and N—H bonds are very polar, and the H atoms are strongly attracted to the F, O, and N atoms in adjoining molecules (Fig. 5-20).

Hydrogen bonds also occur between alcohol molecules (molecules that have an —O—H group) and between important biological molecules that have —N—H groups, including proteins, enzymes, DNA, and RNA. Hydrogen bonds have a profound effect on the properties of the substances that contain them. They account for many of the unusual, often unique, properties of water that make life on earth possible. We will discuss this topic in Chapter 12.

Hydrogen bonds exist *within* molecules as well as *between* molecules. The three-dimensional configuration of many proteins is determined by the formation of hydrogen bonds between different parts of the molecule. The vital role that hydrogen bonding plays in many biological processes will be discussed in more detail in Chapter 10.

EXPLORATIONS

Linus Pauling: The Colorful, Unconventional Genius from California

Linus Pauling (Fig. 5-A) ranks as one of the greatest scientists of the twentieth century. A brilliant, charismatic man of boundless energy, he has had an uncanny genius for intuitively solving many chemical problems. In a career spanning 70 years, he has made important contributions in the fields of mineralogy, physics, biology, and chemistry, and in 1954 he received the Nobel Prize for chemistry in recognition of his work on chemical bonding and protein structure. His interests have not been limited to science. In the 1950s he was an

Figure 5-A Linus Pauling wearing his trademark beret.

outspoken and courageous advocate for banning atmospheric testing of nuclear weapons. For these activities he was both stigmatized and praised. The U. S. government labeled him a communist and traitor; the Nobel Committee awarded him the 1963 Peace Prize.

Linus Pauling was born in a small Oregon town in 1901. He grew up in great poverty in Portland where his father, a druggist, struggled to make a living until his untimely death in 1910. Pauling was a precocious, intensely curious child with a great thirst for knowledge. From an early age, he went his own way with little regard for conventional behavior, a trait that perhaps he inherited from his colorful maternal forebears. A great uncle was a spiritualist who lived in the hope that a dead Indian with whom he kept in touch would lead him to a long-lost gold mine. His aunt, Stella "Fingers" Darling, was renowned for her skill in opening safes.

During his school years, Pauling delivered milk and worked as a photographer's assistant and motion picture projectionist to help pay the family's bills. Despite his strenuous schedule and the chaotic conditions in his mother's rowdy boarding house, he excelled in his studies. He left high school at 16 but was not given a diploma because he had refused to take a civics course, telling the principal he could learn all he needed by reading on his own. A diploma was presented to him after he won his first Nobel prize.

Pauling began experimenting in chemistry when he was very young. As a school boy, he sprinkled explosive chemicals on the trolley tracks and watched them blow up when the cars went over them. And he experimented in his basement with chemicals "borrowed" from a laboratory at an abandoned smelter belonging to the Oregon Iron and Steel Company.

In 1917, over the strong objections of his mother, who wanted him to help support his two younger sisters, Pauling enrolled in the engineering school at tuition-free Oregon Agricultural College, where his exceptional talents were immediately recognized. By keeping his expenses to a minimum and working at a number of odd jobs, he managed to send money home. However, when his mother became seriously ill at the end of his sophomore year he was forced to leave school and obtain full-time work. He was able to return a year later to complete his degree, helped by a paid teaching position. One of the students in the class he taught was Ava Helen Miller, who in 1923 was to become his wife. To Ava Helen's annoyance, Pauling, fearful of being accused of favoritism, always gave her lower grades than she deserved.

In 1922, again in opposition to his mother's wishes, Pauling accepted a fellowship to do graduate work at the California Institute of Technology (Caltech). In his research into the structure of minerals using x-ray crystallography, Pauling demonstrated the talent that was to characterize his career: a remarkable ability to correctly predict structure and then adapt his "guess" to the experimental data. He received his doctorate in 1925, the same year that his first child, Linus Jr., was born. Two other sons and a daughter were to follow during the next 12 years.

Pauling recognized that the revolutionary concept of quantum mechanics then sweeping through Europe had profound implications for chemical bonding, the subject that had become his primary interest. He therefore accepted a one-year Guggenheim fellowship to study in Munich, Germany, with the eminent physicist Arnold Sommerfeld. His stay in Europe was productive, and soon after his return to Caltech, he published a paper on the application of quantum mechanics to chemical bonding that later formed the basis of his classic book, *The Nature of the Chemical Bond*.

145

In the mid-thirties, Pauling turned his attention to the conflicting theories concerning the structure of proteins, the giant biological molecules that form the structural material of the body. Pauling became convinced that the amino acid molecules, the building blocks from which proteins are composed, must be joined together in long coiled chains held together by hydrogen bonds. However, it was not until several years later that he was able to make a model to fit his theory. One day in 1948, sick in bed and bored, he drew a chain of amino acid molecules on paper and began folding and twisting the paper into a spiral resembling a coiled spring. He instantly realized that with this configuration (which he named an alpha-helix), hydrogen bonds could form between opposing amino acids at each turn of the helix. For this work and his work on chemical bonding, he received the 1954 Nobel Prize in chemistry. His confirmation of the helical structure of proteins led directly to Watson and Crick's discovery, in 1956, of the double helix structure of the genetic material DNA.

After the war (1939–1945), during which time he was engaged in military research, Pauling became increasingly concerned about the dangers of radioactive fallout, and he led a crusade to ban atmospheric testing of nuclear weapons. These activities, coupled with his outspoken left-of-center political opinions, were viewed by the government as an effort to undermine the United States' ability to counter a threat from the Soviet Union, and Pauling's loyalty was questioned. Although no proof was ever found, he was labeled a communist, and on two occasions he was denied a passport to travel abroad. The world community viewed his activities in a different light, and in 1963 he was awarded the Noble Prize for Peace.

Pauling's involvement in demonstrations was an embarassment to Caltech, and in 1963, with his best scientific work behind him, he resigned after 40 years at the institute. After his move to Stanford University in 1965, he become increasingly interested in the role of nutrients in human disorders and began his controversial work on vitamin C.

At first, he advocated large doses of vitamin C as a cure for the common cold, and later, in association with the Scots physician Ewan Cameron, as a treatment for cancer. Conclusive proof of the therapeutic value of vitamin C was elusive, and Pauling's work was severely criticized by his peers. Angered by their reaction, Pauling went directly to the public and, despite the disapproval of the medical profession, began promoting vitamin C in popular magazine articles and on television talk shows. (It is a sad irony that his beloved wife, who had been taking massive doses of vitamin C for several years, died of stomach cancer in 1981.)

After he retired from Stanford University in 1973, Pauling established the Linus Pauling Institute of Science and Medicine in Menlo Park, California, where he still remains active. Research on vitamin C continues, and scientists at the institute are also engaged in other aspects of cancer research and in AIDS research.

During his long career, Pauling has often been criticized for his tendency to jump to conclusions before obtaining sufficient evidence to support his theories, and his unconventional behavior and love of the limelight have not always endeared him to his colleagues. But, undoubtedly Linus Pauling will be remembered and honored as one of the most influential and productive scientists of the twentieth century.

Question

What role should scientists play in debates about environmental problems? In your opinion, should they become involved in political issues, or should they devote their energies entirely to their scientific work?

Reference: A. Serafini, *Linus Pauling: A Man and His Science* (New York: Paragon House, 1989).

KEY WORDS AND CONCEPTS

anion	electron configuration	nonbonding electron
cation	energy levels	octet rule
continuous spectrum	hydrogen bonds	orbitals
covalent bond	intermolecular forces	polar covalent bond
dipole	ion	polar molecules
dipole-dipole interactions	ionic bond	polyatomic ions
double bond	Lewis symbols	probability
electron-dot symbols	line spectrum	triple bond
electronegativity	London forces	valence electrons
	metallic bonding	

QUESTIONS AND PROBLEMS

1. **a.** When sunlight is passed through a prism, a _____ spectrum is produced. **b.** When electrons lose energy and fall from an excited state to a lower energy state, they produce _____ spectra.

2. Although an atom of hydrogen has only one electron, its spectrum has many lines. How does the Bohr model of the atom explain this occurrence?

3. Using the periodic table and Figure 5-9 (Aufbau diagram), write the electron configuration for:
   a. silicon **b.** krypton **c.** zinc

4. Using the periodic table and Figure 5-9 (Aufbau diagram), write the electron configuration for:
   a. arsenic **b.** sodium **c.** bromine

5. The atomic numbers of phosphorus, carbon, and potassium are 15, 6, and 19, respectively. Predict:
   a. The number of electrons in each energy level for each element. **b.** The electron configuration of each element.

6. Fill in the blanks.
   a. The fifth energy level can hold a maximum of _____ electrons. **b.** The designations used to describe the first four orbitals are_____. **c.** The maximum number of orbitals in each of the first four energy levels is _____. **d.** The maximum number of electrons that can occupy a d orbital is ___.

7. Fill in the numerical values that correctly complete each of the following statements.
   a. The $2p$ orbital can hold a maximum of ____ electrons.
   b. The $4f$ orbital can hold a maximum of ____ electrons.
   c. The $3d$ orbital can hold a maximum of ____ electrons.
   d. The $2s$ orbital can hold a maximum of ____ electrons.

8. Name the element with the lowest atomic number that has
   a. a completed $3s$ orbital. **b.** one $4s$ electron.
   c. three $3p$ electrons.

9. What is wrong with each of the following attempts to write electron configurations?
   a. $1s^2 1p^6 2s^2 2p^6$ **b.** $1s^2 2s^2 2p^6 3s^2 3p^6 3d^{10}$

10. Does each of the following correctly represent the electron configuration of an element? Explain your answer.
    a. $1s^2 2s^2 3s^2$ **b.** $1s^2 2s^2 2p^6 2d^9$

11. How many valence electrons do atoms of each of the following elements have?
    a. Be **b.** F **c.** Na **d.** K **e.** Se **f.** Sn

12. Each of the following electron dot structures represents a period two element. Identify each.

   a. X • **b.** :X: **c.** :X• **d.** •X•

13. Would you expect each of the following atoms to gain or lose electrons when forming ions? What is the most likely ion each will form? Explain your answer.
    a. Na **b.** Sr **c.** Ba **d.** I

14. The following elements can form ions. Explain why, in doing so, they lose or gain electrons.
    a. Al **b.** S **c.** B **d.** Cs

15. Describe the octet rule in your own words.

16. From the electron configurations listed below, write the element with its electron-dot symbols.
    a. $1s^2 2s^2 2p^5$ **b.** $1s^2 2s^2 2p^6 3s^1$ **c.** $1s^2 2s^2 2p^6 3s^2 3p^3$

17. What elements are represented by the following electron configurations?
    a. $1s^2 2s^2 2p^6 3s^2 3p^6 4s^1$ **b.** $1s^2 2s^2 2p^6 3s^2 3p^6 4s^2$

18. Write the electron-dot symbols of the following elements:
    a. barium **b.** chlorine **c.** aluminum **d.** sulfur

19. The ions Na^+ and Mg^{2+} occur in chemical compounds, but the ions Na^{2+} and Mg^{3+} do not. Explain.

20. Find cesium, strontium, and bromine in the periodic table. Write a formula of the binary ionic compound formed from:
    a. cesium and bromine **b.** strontium and bromine

21. Classify the bonds in the following compounds as ionic or covalent.
    a. NaF **b.** MgS **c.** MgO **d.** $AlCl_3$

22. Write the formula of an ionic compound formed from the O ion and each of the following ions:
    a. magnesium ion **b.** lithium ion **c.** beryllium ion

23. Write the formula of the ionic compounds formed by each of the following with chlorine.
    a. sodium ion **b.** calcium ion **c.** aluminum ion

24. Write the formulas of the ionic compounds formed from the following elements. Name each compound.
    a. sodium and sulfur **b.** potassium and oxygen
    c. chlorine and lithium

25. Write the formulas of the ionic compounds that can be formed from the following pairs of elements. Name the compounds.
    a. beryllium and fluorine **b.** aluminum and phosphorus **c.** bromine and magnesium

26. Describe in your own words:
    a. ionic bond **b.** covalent bond

27. Explain what is meant by the following:
    a. polar covalent bond **b.** polar molecule

28. Using electron-dot symbols, show the bonding for each of the following diatomic molecules:
    a. O_2 **b.** N_2 **c.** H_2

29. One of the major components of marsh gas is methane, CH_4, which is produced by the decay of plant and animal matter. Methane is also produced in the intestinal tract as a byproduct of bacterial metabolism. Write the Lewis structure for CH_4.

30. Draw electron-dot structures to illustrate the covalent bonding found in each of the following molecules.
    a. Br_2 **b.** BrCl **c.** HBr

31. Draw electron-dot structures to show the covalent bonding in the following compounds.
    a. NCl_3 **b.** OF_2 **c.** PH_3

32. What is the difference between a single covalent bond, a double covalent bond, and a triple covalent bond?

33. Draw the Lewis electron-dot structures for the following molecules, each of which contains at least one double or triple bond. (The skeletal arrangement of atoms in each molecule is shown under the formula.)
  a. N_2F_2

  F N N F

  b. C_3H_4

  H   H
  C C C
  H   H

34. Each of the following molecules contains at least one double or triple bond. Draw the electron-dot structure of each. The arrangement of the atoms is shown below each molecular formula.
  a. C_2H_3N

  H
  H C C N
  H

  b. C_2N_2

  N C C N

35. Write formulas for the following compounds:
  a. calcium fluoride   **b.** carbon tetrachloride
  c. magnesium bromide   **d.** nitrogen trichloride

36. Write formulas for the following compounds:
  a. silicon dioxide   **b.** sodium hydroxide   **c.** sulfur hexafluoride   **d.** cesium bromide

37. What type of bond is formed between two elements with the same electronegativity?

38. In each of the following pairs of elements, indicate which element is more electronegative.
  a. H and F   **b.** Be and N   **c.** N and O

39. Which element in the following pairs of elements is the more electronegative?
  a. Cl and Br   **b.** Na and Mg   **c.** O and S

40. Arrange each of the following sets of bonds in order of increasing polarity.
  a. H—Cl, H—O, H—F   **b.** N—O, P—O, Al—O
  c. H—Cl, Br—Br, B—N   **d.** P—N, S—O, Be—F

41. Write formulas (including charges) for the following polyatomic ions:
  a. nitrate   **b.** hydroxide   **c.** sulfate

42. Write the formulas for the following polyatomic ions. Include the charge.
  a. ammonium   **b.** phosphate   **c.** carbonate

43. If potassium perchlorate has the formula $KClO_4$, what formula would you expect for lithium perbromate and sodium periodate?

NUCLEAR CHEMISTRY
The Risks and Benefits of Nuclear Radiation

6

Using the properties of certain atomic nuclei, it has been possible to show that these ancient Indian ruins in Montezuma National Monument in Arizona date from the thirteenth century.

In previous chapters, we examined the electronic configuration of atoms and the way electrons are involved in bonding atoms together to form compounds. In this chapter, we will focus on the nuclei of unstable elements that are the source of nuclear energy and nuclear radiation.

The explosion of the first atomic bomb in 1945 demonstrated the enormous destructive power of the atomic nucleus and the dangers of nuclear radiation. With the end of the Cold War and the collapse of the Soviet Union in 1992, the threat of nuclear war has diminished, but many concerns remain. Radioactive wastes produced by nuclear weapons plants and the nuclear power industry threaten the environment, and safe disposal of these wastes is a major problem. There is also fear that another catastrophic accident, like the one at Chernobyl in 1986, could occur at one of Eastern Europe's aging nuclear power plants.

Nuclear radiation can destroy tissue cells and cause death, and it is particularly frightening because it is invisible and cannot be smelled or felt. But nuclear radiation can also be enormously beneficial. In medicine, it is an extremely useful tool for diagnosis and for cancer therapy. It has applications in biological research, agriculture, archeology, and industry, and many countries depend on nuclear power for most of their energy needs. Nuclear radiation is all around us in the environment. Some of it is artificially produced, but most of it occurs naturally. We need to understand nuclear radiation so that we can better assess both the perils and the benefits.

In this chapter, we will examine the reactions within the atomic nucleus that produce nuclear radiation. We will learn how nuclear radiation was discovered and how and why it occurs. We will study the properties of the different types of nuclear radiation and find out how nuclear reactions can be used to make synthetic isotopes and elements that are unknown in nature. The events that led to the construction and detonation of the first nuclear bomb will also be covered. We will examine the harmful effects of radiation and also the many ways nuclear radiation has been put to use for the benefit of society.

Learning Goals:

In this chapter, you should gain an understanding of:

1. The three types of radiation emitted from atomic nuclei.

2. Nuclear reactions and the way to write balanced nuclear equations.

3. The detection and measurement of radiation.

4. The relative harmfulness of each type of radiation to living things.

5. How the half-lives of radioisotopes are used to date objects of antiquity.

6. Beneficial uses of radioactivity.

7. The process of nuclear fission.

⟶ THE ATOMIC NUCLEUS

In this chapter, we focus on the *nucleus* of the atom, the tiny dense core at the center of every atom. The nucleus of an atom is made up of protons and neutrons packed tightly together. We saw in Chapter 4 that the number of protons in the atoms—the *atomic number*—of a given element is always constant, but that the number of neutrons may vary. The sum of the number of protons and the number of neutrons in the nucleus is the **mass number**. Atoms of an element that have the same number of protons but different numbers of neutrons are called *isotopes*. For example, the symbols $^{235}_{92}\text{U}$ and $^{238}_{92}\text{U}$ identify two different isotopes of uranium.

Number of protons + number of neutrons = mass number.

In ordinary chemical reactions, it is rarely necessary to consider the different isotopes of an element because chemical reactions are concerned with the rearrangement of electrons and do not involve the nucleus. However, when considering **nuclear reactions**—reactions in which changes occur in the nucleus of the atom—it is essential that the specific isotope taking part in the reaction be identified.

Nuclear Stability

The nuclei in the atoms of most—but not all—naturally occurring elements are very stable despite the electrical repulsions between the positively charged protons that tend to pull them apart. Stability is achieved by a powerful, although not well understood, localized force within the nucleus that overcomes this repulsion and allows the nuclei of the majority of natural isotopes to hold firmly together. Nuclei with certain ratios of neutrons to protons and nuclei with very large total numbers of protons and neutrons, however, are not stable and spontaneously emit high-energy radiation as a means of achieving greater stability. These unstable isotopes are called **radioisotopes** because they are **radioactive**. About 25 naturally occurring elements, including all those with atomic numbers greater than 83, have one or more radioisotopes.

Over 1100 unstable radioisotopes are known. About 65 occur naturally; the rest have been made by laboratory-controlled nuclear reactions.

THE DISCOVERY OF RADIOACTIVITY

As we saw in Chapter 4, work with cathode-ray tubes led to the discovery of electrons and protons. It also led to another important discovery, X-rays. In 1895, the German physicist Wilhelm Roentgen (1845–1923) observed that when cathode rays struck certain metals and glass, a new type of ray was emitted. These rays behaved quite differently from cathode rays; they were not deflected by electric or magnetic fields, and they could penetrate deeply into matter, even passing through walls. Roentgen named these extraordinary new rays X-rays. They were later identified as a high-energy short-wavelength form of electromagnetic radiation (Chapter 1).

X-rays, which pass readily through most body tissues except bone, are used to distinguish bruises from broken bones. An image of a bone produced on a photographic plate placed behind the body part being X-rayed will show if there is a break in the bone.

Roentgen's discovery prompted many scientists, including the French physicist Henri Becquerel (1852–1908), to study these new rays. Becquerel was working with substances that, after exposure to sunlight, become luminous and then continue to glow even after being placed in the dark. He wondered if this phenomenon, which is called fluorescence, was related to X-rays. While working with a fluorescent uranium ore, he made an unexpected discovery. He inadvertently placed a sample of an ore on top of an unexposed photographic plate inside a desk drawer. Some days later, he removed the plate from the drawer and was surprised to find that, although the plate had been wrapped in dark paper to protect it from accidental exposure to light, it showed an image of the uranium rock he had placed on it. Becquerel concluded that the uranium ore must have spontaneously emitted some form of radiation that had penetrated the protective wrapping and exposed the photographic plate. He had discovered another type of radiation.

Following this discovery, Becquerel's student, the Polish-born chemist Marie Sklodowska Curie (1867–1934) and her husband, the French physicist Pierre Curie (1859–1906), began a systematic search for other substances that spontaneously emit radiation. They found that all uranium ores, regardless of type and source, emitted radiation, and that this property, which Marie Curie called **radioactivity**, was a characteristic of the element uranium (U, atomic number 92). In the course of much arduous work, the Curies discovered and isolated two new radioactive elements, polonium (Po, atomic number 84), which is 400 times more radioactive than uranium, and radium (Ra, atomic number 88), which is more than one million times more radioactive than uranium. Several tons of uranium ore had to be processed to obtain just 0.1 g of pure radium. In 1903, Becquerel and the Curies were jointly awarded the Nobel Prize in physics for their discovery of radioactivity.

Marie Curie was the first person to win 2 Nobel Prizes. She won her second prize in 1911 for the discovery of polonium and radium.

THE NATURE OF NATURAL RADIOACTIVITY

Soon after Becquerel's discovery that uranium ores were radioactive, Ernest Rutherford (Chapter 4) determined that the radiation emitted by naturally occurring materials is of three distinct types. Rutherford passed a beam of radiation from a radioactive source between electrically charged plates onto a photographic plate, as shown in Figure 6-1, and found that the beam of radiation split into three components. One component was attracted to the negatively charged plate and was therefore positively charged; a second component was attracted to the positively charged plate and was therefore negatively charged; the third component was not deflected from its original path and could be assumed to have no charge. Rutherford named these components alpha (α), beta (β), and gamma (γ) rays, after the first three letters in the Greek alphabet. It was not until some 30 years later that alpha, beta, and gamma rays were completely characterized.

We now know that alpha rays are made up of particles, each particle consisting of two protons and two neutrons. An alpha particle, therefore, has a mass of 4 amu, a 2+ charge, and is identical to a helium nucleus (a helium atom minus its two electrons). Alpha particles are represented as $_2^4\alpha$, or $_2^4$He.

Beta rays are also made up of particles. A beta particle is identical to an electron and therefore it has negligible mass and a charge of 1-. It is usually represented as $_{-1}^0\beta$. Although a beta particle is identical to an electron, it does not come from the electron cloud surrounding an atomic nucleus, as might be expected. It is produced *inside* the atomic nucleus and then ejected. We will discuss this process further in the next section.

Gamma rays, unlike alpha and beta rays, are not made up of particles and therefore have no mass. They are a form of high-energy electromagnetic radiation with very short wavelengths, similar to X-rays, and are usually represented as $_0^0\gamma$. All three types of radiation have sufficient energy to break chemical bonds and disrupt living and nonliving materials upon contact.

Penetrating Power and Speed of Natural Forms of Radiation

Alpha particles, beta particles, and gamma rays are emitted from radioactive nuclei at different speeds and have different penetrating powers. Alpha particles are the slowest. They are emitted at speeds approximately equal to 1/10 the speed of light and can be stopped by a sheet of paper or by the outer layer of skin (Fig. 6-2). Beta particles are emitted at speeds almost equal to the speed of

Rutherford showed that α-particles combine with electrons to form atoms of helium. He therefore concluded that an α-particle must consist of the positively charged nucleus of a helium atom.

Figure 6-1 Behavior of radiation from a radioactive source in an electric field. The radiation splits into three components: α particles, β particles, and γ rays. The positively charged α particles are attracted to the negatively charged plate; the negatively charged β particles are attracted to the positively charged plate; and the uncharged γ rays are not deflected from their original path.

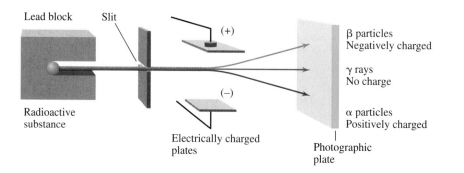

Paper Aluminum (1-mm sheet) Concrete wall

α

β

γ

Figure 6-2 Penetrating abilities of alpha (α), beta (β), and gamma (γ) radiation. Alpha particles are the least penetrating, and are stopped by a sheet of paper or the outer layer of skin; beta particles pass through paper but are stopped by aluminum foil or a block of wood; gamma particles can be stopped only by several centimeters of lead or a concrete wall several meters thick.

light, and, because of their greater velocity and smaller size, their penetrating power is approximately 100 times greater than that of alpha particles. Beta particles can pass through paper and several millimeters of skin but are stopped by aluminum foil. Gamma rays are released from nuclei at the speed of light and are even more penetrating than X-rays. They pass easily into the human body and can be stopped only by several centimeters of lead or several meters of concrete. Properties of the three types of natural radiation are summarized in Table 6-1.

Both γ-rays and X-rays are forms of electromagnetic radiation (Chapter 1). γ-rays are of shorter wavelength and higher energy than X-rays.

NUCLEAR REACTIONS

When a radioactive isotope of an element emits an alpha or a beta particle, a nuclear reaction occurs, and the nucleus of that isotope is changed. The changes that occur during a nuclear reaction can be represented by a nuclear equation.

Let us write nuclear equations to explain the radioactivity that Becquerel and the Curies observed in their studies of uranium ores. The element uranium has several radioactive isotopes, including $^{238}_{92}U$ (also written uranium-238), which spontaneously emit alpha particles. The notation $^{238}_{92}U$ tells us that the atomic number of uranium is 92 and the mass number of this particular isotope is 238. $^{238}_{92}U$ therefore has 92 protons and 146 (238 − 92) neutrons in its nucleus.

When the uranium-238 nucleus emits an alpha particle, $^{4}_{2}\alpha$, it loses 4 atomic mass units (2 protons and 2 neutrons). The resulting atomic nucleus therefore has an atomic mass number of 234 (238 − 4) and, because it has lost 2

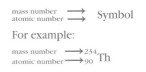

$$\frac{\text{mass number}}{\text{atomic number}} \longrightarrow \text{Symbol}$$

For example:

$$\frac{\text{mass number}}{\text{atomic number}} \longrightarrow {}^{234}_{90}\text{Th}$$

When an α-particle is emitted, the atomic number decreases by two and the mass number decreases by four.

TABLE 6-1 Properties of the Three Types of Radiation Emitted by Radioactive Elements

Name	Symbol	Identity	Charge	Mass (amu)	Velocity	Penetrating Power
Alpha	$^{4}_{2}\alpha$, $^{4}_{2}He$	Helium nucleus	2+	4	1/10 the speed of light	Low, stopped by paper
Beta	$^{0}_{-1}\beta$	Electron	1−	0	Close to the speed of light	Moderate, stopped by aluminum foil
Gamma	$^{0}_{0}\gamma$	High-energy electromagnetic radiation	0	0	Speed of light (3×10^{10} cm/s)	High, stopped by several centimeters of lead

protons, an atomic number of 90 (92 − 2). An atom with an atomic number of 90 is no longer an atom of uranium. If you refer to the periodic table, you will see that the element with atomic number 90 is thorium (Th). Thus, the spontaneous emission of an alpha particle from a uranium atom results in the formation of a completely different element. This **transmutation** of uranium to thorium is represented by the following nuclear equation:

mass number = 238 sum of mass numbers = 234 + 4 = 238

$$^{238}_{92}U \longrightarrow ^{234}_{90}Th + {}^{4}_{2}\alpha$$

atomic number = 92 sum of atomic numbers = 90 + 2 = 92

For the equation to be properly balanced, the mass number on the left side of the equation must equal the sum of the mass numbers on the right side of the equation. Similarly, the atomic number on the left must equal the sum of the atomic numbers on the right.

Thorium-234, the main product in the above equation, is always found in uranium-238 ore deposits. (The Curies isolated it from a uranium ore before it was possible to explain exactly how it had been formed.) It, too, is radioactive, but instead of alpha particles, it emits beta particles. As we mentioned earlier, it may seem improbable that a negatively charged beta particle ($^{0}_{-1}\beta$) identical to an electron can be released from a nucleus made up of positively charged protons ($^{1}_{1}p$) and uncharged neutrons ($^{1}_{0}n$), but this does occur. The accepted explanation is that a neutron in the nucleus changes into a proton and a beta particle according to the following equation

$$^{1}_{0}n \longrightarrow {}^{1}_{1}p + {}^{0}_{-1}\beta + energy$$

Notice that the equation is balanced: The sum of the superscripts on both sides of the equation are equal (1 = 1 + 0); the subscripts are similarly balanced (0 = 1 − 1). Once formed, the beta particle is ejected from the nucleus, leaving the original nucleus with an additional proton but one less neutron. Because the mass of a proton equals that of a neutron and the mass of a beta particle is essentially zero, there is no net change in the mass of the nucleus. But as a result of exchanging a neutron for a proton, the nucleus gains a 1+ charge, and a new element is formed.

The balanced nuclear equation showing the emission of a beta particle from Th-234 and the formation of a new element is given below.

mass number (superscript) = 234 sum of superscripts = 234 + 0 = 234

$$^{234}_{90}Th \longrightarrow {}^{234}_{91}Pa + {}^{0}_{-1}\beta$$

atomic number (subscript) = 90 sum of subscripts = 91 − 1 = 90

When a β-particle is emitted, the atomic number increases by one; there is no change in the mass number.

Because the atomic number has changed, a new element has been formed. This time the atomic number has *increased* by one. If you refer to the periodic table, you can confirm that the element with atomic number 91 is protactinium (Pa).

After emitting an alpha or a beta particle, the nuclei of naturally occurring radioisotopes still have excess energy, and they release this by emitting gamma rays. Since gamma rays are waves and not particles and have no mass and no charge, the identity of the emitting element does not change when a gamma ray is emitted. Gamma rays are not therefore included in nuclear equations. The changes in mass number and atomic number that occur when unstable nuclei emit alpha or beta particles are shown in Table 6-2.

New elements are never produced in a chemical reaction, but new elements are often produced in a nuclear reaction.

TABLE 6-2   Changes in Mass Number and Atomic Number That Occur When Radioactive Elements Decay

Type of Emission*	Symbol	Mass Number (protons plus neutrons)	Charge	New Element	
				Change in Mass Number	Change in Atomic Number
Alpha	$^4_2\alpha$, ^4_2He	4	2+	Decreased by 4	Decreased by 2
Beta	$^0_{-1}\beta$	0	1−	No change	Increased by 1

*Gamma rays are not included because they have no mass and no charge. Therefore, when they are emitted, no new element is formed.

Radioactive Decay Series

The transformation of one element into another element as a result of radioactive emissions is termed **radioactive decay**. The spontaneous decay of uranium-238 to form thorium-234 and of thorium-234 to form protactinium-234 are the first two steps in the uranium **decay series**, which continues through a total of 14 steps until the stable element lead-206 is formed (Fig. 6-3). Of particular interest in this decay series is radon-222 ($^{222}_{86}Rn$), the radioactive gas that (as we will learn in the section "Everyday Exposure to Radiation") accounts for most of the potentially harmful radiation in the environment.

Besides the uranium decay series, two other naturally occurring decay series exist: the **thorium series**, which starts with thorium-232 and ends with lead-208; and the **actinium series**, which starts with uranium-235 and ends with lead-207. All the naturally occurring radioisotopes of high atomic number belong to one of these three decay series.

ARTIFICIAL TRANSMUTATIONS

The transmutation reactions we have described so far occur naturally as a result of the spontaneous emission of alpha or beta particles from unstable nuclei. **Artificial transmutations** of one element into another element can be brought about by bombarding certain stable nuclei with alpha particles, neutrons, or other subatomic particles.

The first artificial transmutation was achieved by Rutherford in 1919. He bombarded stable nitrogen atoms with alpha particles and produced oxygen-17, a stable isotope of oxygen, and protons (^1_1p). This experiment confirmed that, as predicted, protons are fundamental particles that are present in the nuclei of all atoms (Chapter 4). The equation for the nuclear reaction is shown below

$$^{14}_7N + ^4_2\alpha \longrightarrow ^{17}_8O + ^1_1p$$

You will notice that the equation is properly balanced: The sum of the mass numbers on the left of the equation (14 + 4 = 18) equals the sum of the mass numbers on the right (17 + 1 = 18). Similarly, the atomic numbers are balanced (7 + 2 = 8 + 1).

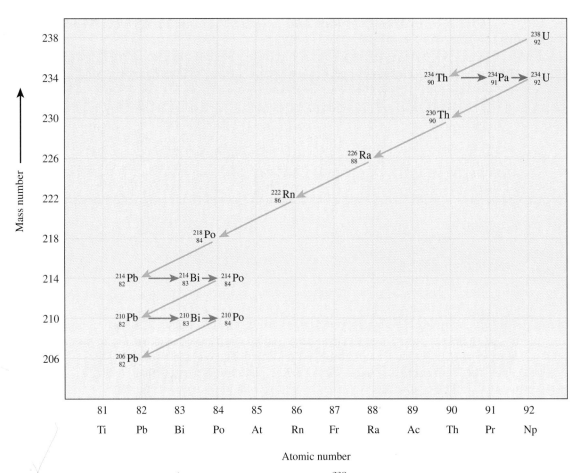

Figure 6-3 The uranium decay series. $^{238}_{92}$U spontaneously decays in a series of steps until the stable element $^{206}_{82}$Pb is formed. Each time an α particle is emitted the mass number decreases by 4 and the atomic number decreases by 2. Each time a β particle is emitted the mass number remains the same but the atomic number increases by 1. The gas radon (Rn), which accounts for most of our everyday exposure to radiation, is formed in the series as a result of the disintegration of $^{226}_{88}$Ra.

During the 1920s and 1930s, many other transmutation reactions were carried out. One of the most important was done by James Chadwick, who bombarded beryllium with alpha particles and obtained carbon atoms and uncharged particles with mass almost identical to the mass of a proton. This reaction, which is shown below, confirmed the existence of the neutron (Chapter 4).

$$^{9}_{4}\text{Be} + ^{4}_{2}\alpha \longrightarrow ^{12}_{6}\text{C} + ^{1}_{0}\text{n}$$

New Isotopes and New Elements

The first radioisotope produced in the laboratory was made by Irene Curie (the daughter of Marie and Pierre Curie) and her husband Frederic Joliot in 1934.

$^{12}_{6}$C and $^{17}_{8}$O, the isotopes of carbon and oxygen produced in the bombardment reactions described above, are stable and occur naturally. Bombardment reactions can also produce isotopes that are unstable and unknown in nature. For example, $^{28}_{13}$Al, a radioactive isotope of aluminum, is produced when magnesium-25 atoms are bombarded with alpha particles.

$$^{25}_{12}\text{Mg} + ^{4}_{2}\alpha \longrightarrow ^{28}_{13}\text{Al} + ^{1}_{1}\text{p}$$

The $^{28}_{13}$Al nucleus emits a beta particle to become a stable isotope of silicon, $^{28}_{14}$Si.

Most of the early bombardment reactions were carried out with slow-moving alpha particles obtained from naturally occurring radioactive materials. It was soon realized that many more nuclear changes could be achieved if nuclei could be bombarded with faster-moving particles. In the 1940s, special instruments called linear accelerators (or cyclotrons) were developed that could generate fast-moving charged particles, including alpha and beta particles, protons, and the nuclei of light elements. The charged particles were shot into target nuclei at high speed and with great force. In this way, thousands of radioisotopes unknown in nature were produced. All the presently known elements not found in nature—mostly those with atomic numbers greater than 92—were formed in bombardment reactions with accelerated particles. Elements with atomic numbers near or above 100 are so unstable and so short-lived that most are no more than scientific curiosities.

Although scientists still cannot transmute base metals, such as lead, into gold—the long-sought goal of the alchemists of the Middle Ages—they are now able to change certain elements into many very useful new isotopes including all those used in medicine.

Linear accelerators and cyclotrons are often called "atom smashers." The National Accelerator Laboratory is located near Chicago in Batavia, Illinois.

EXAMPLE 6-1   Write a balanced nuclear equation for the following transformation: Hafnium-181 undergoes beta decay.

Solution:
1. Look up hafnium (Hf) in the periodic table. Find its atomic number (72).
2. Write the equation with the reactants and products that you know:

$$^{181}_{72}\text{Hf} \longrightarrow \ ^{0}_{-1}\beta \ + \ ?$$

3. Determine the isotope that is formed when hafnium emits a beta particle. Beta particles have almost no mass and a negative charge (−1). Therefore, the isotope formed will have the same mass as hafnium (181) and an atomic number that has increased by one unit to 73. Look in the periodic table for an element with atomic number 73. It is tantalum, Ta.
4. Complete the nuclear equation.

$$^{181}_{72}\text{Hf} \longrightarrow \ ^{0}_{-1}\beta \ + \ ^{181}_{73}\text{Ta}$$

5. To make certain that the equation is correct, check that (a) the sum of the superscripts on the left side of the equation equals the sum of the superscripts on the right side and (b) the sum of the subscripts on the left side of the equation equals the sum of the subscripts on the right side.

superscripts: 181   =   0   +   181
subscripts: 72   =   −1   +   73

The equation is correct.

PRACTICE EXERCISE: Write a balanced nuclear equation for the following: Zirconium-93 undergoes beta decay.

Answer: $^{93}_{40}\text{Zr} \longrightarrow \ ^{0}_{-1}\beta \ + \ ^{93}_{41}\text{Nb}$

$^1_1H + ^{35}_{17}Cl \rightarrow ^4_2He + ^{32}_{16}S$

EXAMPLE 6-2 Write a balanced nuclear equation for the following: Radium-226 decaying to a radon isotope.

Solution:
1. Look up radium (Ra) in the periodic table and find its atomic number (88).
2. Look up radon (Rn) in the periodic table and find its atomic number (86).
3. Write the equation with the reactants and products that you know:

$$^{226}_{88}Ra \longrightarrow \text{?} + ^{?}_{86}Rn$$

4. The atomic number has decreased by two (from 88 to 86). Therefore, Ra must have lost an alpha particle $^4_2\alpha$
5. Write the complete reaction:

$$^{226}_{88}Ra \longrightarrow ^4_2\alpha + ^{222}_{86}Rn$$

6. The radon isotope must be Rn-222 to agree with the requirements that (a) the sum of the superscripts of the reactants equals the sum of the superscripts of the products (226 = 4 + 222) and (b) the sum of the subscripts of the reactants equals the sum of the subscripts of the products (88 = 2 + 86).

PRACTICE EXERCISE: Write a balanced nuclear equation for the following. Hydrogen-1 reacts with chlorine-35 to give an alpha particle.

Answer: $^1_1H + ^{35}_{17}Cl \longrightarrow ^4_2\alpha + ^{32}_{16}S$

THE HALF-LIFE OF RADIOISOTOPES

Different radioisotopes—whether naturally occurring or artificially produced—decay at characteristic rates. The more unstable the isotope is, the more rapidly it will emit alpha or beta particles and change into a new element. The rate at which a particular radioisotope decays is expressed in terms of its **half-life**, *the time required for one-half of any given quantity of the isotope to decay*. There is no way of knowing when any *one* nucleus will disintegrate, but after one half-life, half the nuclei in the original sample will have disintegrated.

Half-lives range from billionths of a second to billions of years. For example, the half-life of boron-9 is only 8×10^{-19} seconds, that of thorium-234 is 24 days, while that of uranium-238 is 4.5 billion years. Half-lives of a number of radioisotopes are given in Table 6-3.

We can construct a **decay curve** (Fig. 6-4) to show graphically what quantity of a given radioisotope will remain after a specific length of time. Let us assume that we start with 16 g of $^{32}_{15}P$, a radioisotope that has a half-life of 14 days and decays by beta emission to form $^{32}_{16}S$. After one half-life (14 days), one-half of the original 16 g of $^{32}_{15}P$ will have decayed and been converted to $^{32}_{16}S$; 8 g of $^{32}_{15}P$ will therefore remain (and 8 g of $^{32}_{16}S$ will have been formed). At the end of two half-lives (28 days), one-half of the 8 g of $^{32}_{15}P$ that was present at the end of one half-life will have decayed; 4 g will remain. At the end of three half-lives (42 days), 2 g will remain, and so on. After six half-lives (84 days), just 0.25 g of the original isotope will remain; the rest of the sample will be the new element, $^{32}_{16}S$. Even after many half-lives, a minute fraction of the original radioisotope will remain.

Many of the radioisotopes in the radioactive wastes generated by the

TABLE 6-3   Half-lives of Some Radio-active Isotopes

Radioactive Isotope	Half-life
Oxygen-13	8.7×10^{-3} seconds
Bromine-80	17.6 minutes
Iodine-132	2.4 hours
Technetium-99	6.0 hours
Radon-222	3.8 days
Barium-140	12.8 days
Hydrogen-3 (tritium)	12.3 years
Strontium-90	28.1 years
Radium-226	1620 years
Carbon-14	5730 years
Plutonium-239	24,400 years
Beryllium-10	4.5 million years
Potassium-40	1.3 billion years
Uranium-238	4.5 billion years

production of nuclear weapons and by nuclear power plants have long half-lives. Long-lived radioisotopes that escape into the environment persist for many years, and there is the possibility that they will be incorporated into food chains. At least 10 half-lives must elapse before a radioisotope has decayed to the point at which it is no longer considered to be a radiation hazard (Chapter 19).

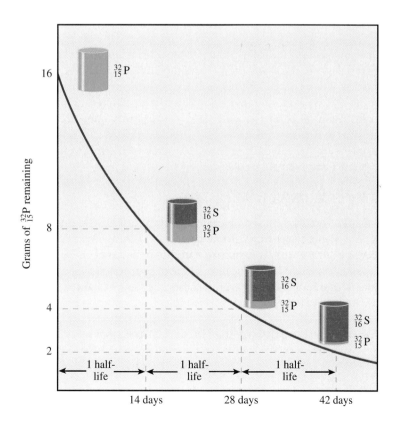

Figure 6-4   Decay of a 16-g sample of $^{32}_{15}$P (half-life, 14 days) to $^{32}_{16}$S by beta emission. After one half-life, one-half of the original 16 g of $^{32}_{15}$P will have been converted to $^{32}_{16}$S; that is, 8 g will remain. After another 14 days, half the remaining 8 g will remain. At the end of 42 days (3 half-lives), 2 g will remain and 14 g of $^{32}_{16}$S will have been formed, and so on.

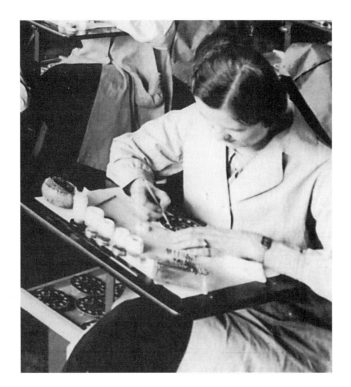

Figure 6-5 Women painting radium onto watch dials to make them glow in the dark.

THE HARMFUL EFFECTS OF RADIATION ON HUMANS

When radioactivity was discovered at the turn of the century, its harmful effects were not recognized. Marie Curie, who worked for many years with radioactive materials, suffered from anemia and died of leukemia; she was probably one of the first victims of radiation poisoning. Other early sufferers were women who, in the 1920s, worked in factories where they painted radium on watch dials to make them glow in the dark (Fig. 6-5). The women frequently licked their brushes to obtain a fine point and often developed cancer of the lips; as a result of ingesting the radioactive material, many of them developed bone cancer or leukemia and died at an early age.

Why Is Radiation Harmful?

Radiation emitted by radioisotopes is harmful because it has sufficient energy to knock electrons from atoms and thus to form positively charged ions. For this reason, it is called **ionizing radiation**. Some of these ions are highly reactive and, by disrupting the normal workings of cells in living tissues, can produce abnormalities in the genetic material DNA (Chapter 10) and increase the risk of cancer (Chapter 16).

Factors Influencing Radiation Damage

The degree of damage caused by ionizing radiation depends on many factors, including (1) the type and penetrating power of the radiation, (2) the location of the radiation—that is, whether it is inside or outside the body, (3) the type of tissue exposed, and (4) the amount and frequency of exposure.

A source of alpha particles outside the body is the least dangerous of the three types of nuclear radiation because alpha particles cannot penetrate the skin and enter the body. However, if an alpha emitter is ingested, for example, in contaminated food, or is inhaled in air containing radon gas, damage to internal body tissues can be severe. Once inside the body, alpha emissions are more damaging than either beta or gamma emissions because they are more effective in forming ions in the surrounding tissues. They travel only short distances through tissues and rapidly transfer the bulk of their energy to a small area. Beta particles and gamma rays travel farther than alpha particles and transfer their energy over a wider area of tissue. As a result, the radiation received per unit area is lower, and the ability to form ions is reduced. The greater the penetrating power of a particular source of radiation, the weaker is its ionizing power.

Beta particles, unlike alpha particles, can penetrate the outer layers of skin and clothing, and produce severe burns. They can also cause skin cancer and cataracts. Trapped inside the body, beta particles are less disruptive to individual cells than are alpha particles, but outside the body, they are more damaging. Gamma rays, because of their great penetrating power (Fig. 6-2), are more dangerous than alpha or beta particles outside the body; inside the body, they are the least dangerous.

X-rays are a fourth form of ionizing radiation. Their penetrating power, and thus their ability to cause tissue damage, lies between that of beta particles and gamma rays.

Different biological tissues vary widely in their sensitivity to ionizing radiation. Rapidly dividing cells are very vulnerable to radiation. Such cells are found in bone marrow, the lining of the gastrointestinal tract, the reproductive organs, the spleen, and the lymph glands. Embryonic tissue is particularly easily damaged, and unless there are compelling medical reasons, pregnant women should avoid all exposure to radiation including X-rays. Cancer cells, which divide very rapidly, are more easily killed by radiation than are healthy cells; this difference explains the success of radiation treatment for some types of cancer.

Red blood cells are formed in bone marrow.

Detection of Radiation

In order to discuss radiation exposure, we must have a means of measuring it. The most common instrument for detecting and measuring radioactivity is the **Geiger counter** (Fig. 6-6), which is essentially a modified cathode-ray tube (Chapter 4). Argon gas is contained in a metal cylinder, which acts as the cathode; a wire anode

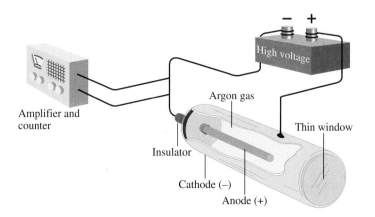

High voltage

Amplifier and counter

Argon gas

Thin window

Insulator

Cathode (−)

Anode (+)

Figure 6-6   Schematic diagram of a Geiger counter. The metal cylinder, which is filled with a gas—usually argon—acts as the cathode; the metal wire projecting into the cylinder acts as the anode. The window in the cylinder is permeable to alpha, beta, and gamma radiation. When radiation enters the cylinder, the gas is ionized and small pulses of electrical current flow between the wire and metal cylinder. The electrical pulses are amplified and counted. The number of pulses per unit time is a measure of the amount of radiation.

runs down the axis of the tube. Radiation from a radioactive source—for example, contaminated soil—enters the tube through a thin mica window, and when it does so, it causes ionization of the argon gas. As a result of the ionization, pulses of electrical current flow between the electrodes, and these pulses are amplified and converted into a series of clicks that are counted automatically.

Units of Radiation

Nuclear disintegrations are measured in **curies** (Ci); 1 Ci is equal to 3.7×10^{10} disintegrations per second. A curie represents a very large dose of radiation. Natural background radiation amounts to only about 2 disintegrations per second. The damage caused by radiation depends not only on the number of disintegrations per second but also on its energy and penetrating power. Another unit of radiation, the **rad,** measures the amount of energy released in tissue (or other medium) when it is struck by radiation. A single medical X-ray is equivalent to 1 rad. The rad is currently being replaced by a new international unit, the **gray** (l gray = 100 rad).

A more useful unit for measuring radiation is the **rem**, which takes into account the potential damage to living tissues caused by the different types of ionizing radiation. For X-rays, gamma rays, and beta particles, 1 rad is essentially equivalent to 1 rem, but for alpha particles, because of their greater ionizing ability, 1 rad is equivalent to 10 to 20 rem. The new international unit to replace the rem is the **sievert** (l sievert = 100 rem).

How Much Radiation Is Harmful?

We still do not know for certain how much radiation is harmful. The study of Japanese survivors of the wartime bombings of Hiroshima and Nagasaki and of people exposed to fallout from nuclear power plant accidents—primarily those at Three Mile Island and Chernobyl—has revealed the probable effects of various large short-term doses of radiation given to the whole body. But there is no agreement on how these numbers should be extrapolated to low dosages. A summary of the effects of large doses is shown in Table 6-4.

TABLE 6-4   Effects on Humans of Short-term Whole-body Exposure to Various Doses of Radiation

Dose (rem)	Effects
<50	Effects inconsistent and difficult to demonstrate.
50–250	Fatigue, nausea, decreased production of white cells and platelets in blood; increased probability of leukemia.
250–500	Same as for 50–250 rem, but more severe; vomiting, diarrhea, damage to intestinal lining; very susceptible to infections because of low white cell count; hemorrhaging because of impaired clotting mechanism; 50% die within months.
500–1000	Damage to cardiovascular system, intestinal tract, and brain; death within weeks.
1000–10,000	Same as for 500–1000 rem, but more severe; coma; death within hours at 10,000 rem.
100,000	Immediate death.

Some scientists believe that because the body has the ability to repair radiation damage, many small doses of radiation over a long period of time produce no lasting effects. Others believe that there is no safe level of radiation exposure. At the present time, there is no agreement on what constitutes a "safe" annual dose of radiation.

EVERYDAY EXPOSURE TO RADIATION

In our daily lives, it is impossible to avoid exposure to every source of radiation. Low-level radiation is all around us in the environment. It has been estimated that the *average* exposure of the U.S. population to ionizing radiation amounts to approximately 360 *mrem* (millirem) per year (note that a millirem is 1/1000 of a rem, the unit used in Table 6-4). By far, the largest contribution (82 percent) comes from natural sources. The remaining 18 percent comes primarily from medical procedures, consumer products, and occupational activities (Table 6-5).

TABLE 6-5   Average Annual Exposure (1990) to Radiation for People in the United States[1]

Source of Radiation	Dose (mrem)	Percent of Total Dose
Natural		
Radon gas	200	55
Cosmic rays	27	8
Terrestrial (radiation from rocks and soil other than radon)	28	8
Inside the body (naturally occurring radioisotopes in food and water)	39	11
Total natural	294	82
Artificial		
Medical		
X-rays	39	11
Nuclear medicine	14	4
Consumer products (building materials, water)	10	3
Other		
Occupational (underground miners, x-ray technicians, nuclear plant workers)	<1	<0.03
Nuclear fuel cycle	<1	<0.03
Fallout from nuclear weapons testing	<1	<0.03
Miscellaneous	<1	<0.03
Total artificial	64	18
Total natural plus artificial	358	100

[1]*Source:* National Council on Radiation Protection and Measurement (NCRP87b), Washington DC: National Academy Press, 1990.

Natural Sources of Radiation

It is now recognized that the major source of natural radiation is **radon** gas, which, on average, contributes 200 mrem to the total annual exposure (Table 6-5). Radon-222 (half-life of 3.8 days), an alpha emitter, is a naturally occurring decay product of uranium-238 (Fig. 6-3), a radioisotope of uranium that is present in widely varying concentrations in most soils and rocks. Radon-222 is an odorless, tasteless, inert gas, and, as it escapes from soil and rocks, it enters the surrounding water and air. The Environmental Protection Agency (EPA) has conducted a nationwide survey of radon in homes to determine where problems are likely to exist (Fig. 6–7). In areas of the country where crustal rocks and soil have high concentrations of uranium-238, radon can seep into homes through basements and, in well-sealed houses, may reach potentially dangerous levels. Inhalation of radon gas can increase the risk of cancer. This effect is due less to radon itself than to its alpha- and beta-emitting decay products (Fig. 6-3), which, when radon is inhaled, become deposited in the respiratory tract.

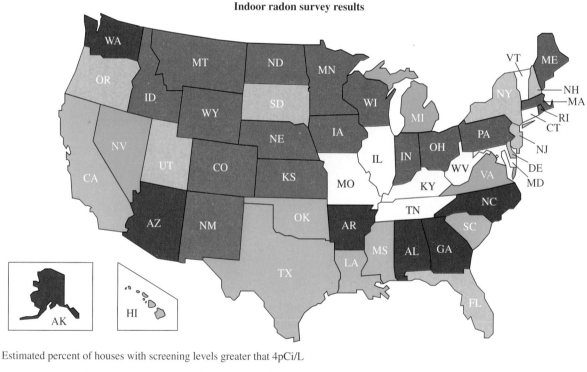

Indoor radon survey results

Estimated percent of houses with screening levels greater that 4pCi/L

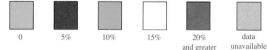

Figure 6-7   Results of a survey of radon in residential dwellings conducted by the Environmental Protection Agency. Exposures to radon below 4 p Ci/L are considered to be average, and generally no action is recommended. If exposures are much greater than 4 p Ci/L, action to reduce the levels is recommended. (p Ci/L = picocuries per liter of air; 1 p Ci = 1×10^{-12} Ci.) (EPA, 1993).

Cosmic rays—a form of short-wavelength electromagnetic radiation that reaches us from outer space—account for 8 percent of natural radiation, or an average dose of 27 mrem/year. Because the dose increases with altitude, residents of mile-high Denver receive about 50 mrem more radiation each year from this source than do people at sea level in New Orleans.

Terrestrial radiation in rocks and soil, other than radon, accounts for another 8 percent of natural radiation (28 mrem/year). Naturally occurring carbon-14 and traces of potassium-40, thorium-223, and uranium-238 are present in food, water, and air and enter the body when ingested or inhaled. This inside-the-body source makes up 11 percent of natural radiation (39 mrem/year). The average dose from all natural sources is about 300 mrem/year.

Radiation from Human Activities

The average annual radiation dose resulting from everyday human activities amounts to about 64 mrem (Table 6-5). Diagnostic X-rays account for 39 mrem/year and nuclear medicine for 14 mrem/year (together accounting for 15 percent of the total annual radiation dose). Consumer products add 10 mrem/year, with most of this radiation coming from radon in domestic water supplies and from building materials—stone, brick, and concrete—which frequently contain uranium ores and thus are sources of radon.

Smokers receive an additional dose of radiation from polonium-210, a naturally occurring alpha emitter present in tobacco. Workers who are potentially at greater risk of exposure to radiation than the average population include underground miners, radiologists, X-ray technicians, and nuclear plant workers. The average additional dose for these workers is less than 1 mrem/year. Although in the past, residents downwind from nuclear weapons testing sites were exposed to radioactive fallout, today, nuclear weapons production and nuclear power plant operations together contribute negligible amounts of radiation. Nuclear wastes, however, from both these activities pose a serious hazard (Chapter 19).

People who work with radiation wear badges containing film that is sensitive to radiation. When developed, the film indicates the extent of the person's exposure to radiation during the monitored period.

USES OF RADIOACTIVE ISOTOPES

Although it presents potential health hazards, radiation from radioisotopes has useful applications in fields as diverse as archeology, medical diagnosis and treatment, agriculture, scientific research, and the power industry.

Determining the Dates of Archeologic and Geologic Events

An ingenious and reliable method based on the half-lives of certain naturally occurring radioisotopes such as carbon-14 can be used to estimate the age of ancient objects (Fig. 6-8). A radioisotope decays at a constant rate that is defined by its half-life, and as it decays, it changes into a new isotope. By measuring the relative amounts of the original radioisotope and the new isotope in an object, the age of the object can be estimated.

Uranium-238, a natural constituent of most rock, has a half-life of 4.5 billion years. It decays to form stable lead-206 and can be used to date very ancient rocks. To take a simple example, let us suppose that equal amounts of uranium-238 and lead-206 are found in a rock we wish to date. This finding tells us that the uranium in the rock must have gone through one half-life (4.5 billion years);

Figure 6-8 Carbon-14 dating showed that this painted bowl from ancient Persia was made more than 5000 years ago.

In 1988, carbon dating showed that the linen cloth known as the Shroud of Turin was made about 1350 AD. This proved that the image could not have been imprinted on the linen by the body of Jesus while it was wrapped in the shroud—as many believed—because Jesus died in about 30 AD.

therefore, the rock is 4.5 billion years old. In coming to this conclusion, we make the assumptions that the rate of decay has always been constant and that no lead-206 was initially present in the sample; that is, all of it came from uranium-238.

Using the decay rate of uranium-238, scientists have dated the oldest rocks on earth at between 3 and 3.5 billion years. Meteorites and moon rocks have been found to be approximately 4.5 billion years old, and, on the basis of this information, geologists conclude that the planets, including our earth, were formed 4.5 to 5.0 billion years ago.

To obtain an accurate date, the half-life of the radioisotope used for dating must be fairly close to the age of the object being studied. Uranium-238 is appropriate for dating objects that are billions of years old, but for objects that are several thousand years old, carbon-14, which has a half-life of 5730 years, is more suitable.

Carbon-14 is formed naturally at a fairly constant rate in the upper atmosphere when neutrons ejected from stable nuclei by cosmic rays interact with ordinary nitrogen nuclei.

$$^{14}_{7}\text{N} + ^{1}_{0}\text{n} \longrightarrow ^{14}_{6}\text{C} + ^{1}_{1}\text{p}$$

The radioactive carbon-14 reacts with oxygen in the atmosphere to form radioactive carbon dioxide ($^{14}_{6}\text{CO}_2$), which together with ordinary nonradioactive carbon dioxide ($^{12}_{6}\text{CO}_2$), is incorporated into plants by photosynthesis and into animals through normal food chains and respiration. While an organism is alive, the ratio of carbon-14 to nonradioactive carbon remains constant, but when the organism dies and incorporation of carbon ceases, the ratio begins to change. The carbon-14 in the tissues slowly disappears as it emits beta particles and becomes nitrogen, but the nonradioactive carbon does not change. The ratio of carbon-14 to carbon-12 decreases.

$$^{14}_{6}\text{C} \longrightarrow ^{14}_{7}\text{N} + ^{0}_{-1}\beta$$

The carbon-14 emits beta particles at a constant rate given by its half-life. Thus, measuring the ratio of carbon-14 to carbon-12 in the plant or animal artifact and comparing it with the ratio in living tissue gives the age of the artifact.

EXAMPLE 6-3 The carbon-14 activity of a piece of ancient wood is one-quarter that of new wood. What is the age of the ancient wood? The half-life of carbon-14 is 5730 years.

Solution:
1. First, represent the carbon-14 activity by a diagram as in Figure 6-4. After one half-life, activity decreases by one-half. After a second half-life, activity is halved again, and one-quarter of the original activity remains.
2. The carbon-14 in the ancient wood is old enough to have gone through two half-lives.

$$\text{age of wood} = 2 \text{ half-lives}$$

The wood is therefore 2(5730 years), or 11,460 years old.

PRACTICE EXERCISE: The carbon-14 activity of an ancient piece of linen cloth is one-eighth that of linen fibers produced today. Given that the half-life of carbon-14 is 5730 years, determine the age of the ancient cloth.

Answer: 17,190 years.

EXAMPLE 6-4   If you have 100 g of radioisotope bromine-80 (half-life, approximately 20 minutes), how many g of the isotope will remain after one hour?

Solution: After 20 minutes (one half-life), one half of the original sample of bromine-80 will remain. 50 g will remain.
After another 20 minutes, 25 g will remain.
After another 20 minutes (that is, after one hour has elapsed) 12.5 g will remain.

PRACTICE EXERCISE: The half-life of tritium (hydrogen-3) is 12.3 years. If 0.005 g of tritium is released from a nuclear power plant during an accident, what mass of this isotope remains after: **a.** 12.3 years, **b.** 49.2 years

Answer: **a.** 0.0025 g   **b.** 0.0003 g

Carbon-14 dating, also known as **radiocarbon dating**, is used to date ancient carbon-containing artifacts such as cloth, wood, leather, and bone that were derived from once-living matter. The method assumes that the flow of carbon-14 into our environment has always been constant. Studies of the annual growth rings in trees have shown that this assumption held over the past 7000 years, but in earlier times, there were some fluctuations. Even so, other dating methods generally agree very well with carbon-14 dating.

Other radioisotopes besides uranium-238 and carbon-14 are used to estimate the age of objects; the choice depends on the nature and the age of the object.

CONSUMER BRIEF　　**Smoke Detectors**

The best protection against fire is early warning. Smoke is nature's way of warning us of danger, but we may not see it or smell it early enough to avoid harm. According to a recent study by the National Fire Protection Association, more than 80 percent of fire fatalities and more than 80 percent of fire injuries occur in homes that have no smoke detector.

Most low-cost smoke detectors are ionization devices that contain the radioisotope americium-241. Americium is an alpha emitter:

$$^{241}_{95}\text{Am} \longrightarrow {}^{4}_{2}\alpha + {}^{237}_{93}\text{Np}$$

The energetic alpha particles released hit air that is flowing through the smoke detector. The air is ionized as electrons are stripped from oxygen and nitrogen molecules in the air.

The electrons that have been stripped from

the ionized air flow to a positive electrode in the smoke detector, setting up a constant flow of electrons, called a standing current. The electronic circuitry of the smoke detector is designed to sense this flow of electrons. As long as the current is flowing, the circuit keeps the alarm shut off.

In the event of a fire, smoke is produced. Smoke consists of small particulate matter that is released by incomplete combustion at the source of the fire. It usually consists of oily, greasy, black carbon particles. When smoke particulates enter the smoke detector, they encounter the electrons. The electron flow is disrupted by the oily particles, and the flow of electrons to the positive electrode decreases. The electronic circuit senses the decrease in standing current and trips the alarm.

The americium-241 in the ionization source has a half-life of 432 years and far outlasts the life

(a)

(b)

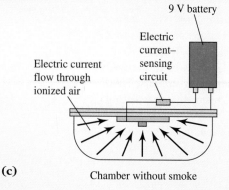

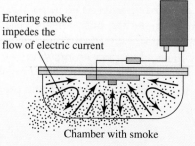

9 V battery

Electric
current–
sensing
circuit

Electric current
flow through
ionized air

Entering smoke
impedes the
flow of electric current

(c)

Chamber without smoke

Chamber with smoke

of the smoke detector. Smoke detectors require a 9-V power source for the electronic circuit that sounds the alarm. The battery lasts for at least a year because the americium source is providing the ionization energy for smoke detection. Because the electronic circuitry consumes an appreciable amount of battery power only when the alarm sounds, the battery life in these units is relatively long.

Question
1. Considering that smoke detectors have a small radioactive source, should mandatory recycling be required?
2. Although some fires give off little smoke, all give off some carbon monoxide, CO. The Consumer Product Safety Commission is currently evaluating carbon monoxide detectors for home use. Where in your home would you place one?

Reference: Consumer Reports (October 1984), pp. 564–567.

Medical Diagnosis and Treatment

Radioisotopes are used in **nuclear medicine** for both diagnosis and therapy. For example, iodine-131, a beta and gamma emitter, is a useful tool for diagnosing thyroid disease and also for treating thyroid cancer.

Nearly all the iodine that we ingest normally in food and water becomes concentrated in the thyroid gland. If a patient drinks a solution containing iodine-131, the radioisotope will behave like normal nonradioactive iodine and accumulate in the thyroid gland. By monitoring the radioactive emissions in the area of the gland, a physician can determine whether the thyroid is functioning normally. Typical scans showing the uptake of iodine-131 in a normal, an enlarged, and a cancerous thyroid gland are shown in Figure 6-9.

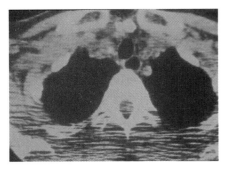

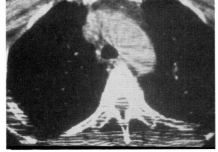

(a)

(b)

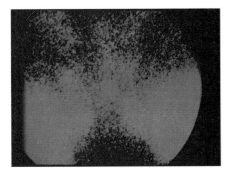

(c)

Figure 6-9 The uptake of iodine-131 by (a) a normal thyroid, (b) an enlarged thyroid, and (c) a cancerous thyroid. The uptake of the radioactive iodine is recorded as a photoscan.

There is obviously some risk attached to introducing radioactive materials into the body. To keep the risk of damaging tissues at a minimum, alpha emitters are never used for diagnosis. Beta or gamma emitters—preferably, the latter—with very short half-lives are used, and the dose is kept as small as possible. Useful diagnostic radioisotopes, besides iodine-131, include gadolinium-153, which is used to evaluate bone mineralization, and sodium-24, which can detect constriction and obstruction in blood vessels.

Like other rapidly dividing cells, cancer cells are very sensitive to radiation. Radiation therapy aims to destroy only cancer cells, but inevitably, some rapidly growing healthy cells—particularly intestinal and blood cells—are also killed. As a result, patients receiving therapy usually experience nausea and vomiting, and their white blood cell counts fall. Aiming radiation, which may amount to 150—200 rem per treatment, as precisely and narrowly as possible at the tumor minimizes damage to normal cells. Thyroid cancer can be treated by ingesting iodine-131; the radioisotope concentrates in the thyroid, where it bombards the cancer cells with radiation and destroys them. For other forms of cancer, an outside source of penetrating radiation is generally used. The most common sources are gamma emitters such as cobalt-60 and cesium-137, and X-rays. Table 6-6 lists a number of radioisotopes and their applications in nuclear medicine.

Applications in Agriculture, Industry, and Scientific Research

Because all isotopes of a given element, whether radioactive or not, behave almost identically in chemical reactions, a radioisotope of an element can be used to label that element for identification. For example, if phosphate labeled with radioactive phosphorus-32 is fed to a plant, the labeled phosphate will be taken

TABLE 6-6   Applications of Radioisotopes in Medicine

Radioisotope	Symbol	Radiation Emitted	Application
Chromium-51	$^{51}_{24}\text{Cr}$	Gamma	Determination of blood flow through heart and of lifetime of red blood cells
Cobalt-57	$^{57}_{27}\text{Co}$	Gamma	Detection of defects in uptake of vitamin B-12
Cobalt-60	$^{60}_{27}\text{Co}$	Beta, gamma	Treatment of cancer
Cesium-137	$^{137}_{55}\text{Cs}$	Gamma	Treatment of cancer
Iodine-131	$^{131}_{53}\text{I}$	Beta, gamma	Determination of activity of thyroid gland; treatment of thyroid cancer
Sodium-24	$^{24}_{11}\text{Na}$	Beta, gamma	Detection of constrictions and obstructions in the circulatory system
Technetium-99m[1]	$^{99\text{m}}_{43}\text{Tc}$	Gamma	Obtaining images of organs, e.g., heart, lungs, liver, kidney
Tritium	$^{3}_{1}\text{H}$	Beta	Determination of total body water

[1]Technetium-99m, with a half-life of about 6 hours, is one of the most useful radioisotopes in medicine. The m indicates that the isotope is metastable; by emitting gamma rays it disintegrates to a more stable form of the same isotope, technetium-99.

up through the plant's roots in the same way as nonradioactive phosphate. Because emissions from radioisotopes are easily detected by Geiger counters and other devices, the movement of phosphorus-32 through the plant can be traced. Such information is valuable in determining how plants grow, how they utilize fertilizers, and how their yields might be improved. In a similar way, radioisotopes, or **tracers**, can be used to trace the movement of pollutants through food chains or through water systems.

A commercially advantageous but still controversial application of nuclear radiation is in food preservation. Irradiation of food with gamma rays destroys insects and the microorganisms that cause spoilage. It also controls mold formation (Fig. 6-10) and retards sprouting in crops such as potatoes and onions. No residual radiation remains in food after irradiation. This procedure is routinely used in Europe, Canada, and Mexico. In the United States, however, because of concern that irradiation may produce as yet undiscovered chemicals capable of causing genetic damage, irradiation is restricted to a few crops. In time, it is likely that irradiation will replace ethylene dibromide (EDB) and other potentially harmful chemicals that are now used to fumigate fruits and vegetables (Chapter 18).

In 1992, the Department of Agriculture proposed the use of irradiation to kill *Salmonella* and other bacteria that, according to the department, contaminate as much as 40 percent of all raw poultry. Irradiation of poultry has been approved by the Food and Drug Administration, but the poultry industry, fearful of a negative response from consumers, is hesitant to adopt the process.

Irradiation offers an effective means of pest control. Irradiated pests become sterile, and, if enough sterile individuals are released into the natural population, most matings will produce no offspring. Unlike chemical pesticides (Chapter 18), this method does not pollute the environment. In another agricultural application, new and improved strains of wheat, corn, and other crops have been produced by irradiating plants to produce intentional mutations.

(a)   (b)

Figure 6-10   Irradiation prevents the growth of mold and keeps fruit fresh. Irradiated strawberries kept at $4°C$ (a) were still fresh after 15 days; those not irradiated (b) had become moldy.

NUCLEAR FISSION

Until about 55 years ago, scientists believed that the only way an unstable nucleus such as uranium-238 could change into one that was more stable was by a series of successive steps, in each of which a small nuclear particle, such as an alpha or beta particle, a proton, or a neutron, was emitted. As a result of these emissions, atomic numbers changed by no more than one or two units at a time (see Fig. 6-3).

In 1938, on the eve of World War II, these beliefs had to be completely revised when two German chemists, Otto Hahn and Fritz Strassman, found that, if they bombarded uranium-238 with neutrons, they obtained not only the expected products—uranium-239 (atomic number 92) and neptunium-239 (atomic number 93)—but also small amounts of barium (atomic number 56), lanthanum (atomic number 57), and cerium (atomic number 58). The scientists were astounded to find elements with such low atomic numbers among the products. They realized that the original uranium nucleus must have been split almost in half, a process they named **nuclear fission**.

The implications of nuclear fission were immediately evident to Hahn's fellow scientist, Lise Meitner, who, because she was Jewish, had fled to Sweden when her homeland, Austria, was annexed by Hitler in 1938. Meitner and her nephew, Otto Frisch, using calculations based on Albert Einstein's famous equation, $E = mc^2$, determined the tremendous amount of energy that would be released if an atomic nucleus was split. They realized that if this energy could be harnessed, it could be used to construct very powerful bombs.

The Energy Source in Nuclear Fission

We mentioned earlier that, despite repulsion between protons, energy within the nucleus holds the nucleus together. This **binding energy** comes from the conversion of a very small amount of mass into energy that occurs when protons and neutrons are packed together to form nuclei. Einstein had discovered that energy and mass are two aspects of the same thing. In his equation, E represents energy, m represents mass, and c is the speed of light (300,000 km/s or 186,000 mi/s). Because c^2 is a very large number, the equation means that even if the mass (m) converted is very small, the amount of energy produced ($m \times c^2$) will be enormous.

The binding energy in the nuclei of different elements is related to the mass of their nuclei as shown in Figure 6-11. Nuclei of iron atoms—and other nuclei of similar mass—are more stable and need less energy to hold them together than either lighter or heavier nuclei. When a heavy nucleus such as a uranium nucleus is split apart to form lighter, more stable nuclei, some of the binding energy is released. It has been calculated that fission of 1 g of uranium releases approximately 10 million times as much energy as burning 1 g of coal.

Figure 6-11 also shows that energy is released when nuclei of very light elements such as hydrogen and helium join, or fuse, together. Nuclear fusion, which occurs in the explosion of a hydrogen bomb, will be discussed in Chapter 15.

Figure 6-11 Intermediate-sized nuclei with mass numbers between about 30 and 65, such as $^{56}_{26}$Fe, need the least amount of energy to hold them together and are the most stable. During fission, elements with high mass numbers, such as $^{235}_{92}$U, split apart to form smaller atoms with more stable nuclei, and energy is released. Energy is also released during fusion when small nuclei such as $^2_1$H and $^4_2$He join to form larger more stable nuclei.

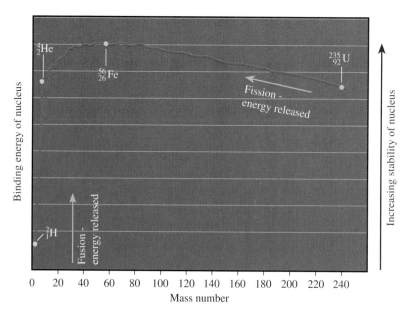

Figure 6-12 Nuclear fission occurs when $^{235}_{92}$U atoms are bombarded with slow-moving neutrons. During fission, $^{235}_{92}$U atoms split apart to form two lighter atoms and from 2 to 4 neutrons. Because the new elements formed are more stable than uranium, a large amount of energy is released. Uranium atoms split in many different ways. As many as 35 different elements have been identified among fission products. $^{146}_{57}$La and $^{87}_{35}$Br are typical of the elements formed.

Fission Reactions

If a fissionable isotope such as uranium-235 or plutonium-239 is bombarded with neutrons, it can split apart in more than one way to produce a variety of lighter elements. Whichever way it splits, between two and four neutrons and a great deal of energy are always released. One way that uranium-235 splits apart is shown in the following equation:

$$^{235}_{92}\text{U} + ^{1}_{0}\text{n} \longrightarrow\ ^{146}_{57}\text{La} + ^{87}_{35}\text{Br} + 3\,^{1}_{0}\text{n} + \text{energy}$$

A slow-moving neutron enters the uranium-235 nucleus and causes fission (Fig. 6-12). Under the right conditions, the neutrons emitted can cause the fission of more uranium-235 nuclei, and thus set off a **chain reaction** as illustrated in Figure 6-13. The reaction is self-perpetuating and can continue, with the release of ever-increasing amounts of energy, until all the uranium nuclei have been split.

A chain reaction is a self-sustaining reaction that continues unchecked once it has begun.

The Atomic Bomb

In August 1939, after President Roosevelt received a letter signed by Einstein warning him that the Germans were already working on a bomb based on nuclear fission, the United States launched the secret Manhattan Project to develop an **atomic bomb**.

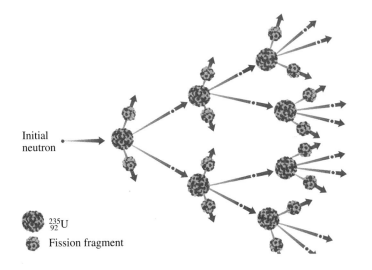

Initial neutron

$^{235}_{92}$U

Fission fragment

Figure 6-13 A nuclear chain reaction. Fission of a $^{235}_{92}$U atom by a single neutron releases from 2 to 4 neutrons from the uranium nucleus. Each neutron released can trigger fission of additional uranium atoms and the release of more neutrons. As a result, an uncontrollable chain reaction occurs, with the release of enormous amounts of energy.

A chain reaction cannot be sustained unless a certain minimum amount of fissionable material, called the **critical mass**, is present. To obtain an adequate supply of uranium-235—which makes up only 0.7 percent of natural uranium—the U.S. government built a top-secret facility for enriching uranium in Oak Ridge, Tennessee. At the same time, a facility was established in Hanford, Washington, to produce plutonium-239, a synthetic fissionable radioisotope.

In 1942, the first sustained chain reaction, using uranium-235 as the starting material, was achieved in a laboratory under the football stadium at the University of Chicago. The buildup of neutrons was carefully controlled to prevent a dangerous explosion. By early 1945, sufficient plutonium-239 had been prepared for the construction of a bomb. To create a bomb, two separate masses of fissionable material, each incapable of sustaining a chain reaction, are brought together to form a critical mass at the moment of detonation.

In July 1945, the first atomic bomb was successfully tested in the desert near Alamogordo, New Mexico. Those who witnessed the explosion were awed by the magnitude of the blast; the brilliance of the light, the tremendous heat generated, and the huge mushroom cloud that formed were overwhelming (Fig. 6-14). On August 6, 1945, President Harry Truman, compelled by his desire to avoid the millions of American and Allied casualties expected in an invasion of Japan, ordered the dropping of a uranium bomb on Hiroshima. Three days later, a plutonium bomb was dropped on Nagasaki. Approximately 200,000 Japanese were either killed or injured by the two bombs. The war ended with the surrender of the Japanese on August 14, 1945.

The Hiroshima bomb was equivalent to about 20,000 tons of TNT.

Figure 6-14   The explosion of an atomic bomb (a bomb based on a nuclear fission reaction) produces a characteristic mushroom-shaped cloud.

Peaceful Uses of Nuclear Fission

Nuclear fission is used to create bombs and other weapons of war, but it can also be harnessed for peaceful purposes. If fission is controlled so that the concentration of neutrons produced is sufficient to maintain the fission process but not great enough to allow an uncontrolled chain reaction, the process can be used as a source of energy for generating electricity in power plants. Nuclear power plants already generate much of the electricity used in Europe, but the United States is still very dependent on fossil fuels to meet its energy needs. Nuclear energy is discussed in more detail in Chapter 15.

EXPLORATIONS

Marie Curie: Wife, Mother, and Two-Time Nobel Prize Winner, Long before Women's Lib

Marie Curie was born Marya Sklodowska in Russian-dominated Warsaw, Poland, in 1867, the youngest of five children—one boy and four girls. She enjoyed a happy childhood until she was nine, when, tragically, her oldest sister died of typhus, and just 2 years later, her mother died of tuberculosis.

After graduating from school at 15 with a gold medal, Marya was anxious to continue her education, but because the Tsar's University of Warsaw was closed to women, her only choice was to study abroad. Her father, after losing his position as a school teacher of mathematics and physics for not being sufficiently pro-Russian, could not afford the expense, and Marya began to save toward her education by working as a tutor. At the same time, ardently patriotic, she joined the illegal "Floating University," well aware that discovery could mean imprisonment or deportation to Siberia. The night school was dedicated to building a core of Polish intellectuals. Teachers gave free instruction to students who, in turn, taught workers.

At 18, to help pay for her sister Bronya's medical studies in Paris, Marya took a position as a governess with the understanding that, when possible, Bronya would in turn help her finance her further education. The bargain was kept, and in 1891, Marya registered for classes at the Sorbonne using the French form of her name, Marie. To be close to the university, she lived in a garret, existing mainly on buttered bread and tea, with an occasional treat of a piece of fruit or an egg. If she wanted heat, she had to carry coal for a small stove up six flights of stairs. Marie was completely absorbed in the joy of learning and, despite the hardships, was later to describe her student days as "a time of charm." In 1893, number one in her class, she became the first woman to receive the *licence es sciences physiques* from the Sorbonne. A year later, second in her class, she received the *licence es sciences mathematiques*.

In the year that she graduated, Marie met Pierre Curie, a shy, reserved man of 35, already recognized for his work on crystal symmetry, magnetism, and piezo electricity. Without any ambition for himself, Pierre was committed to a life devoted entirely to scientific research. Today, many physicists regard him as a greater scientist than his far more famous wife.

There was an immediate rapport between Marie and Pierre, and they were married in July 1895. It was a difficult decision for Marie, because it meant leaving her family and her beloved Poland, but she was persuaded to follow her heart. Always practical, Marie chose for her wedding a plain dark dress that she could wear afterwards in the laboratory. The couple spent their honeymoon bicycling through the French countryside.

Neither marriage, the birth of her daughter, Irene, in 1897, nor the fact that no European woman had ever completed a doctorate in science, deterred Marie from her pursuit of an advanced degree (Fig. 6-A). For her research topic, she chose the rays spontaneously emitted by uranium salts that had recently been discovered by Henri Becquerel. This fateful decision determined the course of her work for the rest of her life.

Becquerel's rays caused air to conduct electricity; using a sensitive electrometer that had been invented by Pierre and his brother, Jacques, Marie tested compounds of all the known elements for this property, which she named radioactivity. Working at her own expense in a small, damp, unheated glassed-in room that Pierre's superior allowed her to use, she soon found that thorium, in addition to uranium, was radioactive. She also found that, regardless of the nature of the radioactive sample, the intensity of its radiation was always proportional to the quantity of uranium or thorium it contained. This finding led her to the then startling conclusion that radiation must come from the atom itself, a fundamental discovery that was to be Marie's most important contribution to science.

Figure 6-A Pierre and Marie Curie with their daughter Irene in 1904.

Marie next discovered that on the basis of its uranium content, the ore pitchblende was much more radioactive than expected. She realized that it must contain a new and extremely radioactive element, present in an amount so small that it had previously escaped detection. The discovery was so important that Pierre decided to abandon his own work temporarily to help Marie isolate the new element and thus prove beyond doubt that it existed. The Curies soon found that two new elements were present in the ore. One, Marie named polonium in honor of her homeland, but it was the second, far more radioactive element, which she named radium, that was to make her famous.

Optimistically, the Curies thought radium might make up 1 percent of pitchblende; in fact, it accounted for one-millionth part. Pitchblende was expensive, and the Curies needed tons of it. Fortunately, they found they could use the cheap residue discarded after the valuable uranium had been extracted. Together, they worked with inadequate equipment under the most primitive conditions in an old shed with a leaky roof. To avoid inhaling noxious fumes, they often worked outside in the yard. Over and over again, Marie took 20-kg samples of pitchblende, the most she could handle at one time, and, using huge cauldrons, separated the different elements following the standard analytical techniques of the day. With each step, the Curies obtained an increasingly pure and more radioactive radium fraction. Finally, in 1902, Marie had a 100-mg sample of pure radium chloride and indisputable proof that she had a new element with an atomic weight of 226.

The work took a heavy toll on the Curies' health, and both Marie and Pierre frequently suffered from extreme fatigue and various other unexplained illnesses. They realized that their red, cracking, permanently damaged fingertips were the result of handling radium. But it never occurred to them—or to anyone else, until many years later—that exposure to radiation, and breathing the radon gas that radium emitted, could be a health hazard, and they took no precautions to protect themselves.

In 1903, with Henri Becquerel, the Curies received the Nobel Prize in physics for the discovery of radioactivity. In that same year, Marie received her doctorate, and 1 year later their second daughter, Eve, was born. Then in 1906, tragedy struck. Pierre Curie, while crossing a street, was hit by a horse-drawn wagon and killed instantly. Marie was completely devastated by the loss of her beloved companion, but in time, she found solace in her work and her children. She was named to Pierre's chair at the Sorbonne and became the first woman in France ever to hold a position in higher education. She continued her work with radium and, in 1911, was awarded the Nobel Prize in chemistry for her isolation of pure radium metal.

Although Marie Curie made no further important discoveries she made possible the work of others, including her daughter, Irene, who with her husband Frederic Joliot won the Nobel Prize in chemistry in 1935 for the synthesis of radioelements. Radium showed great promise as a cure for cancer and was hailed as a wonder drug. As radium's discoverer, Marie was honored around the world, particularly in America, where she raised huge sums of money for research into its applications.

In 1934, at the age of 67, Marie Curie died from leukemia, undoubtedly caused by years of exposure to radiation. Her discovery that radiation had its origin in the atom was a milestone in nuclear physics, and that it was made by a young woman in a field completely dominated by men makes it doubly remarkable. Other scientists elucidated the structure of the atom, but Marie Curie was the first to show that the atom is not immutable, and her seminal discovery ushered in the atomic age.

Question

During her isolation of radium and polonium from uranium ores, and her subsequent work with radium, Marie Curie was exposed to very large doses of radiation. Discuss the type of radiation to which she was exposed. How does her exposure compare with the exposure to naturally occurring radon gas experienced by the average person living in Arkansas today? (The number of nuclear disintegrations per second from 1 g of radium = $3.7 \times 10^{10} = 1$ Ci.)

References: R. Pflaum, *Grand Obsession* (New York: Doubleday, 1989); E. Curie, *Madame Curie: A Biography by Eve Curie*. Translated by V. Sheean (New York: Doubleday, 1937); R. Reid, *Marie Curie* (New York: Saturday Review Press/E. P. Dutton, 1974).

KEY WORDS AND CONCEPTS

alpha particle
artificial transmutation
beta particle
binding energy
chain reaction
cosmic rays
critical mass
curie
decay series

gamma rays
Geiger counter
half-life
ionizing radiation
nuclear fission
nuclear medicine
nuclear reaction
rad (gray)
radioactive decay

radioactivity
radiocarbon dating
radioisotope
radon
rem (sievert)
terrestrial radiation
tracers
transmutation
X-rays

QUESTIONS AND PROBLEMS

1. Write the symbols used to describe the following:
   a. proton **b.** neutron **c.** alpha particle
   d. electron **e.** beta particle **f.** gamma ray

2. Write the number of protons and neutrons in each of the following nuclei:
   a. chlorine-37 **b.** oxygen-17 **c.** molybdenum-99
   d. cadmium-113 **e.** plutonium-234 **f.** silver-115 **g.** cesium-136 **h.** carbon-13 **i.** neon-22
   j. barium-137

3. Write balanced nuclear equations for:
   a. beta emission by magnesium-28 **b.** alpha emission by lawrencium-255 **c.** beta emission by nickel-65

4. Lead-210 is used to prepare eyes for corneal transplants. The product of lead-210 decay is bismuth-210. What particle is emitted as lead-210 decays?

5. Thorium-231 is the product of alpha emission by a radioisotope. Thorium-231 is radioactive and emits a beta particle. Write a reaction that shows from which isotope thorium-231 is produced. Write a second reaction that shows the product of thorium-231 decay.

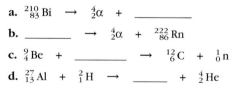

6. Fill in the missing symbol in each of the following nuclear equations.
   a. $^{210}_{83}\text{Bi} \rightarrow {}^{4}_{2}\alpha + \underline{\hspace{2cm}}$
   b. $\underline{\hspace{2cm}} \rightarrow {}^{4}_{2}\alpha + {}^{222}_{86}\text{Rn}$
   c. $^{9}_{4}\text{Be} + \underline{\hspace{2cm}} \rightarrow {}^{12}_{6}\text{C} + {}^{1}_{0}\text{n}$
   d. $^{27}_{13}\text{Al} + {}^{2}_{1}\text{H} \rightarrow \underline{\hspace{1.5cm}} + {}^{4}_{2}\text{He}$

7. What is the effect on the mass number and atomic number of the reacting atom when the following nuclear transmutations occur?
   a. A beta particle is emitted. **b.** An alpha particle is emitted. **c.** A gamma ray is emitted.

8. Write balanced nuclear equations for:
   a. beta emission by nickel-65 **b.** alpha emission by neodymium-150 **c.** beta emission by magnesium-28

9. Sulfur-35 is radioactive and undergoes beta decay. What differences in chemical reactivity would you expect between sulfur-35 and nonradioactive sulfur-32? Explain your answer.

10. Define the following:
    a. alpha particle **b.** mass number **c.** gamma ray
    d. atomic number **e.** beta particle **f.** nuclear transmutation

11. Potassium-42 is a beta-emitting isotope used to locate brain tumors.
    a. Write a nuclear reaction for the decay of potassium-42. **b.** If the half-life of this isotope is 12.4 hours, what fraction of potassium-42 remains after 62 hours? **c.** Give two reasons why you think this particular isotope is chosen for this diagnostic test.

12. The half-life of barium-131 is 12.0 days. How many grams of barium-131 remain after four half-lives if you begin with 100 g? How many days do four half-lives take?

13. The radioisotope americium-241 is used commercially in home smoke detectors. This isotope has a half-life of 433 years.
    a. How long will it take for the activity of this material to drop to less than 1 percent of its original activity? **b.** On the basis of your answer to part a, should a 20-year-old smoke detector be disposed of by placing it in the municipal trash?

14. Describe how a Geiger counter works and how radioactivity is detected.

15. The synthetic radioisotope technetium-99, which is a beta emitter, is a widely used isotope in nuclear medicine. The following data were obtained using a Geiger counter.

Disintegrations of Technetium-99 per Minute	Time (hours)
180	0.0
130	2.5
104	5.0
77	7.5
59	10.0
46	12.5
24	17.5

a. Make a graph of this information similar to Figure 6-4 and determine the half-life of technetium-99. (Remember that disintegrations per minute are directly proportional to the grams of technetium remaining.)   **b.** How long would it take for 99 percent of the technetium-99 to decay after injection into the patient?

16. Chemical reactions can often be used to change a toxic chemical into another compound that is not as toxic. Why can't chemical treatment technology be applied to make nuclear waste harmless?

17. A so-called expert suggested that strontium-90 deposited in the Nevada desert during nuclear testing will undergo radioactive decay more rapidly than strontium-90 elsewhere because of the high desert temperatures. Do you agree with this assessment? Explain.

18. Prior to the Nuclear Test Ban Treaty of 1963, the average exposure to radioactive fallout from nuclear weapons testing amounted to 30 mrem/year per person. In 1990, the average annual exposure was only 4 mrem per person. Estimate the increase in exposure to an average individual during the period 1990–2000, if open-air nuclear bomb testing had not been banned. (Assume the average exposure per year would continue to be 30 mrem per person).

19. An oil painting alleged to be painted by Rembrandt (1606–1669) is subjected to carbon-14 dating. The carbon-14 content of the canvas used is 0.96 times that of a living tree. If the half-life of carbon-14 is 5730 years, could this painting be an original Rembrandt?

20. Would the decay of uranium-238 to lead-208 provide an accurate method for determining the age of a sample thought to be about 500 years old?

21. Define the following:
   a. fission reaction   **b.** nuclear chain reaction
   c. critical mass

22. Describe the sources of everyday radiation to which the general public is exposed and indicate the percentage each source contributes.

23. Answer the following questions.
   a. Why is radioactivity used in treating cancer? How is treatment designed so that damage to healthy cells is minimized?   **b.** Why is a critical mass of radioactive material necessary for an atomic bomb? Could a little bomb be made with less than the critical mass?   **c.** Food is often irradiated before it is sold. Why doesn't the food become radioactive and poisonous?   **d.** Where does radon come from? Where is it most likely to be found in the home? How can you lessen your exposure to it?   **e.** Does our exposure to radiation come from natural or man-made sources? What is the contribution of natural sources relative to our total exposure?

24. Which of the following has the greatest penetrating ability: alpha particle, beta particle, or gamma ray?

25. What type of shield is necessary to stop the following:
   a. X-rays   **b.** alpha particles   **c.** beta particles
   d. gamma rays

26. Explain two ways in which radiation can damage or destroy cells.

27. One gram of uranium-235 produces as much energy as 14 barrels of crude oil. How many barrels of crude oil can be saved for every kilogram of uranium-235 used in a nuclear power plant?

CHEMICAL REACTIONS

In photosynthesis—the series of reactions on which all life depends—green plants, like these in a tropical rain forest in Costa Rica, use energy from the sun to convert carbon dioxide and water to the sugar glucose and release oxygen as a byproduct.

Since ancient times, humans have been initiating chemical reactions to produce new substances from natural materials—pottery from clay, drugs and dyes from plants, and beer and wine from fermenting grain. Many of these early reactions were discovered accidentally. Today, chemists are able to predict the course of a chemical reaction, and they know how to manipulate conditions to increase both the speed of a reaction and the yield of the desired new product.

Can we tell if a chemical reaction will occur when two or more substances are mixed together? Will a reaction occur immediately, or will it need energy to get it started? How fast will a reaction go, and can it be made to go faster or slower? And how much new product can be expected from a given amount of starting materials? In this chapter, we will attempt to answer these questions, and, in the process, we will learn that all chemical reactions, whether occurring in a saucepan on the stove, in a chemical factory, or in the human body, the atmosphere, the ocean, or any other part of the environment, are governed by the same basic principles.

Learning Goals:

In this chapter, you should gain an understanding of:

1. Balanced chemical equations and how to write them.

2. The mole concept.

3. Calculations based on chemical equations.

4. Entropy and the difference between exothermic and endothermic reactions.

5. Factors that influence the rate of a chemical reaction.

6. Reversible reactions and chemical equilibrium.

CHEMICAL EQUATIONS

Chemical symbols are equivalent to the letters of the alphabet, chemical formulas are equivalent to words, and chemical equations are equivalent to sentences.

In any chemical reaction, one or more substances, called **reactants**, are changed into one or more new substances, called **products**. The reaction can be represented in chemical shorthand by a chemical equation. Carbon (C), the main ingredient in coal, burns in atmospheric oxygen to form carbon dioxide (CO_2). The chemical equation for this reaction is:

$$\text{carbon} \qquad \text{oxygen} \qquad \text{carbon dioxide}$$

$$C \quad + \quad O_2 \quad \longrightarrow \quad CO_2$$

The plus sign (+) is read as "and," and indicates that the reactants (C and O_2) are mixed together in some way. The arrow ($\longrightarrow$) means "react to yield."

A chemical equation tells us much more than just what substances react together and what substances are produced. It tells us how many atoms or molecules of each reactant take part in the reaction and how many atoms or molecules of each product are formed. The above equation tells us that one atom of carbon (C) reacts with one molecule of oxygen (O_2) to yield one molecule of carbon dioxide (CO_2).

A chemical equation is also a bookkeeping device: It keeps track of the numbers of atoms of the different elements taking part in a reaction. In accordance with the law of conservation of matter, which states that matter is neither created nor destroyed in a chemical reaction (Chapter 4), the number of atoms of each element present at the start of a chemical reaction will equal the number present at the end of the reaction. Therefore, in a correctly written, or *balanced*, chemical equation for a chemical reaction, the total number of each kind of atom on the left side of the arrow will equal the total number of each kind of atom on the right side of the arrow.

WRITING BALANCED CHEMICAL EQUATIONS

In a balanced equation, the sum of the atoms of each element in the reactants (to the left of the arrow) is equal to the sum of the atoms of each element in the products (to the right of the arrow).

The first step in writing a balanced equation is to write the correct formulas for the reactants and products. The next step is to add the numbers of atoms on both sides of the equation and see if they balance. The equation for the burning of coal, shown above, is properly balanced: The numbers of C and O atoms on the left side of the equation are equal to the number on the right side.

$$\text{carbon} \qquad \text{oxygen} \qquad \text{carbon dioxide}$$

$$C \quad + \quad O_2 \quad \longrightarrow \quad CO_2$$

	reactants	*product*
C	1	1
O	2	2

balanced

Not all chemical reactions can be so easily balanced. For example, it is not as simple to write the correct equation for the explosive reaction between hydrogen (H_2) and oxygen (O_2) to form water (H_2O) that occurs when an electric spark is passed through a mixture of the two gases. As the first step in writing the balanced equation for this reaction, we enter the correct formulas for the reactants and for the product in the equation.

$$\text{hydrogen} \qquad \text{oxygen} \qquad \text{water}$$
$$H_2 \quad + \quad O_2 \quad \longrightarrow \quad H_2O \qquad \text{(unbalanced)}$$

	reactants	product
H	2	2
O	2	1

unbalanced

We can see that this equation is not balanced. There are 2 atoms of oxygen on the left side of the equation and only 1 on the right side. We could balance the oxygen atoms by assuming that 2 molecules of water are formed and putting the coefficient 2 in front of the water molecule (remember that a coefficient in front of a formula multiplies everything in the formula).

The number in front of a formula in a chemical equation is the *coefficient*.

$$H_2 \quad + \quad O_2 \quad \longrightarrow \quad 2\,H_2O \qquad \text{(unbalanced)}$$

	reactants	product
H	2	$2 \times 2 = 4$
O	2	2

unbalanced

But this does not give us a balanced equation. In establishing an oxygen balance, we have created a hydrogen imbalance. This imbalance can be corrected by having 2 molecules of hydrogen react with 1 molecule of oxygen:

$$2\,H_2 \quad + \quad O_2 \quad \longrightarrow \quad 2\,H_2O \qquad \text{(balanced)}$$

	reactants	product
H	$2 \times 2 = 4$	$2 \times 2 = 4$
O	2	2

balanced

Always remember that the *subscripts* in the formulas for the reactants or products cannot be changed to balance an equation because doing so would alter the *identity* of the substances taking part in the reaction. For example, the reaction between hydrogen and oxygen to form water cannot be written as follows:

$$\text{hydrogen} \qquad \text{oxygen} \qquad \text{hydrogen peroxide}$$
$$H_2 \quad + \quad O_2 \quad \longrightarrow \quad H_2O_2 \qquad \text{(incorrect)}$$

This equation is balanced, but it is incorrect because the formula of the product, H_2O_2, represents hydrogen peroxide instead of water, H_2O.

SOME ENVIRONMENTALLY IMPORTANT REACTIONS

For practice in balancing equations, let us look at a number of chemical reactions that are important for the environment.

Oxides of Nitrogen and Smog

Nitrogen and oxygen make up approximately 78 percent and 21 percent by volume of the air we breathe.

Automobiles are a major source of air pollution (Chapter 13). At ordinary temperatures, nitrogen (N_2) and oxygen (O_2) in the atmosphere do not combine, but at the high temperatures that exist inside running automobile engines, they react to produce the colorless gas nitric oxide (NO). As a first step, we can write the equation for this reaction as follows:

$$\underset{N_2}{\text{nitrogen}} + \underset{O_2}{\text{oxygen}} \longrightarrow \underset{NO}{\text{nitric oxide}} \quad \text{(unbalanced)}$$

	reactants	*product*
N	2	1
O	2	1

unbalanced

The equation can be balanced if we have 2 molecules of NO:

$$N_2 + O_2 \longrightarrow 2\ NO \quad \text{(balanced)}$$

	reactants	*product*
N	2	2
O	2	2

balanced

Once in the atmosphere, NO combines rapidly with atmospheric oxygen to form nitrogen dioxide (NO_2), the red-brown toxic gas that is responsible for the color of the yellow-brown smog that often settles over Los Angeles and Denver.

$$\underset{NO}{\text{nitric oxide}} + \underset{O_2}{\text{oxygen}} \longrightarrow \underset{NO_2}{\text{nitrogen dioxide}} \quad \text{(unbalanced)}$$

	reactants	*product*
N	1	1
O	1 + 2 = 3	2

unbalanced

The equation can be balanced if we have 2 molecules of NO and 2 molecules of NO_2.

$$2\ NO + O_2 \longrightarrow 2\ NO_2 \quad \text{(balanced)}$$

	reactants	*product*
N	2	2
O	2 + 2 = 4	2 × 2 = 4

balanced

Oxides of Sulfur and Acid Rain

Much of the coal burned in the United States contains sulfur in the form of the mineral iron pyrite (FeS_2).

One of the main causes of acid rain is the burning of coal that has a high sulfur content (Chapter 13). When the coal is burned, the sulfur (S) is converted to the pungent, choking gas sulfur dioxide (SO_2).

$$\underset{S}{\text{sulfur}} + \underset{O_2}{\text{oxygen}} \longrightarrow \underset{SO_2}{\text{sulfur dioxide}} \quad \text{(balanced)}$$

	reactants	*product*
S	1	1
O	2	2

balanced

If released into the atmosphere, SO_2 combines with atmospheric oxygen to form sulfur trioxide (SO_3).

$$\underset{SO_2}{\text{sulfur dioxide}} \quad + \quad \underset{O_2}{\text{oxygen}} \quad \longrightarrow \quad \underset{SO_3}{\text{sulfur trioxide}} \quad \text{(unbalanced)}$$

	reactants	*product*
S	1	1
O	2 + 2 = 4	3
	unbalanced	

This equation can be balanced by having 2 molecules of SO_2 and 2 molecules of SO_3.

$$2\,SO_2 \quad + \quad O_2 \quad \longrightarrow \quad 2\,SO_3 \quad \text{(balanced)}$$

	reactants	*product*
S	2	2
O	$(2 \times 2) + 2 = 6$	$2 \times 3 = 6$
	unbalanced	

Sulfur trioxide combines very rapidly with water vapor in the atmosphere to form sulfuric acid (H_2SO_4), which reaches the earth in rainfall.

$$\underset{SO_3}{\text{sulfur trioxide}} \quad + \quad \underset{H_2O}{\text{water}} \quad \longrightarrow \quad \underset{H_2SO_4}{\text{sulfuric acid}} \quad \text{(balanced)}$$

	reactants	*product*
S	1	1
O	3 + 1 = 4	4
H	2	2
	balanced	

Photosynthesis

In the process of photosynthesis, green plants absorb solar energy and use it to convert carbon dioxide from the atmosphere and water into glucose, a simple sugar, and oxygen (Fig. 7-1). As we saw in Chapter 3, without the energy stored in green plants and the replenishing of the atmosphere with oxygen, life on earth could not exist. Photosynthesis is a complex process that occurs in many steps. The overall reaction is shown in the following equation:

$$6\,H_2O \quad + \quad 6\,CO_2 \quad \overset{\text{solar energy}}{\longrightarrow} \quad C_6H_{12}O_6 \quad + \quad 6\,O_2 \quad \text{(unbalanced)}$$

	reactants	*products*
H	2	12
O	1 + 2 = 3	6 + 2 = 8
C	1	6
	unbalanced	

To match the 6 C atoms and 12 H atoms in the sugar, 6 molecules of CO_2 and 6 molecules of H_2O are required. To balance the O atoms, 6 molecules of O_2 are needed.

$$6\,H_2O \quad + \quad 6\,CO_2 \quad \longrightarrow \quad C_6H_{12}O_6 \quad + \quad 6\,O_2 \quad \text{(balanced)}$$

	reactants	*products*
H	$6 \times 2 = 12$	12
O	$6 + (6 \times 2) = 18$	$6 + (6 \times 2) = 18$
C	6	6
	balanced	

Directly, or indirectly through food chains, all living creatures are dependent on photosynthesis for their food.

Figure 7-1 Green plants are fundamental to all life on earth. In the process of photosynthesis, they use light energy from the sun to convert carbon dioxide and water to the sugar glucose and release oxygen as a byproduct.

CHEMICAL ARITHMETIC

Chemical equations, in addition to showing the numbers of atoms and molecules that are involved in a reaction, tell us the *relative masses* (weights) of the different reactants and of the different products formed. Such information is essential not only for the laboratory chemist who wants to make a few grams of a particular compound, but also for the plant manager who is planning a large-scale production of an industrial chemical.

Mass Relationships in Chemical Equations

First, let us look at the mass (weight) relationships in the equation for the reaction of carbon with oxygen.

$$C + O_2 \longrightarrow CO_2$$

The equation shows that one atom of carbon reacts with one molecule of oxygen to form one molecule of carbon dioxide.

Recall from Chapter 4 that the atomic mass unit (amu) of the most common isotope of carbon is exactly 12, and the masses of the atoms of all other elements are compared with that value. (The atomic masses, or atomic weights, of the elements are listed inside the back cover of this book.)

Since one atom of carbon weighs 12 amu and one atom of oxygen weighs 16 amu, the equation also shows that, in agreement with the law of conservation of mass,

12 amu of carbon combine with 32 (2×16) amu of oxygen to form 44 (12 + 32) amu of carbon dioxide.

That is, the sum of the formula masses of the reactants equals the formula mass of the product.

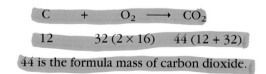

44 is the formula mass of carbon dioxide.

Determining Formula Masses

The **formula mass** of a substance is the sum of the atomic mass units of the constituent atoms in the chemical formula.

The first step in determining mass relationships in any chemical equation is to determine the formula masses of the reactants and products. As we have seen for the reaction of carbon with oxygen, this value is easily obtained for simple molecules like O_2 and CO_2, but it can be more difficult for complex compounds. The following examples describe the steps to be followed.

The formula mass of a molecule such as carbon dioxide (CO_2) or water (H_2O) is often called its molar mass. For an ionic compound such as sodium chloride (NaCl), which does not exist as a discrete molecule, the term formula mass is used.

EXAMPLE 7-1   Calculate the formula mass of silver nitrate, $AgNO_3$

Solution: Formula masses are calculated by adding together the atomic masses of the constituent atoms.
Refer to the table inside the back cover of your textbook to obtain the atomic mass of each type of atom.
The formula of $AgNO_3$ contains

1 atom of Ag	= 108 amu
1 atom of N	= 14 amu
3 atoms of O	$\underline{3 \times 16 = 48 \text{ amu}}$
Formula mass of $AgNO_3$	= 170 amu

PRACTICE EXERCISE: Calculate the formula mass of SO_2.

Answer: 64 amu

EXAMPLE 7-2   Calculate the formula mass of ammonium phosphate, $(NH_4)_3PO_4$

Solution: The formula of $(NH_4)_3PO_4$ contains

3 atoms of N	$3 \times 14 = 42$ amu
$3 \times 4 = 12$ atoms of H	$12 \times 1 = 12$ amu
1 atom of P	= 31 amu
4 atoms of O	$\underline{4 \times 16 = 64 \text{ amu}}$
Formula mass of $(NH_4)_3PO_4$	= 149 amu

PRACTICE EXERCISE   Calculate the formula mass of each of the following: **a.** $KMnO_4$ **b.** $Ca(OH)_2$

Answer: **a.** 158 amu   **b.** 74 amu

Mass relationships in an equation can be expressed in other mass units besides atomic mass units; they can be expressed in grams, kilograms, pounds, tons, and so on. For example, we can express the equation on page 186 as follows:

12 lb of carbon combine with 32 lb of oxygen to form 44 lb of carbon dioxide.

12 g of carbon combine with 32 g of oxygen to form 44 g of carbon dioxide.

Calculations Based on Chemical Equations

If the appropriate equation is known, it is possible to calculate the weight of product that can be obtained from a given amount of reactant or the amount of reactant required to produce a desired amount of product. As we will explain in Examples 7-3 and 7-4, we can, for example, give answers to the following questions:

The law of conservation of mass is the basis for solving mathematical problems based on chemical formulas.

1. How many kilograms of aluminum can be produced from 10 kg of the mineral bauxite, Al_2O_3?

2. How many kilograms of nitrogen are required to react with 5 kg of hydrogen to form ammonia, the starting material for the manufacture of synthetic fertilizers?

Approximately ten times as much energy is required to produce a ton of aluminum as is needed to produce a ton of steel.

Aluminum from Bauxite   As we noted in Chapter 2, aluminum is one of the most essential metals in our modern society. Aluminum is an abundant element and constitutes approximately 8 percent of the earth's crust, but it is tied up in minerals, and until quite recently, there was no feasible cost-effective way to extract it from its ores. Then, in 1886, two young chemists, the American Charles M. Hall (Fig. 7-2) and the Frenchman Paul Heroult simultaneously but independently discovered a way to extract aluminum from bauxite, a mineral composed mainly of Al_2O_3. They discovered that when an electric current was passed through a mixture of purified Al_2O_3 in molten cyrolite (a high-melting aluminum mineral, Na_3AlF_6), molten aluminum metal separated out and settled at the bottom of the reaction vessel (Fig. 7-3).

EXAMPLE 7-3   How many kilograms of aluminum, Al, can be obtained from 10 kg of bauxite, Al_2O_3?

Solution:

1. The first step in solving any problem based on a chemical reaction is to write the balanced equation for the reaction. The balanced equation for the formation of aluminum from bauxite is:

$$2\,Al_2O_3 \longrightarrow 4\,Al + 3\,O_2$$

2. Next, determine the formula masses of Al and Al_2O_3 as shown in example 7-1. (The formula mass of O_2 is not needed because it is not mentioned in the question)

| Atomic mass of Al | | = 27 amu |

Formula mass of Al_2O_3:

2 atoms of Al	2×27 = 54 amu
3 atoms of O	3×16 = 48 amu
Formula mass of Al_2O_3	= 102 amu

3. Substitute these values in the equation:

$$2\,Al_2O_3 \longrightarrow 4\,Al + 3\,O_2$$

| 2×102 amu | 4×27 amu |
| 204 amu | 108 amu |

Figure 7-2   While a student, Charles M. Hall (1863–1914) began looking for a cheap way to obtain aluminum, and at age 22, he invented the electrolytic process that is still used today. He founded the Aluminum Corporation of America (ALCOA) and became a very wealthy man. Generally overlooked is the contribution his elder sister Julia made to the discovery. She, like Charles, studied chemistry at Oberlin College, and brother and sister worked together in the laboratory in the family's woodshed. Julia's records were crucial in establishing that Charles had discovered the aluminum process about a month before Paul Heroult.

4. State the mass (weight) relationship between Al_2O_3 and Al in terms of the given mass unit (kg).

$$204 \text{ kg } Al_2O_3 \text{ yield } 108 \text{ kg Al}$$

5. Solve the problem using dimensional analysis as explained in Appendices C and E.

   a. Given quantity: 10 kg Al_2O_3

   b. Required quantity: kg Al

   c. The conversion factor for solving the problem is the mass relationship between Al_2O_3 and Al as determined from the balanced equation.

   Conversion factor: $\dfrac{108 \text{ kg Al}}{204 \text{ kg } Al_2O_3}$

   d. Solution: $10 \text{ kg } Al_2O_3 \times \dfrac{108 \text{ kg Al}}{204 \text{ kg } Al_2O_3} = 5.29 \text{ kg Al}$

5.29 kg of aluminum can be obtained from 10 kg of bauxite.

PRACTICE EXERCISE: How many kilograms of iron, Fe, can be obtained from 5 kilograms of the mineral hematite, Fe_2O_3?

Answer: 3.5 kg of Fe

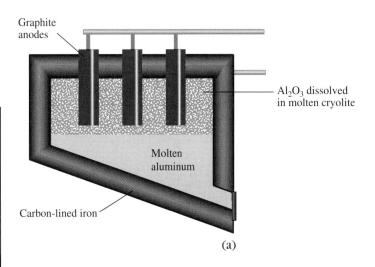

(a)

(b)

Figure 7-3   (a) An electrolysis cell for the production of aluminum from bauxite. Molten aluminum is denser than the molten mixture of Na_3AlF_6 and Al_2O_3 and collects at the bottom of the cell. (b) Molten aluminum flows from furnaces into molds in a factory.

Ammonia from Nitrogen and Hydrogen: The Haber Process In 1908, the German chemist Fritz Haber (Fig. 7-4) established that under certain conditions at high temperature and high pressure, atmospheric nitrogen (N_2) would combine with hydrogen gas (H_2) to produce ammonia (NH_3). Today, in the United States alone, over 10 million tons of ammonia for use as fertilizer are produced by this process annually (Fig. 7-5). Before the development of the Haber process, farmers relied almost entirely on animal manure, crop rotation, and dwindling supplies of natural deposits of sodium nitrate (Chilean saltpeter) to replenish the nitrogen in the soil. Synthetic fertilizers made from ammonia play a crucial role in providing the food needed by the world's ever-growing population.

Figure 7-4 The German chemist Fritz Haber (1868–1934). At the beginning of World War I (1914–1918), Germany relied on nitrates imported from South America to make munitions. When a British Naval blockade cut off this supply, Germany was able to maintain weapons production by industrializing the Haber process and using ammonia to obtain the needed nitrogen compounds. It is ironic that Haber, who contributed so much to Germany's war effort, was dismissed from his position as director of the Kaiser Wilhelm Institute for Physical Chemistry in 1933 because he was a Jew.

EXAMPLE 7-4 Five kilograms of hydrogen react with nitrogen to form ammonia (NH_3). Assuming that the reaction goes to completion: **a.** How many kilograms of nitrogen are required? **b.** How many kilograms of NH_3 are formed?

Solution:
1. Write the balanced chemical equation for the reaction:

$$N_2 + 3\,H_2 \longrightarrow 2\,NH_3$$

2. Determine the formula masses of N_2, H_2, and NH_3:

Formula mass of N_2	$2 \times 14 = 28$ amu
Formula mass of H_2	$2 \times 1 = 2$ amu
Formula mass of NH_3	$14 + (3 \times 1) = 17$ amu

3. Substitute these values in the equation:

$$N_2 \quad + \quad 3\,H_2 \quad \longrightarrow \quad 2\,NH_3$$

$$28 \text{ amu} \quad 3 \times 2 \text{ amu} \quad 2 \times 17 \text{ amu}$$

$$6 \text{ amu} \qquad 34 \text{ amu}$$

4. State the mass relationships between reactants and product in terms of the given mass units (kg).

28 kg of N_2 react with 6 kg of H_2 to give 34 kg of NH_3.

5. **a.** Determine the **number of kilograms of N_2** that will react with 5 kg of H_2 using dimensional analysis:
   (1) Given quantity: 5 kg H_2
   (2) Required quantity: kg N_2

   (3) Conversion factor (from step 4): $\dfrac{28 \text{ kg } N_2}{6 \text{ kg } H_2}$

   (4) Solution: $5 \cancel{\text{ kg } H_2} \times \dfrac{28 \text{ kg } N_2}{6 \cancel{\text{ kg } H_2}} = 23.3 \text{ kg } N_2$

23.3 kg of N_2 are required for the formation of ammonia from 5 kg of hydrogen.

   b. Similarly, determine the **number of kilograms of NH_3** that can be formed.
   (1) Given quantity: 5 kg H_2
   (2) Required quantity: kg NH_3

(3) Conversion factor (from step 4): $\dfrac{34 \text{ kg NH}_3}{6 \text{ kg H}_2}$

(4) Solution: $5 \cancel{\text{ kg H}_2} \times \dfrac{34 \text{ kg NH}_3}{6 \cancel{\text{ kg H}_2}} = 28.3 \text{ kg NH}_3$

28.3 kg of NH_3 are produced from 5 kg of H_2.

PRACTICE EXERCISE: Oxygen masks that produce O_2 in emergencies contain potassium superoxide, KO_2, which reacts with exhaled CO_2 and H_2O to produce oxygen.

$$4 \text{ KO}_2 + 2 \text{ H}_2\text{O} + 4 \text{ CO}_2 \longrightarrow 4 \text{ KHCO}_3 + 3 \text{ O}_2$$

If a person wearing the mask exhales 0.70 g CO_2 per minute, how many grams of KO_2 must the mask contain to remove all exhaled CO_2 for 10 minutes?

Answer: 11.3 g of KO_2

Figure 7-5 A chemical plant for the manufacture of ammonia from nitrogen gas and hydrogen gas by the Haber process. In 1991, ammonia ranked fifth in production of chemicals manufactured in the United States.

CONSUMER BRIEF   **Photography**

Photography is a popular hobby. Although most amateur photographers use color film that requires processing at a commercial laboratory, there are some who prefer to develop their own black-and-white film in their basement photolabs.

Black-and-white photographic film is made by taking a clear strip of plastic and covering it with gelatin in which tiny crystals of an ionic silver compound such as silver bromide (AgBr) are suspended. When the film is exposed to light, an image is recorded on the film through a reaction in which the silver ions and bromide ions are converted to their elemental form:

$$2 \text{ Ag}^+ + 2 \text{ Br}^- \longrightarrow 2 \text{ Ag} + \text{Br}_2$$

The light striking the film converts a small fraction of the silver ions in some of the silver halide crystals to silver atoms, and a latent image is produced. This process is called photosensitization and the crystals that are sensitized in this way are more easily reduced completely to silver than are crystals that have not been sensitized. The reason sensitized crystals are more easily reduced is not well understood, but the process is important in developing black-and-white film.

To convert the latent image to a real image, the film is removed from the camera in the dark, and it is placed in a developer solution, which contains mild reducing agents. The developer causes the Ag^+ to be reduced to Ag. The Ag^+ is transparent, but the Ag is black, so the film produces a dark image wherever it is exposed to light. Less light produces lighter shades of gray, whereas intense light produces a dark spot. The image produced is called a negative because it is the inverse of the final black-and-white photograph. Another step has to be taken before the film can be exposed to light again. The film is placed

in a fixer solution, which removes unexposed AgBr from the film emulsion and creates the film negative. If this is not done, AgBr would be left in the film emulsion and would react with light; the entire film would turn black once the lights were turned on, and the original image would be lost.

After the developed black-and-white film is washed and dried, it is used to make a black-and-white print (photograph). Light is passed through the negative and focused on a sheet of photographic paper coated with a thin gelatin layer that contains AgBr. The same image formation, developing, fixing, washing, and drying process is carried out for the photographic paper as for the exposed film. Because the darkened or exposed areas of the negative allow less light to pass through than do the light areas, they produce light areas on the photographic paper. The positive image produced on the print is the reverse of the image on the negative.

Color film uses a similar process, but it has three emulsion layers. Each layer has a different sensitizer that absorbs a different primary color. Red, green, and blue are primary colors, and any color can be reproduced by combining these three in the proper proportions. Exposed color film is developed just as black-and-white film, except that the developer reacts with compounds called color couplers incorporated into the film next to each emulsion layer. Each color coupler produces a different intensity of one primary color in its emulsion layer. The overlaying of the three primary colors gives a full color image.

The chemical solutions that affect the development of color film must be kept at precise temperatures, and the development time must be carefully controlled to get accurate colors produced on the film. For those reasons, developing color film is too difficult for the amateur.

Questions
1. More than 30 percent of the silver used by industry in the United States goes into the manufacture of photographic film. Should photographic labs be required to collect silver waste for recycling?
2. Could sunglasses be made from glass that contains silver salts?

THE MOLE

Let us return once again to the equation for the burning of coal:

$$C + O_2 \longrightarrow CO_2$$

As we saw earlier, the equation gives the following information:

$$1 \text{ atom C} + 1 \text{ molecule O}_2 \longrightarrow 1 \text{ molecule CO}_2$$

$$12 \text{ g C} + 32 \text{ g O}_2 \longrightarrow 44 \text{ g CO}_2$$

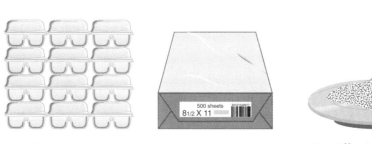

12 eggs — 1 dozen   144 eggs — 1 gross   500 sheets of paper — 1 ream 6.02×10^{23} carbon atoms — 1 mole

Figure 7-6 Convenient counting units for various objects. Chemists use the mole as the counting unit for atoms, molecules, ions, and formula units. In the same way that a dozen eggs means 12 eggs, a mole of carbon atoms means 6.02×10^{23} carbon atoms.

In addition to the above information, it is sometimes necessary to know the quantities of substances that take part in a reaction in terms of *numbers* of atoms and *numbers* of molecules. Because atoms are incredibly small, it is obviously impossible to count out numbers of atoms or molecules. Even to obtain the very smallest amount of carbon visible to the naked eye, billions of atoms of carbon would be required. Instead, chemists use the **mole** as their counting unit to keep track of numbers of atoms and numbers of molecules.

A mole is a specific number of chemical units (atoms, molecules, or formula units).

In our daily lives, we use a variety of counting units. At the grocery store, eggs are purchased by the dozen (12 eggs) and not as individual numbers of eggs. In offices and schools, pencils are often purchased by the gross (144 pencils) and paper by the ream (500 sheets) (Fig. 7-6). In the same way that one dozen is understood to equal 12, one gross 144, and one ream 500, it is understood by chemists that:

one mole = 6.02×10^{23} structural units

The structural units may be atoms, molecules, or formula units.

1 mole of C atoms = 6.02×10^{23} C atoms

1 mole of O_2 molecules = 6.02×10^{23} O_2 molecules

2 moles of CO_2 molecules = $2 \times (6.02 \times 10^{23})$ CO_2 molecules

1 mole of NaCl formula units = 6.02×10^{23} NaCl formula units

The number 6.02×10^{23} is called **Avogadro's number**, in honor of the Italian scientist Amadeo Avogadro (1776–1856), whose pioneering work with gases led to the establishment of this basic number. The theoretical basis for Avogadro's number is beyond the scope of this text and will not be considered.

Because atoms and molecules are incredibly small, Avogadro's number—602,000,000,000,000,000,000,000—is almost incomprehensibly large. If the entire population of the United States (some 250 million people) spent one-dollar bills at the rate of one dollar per second, 12 hours a day, every day of the year, it would take 170 million years for Avogadro's number of one-dollar bills to be used up.

Molar Mass

Although a mole is such a large number, a mole of a chemical is a convenient amount to work with in the laboratory. The mass of a mole, like the mass of a dozen, depends on the identity of the structural units being weighed. In the same way that a dozen grapes do not weigh as much as a dozen apples, a mole of carbon atoms do not weigh the same as a mole of gold atoms. The mass of a mole (or molar mass) of a chemical varies according to the atomic masses of the elements that make up the chemical.

The **molar mass** of a chemical is defined as *the atomic or formula mass expressed in grams*. Thus, for carbon it is 12 g (Fig. 7-7).

A mole of one compound or element contains the same number of chemical units (Avogadro's number) as a mole of another compound or element.

Molar masses of some of the substances taking part in the equations we have just studied are listed in the following table. Molar masses of some common substances are shown in Figure 7-8.

Chemical	Atomic or Formula Mass	Molar Mass
C	12 amu	12 g
O_2	32 amu	32 g
NH_3	17 amu	17 g
Al_2O_3	102 amu	102 g

EXAMPLE 7-5   How much would one mole of calcium chloride, $CaCl_2$, weigh in grams?

Solution: Calculate the formula mass of calcium chloride:

$$1 \text{ atom of Ca} = 1 \times 40.1 \text{ amu} = 40.1 \text{ amu}$$

$$2 \text{ atoms of Cl} = 2 \times 35.5 \text{ amu} = \underline{71.0 \text{ amu}}$$

$$\text{Formula mass of } CaCl_2 = 111.1 \text{ amu}$$

The molar mass of $CaCl_2$ is 111.1 g

One mole of $CaCl_2$ weighs 111.1 g

PRACTICE EXERCISE: How much would one mole of glucose, $C_6H_{12}O_6$, weigh in grams?

Answer: 180 g

EXAMPLE 7-6   How many moles of calcium chloride, $CaCl_2$, are there in 22.2 g of pure $CaCl_2$?

Solution: From the molar mass of $CaCl_2$ calculated in Example 7-5, we know that one mole of calcium chloride weighs 111.1 g.

$$22.2 \text{ g } CaCl_2 \times \frac{1 \text{ mole } CaCl_2}{111.1 \text{ g } CaCl_2} = 0.20 \text{ moles } CaCl_2$$

PRACTICE EXERCISE: Calculate the number of moles in each of the following: **a.** 47.45 g CH_3Br **b.** 16.17 g NaCN.

Answer: **a.** 0.5 mole   **b.** 0.33 mole

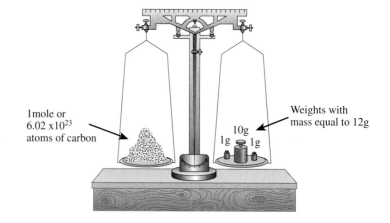

Figure 7-7   The mass of a mole (or molar mass) of a chemical is the atomic mass or formula mass of the chemical expressed in grams. Thus, the mass (or weight) of a mole (6.02×10^{23}) of carbon atoms is 12.0 g.

1 mole or 6.02×10^{23} atoms of carbon

Weights with mass equal to 12 g

10 g

1 g   1 g

Figure 7-8 One-mole amounts of various chemicals. From left to right, top to bottom: Sucrose 342.30g, Lead-shot 207.21g, Potassium Dichromate 294.21g, Mercury 200.61g, Water 18.016g, Copper 63.54g, Salt 58.45g, and Sulfur 32.066g. The mass in grams of a mole of each chemical is calculated from its formula mass.

Calculations Using the Mole and Molar Mass Concepts

A balanced equation shows both the mole relationships and the mass relationships between the substances involved in a particular reaction. For example, from the equation for the reaction of hydrogen with nitrogen to produce ammonia, we know the following:

equation:	N_2	+	$3 H_2$ $\longrightarrow$	$2 NH_3$
number of moles:	1 mole		3 moles	2 moles
molar mass:	28 g		2 g	17 g
molar mass × # of moles:	28 g		6 g	34 g

We can express this information in words as follows:

1. One mole of nitrogen reacts with 3 moles of hydrogen to produce 2 moles of ammonia, and
2. 28 g of nitrogen react with 6 g of hydrogen to produce 34 g of ammonia.

Thus, from the correctly balanced equation for a reaction, we can calculate the number of moles, or grams, of a particular reactant needed to produce a certain number of moles, or grams, of a product.

EXAMPLE 7-7   Ozone, O_3, is formed in the stratosphere from oxygen:

$$3\,O_2 \xrightarrow{\text{sunlight}} 2\,O_3$$

Express this equation in words in terms of moles and grams.

Solution:
1. Three moles of oxygen, O_2, react together to form 2 moles of ozone, O_3.
2. The formula mass of O_2 is $= 2 \times 16.0$ amu $= 32.0$ amu.
Therefore, 1 mole of O_2 weighs 32.0 g:

$$3\ \text{moles}\ O_2 \times \frac{32.0\ \text{g}\ O_2}{1\ \text{mole}\ O_2} = 96.0\ \text{g}\ O_2$$

The formula mass of O_3 is 3×16.0 amu $= 48.0$ amu.

$$2\ \text{moles}\ O_3 \times \frac{48.0\ \text{g}\ O_3}{1\ \text{mole}\ O_3} = 96.0\ \text{g}\ O_3$$

Thus, 96.0 g of oxygen, O_2, react in sunlight to form 96 g of ozone, O_3.

PRACTICE EXERCISE: Iron reacts with oxygen in air to produce iron oxide, Fe_2O_3:

$$4\,Fe + 3\,O_2 \longrightarrow 2\,Fe_2O_3$$

Express the molar and mass relationships in your own words.

Answer:
1. 4 moles of Fe react with 3 moles of O_2 to give 2 moles of Fe_2O_3.
2. 223.2 g of Fe react with 96.0 g of O_2 to give 319.2 g of Fe_2O_3.

EXAMPLE 7-8   How many moles of CO_2 will be produced in the complete combustion of 10 moles of octane, C_8H_{18}, a component of gasoline?

$$2\,C_8H_{18} + 25\,O_2 \longrightarrow 16\,CO_2 + 18\,H_2O$$

Solution:
1. Write in the number of moles of octane and carbon dioxide that take part in the reaction (the number of moles of O_2 and of H_2O need not be included since these substances are not mentioned in the question).

$$2\,C_8H_{18} + 25\,O_2 \longrightarrow 16\,CO_2 + 18\,H_2O$$

$$\text{2 moles} \qquad\qquad\qquad \text{16 moles}$$

2. Use this information to calculate the number of moles of CO_2 produced from 1 mole of C_8H_{18}. You can probably solve this problem by inspection, but for more difficult problems you may want to use dimensional analysis, as follows.
   a. Given quantity: 10 moles of octane
   b. Required quantity: moles of CO_2
   c. The conversion factor is the relationship between moles of octane and moles of CO_2 as shown in the balanced equation.

$$\text{Conversion factor:} \quad \frac{16 \text{ moles of } CO_2}{2 \text{ moles of octane}}$$

d. Solution: $10 \cancel{\text{ moles of octane}} \times \dfrac{16 \text{ moles of } CO_2}{2 \cancel{\text{ moles of octane}}} = 80 \text{ moles of } CO_2$

80 moles of CO_2 will be produced in the complete combustion of 10 moles of octane.

PRACTICE EXERCISE   One way to remove the air pollutant nitric oxide, NO, from the gases discharged from smokestacks is to react it with ammonia, NH_3.

$$NH_3 + NO \longrightarrow N_2 + H_2O \text{ (unbalanced)}$$

How many moles of ammonia need to be added for every six moles of nitric oxide?

Answer: 4 moles ammonia, NH_3

EXAMPLE 7-9   In the reaction shown in Example 7-9, how many moles of water are produced by the complete combustion of 342 g of octane?

Solution: In this problem, you are asked to determine the number of *moles* of water that are produced from 342 *grams* of octane. Therefore, to solve the problem you need a relationship between moles of water and grams of octane.

1. From the balanced equation, we know the following:

$$2\,C_8H_{18} \;+\; 25\,O_2 \;\longrightarrow\; 16\,CO_2 \;+\; 18\,H_2O$$

$$\qquad 2 \text{ moles} \qquad\qquad\qquad\qquad\qquad 18 \text{ moles}$$

2. We also know that 1 mole of a substance is equal to its formula mass in grams. To obtain the relationship between moles of H_2O and grams of octane in the above reaction, convert moles of octane to grams of octane. Calculate the formula mass of octane, C_8H_{18}.

$$8 \text{ atoms of C} \qquad 8 \times 12 = 96 \text{ amu}$$

$$18 \text{ atoms of H} \qquad 18 \times 1 = \underline{18 \text{ amu}}$$

$$\text{formula mass of } C_8H_{18} = 114 \text{ amu}$$

Therefore,

$$1 \text{ mole of } C_8H_{18} \text{ weighs } 114 \text{ g}$$

$$\text{and } 2 \text{ moles of } C_8H_{18} \text{ weigh } 228 \text{ g}$$

Substitute this value in the equation.

$$2\,C_8H_{18} \;+\; 25\,O_2 \;\longrightarrow\; 16\,CO_2 \;+\; 18\,H_2O$$

$$\quad 2 \text{ moles} \qquad\qquad\qquad\qquad\qquad 18 \text{ moles}$$

$$\quad 228 \text{ g}$$

3. Solve the problem using dimensional analysis.
   a. Given quantity: 342 g octane
   b. Required quantity: moles of H_2O
   c. The conversion factor is the relationship between moles of water and grams of octane.

$$\text{Conversion factor: } \frac{18 \text{ moles of } H_2O}{228 \text{ g of octane}}$$

   d. Solution: $342 \text{ g octane} \times \dfrac{18 \text{ moles of } H_2O}{228 \text{ g of octane}} = 27 \text{ moles of } H_2O$

27 moles of water are produced by the complete combustion of 342 g of octane.

PRACTICE EXERCISE: Copper metal can be produced from ores rich in copper sulfide by heating the ore with carbon in the air. How many grams of copper metal can be made by heating 3.0 moles of copper sulfide (CuS)?

$$CuS + C + O_2 \longrightarrow Cu + SO_2 + CO_2 \text{ (unbalanced)}$$

Answer: 190.5 g Cu

CHEMICAL REACTIONS: WHAT MAKES THEM HAPPEN?

Spontaneous Reactions

A spring-wound toy car keeps running until the spring has unwound. Water flows downhill until it reaches level ground. These are **spontaneous** processes; once started, they proceed without the intervention of any outside agency. On the other hand, we cannot make the toy car start to run again or get the water back to the top of the hill without some human (or outside) intervention and expenditure of energy. Rewinding the spring and carrying the water to the top of the hill are both **nonspontaneous** processes.

A spring as it unwinds and water as it flows downhill give up potential energy (Chapter 3) and achieve greater stability. In these and other spontaneous mechanical processes, we recognize that the driving force is the *tendency of the system to move to a lower energy state*. The same principle applies to most—but not all—chemical reactions. As the reaction proceeds, energy in the form of heat is released, and the energy of the products formed is less than the energy of the reactants.

In a chemical substance, potential energy is stored in the chemical bonds that hold the constituent atoms together. In any chemical reaction, bonds between atoms in the reactants break, and the atoms then recombine in a different way to form the products. We will use the following equation to represent a spontaneous chemical reaction:

$$A\text{-}B + C\text{-}D \longrightarrow A\text{-}C + B\text{-}D$$

In the burning of coal, the products (carbon dioxide and ashes) have less energy than the reactants (coal and oxygen) from which they were formed.

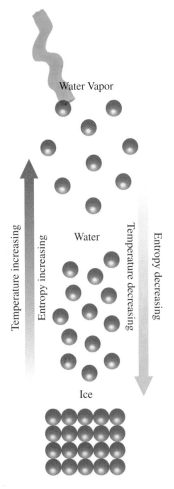

Figure 7-9 An endothermic reaction. When barium hydroxide and ammonium chloride are mixed together, an endothermic reaction occurs. Heat is absorbed from the surroundings, and the temperature of the mixture falls rapidly to well below the freezing point of water.

Figure 7-10 The increase in disorder that occurs as a substance such as ice changes from a solid to a liquid to a vapor. The well-ordered ice structure becomes increasingly disordered, and the entropy of the system increases. When the reverse process occurs and the vapor condenses to a liquid and then freezes to a solid, the system again becomes more ordered and entropy decreases.

If the principle that explains spontaneous mechanical processes also applies to spontaneous chemical reactions, we expect the potential energy in the bonds in the products (A–C and B–D) to be lower than the potential energy in the bonds in the reactants (A–B and C–D). And, in accordance with the first law of thermodynamics (Chapter 3), energy should be released in the reaction. In the great majority of spontaneous chemical changes, energy is indeed released.

In the burning of coal, for example, large amounts of heat energy are produced.

$$C + O_2 \longrightarrow CO_2 + \text{heat energy}$$

Another less obvious example is the rusting of iron. In this process, heat is produced, but the reaction proceeds so slowly that the heat can be detected only with very sensitive instruments. Reactions in which heat is produced are called **exothermic** reactions.

Although the majority of chemical reactions are exothermic, we do not have to look far to find spontaneous reactions in which energy is *absorbed*. For example, when steam is passed over red-hot coals, carbon monoxide and hydrogen gas are produced, and heat is taken up from the surroundings.

$$2\,C + 2\,H_2O + \text{heat energy} \longrightarrow 2\,CO + 2\,H_2$$

If the energy stored in the bonds of the reactants in a particular chemical reaction is *greater* than the energy stored in the bonds in the products, the difference in energy is released as heat (exothermic reaction).

Conversely, if the energy stored in the bonds of the reactants is *less* than the energy stored in the bonds in the products, the difference in energy, in the form of heat, is absorbed from the surroundings (endothermic reaction).

Reactions in which heat is absorbed are called **endothermic** reactions. A dramatic example of a spontaneous endothermic reaction is readily demonstrated in the laboratory. When barium hydroxide and ammonium chloride are mixed together, a precipitous drop in temperature is observed as heat is absorbed from the surroundings (Fig. 7-9). Barium chloride and ammonia are formed according to the following equation:

$$Ba(OH)_2 \cdot 8\,H_2O + 2\,NH_4Cl + energy \longrightarrow BaCl_2 + 2\,NH_3 + 10\,H_2O$$

Obviously, the tendency to achieve a lower energy state is not the sole driving force in a spontaneous chemical reaction.

The melting of ice at room temperature provides a clue to a second driving force. As an ice cube melts, it absorbs heat from its surroundings and, as in all endothermic changes, the energy of the system increases. At the same time, the well-ordered ice-crystal state (see Chapter 12) changes to the less orderly liquid state (Fig. 7-10). This tendency to achieve a more disordered, or random, state is the second driving force in spontaneous chemical reactions.

The term **entropy** is used to describe the degree of disorder, or randomness, of a system. The greater the disorder, the higher the entropy. Liquid water has higher entropy than ice. A shuffled pack of cards has more entropy than one arranged in suits. In our everyday lives, we notice a general tendency toward disorder. An untended garden loses orderliness as weeds and undergrowth take over (Fig. 7-11). From your own experience, you can appreciate that a room becomes increasingly disordered if clothes, books, and other possessions are not tidied away.

We therefore see that natural processes are driven in two ways: toward *lower energy* and toward *higher entropy*. If these two tendencies oppose each other, the dominant tendency determines the course of the reaction. If an ice cube is kept at a temperature below 0°C, it does not melt (Fig. 7-10). In this situation, the tendency to move toward a more disordered state (liquid) is less than the tendency to remain in a lower energy state (ice). In the reaction of steam with coal, the reverse is true.

(a)

(b)

Figure 7-11 In nature there is a spontaneous tendency toward increasing disorder, or entropy, of the system. (a) A neatly tended garden. (b) In an untended garden, weeds grow and the garden becomes increasingly disordered.

Heat Packs and Cold Packs

Athletes use instant heat and cold packs as first-aid devices to treat injuries on the playing field. Skiers and mountain climbers carry these packs as safety equipment: If they are trapped by the snow, the heat packs can hold off frostbite until help arrives. Typically, heat and cold packs contain a dry chemical separated from a pouch of water. When the water pouch is broken by striking the pack, the chemical comes into contact with the water, and a chemical change occurs. Depending on the chemical, the temperature of the water either rises or falls.

Heat packs contain either magnesium sulfate, $MgSO_4$, or calcium chloride, $CaCl_2$. When either of these two solid salts dissolves in water, heat is released. These processes are exothermic. The process can be shown as follows:

$$MgSO_4 \xrightarrow{H_2O} Mg^{2+} + SO_4^{2-} + heat$$

$$CaCl_2 \xrightarrow{H_2O} Ca^{2+} + 2\,Cl^- + heat$$

A typical heat pack contains 40 g of dry $CaCl_2$. A separate compartment contains 100 mL of water. When the pack is activated, the temperature of the water rises from 68°F to 90°F. The pack stays hot for about 15 to 30 minutes.

In cold packs, the dry chemical is solid ammonium nitrate (NH_4NO_3). When ammonium nitrate dissolves in water, it absorbs energy from the surroundings. This process is endothermic. We can express the process as:

$$NH_4NO_3 + heat \xrightarrow{H_2O} NH_4^+ + NO_3^-$$

These packs have 30 g of NH_4NO_3 and a separate compartment with 100 mL of water. When the pack is activated, the temperature of the water

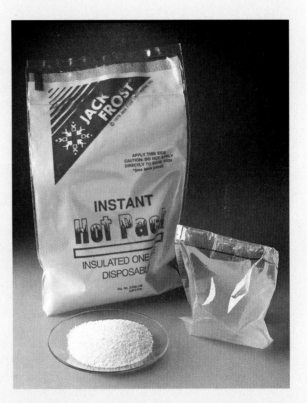

inside is lowered from 75°F to below 32°F. This type of cold pack remains cold for about 15 to 30 minutes.

The process by which these packs increase or decrease in temperature is not a chemical reaction but rather a physical transformation. The heating or cooling effect is caused simply by the solid salt dissolving in water.

Questions
1. Why are heat and cold packs so useful for athletic events? Why not use ice or hot water?
2. You have cut and bruised your arm while playing soccer. You have to cool your injury and you have a choice of a cold pack or ice from a friend's soda. Which should you choose?

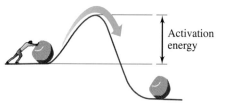

Figure 7-12 Activation energy is comparable to the energy required to raise a boulder over a barrier. Once over the barrier the boulder will roll down the slope of its own accord.

RATES OF CHEMICAL REACTIONS

Some chemical reactions are practically instantaneous; others take months or even years to complete. In the tragic accident in which the space shuttle *Challenger*'s fuel tanks ruptured, a spark caused the escaping hydrogen to react instantly and explosively with oxygen in the air.

$$2\,H_2 + O_2 \longrightarrow 2\,H_2O + energy$$

In contrast, a discarded automobile may stay years at a dump site before it completely rusts away.

The **rate of a chemical reaction** is the rate or speed at which reactants are converted to products. It is measured by determining the amount of reactant used up or product formed in a specified period of time. How fast or how slowly a chemical reaction proceeds depends on a number of factors.

Getting a Reaction Started

Before any reaction between two or more substances can occur, the reactant particles (atoms, molecules, or ions) must make contact or collide with each other. The more frequently collisions occur, the more rapidly product is formed. However, not every collision necessarily produces product. To be successful, collisions must have sufficient energy to break the bonds that hold atoms together in the reactants. Another factor is the orientation of the reactants when they make contact. Head-on collisions are more likely to be effective in breaking bonds than are glancing blows.

The minimum amount of energy that reactant particles must possess in order to react is called the **activation energy**. This amount can be likened to the energy required to lift a boulder over a barrier or hill before it will roll down the other side of its own accord (Fig. 7-12). Lighting a match provides another example. If used as intended, a match will not burst into flame until it is drawn quickly over the rough striking surface. It needs the heat generated by friction to supply the activation energy. Once lit, the match keeps on burning. Similarly, if hydrogen and oxygen are mixed together at room temperature, they do not react. But if a spark is introduced into the mixture, an explosive reaction occurs immediately.

FACTORS THAT INFLUENCE THE RATE OF A REACTION

There are many ways to increase or decrease the rate of a chemical reaction. The chemical industry spends a great deal of money researching ways to make products more quickly. The food industry, on the other hand, is interested in developing ways to slow reactions that lead to spoilage.

The Effect of Temperature

The rate of a chemical reaction can be increased considerably by raising the temperature. A higher temperature gives the reactant particles more energy. They collide more frequently, and more of them have sufficient energy to overcome the activation energy barrier. From our own experience, we know that food cooks more rapidly (if often less satisfactorily) in a hot oven than it does in a cooler one.

Conversely, lowering the temperature decreases the rate of a reaction. Milk turns sour much more slowly in the refrigerator than it does at room temperature, and many foods keep fresh for months if kept frozen. Animals can withstand long periods of hibernation in winter because their metabolic processes operate so slowly in the cold temperature that they can survive on stored fat.

In many reactions, a 10°C increase in temperature will double the rate of the reaction.

The Effect of Concentration

In the same way that increasing the number of cars on the highway increases the chance of collision, an increase in the concentration of reactant particles increases the frequency of collisions, and thus the rate of the reaction. For example, a lighted splint of wood will burn much more vigorously in pure oxygen than in air, which has only a 21 percent concentration of oxygen.

Related to the concentration of the reactants is their state of subdivision. When particle size is very small, reaction rates can increase dramatically. A piece of coal does not ignite very easily, but coal dust ignites explosively, a reaction that has led to many mining accidents. Similarly, grain dust explosions caused by an accidental spark are a constant threat at grain elevators (Fig. 7-13).

When a given amount of a substance is subdivided into small particles the total surface area increases. The chance of ignition is increased because more of the substance is in contact with oxygen in the atmosphere.

The Effect of a Catalyst

The rate of a chemical reaction can often be increased by adding a substance called a catalyst to the reaction mixture. For example, catalysts are the key to the success of automobile emission-control systems (see Chapter 13) in which

Figure 7-13 Explosive ignition of grain dust by an accidental spark destroyed this grain elevator.

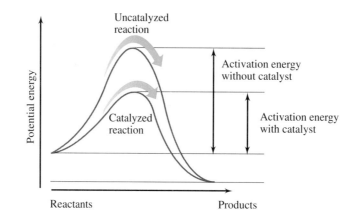

Figure 7-14 A catalyst lowers the activation energy of a reaction. As a result, more reactant particles have sufficient energy to overcome the energy barrier.

A catalyst speeds up a chemical reaction without itself undergoing any permanent change.

carbon monoxide (CO) and nitric oxide (NO)—toxic gases present in engine exhaust—are converted to carbon dioxide and nitrogen gas by passage over the catalysts palladium and rhodium.

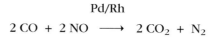

$$\overset{\text{Pd/Rh}}{2\,CO + 2\,NO \longrightarrow 2\,CO_2 + N_2}$$

A **catalyst** is defined as *a substance that increases the rate of a chemical reaction without itself being consumed in the process.* It increases the rate by lowering the activation energy for the reaction (Fig. 7-14), thus increasing the number of collisions that are effective. The working of a catalyst is somewhat analogous to driving through a tunnel in a mountain instead of taking the road over the mountain. The tunnel, like a catalyst, provides a route that saves time and requires less energy.

Catalysts are used extensively in industry. They are important in petroleum refining and in the manufacture of ammonia, sulfuric acid, and many other products. Special kinds of catalysts called **enzymes** exist in all biological systems (Chapter 10). Enzymes, which are complex proteins, make it possible for reactions in living cells to proceed rapidly and at body temperature. Outside the body, in the absence of enzymes, these same reactions occur very slowly, and harsh conditions are needed to make them go faster.

REVERSIBLE REACTIONS AND CHEMICAL EQUILIBRIUM

Reactions That Go In Either Direction

Many chemical reactions are reversible. Depending on conditions, they can be made to proceed in either direction. For example, the following reaction between hydrogen and iodine to form hydrogen iodide:

$$H_2 + I_2 \longrightarrow 2\,HI$$

can be made to go in the opposite direction by applying heat:

$$2\,HI \longrightarrow H_2 + I_2$$
$$\text{heat}$$

Equations describing reversible reactions are written with a double arrow (⇌):

$$H_2 + I_2 \rightleftharpoons 2\,HI$$

Many of the chemical reactions that occur in the body are reversible. Oxygen in the air we breathe combines with a protein called hemoglobin (Hb) in the lungs to form oxyhemoglobin (HbO$_2$). Oxyhemoglobin is carried in the bloodstream to different parts of the body where conditions are such that oxygen is released for use in various metabolic processes.

$$O_2 + Hb \rightleftharpoons HbO_2$$

Reactions That Go Part-Way

In the chemical reactions for which we have written equations, we have assumed that, once started, the reactions go to completion: That is, reactants are completely converted to product. Completion occurs in many reactions. For example, in the presence of sufficient air, coal continues to burn until all the carbon has been converted to carbon dioxide.

$$C + O_2 \longrightarrow CO_2$$

There is no tendency for the carbon dioxide to break down into oxygen and carbon. However, the reaction between nitrogen and hydrogen to produce ammonia does not go to completion at room temperature. At some point, the reverse reaction starts to occur, and ammonia molecules decompose and nitrogen and hydrogen are re-formed.

$$N_2 + 3\,H_2 \rightleftharpoons 2\,NH_3$$

At the point when the rate at which product is being formed equals the rate at which product is being decomposed, the reaction is said to have reached a state of **dynamic equilibrium**. The term *dynamic* is used because even though no further *net* change occurs in the concentrations of reactants and product once equilibrium has been established, molecules are still being changed. Reactant molecules continue to combine and product molecules continue to decompose.

An equilibrium reaction can be forced to go more in one direction than the other by adjusting the conditions (temperature, pressure, or concentration). In the commercial production of ammonia by the Haber process, the reaction is carried out at about 550° C, 250 atm of pressure, in the presence of a catalyst, and the ammonia is removed as it is formed. Under these conditions, the reaction goes almost entirely to the right, and ammonia production is at a maximum.

A great many of the reactions that occur in the natural environment are equilibrium reactions. For example, the concentration of calcium ions in natural waters is largely controlled by an equilibrium reaction between calcium carbonate minerals, atmospheric carbon dioxide, and water.

$$CaCO_3 + CO_2 + H_2O \rightleftharpoons 2\,HCO_3^- + Ca^{2+}$$

Acid rain can seriously upset this and other natural equilibria (Chapter 13). Metabolic processes in living organisms are also controlled primarily by equilibrium reactions (Chapter 10).

If the products of a chemical reaction are much more stable than the reactants, the reaction, once started, will continue until the reactants are used up.

The point at which equilibrium is reached is a characteristic property of a particular reaction under a given set of conditions.

The Man in the Ice: Discovery of a Fifty-Three-Centuries-Old Ancestor

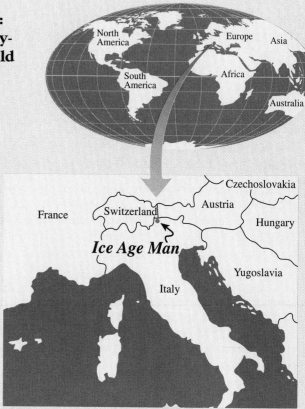

On September 19, 1991, a German couple hiking on an Alpine glacier near a ridge marking the Italian-Austrian border were startled to see the head and shoulders of a human body protruding from the ice in front of them. The Austrian forensic team sent to retrieve the corpse discovered an ax nearby and suspected foul play. As they carelessly hacked the man from the ice, they damaged one hip, tore his clothing, and castrated him. It was not until Innsbruck University archeologist Konrad Spindler saw the body and the axhead that anyone knew that an unprecedented discovery had been made. Spindler immediately realized that the brown, shriveled, remarkably preserved body was that of a man who had lived at least 4000 years ago. Radiocarbon dating proved that the corpse was approximately 5300 years old.

Over the centuries, many people have been lost in the Alps, but most have reappeared within 100 years when glacier flow has transported them to lower altitudes and the ice around them has melted. Usually these bodies had been multilated by the moving ice and had a white waxy appearance as a result of chemical reactions that changed their body fat.

Special circumstances saved "Iceman," as he is now called, from this fate. Experts believe he was camping in the deep ravine where he was found when he was caught in a sudden autumn storm and died of hypothermia while he slept. They judge it was fall because a ripe sloe, a fruit that ripens in the autumn, was found with the body. Cold, dry winds, common at that time of year, quickly desiccated the body before snow and ice buried it. Deep in its frigid grave, the dehydrated body did not undergo the chemical changes that, in other circumstances, would have caused its decomposition. The man and his perishable belongings—things made of wood, leather, and other animal or plant materials—were almost perfectly preserved.

For over 50 centuries, glaciers must have moved above the body without disturbing it. Then in 1991,

freak storms in North Africa blew desert sand onto the Alps. Because a dark layer of sand absorbs more solar energy than pristine white snow, more than the usual amount of snow melted during the summer months. The glacier retreated, and the body was exposed. Had the body not been discovered at that time, it would have been reburied under 2 feet of snow by the end of September.

Archeologists are particularly excited about the find because, unlike most prehistoric remains, this corpse was found not in a grave but at a campsite. He and his extensive possessions provide the most complete picture of the daily life of ancient humans that has ever been revealed.

We now know that Iceman was about 30 years old, 5 foot 2 inches tall, and his features resembled those of the average European alive today (Fig. 7-A). His skin was tattooed with numerous groups of small parallel lines—proof that this practice had begun some 2500 years earlier than was previously believed. His brownish-black hair, which had fallen from his head, was only 3 inches long, evidence that it had been cut. There were no signs of disease or injury.

From the fragments of garments that were retrieved, it appears that he wore a tunic made from pieces of animal skins. The pieces were neatly whip-stitched together with sinew, except in a few places where they had been crudely repaired with twisted dried grass. Iceman wore leather boots stuffed with dried grass for warmth. He had a fur cap and a long woven grass cape, and he probably wore some kind of leggings. Surprisingly, no wool or linen was found, although both were used for clothing at that time.

The man in the ice was well equipped. He had a small, still razor-sharp flint dagger with an ash wood handle that was probably used as a cutting tool rather than a weapon and a leather pouch, most likely worn attached to a belt, that contained several useful items including two flint scrapers, a gouging tool, and some pieces of iron pyrite (FeS_2), a valuable mineral that could be used to strike sparks to kindle a fire. He also had a U-shaped piece of hazel wood and two thin flat pieces of larch that must have formed the frame of a backpack.

Several artifacts had never been seen before. These included two mushrooms of a type known to have antibiotic properties, strung on a leather cord; a pencil-shaped piece of wood with a hard, sharp piece of material inserted in the tip, which may have been used to sharpen flint blades; and two birch bark canisters. One canister contained lumps of charcoal, indicating that the canisters were probably used to carry glowing charcoal embers from one campsite to the next as a means of starting a fire. Still unexplained is a small marble disc with a tasseled leather thong threaded through a hole in its center that may have been a talisman.

The man had a bow and a unique deerskin quiver containing 14 arrows with 2-1/2 foot shafts made of viburnum and dogwood. Strangely, the bow and 12 of the arrows were unfinished and could not have been used. The 6-foot long bow, made of yew, was only roughly whittled, and no grooves had been cut for the bowstring. The two completed arrows had flint heads and feathers arranged at angles that indicate a surprising understanding of ballistic principles.

The ax was particularly intriguing to archeologists. Initially, it was assumed to be bronze but analysis showed that it was pure copper, evidence that Iceman lived during the relatively short transition period when Europeans were emerging from the Stone Age and beginning to work with metals but had not yet entered the Bronze Age. Bronze, an alloy of copper with tin, is a harder substance and easier to work with than copper; it soon became the metal of choice for weapons. Numerous ancient bronze axes have been found in Europe, but copper ones are very rare.

Why was the iceman camping high on a mountain armed with a bow and arrows that could not be used? On the basis of grains of domesticated wheat found in his clothing, experts believe that he came from a farming community in the valley. They speculate that he left, probably with others, to graze animals on the slopes of the mountain. Then, as a result of some accident, he broke or lost his bow and arrows and went up over the Alpine crest and down the

Figure 7A. Drawing of Ice Man's appearance, based on an analysis of artifacts found at the recovery site.

other side to find wood to make new ones. At each stop on his return journey to rejoin his companions, he worked on his weapons but had not completed the task when he was overtaken by a sudden storm. Seeking shelter, he entered the ravine, where, exhausted, he lay down, fell asleep, and froze to death.

When first informed of the body in the ice, the Italian police, assuming it to be that of a lost mountaineer and eager to avoid the task of retrieving it, insisted that it was inside the Austrian border and not their responsibility. Once the corpse's significance became known, however, they had second thoughts, and when a new survey established that the site of the find was 100 yards within their border, the Italian government claimed ownership of the body. The Aus-

trians have agreed to return it in 1994. In the meantime, despite objections from Rome, the Innsbruck team plans to examine Iceman's stomach contents, his brain and other internal organs, and analyze his DNA. Most scientists agree that this incredible relic, even though it is now being kept under very carefully controlled conditions, cannot be preserved indefinitely. It would be a tragedy if petty politics were to prevent a complete study and thus jeopardize a unique opportunity to open a window on our distant past.

References: B. Rensberger, "'Iceman' Yields Details of Stone Age Transition," *Washington Post* (October 15, 1992), p. A1; L. Jaroff. "Iceman." *Time* (October 26, 1992), p. 62.

KEY WORDS AND CONCEPTS

activation energy
Avogadro's number
catalyst
dynamic equilibrium
endothermic

entropy
exothermic
formula mass
molar mass
mole

nonspontaneous reactions
products
rate of reaction
reactants
spontaneous reaction

QUESTIONS AND PROBLEMS

1. Why is it important to make sure that a chemical reaction is properly balanced?

2. Balance the following reactions:
   a. $Al + O_2 \longrightarrow Al_2O_3$
   b. $Fe + O_2 + H_2O \longrightarrow Fe(OH)_3$
   c. $CH_3OH + O_2 \longrightarrow CO_2 + H_2O$
   d. $CH_4 + O_2 \longrightarrow CO_2 + H_2O$

3. Balance the following reactions:
   a. $KClO_3 \longrightarrow KClO_4 + KCl$ **b.** $PbO \longrightarrow Pb + O_2$
   c. $NO_2 + H_2O \longrightarrow HNO_3 + NO$

4. Balance the following reactions:
   a. $Na + Cl_2 \longrightarrow NaCl$ **b.** $H_2 + I_2 \longrightarrow HI$
   c. $Si + Br_2 \longrightarrow SiBr_4$

5. Write balanced equations for:
   a. the reaction of magnesium with solid CO_2 (dry ice) to form MgO and carbon. **b.** the reaction of bromine (Br_2) with potassium (K). **c.** the combustion of silane gas (SiH_4) to produce water vapor and solid silicon dioxide (SiO_2).

6. Write balanced reactions for the reaction of bromine (Br_2) with the following metals to form solids:
   a. potassium (K) **b.** calcium (Ca) **c.** aluminum (Al)

7. One way to remove nitric oxide (NO) from smokestack emissions is to react it with ammonia:
   $$4 NH_3 + 6 NO \longrightarrow 5 N_2 + 6 H_2O.$$

Using this equation, fill in the blanks
a. 10 moles of NO react with _____ moles NH_3.
b. 5 moles of NO produce _____ moles N_2. **c.** the production of 6 moles N_2 requires _____ moles NO.
d. 10 moles of NO produce _____ moles H_2O.
e. _____ grams of H_2O are formed from 3 moles of NH_3.

8. The combustion of hexane (C_6H_{14}) in air yields carbon dioxide (CO_2) and water (H_2O).
   $$2 C_6H_{14} + 19 O_2 \longrightarrow 12 CO_2 + 14 H_2O$$
   a. How many moles of oxygen (O_2) react with 4 moles of hexane? **b.** How many moles of water are produced by burning 4 moles of hexane? **c.** How many moles of carbon dioxide (CO_2), the greenhouse gas, are produced for every mole of hexane burned?

9. What is the molar mass of each of the following in grams?
   a. NH_3 **b.** $MgSO_4$ **c.** C_2H_5OH **d.** Fe_2O_3 **e.** H_2SO_4

10. What is the molar mass of each of the following in grams?
    a. xenon hexafluoride **b.** silicon tetrachloride
    c. nitrogen trifluoride

11. What is the formula of a hydrocarbon that has a molecular mass of 26? Hydrocarbons contain only carbon and hydrogen.

12. Complete the following table for ethyl alcohol, CH_3CH_2OH.

No. of grams	No. of moles
46.0	———
———	0.25
115.0	———
———	0.50

13. How much does 1.0 mole of each of the following substances weigh in grams?
a. NaCl   **b.** $C_{12}H_{22}O_{11}$ **c.** $Pb(OH)_2$
d. $Mg_3(PO_4)_2$

14. Calculate the number of moles of each type of atom present in each of the following.
a. 4 moles of N_2O **b.** 2 moles of $Al(OH)_3$
c. 5 moles of $Na_2S_2O_3$

15. Calculate the number of moles of oxygen (O_2) in each of the following:
a. 3 moles of $C_6H_8O_6$ (vitamin C)   **b.** 5 moles of H_2SO_4 **c.** 2 moles of SO_3

16. The combustion of butane gas, C_4H_{10}, in air yields CO_2 and H_2O.
a. Write a balanced equation for the reaction.
b. How many moles of C_4H_{10} are required to form 10 moles of CO_2?   **c.** How many moles of H_2O are formed from each mole of butane?   **d.** How many grams of H_2O are formed from 4 moles of C_4H_{10}?

17. Arsenic reacts with chlorine (Cl_2) to form a chloride. If 3.17 g of arsenic reacts with 7.51 g of chlorine, what is the formula of the chloride?

18. Iron ore consists mainly of iron oxide (Fe_2O_3). When iron oxide is heated with an excess of coke (carbon), iron metal and carbon monoxide are produced.

a. Write a balanced equation for the reaction.
b. How many moles of iron oxide are required to form 10 moles of iron?   **c.** How many grams of carbon monxide are formed from 12.0 grams of coke?

19. Crude oil burned in an electric generating plant contains about 1.2 percent sulfur by mass. When the oil burns, the sulfur reacts with oxygen and forms sulfur dioxide, a primary air pollutant.

$$S + O_2 \longrightarrow SO_2$$

How many grams of SO_2 are formed when one metric ton (1,000 kg) of coal is burned?

20. The catalytic converter, now required equipment on American automobiles, converts poisonous carbon monoxide (CO) to carbon dioxide (CO_2).

$$2\,CO + O_2 \longrightarrow 2\,CO_2$$

a. What mass of CO_2, in grams, is produced when 100 g of CO react?   **b.** What mass of O_2 is needed to react with 50 g of CO?

21. Which of the following occur spontaneously?
a. iron reacting with oxygen to form Fe_2O_3 (rust)
b. salt dissolving in water   **c.** water vaporizing at 0°C   **d.** making an outline of your class notes

22. Explain the following:
a. not all exothermic reactions are spontaneous
b. not all endothermic reactions are spontaneous

c. if an equilibrium reaction is spontaneous in one direction, it is not spontaneous in the opposite reaction

23. Describe the following in your own words.
a. rate of reaction   **b.** activation energy   **c.** entropy
d. exothermic reaction

24. Write the three physical states of matter, solid, liquid, or gas, in order of increasing entropy. Explain your answer.

25. For each of the following, indicate whether entropy increases or decreases:
a. the freezing of water   **b.** weeding a garden
c. butter melting   **d.** sugar dissolving in coffee

26. In the 1950's, the insecticide DDT was sprayed over wide areas. After application, DDT moves into the environment through soil into plants and water. In lakes it concentrates in the fatty tissue of fish. Consider the environment to be an ecosystem and describe the entropy change associated with each of the processes mentioned. In terms of cleaning DDT from the environment, what entropy change to the ecosystem is required? How can such an entropy change be brought about?

27. Define the term *activation energy*.

28. List three factors that influence the rate of a chemical reaction.

29. What is a catalyst? Explain how a catalyst changes the rate of a chemical reaction. What effect does it have on the activation energy of a reaction?

30. Milk will sour in a day or two when left at room temperature. When it is refrigerated, however, it will keep for two weeks. Explain.

31. Combustible materials burn more rapidly in oxygen than in air. Explain.

32. Explain the following in your own words:
a. a flame lights a cigarette but the cigarette continues to burn after the flame is removed   **b.** a decrease in temperature slows the rate of a reaction   **c.** increasing the concentration of a reactant increases the rate of a reaction

33. Many cold-blooded animals decrease their body temperature in cold weather to match that of their environment. In light of what you have learned in this chapter, explain why they do this and how it affects their need for energy (food).

34. Explain the following:
a. bulk flour is hard to burn, yet grain elevators in which flour is stored have been known to explode
b. solids dissolve in water more rapidly if they are first ground into small particles

35. Describe how the rate of reaction relates to a reaction "going to completion."

36. Fire fighters fighting a natural gas (CH_4) fire try to limit the rate of the following reaction:

$$CH_4 + 2\,O_2 \longrightarrow CO_2 + 2\,H_2O + heat$$

Explain how each of the following will limit the reaction:
a. limiting the fuel supply (CH_4)   **b.** "smothering" the fire   **c.** Applying water   **d.** Using a carbon dioxide fire extinguisher

REACTIONS IN SOLUTION
Acids and Bases, and Oxidation-Reduction Reactions

Acids are present in many natural products, including citrus fruits like these oranges.

Nearly all the chemical reactions that occur in the natural environment occur in aqueous solution, and nearly all these reactions are either acid-base reactions (reactions between acids and bases) or oxidation-reduction reactions. In the living world, photosynthesis, digestion of food, elimination of wastes, and metabolic processes that supply energy to organisms are all brought about by either acid-base or oxidation-reduction reactions. In the nonliving world, these same types of reactions are responsible for the weathering of rocks and soil, the rusting of iron, and the decomposition of dead plants and animals.

Almost everyone is familiar with some acids and bases; we use them every day in our homes, and we have become increasingly aware of the problem of acid rain.

In this chapter, we will learn that certain features common to all acids and other features common to all bases explain the distinctive properties that characterize these two classes of compounds. We will discover that acids and bases are interrelated compounds and that acid-base reactions involve the transfer of protons. We will discuss the causes of acid rain and the impact that increased acidity has on the environment. We will also study some important oxidation-reduction reactions and learn how they can be understood in terms of electron transfer. But first we will look at some general properties of aqueous solutions.

Learning Goals:

In this chapter, you should gain an understanding of:

1. Aqueous solutions and the units used to measure the concentration of solutes in solution.
2. The definitions of acids and bases.
3. The chemical properties that distinguish acids and bases.
4. The difference between a weak acid and a strong acid.
5. Acid-base neutralization reactions.
6. The pH scale and how acidity is measured.
7. Acid rain and its effect on the environment.
8. The definitions of oxidation and reduction.
9. Oxidation-reduction reactions.

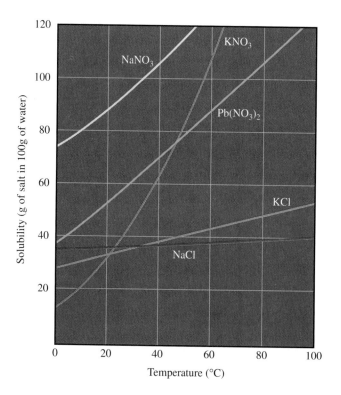

Figure 8-1   The effect of temperature on the solubilities of the following common salts in water: sodium nitrate ($NaNO_3$), potassium nitrate (KNO_3), lead nitrate ($Pb(NO_3)_2$), potassium chloride (KCl), and sodium chloride (NaCl).

AQUEOUS SOLUTIONS

A **solution** is a homogeneous mixture of two or more substances. The substance present in greatest quantity is termed the **solvent**; the substance dissolved in the solvent is called the **solute**. Solutions in which the solvent is water are called **aqueous solutions**.

Water is an exceptionally good solvent and can dissolve a wide variety of ionic and polar substances. In the oceans, sodium chloride is the main solute, and many other solutes are present in smaller quantities. In our bodies and in plants, nutrients are transported in aqueous solution to tissues where they are needed, and waste products of metabolism are carried away in aqueous solution. Oxygen gas in solution in water is essential for the survival of fish and other aquatic organisms.

The maximum quantity of solute that will dissolve in a given quantity of solvent is termed its **solubility**. For most common salts in water, solubility increases with increasing temperature (Fig. 8-1). For many gases, however, the opposite is true. Both oxygen and carbon dioxide are less soluble in hot water than in cold water, which has important implications for aquatic life (Chapter 12).

More than one solute can be present in a solution. For example, instant coffee and sugar can be dissolved in a cup of hot water.

In addition to temperature, solubility may be affected by pressure and the presence of other solutes in solution.

THE FORMATION OF AQUEOUS SOLUTIONS

When any solution is formed, solute particles (molecules or ions) become uniformly dispersed among the solvent particles as shown diagrammatically in Figure 8-2. For this uniform dispersal to occur, attractions between solute particles (in this case, sugar) and between solvent molecules (water) must first be overcome. A rough rule of thumb for determining the solubility of a solute in a particular solvent is "like dissolves like." By like, we mean of like polarity. Ionic and polar covalent solutes tend to dissolve readily in polar solvents such as water but not in nonpolar solvents such as oils and fats. Conversely, nonpolar solutes dissolve in nonpolar solvents but not in polar solvents.

To understand how and why this rule applies, let us consider the dissolution of sodium chloride in water. In a sodium chloride (NaCl) crystal, sodium ions (Na^+) and chloride ions (Cl^-) are held together by electrostatic forces (Chapter 5). These forces must be overcome before dissolution can occur.

When a sodium chloride crystal is added to water, the positive ends of the polar water molecules are attracted toward negatively charged chloride ions at the surface of the crystal (Fig. 8-3). Surface chloride ions, because they are not completely surrounded by positively charged sodium ions, are less strongly held in the crystal than are chloride ions located in the interior. The pull of the water molecules is sufficient to detach Cl^- ions from the crystal, and these ions, surrounded by water molecules, move away from the crystal. In a similar manner, the negative ends of the water molecules are attracted to Na^+ ions at the crystal surface, and Na^+ ions, surrounded by water molecules, leave the crystal. The removal of surface Cl^- and Na^+ ions leaves other ions in the crystal exposed; these, in turn, become hydrated and dispersed. The process continues until the entire crystal has dissolved.

The polar nature of water makes it a good solvent for other polar substances but not for nonpolar substances such as oil (Fig. 8-4). Oil does not dissolve in water because oil molecules have no charges with which to attract

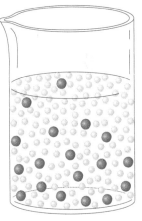

● Sugar molecule
○ Water molecule

Figure 8-2 Diagrammatic representation of sugar molecules dissolved in water. The solute particles (sugar molecules) are uniformly distributed among the solvent particles (water molecules). Not drawn to scale.

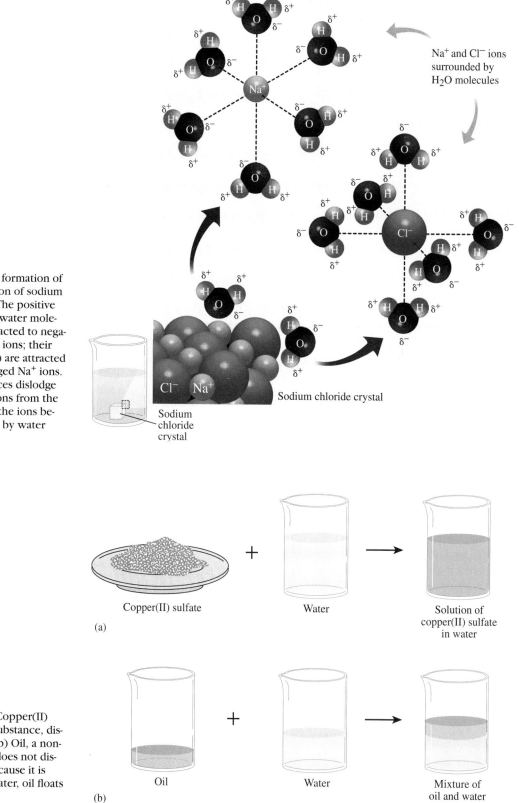

Figure 8-3 The formation of an aqueous solution of sodium chloride (NaCl). The positive ends of the polar water molecules (δ^+) are attracted to negatively charged Cl$^-$ ions; their negative ends (δ^-) are attracted to positively charged Na$^+$ ions. The attractive forces dislodge the Na$^+$ and Cl$^-$ ions from the solid crystal, and the ions become surrounded by water molecules.

Na$^+$ and Cl$^-$ ions surrounded by H$_2$O molecules

Sodium chloride crystal

Sodium chloride crystal

Cl$^-$ Na$^+$

Copper(II) sulfate

Water

Solution of copper(II) sulfate in water

(a)

Oil

Water

Mixture of oil and water

(b)

Figure 8-4 (a) Copper(II) sulfate, an ionic substance, dissolves in water. (b) Oil, a nonpolar substance, does not dissolve in water. Because it is less dense than water, oil floats on the surface.

polar water molecules. Attractive forces between oil and water molecules are therefore very weak and are not sufficient to break the stronger hydrogen bonds between adjacent water molecules.

Dry cleaners apply the "like dissolves like" principle when they use non-polar solvents such as perchloroethylene, rather than water, to remove greasy (nonpolar) dirt from clothes. Grease will not dissolve in water but will dissolve in a nonpolar solvent.

When ions in a solution are surrounded by water molecules, they are said to be hydrated.

CONCENTRATION UNITS

We often need to know the exact concentration of a solute in a given volume of a solvent. We will consider two ways of expressing concentration: (1) molarity and (2) parts per million (ppm) and parts per billion (ppb).

Molarity and Molar Solutions

For aqueous solutions, the unit of concentration most often used by chemists is molarity: the number of moles (Chapter 7) of solute per liter of solution. A molar (1M) solution is one that contains the molecular (or formula) mass of the solute dissolved in one liter (1 L) of solution.

$$\text{molarity} = \text{M} = \frac{\text{moles of solute}}{\text{liters of solution}}$$

Parts per Million (ppm)

Parts per million is a convenient unit for describing very dilute solutions. It is frequently used in discussing concentrations of pollutants in water.

Let us consider a solution of drinking water in which the concentration of lead is 1 ppm. This means there is one part of lead in every one million parts of water. Parts can be expressed in any unit of mass (ounces, tons, micrograms), but the same unit must be used for both solute and solvent. We will express parts in grams.

$$1 \text{ ppm} = \frac{1 \text{ g of solute}}{1 \text{ million g of water}}$$

Because it is more convenient to measure liquids by volume rather than by mass, change mass of water to volume of water:

$$1 \text{ g of water has a volume of } 1 \text{ mL}$$

(The slight variation in weight with temperature is not significant for the present purpose.) Therefore:

$$1 \text{ ppm} = \frac{1 \text{ g solute}}{1 \text{ million } (1,000,000) \text{ mL water}}$$

Change mL to L:

$$1 \text{ ppm} = \frac{1000 \text{ mL}}{1 \text{ L}} \times \frac{1 \text{ g}}{1,000,000 \text{ mL}}$$

$$= \frac{1 \text{ g}}{1000 \text{ L}}$$

Change grams to milligrams:

$$1 \text{ ppm} = \frac{1000 \text{ mg}}{\cancel{1 \text{ g}}} \times \frac{\cancel{1 \text{ g}}}{1000 \text{ L}}$$

$$1 \text{ ppm} = 1 \text{ mg per L}$$

The solution of drinking water contains 1 mg of lead in each liter of water.

Part per million (ppm) means the number of milligrams of solute dissolved in 1 liter of aqueous solution.

It is important to remember that a concentration of 1 ppm is the same as a concentration of 1 mg/L. A solution containing 1 ppm of solute is a very dilute solution. For example, if 1 teaspoon of salt is dropped into a swimming pool, the concentration of salt in the water is approximately 1 ppm.

Parts per Billion

For certain solutions, particularly water samples containing minute traces of contaminants, it is often more convenient to express concentration in parts per billion (ppb) rather than parts per million.

$$1 \text{ ppb} = \frac{\text{ppm}}{1000}$$

$$1 \text{ ppb} = \frac{1 \text{ mg}}{1 \text{ L}} \times \frac{1}{1000}$$

Change milligrams to micrograms:

$$1 \text{ ppb} = \frac{1000 \text{ μg}}{1 \cancel{\text{mg}}} \times \frac{1 \cancel{\text{mg}}}{1 \text{ L}} \times \frac{1}{1000}$$

$$1 \text{ ppb} = 1 \text{ μg per liter}$$

EXAMPLE 8-1   The U.S. Environmental Protection Agency's (EPA) limit for toxic substance A in drinking water is 20 ppb. A laboratory finds the concentration of substance A in a sample taken from a water fountain to be 30 μg/100 mL. Is this above or below the EPA limit? By how much?

Solution:

1. Given:

$$\text{sample A} = \frac{30 \text{ μg}}{100 \text{ mL}}$$

2. To convert concentration to micrograms per liter (ppb), multiply top and bottom by 10 to make the denominator 1000 mL (1 L).

$$A = \frac{30 \text{ μg}}{100 \text{ mL}} \times \frac{10}{10}$$

$$= \frac{300 \text{ μg}}{1000 \text{ mL}}$$

$$= \frac{300 \text{ μg}}{1 \text{ L}}$$

Remember 1 ppb = 1 μg/L. Therefore:

$$\frac{300 \text{ μg}}{1 \text{ L}} = 300 \text{ ppb}$$

3. This concentration is above the EPA limit of 20 ppb. How much above?

$$20 \text{ ppb (x)} = 300 \text{ ppb}$$

$$x = 15$$

The concentration (300 ppb) is 15 times higher than the EPA limit.

PRACTICE EXERCISE: The concentration of a toxic chemical in a stream is found to be 0.06 mg per 100 mL. Express the concentration in ppm and ppb.

Answer: 0.6 ppm or 600 ppb

ACIDS AND BASES: PROPERTIES AND DEFINITIONS

Acids and bases are familiar substances that can be found around the home (Fig. 8-5). Vinegar and citrus fruits (e.g., grapefruit, oranges, and lemons) are sour to the taste because they contain acids. Hydrochloric acid is the main constituent of gastric juice in our stomachs. We clean house with products that contain bases such as ammonia and sodium hydroxide. The bases sodium bicarbonate and calcium carbonate are ingredients in antacids for the relief of an upset stomach.

Properties of Acids and Bases

Long before the currently accepted definitions of acids and bases in terms of proton transfer were established, solutions of acids and bases were identified by their characteristic properties, listed below.

Acids

1. Solutions of acids have a sour taste. We can safely recognize the sour taste of acids in certain everyday foods and beverages. For example, the sour taste of vinegar is caused by acetic acid. Lemons and other citrus fruits taste tart because they contain citric acid. Lactic acid gives yogurt its characteristic taste.

2. When an aqueous solution of an acid is added to certain metals, such as tin, magnesium, or zinc, the acid dissolves the metal, and hydrogen gas is

The word *acid* is derived from *acidus*, the Latin word for sour.

(a)

(b)

Figure 8-5   Acids and bases are present in many consumer products. (a) Acids (b) Bases

The symbols (g), (l), (s), and (aq) are sometimes added after the formula of an element or compound taking part in a chemical reaction, to indicate whether the substance is in a gaseous, liquid, or solid state, or if it is in aqueous solution.

Litmus is extracted from certain lichens containing *erythrolitmin*, a substance that turns red in acid and blue in base.

produced. The equation for the reaction of magnesium with hydrochloric acid is as follows:

$$2 \, HCl(aq) + Mg(s) \longrightarrow MgCl_2(aq) + H_2(g)$$

3. Acids in aqueous solution change the color of litmus from blue to red. Using litmus paper is the simplest and safest way to test for an acid. (Fig. 8-6a).
4. Acids react with bases to form water and ionic compounds called **salts.** (We will consider this reaction in more detail later in the chapter under the heading "Neutralization.")

Bases

1. Bases in solution in water have a bitter taste. Since bases, unlike acids, are not present in common foods or beverages, this property is not easily tested. However, we can recognize the characteristic taste of a base in certain antacids. For example, the bitter taste of Alka Seltzer is due to the base sodium bicarbonate.
2. Bases feel slippery on the skin. Soaps, which contain bases, exhibit this property.
3. Bases in aqueous solution change litmus from red to blue (Fig. 8-6b)— again, using litmus is the simplest and safest way to test for a base.
4. Bases react with acids to form water and salts.

Early Definitions of Acids and Bases

Why do acids and bases have the properties just described? By the end of the nineteenth century, it was generally agreed that the presence of hydrogen ions (H^+) in acid solutions, and of hydroxide ions (OH^-) in basic solutions, accounted for the characteristic properties.

For example, hydrochloric acid (HCl) in water provides hydrogen ions:

$$HCl(g) \xrightarrow{\text{water}} H^+(aq) + Cl^-(aq)$$

and the base sodium hydroxide (NaOH) in water provides hydroxide ions:

$$NaOH(s) \xrightarrow{\text{water}} OH^-(aq) + Na^+(aq)$$

We now know that it is the **hydronium ion** (H_3O^+), rather than the hydrogen ion (H^+), that is responsible for the observed properties of acids in water. The hydroxide ion (OH^-) is responsible for the observed properties of bases.

(a)   (b)

Figure 8-6 (a) Products containing acids turn litmus paper red. (b) Products containing bases turn litmus paper blue.

The Hydronium Ion

A hydrogen ion (H^+) is a hydrogen atom from which the negatively charged electron has been removed. Recall from Chapter 4 that a hydrogen atom—the simplest of all atoms—consists of one electron with a -1 charge and one proton with a +1 charge. When the electron is removed from a hydrogen atom, a bare positively charged nucleus consisting of a single proton is left. For this reason, the hydrogen ion (H^+) is often termed a **proton**.

Free H^+ ions (protons) do not exist in water because they immediately combine with water molecules to form hydronium ions. The slightly negatively charged oxygen atom in a water molecule attracts a positively charged proton. The proton attaches itself to one of the unshared pairs of electrons surrounding the oxygen atom.

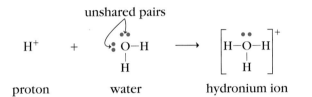

When the hydronium ion is formed, the oxygen atom in the water molecule retains its stable octet of electrons (Chapter 5). All three hydrogen atoms achieve the stable two-electron helium structure. In this way, a stable ion, the hydronium ion, is formed. (In this text, for the sake of simplicity, we will often use H^+ instead of the more accurate H_3O^+. In aqueous solutions, whenever H^+ is written, H_3O^+ should be understood.)

Some common acids are listed in Table 8-1. All of them contain one or more hydrogen atoms that form hydronium ions when the acids are dissolved in water.

TABLE 8-1   Some Common Acids

Name of Acid*	Formula†	Classification
Sulfuric	$\underline{H}_2SO_4$	Strong
Nitric	$\underline{H}NO_3$	Strong
Hydrochloric	$\underline{H}Cl$	Strong
Phosphoric	$\underline{H}_3PO_4$	Weak
Tartaric	CH(OH)COO$\underline{H}$ | CH(OH)COO$\underline{H}$	Weak
Lactic	CH_3CH(OH)COO$\underline{H}$	Weak
Citric	CH_2COO$\underline{H}$ | C(OH)COO$\underline{H}$ | CH_2COO$\underline{H}$	Weak
Acetic	CH_3COO$\underline{H}$	Weak
Carbonic	$\underline{H}_2CO_3$	Weak
Hydrocyanic	$\underline{H}CN$	Weak

*In decreasing order of acid strength.
†The hydrogen atoms that form hydronium ions in solution are underlined.

The Brønsted-Lowry Definitions of Acids and Bases

The definitions of acids and bases just given, although still frequently used, are oversimplified and limited and apply only to acids and bases in solution in water. In 1923, expanded definitions of acids and bases were proposed almost simultaneously by the Danish scientist Johannes Brønsted and the British scientist Thomas Lowry. The Brønsted-Lowry definitions are as follows:

> An **acid** is any substance that can *donate a proton* to some other substance.
>
> A **base** is any substance that can *accept a proton* from some other substance.

The Brønsted-Lowry definitions include all the acids and bases covered by the earlier definitions. For example, in reacting with water to form a hydronium ion, hydrochloric acid donates its proton to the water molecule.

$$\text{proton donor} \qquad \text{proton acceptor}$$

At the same time, the water molecule accepts a proton. In this reaction, therefore, water is acting as a base, according to the Brønsted-Lowry definition.

The base, sodium hydroxide (NaOH), also fits the Brønsted-Lowry definition. As shown in the following equation, the hydroxide ion (OH^-) can accept a proton to form water.

$$OH^- \quad + \quad H^+ \longrightarrow H_2O$$

$$\text{proton acceptor} \quad \text{proton donor}$$

Ammonia

An aqueous solution of ammonia (NH_3) has all the characteristic properties of a typical base. This fact, at first, seems surprising because ammonia does not have an OH group in its formula. However, if we look at the following equation, we can understand what happens:

$$\text{proton acceptor} \qquad \text{proton donor}$$

When NH_3 dissolves in water, it accepts a proton from water; a positively charged ammonium ion (NH_4^+) and a negatively charged hydroxide ion (OH^-) are formed. Ammonia therefore fits both the older definition of a base and the Brønsted-Lowry definition.

Other common bases besides NaOH, KOH, and NH_3 include the ionic compounds sodium carbonate (Na_2CO_3), sodium bicarbonate ($NaHCO_3$), and sodium oxide (Na_2O) and other metal oxides, all of which form hydroxide ions in aqueous solution. In each case, the negative ion in solution accepts a proton from a water molecule, as shown below for bicarbonate (HCO_3^-) and oxide (O^{2-}) ions.

$$HCO_3^- + H-O-H \longrightarrow H_2CO_3 + OH^-$$
$$O^{2-} + H-O-H \longrightarrow 2\,OH^-$$

EXAMPLE 8-2   Show why nitric acid, HNO_3, is an acid.

Solution: When nitric acid is placed in water the following reaction occurs:

$$HNO_3 + H_2O \longrightarrow H_3O^+ + NO_3^-$$

The nitric acid is acting as a **Brønsted-Lowry** acid. It has donated a proton to water, forming the hydronium ion, H_3O^+.

PRACTICE EXERCISE: Show why hydrobromic acid, HBr, is an acid.

Answer: $HBr + H_2O \longrightarrow H_3O^+ + Br^-$

THE STRENGTHS OF ACIDS AND BASES

Acids   In Table 8-1, hydrochloric acid, nitric acid, and sulfuric acid are classified as strong acids, whereas acetic acid, carbonic acid, and citric acids are classified as weak acids. Let us examine the difference between these two groups of acids.

A strong acid is defined as one that completely—or almost completely—**dissociates** into ions in water. For example, hydrochloric acid in water exists almost entirely as H_3O^+ and Cl^- ions. It dissociates almost 100 percent in water. The same is true for nitric acid and sulfuric acid. Using HA to represent the general formula for an acid, we can write the equation for the dissociation of a strong acid as follows:

$$HA + H_2O \longrightarrow H_3O^+ + A^-$$

The boldfaced arrow indicates that the reaction proceeds almost completely to the right.

Weak acids, on the other hand, dissociate very little in aqueous solution—often less than 5 percent. They exist in solution mainly as undissociated molecules (HA). The equation for the dissociation of a typical weak acid can be written as follows:

$$HA + H_2O \rightleftharpoons H_3O^+ + A^-$$

The boldfaced arrow indicates that the reaction lies to the left and that relatively few H_3O^+ and A^- ions are present in solution. The two arrows—one pointing to the left and the other to the right—also indicate that an equilibrium exists in the solution: The rate of formation of ions is equal to the rate at which they recombine to form undissociated molecules. The difference between strong and weak acids is shown diagrammatically in Figure 8-7.

Bases   Bases, like acids, can be classified as strong or weak. Common strong bases are sodium hydroxide and potassium hydroxide.

A **strong base**, like a strong acid, dissociates completely—or almost completely—into ions in water.

$$NaOH \longrightarrow Na^+ + OH^-$$

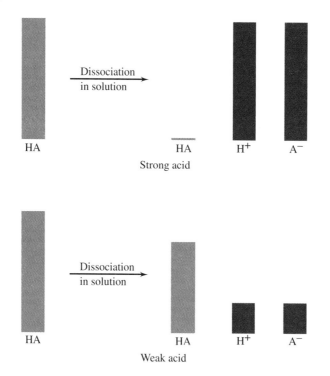

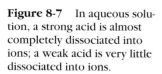

Figure 8-7 In aqueous solution, a strong acid is almost completely dissociated into ions; a weak acid is very little dissociated into ions.

Calcium hydroxide, $Ca(OH)_2$, and magnesium hydroxide, $Mg(OH)_2$, are other examples of strong bases. However, they are only slightly soluble in water. The relatively little that goes into solution dissociates completely, but the total number of ions present is small.

A **weak base**, like a weak acid, dissociates very little in water. A common example of a weak base is ammonia. Although ammonia dissolves readily in water, it dissociates only slightly to give ammonium (NH_4^+) ions and hydroxide (OH^-) ions:

$$NH_3 + H_2O \rightleftharpoons NH_4^+ + OH^-$$

As was the case for a solution of a weak acid, when a weak base dissolves in water, an equilibrium is established.

▶ NEUTRALIZATION

If a solution containing hydronium ions (an acid) is mixed with a solution containing an *equal* number of hydroxide ions (a base), the resulting solution has none of the properties of an acid and none of the properties of a base: It tastes neither sour nor bitter, and it has no effect on litmus paper. The acid and base have **neutralized** each other, and water and a **salt** have been formed.

For example, a solution of hydrochloric acid (HCl) reacts with a solution of sodium hydroxide (NaOH) to produce water and sodium chloride (NaCl), common table salt:

$$\underset{\text{acid}}{HCl} + \underset{\text{base}}{NaOH} \longrightarrow \underset{\text{water}}{H_2O} + \underset{\text{salt}}{NaCl}$$

Or more accurately, remembering that in aqueous solution, hydrogen ions react with water to form hydronium ions:

$$H_3O^+ + Cl^- + Na^+ + OH^- \longrightarrow 2\,H_2O + Na^+ + Cl^-$$

Other examples of neutralization reactions are shown below.

Sodium ions (Na^+) are associated with a salty taste.

HNO_3	+	KOH	$\longrightarrow$	H_2O	+	KNO_3
nitric acid		potassium hydroxide		water		potassium nitrate

CH_3COOH	+	$NaOH$	$\longrightarrow$	H_2O	+	CH_3COONa
acetic acid		sodium hydroxide		water		sodium acetate

In each reaction, a salt is formed, and the hydronium ion from the acid reacts with the hydroxide ion from the base to form water.

H_3O^+	+	OH^-	$\longrightarrow$	$2\,H_2O$
proton donor		proton acceptor		water

EXAMPLE 8-3   Nitric acid, a component of acid rain which is deposited in Adirondack lakes, can be neutralized by adding lime to the lake water. Write the equation for the neutralization of nitric acid, HNO_3, with lime, $Ca(OH)_2$.

Solution: $2\,HNO_3 + Ca(OH)_2 \longrightarrow 2\,H_2O + Ca(NO_3)_2$

PRACTICE EXERCISE: Write the equation for the neutralization reaction between sodium hydroxide, $NaOH$, and hydrobromic acid, HBr.

Answer: $NaOH + HBr \longrightarrow NaBr + H_2O$

THE DISSOCIATION OF WATER

Water is generally thought of as H_2O, a covalent undissociated molecule. In fact, a very small percentage of water molecules dissociates to give hydronium and hydroxide ions:

$$H-O-H + H-O-H \longrightarrow H_3O^+ + OH^-$$

Only about one water molecule in every 500 million (approximately 0.000000002 percent) dissociates. In the dissociation, a proton is transferred from one water molecule to another; thus, water is acting as both an acid and a base. Although the dissociation of water is so slight, it is important for understanding pH.

THE pH SCALE

Most people are familiar with the term pH, and, although they may not know exactly what the term means, they know that it is a measure of acidity. pH is not just a value that is measured by chemists in the laboratory. Gardeners, for example, regularly measure the pH of soil to determine if lime (CaO, a base) should be added to neutralize excess acidity, and the pH of rain and water in lakes is monitored to check for increased acidity.

To understand more precisely what is meant by pH, we need to consider the concentration of H_3O^+ and OH^- ions in aqueous solutions, and in water itself, in terms of molarity. Hydrochloric acid (HCl), a strong acid, dissociates completely into ions in water. Thus, a one-molar (1M) solution of hydrochloric acid contains 1 mole of H_3O^+ ions (and 1 mole of Cl^- ions) per liter of solution.

$$HCl \quad + \quad H_2O \quad \longrightarrow \quad H_3O^+ \quad + \quad Cl^-$$
$$\text{1 mole/L} \qquad\qquad\qquad \text{1 mole/L} \quad \text{1 mole/L}$$

Water, however, dissociates very little, and 1 L of pure water contains just 1.0×10^{-7} (or 0.0000001) moles of H_3O^+ and 1.0×10^{-7} (or 0.0000001) moles of OH^- ions.

A solution that contains more than 1.0×10^{-7} moles of H_3O^+ ions per liter is **acidic**. A solution that contains fewer than 1.0×10^{-7} moles of H_3O^+ ions per liter is **basic**.

The pH scale was introduced to provide a convenient and concise way of expressing small concentrations of H_3O^+ ions in solution. pH can be defined as

$$[\,H_3O^+\,] \; = \; 10^{-pH}$$

It was the Danish biochemist S. P. L. Sørensen who proposed that the number in the exponent be used to express the hydrogen ion concentration.

(The brackets [ ] are a shorthand way of representing concentration in moles per liter.) Thus, on the pH scale, a concentration of 1.0×10^{-7} moles of H_3O^+ ions per liter becomes a pH of 7. Similarly, a concentration of 1.0×10^{-4} moles per liter becomes a pH of 4. It therefore follows that:

A solution with a *pH below 7 is acidic*.
A solution with a *pH of 7 is neutral*.
A solution with a *pH above 7 is basic*.

pH is derived from the French *pouvoir hydrogene* which means hydrogen power.

The relationship between pH and the concentrations of H_3O^+ and OH^- ions in solution is shown in Figure 8-8. In pure water, the numbers of H_3O^+ and OH^- ions are equal, and the pH is 7. The lower the pH of a solution, the more acidic it is and the higher the concentration of H_3O^+ ions in solution; the higher the pH of a solution, the more basic it is and the higher the concentration of OH^- ions in solution. The exponent changes from −1 in a strong acidic solution to −7 in a neutral solution to −14 in a strong basic solution. Figure 8-8 also shows the pH values of some common solutions.

EXAMPLE 8-4   What is the pH of solutions in which the hydronium ion concentration is **a.** 1.0×10^{-6} M and **b.** 0.00001 M?

Solution: **a.** The exponent is −6. Therefore, the pH is 6.
b. Expressing the concentration in scientific notation gives: 1.0×10^{-5} M. The exponent is −5. Therefore, the pH is 5.

PRACTICE EXERCISE: What is the pH of solutions in which the hydronium ion concentration is **a.** 1.0×10^{-4} M and **b.** 0.001 M?

Answer: **a.** 4 **b.** 3

EXAMPLE 8-5 What is the hydronium ion concentration of a solution that has a pH of 2?

Solution: If the pH is 2, the exponent in the expression for the concentration must be −2. Therefore, the hydronium ion concentration is: 1.0×10^{-2} M, or 0.01 M.

PRACTICE EXERCISE: What is the hydronium ion concentration of a solution that has a pH of 1?

Answer: 0.1 M

It is important to note that a change of *one* pH unit corresponds to a *10-fold* change in H_3O^+ concentration. For example, a pH of 3 corresponds to a H_3O^+ concentration of 1.0×10^{-3} (or 0.001) moles per liter, and a pH of 2 corresponds to a H_3O^+ concentration of 1.0×10^{-2} or (0.01) moles per liter. A change of pH from 3 to 2 represents a 10-fold increase in the number of H_3O^+ ions in 1 L of solution since 0.01 is 10 times as great as 0.001.

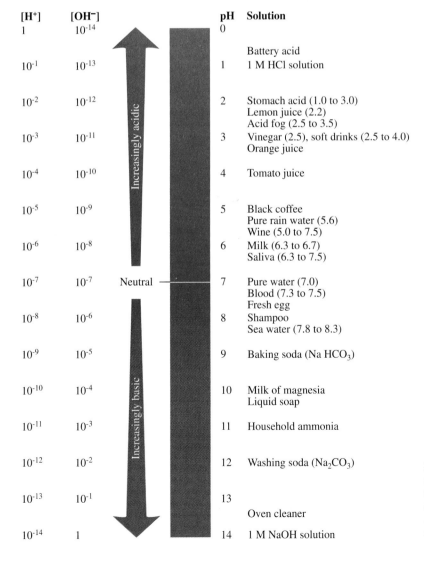

$[H^+]$	$[OH^-]$		pH	Solution
1	10^{-14}		0	
				Battery acid
10^{-1}	10^{-13}		1	1 M HCl solution
10^{-2}	10^{-12}		2	Stomach acid (1.0 to 3.0) Lemon juice (2.2) Acid fog (2.5 to 3.5)
10^{-3}	10^{-11}	Increasingly acidic	3	Vinegar (2.5), soft drinks (2.5 to 4.0) Orange juice
10^{-4}	10^{-10}		4	Tomato juice
10^{-5}	10^{-9}		5	Black coffee Pure rain water (5.6) Wine (5.0 to 7.5)
10^{-6}	10^{-8}		6	Milk (6.3 to 6.7) Saliva (6.3 to 7.5)
10^{-7}	10^{-7}	Neutral	7	Pure water (7.0) Blood (7.3 to 7.5) Fresh egg
10^{-8}	10^{-6}		8	Shampoo Sea water (7.8 to 8.3)
10^{-9}	10^{-5}		9	Baking soda (Na HCO_3)
10^{-10}	10^{-4}		10	Milk of magnesia Liquid soap
10^{-11}	10^{-3}	Increasingly basic	11	Household ammonia
10^{-12}	10^{-2}		12	Washing soda (Na_2CO_3)
10^{-13}	10^{-1}		13	Oven cleaner
10^{-14}	1		14	1 M NaOH solution

Figure 8-8 The pH scale. The figure shows the relationship between pH and the concentration of H^+ and OH^- ions in solution in water at 25°C. It also shows the pH values of some common aqueous solutions.

In the examples above, we considered only solutions with whole-number pH values. Most solutions, as shown in Figure 8-8, have pH values that are not whole numbers. The pH of lemon juice, for example is 2.2, and that of milk lies between 6.3 and 6.7. Solving problems involving nonintegral pH values is more complex than solving problems with integral pH values and involves logarithms. For our present purpose, it will suffice to know that a pH of 7.4, for example, lies between pH 7.0 and pH 8.0, and represents a H_3O^+ concentration between 1.0×10^{-7} M and 1.0×10^{-8} M.

CONSUMER BRIEF   Antacids: How Do You Spell Relief?

Compared with most body fluids such as blood, saliva, spinal fluid, and bile, which have a pH near neutrality (pH of 7), stomach fluid is very acidic. Its pH ranges from about 1 to 3. This acidity is due to hydrochloric acid (HCl) that is constantly secreted in small amounts by the cells of the mucosa, the inner lining of the stomach. When food enters the stomach—or even when food is smelled—acid production increases. The acid assists in the digestion of food and also prevents growth of bacteria. Overeating and emotional factors can cause the stomach to produce excessive amounts of acid, which can result in discomfort as the pH falls. This condition is variously referred to as upset stomach, indigestion, or heartburn. It can be relieved by consuming an antacid, an over-the-counter drug that contains a base capable of neutralizing the excess acid.

Basic substances present in commercial antacids include sodium bicarbonate, calcium carbonate, and magnesium and aluminum hydroxides. Some products contain more than one base. Listed in the table are the active ingredients in some common antacids.

The bases sodium bicarbonate, in Alka-Seltzer, and calcium carbonate, in Tums, react with acid as follows:

$$H^+ + NaHCO_3 \longrightarrow Na^+ + CO_2 + H_2O$$

$$2\,H^+ + CaCO_3 \longrightarrow Ca^{2+} + CO_2 + H_2O$$

In both of these reactions the gas CO_2 is generated in the stomach as the antacids work. The other antacids listed do not generate gas as they work; they react as follows:

$$2\,H^+ + Mg(OH)_2 \longrightarrow Mg^{2+} + 2\,H_2O$$

$$3\,H^+ + Al(OH)_3 \longrightarrow Al^{3+} + 3\,H_2O$$

Brand Name	Active Ingredient	Formula
Alka-Seltzer	Sodium bicarbonate	$NaHCO_3$
DiGel	Aluminum hydroxide	$Al(OH)_3$
Phillip's Milk of Magnesia	Magnesium hydroxide	$Mg(OH)_2$
Maalox	Magnesium hydroxide and aluminum hydroxide	$Mg(OH)_2$, $Al(OH)_3$
Mylanta	Magnesium hydroxide and aluminum hydroxide	$Mg(OH)_2$, $Al(OH)_3$
Rolaids	Dihydroxy aluminum sodium carbonate	$Al(OH)_2NaCO_3$
Tums	Calcium carbonate	$CaCO_3$

An effective antacid should decrease stomach acidity rapidly to give "instant relief." But it should not neutralize it completely. If the stomach acidity were neutralized, digestion would be stopped and the walls of the stomach would begin to secrete even more stomach acid, a condition called acid rebound. Instead, antacids neutralize enough HCl in the stomach to reduce any pain and discomfort, but not enough to stop digestion completely.

Antacids can have side effects. Aluminum and calcium hydroxide products often cause constipation, whereas magnesium hydroxide products can act as laxatives. Sodium bicarbonate dissolves quickly in the stomach and gives very fast relief, which fades within five minutes. And, because of

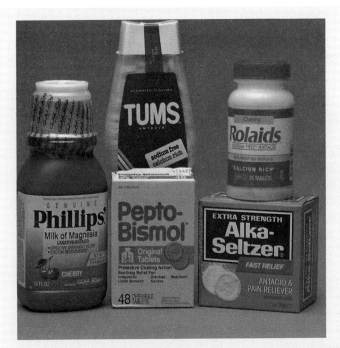

acid rebound, people are often tempted to take more antacid than needed, which only stimulates even more acid production.

When choosing a particular brand of antacid, the consumer should consider the type of base in the formula. People with hypertension should avoid the preparations with sodium bicarbonate, because they need to limit their consumption of sodium. Anyone concerned about osteoporosis, bone degeneration caused by loss of calcium,

would want to consider antacids formulated with calcium carbonate. It is up to the consumer to decide how to spell relief.

Questions
1. Consider the active ingredient in each product listed above. Is there any one base that will consume more stomach acid than the others?
2. Should antacids be sold only in bottles with child-proof caps?

ACID-BASE BUFFERS

In many biological systems, it is important that pH be maintained within a narrow range. You will notice in Figure 8-8 that the pH of blood is slightly basic and lies between 7.3 and 7.5. Even a small deviation below or above these values can be life-threatening.

The pH of blood is maintained within its critical range by **buffers**, which are substances that resist changes in pH. The primary buffer system in blood is the bicarbonate system. Blood contains dissolved carbon dioxide and bicarbonate ions (HCO_3^-). When acid enters the blood as a result of metabolic processes, it reacts with HCO_3^- as shown below: H^+ ions are removed from solution, and a drop in pH is prevented.

$$H^+ + HCO_3^- \rightleftharpoons H_2CO_3 \rightleftharpoons CO_2 + H_2O$$

If base is added, the following reaction occurs: Hydroxide ions are removed, and a rise in pH is prevented.

$$OH^- + H_2CO_3 \rightleftharpoons HCO_3^- + H_2O$$

The bicarbonate buffer system is also important in regulating the pH of oceans. All weak acids and weak bases, in the presence of their salts, form buffer systems.

Citrus fruits, grapes, tomatoes, and rhubarb all contain weak acids.

NATURALLY OCCURRING ACIDS

With few exceptions, naturally occurring acids are weak acids. The most common acid in the environment is carbonic acid, which forms when carbon dioxide gas dissolves in water. Pure water is neutral (pH 7.0), but pure rainwater is slightly acidic because, as it falls, it dissolves carbon dioxide, a natural component of the atmosphere. A dilute solution of carbonic acid is formed, and although carbonic acid dissociates very little, the dissociation is sufficient to provide enough hydrogen ions to lower the pH of rainwater to 5.6.

$$CO_2 + H_2O \rightleftharpoons H_2CO_3 \rightleftharpoons H^+ + HCO_3^-$$

Other weak acids are formed naturally in soil when plant materials decay. These weak organic acids dissociate slightly to give hydrogen ions. (R represents a large organic group.)

$$R-COOH \rightleftharpoons H^+ + R-COO^-$$
organic acid

Complete decomposition of organic materials results in the formation of carbon dioxide, which, as it dissolves, adds additional H^+ ions to soil. Soils rich in organic matter are therefore naturally acidic.

A strong acid that is produced naturally in the environment is sulfuric acid. In many volcanic eruptions—including the eruption of Mount Pinatubo in the Philippines in 1991—large quantities of sulfur dioxide (SO_2) are emitted and carried great distances by the wind. In the atmosphere, SO_2 is converted to sulfur trioxide (SO_3), which dissolves in rainwater, producing H_2SO_4.

USES OF ACIDS AND BASES

Acids and bases have many important industrial uses (Table 8-2) and rank high among those chemicals that are produced in greatest quantity in the United States (Table 8-3).

Sulfuric acid (H_2SO_4), nitric acid (HNO_3), and phosphoric acid (H_3PO_4) are all used in the production of fertilizers. Sulfuric acid, which is produced in the United States in greater quantity than is any other chemical, is also used in the petroleum industry, in metal and ore processing, and in automobile storage batteries. It is important in the manufacture of dyes, drugs, plastics, and many other products. Nitric acid is used in the manufacture of explosives, and in steel refining, and, like sulfuric acid, it plays a role in the manufacture of dyes and plastics. Hydrochloric acid (also known as muriatic acid) is important in the building industry. It is used to remove excess mortar from bricks and scale and

TABLE 8-2   Common Acids and Bases and Their Uses

Acid	Formula	Uses
Sulfuric acid	H_2SO_4	Manufacture of fertilizers, plastics, dyes, paper; petroleum refining; steel production; liquid in automobile batteries
Nitric acid	HNO_3	Manufacture of fertilizers, explosives, plastics, dyes; steel production
Hydrochloric acid	HCl	In building industry for removal of mortar from brick and rust scale from metals
Phosphoric acid	H_3PO_4	Manufacture of fertilizers, detergents, baking powder, fire-resistant fabrics, carbonated drinks
Base		
Sodium hydroxide	NaOH	Petroleum industry; manufacture of aluminum, synthetic fibers, paper, dyes, detergents, soaps; drain cleaners
Potassium hydroxide	KOH	Production of fertilizers, soaps, detergents
Calcium hydroxide	$Ca(OH)_2$	Production of cement, mortar, plaster, paper, bleaching powder; water softening; reducing soil acidity
Ammonia	NH_3	Production of fertilizer, explosives, plastics, synthetic fibers, paper, rubber, detergents; household cleaners

rust from metals. It is also used to clean automobile radiators. Phosphoric acid and its salts are used in the manufacture of many common products including carbonated beverages, detergents, baking powder, and fire-resistant textiles.

The base calcium hydroxide ($Ca(OH)_2$), often called hydrated lime, is produced in very large quantity in the United States. It is made from calcium carbonate ($CaCO_3$), limestone, and is used in manufacturing cement, mortar, and plaster. It is also used for reducing soil acidity and softening hard water. Another base produced in great quantity is sodium hydroxide, which is used in the manufacture of synthetic fibers, dyes, detergents, and household cleaners. Potassium hydroxide is needed for the production of fertilizers and soaps. The base produced in greatest quantity is ammonia. It is essential for the production of fertilizers and is also used in the manufacture of plastics, explosives, paper, and many other products.

TABLE 8-3   Acids and Bases Produced in the United States (1991)

Name	Rank Among All Chemicals Produced	Billions of Pounds
Acid		
Sulfuric acid	1	86.6
Phosphoric acid	7	24.7
Nitric acid	13	15.0
Hydrochloric acid	27	5.6
Base		
Ammonia	5	34.0
Lime	6	33.6
Sodium hydroxide	8	24.4

ACID RAIN

Acid rain is a serious environmental problem. As we have seen, pure rainwater is slightly acidic and has a pH of 5.6. Rainwater with a pH less than 5.6 is described as **acid rain**. In recent years, the average pH of rainfall in many parts of northeastern North America has fallen below 4.6 (Fig. 8-9), and rain with a pH as low as 2.9 has been recorded.

The primary man-made causes of acid rain (more accurately called **acid deposition**) are emissions of sulfur dioxide (SO_2) from coal- and oil-burning power plants and emissions of nitrogen oxide (NO) from automobiles. In the atmosphere, these emissions are chemically converted into H_2SO_4 and HNO_3, which accumulate in cloud droplets and fall to earth in rain and snow (Fig. 8-10). (The sources of acid rain and the steps being taken to control them will be studied in more detail in Chapter 13.)

The Effects of Acid Rain

The natural weathering of rock is in large part caused by the acid contained in pure rainwater and soils. If rain is unusually acidic (pH less than 5.6), the normal weathering process is accelerated, and other chemical reactions also occur. The effects on the environment of both pure rainwater and acid precipitation depend on the type of soil and bedrock that the precipitation encounters.

In the normal process of weathering, when naturally acidic rain falls on soils derived from limestone ($CaCO_3$), some limestone dissolves, and the acid is neutralized:

$$CaCO_3 + H^+ \longrightarrow Ca^{2+} + HCO_3^-$$

Although this reaction occurs very slowly, over thousands of years, great quantities of rock are dissolved. The extensive Carlsbad Caverns in New Mexico and the Caves of Manacor in Spain were formed in this way (Fig. 8-11).

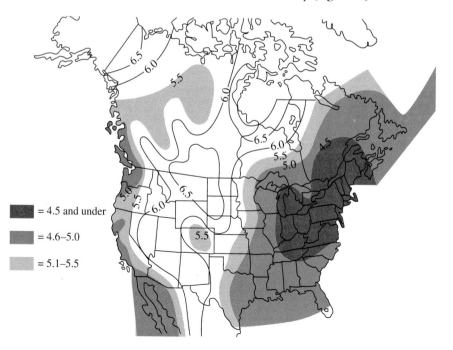

Figure 8-9 Average pH of precipitation in the United States and Canada. In many parts of North America, precipitation is abnormally acid (pH below 5.6). The pH of precipitation is particularly low in the Northeast and along the West Coast.

= 4.5 and under

= 4.6–5.0

= 5.1–5.5

Figure 8-10 The main sources of acid deposition are emissions from oil- and coal-burning power plants and automobiles. Sulfur dioxide and nitrogen oxides from these sources combine with water vapor in the atmosphere to form sulfuric acid (H_2SO_4) and nitric acid (HNO_3). The acids collect in clouds and fall to earth in rain and snow, causing many adverse effects.

When acid rain falls in limestone regions such as the Midwest and the Great Lakes area, the acid is neutralized according to the above reaction, and adverse effects are minimized. However, in many parts of the United States and Canada, the underlying rock is primarily granite and basalt, igneous rock composed of silicate minerals such as feldspar that contain few buffers (Fig. 8-12). In the normal process of weathering, feldspar is partially dissolved by naturally acidic rain and soils and converted to the clay kaolinite and silica. Potassium ions are formed and go into solution:

$$2 \ KAlSi_3O_8 \ + \ 2 \ H^+ \ + \ H_2O \ \longrightarrow \ Al_2Si_2O_5(OH)_4 \ + \ 4 \ SiO_2 \ + \ 2 \ K^+$$

feldspar   rainwater   kaolinite   silica potassium
                                       (clay)   ions

Other silicate minerals weather in the same general way to produce clays and metal ions.

In the presence of an increased concentration of acid, the clay partially dissolves, and aluminum ions, which are toxic to plants and aquatic life, go into solution:

Figure 8-11 Caves of Manacor in Spain. The caverns were formed over thousands of years as a result of the slow dissolution of underground deposits of limestone by naturally acid rain.

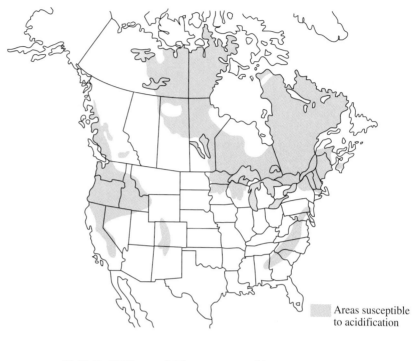

Figure 8-12 The shaded areas of the map are particularly susceptible to acidification because in these regions the underlying rock is mainly granite, which has little buffering capacity.

Areas susceptible to acidification

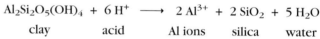

$$Al_2Si_2O_5(OH)_4 + 6\,H^+ \longrightarrow 2\,Al^{3+} + 2\,SiO_2 + 5\,H_2O$$

clay   acid   Al ions   silica   water

Since the 1950s, lakes in many parts of the world have become increasingly acidic, and, at the same time, fish populations have declined. Salmon are no longer found in many of Nova Scotia's rivers, and fish have disappeared from many lakes in eastern Canada, the Adirondacks, and Scandinavia. When pH falls

Figure 8-13 Dead and dying trees in a forest high in the Blue Ridge Mountains. In many parts of the eastern United States, the growth of forests at high elevation has declined, and trees are dying. Acid deposition is blamed for much of the damage.

Figure 8-14 Many ancient limestone and marble monuments in Europe are being severely eroded by acid deposition. The statue shown is in Rome, Italy.

below 5.5, many desirable fish, such as trout and bass, die. At a pH of 5.0, few fish of any kind can survive. At a pH of 4.5, lakes become virtually sterile. The effects on aquatic life are caused both by increased acidity and by the presence of toxic metal ions, particularly Al^{3+}.

Since 1980, trees in many forests in the eastern United States and parts of Europe—including Germany's Black Forest—have suffered severe damage (Fig. 8-13). Acid deposition puts trees under stress, making them unusually susceptible to damage from disease, insects, and cold temperatures. Needed nutrients, such as K^+, Ca^{2+}, and Mg^{2+} ions are leached from the soil by acid, often being replaced by toxic Al^{3+} ions. Unusually acidic soil damages fine root hairs and destroys beneficial microorganisms. Acid rain can damage surface structures on leaves and pine needles, causing them to wither and drop.

Building materials, particularly limestone and marble, which are composed of $CaCO_3$, are readily eroded by acid. As shown in Figure 8-14, many ancient statues in Europe, particularly those on historic cathedrals, have been severely and rapidly eroded over the last 50 years. Acid rain is considered to be responsible for the accelerated weathering of these structures.

Causes of Increased Acidity in the Environment

The damage to lakes and forests in the northeastern United States and in Canada has usually been blamed on emissions from industrial centers in the Midwest, which are carried from their source by the prevailing winds. However, recent work suggests that the natural process of soil formation may be an important factor in causing acidification of lakes and soil. Changes in land use and consequent changes in vegetation on the land, such as have occurred in the northeastern part of the United States and in Scandinavia, are known to acidify soils. Lowering acid-causing industrial emissions may not, therefore, be as effective as hoped in saving lakes and forests.

Industrial centers in Europe are believed to be responsible for increased acidity and environmental damage in Scandinavia.

ACID MINE DRAINAGE

Acid released at certain types of mines can cause local increases in the acidity of soils, streams, and rivers. **Acid mine drainage** is associated primarily with the mining of coal deposits rich in the mineral pyrite (FeS_2). When underground sources of pyrite-rich coal are mined, the coal becomes exposed to water and the atmosphere. The pyrite reacts with oxygen and water, and acid is formed according to the following equations:

$$2\,FeS_2 + 7\,O_2 + 2\,H_2O \longrightarrow 2\,Fe^{2+} + 4\,SO_4^{2-} + 4\,H^+$$
$$4\,Fe^{2+} + O_2 + 10\,H_2O \longrightarrow 4\,Fe(OH)_3 + 8\,H^+$$

Rainwater seeping through sulfur-rich wastes left at mine sites is a further source of acid. Although not widespread, the effects of acid runoff from mines into nearby streams can be severe, and in some instances, runoff has caused sizable fish kills.

Regulations now limit acid discharge from mine sites, and at most active mines, acid is neutralized with lime before it is released into natural water systems. Many abandoned mines have been sealed in an effort to prevent acid formation by excluding water and oxygen. These measures have not always been successful, and acid mine drainage remains a problem in some areas.

OXIDATION AND REDUCTION

Among the most important reactions that occur in the environment, and in industrial processes, are **oxidation-reduction reactions**. In our bodies, cells obtain energy by the oxidation of food. The oxidation of fossil fuels provides 90 percent of the energy used by society. Metals are obtained from their ores by reduction reactions. Photosynthesis by green plants involves reduction of carbon dioxide obtained from the atmosphere. Oxidation and reduction are complementary processes; they always occur together. In any oxidation-reduction reaction, one substance is oxidized, while at the same time, another is reduced.

Reactions were described as oxidation or reduction reactions long before it was possible to explain what was occurring at the molecular level. As was the case with acids and bases, definitions of oxidation and reduction broadened once the structure of the atom was understood. Oxidation and reduction can be defined narrowly in terms of the gain or loss of oxygen and hydrogen atoms, or more generally, in terms of the gain and loss of electrons.

Oxidation-reduction reactions that occur in living organisms, and many that occur in the physical environment, take place in aqueous solution. To obtain a better understanding of oxidation and reduction processes, we will consider some important oxidation-reduction reactions that occur in the absence of water before we discuss oxidation-reduction reactions in aqueous solution.

GAINING AND LOSING OXYGEN ATOMS

The early chemists, who studied many reactions involving oxygen, defined **oxidation** as the combination of oxygen with some other element or compound. The element or compound that gained oxygen was said to be oxidized.

Gaining Oxygen Atoms

Oxygen is a very reactive element, and it combines with every other element except helium, argon, and neon to form compounds called **oxides**. For example, when iron rusts, it combines with oxygen in the air to form iron(III) oxide:

$$4\,Fe + 3\,O_2 \longrightarrow 2\,Fe_2O_3$$

Sulfur combines with oxygen to form sulfur dioxide:

$$S + O_2 \longrightarrow SO_2$$

The reactions in which iron and sulfur combine with oxygen are oxidation reactions; iron and sulfur have been oxidized.

One oxidation reaction that we have discussed many times is combustion (or burning). When wood, coal, or other carbon-containing fuels are burned, the carbon atoms combine with oxygen in the atmosphere to form carbon dioxide:

$$C + O_2 \longrightarrow CO_2$$

If the supply of oxygen is limited, carbon monoxide is formed:

$$2\,C + O_2 \longrightarrow 2\,CO$$

In both reactions, carbon is oxidized. In most oxidation reactions, heat is produced; in combustion reactions, the heat is given off rapidly. We observe this emission of heat when wood is burned in the fireplace or when a forest fire is started.

Many oxidation reactions that occur naturally in the environment proceed very slowly. The breakdown of organic material in dead plants and animals by aerobic bacteria and other decomposers (Chapter 3) is an example. Carbon in the organic matter is gradually oxidized, and the end product, as in combustion, is carbon dioxide.

EXAMPLE 8-6 When chromium oxide, Cr_2O_3, reacts with aluminum metal, Al, the aluminum is oxidized to aluminum oxide, Al_2O_3. Write the equation for the oxidation reaction.

Solution: $Cr_2O_3 + 2\,Al \longrightarrow 2\,Cr + Al_2O_3$

PRACTICE EXERCISE: Boron hydride, BH_3, is pyrophoric, meaning it reacts spontaneously with oxygen in the air, producing a flame. In this oxidation reaction, boric acid, H_3BO_3 is formed. Write an equation for the reaction.

Answer: $2\,BH_3 + 3\,O_2 \longrightarrow 2\,H_3BO_3$

Losing Oxygen Atoms

Originally, the term **reduction** referred to removal of oxygen from a compound. Many of the early reactions that were studied were reduction reactions in which metals were produced from their oxides. One of Lavoisier's experiments (Chapter 4) that led to the formulation of the law of conservation of mass, and the identification of oxygen, was a reduction reaction. Lavoisier heated mercury(II) oxide and obtained the element mercury and oxygen gas:

$$2\,HgO \longrightarrow 2\,Hg + O_2$$

Other reduction reactions in which a metal oxide loses oxygen to yield the metal are shown in the following two equations:

$$CuO + H_2 \longrightarrow Cu + H_2O$$

$$2\,Fe_2O_3 + 3\,C \longrightarrow 4\,Fe + 3\,CO_2$$

In these reduction reactions, the mass of the starting material, the metal oxide, is reduced. This fact explains the origin of the term *reduction* for reactions in which oxygen atoms are lost.

Substances that are in a reduced form, such as coal, have
a high energy content. The energy is released when the substance is oxidized. Conversely,
oxidized substances, such as
carbon dioxide, are low in
energy.

> **EXAMPLE 8-7**   Chlorine dioxide, ClO_2, a very powerful disinfecting agent, is prepared from sodium chlorate, $NaClO_3$, and sulfuric acid. Is the $NaClO_3$ oxidized or reduced?
>
> **Solution:** $NaClO_3 \longrightarrow ClO_2$ (not a complete equation)
> **1.** Even though all the products of the reaction are not given, it can be seen that the $NaClO_3$ has lost oxygen.
> **2.** Since oxygen has been lost, the $NaClO_3$ is reduced.
>
> **PRACTICE EXERCISE:** Potassium permanganate, $KMnO_4$, can be oxidized in some reactions and reduced in others. For the following reaction, is perman-ganate, MnO_4^-, oxidized or reduced?
>
> $$MnO_4^- \longrightarrow MnO \quad \text{(not a complete equation)}$$
>
> **Answer:** Since the MnO_4^- has lost oxygen, it is reduced.

GAINING AND LOSING HYDROGEN ATOMS

In the past, oxidation and reduction were also defined in terms of losing and gaining *hydrogen* atoms. A substance was said to be reduced if it gained hydro-gen atoms, and oxidized if it lost hydrogen atoms.

Gaining Hydrogen Atoms

An important reduction reaction is the combination of carbon monoxide with hydrogen, in the presence of a catalyst, to form methanol:

$$CO + 2\,H_2 \xrightarrow{\text{catalyst}} CH_3OH$$

Carbon monoxide, in gaining four hydrogen atoms, has been reduced. This re-action is important in the development of methanol as a fuel for automobiles (Chapter 15).

Losing Hydrogen Atoms

A reaction in which a substance loses one or more hydrogen atoms is termed an oxidation reaction. Many such reactions occur in metabolic processes in the body and will be discussed in Chapter 10. An example is the enzyme-catalyzed conver-sion of ethanol (C_2H_5OH) to the compound acetaldehyde (CH_3CHO) in the liver:

$$C_2H_5OH \xrightarrow[\text{enzyme}]{-2\,H} CH_3CHO$$

$$(C_2H_6O \longrightarrow C_2H_4O)$$

In this reaction, ethanol has lost two hydrogens and has thus been oxidized.

EXAMPLE 8-8   In which of the following is the reactant undergoing oxidation?

a. $CH_3OH \longrightarrow CH_2O$ (not a complete equation)
b. $C_4H_{10} + H_2 \longrightarrow C_4H_{12}$

Solution: **a.** In this reaction, methyl alcohol, CH_3OH, loses two H atoms, forming CH_2O; therefore, it is oxidized.
b. In this reaction, butene, C_4H_{10}, gains two H atoms, forming C_4H_{12}; therefore, it is reduced.

PRACTICE EXERCISE: When the very reactive chemical reagent, lithium aluminum hydride ($LiAlH_4$), reacts with organic compounds, hydrogen atoms are usually added to the organic compound. When it reacts with acetaldehyde, CH_3CHO, the following occurs:

$$LiAlH_4 + CH_3CHO \longrightarrow CH_3CH_2OH$$

Is acetaldehyde oxidized or reduced?

Answer: Reduced

GAINING AND LOSING ELECTRONS

After the discovery of the electron, oxidation and reduction were defined in much broader terms. According to the newer, more general definition, *oxidation occurs if a substance in a chemical reaction loses one or more electrons. Reduction occurs if a substance gains one or more electrons.* The earlier definitions limited oxidation and reduction to reactions in which oxygen or hydrogen were involved. The newer definition broadens the concept to include many reactions involving compounds that contain neither oxygen nor hydrogen.

Oxidation-Reduction (Redox) Reactions

Oxidation cannot occur without reduction. For simplicity and to emphasize their dual nature, oxidation-reduction reactions are often called *redox reactions.*

   We will consider first an oxidation-reduction reaction in aqueous solution that demonstrates experimentally, and in an obvious way, that electrons are being transferred. If a strip of zinc (Zn) is immersed in a solution of copper(II) sulfate (Fig. 8-15a), a spontaneous redox reaction occurs. Copper(II) sulfate in solution dissociates into Cu^{2+} ions, which impart a blue color to the solution, and colorless SO_4^{2-} ions. The zinc strip in the solution gradually becomes covered with a deposit, which can be shown to be elemental copper (Cu). At the same time, the blue color of the solution fades, indicating a decrease in the number of Cu^{2+} ions in solution (Fig. 8-15b). The Cu^{2+} ions have been reduced. They have gained electrons to become uncharged Cu atoms:

$$Cu^{2+}(aq) + 2e^- \longrightarrow Cu(s)$$

The electrons have been supplied by the zinc, which gradually dissolves in the solution (Fig. 8-15b). In giving up electrons, the elemental zinc (Zn) has been oxidized and converted to colorless Zn^{2+} ions:

(a)

(b)

Figure 8-15   When a strip of zinc is placed in a solution of copper(II) sulfate a spontaneous oxidation-reduction reaction occurs. (a) In solution, copper(II) sulfate dissociates into Cu^{2+} and SO_4^{2-} ions. The Cu^{2+} ions give the solution a blue color. (b) Electrons are transferred from the zinc (Zn) to the Cu^{2+} ions, and Zn^{2+} ions and copper metal (Cu) are formed. The zinc strip gradually dissolves as Zn^{2+} ions go into solution. The blue color of the solution fades as Cu^{2+} ions are removed from solution and deposited as Cu on the zinc strip.

$$Zn(s) \longrightarrow Zn^{2+}(aq) + 2e^-$$

Combining the two reactions, we obtain:

$$Cu^{2+}(aq) + Zn(s) + 2e^- \longrightarrow Cu(s) + Zn^{2+}(aq) + 2e^-$$

The electrons cancel out, and the overall redox reaction becomes:

$$Cu^{2+}(aq) + Zn(s) \longrightarrow Cu(s) + Zn^{2+}(aq)$$

SO_4^{2-} ions take no part in the reaction and are not included in the equation.

The reaction illustrates the complementary nature of oxidation and reduction. Neither can occur without the other. As one substance (in this case, Cu^{2+}) is reduced, another (Zn) is oxidized.

The substance that causes an oxidation by accepting electrons is termed an *oxidizing agent*; in the process of oxidation, the oxidizing agent itself is reduced. Similarly, the substance that causes a reduction by providing electrons is termed a *reducing agent*; in the process of reduction, the reducing agent is oxidized.

EXAMPLE 8-9   In the following reactions, are the metals oxidized or reduced?
a. $Zn + 2HCl \longrightarrow ZnCl_2 + H_2$
b. $Fe^{3+} \longrightarrow Fe^{2+}$ (not a complete equation)

Solution:
a. The metal Zn is converted to its ion, Zn^{2+}. It must be a 2+ ion because we know that each chlorine ion carries a 1– charge, and electrical neutrality must be maintained. To change from Zn to Zn^{2+}, the metal must *lose* 2 electrons. Therefore, zinc is oxidized.
b. To change from a 3+ ion to a 2+ ion, iron must *gain* an electron. Therefore, iron is reduced.

PRACTICE EXERCISE: Are the following reactions oxidation or reduction reactions?
a. $CuSO_4 \longrightarrow Cu$ (not a complete equation)
b. $2Ag + 2HNO_3 \longrightarrow 2AgNO_3 + H_2$

Answer: **a.** reduction   **b.** oxidation

OXIDATION AND REDUCTION: THREE DEFINITIONS

Let us now summarize the three definitions of oxidation and reduction.

	Oxidation	**Reduction**
In terms of oxygen:	gain of oxygen	loss of oxygen
In terms of hydrogen:	loss of hydrogen	gain of hydrogen
In terms of electrons:	loss of electrons	gain of electrons

A simple mnemonic for the definitions of oxidation and reduction is "LEO (the lion) says GER": *L*ose *E*lectrons *O*xidation; *G*ain *E*lectrons *R*eduction.

Although it may not be immediately obvious, in all the reactions used to illustrate oxidation and reduction in terms of oxygen and hydrogen, electrons were transferred.

Figure 8-16   An electrochemical cell uses a redox reaction to produce an electric current. In the cell shown, the solutions are separated by a porous disc made of material that slows mixing of the solutions but permits passage of ions. During the reaction, zinc atoms in the zinc strip (the anode) give up electrons and go into solution as Zn^{2+} ions; Cu^{2+} ions in the copper(II) sulfate solution combine with electrons that have passed through the wire and are deposited as copper atoms on the copper strip (the cathode). The flow of electrons forms an electric current sufficient to light an electric light bulb. Sulfate ions pass through the porous disc to maintain balance between the charges in the solutions.

BATTERIES: ENERGY FROM OXIDATION-REDUCTION REACTIONS

Redox reactions can be used to produce electrical energy. For example, in the spontaneous transfer of electrons between copper ions and zinc atoms, which we discussed previously, chemical energy is converted to electrical energy that can be used to power a battery:

$$Zn(s) + Cu^{2+}(aq) \longrightarrow Zn^{2+}(aq) + Cu(s)$$

If a strip of zinc is placed in a solution containing Cu^{2+} ions, as shown in Figure 8-15, energy is released as heat, which warms the solution. If, however, the apparatus is modified as shown in Figure 8-16, electrical current is produced which can be used to light an electric light bulb. In one compartment of the apparatus, a strip of zinc is immersed in a solution of zinc sulfate (which dissociates to give Zn^{2+} and SO_4^{2-} ions). In the other compartment, a strip of copper is immersed in a solution of copper(II) sulfate (which dissociates to give Cu^{2+} and SO_4^{2-} ions). The two compartments are separated by a porous disc that prevents the two solutions from mixing but allows the passage of ions. The metal strips, which serve as electrodes, are connected through a voltmeter.

As the reaction proceeds, electrons pass through the wire from the zinc electrode to the copper electrode. They pass down through the copper electrode (cathode), and copper is deposited on the electrode as electrons combine with Cu^{2+} ions in solution. Zinc atoms in the zinc electrode (anode) lose electrons, and Zn^{2+} ions are formed and pass into solution. Sulfate ions take no part in the electron transfer, but they migrate through the porous partition to maintain electrical balance between the positive and negative ions in solution. The electric current passing through the wire is sufficient to light a light bulb.

An apparatus like the one just described that produces an electric current is called an **electrochemical cell**. A battery is made up of a series of electrochemical cells.

In any electrical circuit, the anode is the electrode that releases electrons and at which oxidation occurs. The cathode is the electrode that receives electrons and at which reduction occurs. To remember this differentiation, note that the words *a*node and *o*xidation begin with a vowel while *c*athode and *r*eduction begin with a consonant.

Batteries are portable sources of electrical energy.

Dental Fillings: Amalgam, Plastic, Ceramic or Gold?

The material most commonly used by dentists for filling cavities is a composition known as a *dental amalgam*. It is one of the few materials we use that is made up of pure elements without modification. An amalgam is a substance made by combining mercury with the metals silver and tin. It actually consists of three solid phases having the compositions Ag_2Hg_3, Ag_3Sn, and Sn_8Hg. Although the incidence of tooth decay has decreased since the widespread introduction of fluoride toothpaste, as many as 100 million silver-colored fillings are used to repair decayed teeth in the United States each year.

In 1990, the CBS television show "60 Minutes" reported that a minute amount of the mercury in dental amalgam may be slowly released from such fillings and cause adverse health effects. (It has long been known that mercury vapor can have toxic effects, particularly on the brain and nervous system.) Since that report, the Environmental Protection Agency has banned mercury from latex paint but not from dental fillings. The dose of mercury vapor possibly released from dental fillings is considered to be so low as to be an insignificant threat to public health. Still, critics of amalgam want proof that it is safe, while supporters want proof there is a genuine hazard.

Another problem with these fillings is that dental discomfort may occur from the reaction of the dental amalgam with other metals in the mouth. Anyone who bites a piece of aluminum foil in such a way that the aluminum foil presses against a dental filling will probably experience a sharp pain. Contact between the foil and the filling creates a small battery, with the dissolved salts in saliva completing the circuit. The current that flows between the metals is experienced by the nerve of the tooth as an unpleasant pain. The same type of small battery can exist in your mouth if a dental amalgam is placed too close to a gold tooth or a tooth with a gold inlay. This small battery causes the amalgam to corrode, thereby releasing tin from the amalgam. The tin ions have a very unpleasant metallic taste. Eventually, the filling decomposes, and the tooth needs to be refilled.

The main alternatives to amalgam are composite resin fillings, which are tooth-colored plastic materials. These materials look more attractive than the silver-colored amalgam and are preferred for front teeth. Composites placed in back molars are slowly eroded by a combination of chewing force and substances in food that attack them. Composites are also more difficult to place precisely, and a poor fit leaves room for decay-producing bacteria to enter the tooth. Newer formulations that chemically bond the composite to tooth enamel may overcome these problems, but these materials have not yet been perfected.

Other alternatives to dental amalgam are gold, which is very durable, and ceramic, which looks very nice. Both are expensive and require multiple visits for installation. Listed below is a comparison of the cost and durability of different types of dental fillings.

Type	Cost per Filling	Life Span	Visits to Install
Amalgam	$ 50	10-20 years	1
Composite	$ 65	3-10 years	1
Ceramic	$400	10 years	2 or more
Gold	$360	20 or more	2 or more

At this time, the alternatives to dental amalgam do not measure up. Given their solid track record and a risk that has yet to be proven, amalgam fillings are still the best bet.

Questions
1. If mercury from amalgam posed a health risk, would dentists and dental assistants be most at risk? Explain.
2. How would you propose to dispose of the waste generated from dental amalgam?

Reference: Consumer Reports, May 1991, 316–19.

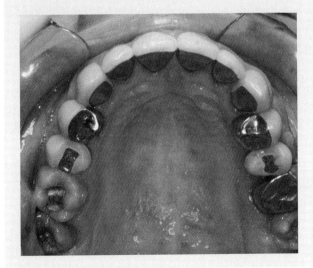

Figure 8-17 Some common commercial batteries. Batteries come in many sizes and depend on a variety of redox reactions to produce electrical current. The large battery is a typical lead-acid automobile battery. Also included are rechargeable nickel-cadmium batteries, zinc-carbon flashlight batteries, alkaline cells, and a number of miniature cells.

Commercial Batteries

Common flashlight batteries, button batteries, and automobile storage batteries all operate on a similar principle (Fig. 8-17).

A cross section of a flashlight battery is shown in Figure 8-18. The outer case, which is made of zinc, serves as the negative electrode (or anode). A carbon rod serves as the positive electrode (or cathode). Surrounding the carbon rod is a moist mixture of ammonium chloride (NH_4Cl), zinc chloride ($ZnCl_2$), and manganese dioxide (MnO_2). The battery is called a dry cell because no free liquid is present.

As current flows through the cell and through the flashlight bulb, the zinc loses electrons and is oxidized:

$$Zn \longrightarrow Zn^{2+} + 2e^-$$

The reaction at the carbon cathode is complex and involves the reduction of manganese dioxide. The overall reaction can be written as follows:

$$Zn + 2\,MnO_2 + 2\,NH_4Cl \longrightarrow ZnCl_2 + Mn_2O_3 + 2\,NH_3 + H_2O$$
$$+ \text{ electrical energy}$$

The dry cell battery was invented by Georges Leclanche in 1866.

Many of the miniature batteries used in watches, hearing aids, cameras, and electronic calculators are mercury cells (Fig. 8-19). In the oxidation-reduction reaction that occurs in these batteries, zinc is oxidized and mercury is reduced:

$$HgO + Zn + H_2O \longrightarrow Hg + Zn(OH)_2 + \text{ electrical energy}$$

Because mercury is toxic, these batteries must be disposed of with care.

The familiar 12-volt automobile storage battery consists of six electrochemical cells joined in series (Fig. 8-20). The electrodes are made of lead and are immersed in dilute sulfuric acid. The negative electrodes (anodes) are impregnated with spongy lead metal; the positive electrodes (cathodes), with lead oxide (PbO_2). As the battery discharges, metallic lead is oxidized to lead

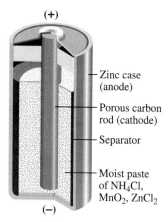

(+)

Zinc case (anode)

Porous carbon rod (cathode)

Separator

Moist paste of NH_4Cl, MnO_2, $ZnCl_2$

(−)

Figure 8-18 Cross section of a zinc-carbon flashlight battery.

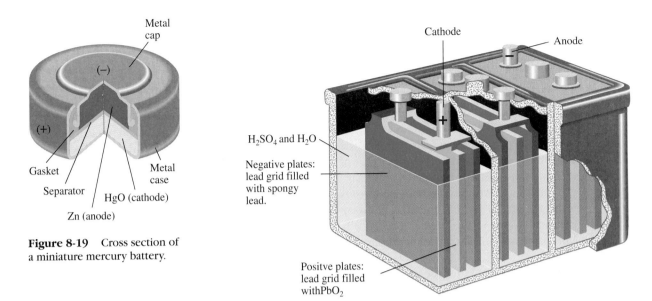

Figure 8-19   Cross section of a miniature mercury battery.

(labels: Metal cap, (−), (+), Gasket, Separator, HgO (cathode), Zn (anode), Metal case)

Figure 8-20   Cross section of lead-acid storage battery.

(labels: Cathode, Anode, H₂SO₄ and H₂O, Negative plates: lead grid filled with spongy lead., Positve plates: lead grid filled withPbO₂)

sulfate ($PbSO_4$) at the anodes, and lead oxide is reduced at the cathode. The net reaction is:

$$\text{discharging}$$
$$Pb + PbO_2 + 2\,H_2SO_4 \;\rightleftharpoons\; 2\,PbSO_4 + 2\,H_2O + \text{electrical energy}$$
$$\text{recharging}$$

A lead-acid battery can be recharged if electrons are forced to flow in the opposite direction by connecting the battery to an external source of electricity. An automobile battery is normally recharged during driving.

Lead storage batteries are reliable and relatively inexpensive. Disadvantages are their weight and the problem of disposal. If carelessly discarded, both the lead and the sulfuric acid can pollute the environment.

The standard carbon-zinc dry cell battery used in flashlights and portable radios is relatively inexpensive. Alkaline, lithium, and mercury batteries are more expensive but last longer.

Lead storage batteries used in automobiles, and nickel-cadmium (Nicad) batteries used in cordless appliances such as calculators, can be recharged.

EXPLORATIONS ▶ Cardiac Pacemakers: Electronic Devices That Save Lives

Today, thousands of people with failing hearts, who in the past would have died, are leading normal lives thanks to tiny artificial cardiac pacemakers implanted in their bodies. These ingenious electronic devices send out electrical impulses that stimulate the failing heart and reestablish its normal pumping action.

The human heart is an amazing organ. Slightly bigger than a clenched fist and weighing less than a pound, it beats continuously, on average, 72 times every minute of our lives. Each year it beats about 38 million times, and in doing so, pumps 700,000 gallons of blood around the body.

The heart is a hollow muscular organ with four chambers: the right and left atria and the right and left ventricles (Fig. 8-Aa). Blood collected into large veins from all parts of the body except the lungs enters the heart through the right atrium, and from there flows into the right ventricle. From the right ventricle, it is pumped to the lungs to be reoxygenated. Oxygenated blood is then returned to the heart through the left atrium and from there enters the left ventricle. From the left ventricle it is pumped to all parts of the body (except the lungs) via the arteries.

The heart has a natural pacemaker that normally initiates and controls its rhythmic pumping action. At regular intervals, this pacemaker, which consists of a group of specialized cells located in the upper part of the right atrium, spontaneously generates an electrical impulse to initiate a heartbeat. The impulse spreads through the heart muscle, causing the atria and ventricles to contract in proper sequence. The coordinated rhythmic contractions keep the blood circulating through the body.

In certain pathological disorders, the heart's own pacemaker is impaired, and the heart ceases to beat with a normal rhythm. For example, in some conditions, the heart beats abnormally slowly. Cardiac output is diminished, and congestive heart failure or other serious problems may result. In other disorders, the heart beats more rapidly than normal or fluctuates between a slow and a fast rate. Most of these potentially fatal disorders can now be treated successfully with an artificial pacemaker.

Electrical impulses have been used for many years to stimulate and control a patient's heartbeat. In early procedures, electrodes placed on a patient's chest were connected to a bulky external pacemaker plugged into an external circuit. Later, battery-powered units that could be attached to patients' clothing were introduced. The first completely implantable pacemaker was placed in a human by a Swedish surgeon in 1958. A year later, an American team headed by Dr. William Chardack implanted a pacemaker designed by the engineer and inventor Wilson Greatbatch.

Wilson Greatbatch had been so convinced of the life-saving potential of an implantable device that he had given up his job so that he could work in a barn behind his house to design one. Within 2 years, he had a working model that he demonstrated to a number of cardiac specialists. Only Dr. Chardack immediately appreciated its possibilities. The Chardack-Greatbatch pacemaker soon became the most widely used pacemaker in the world.

An implantable pacemaker has three main parts: the pulse generator containing the battery and the electronic circuit; the pacing leads, which carry the electrical impulses; and the metal electrodes at the end of the leads, which make contact with the heart muscle (Fig. 8-Ab). Pacemakers are installed in the body in one of two ways. Either the electrodes are attached to the outside surface of the heart and the pulse generator is implanted beneath the soft tissue of the abdominal wall or the electrodes are passed through a vein in the neck directly into the heart and the pulse generator is implanted under the skin just below the collarbone. Depending on the nature of the heart impairment, the electrical stimuli are applied to the ventricle (the most common procedure), the atrium, or the two chambers in sequence. The pacing systems are now programmable: Electrical parameters such as timing, duration, and voltage of the stimuli can be adjusted from outside the body to meet a patient's needs.

The early implantable pacemakers had many problems. They were quite bulky, often causing discomfort to the patient, and, almost invariably, they failed within 1 or 2 years. The greatest problem was the battery. The mercury-zinc batteries used at that time had a life expectancy of about 5 years, but when placed in pacemakers, they failed much sooner, primarily because the pulse generator could not be hermetically sealed. Body fluids seeped in, causing a short circuit. Various alternative power sources were studied, including biological and nuclear sources, but the real breakthrough came with the introduction of lithium batteries in the 1970s.

Lithium makes a good anode for a battery because it loses electrons easily. It is the lightest of all the metals and, weight for weight compared with other metals, it produces a large quantity of energy. Thus, lithium-iodine batteries—the type most commonly used in pacemakers—are very small and lightweight. Today's pacemakers weigh approximately 1 ounce (28 g) (Fig. 8-Ac).

Pulse generators can now be sealed efficiently, and implanted pacemakers generally function reliably for 6 to 10 years, and sometimes longer. Often the cause of failure is not the battery but dislodgment of the leads.

Tremendous advances have been made in electronic circuitry. Developments in microelectronics that revolutionized the computer industry also revolutionized the pacemaker industry. With each advance, pacemaker circuits became smaller, more sophisticated, and more reliable. The first pulse generators delivered stimuli at a fixed rate that was unrelated to the patient's own heart rhythm. The complex circuits in modern units are able to sense a patient's own heart-

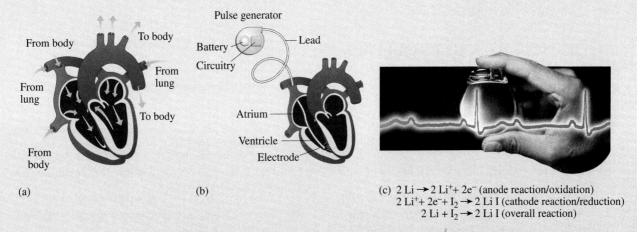

(c) $2 Li \rightarrow 2 Li^+ + 2e^-$ (anode reaction/oxidation)
$2 Li^+ + 2e^- + I_2 \rightarrow 2 Li I$ (cathode reaction/reduction)
$2 Li + I_2 \rightarrow 2 Li I$ (overall reaction)

Figure 8-A. (a) Circulation of blood through the heart. (b) Placement of a pacemaker in the heart. (c) A modern pacemaker. The redox reaction in its lithium-iodine battery occurs as shown under the figure.

beat and adjust the electrical stimuli accordingly; if the circuit senses a natural beat, the generator does not fire, but if there is a longer than normal pause between beats, it fires. Some pacemakers also sense and respond to other body signals. For example, some can sense increases in body temperature and respiratory rate that occur during exercise and then adjust the heat rate to meet the body's changing needs.

Pacemaker patients rely primarily on electrocardiograms to check that their hearts and pacemakers are functioning properly. Using a telephone transmitter, patients in the comfort of their homes can relay tracings to their doctors. Some sophisticated pacemakers are monitored by telemetry; that is, the unit not only receives instructions, it also relays information to an outside receiver. A built-in microprocessor in the unit records and stores data about cardiac function and the condition of the different parts of the pacemaker; it then relays this information to the doctor's office.

Despite all the advances, a pacemaker is still limited by its battery, which inevitably runs down. In the future, it is expected that this problem will be solved by taking advantage of a naturally occurring oxidation-reduction reaction in the body. If such an inexhaustible source of energy can be tapped, a pacemaker will last a patient for as long as it is needed.

Questions
1. Explain why lithium is lighter than other metals.
2. Discuss advances in technology that would improve the efficiency of pacemakers and add to the comfort of patients.

References: D. Sonnenburg, M. Birnbaum, and E. A. Naclerio, *Understanding Pacemakers* (New York: Michael Kesend Publishing, 1982); H. W. Moses, J. A. Schneider, B. D. Miller, G. J. Taylor, *A Practical Guide to Cardiac Pacing* (New York: Little Brown, 3d ed. 1991).

KEY WORDS AND CONCEPTS

acid	hydronium ion	pH
acid rain (deposition)	molar (molarity)	reduction
acid mine drainage	neutralization	solubility
aqueous solution	oxidation	solute
base	oxidation-reduction (redox)	solution
buffer	reactions	solvent
concentration	parts per billion	
dissociation	parts per million	

QUESTIONS AND PROBLEMS

1. Define the following terms:
   a. solution **b.** solvent **c.** solute

2. What ions are present in aqueous solutions of the following salts?
   a. KBr **b.** $MgCO_3$ **c.** Na_2SO_4 **d.** NH_4Cl
   e. $Mg(NO_3)_2$

3. What ions are present in aqueous solutions of the following salts?
   a. $PbCl_2$ **b.** NaF **c.** $Ba(OH)_2$ **d.** AgI **e.** $SrCl_2$

4. A waste discharge stream from a paper mill is sampled for dioxin, a toxic substance that is formed in the bleaching of paper pulp. Analysis of water from the stream shows a dioxin concentration of 0.010 μg/mL. Express this concentration in parts per billion. The EPA standard for dioxin is 1 ppb. Is this wastewater in violation of EPA standards?

5. Which is the higher concentration: 100 ppb or 0.05 ppm?

6. If you dropped 25 g of salt (NaCl) into a 120,000-L swimming pool, what would be the concentration of NaCl in the water in parts per million and parts per billion?

7. Litmus is turned _____ by acids and _____ by bases.

8. Name:
   a. a base present in household products that is used to remove greasy dirt **b.** an acid that is commonly used in food preparation

9. Name and give the formula for the ion that is responsible for:
   a. the properties of an acid in water **b.** the properties of a base in water

10. Identify the Brønsted-Lowry acid and base in the following reactions:
    a. $HF + H_2O \longrightarrow H_3O^+ + F^-$
    b. $S^{2-} + H_2O \longrightarrow HS^- + OH^-$
    c. $H_2CO_3 + H_2O \longrightarrow H_3O^+ + HCO_3^-$
    d. $CH_3NH_2 + H_2O \longrightarrow CH_3NH_3^+ + OH^-$

11. Complete the following reaction:
    $HCl + NaOH \longrightarrow$
    a. Name the products.
    b. This reaction is a _____ reaction.

12. Which of the following are weak acids?
    a. acetic acid **b.** hydrochloric acid **c.** nitric acid

13. Which of the following are strong acids?
    a. hydrocyanic acid **b.** carbonic acid **c.** sulfuric acid

14. Give an equation to show how water dissociates into ions.

15. What is the pH of a neutral solution?

16. Indicate whether solutions having the following pH values are acidic, basic, or neutral.
    a. 3.5 **b.** 7.3 **c.** 9.6

17. Which of the solutions having the following pH values are basic?
    a. 9.4 **b.** 7.0 **c.** 12.5

18. What is the pH of a solution that has a hydrogen ion concentration [H^+] of:
    a. 1.0×10^{-4}M **b.** 1.0×10^{-8}M **c.** 1.0×10^{-1}M

19. What is the pH of a solution that has a hydrogen ion concentration [H^+] of:
    a. 0.001 M **b.** 0.1 M **c.** 0.000000000001 M

20. What is the hydrogen ion concentration [H^+] of a solution that has a pH of:
    a. 10 **b.** 7 **c.** 4 **d.** 2

21. Solution A has a pH of 2. Solution B has a pH of 5.
    a. Which solution is more acidic? **b.** How many times more acidic is the one solution than the other?

22. Describe four general properties of aqueous solutions of:
    a. acids **b.** bases

23. Name acids and bases you would find in the average American home, and explain briefly how they are used.

24. What is the hydronium ion? Explain how it is formed in an aqueous solution of acid. Is it a stable ion? Explain your answer.

25. Explain how ammonia acts as a base in water. Is it a strong or a weak base? Give equations.

26. Using HA to represent an acid, explain and give equations for the difference between a strong acid and a weak acid. Give the names and formulas of two strong acids and two weak acids.

27. Explain how acids and bases neutralize each other. What products are formed? Give two equations to illustrate your answer.

28. Explain, giving an equation, how water acts both as an acid and a base.

29. Why was the pH scale introduced? Explain why the pH of a neutral solution is 7.

30. List three uses of sulfuric acid.

31. Name and give the formula of the most common acid in the natural environment.

32. Give equations showing:
    a. how water dissolves atmospheric carbon dioxide to form an acidic solution **b.** the dissociation of the resulting acid into ions

33. What is the pH of most physiological liquids such as blood?

34. Give equations to show how limestone is weathered by rainfall.

35. What is the pH of pure rain? Why is it acidic? Give equations to explain your answers.

36. Describe two natural sources of sulfuric acid.

37. Describe, giving equations, the natural acid weathering of:
   a. limestone   **b.** feldspar

38. Define *acid rain*.
   a. What are the effects of acid rain on trees and other plant life?   **b.** At what pH will a lake become sterile?

39. Why does acid rain create more-serious problems in lakes in eastern North America than in the Great Lakes? Include chemical equations to illustrate your answer.

40. What is the main cause of acid mine drainage? Give an equation to show how this acid is formed.

41. What steps have been taken to control acid mine drainage? Have these methods been successful?

42. Classify each of the following as either an oxidation or reduction.
   a. $FeO \longrightarrow Fe_2O_3$ **b.** $ClO^- \longrightarrow Cl_2$
   c. $OH^- \longrightarrow O_2$ **d.** $MnO_4^- \longrightarrow MnO_2$

43. Which of the following is an oxidation?
   a. $H_2S \longrightarrow S$ **b.** $Cl_2 \longrightarrow ClO_4^-$
   c. $Fe(OH)_3 \longrightarrow Fe$

44. Which of the following is a reduction?
   a. $NO_2^- \longrightarrow NO_3^-$ **b.** $Cl_2 \longrightarrow ClO_3^-$
   c. $Cr_2O_3 \longrightarrow Cr$

45. Do the following statements describe an oxidation, a reduction, or neither?
   a. hydrogen ion lost   **b.** electrons gained
   c. hydrogen atom lost   **d.** electrons lost

46. Do the following statements describe an oxidation, a reduction, or neither?
   a. hydrogen atom gained   **b.** oxygen atom gained
   c. hydroxide ion gained   **d.** hydroxide ion lost

47. In the following reactions, which metal is oxidized and which is reduced?
   a. $4\,H^+ + Sn^{2+} + O_2 \longrightarrow Sn + 2\,H_2O$
   b. $4\,Zn + NO_3^- + 10\,H^+ \longrightarrow 4\,Zn^{2+} + NH_4^+ + 3\,H_2O$
   c. $6\,Fe^{2+} + Cr_2O_7^{2-} + 14\,H^+ \longrightarrow 6\,Fe^{3+} + 2\,Cr^{3+} + 7\,H_2O$
   d. $MnO_4^- + Cl^- \longrightarrow Mn^{2+} + ClO^-$

CARBON COMPOUNDS
An Introduction to Organic Chemistry

This diamond neck-lace is made of pure carbon. It is resting on coal, which is made up primarily of compounds of carbon with hydrogen.

Compounds of carbon are central to life on earth. Deoxyribonucleic acid (DNA), the huge molecule that carries the genetic code for a given species, is a carbon compound. Skin, muscle, hair, blood, leaves, roots, and all other living tissues are made from carbon compounds. We and the living world around us are not only made from carbon compounds, but we also depend on these compounds to support us in our daily lives and to enhance our standard of living. Our cloth-ing, whether made of natural fibers, such as wool and cotton, or synthetic fibers, such as nylon and polyester, is made from carbon compounds. Numerous other consumer goods—foods, medicines, paper, plastics, pesticides, and tires—are also made from carbon compounds. The combustion of carbon compounds in natural gas, petroleum, and coal generates the energy that fuels our economy.

With the exception of ionic compounds, such as carbonates and cyanides, and oxides of carbon, which are considered to be inorganic, com-pounds of carbon are described as organic compounds (Chapter 2). Over 10 million chemical compounds are known, and thousands more are being synthe-sized each year. Of these, 9 million are organic compounds. The remainder are inorganic compounds. Although there are 92 naturally occurring elements (Chapter 5), only carbon can form so many compounds. Organic chemistry is devoted to the study of carbon compounds; inorganic chemistry covers all the other elements and their compounds.

Learning Goals:

In this chapter, you should gain an understanding of:

1. Why carbon forms so many compounds compared with other elements.

2. The different classes of hydrocarbons.

3. Structural isomerism and cis-trans isomerism.

4. The structure of benzene.

5. Nomenclature of simple organic compounds.

6. Different classes of organic compounds based on functional groups.

THE CARBON ATOM AND CHEMICAL BONDING

Valence electrons are the electrons in the highest energy level of an atom.

The octet rule (or rule of eight) states that in forming compounds, atoms gain, lose, or share electrons until they are surrounded by 8 electrons.

Why does carbon form so many compounds? The answer lies in its atomic structure. Carbon, atomic number 6, has four valence electrons, and when it forms compounds by sharing these valence electrons (Chapter 5) with other carbon atoms, or atoms of other elements, it obeys the octet rule. It forms carbon-carbon single bonds by sharing pairs of electrons:

$$\bullet \overset{\bullet}{\underset{\bullet}{C}} \bullet \; + \; \bullet \overset{\bullet}{\underset{\bullet}{C}} \bullet \; + \; \bullet \overset{\bullet}{\underset{\bullet}{C}} \bullet \; \longrightarrow \; \bullet \overset{\bullet}{\underset{\bullet}{C}} \overset{\bullet}{\underset{\bullet}{:}} \overset{\bullet}{\underset{\bullet}{C}} \overset{\bullet}{\underset{\bullet}{:}} \overset{\bullet}{\underset{\bullet}{C}} \bullet \quad (-\overset{|}{\underset{|}{C}}-\overset{|}{\underset{|}{C}}-\overset{|}{\underset{|}{C}}-)$$

When it combines with four hydrogen atoms to form methane, the carbon atom shares its four valence electrons with the hydrogen atoms, thus forming four stable covalent bonds:

Remember that although we have represented the electrons in the carbon atom and in the hydrogen atoms differently, all electrons are identical regardless of source.

$$\bullet \overset{\bullet}{C} \bullet \; + \; 4\,H \; \longrightarrow \; H \overset{\overset{H}{\bullet\,\circ}}{\underset{\underset{H}{\circ\,\bullet}}{C}} H \quad \text{or} \quad H-\overset{\overset{\displaystyle H}{|}}{\underset{\underset{\displaystyle H}{|}}{C}}-H \quad \text{or} \quad CH_4$$

methane

Carbon forms more compounds than any other element primarily because carbon atoms link up with each other in so many different ways. A molecule of a carbon compound may contain a single carbon-carbon bond or thousands of such bonds. The carbon atoms can link together in straight chains, branched chains, or in rings (Fig. 9-1). In addition to carbon-carbon single bonds, carbon can form carbon-carbon double and triple bonds. In each of these different bonding patterns, the carbon atoms form four covalent bonds.

Only the element carbon is able to form long chains between like atoms. Some elements, including oxygen (O), nitrogen (N), and chlorine (Cl), form stable two-atom molecules; sulfur (S), silicon (Si), and phosphorus (P) form unstable chains of four to eight like atoms, but no element can form chains as long as carbon. Even silicon and germanium (Ge), which are in the same group of the periodic table as carbon, do not form long like-atom chains.

Different Forms of Carbon

Elemental carbon exists in two very different crystalline forms: diamond and graphite. In diamond, each carbon atom is joined by strong covalent bonds to four other carbon atoms. Each of these carbon atoms is also joined to four more carbon atoms, and so on, until a huge three-dimensional interlocking network of carbon atoms is formed (Fig. 9-2a). This carbon-carbon bonding pattern accounts for the

methane: saturated & CH4 b/c its filled name b/c its filled

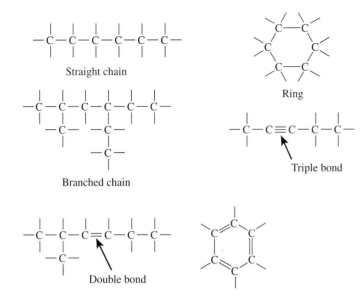

Straight chain

Branched chain

Double bond

Ring

Triple bond

Figure 9-1 Carbon atoms can join to form straight chains, branched chains, and rings. In addition to carbon-carbon single bonds, double and triple bonds are also formed.

stability and extreme hardness of diamond. For a diamond to undergo a chemical change, many strong bonds within the crystalline structure must be broken.

Graphite is completely different from diamond. It is a soft, black, slippery material made up of hexagonal arrays of carbon atoms arranged in sheets (Fig. 9-2b). Each carbon atom in the array is joined to three other carbon atoms by forming two single bonds and one double bond. Graphite is slippery because the intermolecular attractive forces holding the sheets together are relatively weak, and the sheets easily slide past each other. Graphite is used as a dry lubricant and, combined with a binder, forms the "lead" in pencils.

Two or more forms of an element that differ in their bonding or molecular structure are called allotropes. Diamond and graphite are all allotropes of carbon.

Macroscopic

Macroscopic

Submicroscopic

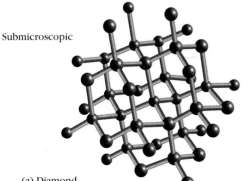

Submicroscopic

(a) Diamond

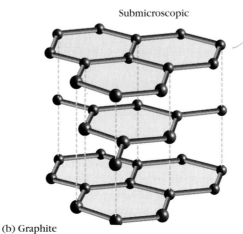

(b) Graphite

Figure 9-2 Two very different forms of carbon. **(a)** In diamond, the carbon atoms are joined to form a strong three-dimensional network. **(b)** In graphite, the carbon atoms are joined to form sheets held together by weak attractive forces.

Ethane

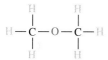

Dimethyl ether

Chloroform

Methylamine

Figure 9-3 Examples of compounds in which carbon atoms are covalently bonded to hydrogen, oxygen, chlorine, and nitrogen atoms.

Compounds of Carbon with Other Elements

In addition to forming carbon-carbon bonds, carbon atoms form strong covalent bonds with other elements, particularly with the nonmetal elements hydrogen (H), oxygen (O), nitrogen (N), fluorine (F), and chlorine (Cl). Simple compounds in which carbon is combined with other elements are shown in Figure 9-3. Carbon also bonds with phosphorus (P), sulfur (S), silicon (Si), boron (B), and the metals sodium (Na), potassium (K), calcium (Ca), and magnesium (Mg).

The enormous number of carbon compounds that exist can be divided into a relatively small number of classes according to the functional group they contain. A **functional group** is a particular arrangement of atoms that is present in each molecule of the class and that largely determines the chemical behavior of the class. Examples of functional groups are −Cl, −OH, −COOH, and −NH_2. Before we examine the different functional groups, we will study the hydrocarbons, the parent compounds from which all carbon compounds are derived.

HYDROCARBONS

Hydrocarbons are composed of the two elements carbon and hydrogen. There are four classes of hydrocarbons: **alkanes**, which contain −C−C− single bonds; **alkenes**, which contain one or more −C=C− double bonds; **alkynes**, which contain one or more −C≡C− triple bonds; and **aromatic hydrocarbons**, which contain one or more benzene rings. Complex mixtures of hydrocarbons, which are present in enormous quantities in natural gas and oil (petroleum), are the source of many of the organic compounds used by industry.

Alkanes

The simplest alkane is **methane**, CH_4, the major component of natural gas. The structure of the molecule can be represented in a number of different ways as illustrated in Figure 9-4. The expanded structural formula includes the four covalent bonds; the ball-and-stick and space-filling models show the spatial arrangement of the atoms. As the ball-and-stick model indicates, the methane molecule is in the shape of a tetrahedron. The space-filling model gives the most accurate representation of the actual shape of the molecule. For simplicity, the condensed structural formula, which does not show the bonds or the bond angles, is usually used to represent the molecule.

The two alkanes that follow methane are **ethane**, C_2H_6, and **propane**, C_3H_8 (Fig. 9-4). Propane is a major component of bottled gas. The next member of the series is **butane**, C_4H_{10}, and for this alkane, two structures are possible (Fig. 9-5). The four carbon atoms can be joined in a straight line (*n*-butane), or the fourth carbon can be added to the middle carbon atom in the −C−C−C− chain to form a branch (isobutane).

Alkanes are known as **saturated hydrocarbons**, because, for a given number of carbon atoms, they contain the largest possible number of hydrogen atoms. The first 10 straight-chain alkanes are shown in Table 9-1. The *n*− stands for normal and signifies straight-chain. Notice that each alkane differs from the one preceding it by the addition of a −CH_2 group. A series of compounds in which each member differs from the next member by a constant increment is called a **homologous series**. The general formula for the alkane series is C_nH_{2n+2} where *n* is the number of carbon atoms in a member of the series. In a homologous series, the properties of the members

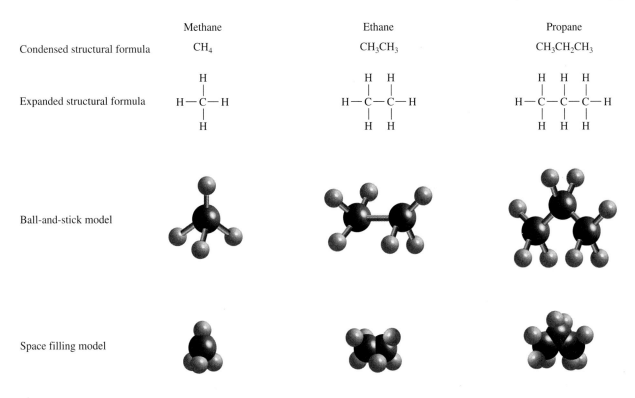

	Methane	Ethane	Propane
Condensed structural formula	CH_4	CH_3CH_3	$CH_3CH_2CH_3$

Figure 9-4   Organic molecules can be represented in several different ways, as shown for the first three members of the alkane series.

change systematically with increasing molecular weight. As Table 9-1 shows, in the alkane series the boiling points of the straight-chain alkanes rise quite regularly as the number of carbon atoms increases. 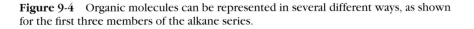 The first four alkanes are gases, and the remainder of those shown are liquids.

EXAMPLE 9-1   Write the structural formula for the straight-chain alkane C_5H_{12}.

Solution: Write five carbon atoms linked together to form a chain,
$$C-C-C-C-C$$

Attach hydrogen atoms to the carbon atoms so that each carbon atom forms four covalent bonds.

$$H-\overset{\displaystyle H}{\underset{\displaystyle H}{C}}-\overset{\displaystyle H}{\underset{\displaystyle H}{C}}-\overset{\displaystyle H}{\underset{\displaystyle H}{C}}-\overset{\displaystyle H}{\underset{\displaystyle H}{C}}-\overset{\displaystyle H}{\underset{\displaystyle H}{C}}-H$$

PRACTICE EXERCISE: Write the structural formula for the straight-chain hydrocarbon represented by C_nH_{2n+2}, where $n = 7$.

Answer:

$$H-\overset{\displaystyle H}{\underset{\displaystyle H}{C}}-\overset{\displaystyle H}{\underset{\displaystyle H}{C}}-\overset{\displaystyle H}{\underset{\displaystyle H}{C}}-\overset{\displaystyle H}{\underset{\displaystyle H}{C}}-\overset{\displaystyle H}{\underset{\displaystyle H}{C}}-\overset{\displaystyle H}{\underset{\displaystyle H}{C}}-\overset{\displaystyle H}{\underset{\displaystyle H}{C}}-H$$

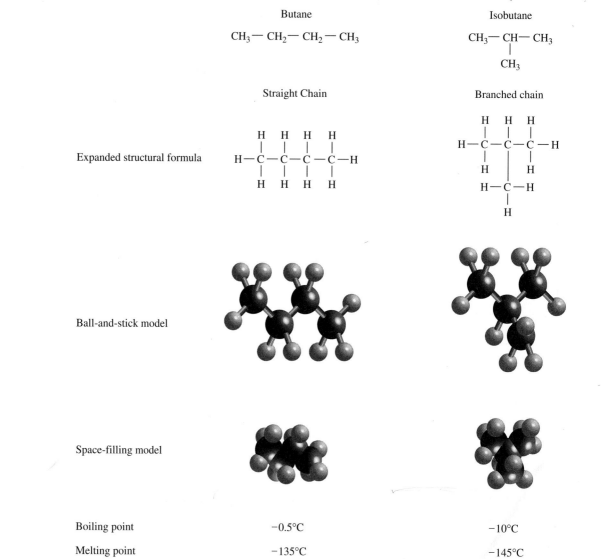

Figure 9-5 The two structural isomers of butane (C_4H_{10}).

We have described unbranched chains as straight chains. In fact, because of tetrahedral bonding, the carbon atoms are not in a straight line as shown in Table 9-1 but are staggered as shown below and in the ball-and-stick models for propane and *n*-butane in Figures 9-4 and 9-5.

Structural Isomerism

Compounds such as butane and isobutane that have the same molecular formula (C_4H_{10}) but different structures are called **structural isomers**. Because their structures are different (Fig. 9-5), their properties—for example, boiling point and melting point—are also different.

All alkanes containing four or more carbon atoms form structural isomers. The predicted number of possible isomers increases rapidly as the number of carbon atoms in the molecule increases. For example, butane, C_4H_{10}, forms two isomers; decane, $C_{10}H_{22}$, theoretically forms 75; and $C_{30}H_{62}$ forms over 400 million (Table 9-2). Most of these isomers do not exist naturally and have not been synthesized, but the large number of possibilities helps explain the abundance of carbon compounds. In isomers with large numbers of carbon atoms, crowding of atoms makes some structures too unstable to exist.

TABLE 9-1   The First Ten Straight-Chain Alkanes

Name	Formula	Boiling Point (°C)	Structural Formula
Methane	CH_4	-162	
Ethane	C_2H_6	-88.5	
Propane	C_3H_8	-42	
n-Butane	C_4H_{10}	0	
n-Pentane	C_5H_{12}	36	
n-Hexane	C_6H_{14}	69	
n-Heptane	C_7H_{16}	98	
n-Octane	C_8H_{18}	126	
n-Nonane	C_9H_{20}	151	
n-Decane	$C_{10}H_{22}$	174	

TABLE 9-2   Number of Possible Isomers for Selected Alkanes

Molecular Formula	Number of Possible Isomers
C_4H_{10}	2
C_5H_{12}	3
C_6H_{14}	5
C_7H_{16}	9
C_8H_{18}	18
C_9H_{20}	35
$C_{10}H_{22}$	75
$C_{15}H_{32}$	4,347
$C_{20}H_{42}$	336,319
$C_{30}H_{62}$	4,111,846,763

EXAMPLE 9-2   Give the structural and condensed formulas of the isomers of pentane.

Solution: Hexane has five carbon atoms. The carbon skeletons of the three possible isomers are:

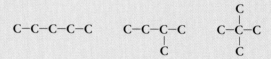

Attach hydrogen atoms so that each carbon atom forms 4 bonds. Each isomer has the same condensed formula, C_5H_{12}.

PRACTICE EXERCISE: Give the structural formulas of the possible isomers of hexane C_6H_{14}.

Answer: The carbon skeletons are as follows. Add the hydrogen atoms and check that each isomer has 14 hydrogen atoms.

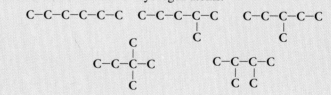

Nomenclature of Alkanes

Organic compounds are named according to rules established by the International Union of Pure and Applied Chemistry (IUPAC). As shown in Table 9-1, except for the first four members of the family, the first part of the name of an alkane is derived from the Greek name for the number of carbon atoms in the molecule. The suffix *-ane* means that the compound is an alkane.

Before we can follow the IUPAC rules for naming branched-chain alkanes, we must consider the names of the groups that are formed when one hydrogen atom is removed from the formula of an alkane. For example, when an H atom is removed from methane, CH_4, a $-CH_3$, or **methyl**, group is formed. Similarly,

Pent-, hex-, hept-, oct-, non-, and dec- are derived from the Greek (and Latin) words for 5, 6, 7, 8, 9, and 10.

removal of an H atom from ethane, C_2H_6, gives $-C_2H_5$, or an **ethyl** group. Because the groups are derived from alkanes, they are called **alkyl** groups. Examples of some common alkyl groups are given in Table 9-3.

Branched-chain alkanes are named by applying the following rules:

1. *Determine the name of the parent compound by finding the longest continuous chain of carbon atoms.* Consider the following example:

(a)
$$H_3C-CH_2-CH_2-\underset{\underset{\displaystyle CH_3}{|}}{CH}-CH_3$$

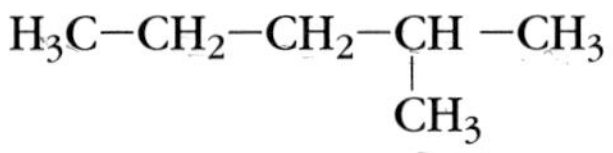

The longest chain contains **5** carbon atoms; therefore, the parent name is *pentane*.

Sometimes, because of the way in which a formula is written, it is not easy to recognize the longest chain. For example, the alkane shown below is not a hexane (6 carbon atoms in the longest chain) as at first might be supposed, but a *heptane* (7 carbon atoms in the longest chain).

(b)
$$H_3C-CH_2-CH_2-CH_2-\underset{\underset{\displaystyle CH_3}{|}}{\underset{\underset{\displaystyle |}{CH_2}}{CH}}-CH_3$$

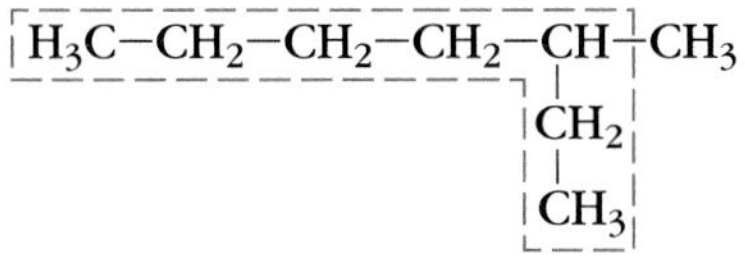

2. *Number the longest chain beginning with the end closest to the branch. Use these numbers to designate the location of the groups (or **substituents**) at the branch.*

Applying this rule to the above examples, the carbon atoms are numbered and named as follows:

(a)
$$\overset{5}{H_3C}-\overset{4}{CH_2}-\overset{3}{CH_2}-\underset{\underset{\displaystyle CH_3}{|}}{\overset{2}{CH}}-\overset{1}{CH_3}$$

CH₃ ⟵ methyl group (substituent)

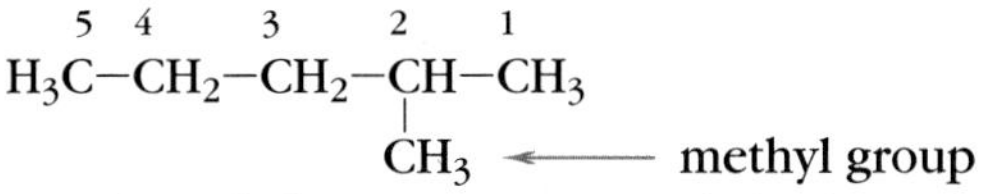

2-methylpentane

(b)
$$\overset{7}{H_3C}-\overset{6}{CH_2}-\overset{5}{CH_2}-\overset{4}{CH_2}-\underset{\underset{\displaystyle 1\ CH_3}{\underset{\displaystyle |}{2\ CH_2}}}{\overset{3}{CH}}-CH_3$$

⟵ methyl group (substituent)

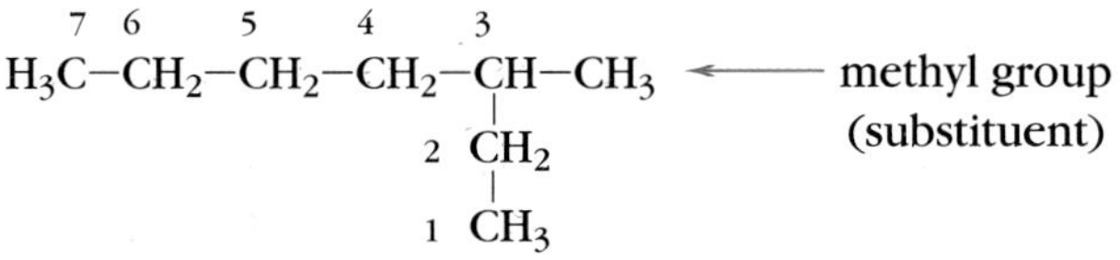

3-methylheptane

TABLE 9-3   Some Common Alkyl Groups

Name	Group
Methyl	$-CH_3$
Ethyl	$-CH_2-CH_3$
n-Propyl	$-CH_2-CH_2-CH_3$
n-Butyl	$-CH_2-CH_2-CH_2-CH_3$
Isopropyl	$-\underset{\underset{\displaystyle CH_3}{|}}{\overset{\overset{\displaystyle CH_3}{|}}{C}}-H$

In the names of compounds, the numbers are separated from words by a hyphen, and the parent name is placed last. In compound (a) a methyl group is attached to carbon number 2; in compound (b) a methyl group is attached to carbon number 3. Thus the names of the compounds are (a) 2-methylpentane and (b) 3-methylheptane.

EXAMPLE 9-3   Name the following compound:

$$CH_3-CH-CH_2-CH_2-CH_3$$
$$\quad\quad\;\; | $$
$$\quad\quad\; CH_3$$

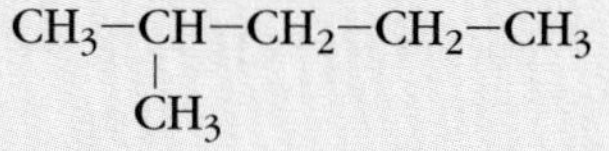

Solution:

1. Determine the name of the parent compound by finding the longest continuous chain of carbon atoms.

The longest chain is five carbons long. Therefore, the parent compound is a pentane.

2. Number the longest chain beginning with the end closest to a branch. Use these numbers to designate the location of the groups (or substituents) at the branch.

The methyl is on the 2-position.

3. The compound is 2-methylpentane.

PRACTICE EXERCISE: Name the following:

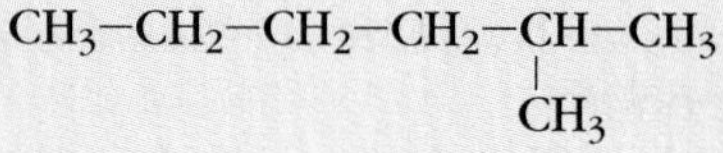

$$CH_3-CH_2-CH_2-CH_2-CH-CH_3$$
$$\quad\quad\quad\quad\quad\quad\quad\quad\quad\quad | $$
$$\quad\quad\quad\quad\quad\quad\quad\quad\quad CH_3$$

Answer: 2-methylhexane

---

3. *When two or more substituents are present on the same carbon atom, use the number of that carbon atom twice. List the substituents alphabetically.*

   Thus, the compound shown below is 3-ethyl-3-methylhexane.

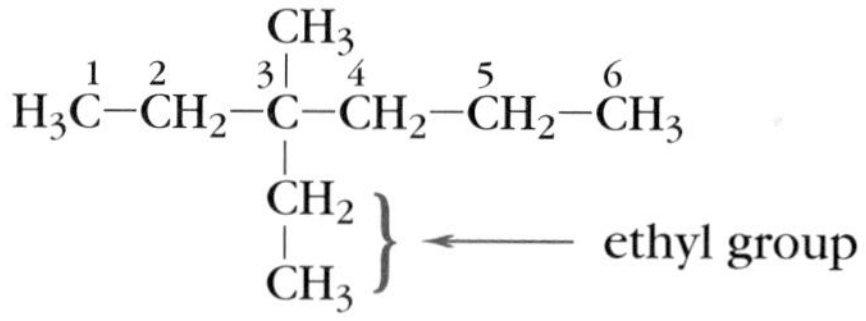

$$\begin{array}{c} CH_3 \\ \,\,\,\,1\,\,\,\,\,2\quad 3|\;\;4\;\;\;\;5\;\;\;\;6 \\ H_3C-CH_2-C-CH_2-CH_2-CH_3 \\ | \\ CH_2 \\ | \\ CH_3 \end{array} \Big\} \longleftarrow \text{ethyl group}$$

3-ethyl-3-methylhexane

4. *When two or more substituents are identical, indicate this fact by the use of the prefixes* di-, tri-, tetra-, *and so on. Commas are used to separate numbers from each other.*

   Thus the following compounds are 2,2,4-trimethylhexane and 3-ethyl-2,4-dimethyloctane:

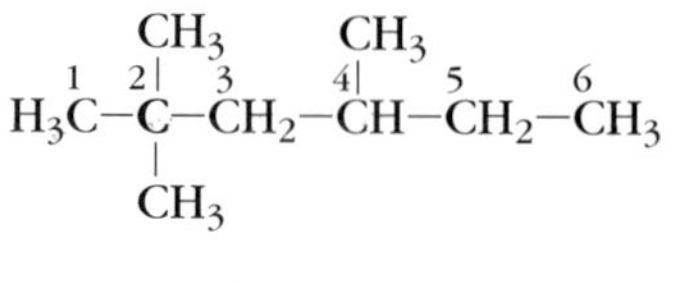

$$\begin{array}{c} CH_3\quad\;\; CH_3 \\ 1\;\;\;2|\;\;\;3\;\;\;\;4|\;\;\;5\;\;\;\;6 \\ H_3C-C-CH_2-CH-CH_2-CH_3 \\ | \\ CH_3 \end{array}$$

2,2,4-trimethylhexane

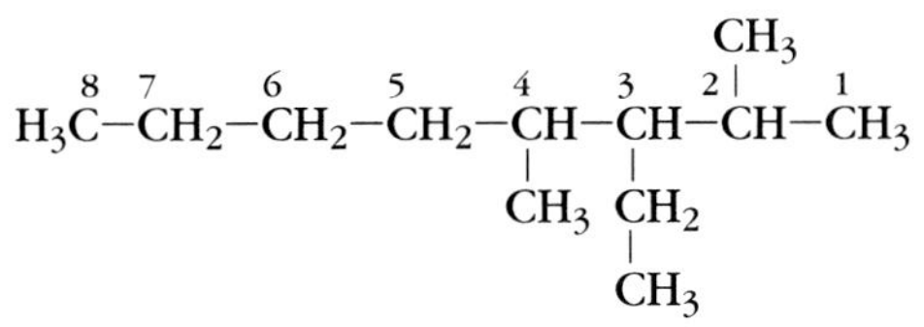

$$\underset{8}{H_3C}-\underset{7}{CH_2}-\underset{6}{CH_2}-\underset{5}{CH_2}-\underset{4}{CH}-\underset{3}{CH}-\underset{2}{\overset{\displaystyle CH_3}{|}}{CH}-\underset{1}{CH_3}$$

with CH₃ and CH₂ (CH₃) branches

3-ethyl-2,4-dimethyloctane

In the examples in this section, the left-hand terminal methyl group has been written H_3C-; when writing condensed formulas, it is written CH_3.

EXAMPLE 9-4 Name the following compound:

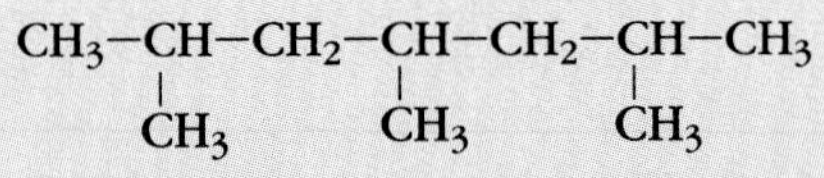

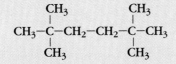

Solution:

1. Determine the name of the parent compound by finding the longest continuous chain of carbon atoms.

The longest chain is seven carbon atoms long. Therefore, the parent compound is a heptane.

2. Number the longest chain beginning with the end closest to a branch. Use these numbers to designate the location of the groups (or substituents) at the branch.

There are methyl groups at positions 2, 4, and 6.

Trimethyl- indicates three methyl groups.

3. The name of the compound is 2,4,6-trimethylheptane.

PRACTICE EXERCISE: Name the following compound:

$$CH_3-\overset{\displaystyle CH_3}{\underset{\displaystyle CH_3}{\overset{|}{\underset{|}{C}}}}-CH_2-CH_2-\overset{\displaystyle CH_3}{\underset{\displaystyle CH_3}{\overset{|}{\underset{|}{C}}}}-CH_3$$

Answer: 2,2,5,5-tetramethylhexane

Reactions of Alkanes

Alkanes are relatively unreactive. Like all hydrocarbons, alkanes undergo **combustion**, but for this to occur, a flame or spark is required to initiate bond cleavage. When burned in air, alkanes form carbon dioxide and water as shown below for methane and butane.

$$CH_4 + 2\,O_2 \longrightarrow CO_2 + 2\,H_2O + heat$$
$$2\,C_4H_{10} + 13\,O_2 \longrightarrow 8\,CO_2 + 10\,H_2O + heat$$

Once initiated, the reactions are exothermic (Chapter 7) and produce a considerable amount of heat. They are the source of the energy we obtain when natural gas, gasoline, and fuel oil are burned (Chapter 14).

Combustion is a chemical reaction in which heat is produced. It usually refers to the combination of a fuel with atmospheric oxygen.

Alkenes

Alkenes are hydrocarbons that have one or more carbon-carbon double bonds ($-C=C-$). The two simplest are **ethene** (C_2H_4), commonly called ethylene, and **propene** (C_3H_6), commonly called propylene. In forming the double bond, the two carbon atoms share two pairs of electrons to acquire the stable octet configuration.

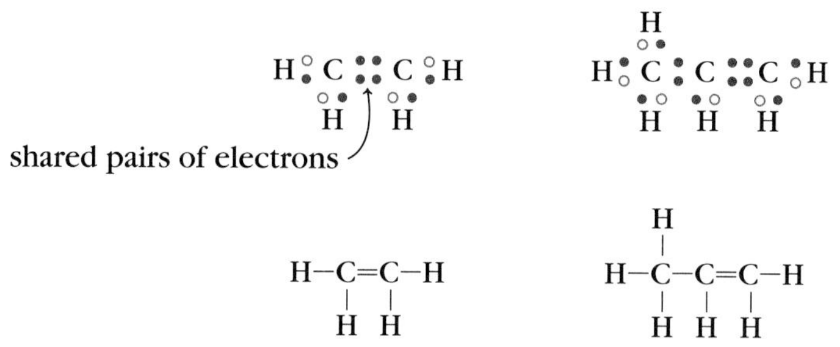

H°C ⦂⦂ C°H   H°C⦂C⦂⦂C°H (with H above center C)

shared pairs of electrons

$$H-C=C-H$$
$$H \; H$$
ethene
(ethylene)

$$H-C-C=C-H$$
propene
(propylene)

Different ways of representing the structure of alkenes are shown in Figure 9-6. Alkenes are identified by the suffix -*ene*. Like alkanes, they form a homologous series. The general formula is C_nH_{2n}.

Alkenes are called **unsaturated hydrocarbons** because they do not contain the maximum possible number of hydrogen atoms. In the presence of a catalyst such as platinum (Pt), they react with hydrogen to form the saturated parent alkane.

$$H_2C=CH_2 + H_2 \xrightarrow{\text{Pt}} H_3C-CH_3$$

ethene   ethane

	Ethene (Ethylene)	Propene (Propylene)
Condensed structural formula	C_2H_4 $CH_2=CH_2$	C_3H_6 $CH_3CH=CH_2$
Expanded structural formula		
Ball-and-stick model		
Space filling model		

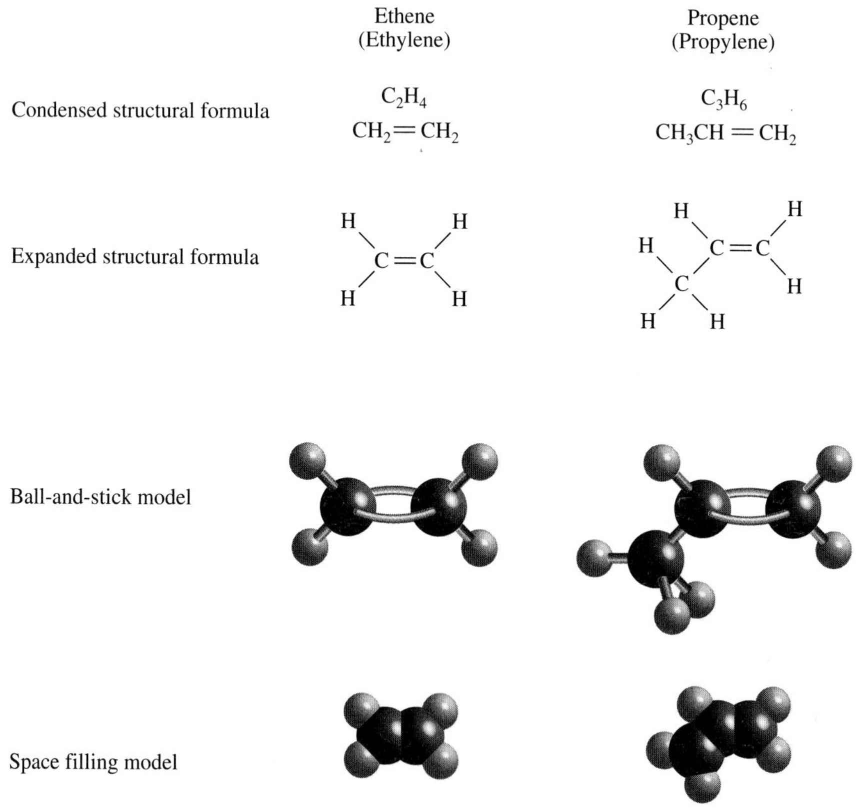

Figure 9-6 The first two members of the alkene series.

The third member of the alkene series is **butene** (C_4H_8). There are two possible positions for the double bond, and two isomers exist:

$$\overset{1}{H_2}C=\overset{2}{C}H-\overset{3}{C}H_2-\overset{4}{C}H_3 \qquad \overset{1}{H_3}C-\overset{2}{C}H=\overset{3}{C}H-\overset{4}{C}H_3$$

1−butene   2−butene

In naming alkenes, the first rule used for naming alkanes is modified: *The longest hydrocarbon chain is numbered from the end that will give the carbon-carbon double bond (−C=C−) the lowest number. Substituents are then numbered and named as for alkanes except that the ending -ene is used instead of -ane.* Thus the compounds shown below are 7-chloro-3-heptene and 4-ethyl-5-methyl-2-hexene.

$$\overset{1}{H_3}C-\overset{2}{C}H_2-\overset{3}{C}H=\overset{4}{C}H-\overset{5}{C}H_2-\overset{6}{C}H_2-\overset{7}{C}H_2-Cl$$

7-chloro-3-heptene

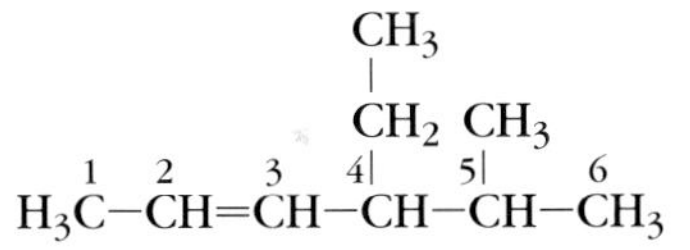

$$\begin{array}{c} CH_3 \\ | \\ CH_2 \quad CH_3 \\ \overset{1}{H_3}C-\overset{2}{C}H=\overset{3}{C}H-\overset{4}{\underset{|}{C}}H-\overset{5}{\underset{|}{C}}H-\overset{6}{C}H_3 \end{array}$$

4-ethyl-5-methyl-2-hexene

Ethene is the most important raw material in the organic chemical industry. It is produced in enormous quantity by the "cracking" of petroleum (Chapter 14), and it is used primarily to make the plastic polyethylene (Chapter 11). It is also the starting material for the manufacture of ethylene glycol (the major component of antifreeze in automobile radiators), other plastics, and many chemicals. Propene is important in the production of polypropylene and other plastics.

Ethene has a natural source: It is produced by fruits and causes the change in skin color that occurs as fruits ripen, for example, when tomatoes change from green to red. To avoid damage in transit, many producers pick fruit when it is green, ship it to its destination, and then treat it with ethene gas to produce the proper ripe color. The flavor, however, is generally considered to be inferior to naturally ripened fruit.

Notice that in all the formulas we have written, every carbon atom has 4 bonds.

In the United States, ethene is produced in greater quantity than any other organic chemical. Approximately 8 million tons are produced annually.

EXAMPLE 9-5   Name the following compound:

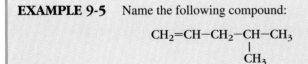

Solution:
1. The longest chain has five carbon atoms. The compound is a pentene.
2. The double bond is between carbon 1 and carbon 2.
3. The methyl group is on carbon 4.
4. The name of the compound is 4-methyl-1-pentene.

PRACTICE EXERCISE: Name the following compound:

$$CH_2=CH-CH_2-CH_2-CH_2-CH_2-CH_2-CH_3$$

Answer: 1-octene

Cis-Trans Isomerism

Groups joined by a single carbon-carbon bond can rotate freely about the bond with respect to each other, at ordinary temperatures. For example, the two methyl groups in ethane (Fig. 9-4) can rotate freely about the single C—C bond. Groups joined by a C=C double bond, however, are restricted and cannot rotate. The two CH_2 groups in ethene (Fig. 9-6) are locked together so that they lie in the same plane.

If one of the hydrogen atoms attached to each of the two carbon atoms in ethene is replaced by a chlorine atom, two distinct compounds are formed:

cis-1,2-dichloroethene   trans-1,2-dichloroethene

The prefixes *cis* and *trans* are derived from the Latin for "on this side" and "across."

The two compounds have the same name—dichloroethene—but are distinguished from each other by the prefixes *cis* and *trans*.

Although they have the same formula, they are not identical (that is, they cannot be superimposed one upon the other so that all atoms and bonds coincide), and their physical properties are different.

	cis-1,2-dichloroethene	trans-1,2-dichloroethene
Boiling point	60.3°C	47.5°C
Melting point	−80.5°C	−50°C

The two compounds are not structural isomers because the atoms attached to each atom in the two compounds are the same in both cases. For example, in both isomers, each carbon atom is joined to another carbon atom, a chlorine atom, and a hydrogen atom. Only the spatial arrangement is different. This type of isomerism is called **cis-trans isomerism**.

If two identical groups are attached to the same carbon atom in the double bond, cis-trans isomerism is not possible. For example, 1,1-dichloroethene—the *structural* isomer of the cis- and trans- isomers shown above—exists in just one form and cannot form cis- and trans- isomers.

Ethyne
(Acetylene)

C_2H_2

Condensed structural formula $CH \equiv CH$

Expanded structural formula $H—C \equiv C—H$

Ball-and-stick model

Space filling model

Figure 9-7 Different ways of representing the alkyne ethyne, better known as acetylene.

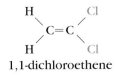

1,1-dichloroethene

Alkynes

Alkynes are hydrocarbons that have one or more triple bonds ($-C\equiv C-$). The simplest alkyne is **ethyne** (C_2H_2), better known by its common name, **acetylene**. The next member of the series is **propyne** (C_3H_4). In forming the triple bond, the two carbon atoms share three pairs of electrons to acquire the stable octet configuration:

Figure 9-8 The high temperature of an oxyacetylene torch can be used to cut steel.

$$HC\equiv CH \qquad CH_3-C\equiv CH$$

ethyne (acetylene)   propyne

Different ways of representing ethyne are shown in Figure 9-7.

Acetylene has many industrial uses. When burned in oxyacetylene torches it produces a very hot flame (about 3000°C, or 5400°F) that is used for cutting and welding metals (Fig. 9-8). It is also an important starting material for the production of plastics and synthetic rubber.

Saturated Hydrocarbon Rings (Cycloalkanes)

Carbon atoms can join together in rings as well as in straight chains and branched chains. Alkanes joined in rings are called **cycloalkanes**. The structures of the first four are shown in Figure 9-9. In the simplified depictions of the formulas, each corner represents a carbon atom attached to two hydrogen atoms, and the lines represent C–C single bonds.

Stability increases from cyclopropane to cyclohexane as the bond angles increase. Cyclohexane is important as the parent compound of glucose and many other sugars and carbohydrates that are vital for animal metabolism (Chapter 10).

AROMATIC HYDROCARBONS

Hydrocarbons can form rings in which some of the carbon-carbon bonds are double bonds as shown in the following 6-carbon ring compounds.

cyclohexene   1,3-cyclohexadiene   benzene

By far the most important unsaturated ring hydrocarbon compound is **benzene**, C_6H_6. Compounds derived from benzene are called **aromatic compounds** because many of the first ones described had distinctive pleasant scents (or aromas).

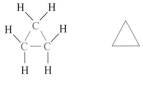

Cyclopropane

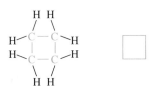

Cyclobutane

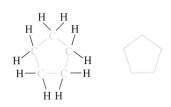

Cyclopentane

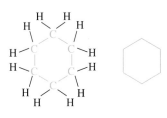

Cyclohexane

Figure 9-9 Expanded and simplified structures of the first four cycloalkanes.

The Structure of Benzene

Benzene was discovered by Michael Faraday in 1826, but its structure remained a mystery for 40 years. The molecular formula, C_6H_6, indicated a very unsaturated hydrocarbon, but benzene did not behave like the unsaturated alkenes or alkynes. Then in 1865, the German chemist August Kekulé proposed a ring of six carbon atoms joined by alternating single and double bonds, with one hydrogen atom bonded to each carbon.

Kekulé said that he solved the riddle of the structure of benzene after he had a dream in which he saw rows of atoms twisting and turning in snake-like motion. Suddenly, he saw one snake whirling with its tail in its mouth, giving him the idea that the carbon atoms in benzene must be joined in a ring.

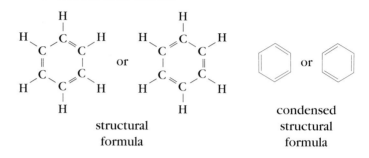

structural
formula

condensed
structural
formula

Although these structures explain much of the chemistry of benzene, they cannot accurately explain all the established experimental facts. We now know that the bonds joining the carbon atoms in the ring are neither single bonds nor double bonds. They are shorter than normal single bonds, but longer

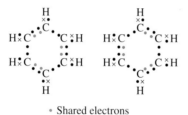

(a)   • Shared electrons

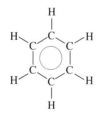

(b)

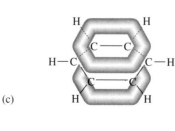

(c)

(d)

Figure 9-10 The benzene molecule. **(a)** The six carbon-carbon bonds in benzene are identical. The six electrons associated with the three double bonds are shared equally among the six atoms in the ring. **(b)** The circle in the center of the ring represents the electrons that are shared equally among the six carbon atoms. **(c)** A representation of the delocalized electrons above and below the plane of the benzene ring. **(d)** The simplified form in which the benzene molecule is usually written.

than normal double bonds, and all are identical. The six electrons associated with three double bonds are shared equally among all the carbon atoms in the ring as depicted in Figure 9-10a or more simply in Figure 9-10b. The shared electrons are said to be **delocalized**. They can be visualized as moving above and below the flat plane of the carbon ring as shown in Figure 9-10c. This arrangement leads to increased stability, and aromatic compounds are more stable than compounds with regular carbon-carbon double bonds. The formula for benzene is usually written in condensed form without the carbon and hydrogen atoms, and with a circle in the center of the ring to represent the delocalized electrons (Fig. 9-10d).

Alkyl Derivatives of Benzene

Replacement of hydrogen atoms in benzene with alkyl groups gives rise to numerous other aromatic hydrocarbons. The simplest member of the family, methyl benzene (common name, toluene), is formed when one hydrogen atom is replaced with a methyl ($-CH_3$) group (Fig. 9-11). If two hydrogens are replaced by methyl groups, three different dimethylbenzenes (commonly called xylenes) can be formed.

The location of a second substituent relative to the first is indicated by numbering the ring carbons. If two substituents are present, as in the xylenes, the prefixes **ortho-** (*o-*), **meta-** (*m-*), and **para** (*p-*) are often used instead of numbers. For more than two substituents, numbers must be used. The numbers are arranged to give the lowest possible numbers to the substituents.

Benzene, toluene, and the xylenes are used extensively as raw materials for the manufacture of plastics, pesticides, drugs, and hundreds of other organic chemicals. The use of benzene as a laboratory solvent was discontinued in the 1970s after it was discovered that inhalation of its fumes lowered the white blood cell count in humans and caused leukemia in laboratory rats.

Polycyclic Aromatic Compounds

Benzene rings can fuse, or join together, to form polycyclic aromatic compounds (Fig. 9-12). The simplest one is naphthalene, which is used as a moth repellent. Additional benzene rings can be added to form a wide variety of fused-ring compounds including benzo (α) pyrene, a known carcinogen that is present in tobacco smoke.

▶ FUNCTIONAL GROUPS

The majority of organic compounds are derivatives of hydrocarbons, obtained by replacing one or more hydrogen atoms in the parent hydrocarbon molecule with a different atom or group of atoms. These atoms or groups of atoms, called functional groups, form the basis for classifying organic compounds. Most chemical reactions involving an organic molecule take place at the site of the functional group. The functional group, as the name implies, determines how the molecule functions as a whole. Molecules containing the same functional groups exhibit similar chemical behavior. The major classes of organic compounds that result from replacing hydrogen atoms in hydrocarbons with functional groups are shown in Table 9-4, together with examples from each class.

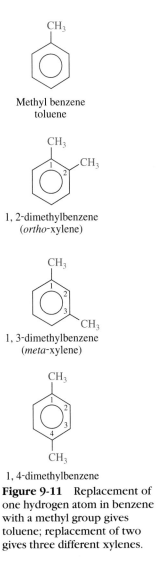

Methyl benzene
toluene

1, 2-dimethylbenzene
(*ortho*-xylene)

1, 3-dimethylbenzene
(*meta*-xylene)

1, 4-dimethylbenzene

Figure 9-11 Replacement of one hydrogen atom in benzene with a methyl group gives toluene; replacement of two gives three different xylenes.

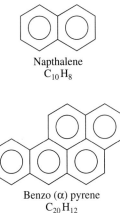

Napthalene
$C_{10}H_8$

Benzo (α) pyrene
$C_{20}H_{12}$

Figure 9-12 Two polycyclic aromatic compounds.

TABLE 9-4 Classes of Organic Compounds Based on Functional Groups

General Formula of Class*	Name of Class	Example	Name of Compound
R—X	Halide	CH_3—Cl	Chloromethane (methyl chloride)
R—OH	Alcohol	C_2H_5—OH	Ethanol (ethyl alcohol)
R—O—R'	Ether	C_2H_5—O—C_2H_5	Diethyl ether (ethyl ether)
$R—\overset{\overset{\textstyle O}{\|\|}}{C}—H$	Aldehyde	$H—\overset{\overset{\textstyle O}{\|\|}}{C}—H$	Methanal (formaldehyde)
$R—\overset{\overset{\textstyle O}{\|\|}}{C}—R'$	Ketone	$CH_3—\overset{\overset{\textstyle O}{\|\|}}{C}—CH_3$	Propanone (acetone)
$R—\overset{\overset{\textstyle O}{\|\|}}{C}—OH$	Carboxylic acid	$CH_3—\overset{\overset{\textstyle O}{\|\|}}{C}—OH$	Ethanoic acid (acetic acid)
$R—O—\overset{\overset{\textstyle O}{\|\|}}{C}—R'$	Ester	$C_2H_5—O—\overset{\overset{\textstyle O}{\|\|}}{C}—CH_3$	Ethyl ethanoate (ethyl acetate)
$R—N{\overset{\textstyle H}{\diagdown}}_{\diagup H}$	Primary amine	$C_2H_5—N{\overset{\textstyle H}{\diagdown}}_{\diagup H}$	Ethylamine
$R—\overset{\overset{\textstyle O}{\|\|}}{C}—N{\overset{\textstyle H}{\diagup}}_{\diagdown H}$	Simple amide	$CH_3—\overset{\overset{\textstyle O}{\|\|}}{C}—N{\overset{\textstyle H}{\diagup}}_{\diagdown H}$	Acetamide

*R represents an H atom or a carbon-containing group, often an alkyl group such as –CH_3 or –C_2H_5. R' may be the same as R or different.

In the remainder of this chapter, we will examine important examples from these major functional group classes.

Organic Halides

Organic halides are hydrocarbon derivatives in which hydrogen atoms in the parent hydrocarbon have been replaced by one or more halogen atoms—fluorine, chlorine, bromine, or iodine. Halogens, which are in Group VIIA of the periodic table, form a stable covalent bond by sharing one of their seven valence electrons, as shown below for chloroethane:

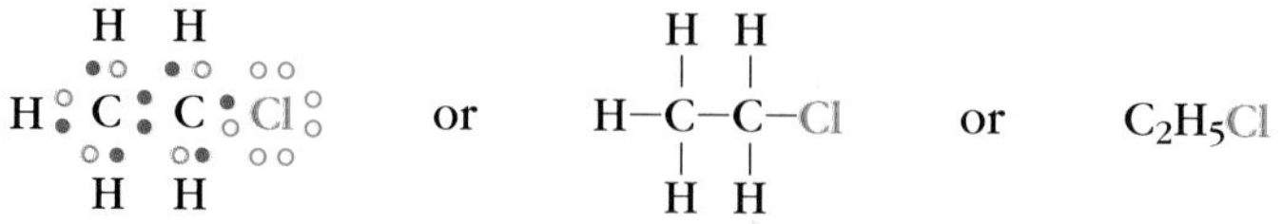

$$H\overset{H}{\underset{H}{\overset{\bullet\circ}{\underset{\circ\bullet}{C}}}}\overset{H}{\underset{H}{\overset{\bullet\circ}{\underset{\circ\bullet}{C}}}}\overset{\circ\circ}{\underset{\circ\circ}{Cl}}\overset{\circ}{\circ} \quad \text{or} \quad H—\overset{\overset{\textstyle H}{\|}}{\underset{\underset{\textstyle H}{\|}}{C}}—\overset{\overset{\textstyle H}{\|}}{\underset{\underset{\textstyle H}{\|}}{C}}—Cl \quad \text{or} \quad C_2H_5Cl$$

1-chloroethane (or *n*-ethyl chloride)

According to the IUPAC system, organic halides are named by writing the name of the parent hydrocarbon preceded by the appropriate designation for the halogen—fluoro-, chloro-, bromo-, or iodo-. In many cases, common names are often used. For example, trichloromethane is usually called chloroform (Fig. 9-13).

Organic halides are very versatile compounds and have been used in numerous ways. Some examples are shown in Figure 9-13. In recent years, some of these halides have been shown to be health hazards, and their use has been discontinued. For years, chloroform, $CHCl_3$, and carbon tetrachloride, CCl_4, both excellent solvents for grease and oils, were used in the dry-cleaning industry and in the laboratory. Since they were shown to be toxic and carcinogenic, they have been replaced by safer solvents. We will see in Chapter 13 that chlorofluorocarbons (CFCs)—substituted alkanes, such as Freon-12, that contain both fluorine and chlorine—were widely used as coolants and aerosol propellants before it was discovered that their release into the atmosphere was a factor in the depletion of the ozone layer. Other organic halides that have caused environmental problems include DDT and the polychlorinated biphenyls (PCBs) (Chapter 12). The two plastics polyvinyl chloride (PVC) and Teflon (the nonstick coating on cookware) (Chapter 11) are made from the alkyl halide derivatives of chloroethene and tetrafluoroethene, respectively.

EXAMPLE 9-6   Give the IUPAC name of the following alkyl halide.

$$CH_3-CH_2-CH_2-Cl$$

Solution:
1. The parent hydrocarbon has three carbon atoms. The compound is a propane.
2. The chlorine atom is on a terminal carbon atom.
3. Carbon atoms are numbered so that the carbon attached to the substituent (Cl) is given the lowest possible number.
4. The compound is 1-chloropropane.

PRACTICE EXERCISE: Write the formula of bromomethane (methyl bromide), CH_3Br:

Answer:

H
|
H—C—Br
|
H

Trichloromethane
(chloroform)

Cl—C—Cl with Cl

Tetrachloromethane
(carbon tetrachloride)

Dichlorodiflouromethane
(Freon-12)

$$CH_2=C-Cl$$ with H

Chloroethane
(vinyl chloride)

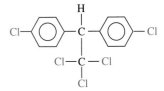

Dichlorodiphenyl trichloroethane
(DDT)

Figure 9-13   Examples of organic halides.

Alcohols

Alcohols are hydrocarbon derivatives in which hydrogen atoms in the parent hydrocarbon have been replaced by one or more hydroxyl groups (–OH) (Fig. 9-14). Oxygen has six valence electrons, and in forming an alcohol, it acquires a stable octet of electrons by forming one covalent bond with a carbon atom and one with a hydrogen atom:

H
|
H : C : O : H
|
H

methanol (methyl alcohol)

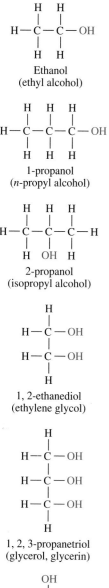

Ethanol
(ethyl alcohol)

1-propanol
(*n*-propyl alcohol)

2-propanol
(isopropyl alcohol)

1, 2-ethanediol
(ethylene glycol)

1, 2, 3-propanetriol
(glycerol, glycerin)

Phenol

2-methylphenol
(o-cresol)

Figure 9-14 Examples of alcohols.

The IUPAC name of an alcohol is obtained from the parent alkane by replacing the final -*e* with the suffix *ol*. Thus, methane becomes methanol, ethane becomes ethanol, and propane becomes propanol.

Alcohols, like water, contain a polar –OH group, and as a result, hydrogen bonding (Chapter 5) can occur between hydrogen and oxygen atoms in adjacent molecules, as shown for methanol in Figure 9-15. When alcohols are converted from the liquid state to the gaseous state, energy is needed to break the hydrogen bonds. As a result, the boiling points of alcohols are higher than are the boiling points of the alkanes from which they are derived (Table 9-5).

Alcohols with low molecular weights such as methanol and ethanol are very soluble in water because hydrogen bonding occurs between water and alcohol molecules. Alcohols with higher molecular weights are less soluble because the increasing size of the nonpolar part of the molecule disrupts the hydrogen bonding network.

Methanol is known as **wood alcohol** because for many years it was produced by heating hardwoods such as maple, birch, and hickory to high temperatures in the absence of air. The present methods of production from coal and coal gas will be examined in Chapter 14. Methanol has many industrial uses. It is used in the manufacture of plastics and other polymers, as an antifreeze in windshield washer fluid, and as a gasoline additive. Its use as a fuel to replace gasoline will be discussed in Chapter 15. Methanol is extremely toxic. Drinking quite small quantities can result in temporary blindness, permanent blindness, or even death.

Ethanol, ethyl alcohol (Fig. 9-14), is the active ingredient in alcoholic beverages. It is produced by the **fermentation** of carbohydrates (sugars and starches) in fruits and other plant sources, using a process that was discovered thousands of years ago. During fermentation, microorganisms such as yeasts convert sugar to alcohol and carbon dioxide:

$$C_6H_{12}O_6 \longrightarrow 2\ C_2H_5OH + 2\ CO_2$$

Ethanol is called **grain alcohol** because it can be produced from grains such as barley, wheat, corn, and rice. Beer is usually made from barley with the addition of hops (Chapter 10). Wine is made from grapes and other fruits.

The percentage of alcohol in the fermentation product is determined by the enzymes present. At alcohol concentrations above about 15 percent, the enzymes begin to be destroyed, and the reaction can proceed no further. The

TABLE 9-5 Comparison of Boiling Points of Alcohols and Their Corresponding Alkanes

Alcohol	Bp (°C)	Alkane	Bp (°C)
Methanol	65.0	Methane	-162
Ethanol	78.5	Ethane	- 88.5
1-Propanol	97.4	Propane	- 42
2-Propanol	82.4		

alcohol content can be increased by distillation. For example, whiskey is made by distilling fermented grains. The *proof* of an alcoholic beverage is equal to twice the percentage (by volume) of the ethanol. Thus, 90-proof rum is 45 percent ethyl alcohol.

Ethanol is not as toxic as methanol, but one pint of pure ethanol, if ingested rapidly, is likely to prove fatal for most people. Ethanol is quickly absorbed into the bloodstream; excessive use causes damage to the liver and the neurological system and leads to physiological addiction.

In addition to being a constituent of alcoholic beverages, ethanol has many industrial uses. It is used in the preparation of many organic chemicals and is an important industrial solvent. It is also used by the pharmaceutical industry as an ingredient in cough syrups and other medications and as an antiseptic for skin disinfection.

Ethanol used in industry is produced by the high-pressure hydration of ethylene, not by fermentation:

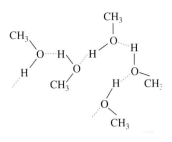

---- Hydrogen bonds

Figure 9-15 Hydrogen bonding in the alcohol methanol.

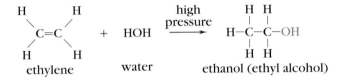

Alcoholic beverages are heavily taxed, but alcohol intended for industrial use is not. To avoid the tax, manufacturers must **denature** the alcohol; that is, render it unfit for human consumption by the addition of a small amount of a toxic substance. Methanol or benzene is often added, because these substances do not alter the solvent properties of the alcohol.

Propanol, the next member of the series following ethanol, can exist in two forms: 1-propanol (*n*-propyl alcohol) and 2-propanol (isopropyl alcohol) (Fig. 9-14). Drugstore rubbing alcohol is a 70 percent solution of 2-propanol and is an effective disinfectant.

Important alcohols with more than one hydroxyl group in the molecule include **ethylene glycol** (1,2-ethanediol) and **glycerol** (glycerin or 1,2,3-propanetriol) (Fig. 9-14). Ethylene glycol, with its two polar −OH groups, is very soluble in water and can be mixed with it in any desired proportion. Because of hydrogen bonding, it has a relatively high boiling point. These properties make it an ideal choice for use as the main ingredient in permanent antifreeze. Glycerol is a nontoxic, clear, syrupy liquid. With its three −OH groups, it has an affinity for moisture. For this reason, it is often added to pharmaceutical preparations such as hand lotions, shaving creams, and soaps to help keep skin moist and soft. Glycerol is also used as a lubricant and in the manufacture of explosives and a variety of other chemicals.

The simplest aromatic alcohol is **phenol** (Fig. 9-14), which was the first widely used disinfectant. It was introduced in 1867 by the British surgeon Joseph Lister. However, because it could burn skin if not in very dilute solution, it was soon replaced by safer related compounds. A methyl derivative of phenol, o-cresol, is the main ingredient in the wood preservative creosote.

The term proof orginated in England in the seventeenth century. To prove to their customers that their product was not diluted, whiskey dealers would pour a small amount of it onto gunpowder. If the mixture could be ignited, it was proof that water had not been added to thew whiskey.

CONSUMER BRIEF

Aspirin: Wonder Drug

Since ancient times humans have sought substances that would relieve pain. Alcohol, opium, cocaine, and marijuana have all been used for this purpose.

The ancient Greeks knew that fever could be lowered by chewing willow bark. In the early 1800's chemists successfully obtained an active extract from the tree's bark, and by 1860, the active ingredient had been identified as salicylic acid. Salicylic acid began to be used as a painkiller, but it had a very sour taste and it damaged tissues in the throat and mouth. In the 1890's, the Bayer Company found that by reacting salicylic acid with acetic anhydride, acetylsalicylic acid, the compound we know as aspirin, was formed. Aspirin was as effective as salicylic acid in relieving pain, but did not have its disadvantages.

Pain-relieving substances are known as *analgesics*, those that lower fever are called *antipyretics*; *anti-inflammatory* agents reduce swelling. Aspirin, which is an analgesic, an antipyretic, and an anti-inflammatory agent, has been taken by more people around the world than any other drug.

Aspirin reduces pain in three ways. First, it reduces the swelling that occurs at the site of an injury or infection. This reduces pressure on the nerve, and thus relieves pain. Second, it reduces the heat generated by inflammation and by the flow of blood in constricted blood vessels at the injury site. Third, it prevents the formation of prostaglandins, substances that play a part in the sensation of pain and its transmission through the nervous system.

In the United States, more than $100 million is spent annually on thousands of tons of aspirin. An average aspirin tablet contains 300-350 mg of acetylsalicylic acid that, together with a binder, is pressed into the common tablet shape with which we are familiar. "Extra strength" aspirin tablets contain more acetylsalicylic acid, usually about 500 mg, but otherwise are no different from ordinary tablets. Aspirin is often formulated in combination with other substances. For example, Anacin tablets that advertisements claim contain "more medicine", contain caffeine in addition to acetylsalicylic acid. Products with additives are considerably more expensive than regular aspirin, and most are no more effective. One could get the same effect obtained from an Anacin tablet by drinking a cup of coffee after taking an aspirin.

Aspirin cannot act to relieve pain until it enters the bloodstream. The acetylsalicylic acid in an aspirin tablet is released as the tablet dissolves in the stomach. Dissolution, which is the slowest step in getting relief to the body, is controlled by the acidity, or pH, of the stomach. Many companies produce "buffered" aspirin to accelerate the dissolution of the aspirin tablet in the stomach. Bufferin tablets, for instance, contain acetylsalicylic acid and the base magnesium carbonate ($MgCO_3$). The base neutralizes the hydrochloric acid in the stomach, which causes the tablets to disintegrate and the acetylsalicylic acid to dissolve. Other manufacturers use other bases, including magnesium hydroxide, $Mg(OH)_2$, and aluminum hydroxide, $Al(OH)_3$. Once the aspirin has dissolved, it is absorbed into the bloodstream.

Although buffered aspirin dissolves faster than nonbuffered aspirin, clinical tests have failed to show that it relieves pain more quickly. The buffer, however, helps prevent gastrointestinal bleeding, aspirin's only undesirable side effect. Although bleeding is usually slight and unimportant, in some people it may be a serious problem. Those who suffer stomach upset from aspirin prefer the buffered product.

Although aspirin is the the most likely drug to be found in the home, it can be toxic. As few as 50 aspirin tablets can kill a 20 pound child, and prior to 1970, aspirin was responsible for poisoning more children under five than any other chemical. Enactment of the Poison Prevention Packaging Act of 1970, which mandated the introduction of child-proof containers, has had a dramatic effect in reducing the number of children poisoned with aspirin.

Questions

1. The "extra medicine' that is advertised on Anacin bottles is caffeine. Would it be better to take aspirin or Anacin for a headache just before going to bed?

2. Should people with stomach ulcers take aspirin?

Ethers

Compounds that have two hydrocarbon groups attached to a single oxygen atom are called **ethers** (Table 9-4). They have the general formula, R—O—R', where R and R' may or may not be the same (Fig. 9-16). The best known ether is diethyl ether, CH_3CH_2—O—CH_2CH_3, which is commonly called ether.

Ether acts as a general anesthetic; its introduction for this purpose in the mid-1840s revolutionized medicine (Fig. 9-17). Surgical procedures that previously could not be performed because of the pain and shock involved became possible. General anesthetics cause unconsciousness and insensitivity to pain by temporarily depressing the central nervous system. Ether is not an ideal anesthetic; it is highly flammable, and inhalation produces some unpleasant side effects, such as nausea and irritation of the respiratory tract. It has now been replaced by other more efficient anesthetics including a related ether, neothyl (Fig. 9-16).

A commercially important ether is methyl tertiary-butyl ether (MTBE) (Fig. 9-16). Since the phaseout of the antiknock agent tetraethyl lead, various other agents, including MTBE, have been investigated as gasoline additives to increase octane rating (Chapter 14).

The word ether is derived from the Greek word *aither* meaning "to ignite".

EXAMPLE 9-7   Indicate the functional group in each of the following. Identify the compounds as an ether, alcohol or phenol.

a. OH
Cl

b. CH_3OCH_3

c. OH
CH_3CHCH_3

Solution:
a. The OH functional group is attached to a benzene ring. This compound is a phenol. It also has a chlorine. It is ortho-chlorophenol.
b. The O functional group is between two carbon atoms. This compound is an ether. It is dimethylether.
c. The OH functional group is attached to a carbon atom. This compound is an alcohol. The OH is attached to carbon 2. The compound is 2-propanol.

PRACTICE EXERCISE: Draw the structure of 1-propanol.

Answer: HO—CH_2—CH_2—CH_3

Aldehydes and Ketones

Aldehydes and **ketones** are related families of compounds (Table 9-4). Both contain a **carbonyl group** in which a carbon atom is joined by a double bond to an oxygen atom. In aldehydes, the carbonyl carbon is attached to at least one hydrogen atom; in ketones, it is attached to two carbon atoms:

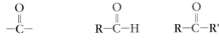

carbonyl group   aldehyde   ketone

CH_3—CH_2—O—CH_2—CH_3

Ether
(diethyl ether)

CH_3—O—CH_2—CH_2—CH_3

Methyl propyl ether
(neothyl)

CH_3—O—C—CH_3 with CH_3 groups above and below

MTBE
(methyl tertiary-butyl ether)

Figure 9-16   Examples of ethers.

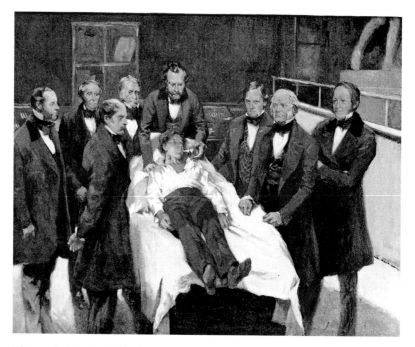

Figure 9-17   In 1846, the Boston dentist William Morton gave a public demonstration of the use of ether as an anesthetic.

The simplest aldehyde is **formaldehyde** (Fig. 9-18). As a 40 percent solution in water, called *formalin*, formaldehyde is used as a preservative for biological specimens and as an embalming fluid. It is also a starting material for the manufacture of certain polymers. Formaldehyde has a disagreeable odor and is a suspected carcinogen.

In the IUPAC naming system, the suffixes *-al* and *-one* are used to identify aldehydes and ketones, respectively, but many are better known by their common names.

Aldehydes and ketones are made by oxidizing the appropriate alcohols, as shown below for the preparation of propanal and propanone (acetone).

$$CH_3-CH_2-CH_2-OH \quad \xrightarrow{\text{oxidation}} \quad CH_3-CH_2-\overset{\overset{\displaystyle O}{\|}}{C}-H$$

propanol   propanal
(propyl alcohol)

$$\overset{\overset{\displaystyle OH}{|}}{CH_3-CH-CH_3} \quad \xrightarrow{\text{oxidation}} \quad CH_3-\overset{\overset{\displaystyle O}{\|}}{C}-CH_3$$

isopropanol   propanone
(isopropyl alcohol)   (acetone)

Acetone, the simplest ketone, is an excellent solvent. It is miscible (soluble in all proportions) with both water and nonpolar liquids and is used to remove varnish, paint, and fingernail polish. Many aromatic aldehydes and ketones have pleasant aromas (Fig. 9-18), and some, including almond and cinnamon, are used as food flavorings. The perfumes of many flowers—violets, for example—are due to the release of small amounts of ketones.

EXAMPLE 9-8   Identify each of the following as an aldehyde or ketone.

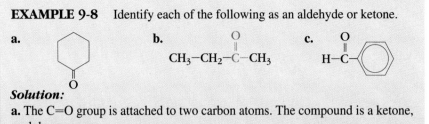

a.   b.   c.

Solution:
a. The C=O group is attached to two carbon atoms. The compound is a ketone, cyclohexanone.
b. The C=O group is attached to two carbon atoms. The compound is a ketone, 2-butanone.
c. A hydrogen atom is attached to the C=O group. The compound is an aldehyde, benzaldehyde.

PRACTICE EXERCISE: Identify the following as ketones or aldehydes:

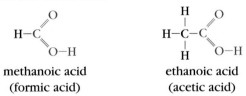

Answer: **a.** ketone **b.** aldehyde **c.** ketone

Carboxylic Acids

Organic acids contain the carboxyl group, a group containing a carbon atom that is double bonded to an oxygen atom and also bonded to a hydroxyl group:

$$-\overset{\overset{\text{O}}{\|}}{\text{C}}-\text{OH} \quad\quad \text{or} \quad\quad -\text{COOH}$$
carboxyl group

Examples of carboxylic acids are shown in Fig. 9-19.

In naming carboxylic acids, the final -e in the parent alkane is replaced with the suffix *-oic*, and the word *acid* is added. The two simplest carboxylic acids are methanoic acid and ethanoic acid, better known by their common names, formic acid and acetic acid.

methanoic acid   ethanoic acid
(formic acid)   (acetic acid)

Compared with inorganic acids such as hydrochloric acid (HCl) and nitric acid (HNO_3), carboxylic acids are weak acids (Chapter 8) and dissociate very little to form ions. Most are less than 2 percent ionized. The acidic hydrogen is the one in the −OH of the carboxyl group.

$$R-\overset{\overset{\text{O}}{\|}}{\text{C}}-\text{OH} \;\rightleftharpoons\; R-\overset{\overset{\text{O}}{\|}}{\text{C}}-\text{O}^-$$

The arrows $\rightleftharpoons$ indicate that the equilibrium favors the undissociated form. Like inorganic acids, carboxylic acids are neutralized by bases and form salts:

Formaldehyde
(methanal)

Acetaldehyde
(ethanal)

Benzaldehyde
(bitter almonds)

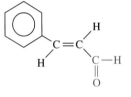

Cinnamaldehyde
(cinnamon)

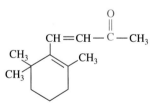

β-ionone
(scent of violets)

Figure 9-18 Examples of aldehydes and ketones.

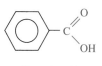

Benzoic acid

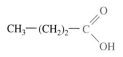

Butyric acid

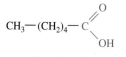

Caproic acid

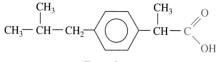

Ibuprofen

Oxalic acid

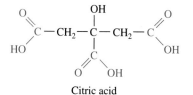

Citric acid

Figure 9-19   Examples of carboxylic acids.

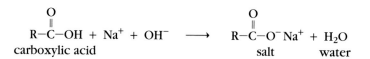

Sodium benzoate, the sodium salt of benzoic acid, is used as a food preservative.

Carboxylic acids have many natural sources. Formic acid is present in ants, various other insects, and nettles. It is the irritant injected under the skin by the sting of red ants, bees, and other insects.

The Latin word for ant is *formica*.

Vinegar, which is a 5 percent solution of acetic acid, is the most commonly encountered organic acid. It can be prepared by the aerobic fermentation of ethyl alcohol. This is an oxidation process in which acetaldehyde is formed as an intermediate:

$$\text{CH}_3\text{CH}_2\text{OH} \xrightarrow{\text{oxidation}} \text{CH}_3\text{CHO} \xrightarrow{\text{oxidation}} \text{CH}_3\text{COOH}$$

ethyl alcohol   acetaldehyde   acetic acid

Acetic acid is used as a starting material in the manufacture of textiles and plastics.

Some of the simple carboxylic acids have very unpleasant odors. Butyric acid is responsible for the smell of rancid butter and, in part, for body odor. The powerful odors of certain cheeses and of goats are due to longer-chain carboxylic acids such as caproic acid. Carboxylic acids with 12 or more carbon atoms are called **fatty acids** and will be studied in Chapters 10 and 17.

Many naturally occurring carboxylic acids contain more than one carboxyl group, and some also contain hydroxyl groups. Citric acid, which is found in all citrus fruits and gives them their tart flavor, contains three carboxyl groups. Oxalic acid, with two carboxyl groups, is toxic in large doses and is present in rhubarb, tomatoes, and other vegetables. It is useful for removing rust and ink stains.

The Latin words for butter and goat are butyrum and caprer.

Esters

Esters (Table 9-4) are derivatives of carboxylic acids in which the −OH of the carboxyl group is replaced by an -OR' group.

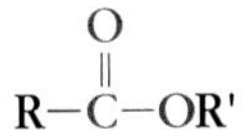

$$\begin{array}{c} \text{O} \\ \| \\ \text{R}-\text{C}-\text{OR'} \end{array}$$

For example, the ester ethyl acetate is formed by replacing the −OH group in acetic acid with an ethyl group.

The reaction between an alcohol and a carboxylic acid to yield an ester and water is called an esterification reaction.

$$\text{CH}_3-\text{COOH} \qquad\qquad \text{CH}_3-\text{COOCH}_2\text{CH}_5$$

acetic acid   ethyl acetate

The names of esters are based on the acids from which they are derived.

Esters are widely distributed in nature and are responsible for the tastes of many fruits and the perfumes of many flowers. Many esters have been synthesized for use as flavorings in foods and beverages. It is interesting to note that just a small change in an R group can significantly alter the flavor sensation of an ester. Apple and pineapple flavors, for example, differ from each other by a single −CH_2 group (Fig. 9-20).

Many fragrances in perfumes are esters. For example, benzyl acetate is the main constituent of oil of jasmine. Ethyl acetate is a useful solvent for removing lacquers, paints, and nail polish.

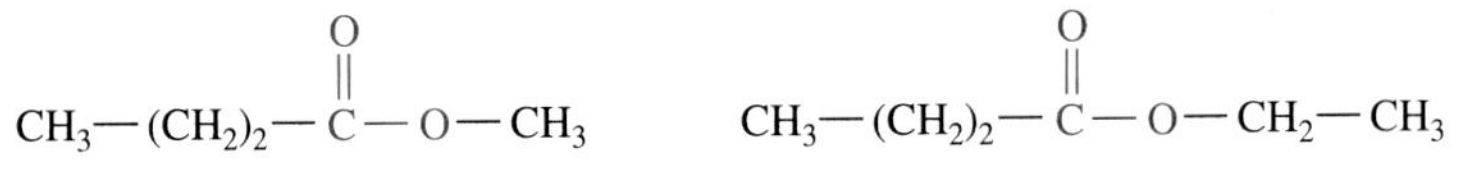

$$\begin{array}{cc} \begin{array}{c} \text{O} \\ \| \\ \text{CH}_3-(\text{CH}_2)_2-\text{C}-\text{O}-\text{CH}_3 \end{array} & \begin{array}{c} \text{O} \\ \| \\ \text{CH}_3-(\text{CH}_2)_2-\text{C}-\text{O}-\text{CH}_2-\text{CH}_3 \end{array} \end{array}$$

Methyl butanoate   Ethyl butanoate
(apple)   (pineapple)

Figure 9-20 The esters that give apples and pineapples their characteristic flavors differ by a single CH_2 group.

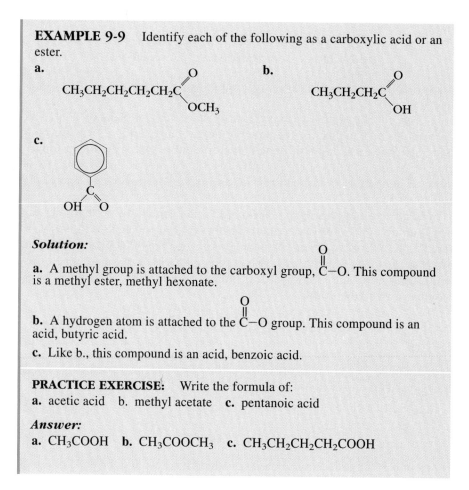

EXAMPLE 9-9   Identify each of the following as a carboxylic acid or an ester.

a.

$$CH_3CH_2CH_2CH_2CH_2C\overset{O}{\underset{OCH_3}{\diagup\diagdown}}$$

b.

$$CH_3CH_2CH_2C\overset{O}{\underset{OH}{\diagup\diagdown}}$$

c.

Solution:

a. A methyl group is attached to the carboxyl group, $C{-}O$. This compound is a methyl ester, methyl hexonate.

b. A hydrogen atom is attached to the $C{-}O$ group. This compound is an acid, butyric acid.

c. Like b., this compound is an acid, benzoic acid.

PRACTICE EXERCISE: Write the formula of:
a. acetic acid   b. methyl acetate   c. pentanoic acid

Answer:
a. CH_3COOH b. CH_3COOCH_3 c. $CH_3CH_2CH_2CH_2COOH$

CONSUMER BRIEF

Anti-Inflammatories: Use Aspirin or Nupe-It?

Inflammation is a swelling that occurs as a result of an overproduction of prostaglandins. These compounds, which were named for the prostate gland—the first organ from which they were isolated—occur in almost every organ of the body. They play important roles in the transmission of pain, the swelling of inflammations, and the onset of fever. Anti-inflammatory drugs relieve these conditions by interfering with the production of prostagladins.

In the past, aspirin was the drug of choice for the treatment of inflammation, and it remains one of the most useful drugs for treating arthritis, a disease characterized by inflamma-

tion of joints and connective tissue. Some people, however, are allergic to aspirin. Another problem is that prolonged use, as needed by people with arthritis, can lead to stomach disorders. For these people alternatives are now available.

Until 1984, when it was approved by the the Food and Drug Administration (FDA) for over-the-counter sales, the anti-inflammatory drug ibuprofen was only available as the prescription drug Motrin. Now it is available in generic form and under the trade names Advil and Nuprin, in addition to Motrin. Ibuprofen appears to be superior to aspirin in its action against inflammation, but it is no more effective than aspirin in relieving pain or reducing fever. Acetaminophen (Tylenol), which is a useful alternative to aspirin for relieving pain and reducing fever, is not, however, an anti-inflammatory drug.

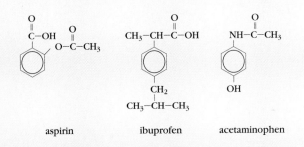

| aspirin | ibuprofen | acetaminophen |

Nolan Ryan, the 46-year old pitching ace, and Jimmy Connors, the 40-year-old tennis star, are both promoting ibuprofen's powerful anti-inflammatory action. Jimmy Connors tells his audience that when they hurt from inflammation, they should "Nupe-it". A new study indicates that there may be some truth in what he says.

Delayed onset muscle soreness (DOMS) is a sore ache that differs from the sharp pain of most sports injuries because this pain is not felt immediately. Pain is not experienced until the athlete wakes on the morning following strenuous exercise, and begins to move.

Studies indicate that taking ibuprofen before unaccustomed vigorous activity may lessen the soreness that is likely to occur 24 to 48 hours later. Usually, an anti-inflammatory medicine is administered when pain begins. But by the time someone with DOMS feels pain, the initial inflammatory stage may be completed and anti-inflammatory agents may not be as effective. No one should routinely take medication just because they are going to be active. But on rare occasions when a person is going to perform unusually strenuous exercise, taking 400 mg of ibuprofen before, and every eight hours after the activity over the next one or two days can minimize soreness.

Doctors warn against abusing painkillers because even over-the-counter drugs can have serious side effects. An extreme example of ibuprofen abuse is the case of a former Seattle Seahawks star who says he developed kidney failure from taking 16-20 tablets of ibuprofen daily for at least three months to help reduce the swelling of an injured ankle. More commonly, athletes with minor injuries take a couple of ibuprofen or aspirin tablets before a game so they can play without pain.

Questions
1. Aspirin can cause stomach bleeding because of its acidity. Is acetaminophen or ibuprofen an acid? Would you expect acetaminophen to cause less stomach upset?
2. Are you willing to pay a premium price for a brand name anti-inflammmatory? If so, why?

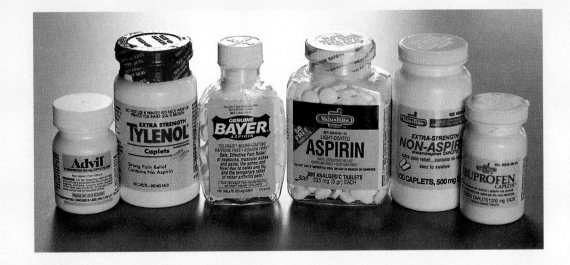

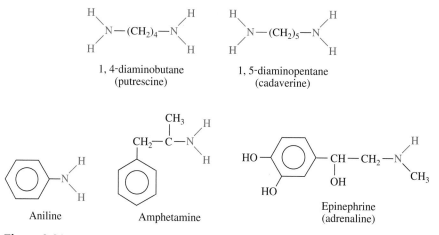

Figure 9-21 Examples of amines.

Amines and Amides

Many important organic compounds—both naturally occurring and synthetic—contain nitrogen. Two important families of nitrogen-containing compounds are amines and amides.

Amines can be thought of as being derived from ammonia, NH_3, by replacing one, two, or all three of the hydrogen atoms with alkyl groups. In this way, primary, secondary, and tertiary amines are formed. The general formulas for these three types of amines are shown below:

$$\underset{\text{ammonia}}{H-\underset{\overset{|}{H}}{N}-H} \qquad \underset{\text{primary amine}}{R-\underset{\overset{|}{H}}{N}-H} \qquad \underset{\text{secondary amine}}{R-\underset{\overset{|}{H}}{N}-R'} \qquad \underset{\text{tertiary amine}}{R-\underset{\overset{|}{R''}}{N}-R'}$$

The two simplest amines are methylamine and ethylamine.

$$\underset{\text{methylamine}}{CH_3-\underset{\overset{|}{H}}{N}-H} \quad (CH_3NH_2) \qquad \underset{\text{ethylamine}}{CH_3-CH_2-\underset{\overset{|}{H}}{N}-H} \quad (C_2H_5NH_2)$$

Most decaying organic matter produces amines, many of which have disagreeable fishy odors (Chapter 3). The smells associated with sewage-treatment plants and meat-packing plants are due primarily to amines. Two particularly unpleasant amines are cadaverine and putrescine, which are products of the bacterial decomposition of proteins and are partly responsible for the smell of decaying flesh (Fig. 9-21).

Important amines that include a benzene ring in their structure are aniline, amphetamine, and epinephrine (commonly called adrenaline) (Fig. 9-21). Aniline is used as a starting material for the manufacture of dyes and certain drugs. Amphetamines are stimulants and have been widely used by people who need to stay alert. They are also used to treat mild depression and as appetite suppressants to treat obesity. Epinephrine and the related compound norepinephrine are secreted by the adrenal gland and play an important role in controlling blood pressure and cardiac output.

EXAMPLE 9-10   Draw the structures of the following and indicate if they are a primary, secondary or tertiary amine:
a. propylamine **b.** methylbutylamine **c.** trimethylamine

Solution:
a. This compound has a propyl group attached to the nitrogen atom in the amine. Nitrogen forms three bonds. The other two bonds must be attached to hydrogen atoms. Since there is only one alkyl group, the compound is a primary amine.

$$CH_3CH_2CH_2{-}\overset{\displaystyle |}{\underset{\displaystyle H}{N}}{-}H$$

b. This compound has a methyl group and a butyl group attached to the nitrogen atom. It is a secondary amine.

$$CH_3{-}\overset{\displaystyle |}{\underset{\displaystyle H}{N}}{-}CH_2CH_2CH_2CH_3$$

c. This compound has three methyl groups attached to the nitrogen atom. It is a tertiary amine.

$$CH_3{-}\overset{\displaystyle |}{\underset{\displaystyle CH_3}{N}}{-}CH_3$$

PRACTICE EXERCISE: Are the following primary, secondary, or tertiary amines?
a. trioctylamine **b.** dibutylamine c. isopropylamine

Answer: **a.** tertiary **b.** secondary **c.** primary

Amides are derivatives of carboxylic acids in which the −OH of the carboxyl group is replaced with an amino group (−NH$_2$), or a substituted amino group (−NHR or −NRR'). Unlike amines, the nitrogen atom in amides is attached to a −C=O group. The general formulas for the three types of amides are shown below:

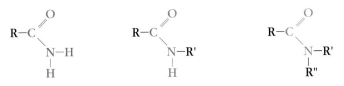

simple amide   monosubstituted amide   disubstituted amide

The simplest amide is acetamide. Amides are named by replacing the ending *-ic* (or *-oic*) and the word *acid* in the name of the corresponding acid with *-amide*.

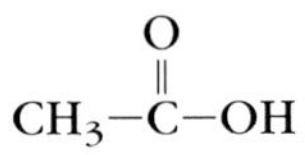

acetic acid 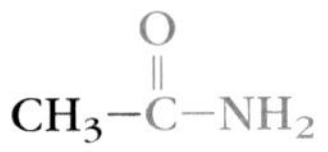   acetamide

The amide functional group is present in many important biological compounds, including proteins and urea, the end product of human metabolism of proteins in food (Fig. 9-22). These compounds will be studied in more detail in Chapters 10 and 17.

Figure 9-22
Examples of amides.

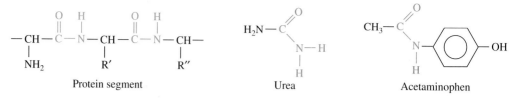

Protein segment   Urea   Acetaminophen

Many amides have been synthesized for use as drugs. For example, acetaminophen, the active ingredient in the pain reliever Tylenol, is an amide (Fig. 9-22). Amides are also important intermediates in the manufacture of nylon and other polymers.

EXAMPLE 9-11   Draw the structure for each of the following compounds: **a.** methylethylamine **b.** 3-pentanone **c.** butyric acid **d.** methylacetamide.

Solution:
a. This compound has a methyl group and an ethyl group attached to the nitrogen atom of an amine. Nitrogen forms three bonds, therefore the third bond is attached to a hydrogen atom.

$$\overset{\displaystyle H}{\underset{}{CH_3-\overset{|}{N}-CH_2CH_3}}$$

b. This compound contains five carbon atoms (penta) and the oxygen of the ketone is attached to carbon 3.

$$CH_3CH_2-\overset{\displaystyle O}{\overset{\|}{C}}-CH_2CH_3$$

c. This compound contains four carbon atoms (butyric) and is a carboxylic acid.

$$CH_3-CH_2-CH_2-\overset{\displaystyle O}{\overset{\|}{C}}-OH$$

d. This compound is formed from acetamide by the addition of a methyl group.

$$CH_3-\overset{\displaystyle O}{\overset{\|}{C}}-\underset{\displaystyle H}{\overset{|}{N}}-CH_3$$

PRACTICE EXERCISE: Identify all the functional groups in epinephrine, a compound used to control allergic reactions.

$$HOCHCH_2NHCH_3$$

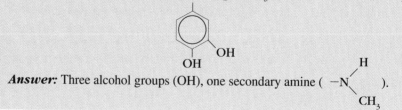

OH

Answer: Three alcohol groups (OH), one secondary amine ($-N\overset{\displaystyle H}{\underset{\displaystyle CH_3}{\big\langle}}$).

This chapter has provided a brief introduction to organic chemistry. We have discussed the major functional group classes but there are many others that have not been mentioned. In Chapters 10 and 17, we will examine in more detail the naturally occurring organic compounds that constitute living tissues and also make up the food on which we depend for energy and survival. Fossil fuels, the starting materials for the synthesis of a great many of the organic compounds produced by industry, will be studied in Chapter 14.

Tyrian Purple: A Color Fit for a King

In imperial Rome, wearing a toga dyed with Tyrian purple could cost you your life—unless you were the emperor. The royal purple dye, obtained at enormous expense from shellfish, was the most renowned dye in the ancient world (Fig. 9-A).

In Republican Rome, purple robes were a sign of rank and prestige. Only military commanders triumphant in battle and the two magistrates (or censors) who supervised public morals and kept the census were permitted to wear garments dyed with Tyrian purple. Consuls and praetors were allowed to wear clothing trimmed with purple, but for lesser mortals, the dye was prohibited. Under the emperors, wearing purple was further restricted. A decree issued by Nero in the first century A.D. gave to the emperor the exclusive right to wear the royal color. Purple dyeing became a state monopoly, and the method of production as well as the workers were tightly controlled. By imperial edict, it became a crime punishable by death to manufacture Tyrian purple anywhere but in the imperial dyeworks. The association of purple with emperors and kings is the origin of the phrase "born in the purple" used to describe infants of royal birth.

Until the middle of the nineteenth century, when the explosive growth in the synthetic dye industry began, the only dyes available were of natural origin. The yellow, red, and brown earth pigments were generally quite stable, but the colors derived from plants and animals were more elusive. Most were not true dyes, and the colors washed out or faded in sunlight. What made Tyrian purple so desirable was that, in addition to being a brilliant color, it was a permanent dye. According to Ronulph of Chester, writing in the fourteenth century, Tyrian purple was "wonder fair and stable, staineth never with cold or with heat, nor with wet, nor with dry, but ever the older the color is fairer."

Legend attributes the discovery of Tyrian purple to the Phoenician god Melkarth, known to the Romans as Hercules. According to the ancient story, Hercules was walking along the seashore one day with his mistress, the nymph Tyros, and his sheep dog. The dog, nosing about among the rocks, found a shellfish and crushed it between his teeth. Juice from the shellfish stained the dog's jaw a bright purplish red. Tyros admired the color, and, to please his mistress, Hercules dyed a gown for her in the new color. The legend lends support to the belief that the dye originated with the Phoenicians. Certainly we know that these enterprising seafarers understood the art of purple dyeing and set up many dye factories in the colonies they established along the Mediterranean coast.

According to some authorities, a dye made from shellfish was used in Crete as early as 1600 B.C., and shellfish were the source of the purple mentioned in the Bible. Ezekiel (27:16), describing trade with Syria in the eighth century B.C., writes "they exchanged for your wares emeralds, purple, embroidered work, fine linen, coral, and agate." Evidence that the Greeks and Romans carried on the art of purple dyeing is found in the huge piles of mollusk shells that have been found close to sites of ancient dyeworks in Athens and Pompeii.

The dye used by the ancients was extracted from two species of mollusks, *Murex brandaris* and *Thais haemastoma*, that were common along the shores of the Mediterranean (Fig. 9-B). A small gland close to the mollusk's head secretes a whitish fluid containing the precursors of the dye. When the

Figure 9-A A toga dyed with Tyrian purple.

Figure 9-B Mediterranean mollusks from which Tyrian purple can be extracted. *Murex brandaris* (left). *Thais haemastoma* (right).

fluid is exposed to the oxygen of the atmosphere and bright sunlight, a series of reactions occurs, and the color of the fluid changes gradually from white to yellow and then to green, blue, and finally to a deep purplish red.

The origin of the name, Tyrian purple, is uncertain but probably derives from the city of Tyre, where, according to the Roman scholar Pliny the Elder, the best dye was made. The word *purple* is derived from the Latin word *purpura*, the name by which the mollusk was known in classical Rome. Our perception of the meaning of purple has changed over the centuries, and today we would describe the famous color as crimson rather than purple. In fourteenth-century England, the color was described simply as red.

Pliny, in his *Historia Naturalis*, written in the mid-first century A.D., gives a detailed account of the dyeing process. According to Pliny, the best time to collect the mollusks was in the spring after the dog star, Sirius, had risen. Fishermen captured the creatures in baited wicker baskets, and when enough had been collected, the glands were carefully removed. After they were crushed, the glands were left to stand over salt for several days before being boiled with water in a large vat. Boiling continued for several weeks, during which time debris rising to the surface was removed, and, at intervals, the dyeing properties of the liquid were tested (Fig.

9-C). When the liquid was of the right strength to produce the desired deep color, fabric was added. In the vat, the fabric acquired a yellow color, but when it was removed and spread out to dry in the sun, like magic, the brilliant royal color appeared. A second dipping in the vat produced an even deeper, richer color.

Because thousands of mollusks were needed to obtain enough dye to color a single garment, Tyrian purple was extremely costly. It has been estimated that in 301 A.D. a pound of dyed wool cost about three times as much as a baker would earn in a year, and a pound of dyed silk was several times more expensive. At certain times, Tyrian purple was worth as much as 10–20 times its weight in gold.

Mysteriously, by the end of the fifteenth century, the art of making Tyrian purple appears to have been forgotten and the dye is rarely mentioned again. One probable reason was the fall of Constantinople (Istanbul) to the Turks in 1453, which ended the rule of the Byzantine emperors. No doubt, in the resulting turmoil, the imperial dyeworks ceased to function. From time to time during the ensuing centuries, the dyeing properties of the mollusks were accidentally rediscovered. There are occasional reports of local fishermen dyeing their shirts red, but the royal dye never again attained the prestige it had known in classical times.

Figure 9-C The dyeing process in ancient times.

In 1908, P. Friedlander made a scientific study of the dye. He extracted 1.4 g of dye from 12,000 mollusks of the species *Murex brandaris* and determined that the purple dye so prized in antiquity was the organic compound 6,6'-dibromoindigo:

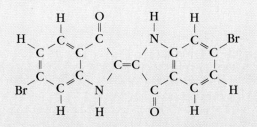

It is not surprising that the pigment contains bromine when we remember that there is a consid-erable amount of bromide in seawater (67 ppm). The crimson color is explained by the large number of alternating double bonds in the molecule, but that is another story.

Question

How many different functional groups are present in the formula given for the purple dye? Name the groups.

References: M. R. Fox, *Vat Dyestuffs and Vat Dyeing*, (New York: John Wiley, 1947); Lyde S. Pratt, *The Chemistry and Physics of Organic Pigments*, (New York: John Wiley, 1947); Stuart Robinson, *A History of Dyed Textiles*, (Cambridge, MA: MIT Press, 1969); P. E. McGovern and R. H. Michel, "Royal Purple Dye: The Chemical Reconstruction of the Ancient Mediterranean Industry," American Chemical Society, *Accounts of Chemical Research*, (1990), *23*, 152.

KEY WORDS AND CONCEPTS

alcohol
aldehyde
alkane
alkene
alkyl group
alkyne
amide
amine
aromatic compound
benzene

carbonyl group
carboxyl group
carboxylic acid
cis-trans isomerism
cycloalkane
delocalized electrons
ester
ether
fatty acid
functional groups

homologous series
hydrocarbon
ketone
organic halide
ortho-, meta-, para-substitution
polycyclic aromatic compounds
substitution
saturated hydrocarbon
structural isomers
unsaturated hydrocarbon

QUESTIONS AND PROBLEMS

1. Write the names of the following alkanes:
   a. $CH_3CH_2CH_2CH_2CH_3$
   b. $CH_3CH_2CH_2CH_2CH_2CH_2CH_2CH_3$ c. $CH_3CH_2CH_3$
   d. $CH_3CH_2CH_2CH_3$

2. Write the names of the following branched alkanes:
   a. $CH_3-CH-CH-CH_3$
       | |
      CH_3 CH_3
   b. $CH_3CH_2CHCH_2CH_3$
         |
        CH_2
        |
        CH_3
   c. $CH_3CHCH_2CH_3$
     |
    CH_3
   d. $CH_3CH_2CH_2CHCH_2CH_2CH_3$
         |
        CH_3

3. Why does carbon form so many compounds with oxygen, nitrogen, and hydrogen?

4. Write the chemical structures and names for the members of the homologous series of saturated hydrocarbons starting with two carbon atoms and ending with six carbon atoms.

5. Draw the chemical structures of the following compounds:
   a. cis-2-butene b. 1-hexene c. propene
   d. 1-pentene

6. Draw the chemical structures of the following compounds:
   a. cyclohexene b. 3-methyl-1-butene
   c. 3,6-dimethyl-2-octene d. 2-methyl-2-butene

7. Draw five different structural isomers for six-carbon hydrocarbons with the formula C_6H_{14}.

8. Which of the following alkenes show geometric (cis-trans) isomerism? Draw the isomeric structures.
   a. 1-butene **b.** 2-methyl-2-butene **c.** 2-butene
   d. 1-chloropropene

9. Which of the following alkenes show geometric (cis-trans) isomerism? Draw the isomeric structures.
   a. 3-methyl-4-ethyl-3-hexene **b.** 2-pentene
   c. 2,3-dichloro-2-butene **d.** 1-pentene

10. Write structural formulas for each of the following compounds:
    a. 2,3-dichloropentane **b.** 1,4-dichlorocyclohexane
    c. 3-ethylpentane **d.** 1-bromobutane

11. Write structural formulas for each of the following compounds:
    a. 2,3-dimethyl-2-butene **b.** cis-2-butene
    c. 1,1-dimethylcyclopentane **d.** 2,3-dichlorobutane

12. Identify the functional group in each of the following compounds:
    a. acetone, CH_3COCH_3 **b.** butyric acid, $CH_3CH_2CH_2COOH$ **c.** methyl acetate, CH_3COOCH_3

13. Identify all the functional groups in each of the following compounds.
    a. glutaraldehyde, $OHCCH_2CH_2CH_2CHO$
    b. glycine, H_2NCH_2COOH **c.** oxalic acid, $HOOCCOOH$

14. Identify all the functional groups in each of the following compounds.
    a. a sex attractant of the female tiger moth
    $$CH_3CH(CH_2)_{12}CH_3$$
    $$|$$
    $$CH_3$$

    b. component of peppermint oil

    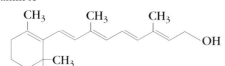

    c. vitamin A

    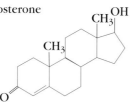

    d. testosterone

    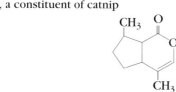

    e. nepetalactone, a constituent of catnip

    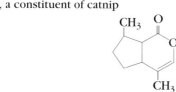

    f. an anesthetic $CH_2=CH-O-CH=CH_2$

15. Cyclohexane and 1-hexene have the same molecular formula, C_6H_{12}. Draw each structure. Explain why they both have the same formula.

16. Name the following:
    a. $(CH_3)_3CCH=CHCH_2CH_3$
    b. $CH_3CH_2CH=CH(CH_2)_4CH_3$
    c. $CH_3CH_2CH=CHCH_2CH_3$

17. Write formulas for:
    a. methylacetylene **b.** methylpropylacetylene
    c. methylbutylacetylene

18. Write the general formula for:
    a. a carboxylic acid **b.** a ketone **c.** an alcohol **d.** an ester

19. Write the general formula for:
    a. an amine **b.** an amide **c.** an ether **d.** an aldehyde

20. What are the chemical names for the following materials?
    a. rubbing alcohol **b.** wood alcohol
    c. grain alcohol **d.** antifreeze

21. Draw the chemical structure for the aromatic alcohol, phenol.

22. Draw the chemical structure for the aromatic carboxylic acid, benzoic acid.

23. What is the percent alcohol by volume in 96-proof whiskey?

24. Why is ethyl alcohol used in cough syrup preparations and not methyl alcohol?

25. What is the product of the oxidation of ethyl alcohol?

26. Write a chemical equation that shows why acetic acid is an acid as defined in Chapter 8.

27. What is the difference between an aldehyde and a ketone?

28. Draw the structure of:
    a. formaldehyde **b.** ethyl acetate **c.** ethyl alcohol
    d. acetone

29. Draw the structure of:
    a. acetic acid **b.** methylamine **c.** methyl alcohol
    d. phenol

30. Compare the odor of butyric acid with that of the ester methyl butyrate.

31. Identify each of the following as a primary, secondary, or tertiary amine.
    a. ethylamine **b.** methylamine **c.** dimethylamine
    d. diethylamine **e.** trimethylamine **f.** triethylamine

32. Name the following functional groups:
    a. $-NH_2$ **b.** $-CHO$ **c.** $-COOH$ **d.** $-COOCH_3$

33. Name the following functional groups:
    a. $-CONH_2$ **b.** $-OH$ **c.** $-CO-$ **d.** $-O-$

34. Draw the structural formula for each of the following molecules:
    a. four-carbon primary alcohol with a double bond in the two position. **b.** ketone containing three carbon atoms **c.** 12-carbon carboxylic acid with a double bond in the six position.

35. Draw the structural formula for each of the following molecules:
    a. methyl ester of a 4-carbon, straight-chain, carboxylic acid. **b.** tertiary amine with three ethyl groups.
    c. 2-carbon aldehyde. **d.** the methyl amide of propionic acid.

36. Draw structural formulas for the following aromatic compounds:
   a. toluene **b.** 1,3-dimethylbenzene (Are the methyl groups in the ortho-, meta- or para- positions?)
   c. para-dimethylbenzene
37. Draw structural formulas for the following aromatic compounds:

   a. meta-dichlorobenzene **b.** 1,2-dimethylbenzene
   c. 3-bromotoluene
38. How many structural formulas can be written for tetrachlorobenzene?
39. How does benzene differ from cyclohexane?

BIOCHEMISTRY
The Molecules of Life

Biochemicals—carbohydrates, fats, proteins, and nucleic acids—are essential ingredients of life. They make up the structural materials of the body, and provide the energy being used by these people running in the New York marathon.

Biochemistry is the study of the living part of our environment. It deals with the substances found in living organisms and the changes those substances undergo during the life of the organism. Living things come in many forms, from bacteria and plants to fish and mammals, but no matter how different these forms are, all are made from the same basic organic components: carbohydrates, fats, proteins, and nucleic acids. In addition to these four biochemicals, living organisms contain water, various minerals and salts, and a number of miscellaneous organic substances, including vitamins.

The four major groups of biochemicals provide a living organism with energy and with the components from which the complex molecules needed for life processes are built. They provide the biological catalysts that control the chemical reactions occurring within an organism and the molecules responsible for the transmission of genetic information from one generation to the next.

Until the early part of the nineteenth century, it was generally believed that living organisms contained a "vital force" that enabled them to synthesize the biochemicals found in living matter and that it would be impossible to synthesize these biochemicals in the laboratory in the absence of this force. This belief was shaken in 1828 when the German chemist Friedrich Wohler synthesized urea (Chapter 9)—an organic compound found in the urine of most mammals—from inorganic starting materials. Since then, scientists have synthesized numerous biological compounds, including carbohydrates, lipids, proteins, hormones, and vitamins. We still may not understand the secret of life, but we do know that biochemicals synthesized in the laboratory have exactly the same structures as the naturally occurring biochemicals and function in the same way when introduced into an organism.

In this chapter, we will study the chemical structures and functions of the four important classes of biological compounds: carbohydrates, lipids, proteins, and nucleic acids. Before we begin this study, however, we will consider a structural feature that is common to many biological molecules: stereoisomerism.

Learning Goals:

In this chapter, you should gain an understanding of:

1. Left-handed and right-handed molecules called stereoisomers.
2. The classification of naturally occurring carbohydrates.
3. The molecular structure of mono-, di-, and polysaccharides.
4. The properties of saturated and unsaturated fatty acids.
5. Proteins and the amino acids from which they are made.
6. Nucleic acids and the nucleotides from which they are made.

STEREOISOMERISM: RIGHT-HANDED AND LEFT-HANDED MOLECULES

Many organic molecules exist in two forms that are related to each other in the same way that a right-hand glove is related to a left-hand glove. Two gloves in a pair are alike in every respect, but they cannot be superimposed, one upon the other. You cannot fit a right-hand glove on your left hand, or vice versa. The two gloves—as well as your hands—are mirror images of each other (Fig. 10-1). The mirror reflection of a left hand looks like a right hand.

Recall from Chapter 9 that *isomers* are *different compounds that have the same molecular formula*. Isomers can be divided into structural isomers and stereoisomers. **Structural isomers** have the *same molecular formula but different structures*. Examples of structural isomers are shown in Figure 10-2. In these pairs of isomers, the atoms are joined in a different order. For example, in ethanol, the oxygen atom is bonded to a carbon atom and a hydrogen atom; while in dimethyl ether, the oxygen atom is bonded to two carbon atoms.

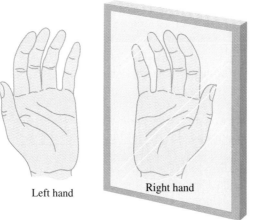

Left hand   Right hand

Mirror

Left and right hands are not superimposable

Figure 10-1   The mirror image of your left hand is your right hand, but one hand cannot be superimposed upon the other.

Molecular Formula   Structural Isomers

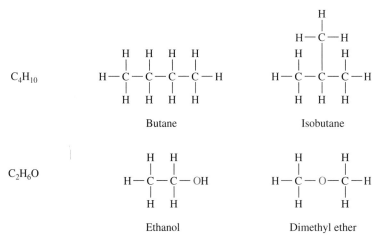

C_4H_{10}

Butane   Isobutane

C_2H_6O

Ethanol   Dimethyl ether

Figure 10-2   The pairs of structural isomers have the same molecular formula but differ in the way the atoms in their molecules are bonded together.

The part of chemistry that is concerned with the structure of molecules in three dimensions is called stereochemistry (from the Greek word *stereos*, meaning solid).

Stereoisomers have the same formula, and their atoms are joined in the same order; *they differ only in the arrangement of the atoms in space*. There are two types of stereoisomers: enantiomers and diastereomers. **Enantiomers** are mirror images of each other. **Diastereomers** are *not* mirror images of each other. We will examine molecules that are diastereomers, but not geometric isomers, in the section on monosaccharides.

Enantiomers

Enantiomers are possible when a molecule is asymmetric. A simple asymmetric molecule is one consisting of a tetrahedral carbon atom attached to four *different* atoms or groups of atoms. Such a molecule is represented by the ball-and-stick model, A, shown in Figure 10-3a. Four different-colored balls represent four different atoms, or groups of atoms, attached to the carbon atom. A second model, B, can also be constructed. The only difference between models A and B is that the red and green balls are switched. As can be seen in Figure 10-3b, model B is the mirror image of model A. Models A and B, therefore, represent enantiomers.

The models are not identical because no matter how they are turned, model A cannot be superimposed on model B (Fig. 10-3c). Models A and B represent two distinct molecules. They differ from each other much as the left hand does from the right hand. Because of this "handedness," asymmetric molecules are called **chiral molecules** (from the Greek word *cheir*, meaning "hand").

A chiral carbon atom has four different atoms or groups attached to it.

Thousands of enantiomers exist. Two simple examples, lactic acid and 2-butanol, are shown below.

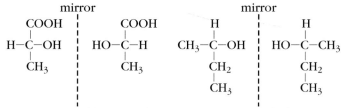

enantiomers of lactic acid   enantiomers of 2-butanol

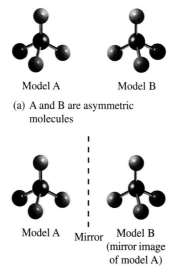

Model A   Model B

(a) A and B are asymmetric
     molecules

Model A   Mirror   Model B
                   (mirror image
                    of model A)

(b) A and B are enantiomers

Figure 10-3   Pairs of enantiomers exist for molecules
that contain one asymmetric
carbon atom. **(a)** Model A represents a simple molecule containing an asymmetric carbon
atom. Model B resembles
model A except that the groups
represented by the red and
green balls have changed positions. **(b)** A and B are related
as an object and its mirror reflection; A and B are enantiomers. **(c)** Model A cannot be
superimposed on model B.

(c) A and B are not superimposable

The different classes into which isomers can be subdivided are summarized in Figure 10-4.

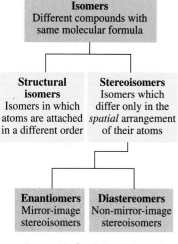

Isomers
Different compounds with
same molecular formula

**Structural
isomers**
Isomers in which
atoms are attached
in a different order

Stereoisomers
Isomers which
differ only in the
spatial arrangement
of their atoms

Enantiomers
Mirror-image
stereoisomers

Diastereomers
Non-mirror-image
stereoisomers

Figure 10-4   Subdivision of
isomers into classes.

EXAMPLE 10-1   2-chlorobutane is a chiral molecule. Draw the two enantiomers of this molecule.

Solution:
1. Draw the condensed formula of butane. Add the chlorine atom at carbon 2.

$$CH_3-CH_2-CH_2-CH_3$$
butane

$$\overset{4}{C}H_3-\overset{3}{C}H_2-\overset{2}{C}H-\overset{1}{C}H_3$$
$$|$$
$$Cl$$

2. Identify the asymmetric carbon: carbon 2 is asymmetric because four different groups are attached to it (methyl, ethyl, chloro, hydrogen).
3. Draw the first enantiomer with carbon 2 in the center.
4. Draw the second enantiomer as a mirror image of the first.

$$
\begin{array}{c}
CH_3 \\
| \\
H-C-Cl \\
| \\
C_2H_5
\end{array}
\qquad
\begin{array}{c}
CH_3 \\
| \\
Cl-C-H \\
| \\
C_2H_5
\end{array}
$$

PRACTICE EXERCISE: 2-bromo-2-chlorobutane is a chiral molecule.
Draw the two enantiomers.

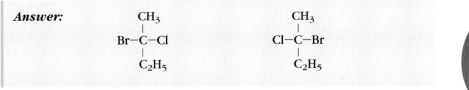

Answer:

(a)

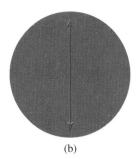

(b)

Figure 10-5 **(a)** Ordinary light vibrates in all possible planes perpendicular to the path of the light. **(b)** Plane-polarized light vibrates in one plane only.

Optical Activity and Enantiomers

The properties of pairs of enantiomers are almost identical. Unlike diastereomers, enantiomers have *identical* boiling points and melting points. Many other physical properties and many chemical properties are also identical. However, enantiomers can always be distinguished from each other by one property: their behavior toward **plane-polarized light**.

According to the wave theory of light, light waves oscillate (vibrate) at right angles to the direction in which the light is traveling through space. The oscillations occur in all planes that are perpendicular to the path traveled by the light. Figure 10-5a schematically shows oscillations present in a flashlight beam as viewed by an observer looking directly into the beam. If the beam of light is passed through a sheet of Polaroid material, the light is changed and emerges from the Polaroid sheet oscillating in *one* plane only (Fig. 10-5b). Such light is called plane-polarized light. If plane-polarized light is passed through a solution containing a single enantiomer, it interacts with the chiral molecules and emerges from the solution oscillating in a different plane. The enantiomer has *rotated* the plane of the polarized light. Because of this ability to rotate the plane of polarized light, enantiomers are said to be **optically active**.

Separate enantiomers of a pair rotate the plane of polarized light by an equal amount, but in opposite directions. For example, if the compound represented by model A in Figure 10-3 rotates the plane by 5 degrees in one direction, the compound represented by model B, its mirror image, would rotate the plane 5 degrees in the opposite direction.

Figure 10-6 explains how a polarimeter can be used to observe and measure the degree of rotation produced by an optically active compound. If an enantiometer rotates the beam of light to the right in a clockwise direction (as the observer looks toward the light source), the enantiomer is said to be *dextrorotary* and is designated (+). If the enantiomer rotates the beam of light to the left in a counterclockwise direction, the enantiomer is said to be *levorotary* and is designated (−). A solution containing equal amounts of the (+) and (−) isomers does not affect the plane of polarized light because the tendencies to rotate in opposite directions cancel each other out.

The terms dextro and levo are derived from the Latin words *dexter*, meaning right, and *laevus*, meaning left.

D- and L- Isomers

The designation of individual enantiomers as (+) or (−) defines how the isomer behaves with respect to plane-polarized light but tells nothing about the *molecular* structure of the isomer. "Handedness" in terms of molecular structure is defined by D- (for right-handed) and L- (for left-handed). The D- and L- forms of glucose and the amino acid alanine are shown in Figure 10-7.

Placing a compound in the D- or L- series defines its molecular structure but does not describe its optical activity. Some D- sugars, for example, have a (+) rotation and some have a (−) rotation.

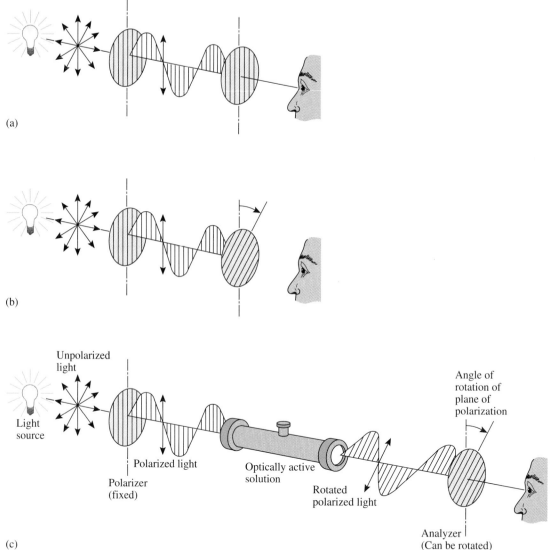

(a)

(b)

Unpolarized light

Light source

Polarized light

Polarizer (fixed)

Optically active solution

Rotated polarized light

Angle of rotation of plane of polarization

Analyzer (Can be rotated)

(c)

Figure 10-6   A polarimeter consists of **(1)** a light source, **(2)** two Polaroid lenses (polarizer and analyzer), and **(3)** a tube between the lenses to hold the sample. The light passes through the polarizer, the tube, and the analyzer before reaching the eye. **(a)** Without a tube, the maximum amount of light reaches the eye when the polarizer and analyzer are parallel, that is, when the analyzer is rotated so that both lenses pass light vibrating in the same plane. **(b)** If the analyzer is rotated so that the lenses are crossed, no light reaches the eye. **(c)** If a tube containing an optically active substance is placed between the polarizer and analyzer, the plane of the light will be rotated and light will emerge. The angle through which the analyzer must be rotated so that light no longer emerges is a measure of the optical activity of the substance in the tube.

Glucose enantiomers

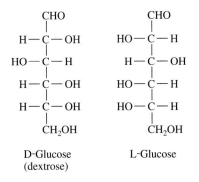

D-Glucose
(dextrose)

L-Glucose

Alanine enantiomers

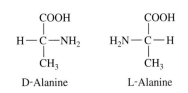

D-Alanine   L-Alanine

Figure 10-7 The D- and L-isomers of glucose and alanine. The D- and L- isomers are enantiomers (mirror images).

EXAMPLE 10-2   2-methyl-1-butanol is a chiral molecule. Draw its enantiomers.

Solution:

1. Write the condensed formula of 1-butanol (n-butyl alcohol). Add the methyl group at carbon 2.

$$\overset{4}{CH_3}-\overset{3}{CH_2}-\overset{2}{CH_2}-\overset{1}{CH_2}-OH \qquad \overset{4}{CH_3}-\overset{3}{CH_2}-\overset{2}{CH}-\overset{1}{CH_2}-OH$$
$$| $$
$$CH_3$$

2. Identify the asymmetric carbon: carbon 2 is asymmetric because four different groups are attached to it (methyl, ethyl, hydrogen, and a CH_2OH).

3. Carbon 1 is not chiral. It does not have four different groups attached (there are two H atoms).

4. Draw the first enantiomer with carbon 2 in the center.

5. Draw the second enantiomer as a mirror image of the first.

$$\begin{array}{c} CH_2OH \\ | \\ H-C-CH_3 \\ | \\ C_2H_5 \end{array} \qquad \begin{array}{c} CH_2OH \\ | \\ CH_3-C-H \\ | \\ C_2H_5 \end{array}$$

PRACTICE EXERCISE: 2-butanol is a chiral molecule. Draw the two enantiomers.

Answer:

$$\begin{array}{c} CH_3 \\ | \\ H-C-OH \\ | \\ C_2H_5 \end{array} \qquad \begin{array}{c} CH_3 \\ | \\ HO-C-H \\ | \\ C_2H_5 \end{array}$$

Enantiomers and Living Processes

Nearly all organic molecules in living organisms and in most of the foods we eat are chiral molecules and have the property of "handedness." Theoretically, these molecules can exist as D- and L- isomers. However, in nature, chiral molecules exist almost exclusively in one form or the other, but not both. For example, naturally occurring carbohydrates exist as D-isomers, while amino acids exist as L-isomers. How and why this happened as life developed on earth is one of the mysteries of nature. As a result of this split, living organisms are capable of metabolizing, and making use of, only those molecules that are in the naturally occurring form. If a person were fed a synthetic diet composed of L-carbohydrates and proteins made from D-amino acids, that person would eventually die of starvation.

CARBOHYDRATES

Carbohydrates are compounds of the elements carbon, hydrogen, and oxygen. Common carbohydrates include glucose, sucrose (table sugar), starch, and cellulose. Carbohydrates are vitally important to both plants and animals. Plants are composed almost entirely of carbohydrates and water. Typically, two thirds of the human diet is made up of carbohydrates. Rice, flour, and corn meal, which are the basic foods of most people in the world, are composed of carbohydrates.

Almost all carbohydrates are produced as a result of photosynthesis in green plants. As we saw earlier (Chapters 3 and 7), in a series of complex steps, green plants use energy from the sun to convert carbon dioxide in the air and water in the soil into simple carbohydrates, mainly glucose. The overall reaction is as follows:

$$6\,CO_2 + 6\,H_2O \xrightarrow{\text{sunlight}} C_6H_{12}O_6 + 6\,O_2$$

In plants, glucose is converted to cellulose and starch. In animals and humans, the oxidation of plant carbohydrates provides both energy and carbon, hydrogen, and oxygen atoms needed to synthesize lipids, proteins, and other biological molecules.

As their name implies, carbohydrates were once regarded as hydrates of carbon: that is, compounds in which carbon is bound to water molecules. Glucose, $C_6H_{12}O_6$, for example, can be written $C_6(H_2O)_6$ and sucrose, $C_{12}H_{22}O_{11}$, can be written $C_{12}(H_2O)_{11}$. We now know that the three elements, C, H, and O are arranged primarily as hydroxyl ($-OH$), and aldehyde ($-\overset{\overset{O}{\|}}{C}-H$) or ketone ($-\overset{\overset{O}{\|}}{C}-$) groups (Chapter 9).

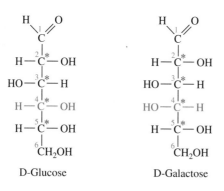

Figure 10-8   Glucose has four asymmetric carbon atoms. D-glucose and D-galactose are two of the 16 possible diastereomers of glucose. D-galactose differs from D-glucose in the orientation of the −OH attached to carbon 4.

* Asymmetric carbon atom

Naturally occurring carbohydrates can be divided into three main classes: monosaccharides, disaccharides, and polysaccharides. **Monosaccharides,** or simple sugars, cannot be broken down into smaller carbohydrate units. **Disaccharides** are made up of two monosaccharide units bonded together. **Polysaccharides** are giant molecules, made up of hundreds, sometimes thousands, of monosaccharide units linked together in a way analogous to how identical railroad cars are coupled to form long freight trains. In biologically important polysaccharides, the monosaccharide units are glucose.

The term saccharide is derived from the Latin word *saccharum*, meaning sugar.

MONOSACCHARIDES

Important monosaccharides include glucose, galactose, fructose, ribose, and deoxyribose. Glucose, galactose, and fructose all have the same formula: $C_6H_{12}O_6$. These six-carbon carbohydrates are known as hexoses. Ribose, $C_5H_{10}O_5$, and deoxyribose, $C_5H_{10}O_4$, are five-carbon carbohydrates and are known as pentoses. Ribose is a component of RNA, and deoxyribose is a component of DNA; they will be considered in the section devoted to nucleic acids.

Molecular Structures of Monosaccharides

Many organic molecules, including monosaccharides, have more than one asymmetric carbon in their molecule. As a result, many different stereoisomers are possible. For any molecule, the maximum possible number of stereoisomers that can exist is equal to 2^n, where n is the number of asymmetric carbon atoms in the molecule. As shown in Figure 10-8, glucose has four asymmetric carbons, and, therefore, 2^4 (16) stereoisomers are possible: eight D-isomers and eight L-isomers. Of these, D-glucose and D-galactose are the most important in the natural world.

Mannose, a stereoisomer of glucose, is found in many plants. Its name is derived from the Biblical word *manna*.

D-glucose and D-galactose (Fig. 10-8) are diastereomers. They have the same formulas and the same arrangement of atoms. However, because they are not mirror images of each other, they are not enantiomers.

D- and L-isomers are distinguished by the arrangement of the −OH group on the highest numbered asymmetric carbon atom in the formula (carbon 5 for the hexoses shown in Fig. 10-8). By convention, a D- sugar is written with the −OH on this carbon pointing to the right; an L- sugar is written with the −OH pointing to the left. The D- and L- forms of glucose are shown in Figure 10-7. In future discussions, it can be assumed, unless stated otherwise, that all sugars mentioned are D- sugars.

The open-chain structures shown for glucose and galactose do not adequately represent the actual molecular structures of hexoses. In fact, the open-chain structures exist in dynamic equilibrium with two cyclic structures. As shown for glucose (Fig. 10-9), a reaction between carbon 1 and carbon 5 in the open-chain structure results in a bond forming through the oxygen atom of the −OH group on carbon 5. Two 6-membered cyclic structures, α and β, which differ only in the spatial arrangement of the −OH group attached to carbon 1, are formed. In aqueous solutions of glucose, the cyclic forms predominate. As we shall see later, the existence of two cyclic forms (α and β) influences the properties of biologically important polysaccharides.

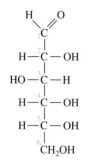

Open-chain glucose

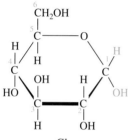

α-Glucose

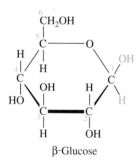

β-Glucose

Figure 10-9 The open-chain and α- and β-cyclic structures of glucose. The α and β structures differ only in the orientation of the −OH attached to carbon 1.

Glucose

By far the most biologically important monosaccharide is **glucose** (also known as dextrose, grape sugar, and blood sugar). Glucose is a moderately sweet sugar. It occurs naturally in honey, grapes, dates, and other fruits. Because glucose contains numerous −OH groups that can form hydrogen bonds with water, it is very soluble in water.

In the human body, glucose is formed from disaccharides and polysaccharides in foods when the foods are broken down during the digestive process. Glucose enters the bloodstream and is carried to all the tissues of the body. By means of a complex series of steps, glucose in body cells is oxidized to carbon dioxide and water, and energy is released.

$$C_6H_{12}O_6 + 6\,O_2 \longrightarrow 6\,CO_2 + 6\,H_2O + energy$$

Glucose is the only carbohydrate that is used directly by the body as a source of quick energy. For this reason, patients in hospitals who are unable to take food by mouth are given a glucose (dextrose) solution intravenously.

Galactose

Galactose is a constituent of many plant gums and pectins. It is also a component of the disaccharide lactose. Although galactose differs from glucose only in the orientation of the −OH group on carbon 4 (Fig. 10-8), galactose cannot be used by the body unless the orientation is changed so that it becomes glucose. Galactose, like glucose, exists in two cyclic forms.

Fructose

Fructose is the sweetest of all the naturally occurring sugars. It is found in honey and many fruits. The open-chain structural formula of fructose is shown in Figure 10-10a, together with that of glucose for comparison. The two formulas are identical for carbons 3 through 6. They differ at carbons 1 and 2 because

fructose contains a ketone group $(-\overset{O}{\underset{\|}{C}}-)$ at carbon 2 while glucose contains an

aldehyde group $(-\overset{O}{\underset{\|}{C}}-H)$ at carbon 1.

Fructose, like glucose, exists in cyclic forms (Fig. 10-10b). Two *5-membered* rings are formed by closure between carbons 2 and 5. Two 6-membered rings, formed by closure between carbons 2 and 6 (formulas not shown) also exist. The straight chain and ring structures exist in equilibrium with each other. We will use the 5-membered ring formula in future discussions.

DISACCHARIDES

Disaccharides consist of two monosaccharides joined together by the elimination of a molecule of water. The three most common disaccharides are:

1. sucrose (one glucose unit and one fructose unit)
2. lactose (one glucose unit and one galactose unit)
3. maltose (two glucose units)

The formation of maltose from two glucose molecules with the elimination of a molecule of water is shown in Figure 10-11a. Sucrose and lactose are formed in a similar manner. The molecular formula of these disaccharides, therefore, is

$$C_{12}H_{22}O_{11} \quad (C_6H_{12}O_6 + C_6H_{12}O_6 - H_2O)$$

Sucrose

Sucrose (common table sugar) occurs naturally in sugarcane and sugar beets. Over 80 million tons of pure sucrose are produced annually from these sources (Fig. 10-11b). On average, in the United States, sucrose provides more than one-quarter of the calories consumed daily. Because it contains numerous hydroxyl groups, which form hydrogen bonds with water molecules, sucrose dissolves readily in tea, coffee, and other aqueous solutions. During digestion (Chapter 17), enzyme action converts sucrose to glucose and fructose.

$$\text{sucrose} \longrightarrow \text{glucose} + \text{fructose}$$

Lactose

Lactose, sometimes called milk sugar, is the major sugar in milk. In the mammary glands, glucose obtained from the blood supply is converted by enzyme action into lactose. When milk is drunk, enzyme action in the digestive system converts the lactose to glucose and galactose.

$$\text{lactose} \longrightarrow \text{glucose} + \text{galactose}$$

Further enzyme action changes the galactose to glucose.

Maltose

Maltose, which is present in germinating grains, is found in beer, malted milk, and corn syrup. Maltose is important primarily as the main product of the digestion of starch. During digestion, maltose breaks down to yield two molecules of glucose.

$$\text{maltose} \longrightarrow \text{glucose} + \text{glucose}$$

POLYSACCHARIDES

The three most important polysaccharides are starch, glycogen, and cellulose. When completely broken down into monosaccharide units, all three yield glucose as the only product. Unlike monosaccharides, and most disaccharides, polysaccharides are not sweet.

Starch and Glycogen

Starch and glycogen are composed entirely of α-glucose molecules. Starch is stored in plants as a food reserve and source of energy. Glycogen plays this role in animals.

Structurally, **starch** is a mixture of two types of polysaccharides: amylose (20–30 percent) and amylopectin (70–80 percent). **Amylose** is a straight-chain molecule (Fig. 10-12a) consisting of about 200 α-glucose molecules joined together by 1–4 linkages as was shown for the formation of maltose in Figure

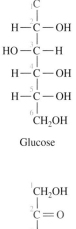

Glucose

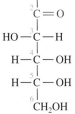

Fructose

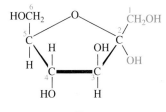

α-Fructose

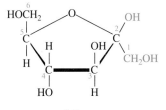

β-Fructose

Figure 10-10   Fructose differs from glucose at carbons 1 and 2. Fructose contains a ke-

tone group (−C−); glucose contains an aldehyde group

(−C−H).

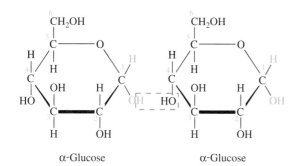

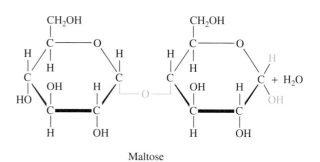

α-Glucose   α-Glucose   Maltose

(a)

(b)

(a)

Figure 10-11 **(a)** The disaccharide maltose is formed when two glucose molecules link together. Linkage occurs between the −OH on carbon 1 on one molecule and the −OH on carbon 4 on the second molecule and a molecule of water is eliminated. **(b)** Cutting sugarcane, the source of much of the sucrose (table sugar) consumed in the United States.

10-11. A typical **amylopectin** molecule consists of about 1000 α-glucose molecules, joined together to form branched chains. The branching occurs through 1–6 linkages (Fig. 10-12b), approximately once every 25 glucose units.

Starch present in potatoes, rice, wheat, and other cereal grains accounts for about two-thirds of the average human diet. During digestion in the body, starch is gradually broken down by enzyme action to yield glucose.

When the starch we eat provides more glucose than the body requires for its immediate needs, some of the excess starch is converted to **glycogen** and stored in the liver and muscles. Large excesses are converted to fats and stored in special fatty tissue called adipose tissue. Glycogen forms the storage polysaccharide in animals in the same way that starch forms the storage polysaccharide in plants. When needed by the body, glycogen is converted back to glucose. Structurally, glycogen is similar to amylopectin except that its chains are more highly branched, with a branch occurring at approximately every tenth glucose unit.

Cellulose

Cellulose forms the structural component of the cell walls of plants. It is by far the most abundant polysaccharide in the natural world. Approximately 50 percent of the mass of a tree is cellulose; cotton is almost 100 percent cellulose. Cellulose provides us with many manufactured products, including paper and paper products, rayon, linen, and cellulose nitrate (a constituent of nail polish).

Cellulose is composed of long unbranched chains of from 100 to 10,000 β-glucose molecules (Fig. 10-13a). Groups of chains, held together by hydrogen bonding between −OH groups on adjacent chains, are twisted into ropelike structures that make cellulose tough and fibrous (Fig. 10-13b). The absorbent properties of cotton and paper towels are due to capillary action and the formation of hydrogen bonds between water molecules and −OH groups in cellulose.

Cellulose is the main food source for horses, cows, sheep, termites, and other herbivores. In the intestinal tracts of these animals, microorganisms produce enzymes that break down cellulose into glucose units. Humans and carnivores, however, cannot use cellulose as a food source because they lack the enzymes needed to break the linkages between the β-glucose units. Although humans cannot digest the cellulose in celery, lettuce, and other green vegetables, these products provide fiber that is needed to facilitate the excretion of solid wastes (Chapter 17).

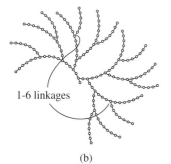

1-6 linkages

(b)

Figure 10-12 Starch is a mixture of amylose and amylopectin molecules. **(a)** In amylose, α-glucose units are joined together in long straight chains through 1-4 linkages. **(b)** In amylopectin the chains are branched; branching occurs by linkage through the −OH on carbon 1 of one glucose molecule and the −OH on carbon 6 of a second molecule (a 1-6 linkage).

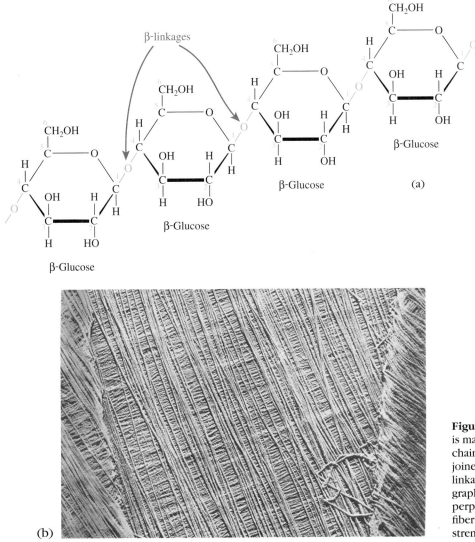

β-linkages

CH₂OH
β-Glucose

β-Glucose

β-Glucose

β-Glucose

β-Glucose

(a)

(b)

Figure 10-13 **(a)** Cellulose is made up of long unbranched chains of β -glucose units joined together through 1-4 linkages. **(b)** Electron micrograph of cellulose fibers. The perpendicular arrangement of fibers in alternate layers gives strength to the cellulose.

LIPIDS

Lipids, like carbohydrates, are compounds of carbon, hydrogen, and oxygen, but unlike carbohydrates (and proteins), lipids cannot be defined on the basis of their chemical structure. They are generally defined as the greasy materials in plants and animals that are insoluble in water but soluble in nonpolar solvents such as benzene, chloroform, and ether. Lipids include compounds of widely differing chemical composition, including fatty acids, triglycerides (or neutral fats), waxes, terpenes, steroids, fat-soluble vitamins, phospholipids, and prostaglandins.

The most abundant lipids in nature are fats and oils, which belong to the class of lipids called **triglycerides**. Triglycerides that are solids at room temperature, such as lard, butter, suet, and tallow, are generally known as **fats**; those that are liquids, such as corn oil, olive oil, soybean oil, and linseed oil are usually called **oils**.

The term lipid is derived from the Greek word *lipos*, meaning fat or lard.

Fats and oils are lighter than water, and float on its surface.

In this chapter, we will limit our discussion of lipids to fatty acids, triglycerides, and steroids. Vitamins will be considered in Chapter 17.

FATTY ACIDS AND TRIGLYCERIDES

Fatty Acids

Monocarboxylic acids contain one carboxyl group, COOH.

Fatty acids are long-chain monocarboxylic acids (Chapter 9) with an *even* number of carbon atoms. Common fatty acids have between 12 and 26 carbon atoms. An example is stearic acid, which contains 18 carbon atoms.

$$CH_3-CH_2-CH_2-CH_2-CH_2-CH_2-CH_2-CH_2-CH_2-CH_2-CH_2-CH_2-CH_2-CH_2-CH_2-CH_2-CH_2-C\overset{\displaystyle O}{\underset{OH}{\big\|}}$$

stearic acid

The formula of stearic acid, and of any other fatty acid, is usually written in condensed form:

$$CH_3-(CH_2)_{16}-COOH \quad \text{stearic acid}$$

As was the case for the straight-chain alkanes (Chapter 9), the carbon atoms in a fatty acid are not in a straight line but are staggered:

The long chains of $-CH_2$ groups make fatty acids and most of their derivatives nonpolar and thus insoluble in water.

Stearic acid is a **saturated fatty acid**: one in which all the carbon-carbon bonds are single bonds. Other fatty acids, such as oleic acid, are **unsaturated** and contain one or more carbon-carbon double bonds.

$$CH_3-(CH_2)_7-CH=CH-(CH_2)_7-COOH$$

Oleic acid

Each double bond causes a bend in the molecule as illustrated in skeleton form for oleic acid.

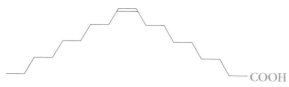

Fatty acids containing more than one double bond are often called **polyunsaturated** fatty acids. Biologically important unsaturated fatty acids contain up to four double bonds.

Important saturated and unsaturated fatty acids are shown in Table 10-1, along with their melting points and natural sources. The uniform straight-chain structure of saturated fatty acid molecules allows them to pack together in an orderly fashion. However, unsaturated fatty acid molecules, because of their bent structure, cannot pack together as closely. As a result, unsaturated fatty acids are less dense than saturated fatty acids and attractions between adjacent molecules are weaker; melting points decrease as the number of double bonds per molecule increases and also as the length of the hydrocarbon chain decreases.

TABLE 10-1   Structures, Melting Points, and Sources of Some Common Fatty Acids

Name	Number of Carbon Atoms	Formula	Melting Point (°C)	Sources
Saturated Fatty Acids				
Lauric	12	$CH_3 (CH_2)_{10} COOH$	44	Coconut oil
Myristic	14	$CH_3 (CH_2)_{12} COOH$	54	Butterfat, coconut oil
Palmitic	16	$CH_3 (CH_2)_{14} COOH$	63	Beef fat, butterfat, cottonseed oil, lard
Stearic	18	$CH_3 (CH_2)_{16} COOH$	70	Beef fat, butterfat, cottonseed oil, lard
Arachidic	20	$CH_3 (CH_2)_{18} COOH$	76	Peanut oil
Unsaturated Fatty Acids				
Oleic	18	$CH_3 (CH_2)_7 CH=CH (CH_2)_7 COOH$	13	Beef fat, lard, olive oil, peanut oil
Linoleic	18	$CH_3 (CH_2)_4 (CH=CHCH_2)_2 (CH_2)_6 COOH$	−5	Corn oil, cottonseed oil, linseed oil, soybean oil
Linolenic	18	$CH_3CH_2 (CH=CHCH_2)_3 (CH_2)_6 COOH$	−11	Corn oil, linseed oil
Arachidonic	20	$CH_3 (CH_2)_4 (CH=CHCH_2)_4 (CH_2)_2COOH$	−50	Animal tissues, corn oil, linseed oil

Triglycerides

Free fatty acids make up a small fraction of natural lipids. Fatty acids are important primarily as the building blocks from which the triglycerides in fat and oils are formed.

We encountered esters and the polyhydroxy alcohol glycerol in Chapter 9. Triglycerides are triesters formed by the reaction of three fatty acid molecules with a glycerol molecule (Fig. 10-14). During the reaction, three molecules of water are eliminated. The three fatty acid molecules may be identical or they may be different. Natural fats and oils are complex mixtures of many different triglycerides, the exact composition varying with the source.

Animal fats, such as lard, suet, and butter, are composed of triglycerides rich in saturated fatty acids, and in general they are solids at room temperature. Vegetable oils are liquids because of their high content of triglycerides composed of unsaturated fatty acids.

An organic ester, with the general formula

$$R-O-\overset{\overset{\displaystyle O}{\|}}{C}-R'$$

is formed from the reaction of an alcohol with an organic acid.

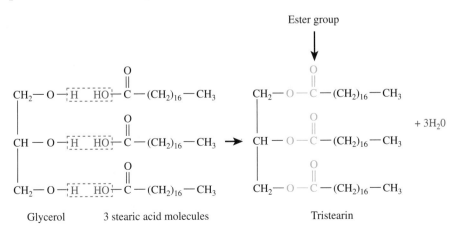

Figure 10-14   The formation of the triglyceride tristearin from glycerol and 3 molecules of stearic acid. During the reaction, three molecules of water are eliminated.

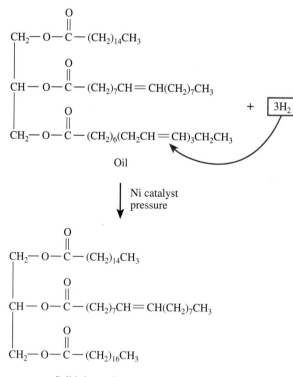

Figure 10-15 Margarine is made from vegetable oils by the hydrogenation of double bonds in the oils. Hydrogenation converts liquid oils (polyunsaturated fats) into semi-solid fats (partially saturated fats).

Studies have shown that a diet high in saturated fats increases the risk of developing atheroschlerosis, the condition in which fatty deposits accumulate in the arteries, and many people now avoid eating animal fats. To encourage consumers to cook with vegetable oils instead of butter or lard, advertisers of oils emphasize their high content of polyunsaturated fats.

Hydrogenation of Oils

An oil can be converted to a semisolid fat by adding hydrogen to some of the fatty acid double bonds, thus decreasing the degree of unsaturation. Soft-spread margarines are prepared by the catalytic partial hydrogenation of vegetable oils, such as corn oil, soybean oil, and cottonseed oil (Fig. 10-15). The reaction is carefully controlled to achieve the desired consistency for the product. Vegetable oils are relatively inexpensive, and margarines can be produced more cheaply than butter.

Soaps from Fats

Soaps have been made from fats for thousands of years. If fats are treated with a strong base, typically NaOH, glycerol and the sodium salts of the fatty acids are formed:

$$
\begin{matrix}
& \overset{O}{\overset{\|}{}} \\
CH_2-O-C-(CH_2)_{16}CH_3 & \\
& \overset{O}{\overset{\|}{}} \\
CH-O-C-(CH_2)_{16}CH_3 & + \ 3 \ NaOH \\
& \overset{O}{\overset{\|}{}} \\
CH_2-O-C-(CH_2)_{16}CH_3 &
\end{matrix}
\longrightarrow
\begin{matrix}
CH_2-OH \\
| \\
CH-OH \\
| \\
CH_2-OH
\end{matrix}
+ \ 3 \ CH_3(CH_2)_{16}COO^- \ Na^+
$$

Tristearin	glycerol	Sodium stearate
(glycerol tristearate)		(a soap)

This reaction is called **saponification**, and the sodium salts of the fatty acids are soaps.

Fats that are used to make soaps include tallow from beef and mutton, palm oil, cottonseed oil, and coconut oil. Bar soaps may contain any or all of the sodium salts of the fatty acids shown in Table 10-1. Potassium hydroxide is used to make the softer potassium soaps from which shaving cream and liquid soaps are made. Toilet soaps contain additives such as perfumes, dyes, and oils.

The cleansing action of soap is explained by the structure of the soap molecule:

$$
CH_3-CH_2-CH_2-CH_2-CH_2-CH_2-CH_2-CH_2-CH_2-CH_2-CH_2-CH_2-CH_2-CH_2-CH_2-CH_2-CH_2-C\overset{\displaystyle O}{\underset{\displaystyle O^- \ Na^+}{\diagdown}}
$$

nonpolar

Sodium stearate

polar

The hydrocarbon end of a long soap molecule, such as sodium stearate, is nonpolar, while the ionized carboxylate end is highly polar. The nonpolar end dissolves in oily materials; the polar end dissolves readily in water. When greasy, oily dirt is vigorously mixed with soapy water, oily particles become surrounded by soap molecules (Fig. 10-16). A grease-soap droplet called a **micelle** is formed as the nonpolar end of the soap enters the oil and the polar end remains dissolved in the water. The micelles cannot coalesce into larger drops because the negative charges on their surfaces repel. They are washed away during rinsing, leaving behind a clean, grease-free surface.

The sulfonic acid salts present as surfactants in synthetic detergents (Chapter 12) function on the same principle. Like soaps, these synthetic molecules have a polar end and a nonpolar end.

The Role of Fats in Living Creatures

In humans and other animals, any excess glucose formed in the body from carbohydrates in foods is converted to fats and stored in the adipose tissue. In this process, glucose is broken down into two-carbon units, which are then built up into fatty acids and finally triglycerides. Because they are built from two-carbon units, fatty acids always have an even number of carbon atoms.

Fats insulate the body against changes in temperature and also protect vital internal organs from mechanical damage. Fat deposits represent the most efficient form in which energy is stored in the body. Triglycerides not only pack together more compactly in muscles and the liver than glycogen, they also yield more energy per gram. A typical triglyceride breaks down as follows:

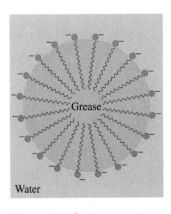

Figure 10-16 Cross section of a micelle. When soapy water is mixed with greasy dirt, nonpolar hydrocarbon ends of the soap molecules encapsulate and dissolve the grease; the polar $-COO^-$ ends of the soap molecules dissolve in the water. The micelles, which repel each other because of their like charges, are washed away during rinsing.

Adipose tissue, which is found primarily under the skin, around deep blood vessels, and in the abdominal cavity, represents carbon compounds consumed in excess of the body's need for energy and for the replenishment of glycogen reserves

$$2\ C_{55}H_{100}O_6\ +\ 154\ O_2\ \longrightarrow\ 110\ CO_2\ +\ 100\ H_2O\ +\ energy$$

The gradual breakdown of stored fat supplies hibernating animals with the water and energy they need to survive the months they spend without food. Similarly, camels in the desert survive on fats stored in their humps.

Fat Absorption of Toxic Substances

Because they are nonpolar, lipids in animal tissues absorb and dissolve other nonpolar substances. The damaging effects of polychlorinated hydrocarbons, such as PCBs and the insecticide DDT, occur when these substances enter natural waters and then become concentrated in the fatty tissues of fish and fish-eating birds (Chapter 12).

▶ WAXES

Waxes are monoesters of fatty acids and long-chain monohydroxy alcohols.

Glands in the human ear produce a wax that protects against infection by trapping airborne particles.

A great variety of waxes are found in the natural environment. All are mixtures made up from many different fatty acids. Waxes function as protective coatings on the leaves and fruits of plants, on the fur and skin of animals, and on the feathers of birds. Beeswax forms the structural part of a beehive. Commercially, waxes are used in making candles and protective coatings for floors, furniture, and automobiles.

Wax coatings on feathers are particularly vital for the survival of aquatic birds. The waxes waterproof the birds, insulating them from cold water. If the waxes are removed, feathers become wet and lose their shape (Fig. 10-17), and the birds lose their insulation and their buoyancy. This damage occurs when birds become covered in oil from an oil spill (Chapter 14). The nonpolar hydrocarbons in the crude oil dissolve the waxes on feathers and the birds cannot stay afloat; insulation is lost, and in cold waters the birds die quickly from hypothermia.

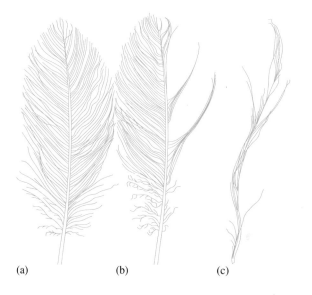

Figure 10-17 Feathers **(a)** form a watertight seal on a bird's body. Oiled feathers **(b, c)** lose their shape and thus their insulating ability.

(a)   (b)   (c)

STEROIDS

The important group of lipids called **steroids** includes cholesterol, bile salts, vitamin D, and many hormones. These lipids, which are present in both plants and animals, are completely different in chemical structure from the lipids we have discussed so far. Their basic structure, which is shared by all steroids, consists of three 6-membered rings and one 5-membered ring fused together. Steroids are distinguished from one another by the location of double bonds and the nature and location of substituent groups in the skeleton. The steroid skeleton and the formulas of cholesterol and some important hormones are shown in Figure 10-18.

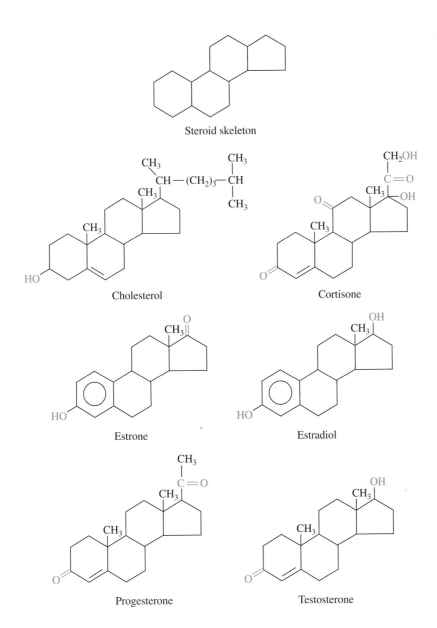

Steroid skeleton

Cholesterol

Cortisone

Estrone

Estradiol

Progesterone

Testosterone

Figure 10-18 Steroids all share the same basic 4-ring structure (three 6-membered rings and one 5-membered ring joined together). Cholesterol is the starting material for the synthesis of all other steroids in the body. Cortisone is used as an anti-inflammatory agent. Estrone, estradiol, and progesterone are female sex hormones; testosterone is the most important male sex hormone. (Functional groups are colored green.)

Cholesterol

Cholesterol is the most abundant steroid in the human body, and it is the starting material for the synthesis of all the other steroids in the body. It is found in brain and nerve tissues and in blood; it is the main component of gallstones and is present in all cell membranes.

Cholesterol is present in animal fats; therefore, milk, butter, eggs, and meat are all sources of dietary cholesterol. Plants, however, do not contain cholesterol, and foods such as vegetable oils, peanut butter, and margarine are cholesterol-free.

Although some of the body's cholesterol is obtained directly from cholesterol-containing foods, most is synthesized in the liver from carbohydrates, fats, and proteins. Reducing the intake of cholesterol-containing foods will not, therefore, automatically lower blood cholesterol levels. To achieve this result, the entire diet must be modified.

Cholesterol is a major factor in the development of atherosclerosis, the condition in which fatty deposits called **plaque** build up on the inner walls of arteries (Fig. 10-19). Plaque is composed largely of cholesterol, with some triglycerides, and its formation is associated with high levels of cholesterol and triglycerides in the blood. As plaque accumulates, the arteries become narrower and the heart must work harder to pump blood through them. If blood vessels to the brain or heart become blocked, the flow of blood is reduced, and the resulting lack of oxygen can cause a stroke or a heart attack.

Cholesterol is the starting material for the synthesis of the related steroids, the bile salts, which are stored in the gallbladder. Bile salts are released into the intestine after a meal to aid in the digestion of fats and oils (Chapter 17). In some conditions, cholesterol in the gallbladder forms into stones, sometimes as large as marbles, a condition that can cause nausea and severe abdominal pain.

Steroid Hormones

Hormones are substances produced in specialized glands in the body: the pituitary, thymus, hypothalamus, thyroid, parathyroid, adrenals, pancreas, and gonads (testes and ovaries). Hormones are secreted directly into the bloodstream and carried to different locations in the body where they regulate a variety of

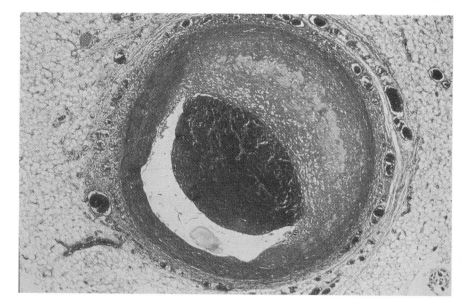

Figure 10-19 In atherosclerosis, plaque buildup (pink) can severely narrow coronary arteries and cause a heart attack.

biological processes. Many, but not all, hormones are steroids and are formed from cholesterol.

An important steroid hormone secreted by the adrenal gland is **cortisone**. When applied to the skin or injected, synthetic cortisone, which is identical to the natural hormone, acts to reduce inflammation. It is also used to treat arthritis.

The sex hormones produced by the ovaries in women and the testes in men are structurally related to cholesterol and cortisone. Two important female hormones are estradiol and estrone, which together are called **estrogens**. Unlike cholesterol and the other steroids shown in Figure 10-18, estrogens include an aromatic ring in their structure. Estrogens control the menstrual cycle, breast development, and other female sex characteristics. Another female hormone is **progesterone**, which causes changes in the uterine wall that prepare the uterus for pregnancy and which also prevents the further release of eggs from the ovary during a pregnancy.

Male sex hormones are called **androgens**. They are responsible for the development of male sex organs and masculine sex characteristics, such as facial hair and a deep voice. The most important male sex hormone is **testosterone**, which has both masculinizing (androgenic) and muscle-building (anabolic) effects. Note how little the formulas of the male sex hormone testosterone and the female hormone progesterone differ.

Synthetic anabolic steroids that some athletes use to improve performance are now banned by the world's major amateur athletic organizations. These steroids can cause sterility in men and masculinization in women, and other adverse effects, but they are useful in preventing muscle wasting in patients recovering from surgery.

CONSUMER BRIEF   Testosterone and Anabolic Steroids

As we have just seen, the structure of steroids is based on a tetracyclic ring system consisting of three 6-membered rings and one 5-membered ring (Fig. 10-18). As functional groups are added to the basic ring system, the identity and function of the steroid is defined. Cholesterol, in addition to being a steroid, is an unsaturated alcohol with a side chain (Fig. 10-18). It is the most abundant steroid in the body. An average 130 pound person has a total of 175 g of cholesterol distributed throughout the body.

There are two major classes of steroid hormones: the sex hormones, which control maturation and reproduction, and the adrenocortical hormones, such as cortisone, which regulate a variety of metabolic processes. The male sex hormone, testosterone, is responsible for the development of male secondary characteristics during puberty, and for promoting tissue and muscle growth.

Testosterone is synthesized in the testes from cholesterol. During synthesis, the alcohol group (-OH) in cholesterol is oxidized to a ketone (-C=O), the double bond is moved, and the side chain is replaced with an alcohol group.

In addition to the several hundred known natural steroids that have been isolated from plants and animals, a large number of steroids have been synthesized in the laboratory in a search for new drugs. The best known synthetic steroids are the oral contraceptives and the anabolic agents. Because anabolic steroids cause the body to build new tissue, they are sometimes taken by athletes seeking to build muscle mass.

The abuse of anabolic steroids by athletes became front page news during the 1988 Olympic games when they were detected in the urine of several athletes. Ben Johnson, the Canadian sprinter, was stripped of his gold medal in the 100 yard dash when his urine tested positive for the steroid stanozolol. In the body, stanozolol's five-membered ring containing two nitrogens, is opened and removed from the molecule. This produces a metabolite with a structure almost identical to that of testosterone. Like all anabolic steroids, Stanozolol has serious side effects. It can cause unwanted hair

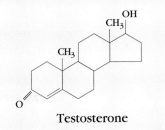

Testosterone

growth, liver dysfunction, acne, and increase the risk of atherosclerosis.

Another anabolic steroid taken by body-builders is nandrolone.

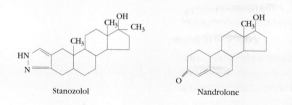

Stanozolol

Nandrolone

Notice that the only structural difference between nandrolone and testosterone is the extra methyl group on the first 6-membered ring in testosterone. Because the structure of this drug is so sim-

ilar to that of testosterone, and testosterone is naturally present in the urine of adult males, it is especially difficult for laboratories to detect it in urine samples. For this reason, nandrolone is in great demand by those abusing steroids.

Questions
1. Since steroids are not addictive, do you think they should be considered to be as dangerous as addictive drugs such as cocaine?
2. Steroids can be given to farm animals to increase their bulk. Does the meat from these animals pose a risk to children?

PROTEINS

Proteins are essential components of all living cells. Their vital importance is indicated in their name, **protein**, which is derived from the Greek word *proteios*, meaning "first" or "foremost." Proteins are present in the smallest cellular organisms as well as in more complex life forms. They are involved in practically every process that occurs in living tissues.

Proteins form the structural material of animals in much the same way that cellulose forms the structural material of plants. Skin, hair, wool, nails, tendons, and muscles are all made of proteins. As antibodies, proteins protect us from disease; as enzymes, they catalyze numerous metabolic processes in the body. The blood protein hemoglobin transports oxygen throughout the body; the protein insulin is a hormone that regulates glucose metabolism. Proteins are an important source of energy for animals. Foods that are high in proteins include meat, poultry, fish, cheese, eggs, beans, and nuts.

All proteins contain the elements carbon, oxygen, hydrogen, nitrogen, and sulfur. Some also contain phosphorus, iodine, iron, magnesium, and a number of other elements. The body contains many thousands of different proteins, and each type of cell in the body makes its own particular kinds of proteins. The building blocks of proteins are the amino acids.

Amino Acids

Proteins are synthesized in body cells by enzyme-catalyzed reactions between amino acids. **Amino acids**, as their name implies, contain an amino group ($-NH_2$) and a carboxylic acid group ($-COOH$) (Chapter 9). In nearly all the amino acids that join together to form proteins, the amino group and the carboxyl group are bonded to the same carbon atom, as shown below for a representative amino acid.

Proteins are very large molecules with formula masses ranging from 5000 amu to several million amu.

$$H_2N-\underset{\underset{R}{|}}{\overset{\overset{H}{|}}{C}}-C\underset{OH}{\overset{O}{<}}$$

amino group   characteristic side chain   ⟵ carboxylic acid group

The R-group side chain varies with each amino acid and takes many forms. In the simplest amino acid, glycine, R is a hydrogen atom. In the next simplest, alanine, R is a methyl group.

$$H_2N-\underset{\underset{H}{|}}{\overset{\overset{H}{|}}{C}}-COOH \qquad H_2N-\overset{\overset{H}{|}}{\underset{\underset{CH_3}{|}}{\overset{*}{C}}}-COOH$$

glycine   alanine   *asymmetric carbon atom

With the exception of glycine, all amino acids have one or more asymmetric carbon atoms in their structures, and, therefore, they can exist as enantiomers. As we noted earlier in the chapter, nature prefers the L-amino acid isomers. In humans and most other animals, all proteins are made from L-amino acids.

Each unique protein is composed of varying numbers of different amino acids that are linked together in a definite sequence. Virtually all proteins in our bodies are made up from the 20 amino acids listed in Table 10-2. All these amino acids are necessary constituents of human proteins. The 10 with their names written in blue are called **essential amino acids**, because they cannot be synthesized in the body from other chemicals and must be obtained from food. The remainder—the nonessential amino acids—can be synthesized in the body. The names of amino acids are abbreviated by using the first three letters of the amino acid as indicated in Table 10-2.

TABLE 10-2 Amino Acids Found in Human Proteins*

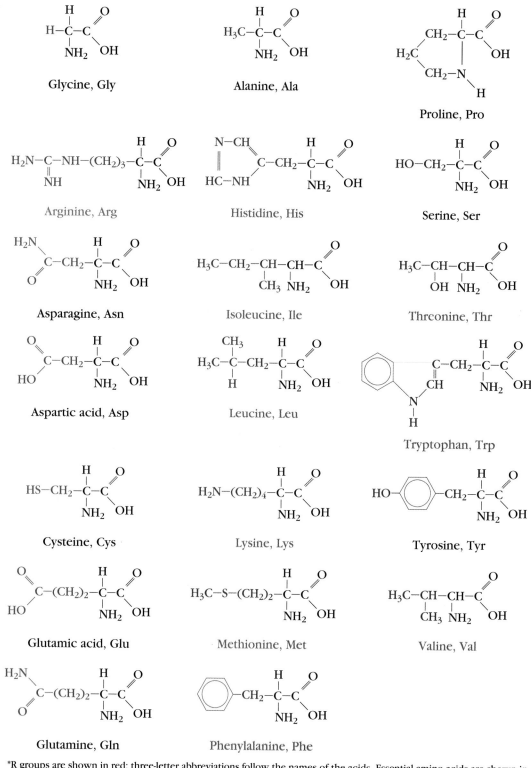

*R groups are shown in red; three-letter abbreviations follow the names of the acids. Essential amino acids are shown in blue (histidine and arginine are essential for infants only).

Peptide Bonds

Amino acids become linked together in proteins by forming amide (Chapter 9) or **peptide bonds**. The amino end of one amino acid reacts with the carboxylic acid end of a second amino acid. A molecule of water is eliminated, and a **dipeptide** is formed.

Peptides with molecular weights up to 10,000 are called polypeptides; those with higher molecular weights are called proteins.

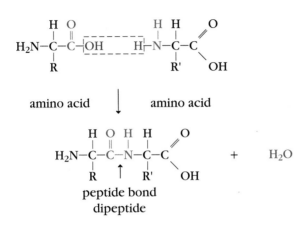

Two different amino acids can bond to form two distinct dipeptides. For example, depending on which amino group bonds with which carboxylic acid group, glycine and alanine can join to form both Gly-Ala and Ala-Gly:

$$H_2N-\overset{\overset{\displaystyle H}{|}}{CH}-\overset{\overset{\displaystyle O}{\|}}{C}-\overset{\overset{\displaystyle H}{|}}{N}-\overset{\overset{\displaystyle CH_3}{|}}{CH}-COOH \qquad H_2N-\overset{\overset{\displaystyle CH_3}{|}}{CH}-\overset{\overset{\displaystyle O}{\|}}{C}-\overset{\overset{\displaystyle H}{|}}{N}-\overset{\overset{\displaystyle H}{|}}{CH}-COOH$$

$$\text{Gly-Ala} \qquad\qquad\qquad \text{Ala-Gly}$$

These two isomers have different physical and chemical properties.

The number of possible isomers increases very rapidly as the length of the peptide chain increases. If only one molecule of each different amino acid is present, three different amino acid molecules can bond to form six distinct tripeptides:

Gly–Ala–Val	Gly–Val–Ala	Ala–Gly–Val
Ala–Val–Gly	Val–Gly–Ala	Val–Ala–Gly

Four amino acids can form 24 tetrapeptides. Eight can be sequenced in over 40,000 different ways. Sequences of up to 50 amino acids are generally called polypeptides. Longer sequences are called proteins.

Since proteins typically contain from 50 to well over a thousand amino acids in sequence, the number of possible combinations that can be made from the 20 amino acids listed in Table 10-2 is almost endless. However, nature is very selective, and only those proteins required for specific functions are actually made. As we shall see in the next section, if just one amino acid in the sequence is altered, the character of a protein is changed, and it may then be unable to perform its normal function. The human body may contain as many as 30,000 distinct proteins.

EXAMPLE 10-3   Write the structures of the dipeptides that can be formed from the amino acids alanine and phenylalanine.

Solution:

1. Two dipeptides can be formed:
   a. The amino group of alanine can bond with the carboxyl group of phenyl-alanine.
   b. The amino group of phenylalanine can bond with the carboxyl group of alanine.
2. Draw the structures of the two amino acids.
3. Show how the amino end of one combines with the carboxyl end of the other with the elimination of a molecule of water.

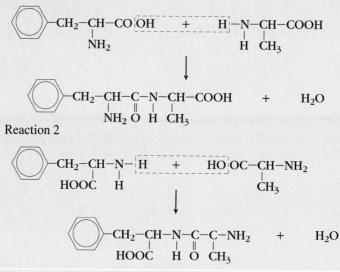

Reaction 1

Reaction 2

PRACTICE EXERCISE: Write the structures of the dipeptides that can be formed from the amino acids alanine and valine.

Answer:

The Structure of Proteins

The sequence of amino acids in a protein, its *primary structure*, determines the role the protein plays in the body. Also important is the *secondary structure*, the well-defined shape that a protein adopts because of hydrogen bonding between peptide linkages in the amino acid chains. In many proteins, including those in hair, wool, and nails, hydrogen bonding causes the amino acid chain to become twisted into a tightly coiled **helix** (Fig. 10-20a), a structure that was first proposed by Linus Pauling (Chapter 5). In the helix, hydrogen bonding occurs between the carbonyl oxygen of one peptide group and the amide hydrogen of another peptide group four amino acids farther along the chain. (Fig. 10-20b).

Another, less common secondary structure is a **pleated sheet** structure in which amino acids chains are held together, side by side, by hydrogen bonds. This type of structure is found in silk and in myosin, a muscle protein. A third secondary structure, a **triple-helix**, is found in collagen, the fibrous protein of connective tissues such as cartilage, tendon, and skin. In these proteins, three helical amino acid chains are braided together to form strong ropelike structures.

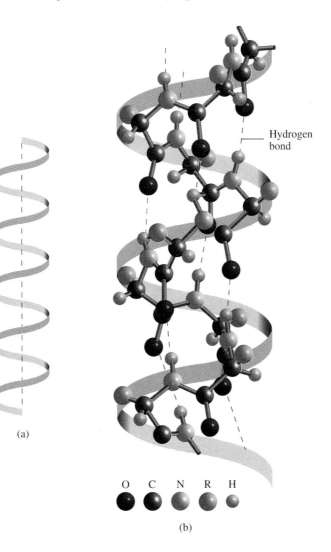

(a)

(b)

O C N R H

Hydrogen bond

Figure 10-20   The helical structure proposed by Linus Pauling. **(a)** A right-handed helix. **(b)** Schematic diagram showing how hydrogen bonding between different parts of the amino acid chain holds the helix together. R represents any of the side chains listed in Table 10-2.

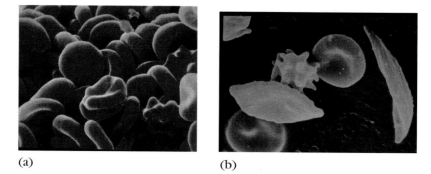

(a)   (b)

Figure 10-21 Electron micrograph showing **(a)** normal red blood cells and **(b)** sickled red blood cells.

Proteins also have a *tertiary structure*. This is the three-dimensional shape that results from further folding, cross-linkings, and coiling of the amino acid chains caused by interactions between the R groups of the amino acids. Four major types of interactions are: (1) the formation of -S-S- covalent bonds (disulfide linkages) between pairs of cysteine side chains (Table 10-2); (2) electrostatic attractions between amino acids with polar side chains such as aspartic acid and lysine; (3) hydrogen bonding between amino acids with $-OH$, $-NH_2$, $-COOH$, and $-CONH_2$ groups in their side chains; and (4) weak London forces (Chapter 7) between nonpolar side chains. As a result of these interactions, most proteins adopt a **globular** shape; others have a more linear, or **fibrous**, configuration.

The vital importance of correct protein structure is demonstrated in the hereditary disease **sickle cell anemia**. In this disorder, the nonpolar amino acid valine is substituted for one of the polar glutamic acid molecules in one of the polypeptide chains of hemoglobin. This alteration of one specific amino acid in the sequence of 146 in the chain alters the tertiary structure of hemoglobin. The protein assumes an abnormal three-dimensional shape (Fig. 10-21), the red blood cells become distorted, and the blood's ability to transport oxygen is reduced. The cells break up easily and block capillaries, causing severe pain.

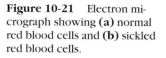

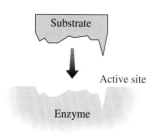

Substrate-enzyme complex

ENZYMES

Enzymes are biological catalysts. They are synthesized by all living organisms, and, with very few exceptions, all are globular proteins. Enzymes are essential for nearly all the chemical reactions that occur in living systems, and thousands of them are present in the human body. They take part in the digestion of food; the synthesis of complex carbohydrates, fats, and proteins; the process of photosynthesis; and the reactions that release energy to tissues. Like all catalysts, enzymes increase the rate of a reaction by lowering the activation energy (Chapter 7).

Enzymes are extraordinarily efficient catalysts. In the body, enzyme-catalyzed reactions occur extremely rapidly under very mild conditions, that is, at neutral pH and a temperature of 37°C (98.6°F). In the laboratory, in the absence of the enzyme, these same reactions occur very slowly unless accelerated by raising the temperature and adding an acid or a base.

The action of an enzyme can be compared to a key opening a lock (Fig. 10-22). Just as a key must have the right shape to open a particular lock, an enzyme for a particular reaction must have the right shape to fit the molecule on which it is acting. The substance with which an enzyme interacts is called the **substrate**; the part of the protein where the interaction occurs is the **active site**. The active

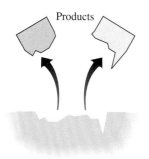

Figure 10-22 The lock-and-key model for enzyme action. The substrate fits the active site on the enzyme in the same way that a key fits a lock. The separation of products from the substrate-enzyme complex represents the opening of the lock.

site occupies a very small part of the total enzyme. A substrate-enzyme complex is formed during an enzyme reaction. The complex breaks apart as the product(s) is formed, leaving the enzyme free to catalyze further reactions.

Some enzymes are simple proteins. Others consist of a protein plus an additional nonprotein group called a **coenzyme**. The coenzyme may be a metal ion (for example, Fe^{3+}, Mg^{2+}, or Cu^{2+}) or one of the water-soluble vitamins (Chapter 17). Without its coenzyme, an enzyme has no activity.

Most enzymes are very specific in their action and catalyze just one reaction or a few related reactions. The group of intestinal enzymes called **amylases** catalyze the breaking of bonds between the α-glucose units in starch in foods, but they do not catalyze the breaking of bonds between the β-glucose units in cellulose. The enzyme **sucrase** catalyzes a single reaction: the breakdown of sucrose to glucose and fructose. The enzyme **trypsin** is less specific. It aids in the digestion of all proteins by breaking peptide bonds. The common names of most enzymes, and all their systematic names, end in the suffix *-ase* (trypsin is an exception).

The lack of a particular enzyme can have serious consequences and is the cause of a number of hereditary conditions including **lactose intolerance**. This condition affects many Asians and Africans, who in their own countries traditionally drink little milk after they are weaned. When they become adults, their bodies stop producing **lactase**, the enzyme needed to digest the lactose in milk. As a result, if they consume milk or milk products, they suffer stomach pain and diarrhea.

We noted at the beginning of the chapter that a person fed a diet composed of L-sugars and D-amino acids would starve. The main reason is that the active sites on enzymes have a particular "handedness." Only molecules with the correct handedness can fit into the active sites on the body's enzymes and form products.

One reason that vitamins and minerals are an important part of our diet is that they often act as coenzymes in vital metabolic processes.

CONSUMER BRIEF   **Hair Dyes**

The natural pigment melanin, which gives skin its color, also colors hair. Blondes have a little melanin in their hair, brunettes a moderate amount, and those with jet black hair have large quantities. The loss of melanin is responsible for gray and white hair. Only redheads are different; their hair contains an additional unique iron-based pigment, phaeomelanin, which makes "rusty" an especially accurate description of a redhead.

Bleaching hair to a blond color can be accomplished easily by applying hydrogen peroxide. Hydrogen peroxide is an oxidizing agent that reacts with melanin and removes it from the cortex of the hair, producing the bleached blond color. Converting dark hair to a lighter shade of brown involves bleaching with hydrogen peroxide followed by

dyeing with a light pigment. The addition of dyes to unbleached hair darkens the color.

Hair dyeing can be accomplished with temporary or permanent dyes. Temporary dyes color the hair with chemicals that are water-soluble and wash out easily when the hair is shampooed. More permanent dyes penetrate the hair and remain fixed there.

Temporary dyes such as FD&C blue, yellow, and red are mixed together to give a pleasing color when coated on hair. Not only are the molecules in these dyes water-soluble, but they are also so large that they don't penetrate into the cortex of the hair. They are deposited on the cuticle and are easily removed by repeated washing.

Permanent dyes last as long as the strand of hair. They are a three-component system that consists of a primary intermediate, a coupler, and an oxidizing agent. The primary intermediate molecules are small mobile molecules, such as para-phenylenediamine or para-aminophenol, that can migrate into the keratin.

NH_2 —⬡— NH_2 NH_2 —⬡— OH

p-phenylenediamine   p-aminophenol

These molecules have reactive functional groups on both ends. One end can react with the shaft of the hair. The other end is attached to the coupler molecule by the oxidizing agent in the multistep process of dyeing. Para-phenylenediamine produces a black color, but variations in hair color can be achieved by placing a variety of substituents on this molecule. In this process, a molecule of dye is synthesized and permanently attached to the hair. It should be noted that even permanent dyes affect only the exposed hair. As new hair grows from the scalp, it has its natural color and "roots" begin to show.

There are some risks associated with using permanent dyes. One derivative of p-phenylenediamine, p-methoxy-m-phenylenediamine (MMPD) has been shown to be carcinogenic when fed to laboratory mice.

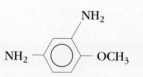

p-methoxy-m-phenylenediamine, MMPD

Because of its cancer risk, it has been removed from most commercial hair dyes. Those dyes that still use it must carry a warning on the label. Because MMPD is structurally similar to many of the other chemicals used for permanently dyeing hair, there is concern about the risk posed by these chemicals, as well. None of them are completely harmless.

Inorganic hair dyes, such as Grecian Formula darken the hair slowly and are often used daily. Grecian Formula gets its name from the aqueducts used by the ancient Greeks, who made their aqueducts watertight by lining them with a lead sheet. The soft, low-melting metal was easy to fabricate and formed an effective seal, but the flowing water slowly dissolved the lead and contaminated the water. It was noticed that people who bathed often in this water had very dark hair. Today, Grecian Formula contains colorless lead acetate, $Pb(CH_3COO)_2$, and is formulated as a shampoo or cream. As the lead ion in the solution penetrates the shaft of the hair, it reacts with sulfur in the amino acid cysteine in the hair protein and forms lead sulfide, PbS, which is black. Repeated application deepens the color as more lead reacts. Depending on genetics, individuals who have a high percentage of cysteine in their hair protein will achieve dark hair more rapidly than people with little cysteine.

Questions
1. Should consumers be concened about cancer-causing chemicals in hair dyes?
2. Does using a hair dye that contains lead have any adverse effects on the environment?

NUCLEIC ACIDS

The complex compounds called **nucleic acids** are found in all living cells except mammalian red blood cells. There are two kinds of nucleic acids: **deoxyribonucleic acid (DNA)** and **ribonucleic acid (RNA)**. DNA is found primarily in the nucleus of cells, while RNA is found primarily in the cytoplasm, the part of the cell surrounding the nucleus.

Functions of Nucleic Acids

DNA and RNA are responsible for the storage and transmission of genetic information. In the cell nucleus, DNA is bonded to protein to form **chromosomes**. The **genes**, which are segments of the DNA molecule, are the basic units of heredity. DNA molecules have the ability to replicate, that is, to produce exact copies of themselves. This ability makes it possible for genetic information to be transferred from one cell to another and for genetic traits to be transmitted, via sperm and eggs, from parents to offspring. A major function of DNA, and one in which RNA is involved, is the control and direction of protein synthesis in body cells. Chemical information stored in the DNA of genes specifies the exact nature of the protein to be made and thus determines the character of the organism.

The Primary Structure of Nucleic Acids

In much the same way that polysaccharides are composed of monosaccharides and proteins are composed of amino acids, nucleic acids are composed of long chains of repeating units called **nucleotides**. DNA molecules are the largest of the naturally occurring organic molecules.

Each nucleotide in a chain is made up from three components: (1) a **sugar**, (2) a **nitrogen-containing heterocyclic base**, and (3) a **phosphoric acid** unit (Fig. 10-23).

The sugar is a pentose, either ribose or deoxyribose (Fig. 10-24). The only difference between these two sugars is at carbon 2, where ribose has a hydrogen atom and an −OH group, and deoxyribose has two hydrogen atoms. As the names indicate, the sugar in DNA is deoxyribose, while that in RNA is ribose.

The five different bases found in nucleic acids are shown in Figure 10-25. Two—**adenine** and **guanine**—are double-ring bases and are classified as **purines**. The other three—**cytosine**, **thymine**, and **uracil**—are single-ring bases and belong to the class of compounds called **pyrimidines**. Purines and pyrimidines are bases because their nitrogen atoms can accept protons (Chapter 8). Adenine, guanine, and cytosine are found in both DNA and RNA. Thymine is found only in DNA, while uracil is normally found only in RNA.

The formation of a representative nucleotide from its three components is shown in Figure 10-26. The pentose forms an ester bond with a phosphoric acid unit through the −OH on carbon 5 and forms another bond with the base through the −OH group on carbon 1. Water is eliminated as the bonds are formed. When successive nucleotides bond together to form long-chain nucleic acids, the phosphoric acid units form a second ester bond through the −OH group on carbon 3 of a second pentose molecule (Fig. 10-27). The alternating sugar-phosphoric acid units form the backbone of the nucleic acids.

The Double Helix

The sequence of the four bases in a nucleic acid, like the sequence of amino acids in proteins, is its primary structure. Like proteins, nucleic acids also have a secondary structure. The secondary structure of DNA was determined in

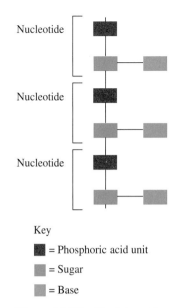

Key

■ = Phosphoric acid unit

▬ = Sugar

▬ = Base

Figure 10-23   Nucleic acids consist of long chains of nucleotides. Each nucleotide in the chain is composed of a phosphoric acid unit, a sugar, and a base.

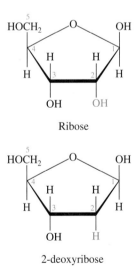

Ribose

2-deoxyribose

Figure 10-24   The pentoses ribose and deoxyribose differ at carbon 2 where ribose has a hydrogen atom and an −OH group, and deoxyribose has 2 hydrogen atoms. Ribose is present in RNA; deoxyribose is present in DNA.

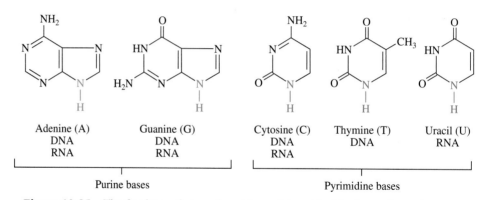

Figure 10-25 The five bases that are found in nucleic acids. Adenine and guanine are purines; cytosine, thymine, and uracil are pyrimidines.

1953, at Cambridge University, by the American James D. Watson and the Englishman Francis H. C. Crick (Fig. 10-28). Their discovery, with all its biological implications, is one of the most significant scientific achievements of the twentieth century.

In the early 1950s, many facts were known about DNA. One particularly important observation was that when DNA from many different organisms was analyzed, the amount of adenine (A) always equaled the amount of thymine (T), and the amount of cytosine (C) always equaled the amount of guanine (G). X-ray studies by Maurice H. F. Wilkins and Rosalind Franklin, who were working at King's College (London), revealed the distances and angles between atoms in the DNA molecule, but the precise three-dimensional configuration of the nucleotide units in the molecule eluded scientists until 1953.

On the basis of the available information, Watson and Crick concluded that a DNA molecule must consist of two polynucleotide chains wound around one another to form a **double helix**, a structure that can be compared to a spiral staircase (Fig. 10-29a). The phosphate-sugar backbone represents the handrails, and the pairs of bases linked together by hydrogen bonds represent the steps (Fig. 10-29b). Working with molecular scale models, Watson and

> Rosalind Franklin did not share the Nobel prize with Watson, Crick, and Wilkins. She died of cancer at the age of 37, four years before it was awarded.

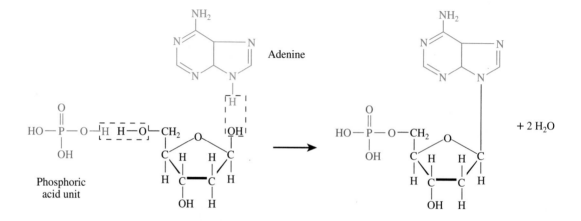

Figure 10-26 The formation of a nucleotide from its three components.

Crick realized that, because of the dimensions of the space available inside the two strands of the helix and the relative sizes of the bases, base G on one chain must always be bonded to base C on the other chain, and base A must always be bonded to base T. This arrangement gives "steps" of almost equal lengths and explains why amounts of C and G and amounts of A and T in DNA molecules are always equal (Fig. 10-29c). Although hydrogen bonds are relatively weak, there are so many of them in a DNA molecule that, under normal physiological conditions, they hold the two chains together. The C-G (and G-C) base pairs are held together by three hydrogen bonds, the A-T (and T-A) base pairs by two (Fig. 10-29c).

For their work in solving the DNA puzzle, Watson, Crick, and Wilkins were awarded a Nobel Prize in 1962.

DNA molecules vary in the number and sequence of the base pairs they contain. In some molecules, there may be as many as 30,000 base pairs. The precise sequence of base pairs in the DNA molecule is the key to the genetic information that is passed on from one generation to the next; it is this sequence that directs and controls protein synthesis in all living cells.

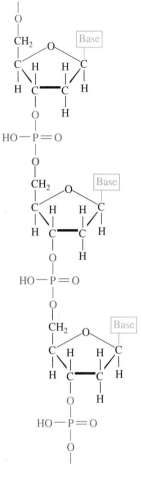

Figure 10-27 The structure of a 3-nucleotide section of a DNA molecule. The alternating sugar and phosphate groups form the backbone of the molecule.

Figure 10-28 James D. Watson (left) and Francis H. C. Crick at the Cavendish Laboratory, Cambridge University, with their model of DNA.

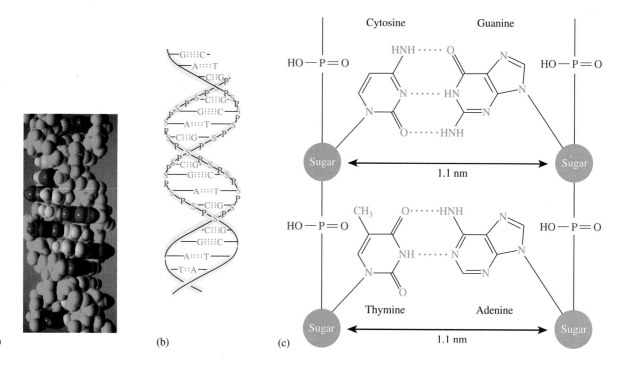

Figure 10-29 **(a)** A computer-generated model of the DNA double helix proposed by Watson and Crick. The blue atoms represent the sugar-phosphate chains. The red and yellow atoms represent the bases. **(b)** Schematic representation of the double helix model of DNA. A = adenine, C = cytosine, G = guanine, T = thymine, S = sugar, P = phosphate, ⁝⁝⁝ and ⁝⁝⁝ represent hydrogen bonds. **(c)** In the DNA molecule, thymine (a pyrimidine) pairs with adenine (a purine) and cytosine (a pyrimidine) pairs with guanine (a purine) to form the steps in the spiral staircase structure. This pairing gives steps of equal length (1.1 nm). Each thymine-adenine pair is held together by two hydrogen bonds; each cytosine-guanine pair by three.

Cell Replication

The double helix structure of DNA proposed by Crick and Watson explains very simply and elegantly how cells in the body divide to form exact copies of themselves. The two intertwined chains of the double helix are complementary. An A on one always pairs with a T on the other, and similarly a C pairs with a G. Just before a cell divides, the double strand begins to unwind (Fig. 10-30). Each unwinding chain serves as a pattern, or template, for the formation of a new complementary chain. Nucleotides, which are always present in the cell fluid surrounding DNA, are attracted to the exposed bases and become bonded to them: A to T, T to A, C to G, and G to C. In this way, two identical DNA molecules are formed, one for each of the two daughter cells.

The Structure and Functions of RNA

As we noted earlier, the sugar in RNA is ribose, and the bases are uracil (U), adenine (A), guanine (G), and cytosine (C) (Fig. 10-25). The primary structure of RNA is similar to that of DNA (Fig. 10-23). Ribose-phosphoric acid units form the backbone, and each ribose unit is bonded to one of the four bases.

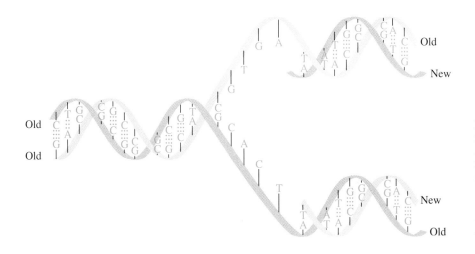

Figure 10-30 Schematic representation of DNA replication. As the original DNA double helix gradually unwinds, new complementary chains of nucleotides are attached to the single strands and two new identical DNA molecules are formed.

RNA molecules are much smaller than DNA molecules and contain from 75 to a few thousand base pairs. RNA molecules exist primarily as single strands rather than in a double helix form, and therefore base pairing is not part of their secondary structure. However, during their involvement in protein synthesis, RNA molecules form base pairs with other RNA molecules; G bonds with C as in DNA, and U bonds with A.

Three types of RNA have been identified, each with a different role in the complex process of protein synthesis. This process proceeds in two main steps:

$$\text{DNA} \xrightarrow{\text{transcription}} \text{RNA} \xrightarrow{\text{translation}} \text{protein}$$

In the first step, the genetic message in a gene is transferred (or transcribed) to one form of RNA, which acts as a template for protein synthesis. The genetic message in RNA is then deciphered (or translated), and, with the intervention of the other forms of RNA, a protein is formed.

EXPLORATIONS

Beer: America's Favorite Alcoholic Beverage

Alcoholic beverages have been consumed since before the dawn of recorded history; long before there was Budweiser or Miller, there was beer. Ancient records reveal that by 3000 B.C., beer was already an important commercial product in Babylon and a common dietary item in Egypt. Greek and Roman writers describe exuberant consumption of wine and beer in classical times by the gods and by mortals. In the first few centuries A.D., beer drinking became particularly popular among Germanic and Nordic tribes and among the Saxons and Celts of Northern Europe where the climate was unsuitable for cultivating grapes for making wine. By the Middle Ages, most larger households made beer for their own consumption. It was in the monasteries, however, that the real advances in the art of brewing occurred. Most monasteries maintained at least one brewhouse, and monks received a daily ration of about a gallon of beer. Gradually, as towns grew, local commercial breweries were established, each one producing its own unique brew (Fig. 10-A).

Figure 10-A   A seventeenth century brewhouse.

Beer came to North America with the early English settlers, and in the years leading up to the American Revolution, breweries became an important part of the economy. Homebrewing remained popular, and we know that George Washington had his own recipe for making beer on his Mount Vernon estate. The Pilsner-type beers, so popular today, came to the United States when the German immigrants began arriving early in the nineteenth century.

Americans love their beer and today drink more of it than either wine or hard liquor. Wine, the second most popular alcoholic beverage, is made primarily from grapes, while beer is made from sprouted grain, called malt. The principal ingredient in all these alcoholic beverages, and the cause of the intoxicating effects they produce, is the simple but mind-altering chemical ethanol.

The term *beer* usually implies a beverage of Western culture made principally from fermented barley and flavored with hops. But nearly all other cultures have a long history of beer making as well. Beer made from rice has been made in Asia for thousands of years, and long before the discovery of the New World by Europeans, the Aztecs and Incas were making beer from fermented corn. In Africa, excellent beers are made from sorghum seeds.

Hops have been used as bittering agents to counteract the sweet taste of sugar in fermented grain since about 3000 B.C., but were probably not cultivated for this purpose until the ninth century A.D. The hop (*Humulus lupus*) is a perennial vine common in Europe, Asia, and North America. Male and female flowers grow on separate plants, and it is the cone-shaped flower clusters of the female plant that contain the oils and resin that give beer its characteristic bitter taste. Male plants are excluded from hop fields as far as possible to prevent fertilization, which results in the formation of unpleasant-tasting seeds. Hops require a mild climate and an abundant supply of water, and in the United States, they are cultivated primarily along the Pacific coast.

Today, the complex chemical process of brewing is fully automated, but it still follows the four basic steps established in the Middle Ages: malting, mashing, wort boiling, and fermentation. To make hard liquor, such as vodka, gin, whiskey, rum, and brandy, an extra distillation step is required.

In malting, the grain is steeped in water, which softens it and initiates germination. As the grain sprouts, enzymes are formed and begin the conversion of starch to dextrins, malt sugar (maltose), and other fermentable sugars. At the appropriate stage in germination, the grain is kiln-dried to stop further growth. The kilning temperature is critical for the development of the desired amount of color in the final product. The higher the temperature, the darker the beer.

In the next step, mashing, dried sprouted grain, now called malt, is ground and mixed with hot water. Enzymes continue to work, and residual starch is broken down into simpler carbohydrates. If a pale light American beer is being brewed, corn or rice, which are cheaper than malt, is added at this stage. Enzymes in the malt act on starch in the added grains, thus increasing the production of fermentable sugars. Soluble carbohydrates, together with various other organic compounds and minerals, dissolve in the water, which is then separated from the spent grain. (Rich in protein, the spent grain is sold as animal feed.) The liquid, called wort, is boiled with hop cones, or an extract of hops, to acquire the desired bitter flavor. After the wort has been cooled and strained, yeast is added, and fermentation, the most critical step in the brewing process, begins.

During fermentation, enzymes produced by yeast convert sugars to ethanol (ethyl alcohol) and carbon dioxide.

$$C_6H_{12}O_6 \longrightarrow 2\ C_2H_5OH + 2\ CO_2$$

Since the yeast greatly affects the flavor of the final product, each brewery has its own special strain. To make a typical American, or Pilsner-type, lager beer, the brewer uses a bottom-fermenting yeast, which sinks to the bottom when spent. When fermentation is complete, the beer is stored for several months in a cold cellar (the German word *lager* means "to store"), where it becomes clear and acquires its mature flavor. To produce the slightly stronger, more highly flavored ales popular in the United Kingdom, a higher temperature and a top-fermenting yeast, which rises to the surface when spent, are used. Cask beers acquire their unique flavor as a result of residual fermentation that is allowed to occur in the cask. Before beer is bottled or canned, it is filtered, carbonated, and usually pasteurized.

The main ingredients in beer are ethanol and carbon dioxide, but trace amounts of many other compounds, including other alcohols, aldehydes, ketones, acids, esters, vitamins, and minerals, are also present. These ingredients, derived in varying amounts from hops, grain, yeast, and water, are responsible for subtle nuances of flavor in the final product. Although many beer drinkers often express a preference for a particular brand of beer—Budweiser, Michelob, Miller, or Schlitz, for example—in blind tests, very few are ever able to recognize their proclaimed favorite.

When a person drinks a can of beer, most of the alcohol in it (4.0–5.0% by volume in the average American beer) is absorbed directly from the intestines into the bloodstream and carried to the liver. In the liver, enzymes begin the oxidation of the alcohol to acetaldehyde, then to acetic acid, and finally to carbon dioxide and water.

Ethanol   Acetaldehyde   Acetic acid

$$C_2H_5OH \longrightarrow CH_3CHO \longrightarrow CH_3COOH$$
$$\longrightarrow CO_2 + H_2O$$

If alcohol is consumed faster than it can be oxidized in the liver (more than about 1 oz of alcohol per hour), blood going to the brain contains alcohol, which has an immediate effect on mental processes. At low blood concentration (slow consumption of a 12-oz can of beer), alcohol relaxes tension and acts as a stimulant but, as the alcohol concentration in blood increases, the brain begins to process information in abnormal ways. Learning skills deteriorate, and the effects are increasingly depressant. If three or four beers are consumed rapidly, judgment is measurably impaired. After about six beers, a person is visibly drunk, with slurred speech, unsteady gait, and confused thinking. After nine or ten beers in quick succession, most people lose consciousness.

The effects of alcohol vary considerably depending on a person's sex, size, genetic makeup, previous drinking habits, and many other factors. Women, because they lack a stomach enzyme that in men destroys some alcohol before it reaches the intestines, are approximately twice as vulnerable to alcohol as men. And they face another danger if they are pregnant. Any alcohol in their blood will reach the fetus, where it can cause stunted growth, disfigurement, and mental retardation in the unborn child, a condition known as fetal alcohol syndrome.

In moderate quantity, alcohol produces pleasurable effects and may even be beneficial. Research has shown that a daily glass of wine or can of beer can lessen the chance of developing heart disease. Beer drinking can be a civilized activity. The convivial Sunday afternoon spent watching football with friends and drinking a few beers has become a popular American ritual. However, irresponsible drinking can have devastating effects—destroying lives, ruining health, and causing thousands of highway accidents. Beer is great stuff, but a beer binge can be a killer. The philosophers made sense when they advised "moderation in all things."

Question

Irresponsible consumption of alcohol causes many health problems and is responsible for numerous automobile accidents. In your opinion, should alcoholic beverages be taxed to help pay for a national health plan?

Reference: *Encyclopedia Brittanica*, 1974 ed. See "beer" and "brewing."

⏩ KEY WORDS AND CONCEPTS

amino acid	fatty acid	peptide bond
androgens	fructose	plane-polarized light
carbohydrate	galactose	polysaccharide
cellulose	glucose	progesterone
chiral molecules	glycogen	protein
cholesterol	hexose	ribonucleic acid (RNA)
coenzyme	lactose	saturated fatty acid
cortisone	lipid	starch
deoxyribonucleic acid	maltose	stereoisomers
(DNA)	micelle	steroid
diastereomers	monosaccharide	structural isomers
disaccharide	nucleic acid	substrate
enantiomers	nucleotide	sucrose
enzyme	oil	testosterone
estrogens	optically active	triglycerides
fat	pentose	unsaturated fatty acid

⏩ QUESTIONS AND PROBLEMS

1. Which of the following is a chiral molecule? Explain and draw the enantiomer of the chiral molecule.

$$
\begin{array}{ccc}
& H & & H \\
& | & & | \\
R-&C&-R & R''-C-R \\
& | & & | \\
& R' & & R'
\end{array}
$$

2. What is a chiral molecule, and how do you determine how many asymmetric carbon atoms are present?

3. What are the three most important classes of carbohydrates?

4. Glucose and fructose have the same molecular formula: $C_6H_{12}O_6$. Draw their structural formulas. How do they differ?

5. Cellulose is made from glucose, but humans cannot digest cellulose. Explain.

6. How does your body store carbohydrates?

7. What similarities are there between starch and cellulose?

8. Glucose molecules exist mostly as cyclic rings, rather than as straight chains. Draw the straight-chain and cyclic structures. Number the carbon atoms in each structure.

9. Give chemical names for each of the following:
   a. cane sugar **b.** grape sugar **c.** milk sugar **d.** sugar found in cola

10. Identify whether the following are monosaccharides, disaccharides, or polysaccharides:
    a. sucrose **b.** glucose **c.** cellulose **d.** lactose

11. Name the two monosaccharides in
    a. sucrose **b.** maltose **c.** lactose.

12. What is glycogen? How is it formed, and what is its role in the body?

13. Is (+) glucose an enantiomer of (−) glucose?

14. Name the enzyme our bodies use to break down table sugar. What are the products of this breakdown?

15. What is the difference between starch, amylopectin, and glycogen?

16. Why are carbohydrates considered to be a source of energy?

17. How are lipids classified? List two physical properties of lipids.

18. What is the difference between unsaturated and saturated fatty acids? Give an example of each.

19. What is a polyunsaturated fat?

20. What happens when a polyunsaturated fat is hydrogenated?

21. Name several foods that contain unsaturated fats.

22. What molecules react to form triglycerides? Write an equation showing how a triglyceride is formed.

23. Name an anti-inflammatory steroid produced naturally in our bodies. Draw the skeletal structure of a steroid.

24. In what organ is cholesterol converted to bile salts?

25. What two functional groups are present in an amino acid molecule? Draw the structure of a representative amino acid.

26. Name the simplest amino acid and draw its structure. Is there an asymmetric carbon in the molecule? Explain.

27. Approximately how many amino acids are needed to make the proteins in our bodies? How many are essential amino acids, and what does that term mean?

28. What element is always present in amino acids but is not present in sugars or fats?

29. Show how two amino acids join to form a dipeptide. Outline the peptide bond.

30. If three different amino acids (leucine, glycine, histidine) formed all possible different dipeptides, how many would there be? List them, using the appropriate three-letter abbreviation to represent each amino acid.

31. What is meant by the terms *primary*, *secondary*, and *tertiary structures* of proteins? Draw a diagram showing the outline of the secondary structure of the protein in hair. Name this type of structure.

32. What type of bonds hold the helix structure of proteins in place?

33. Describe the lock-and-key theory of enzymes. Include a diagram.

34. What are the functions of nucleic acids?

35. Nucleotides are made up of what three components? Use a simple diagram to show how the components are bonded together.

36. Name two types of nucleic acids. How do the sugar molecules in these two types differ from each other?

37. Write the base that forms a pair with the following bases:
a. uracil **b.** guanine **c.** adenine **d.** cytosine

38. The base sequence along one strand of DNA is ACTGT. What would be the sequence on the complementary strand of DNA?

39. Describe how DNA replication occurs.

SYNTHETIC POLYMERS
Materials for the Modern Age

Synthetic polymers are made into numerous consumer products including the thin film used in food packaging shown being manufactured here.

During the last 75 years in the United States, the production of synthetic polymers, which are best known to the general public as plastics, has developed into a multibillion dollar industry that has revolutionized the way we live. Polyester, nylon, Dacron, Styrofoam, Teflon, Formica, and Saran are all familiar synthetic polymers. We use them, and many other polymers, in almost every aspect of our daily lives. In the home, synthetic polymers are used in roofing, wall paneling, carpeting, furniture, paint, appliances, dishes, and food packaging. The steering wheel, instrument panel, upholstery, tires, and other parts of our cars are made of polymers; many of the clothes we wear are made from synthetic polymer fibers. In the United States today, production of synthetic polymers exceeds 90 kg (200 lb) per person, an annual amount that has exceeded that of steel since 1976. We are indeed living in the Age of Polymers.

The great surge in the production of polymers occurred in response to several factors. These included the need to find alternatives for dwindling supplies of natural raw materials, such as wood, rubber, and some metal ores, and the increasing ability of chemists to make polymers that, for many purposes, were superior to materials normally used. Certain polymers, for example, are stronger than steel, and many have specialized properties that natural materials do not possess. The cheap and plentiful supply of petroleum, which is the main source of raw materials for polymer manufacturing, made it less costly to make products by molding plastics than by shaping and joining wood or metals.

In this chapter, we will examine the basic chemical reactions that lead to the formation of different types of polymers. We will study the structural differences that account for the diverse properties of polymers, and we will consider some of the ways in which common polymers are used. We will also discuss the problem of disposing of the mounting volume of discarded plastic products.

Learning Goals:

In this chapter, you should gain an understanding of:

1. The synthesis of macromolecules by addition, condensation, and rearrangement polymerization.

2. Similarities and differences between natural and synthetic polymers.

3. The structures and properties of common synthetic polymers.

4. Differences between silicones and carbon-based polymers.

5. The problems associated with the disposal of plastics.

POLYMERS: CHARACTERISTICS AND DIFFERENT TYPES

Polymers are giant molecules with molecular weights ranging from several thousand to over a million. These macromolecules are composed of thousands of small molecular repeating units called **monomers** that are linked together to form long chains and, in some cases, complex three-dimensional networks. By changing the building blocks (the monomers), an incredible number of polymers with a wide range of properties can be synthesized.

The word polymer is derived from the Greek words *poly*, meaning "many," and *meros*, meaning "parts."

Homopolymers and Copolymers

Polymers made from identical monomers are called **homopolymers** (Fig. 11-1a). Those made from two or more different monomers are called **copolymers**. The monomers in copolymers may be linked in both regular and irregular patterns. For example, depending on the reaction conditions for polymerization, two different monomers can alternate regularly or they can link together randomly. In still other arrangements, blocks of one monomer can be introduced at intervals along chains of the other monomer, or they can be grafted on as side chains (Fig. 11-1b).

Plastics

Polymers are frequently—but often incorrectly—called plastics. While all plastics are polymers, not all polymers are plastics. A **plastic**, in polymer terminology, is a material that, at some point in its manufacture, is in a sufficiently fluid state to allow it to be molded into various shapes.

The word plastic is derived from the Greek word *plastikos*, meaning "capable of being molded or shaped."

Plastics can be classified as either thermoplastic or thermosetting. **Thermoplastics** can be repeatedly softened by heating and molded into shapes that harden on cooling. **Thermosetting** plastics are also soft when first heated, and they can be molded into shapes that they retain on cooling. When reheated, however, they do not again become soft but degrade or decompose.

Natural Polymers

Many polymers, both inorganic and organic, are found in the natural world. Silicates (Chapter 2), which make up so much of the earth's rocky surface, are polymers constructed from silicon dioxide (SiO_2) monomers. Diamond and graphite (Chapter 9) are polymers in which the monomer is a carbon atom. Other naturally occurring polymers include starches, cellulose, proteins, the nucleic acids DNA and RNA (Chapter 10), and natural rubber. The monomer in starches is α-D-glucose, while in cellulose it is β-D-glucose. In proteins, the monomers are amino acids. In nucleic acids, they are simple sugars, nitrogen bases, and phosphoric acid species.

Until the advent of synthetic polymers, building construction, furnishings, and clothing relied heavily on natural materials. These materials were primarily wood, cotton, and linen (cellulose polymers); wool, silk, and leather (protein polymers); and natural rubber.

Early Synthetic Polymers

The early synthetic polymers were designed as substitutes for natural materials and were manufactured using natural starting materials. The first synthetic plastic was developed in 1869 by the American inventor John Wesley Hyatt. His

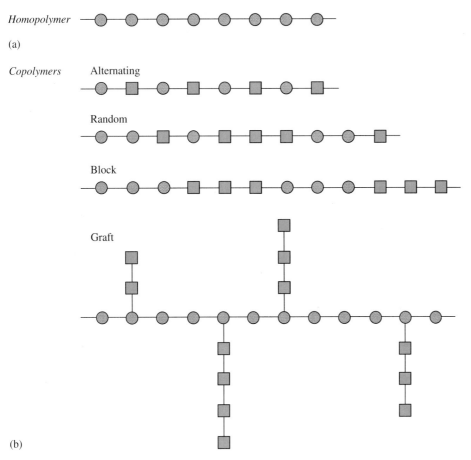

Figure 11-1 The formation of homopolymers and copolymers.

By the latter half of the nineteenth century, huge numbers of elephants were being slaughtered to meet the enormous demand for ivory, and in 1863, a New England billiard-ball manufacturer offered a $10,000 reward to the person who could develop an ivory substitute. Hyatt developed celluloid in pursuit of this prize, but ironically never won it.

product, which he called **celluloid**, was made from cellulose nitrate obtained by treating cellulose with nitric acid. Celluloid was used to make billiard balls, piano keys, combs, stiff collars, toys, and early motion picture film. It was regarded primarily as a substitute for ivory and tortoise shell. Despite its flammability, celluloid remained popular until the end of the nineteenth century, when it was gradually replaced with new, safer synthetic materials.

The first artificial fibers that could be woven into fabric were also made from cellulose nitrate. These shiny fibers were developed as a replacement for silk and given the name **rayon**. In modified form, rayon is still produced today.

By the early 1900s, the chemical industry had discovered that synthetic polymers could be made more efficiently by using small molecules as starting materials than by altering natural polymers. The first completely synthetic plastic, Bakelite, was made in 1909. It will be considered in the section on condensation polymers.

The Formation of Different Types of Polymers

There are three general types of polymers: (1) addition polymers, (2) condensation polymers, and (3) rearrangement polymers. **Addition polymers** are formed when monomers link together without the loss of any atoms from the monomers. Important addition polymers that we will consider are polyethylene, poly(vinyl chloride) (PVC), polystyrene (Styrofoam), polyacrylonitrile (Dacron, Acrilan), polytetrafluoroethylene (Teflon), and synthetic rubber. **Condensation polymers** are formed when monomer units, in linking together, eliminate a small molecule, usually water. Examples of condensation polymers are polyesters, polyamides (nylon), and formaldehyde resins (Bakelite, Formica). In **rearrangement polymers**, monomers link together without the loss of any atoms, but, unlike addition polymers, atoms are rearranged in the process of linking. The polyurethane known as "foam rubber" is the most common example of a rearrangement polymer. We will consider each of the three general types of polymers in turn.

ADDITION POLYMERS

Polyethylene

In 1991, about 18 billion kg (40 billion lb) of ethylene were produced in the United States—more than any other chemical except sulfuric acid and nitrogen.

The simplest addition polymer is **polyethylene**, which is produced in greater quantity than any other polymer. In 1991, more than 9 billion kg (20 billion lb) were produced in the United States, which amounts to approximately 75 lb per person annually. Polyethylene is a very stable material, it is resistant to attack by chemicals, and it is cheap to produce.

Polyethylene is made from ethylene (ethene) (Chapter 9), which is obtained from the cracking of petroleum (Chapter 14). At high temperature and pressure, and in the presence of a catalyst, ethylene monomers link together to form long single-bonded carbon-carbon chains. No atoms in the ethylene molecules are lost as polymerization occurs:

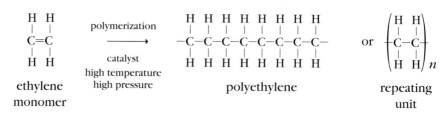

The double bond in ethylene contains two pairs of electrons. As polymerization occurs, one pair bonds the two carbon atoms in the monomer together, thus forming a single bond; the other pair is used to link successive monomers together.

It is not possible to write an exact formula for polyethylene (or any other polymer) because the number of carbon atoms in the chain can vary from several hundred to as many as 50,000. The notation $-(CH_2-CH_2)_n$ indicates that the unit inside the parentheses is repeated n times, where n is a very large number.

Polyethylene can be processed into many forms. Varying the catalyst and altering the pressure or temperature allow polyethylene to be converted into long filaments that can be made into rope, thin flexible sheets, or solid shapes. The very different properties of these various forms of polyethylene reflect structural differences in the polymer chains. At relatively low temperature and pressure and in the presence of a chromium oxide catalyst, long, linear, unbranched chains (n = approximately 10,000) predominate. The chains are angled as shown in Figure 11-2a and can pack closely together to give an orderly, uniform arrangement. The product is **high-density polyethylene (HDPE)**, which is hard, tough, and rigid.

HDPE accounts for one-third of all polyethylene produced in the United States. It is used to make toys, barrels and plungers for syringes, pipe, bottles, and containers for many liquid household products such as bleach and detergent. Nearly all one-gallon beverage containers in the United States are made of HDPE (Fig. 11-2b). HDPE is quite strong and resistant to breakage, but it can be crushed fairly easily.

At high temperature and pressure with a different catalyst, shorter chains (n = approximately 500) with irregular branching are formed (Fig. 11-3a). Because of the branching, the chains cannot pack as closely together as HDPE chains, and the product is **low-density polyethylene (LDPE)**, a material that is waxy, transparent (or translucent), and very flexible. It can be processed into

During World War II, polyethylene was used to insulate electrical wiring in radar sets. The invention of radar enabled British pilots to detect enemy aircraft before they became visible and was a major factor in the defeat of Germany.

(b)

(a)

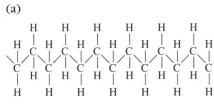

Part of a polyethylene molecule

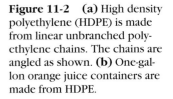

Figure 11-2 **(a)** High density polyethylene (HDPE) is made from linear unbranched polyethylene chains. The chains are angled as shown. **(b)** One-gallon orange juice containers are made from HDPE.

clear thin sheets that are used as food wrap. It is also made into trash bags, grocery bags (Fig. 11-3c), shopping bags of various kinds, and into squeeze bottles and flexible tubing. When this material is stretched, the short polymer chains move easily past each other; as they are pulled apart, the sheets tear.

Reaction conditions can be adjusted to give an LDPE in which the carbon chains are frequently **cross-linked** (Fig. 11-3b). This weblike structure makes the material extremely strong and very difficult to tear. It is used to make threaded bottle caps and plastic ice at skating rinks.

Substituted Ethylenes

We have seen that changing structural features such as chain length and degree of branching and cross-linking produces polymers with different properties. Polymers with an even wider range of properties can be made if one or more of the hydrogen atoms in the ethylene monomer are replaced by other atoms or

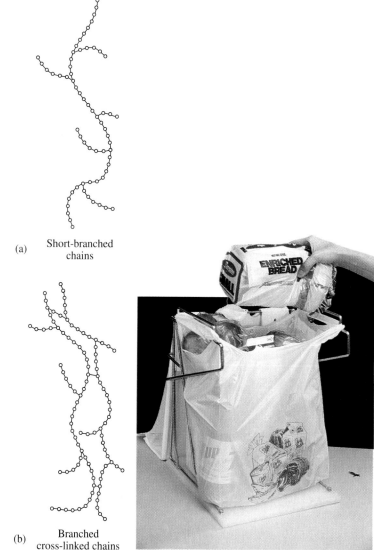

(a)   Short-branched
chains

Figure 11-3 Low-density polyethylene (LDPE) is formed from branched polyethylene chains. **(a)** LDPE made from short branched chains is very flexible. **(b)** LDPE made from cross-linked branched chains is extremely hard and strong. **(c)** Grocery bags made of LDPE.

(b)   Branched
cross-linked chains

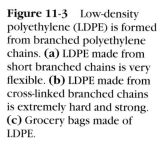

(c)

groups of atoms. As is the case for ethylene-based polymers, the carbon-carbon double bond ($-C=C-$) is the key to the formation of substituted-ethylene polymers.

The three substituted polyethylenes produced in greatest quantity are (1) poly(vinyl chloride) (PVC), (2) polypropylene, and (3) polystyrene. Others, produced in smaller quantities, are best known by their trade names: Saran, Teflon and Acrilan.

Poly(vinyl chloride) is produced from vinyl chloride, the monochloro derivative of ethylene.

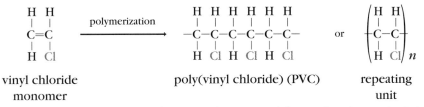

vinyl chloride monomer — poly(vinyl chloride) (PVC) — repeating unit

In the same way that $-CH_3$ represents a methyl group, $CH_2=CH-$ represents a vinyl group.

Poly(vinyl chloride) (PVC), the second most widely used polymer, is lightweight, hard, and resistant to chemicals. It is used to make plumbing pipes, floor tiles, credit cards, phonograph records, garden hoses, trash bags, and imitation leather (Fig. 11-4).

If vinyl chloride is polymerized with an excess of the dichloro-substituted derivative of ethylene, vinylidene chloride (1,1-dichloroethene), a copolymer, known as **Saran**, is produced.

The carcinogenicity of vinyl chloride was first suspected in the 1960s when a high incidence of cancer among workers exposed to it was reported. The risk increased with the length of time of exposure.

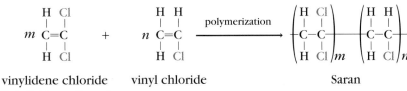

vinylidene chloride (1,1-dichloroethene) — vinyl chloride — Saran

The two monomers alternate randomly along the polymer chain. Saran is used extensively for food packaging (Fig. 11-5).

Polypropylene, the methyl derivative of polyethylene, is an exceptionally strong material.

Figure 11-4 Items made from poly(vinyl chloride) (PVC).

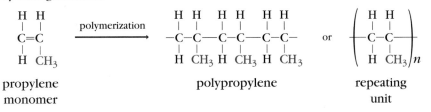

propylene monomer — polypropylene — repeating unit

It is used in the transport industry for making battery cases (Fig. 11-6) and automobile trim, and, because of its strength, fibers made from it are used to make outdoor carpeting, fishing nets, and rope. In 1991, propylene ranked tenth among chemicals produced in the United States, with production at over 20 billion pounds.

Polystyrene, which was first prepared as early as 1920 by the German chemist Hermann Staudinger, is made by the polymerization of styrene. Styrene is a derivative of ethylene in which one hydrogen atom is replaced by a benzene ring.

Figure 11-5 The plastic film Saran is used extensively for wrapping food. The principal monomer in Saran is vinylidene chloride.

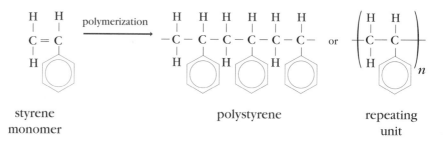

styrene
monomer

polystyrene

repeating
unit

Figure 11-6 The exceptionally strong polymer polypropylene is used to make automobile battery casings and other items.

The benzene rings in polystyrene make it more rigid than polyethylene. As a result, polystyrene tends to be brittle and easily shattered. Clear, hard polystyrene is used to make glasses, containers, wall tile, and appliance parts (Fig. 11-7a).

Polystyrene can also be made into a lightweight solid foams by adding a chemical agent that produces gas bubbles in molten styrene. The foams make good insulating and packing materials. They are used to make picnic coolers, egg cartons, and cups for keeping drinks hot and cold (Fig. 11-7b). Foam sold under the trade name Styrofoam is used as a cushioning material for shipping containers, and for insulating buildings.

Polytetrafluoroethylene (Teflon) is formed from the derivative of ethylene in which all four hydrogen atoms are replaced with fluorine atoms:

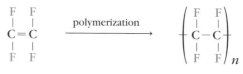

tetrafluoroethylene
monomer

polytetrafluoroethylene
(Teflon)

(a)

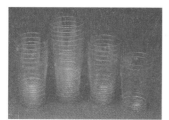

(b)

The presence of the fluorine atoms makes the polymer much more resistant to heat and to attack by chemicals than polyethylene is. These properties make Teflon very suitable for use in gaskets, bearings, and machine parts that must operate in corrosive environments. Its best known use is as a nonstick coating for cooking utensils.

Polyacrylonitrile is a polymer of acrylonitrile, a molecule in which a nitrile ($-C\equiv N$) group is substituted for one of the hydrogen atoms in ethylene.

acrylonitrile
monomer

polyacrylonitrile

Figure 11-7 Items made from polystyrene. **(a)** Clear polystyrene glasses. **(b)** Items made of polystyrene foam.

Polyacrylonitriles are processed into tough fibers for carpeting and fabrics. Because the nitrile group is polar, polymer chains are attracted to one another; this attraction makes the fibers stronger than those formed from nonpolar ethylene. The acrylic fibers sold under the trade names Orlon, Acrilan, and

TABLE 11-1   Some Addition Polymers

Monomer	Polymer	Typical Uses
$CH_2=CH_2$ Ethene	$(CH_2-CH_2)_n$ Polyethylene	Containers, pipes, bags, toys, wire insulation, bottle caps
$CH_2=CHCH_3$ Propene	$\left(CH_2-CH\atop\ \ \ \ CH_3\right)_n$ Polypropylene	Fibers for carpets, artificial turf, rope, fishing nets, automobile trim
$CH_2=CHCl$ Chloroethene (vinyl chloride)	$\left(CH_2-CH\atop\ \ \ \ Cl\right)_n$ Poly(vinyl chloride) (PVC)	Garden hoses, floor tiles, plumbing, phonograph records, artificial leather, food wrap
$CH_2=CHCN$ Acrylonitrile	$\left(CH_2-CH\atop\ \ \ \ CN\right)_n$ Polyacrylonitrile (Orlon, Acrilan)	Fibers for cloth, carpets, upholstery
$CH_2=CH-$⬡ Styrene	$\left(CH_2-CH\atop\ \ \ \ \text{⬡}\right)_n$ Polystyrene	Styrofoam, hot-drink cups, insulation, packaging
$CF_2=CF_2$ Tetrafluoroethene	$(CF_2-CF_2)_n$ Teflon	Nonstick coating for kitchen utensils

Creslan are copolymers of acrylonitrile with varying percentages (25 to 85 percent) of vinyl chloride.

Examples of some important addition polymers are listed in Table 11-1, which shows the monomers from which they are formed and some of their uses.

EXAMPLE 11-1   Write the structure of the addition polymer formed from the monomer CH_2CH-R (R = an atom or group, e.g., Cl or CH_3). Show three segments.

Solution: The structural formula of the monomer is:

$$\begin{array}{ccc} H & & H \\ | & & | \\ C & = & C \\ | & & | \\ H & & R \end{array}$$

No atoms are lost in the formation of an addition polymer. Each C atom forms 4 bonds. Two of the electrons in the C=C bond are used to join adjacent monomers to give a polymer with only single bonds. The polymer is:

$$\begin{array}{ccccccc} & H & H & H & H & H & H \\ & | & | & | & | & | & | \\ -C & -C & -C & -C & -C & -C- \\ & | & | & | & | & | & | \\ & H & R & H & R & H & R \end{array} \qquad \text{or} \qquad \left(\!\!\begin{array}{c} H \ H \\ | \ \ | \\ -C-C- \\ | \ \ | \\ H \ R \end{array}\!\!\right)_{\!n}$$

PRACTICE EXERCISE: Show four segments of the addition polymer that can be constructed from CH_2CCl_2.

Answer:

$$\begin{array}{ccccccccc} & H & Cl & H & Cl & H & Cl & H & Cl \\ & | & | & | & | & | & | & | & | \\ -C & -C & -C & -C & -C & -C & -C & -C- \\ & | & | & | & | & | & | & | & | \\ & H & Cl & H & Cl & H & Cl & H & Cl \end{array}$$

⬤ ELASTOMERS

Elastomers are synthetic polymers with properties similar to those of natural rubber. As their name implies, they are elastic; like rubber they have the property of immediately returning to their original form after being stretched.

Natural Rubber

The people of Central and South America were using rubber balls for games long before Christopher Columbus observed the practice in 1496. According to one astonished witness, the balls rebounded so much that they appeared to be alive.

Natural **rubber** is an addition polymer that is obtained as a milky white fluid called latex from the tropical rubber tree *Hevea brasiliensis*. The latex, which oozes out when the tree is cut, is an emulsion of rubber particles in water. The rubber can be precipitated from the latex and then processed.

Natural rubber is a polymer of the monomer **isoprene** (2-methyl-1,3-butadiene).

isoprene

Isoprene has two double bonds. When it polymerizes, the repeating units that are formed have one double bond. The monomers can link together to form *cis-* or *trans-* arrangements (Chapter 9) with the linked $-CH_2-$ groups on the same or on opposite sides of the double bond.

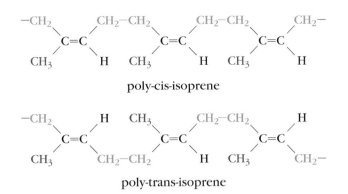

poly-cis-isoprene

poly-trans-isoprene

Poly-*trans*-isoprene is present in *gutta percha*, a natural material found in certain trees in Malaysia and South America. It is not as elastic as rubber, and is used to make golfball covers, and insulation for underwater electrical equipment and cables.

In natural rubber, the linkages are all cis- arrangements.

Natural rubber is a soft, water-repellent, elastic material that becomes sticky when warmed and stiff when cooled. If untreated, it is of limited use, and until the middle of the nineteenth century, it was used primarily in pencil erasers.

In 1839, Charles Goodyear (1800–1860), who for many years had been working to find ways to improve the qualities of rubber, accidentally spilled a mixture of rubber and sulfur on a hot surface. To his surprise, he found that the rubber had become tougher and more resistant to heat and cold, and its elasticity had improved. Later, the process was named **vulcanization** after Vulcan, the Roman god of fire. Vulcanization made natural rubber suitable for numerous applications including automobile tires, and was a major factor in the development of the automobile industry.

Rubber owes its elasticity to the flexibility of its randomly intertwined long polymer chains. When rubber is stretched, the chains straighten; when the tension is released, the chains recoil. If the rubber is stretched strongly enough, adjacent chains slip past each other, and the rubber loses its elasticity. During vulcanization, short chains of sulfur atoms of varying lengths form links between sections of the polymer chains (Fig. 11-8a). When stretched, the chains can still untwine and straighten (Fig. 11-8b), but there is the advantage that, when the tension is released, the sulfur cross-links ensure that the chains spring back to their original positions. The sulfur linkages hold the chains together and prevent them from slipping past each other when stretched.

Rubber is said to have been given its name by Joseph Priestley (Chapter 4) who found that it could be used to rub out (erase) lead pencil marks.

Synthetic Rubber

The first commercially successful synthetic elastomer to be produced in the United States was **neoprene**. Neoprene is an addition polymer of 2-chlorobutadiene, an isoprene derivative in which a chlorine atom is substituted for the methyl group.

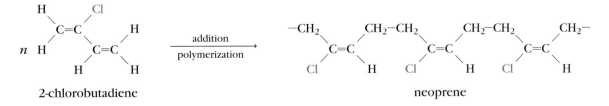

2-chlorobutadiene   neoprene

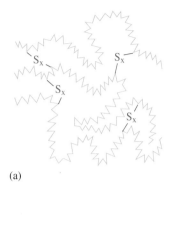

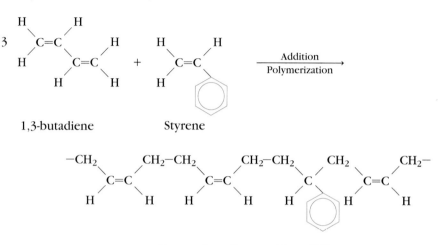

Neoprene is resistant to abrasion and to oils, gasoline, and other solvents. It is a useful material for making gasoline hoses, garden hoses, gaskets, shoe soles, adhesives, and conveyor belts, but it is not suitable for automobile tires.

The most important synthetic rubber is **styrene-butadiene rubber (SBR)**, which was developed during World War II when supplies of imported natural rubber were cut off. SBR is a block copolymer made by mixing styrene and 1,3-butadiene in a 1-to-3 ratio.

1,3-butadiene   Styrene

Styrene-butadiene rubber (SBR)

Figure 11-8 **(a)** When rubber is vulcanized, short chains of sulfur atoms of varying length (indicated by the subscript x) form cross-links between adjacent hydrocarbon chains. **(b)** When tension is applied, the chains straighten out but cannot slip past each other because of the sulfur cross-links. When the tension is released, the chains become coiled again, and the rubber resumes its original shape.

SBR is more resistant to abrasion and oxidation than natural rubber. Like natural rubber, it can be vulcanized to improve its properties. SBR, which accounts for more than 40 percent of synthetic rubber production, is used primarily for the manufacture of tires.

Worldwide, tires and tire products account for more than half the combined production of natural and synthetic rubber. A typical tire is only about 60 percent rubber. The other main ingredient is carbon black, which is added to increase strength and resistance to abrasion. Steel belts, nylon cord, and fiberglass give tires added strength.

Synthetic poly-cis-isoprene, structurally identical to natural rubber, can now be produced on an industrial scale. However, because of the cost of the raw materials, which are obtained from petroleum, this synthetic material is more expensive to produce than is natural rubber.

▓ PAINTS AND POLYMERS

An important result of the tremendous growth of the plastics industry after World War II was the production of water-based paints, a development that revolutionized the paint industry. Water-based and oil-based paints are composed of three major ingredients: (1) a pigment, (2) a solvent, and (3) a binder. The **pigment** provides the desired color. The primary pigment in a paint is usually titanium dioxide (TiO_2) or zinc oxide (ZnO), compounds that are white, opaque, and serve to cover any previous color on the surface to be painted. White lead (2 $PbCO_3 \cdot Pb(OH)_2$) had been used extensively as a primary pigment until 1977, when it was banned for interior use because of its toxicity (Chapters 12, 16). Pigments such as carbon black, iron oxide (Fe_2O_3, brown and red), cadmium sulfide (CdS, orange), chromium oxide (Cr_2O_3, green), and various organic compounds are mixed with the primary pigment to obtain the desired color.

The **solvent** in a paint is the liquid in which the pigment, binder, and other ingredients are suspended, usually in the form of an emulsion. Once the paint has been spread, the solvent evaporates, and the paint dries. In oil-based paints, the solvent is usually a mixture of hydrocarbons obtained from petroleum. A compatible organic solvent must be used for thinning the paint and cleaning up after painting. This is usually turpentine, which, like most other suitable solvents, gives off hazardous fumes. In water-based paints, the solvent is water, which can be used both for thinning and for cleaning up.

The **binder** is a material that polymerizes and hardens as the paint dries, forming a continuous film that holds the paint to the painted surface. The pigment becomes trapped within the polymer network. In oil-based paints, the binder is usually linseed oil, but coconut, soybean, and castor oils are also used. The oils are mixtures of triglycerides of several long-chain fatty acids (Chapter 10). As the solvent evaporates and the paint hardens, the oil reacts with atmospheric oxygen to form a cross-linked polymeric material.

The binder in water-based paints is made by mixing a partially polymerized material with some of its unreacted monomer. The slightly rubbery nature of the partially polymerized material explains why water-based paints are called latex paints. As the water evaporates, further polymerization occurs, and the paint hardens.

Several synthetic polymers are used as binders in latex and acrylic paints. Paints with poly(vinyl acetate) binders now account for about 50 percent of the paint used for interior work. Those with poly(styrene-butadiene) binders are less expensive but do not adhere quite as well and have a tendency to yellow. Acrylic paints, made with acrylonitrile binders, are considerably more expensive than other types of paints, but they are excellent for exterior work. They adhere well and are washable, very durable, and resistant to damage by sunlight. In most respects, they are superior to oil-based paints, which they have now largely replaced.

CONSUMER BRIEF   **Hair Permanents: Straighten Then Curl**

Hair is composed primarily of the protein keratin. In hair, adjacent protein strands are held together by covalent disulfide bonds and by weaker, but numerous, hydrogen bonds. Hair can be curled to produce temporary or permanent waves by breaking and rearranging some of these bonds. The simplest and least permanent procedure is to wet the hair, which disrupts the hydrogen bonds holding the proteins in shape. If the wet hair is allowed to dry while wound on rollers, hydrogen bonds reform in new positions, and the hair is set temporarily in waves. Curling irons also induce a temporary set. The change in shape achieved in these two procedures is quickly reversed if there is an increase in humidity. Flowing waves that have taken considerable time to create can disappear in minutes if there is a sudden change in weather.

Longer-lasting waves can be obtained by spraying the hair after it has been set, with a resin-type polymer dispersed in a solvent. The polymer can also be applied as a mousse. After the polymer is sprayed or rubbed on the hair, the solvent evaporates, leaving the hair coated with the polymer. Commonly used products contain poly(vinylpyrolidine) (PVP) or poly(vinylacetate).

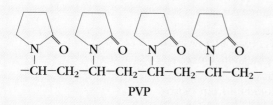

PVP

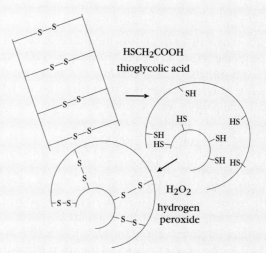

HSCH$_2$COOH
thioglycolic acid

H$_2$O$_2$
hydrogen
peroxide

The plastic coating holds the hair in place longer than does a heat-induced set, but the effect lasts only for a matter of hours.

A permanent wave can only be achieved by initiating a chemical reaction that breaks the disulfide linkages that hold the protein strands together. During World War II, research at the National Bureau of Standards into the structures of natural fibers, such as cotton and wool, showed that wool's natural resistance to wrinkling was due to its three dimensional protein structure. After the war, Dr. Milton Harris determined that hair, which has a structure similar to that of wool, could be curled by altering the way the protein strands in hair are linked together. He sold his chemical process to a little company called Toni, and the permanent wave was born.

A permanent wave is set by applying a lotion containing the oxidizing agent thioglycolic acid to the hair. The oxidizing agent ruptures disulfide linkages in the hair protein, a step that is accompanied by the release of hydrogen sulfide (H$_2$S), and the smell of rotten eggs. Next, the hair is set on rollers and treated with a mild oxidizing agent, such as hydrogen peroxide (H$_2$O$_2$). This agent causes disulfide linkages to reform in new positions. The strands of protein lock together in the shape of the roller and leave the hair with a permanent curl.

The same process can be used to straighten moderately curly hair. Instead of being curled on rollers after treatment with thioglycolic acid, the hair is rinsed and combed straight. For very curly hair, "relaxers" such as sodium or lithium hydroxide must be used. Very curly hair has so many disulfide bonds holding the hair in tight curls, that a strong base is needed to complete the bond rupture in a reasonable time. Care must be taken not to burn the scalp with the strong base.

A permanent wave lasts the lifetime of the hair. As new hair grows, it assumes its natural degree of straightness or curliness, and permanent-wave treatment must be applied to the new hair.

Questions
1. Why do you smell rotten-eggs when applying a permanent?
2. You are having your very curly hair straightened. The hair stylist suggests that you sit under the hair dryer while the chemical treatment is working. Should you agree?

CONDENSATION POLYMERS

A **condensation reaction** is a reaction in which two molecules react to form a larger molecule with the elimination of a small molecule such as water. An example is the reaction of a carboxylic acid and an alcohol, with the elimination of water, to form an ester.

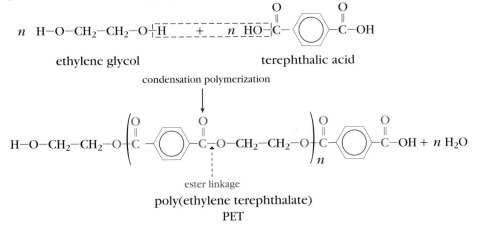

This type of reaction depends on the presence of two different functional groups on the two reacting molecules—a carboxylic acid group (–COOH) and a hydroxyl group (–OH), in the above example. If the reacting molecules are bi-functional (if they have a chemically reactive group on *both* ends of their molecules), they can polymerize and become linked end-to-end to form long chains. Both polyesters and polyamides are formed in this way.

Polyesters

The most important **polyester** is **poly(ethylene terephthalate) (PET)**, which is an alternating copolymer of ethylene glycol (an alcohol with two –OH groups; Chapter 9) and terephthalic acid (a carboxylic acid with two –COOH groups). The formation of the polymer occurs as follows:

As each linkage is formed, a water molecule is eliminated.

PET can be processed into fibers, rigid material, and film. The textile fibers marketed as **Dacron** account for 50 percent of all synthetic fibers and are used primarily in clothing. Blended with cotton, Dacron forms a fabric that requires no ironing after washing. Formulated into tubing, it can be used to replace diseased blood vessels in open-heart surgery.

Cross-linked PET is used in making soft-drink bottles. The cross-links make the polymer strong enough to withstand the pressure exerted by the carbon dioxide gas in the bottle. PET can also be made into a very strong, very thin film known as **Mylar**, which is used for packaging frozen foods and as a skin substitute in the treatment of burn victims. When magnetized, Mylar is used for recording tape.

> The term "polyester" is commonly used to describe fabric made from PET fibers.

Polyamides

Polyamides, which are better known as **nylons**, are condensation copolymers of diamines and dicarboxylic acids. One of the most important is the copolymer nylon 66, which was the first completely synthetic fiber to be produced. It was discovered in 1931 by Wallace H. Carothers, a chemist working for DuPont.

Nylon 66 is prepared from the diamine 1, 6-diaminohexane and the dicarboxylic acid adipic acid. As the two monomers react, a molecule of water is eliminated, and an amide linkage is formed.

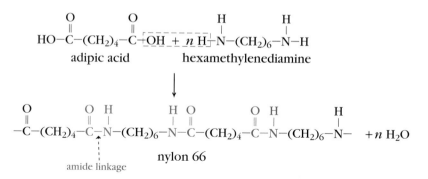

With some modifications, the polymerization reaction can be carried out in the laboratory as shown in Figure 11-9. Nylon 66 was given the number 66 because each of the two monomer molecules from which it is made has 6 carbon atoms.

The second most important polyamide is nylon 6, a polymer made from a single repeating monomer: aminocaproic acid (which has 6 carbon atoms). The amino end of one molecule reacts with the carboxylic acid end of the next monomer:

Figure 11-9 When a solution of adipoyl chloride (a derivative of adipic acid) in hexane is poured gently onto a solution of hexamethylenediamine in water, a white film of nylon forms at the interface between the two layers. The film can be pulled up as a string and wound onto a stirring rod.

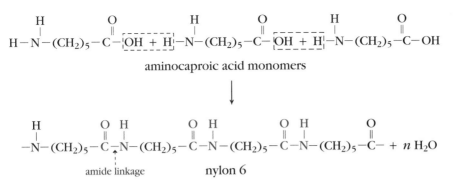

aminocaproic acid monomers

nylon 6

Nylons can be made into fibers that are stronger, more durable, more chemically inert, and cheaper to produce than wool, silk, or cotton. Nylon is used for carpeting, clothing (particularly hosiery), tire cord, rope, surgical sutures, and parachutes. Nylons can also be molded to make electrical parts, valves, and various machine parts.

When first introduced to the public in 1939, nylon hose were an immediate success. However, production of hosiery had to be discontinued during World War II, when all nylon was reserved for military use, particularly for making parachutes. Once the war ended, the industry had difficulty keeping up with the enormous demand for nylon. Nylon hose were sheer and more durable and less expensive than silk hose; after the restrictions of the war years, women were eager to buy them.

The natural protein fibers silk and wool, like the synthetic nylons, are polyamides and are chemically very similar to nylon 6 (Fig. 11-10). As we saw in Chapter 10, proteins are formed from amino acids by the formation of amide bonds, which in protein chemistry are called peptide bonds. Thus all proteins are polyamides.

The extensively cross-linked polyamide Kevlar is 20 times stronger than steel, and Kevlar ropes have replaced steel ropes in many applications.

Unlike natural fibers, nylon, Dacron, Orlon, and other synthetic fibers are resistant to moths and mildew, and are wrinkle-resistant.

EXAMPLE 11-2   Dacron, a polyester, is a condensation polymer formed from a diol and a dicarboxylic acid. Its structure is:

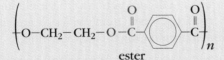

ester

What two monomers are used to produce Dacron?

Solution: A diol has −OH groups at both ends of the molecule; a dicarboxylic acid has −COOH groups at both ends of the molecule. A water molecule is lost in the formation of the condensation polymer. To obtain the monomers add −H and −OH at the ester linkage where the monomers are joined. The monomers are:

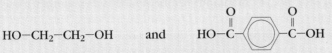

PRACTICE EXERCISE: What monomers are used to make the following polyamide condensation polymer?

$$
\begin{array}{c}
| \\
C=O \\
| \\
NH \\
| \\
CH-R \\
| \\
C=O \\
| \\
NH \\
| \\
CH-R \\
| \\
C=O
\end{array}
$$

Figure 11-10   The protein keratin in wool is a natural polymer. Like nylon, it is formed from monomers that join through amide (peptide) bonds.

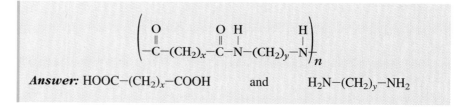

Answer: $HOOC-(CH_2)_x-COOH$ and $H_2N-(CH_2)_y-NH_2$

Formaldehyde-Based Network Polymers

Most of the polymers we have considered so far, including polyethylene, the substituted-ethylene polymers, polyesters, and nylon, have been thermoplastics (plastics that can be softened by heat and remolded). Formaldehyde-based polymers, however, are thermosetting plastics and cannot be remolded once hardened. They form as sticky, viscous liquids called **resins** that become hard on exposure to air.

The first thermosetting plastic was synthesized in 1909 by Leo Baekeland and marketed under the trade name **Bakelite**. Bakelite is produced from formaldehyde and phenol monomers by means of a two-step polymerization reaction. In the first step, formaldehyde is added to the ortho- and para- positions of phenol.

A formaldehyde-amine resin is used to make almost indestructible dinnerware (Melmac), and counter tops (Formica). A formaldehyde-urea resin is used to make particle-board binder.

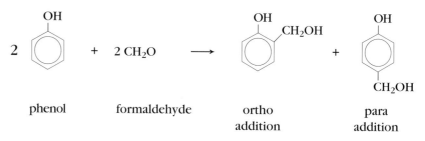

| phenol | formaldehyde | ortho addition | para addition |

In the second step, which occurs during hardening, the substituted monomers undergo a condensation reaction with the elimination of water. (Recall from Chapter 9 that a hydrogen atom, which is not shown, is present at each of the unsubstituted positions on the benzene ring.)

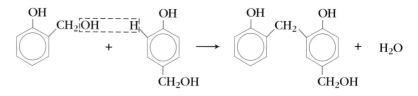

Unlike the bifunctional monomers used to make polyesters and polyamides, each phenol monomer can be linked to *three* other monomers to form a complex, three-dimensional, cross-linked network (Fig. 11-11). The benzene rings are joined through $-CH_2-$ linkages derived from formaldehyde. The numerous cross-links make the polymer very hard and rigid.

Bakelite has excellent insulating properties and is used to make housings for radios and other electrical circuits. It is also used for buttons and bottle caps, but its primary use is as an adhesive in the production of plywood and particle board.

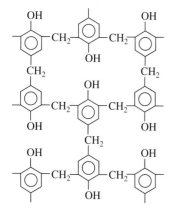

Figure 11-11 The resin formed from phenol and formaldehyde has a three-dimensional network structure.

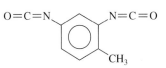

Figure 11-12 Toluene-2,4-diisocyanate.

REARRANGEMENT POLYMERS: POLYURETHANES

The most important rearrangement polymers are the **polyurethanes**, which are made from dialcohol and diisocyanate monomers. Isocyanates are compounds that contain the O=C=N– functional group. The most commonly used diisocyanate monomer is toluene-2,4-diisocyanate (Fig. 11-12). The rearrangement reaction that leads to the formation of a urethane linkage is shown below, using general formulas for the dialcohol and the diisocyanate monomers.

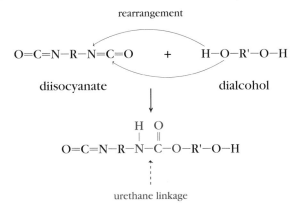

The reaction is not a condensation reaction; there is a *rearrangement* of atoms, but no atoms are lost. As the urethane linkage forms, a hydrogen atom on the alcohol moves to a nitrogen atom on the isocyanate, and the nitrogen-carbon double bond is converted to a nitrogen-carbon single bond. The carbon atom in the isocyanate bonds to the alcohol through an oxygen atom.

Further rearrangement reactions lead to the formation of a long-chain polyurethane polymer.

O H   H O   O H   H O
‖ | | ‖   ‖ | | ‖
–O–C–N–R–N–C–O–R'–O–C–N–R–N–C–O–R'–O–

polyurethane

A urethane linkage is similar (although not identical) to an amide linkage, and, as a result, polyurethanes are similar in many ways to nylons. They can be drawn into fibers, which are used primarily for upholstery, and they can be processed into film. Some polyurethanes are rubbery materials that have found uses as skin replacements for burn victims and as diaphragms in artificial hearts (Fig. 11-13a,b).

Polyurethane foam is often called foam rubber.

(a)   (b)

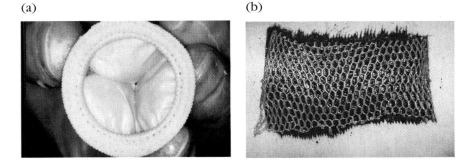

Figure 11-13
(**a**) Polyurethanes are used to make diaphragms in artificial hearts. (**b**) Thin sheets are used as skin replacements for burn patients.

Until quite recently, **polyurethane foam** was used extensively for building insulation. It was easily made at building sites by adding water to a mixture of the dialcohol and diisocyanate monomers. The water reacted with the diisocyanate to produce carbon dioxide, which acted as the foaming agent.

$$O=C=N-R-N=C=O \ + \ H_2O \longrightarrow \ O=C=N-R-NH_2 \ + \ CO_2$$
diisocyanate

The foam was then injected directly into attics and behind walls to improve insulation. This procedure offered a useful way to reduce the amount of energy used for heating and cooling, but it had to be discontinued when some unreacted diisocyanate was found to remain in the foam. The diisocyanate most frequently used for foaming was toluene-2,4-diisocyanate (Fig. 11-12), which is toxic and can cause respiratory problems.

⬤ SILICONES

All the polymers we have discussed so far have been organic polymers with carbon atom backbones. Chemists have also made very useful polymers with silicon atom backbones.

Silicon is in the same group of the periodic table as carbon, and it forms a few low-molecular-weight compounds with hydrogen called silanes that are analogous to simple hydrocarbons. These include SiH_4, Si_2H_6, and Si_3H_8. Si–Si and Si–H covalent bonds are not as stable as C–C and C–H covalent bonds, and silicon does not form stable long-chain molecules analogous to long-chain hydrocarbons. But the hydrogen atoms in silanes can be replaced quite easily by other atoms, or groups of atoms, to give compounds that can readily polymerize. For example, dihydroxy silicon monomers with the general formula $R_2Si(OH)_2$ (where R is a CH_3, C_2H_5, C_3H_7, or some other hydrocarbon group) can undergo condensation reactions to form both linear and cross-linked network polymers known as **silicones**, as shown in Figure 11-14a.

Silicones are water-repellent, heat-stable, and very resistant to chemical attack. Depending on the nature of the $R_2Si(OH)_2$ monomer and the degree of polymerization, silicones are obtained as either oils, greases, rubberlike materials, or resins (Fig. 11-14b). Silicone oils, unlike hydrocarbon oils, do not decompose at high temperatures and do not become viscous. They are particularly valuable as lubricants for machinery and engines that must operate at extremes of temperature.

Silicone rubbers (Fig. 11-14c), which are produced when very long chains are extensively cross-linked with –Si–O–Si– bonds, are less subject to oxidation and attack by chemicals than are either natural rubber or synthetic carbon-based rubbers. Another advantage is that they retain their elasticity at very low temperatures, which makes them valuable as sealants for aircraft and space vehicles that must operate at low temperatures. Silicone rubber is easily molded, and, because it is so unreactive, it is an excellent replacement material for body parts. It is used for repairing the nose and ear and in the construction of artificial joints. Other silicones are used as hydraulic fluids, electrical insulators, and moisture-proofing agents in fabrics.

Until recently, liquid silicone contained in urethane bags was used extensively for breast augmentation. In 1992, reports that implants increased the risk of developing cancer and that leaking silicone might cause a severe immune response prompted the Food and Drug Administration to limit the use of silicone to breast augmentation in mastectomy patients. The controversial decision has since been criticized by other researchers who oppose the ban.

Silly Putty owes its dual elastic and liquid properties to its structure which is between that of a silicone rubber and a silicone oil.

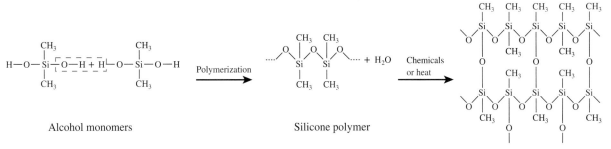

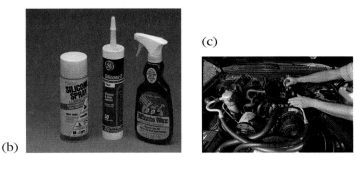

(a)

Alcohol monomers   Silicone polymer

Methyl silicone cross-linked polymer

(c)

(b)

Figure 11-14 **(a)** Dihydroxy silicon monomers can form long-chain condensation polymers called silicones. Under certain conditions cross-links can be formed between adjacent chains to give silicone rubbers. **(b)** Items containing silicone polymers. **(c)** Items made of silicone rubber.

POLYMERS FOR THE FUTURE

Polymer research is a fertile field. Many of the newest polymers are **composites**, exceptionally strong materials in which fibers, metals, and other additives are embedded in the polymer matrix. Examples are glass-reinforced nylons and polymers reinforced with graphite. Other promising developments include strong materials made from polycarbonates.

     Considerable progress has been made in the development of stable polymers that conduct electricity. These materials, which in most cases are based on acetylene polymers, are lightweight and flexible. They are expected to have applications in batteries, semiconductor circuits, radar dishes, and electrical wiring. New thermoplastic elastomers (TPEs) may soon replace natural rubber and traditional synthetic thermoset rubber in many applications. TPEs are easier and less expensive to produce, and they present fewer disposal problems because they can be ground up and reused. TPEs are not suitable for tires, but they are ideal for most other automotive rubber parts, including bumpers, belts, drive boots, and air-conditioning hoses.

PROBLEMS WITH POLYMERS

The major problem with synthetic polymers is disposing of the discarded items made from them. There is also concern about the hazardous chemicals that some polymer materials release into the environment. Another problem is the dwindling supply of petroleum, the primary source of the raw materials from which synthetic polymers are made.

Disposal of Plastics

The very properties that make plastics so useful—their stability and resistance to oxidation and attack by chemicals and bacteria—make them almost indestructible.

They have become unsightly litter along roadsides and wherever else they are discarded, and they are found floating on the surfaces of all oceans of the world.

Plastics account for only 7 percent of the total weight of solid waste generated in the United States, but they make up approximately 20 percent of its volume. A 1990 report by the EPA predicted that by the turn of the century, the weight of plastic wastes will increase by 50 percent.

There is no single way to solve the problem. It will have to be tackled with a combination of several procedures, including recycling, incinerating, making plastics more degradable, and reducing consumption.

Recycling About 87 percent of all plastics are thermoplastic and can be recycled by melting and remolding. At present, however, only about 2 percent of plastics is being recycled, compared with 20 percent of paper and 30 percent of aluminum (Chapter 19). As explained in the following Consumer Brief, for recycling to be successful, plastics must first be separated according to type.

Incineration Incineration of plastics to produce energy is another approach to disposal, but its drawback is that certain plastics release toxic gases when burned. Vinyl chloride polymers, for example, release toxic hydrogen chloride gas; polyurethanes and polyacrylonitriles produce deadly hydrogen cyanide. Even if a polymer does not release a toxic gas when burned, it will produce carbon dioxide, which contributes to the greenhouse effect (Chapter 13). The formation of toxic gases on burning points up one of the dangers often associated with accidental fires in homes, other buildings, and in aircraft, where many furnishings and other fittings are made of plastics.

Degradable Plastics Manufacturers are increasingly seeking ways to make plastics more degradable. Photodegradable plastics, which are broken down by ultraviolet light, can be made by incorporating light-sensitive carbonyl molecules (molecules containing a —C=O group) into the polymer chains. Biodegradable plastics that can be digested by microorganisms are made by incorporating starch or cellulose into the polymer during production. As bacteria consume the starch or cellulose, the plastic is broken down into small pieces. This approach is particularly useful for thermosetting plastics, which are not readily recycled.

The major value of both types of degradable plastics is their reduction of litter. If they are buried in a landfill, they degrade very slowly, if at all, because of lack of light, oxygen, and moisture. Even naturally biodegradable materials such as newspapers have been found almost unchanged after 20 years in landfills. Degradable plastics are used primarily for six-pack beverage rings, trash bags, and disposable diapers. These products are chosen because they entangle wildlife, make up a high percentage of litter, or cause pollution.

In a recent development that holds promise for the future, biodegradable polymers based on carbohydrates, proteins, and carboxylic acids have been made by bacteria in fermentation processes. The same bacteria that produce the polymers will decompose them under the anaerobic condition of a landfill. Another degradable polymer was recently discovered accidentally by researchers studying the loss of crunchiness of cereals in milk. In the course of their work, the scientists produced a cornstarch-based polymer. Called Eco-foam, this water-soluble material is now marketed as a packing material alternative to Styrofoam. At present, these new degradable materials have a limited market because they are considerably more expensive to produce than nondegradable alternatives.

Reducing Plastic Usage Nearly one-third of all plastics production goes into making packaging materials and other quickly discarded items. Plastic

Before plastic yokes are discarded, each ring should be cut to prevent animals from becoming caught in them.

CONSUMER BRIEF

Recycling Plastics: Which Is Which?

In November, 1992, voters in Massachusetts were faced with a ballot initiative known as Question #3. This initiative, if passed, would have required all materials used for packaging to be reusable or be made partly from recycled materials. Although Question #3 was defeated, the proposal reflects the public's concern about plastic waste.

Plastics are not biodegradable, and when they are buried in a landfill, they remain there unchanged, almost indefinitely. One solution to the landfill problem is to recycle plastic bottles and packaging materials. Recycling, however, presents many problems: there are many types of plastics; the different types have different physical properties; and it is not easy to tell one plastic from another.

Recycling thermosetting polymers is difficult because these polymers decompose when heated and cannot be re-formed into new products. Thermoplastic polymers, however, are more amenable to recycling. They can be softened by heat and then molded into a new shape. There is a wide variety of thermoplastic polymers, and each must be softened and remolded at a different temperature.

Unseparated plastic refuse, which is called comingled plastic, has little value because its properties depend on the type of plastic containers that make up the waste mixture. Industry does not want to recycle material that may plug up its machinery because it melts at a too-high or too-low temperature. Only plastic that has been sorted by polymer type has value to industry. Separating plastics by polymer type, however, is very costly because it cannot be done by machines. Before a large-scale recycling process could be implemented, an easy way of sorting plastic waste had to be developed.

In 1988, the Society of the Plastics Industry (SPI) developed a uniform coding system that makes it possible to sort waste plastic. The SPI coding symbol consists of three triangulated arrows surrounding a single-digit code number, with the abbreviation of the plastic underneath. There are seven code numbers: 1 through 6 refer to specific polymers; 7 refers to all other types.

Today, more than 35 states mandate the use of this coding system on containers with capacities of 8 or more ounces and on bottles of 16 or more

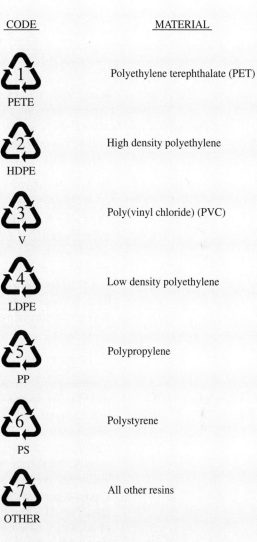

CODE	MATERIAL
1 PETE	Polyethylene terephthalate (PET)
2 HDPE	High density polyethylene
3 V	Poly(vinyl chloride) (PVC)
4 LDPE	Low density polyethylene
5 PP	Polypropylene
6 PS	Polystyrene
7 OTHER	All other resins

ounces. Smaller containers take longer to sort and cannot be recycled profitably.

Uses are being developed for recycled plastic that take advantage of its low cost and durability. One effort is focusing on converting plastic products with short service lives, such as foam, wrap, and containers, into products with long service lives, such as construction materials and plastic pipe. Recycled plastic soft drink bottles, which are made of poly(ethylene terephthalate) (code, PETE), are being used to make carpeting and insulation for ski jackets and parkas. In 1989, 20 percent of all PETE was being recycled. Used polystyrene (PS) from coffee cups and throw-away plates and food trays, is being converted to plastic "lumber." This lumber can be drilled, cut, or nailed, just as real lumber can, and it has the advantage that it is not eaten by termites or rotted by water. Recycled poly(vinyl chloride) (V) from automobile seats and luggage is being converted into sewer drainpipes.

Dow Plastics and Advanced Recycling Technologies, Inc. have developed a process that increases the quality and consistency of recycled polyethylene used to make grocery bags. In the past, polyethylene that had been recycled was not of high enough quality to be stretched back into thin sheets and made into bags again. The new process removes food debris and paper—including the receipt—from the grocery bag. When recycled, the polyethylene is of high enough quality to be used to make new bags without the need to add large quantities of virgin polyethylene.

Questions
1. Explain why plastics are not consumed like gasoline and many other products. Do we really dispose of plastics?
2. Should a federal law be passed requiring deposits on certain plastic bottles?

cups, plates, forks, carrying bags, and numerous other products are discarded after one use. In many instances, washable, reusable utensils and reusable string or cloth bags could be used instead of the throw-away items. Fast-food establishments are discovering that it is good public relations to cut back on the amount of plastic materials they use for food packaging. Some local governments have banned the use of certain types of plastic containers. However, to have any significant impact on the plastic waste problem, such measures will have to be adopted much more widely.

Polymer Additives: Plasticizers

Over time, plasticizers in plastic materials evaporate, and the material loses its flexibility and tends to crack.

Another problem associated with polymers is the chemicals that are frequently added to them during production to improve or alter their properties and that may contaminate the environment when the plastic is discarded. Polymers that are brittle and hard can often be made soft and pliable by the addition of compounds called **plasticizers**. For example, PVC treated in this way can be used to make raincoats, seat coverings, and other similar materials. For many years until they were banned, PCBs (Chapter 12) were widely used as plasticizers. Today, the most commonly used plasticizers are phthalates, esters of phthalic acid (Fig. 11-15). Phthalates, as a result of their widespread use in plastics (approximately 8 billion pounds were produced in 1991), have now been detected in virtually all soil and water ecosystems, including the open ocean. The levels found are very low and the toxicity of phthalates is also low, but because of the widespread distribution, the use of phthalates is a cause of concern.

The Raw Materials for Polymer Production

Ethylene, propylene, butylene, styrene, and phenol, the monomers from which the giant polymer molecules are made, are derived mainly from petroleum (Chapter 14). As petroleum reserves are depleted, more of the raw materials for making synthetic polymers are expected to be obtained from coal. Another possibility is biomass conversion (Chapter 15) to produce small molecules such as methyl and ethyl alcohol and methane, from which the necessary monomers could be synthesized.

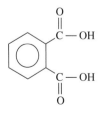

Figure 11-15 Phthalic acid.

EXPLORATIONS

The Electric Car: Up-scale Automobile for the Twenty-first Century

The gasoline engine has dominated road transportation for almost a hundred years. By the beginning of the next century, it may be facing stiff competition from a battery-powered motor. As petroleum supplies inevitably dwindle, gasoline prices rise, and the demand for vehicles emitting fewer pollutants increases, the motoring public may be ready for an electric car. To date, electric vehicles (EVs) have had little success because of their relatively poor performance and limited range between battery charges. Now this situation may be about to change. The big automakers in the United States and abroad are devoting large sums of money on research to develop a viable electric car.

At the end of the nineteenth century, when the production of a horseless carriage was still little more than a dream, three technologies were being explored: gasoline engines, steam engines, and battery-powered motors. By the early 1900s, 40 percent of the approximately 8000 automobiles in the United States were powered by steam, 38 percent by batteries, and only 22 percent by gasoline. At that time, gasoline engines were difficult to start, noisy, and unreliable. Steam-powered cars were easier to operate, but they needed a constant supply of water and high steam pressure, which made them expensive and difficult to maintain; most were gone by 1910. Early EVs had many advantages over their competitors: they started up instantly, were easy to operate, required minimum maintenance,

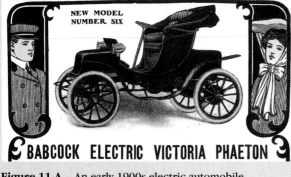

Figure 11-A An early 1900s electric automobile.

and ran silently—so silently that they were judged dangerous because pedestrians could not hear them coming! At the height of their popularity in 1912, EVs had a cruising speed between 15 and 20 mph (miles per hour) and could travel 30 to 40 miles between charges (Fig. 11-A).

The popularity of EVs was short-lived, and by the 1920s, gasoline-powered vehicles were firmly in the lead. Two factors doomed the electric car: the limited capacity of the battery, which restricted speed and range, and, ironically, replacement of the gasoline motor's dangerous hand crank with a battery-powered electric starter.

Despite their limitations, EVs never completely disappeared. In the United Kingdom, in particular, they have been in continuous use for over 70 years as neighborhood delivery vehicles. In this role their modest speed and range are not a handicap, and their low operating costs and few repair needs make them more economical than gas-powered vehicles. However, it was not until the oil crisis in the 1970s, when gasoline prices soared and long lines formed at gas stations, that automakers made a serious commitment to market electric cars for general use. By 1979, large and small companies in the United States and abroad were producing an assortment of often strange-looking vehicles, most with cruising speeds between 30 and 50 mph. Once the oil crisis ended, interest in EVs again waned.

Today, as concern for the environment increases, EVs look attractive again. A significant spur to development is a California law requiring that by 2003, 10 percent of new cars sold in the state operate without emitting any pollutants. Currently, only EVs can meet this standard. Even when the power-plant emissions associated with generating the electricity needed to power the batteries are taken into account, EVs emit 99 percent less carbon monoxide, 98 percent less hydrocarbons, 89 percent less nitrogen oxides, and more than 50 percent less carbon dioxide per mile than conventional automobiles.

General Motors, Ford, and Chrysler, which all now have EV programs, are well aware that to be acceptable to the general public, an EV must be comparable in cost and performance to a gasoline-powered car. GM's test model, the sleek high-performance Impact (Fig. 11-B), created a sensation when it was first displayed in 1990. The vehicle had a top speed of 110 mph and a range of 120 miles between charges. Under the hood, electricity was generated by a row of advanced-design lead-acid batteries, and an inverter converted the direct current to alternating current and transferred it to two induction motors—one for each front wheel. Start-up was soundless, and as the accelerator pedal was pressed, the car hummed into motion and accelerated rapidly and smoothly to 70 mph.

Unfortunately, despite its great looks and impressive performance, the Impact was not a prototype for a marketable vehicle. Not only was it prohibitively expensive, but it had numerous safety problems as well. It was difficult to handle in turns, and its fiberglass frame could not withstand even a minor collision. The 320-volt battery pack could deliver a lethal shock, and if one of the induction motors failed, the car would swerve out of control. Despite these shortcomings, however, the test model had tremendous potential. The challenge was to modify the design so as to reduce cost and improve safety while maintaining acceptable performance.

The main difficulty with EVs has always been the battery. In 1991, the three big automobile companies, the Department of Energy, and the electric utilities began a joint effort to develop promising battery technologies for future EVs. Chrysler is focusing its research on a nickel-iron system, while Ford favors a sodium-sulfur system that it believes could extend an EV's range to over 200 miles. For the long term—the year 2010 and beyond—lithium-iron disulfide and lithium-polymer systems are expected to give even better performance. Fuel cells are another promising option. But for the present, no battery system can compete with an optimally efficient lead-acid battery.

Lead-acid batteries are heavy. The pack required to operate GM's Impact weighs 870 pounds. Since every pound reduces efficiency, the frame of the EV that GM hopes to market by 1994 will be made of lightweight aluminum instead of steel. Body parts will be made of the most advanced lightweight but strong plastics; new synthetic rubber that has less rolling resistance than conventional rubber will be used to make the tires. Another factor influencing a vehicle's efficiency is its drag coefficient, that is, its resistance to air. The Impact's sleek sporty body gave it an impressively low drag coefficient but did not provide sufficient strength and bulk to allow the vehicle to withstand a crash; the shape will have to be altered. Each design modification is a compromise.

GM expects its modified Impact to sell in the $20,000 to $30,000 range. Even at that price, and allowing for the need to replace the entire battery pack every 20,000 miles at a cost of $1500, they estimate that the EV would be more economical than a gas-powered car over the lifetime of the vehicle. EVs last longer, and maintenance costs are much lower. There are few parts to wear out, no oil to change, and the per-mile cost of charging the batteries is one-fifth that of gasoline. If fully discharged, batteries require 2 to 3 hours to recharge; if 15 to 20 percent discharged, only 20 to 30 minutes are needed. An EV would be ideal for the daily commute to work, which for most people rarely exceeds 10 miles.

Despite the environmental and potential cost advantages of EVs, motorists are likely to balk at the high purchase price. To encourage sales, the government may have to provide incentives such as the right to drive in high-occupancy-vehicle (HOV) lanes, tax breaks, reduced electric rates, and easy access to cheap public battery chargers.

Obviously, many problems remain to be solved before we switch from "filling up" to "plugging in," but it appears likely that to those born in the twenty-first century, our present gas-guzzlers will be as old-fashioned as the phonograph record is to today's teenagers.

Question
Discuss the environmental advantages of an electric vehicle as a replacement for a standard gasoline-powered automobile.

References: M. Fischetti, "Here comes the electric car—It's Sporty, Aggressive, and Clean." *Smithsonian* (April 1992), p. 34; *World Guide to Battery-Powered Road Transportation: Comparative Technical and Performance Specifications*, compiled by Jeffrey M. Christian. (New York: McGraw Hill, 1980).

Figure 11-B   General Motors' Impact.

KEY WORDS AND CONCEPTS

addition polymer	plasticizer	poly(vinyl chloride) (PVC)
Bakelite	polyacrylonitrile	rearrangement polymer
condensation polymer	polyamide	resin
copolymer	polyester	rubber
cross-linked	polyethylene	silicone
elastomer	poly(ethylene terephthalate)(PET)	styrene-butadiene rubber (SBR)
homopolymer	polymer	thermoplastic
isoprene	polypropylene	thermosetting
monomer	polystyrene	vulcanization
neoprene	polytetrafluoroethylene	
plastic	polyurethane	

QUESTIONS AND PROBLEMS

1. Define the following:
   a. HDPE **b.** copolymer **c.** thermoplastic

2. Define the following:
   a. monomer **b.** vulcanization **c.** addition polymer

3. Nylon is a copolymer. Orlon and polyethylene are not. Explain.

4. The first man-made polymer was celluloid. What was it made from?

5. Write a chemical equation that shows how polyethylene is formed from ethylene.

6. Write a chemical equation that shows how poly(vinylchloride) is formed from vinyl chloride.

7. Describe the addition polymerization process.

8. Give the structure and name of the monomer used to make the polymer $-(C_2-F_4)_{\overline{n}}$. Is this an addition or condensation polymer?

9. Polyethylene can be manufactured in two different forms, high-density polyethylene (HDPE) and low-density polyethylene (LDPE).
   a. Is there a difference in molecular structure between HDPE and LDPE? **b.** List the properties and uses of HDPE and LDPE.

10. Explain what is meant by vulcanization.

11. Why is rubber elastic?

12. Because of its low reactivity, Teflon is used to line pots and pans. Explain why this polymer is unreactive. Why is polystyrene unsuitable for lining kitchenware?

13. The very strong condensation polymer Kevlar, which is used to make bullet-proof vests, has the following formula. What monomers are used to make Kevlar?

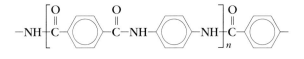

14. Burning plastic can give off choking and even poisonous gas. Name the plastic that would give off the following on burning.
    a. HCl **b.** HCN **c.** HCHO **d.** C_2H_3Cl

15. Draw a polymer made from five styrene monomers.

16. Draw a polymer made from four tetrafluoroethylene monomers.

17. Name a commercial plastic that contains the following functional group:
    a. chlorine **b.** fluorine **c.** nitrogen **d.** phenyl

18. Natural rubber is made from isoprene.
    a. Draw the structure of isoprene. **b.** Draw a polymer made from three isoprene monomers that polymerize in a cis- arrangement. **c.** On the basis of your drawing for part b, would you expect this polymer to be easily stretched?

19. Draw the structures of phenol and formaldehyde and the polymer they form when they react together.

20. List five paint pigments and the color each produces. Which white pigment has been banned?

21. What feature do all condensation polymerization reactions have in common?

22. What are the main ingredients in a paint? What is the purpose of each ingredient?

23. Describe the changes that take place when an oil-based paint dries. Does the wet paint react with any atmospheric gases? Does it add anything to the atmosphere?

24. Nylon 46 is made from a diamine with four carbon atoms and a diacid with six carbon atoms. Draw the structure of nylon 46.

25. What is the raw material from which the monomers used in the synthesis of most polymers are made?

26. What properties make polyester a good material for clothing? Would HDPE be a good polymer for making clothing? Explain.

27. Draw the structure and give the trademark name of the following:

    a. a thermosetting polymer **b.** a thermoplastic polymer **c.** a silicone polymer

28. List three factors that make silicones superior to some carbon-based rubbers.

29. What is the main environmental issue that needs to be addressed concerning the continued use of plastics?

30. What percentage of plastic waste is recycled? How does this amount compare with the amounts of aluminum and paper that are recycled?

31. Describe two substances that can be added to a plastic that will make it biodegradable.

32. Why is a plasticizer added to a polymer? Name a commonly used plasticizer.

33. List three ways to dispose of used plastics. Give two problems associated with each.

34. Would you support a national container deposit law? Explain.

35. If you were going to recycle plastics from the trash in your home town, describe how you would separate the plastics and what you would do with them once separated.

WATER RESOURCES AND WATER POLLUTION

A clean source of water is essential for all life. Waterfalls in the Deschutes National Forest in Oregon.

Water is one of our most precious resources; without it, there would be no life on earth. We use water in the home and for recreation, and water is crucial for agriculture and industry. Although the total amount of water on earth is fixed and cannot be increased, we are in no danger of running out of it. Water is constantly being recycled and replenished by rainfall; there is plenty of fresh water to meet the needs of everyone on earth. However, because of the uneven distribution of rainfall and the heavy use of water in certain areas, many regions in the United States and other parts of the world are experiencing severe water shortages.

Water in all its forms—ice, liquid water, and water vapor—is very familiar to us, and yet, compared with most other substances, water has many unusual properties. For example, its exceptional ability to store heat modifies the earth's climate, and ice's ability to float on water allows aquatic creatures to survive in winter.

Water is an excellent solvent and can dissolve a wide variety of ionic and polar substances. Thus, it is an excellent medium for carrying nutrients to plants and animals. It is also a good medium for carrying toxic substances and other pollutants.

355

In this chapter, we will examine the composition of natural waters, the way water is distributed on the earth, and how it is recycled. We will learn how the molecular structure of water accounts for its extraordinary properties, and we will consider the implications of those properties for life on earth. We will discuss the way we use our water supplies and the measures that can be taken to better manage and conserve them. Last, we will study water pollution and examine the main sources of pollution, the effects of pollutants on human health and the environment, and the steps that can be taken to preserve water quality.

Learning Goals:

In this chapter, you should gain an understanding of:

1. The distribution of water resources on the surface of the earth and how water is recycled in the hydrologic cycle.
2. The unique physical properties of water and why these properties are so important.
3. Water management and conservation.
4. The different types and sources of water pollutants.
5. Sewage treatment.

 ## DISTRIBUTION OF WATER ON THE EARTH

Water is the most abundant compound on the earth, covering nearly three-quarters of its surface. Although the total amount of water on earth is enormous, only a small percentage of it is fresh water—much of which is not readily available for human use. Over 97 percent of the world's total reservoir of water is found in oceans and is too salty for drinking, irrigation, and most industrial and household needs. Of the remaining approximately 3 percent, about 2 percent is in the form of glaciers and polar ice caps. Most of the rest is found underground (groundwater), and much of it is too difficult or too expensive to tap. Lakes and rivers, which are the major sources of the world's drinking water, account for just over 0.1 percent of the earth's total water (Fig. 12-1).

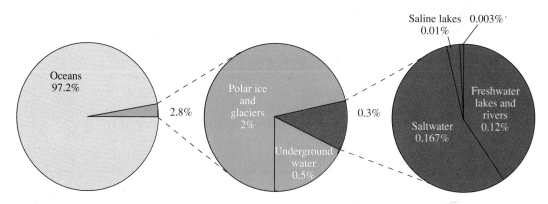

Figure 12-1   The distribution of water on the earth. Most of the water on earth is salt water in the oceans. The less than 3 percent that is fresh water exists mainly as inaccessible ice in glaciers and the polar ice caps. For fresh water we must depend on the 0.7 percent that is found underground and in inland seas, lakes, and rivers.

TABLE 12-1 Water Content of Selected Organisms and Foods

	Percent of Mass
Marine invertebrates	97
Human fetus (1 month)	93
Fish	82
Human adult	70
Broccoli	90
Milk	88
Apples	85
Grapes	80
Eggs	75
Potatoes	75
Steak	73
Cheese	35

Water is the major component of all living things (Table 12-1). It constitutes 70 percent of the human adult body and from 50 to 90 percent of all plants and animals. Water enters the body in the liquids we drink and the foods we eat; it leaves the body in urine, feces, sweat, and the air we exhale. Humans can survive only a few days without water.

THE COMPOSITION OF NATURAL WATERS

All bodies of natural water are solutions and contain varying concentrations of dissolved solids. Water containing up to 0.1 percent (1000 ppm) of dissolved solids is generally termed **fresh water**, but, assuming it is safe in other respects, such water is not necessarily suitable for drinking. For drinking, the U.S. Public Health Service recommends a solid content of no more than 0.05 percent (500 ppm). The concentration of dissolved solids in seawater averages 3.5 percent (35,000 ppm). Water is termed **brackish** when the solid content lies somewhere between that of fresh water and seawater.

The ions present in seawater at 1 ppm or more are shown in Table 12-2. Present at lower concentrations are at least 50 other elements. The high concentration of salts in oceans comes primarily from the constant evaporation of water from their surfaces.

The concentration of ions in fresh water is much lower than that in seawater, and the distribution of ions is quite different (Table 12-3). In seawater, the main cation is sodium (Na^+), and the main anion is chloride (Cl^-). In fresh water, on the other hand, calcium (Ca^{2+}) and magnesium (Mg^{2+}) are the dominant cations, and bicarbonate (HCO_3^-) is the dominant anion. The ions in fresh water come from the weathering of rocks and soil. A main source of the Na^+ and Cl^- in river water, particularly water near a shore, is salt spray that is thrown up into the atmosphere from the ocean and then deposited on land and carried, by runoff, into rivers.

In the Great Salt Lake in Utah and the Dead Sea in Israel, the concentration of dissolved solids is as high as 25 percent.

TABLE 12-2 Major Constituents of Seawater

Ion	ppm
Chloride, Cl^-	19,000
Sodium, Na^+	10,600
Sulfate, SO_4^{2-}	2,600
Magnesium, Mg^{2+}	1,300
Calcium, Ca^{2+}	400
Potassium, K^+	380
Bicarbonate, HCO_3^-	140
Bromide, Br^-	65
Other substances	34
Total	34,519

TABLE 12-3 Comparison of the Concentrations of Major Ions in Fresh Water and Seawater (percent of total ionic concentration)

Ion	Fresh Water	Seawater
HCO_3^-	41.0	0.2
Ca^{2+}	16.0	0.9
Mg^{2+}	14.0	4.9
Na^+	11.0	41.0
Cl^-	8.5	49.0

THE HYDROLOGIC CYCLE: RECYCLING AND PURIFICATION

The earth's supply of water is continually being purified and recycled in the **hydrologic**, or **water**, **cycle**, illustrated in Figure 12-2. Through the processes of evaporation, transpiration, condensation, and precipitation, solar energy and gravity are responsible for the ceaseless redistribution of water among oceans, land, air, and living organisms. The heat of the sun warms the surface of the earth, causing enormous quantities of water to evaporate from the oceans. On

The term vapor, rather than gas, is generally used to describe a substance that is in a gaseous form below its normal boiling point.

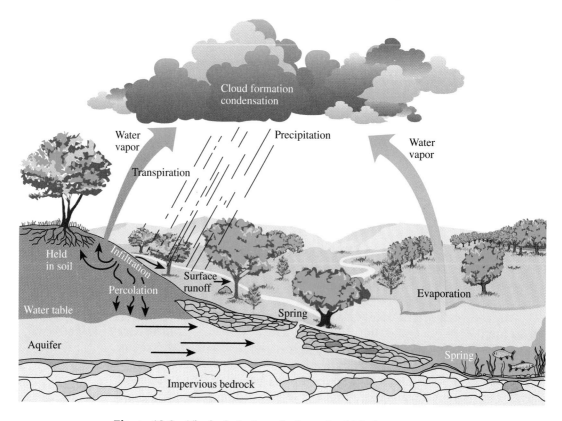

Figure 12-2 The hydrologic cycle through which the water on earth is continuously recycled. Powered by the sun, evaporation, condensation, and precipitation redistribute water among the oceans, the land, the atmosphere, and living organisms.

land, water vapor enters the atmosphere from plants by the process of transpiration, in which water escapes through the pores on leaf surfaces and evaporates. Additional water evaporates from lakes, rivers, and wet soil.

In the process of evaporation, water is purified. Dissolved substances are retained in the oceans, land, and plants, and purified water vapor enters the atmosphere. As the water vapor rises, it cools and condenses into fine droplets that form into clouds. Prevailing winds carry moist air and clouds across the surface of the earth. If the air cools sufficiently, water droplets or ice crystals fall to earth as precipitation (rain, sleet, hail, or snow). Precipitation that falls through clean air is pure and contains only gases dissolved from the atmosphere (primarily carbon dioxide) and traces of dissolved salts. Some precipitation becomes locked in glaciers, but most of it either sinks into the soil or flows downhill, as runoff, into nearby streams, eventually making its way through rivers, lakes, and wetlands to the ocean.

Precipitation that seeps into the soil is either taken up by plant roots or continues to percolate into the ground until it reaches an impermeable layer of rock that stops its downward progress. Water collects in the porous rock above this impermeable layer, forming a reservoir of groundwater termed an **aquifer** (Fig. 12-2). Groundwater is generally of a high quality because the porous rock acts as a filter and retains suspended particles and bacteria. Groundwater is replenished much more slowly than is water at the surface of the earth; it flows slowly (a speed of 15 m [50 ft] per year is typical) through an aquifer until it reaches an exit to the surface, where it either emerges as a spring or seeps out over a relatively wide area. These springs and seeps feed lakes, streams, and rivers—and ultimately, the oceans.

> On land, transpiration often accounts for more water vapor entering the atmosphere than direct evaporation from lakes, rivers, and soil. Worldwide, direct evaporation from the oceans is the major source of atmospheric water vapor.

CONSUMER BRIEF **Sport Drinks: Electrolyte and Energy Replacement**

Sport drinks have become very popular among young, active people for the replacement of minerals and energy lost in exercise.

The most prominent of them is Gatorade, which was developed at the University of Florida, and is named after the team mascot, the Gator.

The ingredients of a typical sport drink fall into three groups: sugars, salts and acids. The drinks, which also contain preservatives, flavorings, gums, and artifical colors are made up in water.

As you exercise, sweat is produced to cool your body. Burning calories while exercising raises your body heat, and sweating cools the body. The harder you exercise, the more you sweat and you need to replenish the lost fluid before you become dehydrated.

Blood contains many components, including sugars, salt, and blood cells. The membranes of red blood cells are semipermeable, meaning that they allow some molecules and ions to pass through but not others. By the process of osmosis water molecules pass through the membrane, but many of the dissolved chemicals are stopped. If a red blood cell is placed in a solution that has a lower concentration of dissolved substances relative to the intracellular fluid, water will start to pass into the cell causing it to swell. This eventually leads to rupture of the cell, a process called hemolysis. If the red blood cells are placed in a solution with a salt concentration greater than that of the intracellular fluid, water passes out of the cell into the surrounding solution, causing the cell to shrivel, a process called crenation. In blood, the fluid surrounding red blood cells has a concentration of dissolved substances equal to that in the intracellular fluid, and is described as being "isotonic." A solution of water containing 0.9 percent (0.9 g/100ml) of sodium chloride approximates the normal osmotic concentration of the extracellular fluid.

Sport drinks contain sodium chloride, potassium citrate, and phosphate. Although they attempt to be isotonic, there is a practical limit to how much salt can be added to these drinks before they taste salty. Sodium and potassium ions both have a charge of 1+, but a sodium ion has only about one-half the mass of a potassium ion. Thus, less sodium than potassium has to be added to the solution to raise the osmotic concentration. Most natural juices have a high content of dissolved salts, especially those such as orange juice, which contain quite large amounts of potassium.

Sport drinks provide "energy replacement" in the same way that other drinks do. They contain the sugars sucrose and fructose that provide fuel for metabolism, but the concentration of sugar in these drinks is about half that in a soft drink. Because the concentration of sugar is lower, sport drinks contain only about half the calories of soft drinks.

Questions
1. Why is it important to know the concentrations of sodium and potassium in fluids we drink?
2. What are the main differences between sport drinks and mineral water?

Comparison of Sport Drinks and Other Common Drinks

	Calories	Carbohydrates (g)	Sodium (mg)	Potassium (mg)	Vitamin C (mg)
Orange juice	110	25	2	470	80
Lemonade	110	28	1	40	17
Cola	105	28	12	5	0
Gatorade	50	14	110	25	0
10-K	60	15	54	29	30

Per 8 oz (240 ml) serving.
Reference: Consumer Reports (August 1993), pp. 491–494.

THE UNIQUE PROPERTIES OF WATER

Water is the only common pure compound on earth that is a liquid, and it possesses many unusual—often unique—properties. If it behaved like other chemical compounds of similar molecular weight and structure, life on earth could not exist. The reason for most of water's extraordinary properties is hydrogen bonding (Chapter 5). Although hydrogen bonds are much weaker than ionic or covalent bonds, they have a profound effect on the physical properties of water in both its liquid and solid states.

The Water Molecule and Hydrogen Bonding

A water molecule can form four hydrogen bonds (Fig. 12-3). The oxygen atom in the molecule can bond to two hydrogen atoms because it has two pairs of unshared electrons. Each hydrogen atom can bond to an oxygen atom in another water molecule. The bonds are directional, which means they can form only when the molecules are correctly oriented toward each other.

The bond angle in the water molecule is 104.5°.

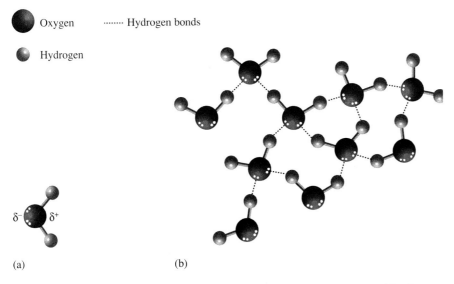

Oxygen ········ Hydrogen bonds

Hydrogen

δ^- δ^+

(a)　　　　　　　　(b)

Figure 12-3 Hydrogen bonding in water. (a) Water is an angular polar molecule. Because oxygen is more elecronegative than hydrogen, it has a partial negative charge (δ^-), and the two hydrogen atoms have partial positive charges (δ^+). (b) The slightly negatively charged oxygen atoms are attracted to the slightly positively charged hydrogen atoms, and hydrogen bonds form between adjacent water molecules. Each oxygen atom has two pairs of unshared electrons and can bond to two hydrogen atoms. Thus each water molecule can form four hydrogen bonds.

In the liquid state, water molecules are in constant motion, and hydrogen bonds are continually being formed and broken. The arrangement of molecules is random, and not all possible hydrogen bonds are formed. In solid water (ice), molecular motion is at a minimum, and molecules become oriented so that maximum hydrogen bonding occurs. An ordered and strong, extended three-dimensional open lattice structure is formed (Fig. 12-4a). The size of the "holes" in the lattice is dictated by the bond angle in the water molecule, which determines how close adjacent molecules can come to each other. As a result, adjacent water molecules in ice are not as close to each other as they can be in liquid water. The ordered arrangement of atoms in ice accounts for the symmetry of ice crystals in snowflakes (Fig. 12-4b).

Boiling Point and Melting Point

Compared with hydrogen compounds of other elements in Group VIA of the periodic table—H_2S, H_2Se, and H_2Te—water (H_2O) has an unexpectedly high boiling point (Fig. 12-5). Normally, boiling points in a series of compounds of elements in the same group increase regularly with increasing molecular weight, as occurs for H_2S, H_2Se, and H_2Te. The unexpectedly high boiling point of H_2O is caused by the hydrogen bonding that exists in water but does not exist, to any significant extent, in the other compounds in the series. When water is converted to vapor, additional energy in the form of heat is required to break its hydrogen bonds; consequently, its boiling point is higher than would be expected. If water boiled at the predicted temperature of $-80°C$ ($-112°F$) it would be a gas at the temperatures found on earth, and our form of life and environment would not be possible.

Water also has an exceptionally high melting point because of the large quantity of heat energy required to break its hydrogen bonds. When ice melts, about 15 percent of its hydrogen bonds are broken, the three-dimensional structure collapses, and water forms.

Heat Capacity

Heat capacity is the quantity of heat required to raise the temperature of a given mass of substance by 1°C. It takes 1 calorie of heat to raise the temperature of 1 g of liquid water by 1°C. Water has the highest heat capacity of any common liquid or solid. From the definition of heat capacity, it follows that the

The high heat capacity of water helps organisms maintain a constant internal temperature when outside temperatures fluctuate.

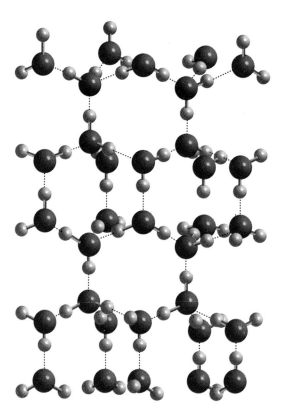

Hydrogen

Oxygen (a)

(b)

Figure 12-4 (a) The well-ordered structure of ice. In the three-dimensional structure, each water molecule is hydrogen-bonded to four other water molecules. (b) The ordered arrangement of H_2O molecules in ice accounts for the intricate hexagonal shapes of snowflakes.

362

higher the heat capacity of a substance, the less its temperature will rise when it absorbs a given amount of heat, and conversely, the less its temperature will fall when the same amount of heat is taken away (released) from it.

The high heat capacity of water has enormous implications for earth's climate. Because of this capacity, the oceans of the world can absorb very large amounts of heat without having a corresponding rise in their temperature. The oceans absorb heat from the sun on warm days, primarily in the summer, and release it in the winter. If there were no water to absorb and release heat, daily and seasonal temperatures on the earth would fluctuate as drastically as they do on the waterless moon and the planet Mercury, where temperatures vary by hundreds of degrees during the light-dark cycle.

Heat of Fusion and Heat of Vaporization

Heat of fusion is the amount of heat required to convert 1 g of a solid to a liquid at its melting point; the same amount of heat is released when 1 g of the liquid is converted to the solid. **Heat of vaporization** is the amount of heat required to convert 1 g of a liquid to a vapor at its boiling point; the same amount of heat is released when 1 g of the vapor condenses to its liquid.

During the process of fusion (melting), heat energy is absorbed but the temperature of the solid-liquid mixture does not begin to rise above the melting point until melting is complete. During the reverse process, freezing, heat energy is released to the surroundings; the temperature does not begin to fall until freezing is complete. Similarly, there is no change in temperature during vaporization and condensation until each process is complete. The changes in energy and temperature that occur when water changes successively from a solid, to a liquid, to a gas, and vice versa, are shown diagrammatically in Figure 12-6.

Since heats of fusion and vaporization are related to heat capacity, it is not surprising that the values for water are higher than for practically any other substance. Again, the explanation is hydrogen bonding: For ice to melt and for water to vaporize, hydrogen bonds must be broken, and breaking them requires a considerable input of energy in the form of heat.

The fact that a relatively large amount of heat is required to evaporate a small volume of water has important consequences. It means that the human body can be cooled efficiently by the evaporation of a small amount of water (perspiration) from the skin. Extensive water loss from the body, which could upset fluid balance within the body, is thus kept at a minimum.

Anyone who has accidentally put his or her hand in the steam coming from a kettle of boiling water knows what a painful experience it can be. The hand is severely burned because as the steam condenses it releases heat, damaging the skin and causing pain. In contrast, a burn from water at the same temperature (100°C, 212°F) is far less severe.

Water's high heat of vaporization affects the earth's climate. In summer, water evaporates from the surfaces of oceans and lakes. The heat energy needed for evaporation is drawn from the surroundings and, in consequence, nearby land masses are cooled. On a hot day, land close to a large body of water is always cooler than land in the interior. At night, when air cools, water vapor condenses, heat is released, and the temperature of the surroundings is raised. In this way, temperature variations between day and night are minimized. A similar modifying effect occurs in winter. When water freezes, heat energy is released and the surroundings are warmed.

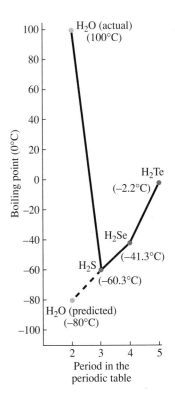

Figure 12-5 With the exception of oxygen, the boiling points of the hydrogen compounds of the group VIA elements increase regularly as one goes down the group. The unexpectedly high boiling point of water is due to hydrogen bonding, which occurs in H_2O but not to any extent in H_2S, H_2Se, and H_2Te.

540 cal are needed to vaporize 1 g of water at its boiling point compared with 73 cal/g for mercury, and 204 cal/g for ethyl alcohol at their boiling points.

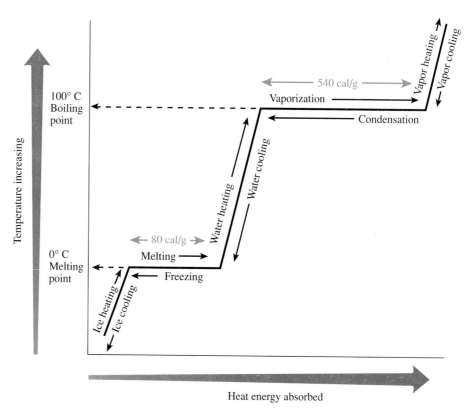

Figure 12-6 Changes in state as water is heated and cooled. There is no change in temperature as water melts, vaporizes, condenses, or freezes until the process is completed. During melting and vaporization, 80 and 540 cal/g of water, respectively, are absorbed; during condensation and freezing, 540 and 80 cal/g, respectively, are released.

Temperature-Density Relationship

Density is defined as mass per unit volume. We often say that one substance is heavier or lighter than another. What we actually mean is that the two substances have different densities: A particular volume of one substance weighs more or less than the same volume of the other substance.

The density of most liquids increases with decreasing temperature and reaches a maximum at the freezing point. The density of water does not. On cooling, the density of water reaches a maximum at 4°C—four degrees *above* its freezing point—and then decreases until the freezing point is reached at 0°C (32°F). The fortunate consequence of this property is that ice floats on the surface of water. This behavior is so familiar that we tend to forget that it is not typical of most liquids. For example, if pieces of solid paraffin are put into a container of liquid paraffin, the solid pieces sink to the bottom of the container because they are denser than liquid paraffin (Fig. 12-7).

The unusual behavior of ice is the result of the open lattice arrangement of hydrogen and oxygen atoms that forms when water freezes (see Fig. 12-4). As noted earlier, molecules of water are farther apart in ice than they are in liquid water. As a result, when water starts to freeze, the number of molecules per unit volume (the density) decreases.

The fact that ice is less dense than water has important consequences for aquatic life. When air temperature falls below freezing in winter, surface water in lakes begins to freeze, and a layer of ice forms and floats on the surface. The ice covering the surface acts as an insulating layer, reducing heat loss from the water under the ice. As a result, most lakes in temperate climates never freeze to a depth of more than a few feet, and fish and other aquatic organisms are able to survive the winter in the water under the ice. If water behaved like most other liquids, ice

Wood floats on water because its density is less than that of water; lead sinks because it is denser than water.

(a) (b)

(a) (b)

Figure 12-7 (a) A solid piece of paraffin sinks in liquid paraffin. (b) Ice cubes float on water. The behavior of ice is exceptional; most chemical compounds behave like paraffin.

would sink to the bottom as it formed, and lakes would freeze from the bottom up. Even deep lakes would freeze solid in winter, and aquatic life would be killed.

The unusual behavior of water when freezing has other consequences for the environment. If water trapped in cracks in rocks freezes, the force of expansion is so powerful that the rock may split, an important factor in the weathering of rock (Chapter 2).

WATER USE AND WATER SHORTAGES

Throughout the world, as populations have grown, the demand for water has increased tremendously; many regions, including parts of the United States, are now experiencing serious water shortages. Water pollution has added to the problem by reducing the amount of drinkable (potable) water.

There are two sources of usable water: (1) **surface water** in lakes and rivers and (2) **groundwater** from wells drilled into aquifers (see Fig. 12-2). At the present time, only groundwater that lies within about 1000 m (3000 ft) of the earth's surface can be tapped economically.

Figure 12-8 The vast Ogallala aquifer lies beneath the Great Plains stretching from South Dakota to Texas. It was formed more than 2 million years ago from melting glaciers. Today, water is being withdrawn from the aquifer much more rapidly than it is being returned.

Figure 12-9 In 1991, excessive withdrawal of groundwater caused this home in Frostproof, Florida, to collapse into a sinkhole.

mainly used for cooling

Industry and agriculture are the major users of water in the United States. Oil refining, steel and paper manufacture, and the electric power industry, all require large quantities of water. However, most of the water industry uses is for cooling purposes and is recycled at the plant site to be reused. On the other hand, water for irrigation, which accounts for 70 percent of all groundwater withdrawals, is consumed and must be continually withdrawn from the ground.

The arid western states of the United States depend on the water in the vast Ogallala aquifer to irrigate over 10 million acres of cropland (Fig. 12-8). Although the aquifer is enormous, the rate at which water is being withdrawn today greatly exceeds its recharge rate, which is extremely slow because the aquifer underlies a region of low rainfall. It has been estimated that, at the present rate of withdrawal, much of the aquifer could be dry by the year 2020.

When withdrawal of water from an aquifer is excessive, land above it may collapse, producing a large sinkhole (Fig. 12-9). Depletion of groundwater near coastal areas can cause seawater to flow into the aquifer, making the water unfit for drinking. Regions of the United States where groundwater has been depleted are shown in Figure 12-10.

TABLE 12-4 Average Per Capita Use of Water Per Day in the United States

Water Use	Gallons
Flushing toilets	30
Bathing	23
Laundry	11
Drinking and cooking	10
Washing dishes	6
Miscellaneous	10
Total	90

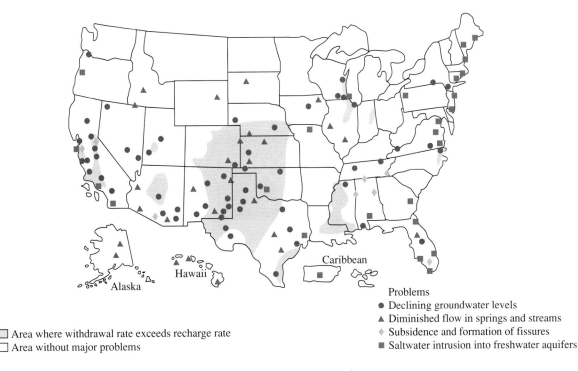

Problems
- ● Declining groundwater levels
- ▲ Diminished flow in springs and streams
- ◆ Subsidence and formation of fissures
- ■ Saltwater intrusion into freshwater aquifers

Area where withdrawal rate exceeds recharge rate
Area without major problems

Figure 12-10   In many parts of the United States groundwater withdrawals exceed the recharge rate. Lowered water tables cause subsidence, saltwater intrusions, and other problems.

Potable water for domestic purposes, which is obtained almost equally from surface water and groundwater, accounts for approximately 10 percent of our total water use. Although we need only about one-half gallon of water per day for drinking, average per capita daily use in the United States is 90 gallons, divided up as shown in Table 12-4. This amount is nearly 3 times as much as the average per capita use worldwide and up to 20 times as much as the per capita use in less developed countries. In some arid western areas of the United States, millions of gallons of water are used daily to water lawns and golf courses, and domestic water consumption is approaching 400 gallons per person per day.

Because of our extravagant use of water, we are now facing a serious water crisis. While we cannot increase the earth's supply of water, we can take steps to manage the water that we have more efficiently.

WATER MANAGEMENT AND CONSERVATION

Water Transfer

Limited water supplies in one region can often be increased by diverting river water to the needy area or by building dams along rivers to collect water in huge reservoirs that can then feed water into depleted areas. However, these undertakings are always controversial. When dams are built, agricultural land and wildlife habitat are lost, and nutrients in water downriver from the dam decrease. River diversion often leads to acrimonious debate between neighbors who are competing for the same water supplies. For example, disputes between California and Arizona over diverting water from the Colorado River went all the way to the U.S. Supreme Court before a settlement was reached.

Conservation

Conservation of water is essential if serious water shortages are to be avoided. Farmers could reduce their use of water by lining irrigation canals with plastic to reduce loss by seepage and by burying pipes to irrigate from below ground. The commonly used method of spraying water over the land is inefficient because much water is lost by evaporation before it can sink into the soil and reach plant roots.

Significant savings could also be made in the home. The average toilet, which with each flush converts 19 L (5 gal) of drinkable water into wastewater, could be replaced with a watersaving model that works effectively with as little as 2 L (one-half gallon) of water. The excessive amount of water released by a conventional shower (almost 40 L [10 gal] of water per minute) could be reduced very cheaply by switching to a watersaving showerhead.

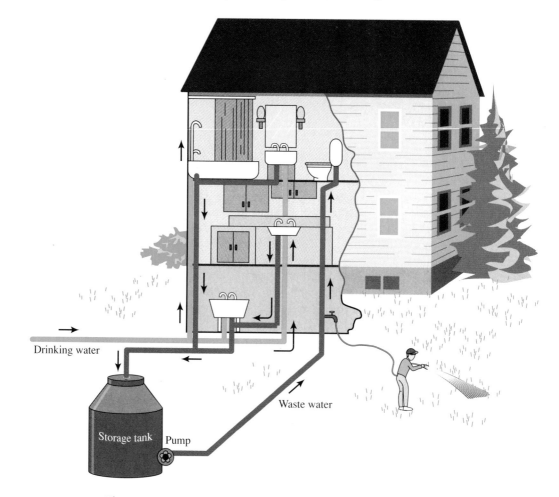

Figure 12-11 Water could be saved in the home if wastewater from sinks and tubs were piped into a storage tank and reused for flushing toilets, washing cars, and other activities that do not require drinking-quality water.

Water could also be saved by recycling wastewater from kitchen sinks and laundry tubs and using it for such purposes as flushing toilets and watering lawns (Fig. 12-11). Another idea is for public water systems to provide two kinds of water: one of drinking quality and a second that would not meet all drinking-water standards but would be free of bacteria and suitable for many other household purposes.

Potable Water from Wastewater

In most urban areas in developed countries, wastewater from homes, businesses, and industrial sites flows through a network of sewer pipes to sewage-treatment plants for purification. After it has been treated, the water is discharged into rivers or oceans. Water could be conserved, and money saved, if purified water from sewage plants were recycled instead into public water systems. Although the idea may not be attractive to many people, the technology is available for obtaining drinking-quality water from sewage effluent. We will consider wastewater treatment in detail later in the chapter.

Desalination

One approach to increasing the supply of fresh water is **desalination**, the removal of salts from seawater and brackish water. Although desalination technology exists, it is very expensive because it requires a large input of energy. Consequently it is used only in regions where there is no alternative source of fresh water. Distillation and reverse osmosis are the most widely used methods for converting salt water to fresh water.

In **distillation**, salt water is heated to evaporate off the water; water vapor is condensed to pure water, leaving the salts behind. Since the need for water is greatest in hot arid lands where sunlight is plentiful, energy costs can be reduced by using solar radiation to heat the water.

The process of **osmosis** is illustrated in Figure 12-12. Two compartments are separated by a **semipermeable membrane**, a thin membrane that contains minute pores through which water molecules, but not ions or molecules larger than water, can pass. When pure water is placed on one side of the membrane and salt water on the other, water molecules pass through the

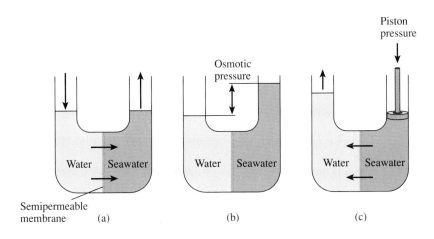

Figure 12-12 Desalination of seawater by reverse osmosis. (a) Water and a salt solution separated by a semipermeable membrane. (b) Water molecules pass through the membrane into the salt solution, raising the level of liquid in the salt solution compartment. The difference in levels between the two compartments represents the osmotic pressure. (c) In reverse osmosis, pressure in excess of the osmotic pressure is applied to force water molecules from the salt solution into the water compartment.

membrane from the water compartment to the salt compartment by the process of osmosis. The level of liquid in the salt compartment rises and reaches a maximum at equilibrium (when the movement of water molecules back and forth between the two compartments is equal). The pressure that must be exerted to prevent osmosis is termed the osmotic pressure. When pressure in excess of the osmotic pressure is exerted on the liquid in the salt compartment, pure water will flow from the salt compartment through the membrane into the water compartment (Fig. 12-12c). This process is **reverse osmosis**.

Reverse osmosis plants are used in California, Texas, and the Florida Keys. They are a major source of fresh water in Israel, Saudi Arabia, and the Mediterranean island of Malta, all arid countries where other sources of water are in very short supply. A problem with desalination, apart from its cost, is disposing of the mountains of salt it produces. If the salt is returned to the ocean near a coast, the increased salt content of the coastal water could have an adverse effect on wildlife.

WATER POLLUTION: A HISTORICAL PERSPECTIVE

Human waste was the first pollution problem. In ancient times, people naturally settled near a source of water, and communities grew up beside lakes, along rivers, and in areas where well water was available. People often drank water from the same river in which they disposed of their wastes and in which they washed themselves and their clothes. As a result, the water often became contaminated with disease-causing microorganisms from human and animal wastes.

During the Industrial Revolution in the nineteenth century, cities in the United States and in Western Europe grew at a tremendous rate. Refuse of all kinds—including great quantities of horse manure—ended up in the streets, in open sewers, and in nearby rivers. Devastating epidemics of waterborne diseases, such as cholera, typhoid, and dysentery, were common in large cities.

By the early part of the twentieth century, the connection between disease and sewage-borne microorganisms had been recognized, and safe water supplies were established in most industrialized nations. As a result, waterborne diseases have been virtually eliminated in the developed countries. However, they are still common in less developed countries, where sewer systems are often inadequate or nonexistent. As recently as 1991, nearly 140,000 cases of cholera were reported in an epidemic that swept through South America. In industrialized nations, contamination with hazardous chemicals has now become the main threat to water supplies.

TYPES OF WATER POLLUTANTS

Water pollutants can be divided into the following broad categories: (1) disease-causing agents, (2) oxygen-consuming wastes, (3) plant nutrients, (4) suspended solids and sediments, (5) dissolved solids, (6) toxic substances, (7) heat (thermal pollution), (8) radioactive substances, (9) oil, and (10) acids. In this

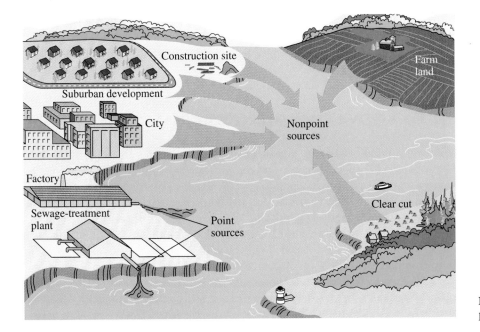

Figure 12-13 Point and nonpoint sources of pollution.

chapter, we will consider the sources, and effects on the environment, of the first seven types of pollutants and the steps that can be taken to control them. Radioactive substances will be discussed in Chapter 19 (The Disposal of Dangerous Wastes) and oil in Chapter 14 (Fossil Fuels). Acids were considered in Chapter 8.

Point and Nonpoint Sources of Water Pollutants

Pollutants enter waterways from both point and nonpoint sources (Fig. 12-13). **Point sources** include sewage-treatment plants, factories, electric power plants, mines, and offshore oil-drilling rigs—in other words, sources that discharge wastes at specific locations, usually through pipes. **Nonpoint sources** discharge pollutants over a wide area and include general runoff from the land, for example, from feedlots, cultivated land, logged forests, and construction sites. Discharges from nonpoint sources are generally more dilute than are those from point sources, but they are more difficult to identify and regulate. They occur irregularly, usually as the result of severe storms.

DISEASE-CAUSING AGENTS

Outbreaks of disease, often of epidemic proportions, can occur if feces from people infected with **pathogens** (disease-causing agents) enter water supplies. Diseases that are transmitted when people drink infected water, swim in it, or eat contaminated food include cholera, typhoid fever, dysentery, infectious hepatitis, and polio.

The coliform bacteria count is used to test water for contamination by microorganisms. **Coliform bacteria** live naturally in the human intestinal tract, and the average person excretes billions of them in feces each day. Coliform bacteria are harmless and cause no diseases, but their presence in water is an indication of fecal contamination. If none are found, the water is free from fecal contamination and can be assumed to be free from pathogenic organisms.

OXYGEN-CONSUMING WASTES

Animals and plants that live in an aquatic habitat depend on oxygen dissolved in the water for their survival. Oxygen is not very soluble in water, and the amount that does dissolve depends on both the temperature and the altitude of the water. As shown in Figure 12-14, solubility decreases with increasing temperature and with increasing altitude. At sea level at 20°C (68°F), the concentration of oxygen in water saturated with oxygen (water that contains the maximum amount of oxygen it can dissolve) will be about 9 ppm as opposed to only 6 ppm at 30°C (86°F) at 2000 m (6600 ft).

Dissolved oxygen (often abbreviated DO) in lakes and rivers is rapidly depleted if organic waste materials are released into the water. Typical sources of such wastes—collectively called **oxygen-consuming wastes**—are human and animal sewage and industrial wastes from paper mills, tanneries, and food-processing plants. Wastes from slaughterhouses and meat-packing plants are a particularly concentrated source of oxygen-consuming wastes.

Organic detritus in aquatic ecosystems ordinarily is decomposed by aerobic (oxygen-consuming) decomposers, primarily bacteria and fungi (Chapter 3). If the water is overloaded with organic wastes, aerobic decomposers proliferate and dissolved oxygen is consumed more rapidly than it can be replaced from the atmosphere. If the level of dissolved oxygen falls below 5 ppm, fish—particularly desirable game fish—start to die (Fig. 12-15). If the concentration of dissolved oxygen continues to fall, invertebrates and aerobic bacteria will be unable to survive.

In the complete absence of dissolved oxygen, decomposition of organic matter continues but it is taken over by anaerobic (non-oxygen-requiring) bacteria. If this stage is reached, the water begins to smell unpleasant because different pathways of decomposition are being followed (Table 12-5). Many of the end products of anaerobic decomposition, including hydrogen sulfide, ammonia, amines, and phosphorus compounds, have disagreeable odors.

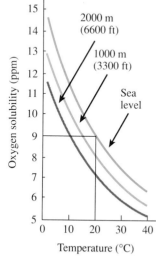

Figure 12-14   The solubility of oxygen in water at different temperatures and altitudes. Solubility decreases as temperature rises and as altitude increases.

TABLE 12-5   End Products of Decomposition of Organic Compounds Under Aerobic and Anaerobic Conditions

Element in Organic Compound	End Product(s) of Decomposition	
	Aerobic conditions	Anaerobic conditions
Carbon, C	CO_2	CH_4
Nitrogen, N	NO_3^-	NH_3 and amines
Sulfur, S	SO_4^{2-}	H_2S
Phosphorus, P	PO_4^{3-}	PH_3 and other P compounds

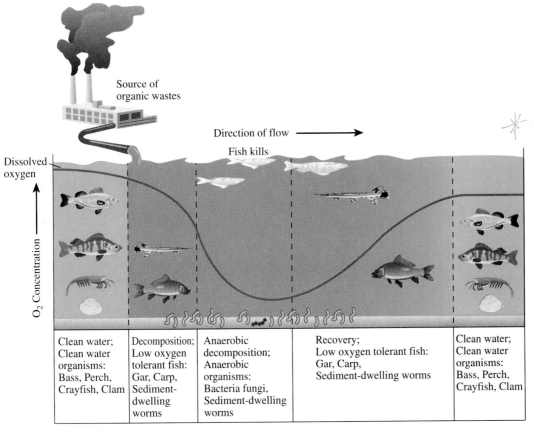

Clean water; Clean water organisms: Bass, Perch, Crayfish, Clam	Decomposition; Low oxygen tolerant fish: Gar, Carp, Sediment-dwelling worms	Anaerobic decomposition; Anaerobic organisms: Bacteria fungi, Sediment-dwelling worms	Recovery; Low oxygen tolerant fish: Gar, Carp, Sediment-dwelling worms	Clean water; Clean water organisms: Bass, Perch, Crayfish, Clam

Distance downstream ⟶

Figure 12-15 If organic wastes are discharged into a river, the level of dissolved oxygen in the water falls, and aquatic organisms begin to die. First to be affected by the decrease in dissolved oxygen are crayfish, clams, and game fish such as bass and perch. If the level of dissolved oxygen continues to fall, aerobic bacteria disappear, other fish such as gar and carp die, and decomposition is taken over by anaerobic bacteria. Once the waste discharge ceases, the river recovers.

In a river, water downstream recovers quite quickly once the discharge ceases, particularly if the water is free-flowing and turbulent. Organic material is diluted, and the water is reoxygenated as fresh water is brought to the surface. Lakes, which have little flow of water, take much longer to recover.

PLANT NUTRIENTS

Aquatic plants need the elements nitrogen (as nitrates) and phosphorus (as phosphates) for proper growth. These elements are usually scarce in most natural bodies of water, and this shortage normally keeps the spread of vegetation under control. However, if a new source of nitrogen or phosphorus is introduced into the water, excessive plant growth occurs, and the algae population explodes.

Figure 12-16 Eutrophication of a lake. If wastewater that contains plant nutrients is discharged into natural bodies of water, excessive growth of algae and aquatic plants occurs.

In an undisturbed environment, eutrofication occurs naturally over thousands of years as lakes age and nutrients are slowly enriched. When human activities add nutrients, the process is greatly accelerated.

The effects of eutrophication are most severe in reservoirs, lakes, estuaries, and water near shorelines where there is little or no flow of water.

As a result of this phenomenon, which is termed **eutrophication**, dense mats of rooted and floating plants are formed. Waterways become clogged, and boat propellers are fouled. The cost of removing the unwanted vegetation is usually prohibitive. Blue-green algae blooms that appear on the water's surface (Fig. 12-16) release unpleasant-smelling, bad-tasting substances. The water becomes turbid, and once-attractive recreation areas are spoiled. As the vegetation and algae decay, they consume oxygen dissolved in the water, thereby producing the results described in the previous section.

A main cause of eutrophication is domestic sewage, which is made up largely of nitrogen-containing human wastes and, except where their use is prohibited by law, phosphorous-containing detergents. Other causes are runoff from agricultural land that has been treated with nitrate and phosphate fertilizers, discharges from industries, and phosphate mining.

Control of Eutrophication

Action to control eutrophication has been concentrated on limiting discharges of phosphorus-containing wastes, which come mainly from point sources: sewage-treatment plants, industrial plants that use large amounts of phosphorus-containing cleaning agents, and phosphate mines. Nitrogen-containing wastes are discharged primarily from nonpoint sources that are difficult to control: agricultural land treated with manure or nitrate fertilizers, slaughterhouses, and stockyards. Nitrogen is usually present in wastes as nitrates, substances that are very soluble in water and therefore difficult to remove. Phosphates, on the other hand, are much less soluble and can be removed by precipitation before the water is discharged (see "Sewage Treatment," later in the chapter).

Limiting phosphorus discharges is effective in controlling eutrophication because, without an adequate supply of phosphorus, excessive plant growth is prevented even if sources of nitrogen remain abundant.

Laundry Soap: Superconcentrates

The average washing machine consumes about 20 pounds of laundry detergent per year. The two main ingredients in synthetic detergents are a surfactant and a builder. A surfactant (or wetting agent) is a substance that makes it possible for water—a polar substance—to dissolve non-polar greasy dirt. Commonly used surfactants are alkyl sulfonates, long-chain substances that have a non-polar end to attract grease, and a polar end that dissolves in water. Forty-five percent of all laundry detergents contain linear alkylbenzene sulfonate (LAS).

$$CH_3CH_2CH_2CH_2CH_2CH_2CH_2CH_2CH_2CH_2CH_2CH_2 - \bigcirc - SO_3^- Na^+$$

nonpolar   polar

Linear alkylbenzene sulfonate (LAS)

The hydrocarbon chain in LAS and in other modern surfactants is unbranched. Some of the first synthetic detergents introduced in the 1950's contained surfactants with branched-chain hydrocarbons, but it was soon found that these detergents caused problems. Bacteria in natural waters and sewage treatment plants were slow to degrade the branched chains, and the surfactants passed through sewage plants almost unchanged. When discharged into streams and lakes, they produced suds.

The function of the builder is to soften water by tying up hard-water ions (Ca^{2+}, Mg^{2+}, and Fe^{2+}) in large water-soluble ions. Common builders (or sequestering agents) are polyphosphates such as tripolyphosphate ($Na_5P_3O_{10}$). In addition, phosphates furnish the necessary alkalinity for cleaning. They keep the dirt particles in suspension, thus helping prevent them from being redeposited on garments or dishes as they emulsify oily, greasy dirt. When first introduced, polyphosphates appeared to have no drawbacks. They were safe for all kinds of fabrics (white and colored), noncorrosive to washing machines, and they biodegraded to simple phosphates, compounds that are harmless to humans and aquatic life. However, it soon became evident that the release of detergents containing phosphate builders into natural waterways caused eutrophication. Twenty three states now limit the discharge of phosphates, and seven more are considering new restrictions.

Many detergent manufacturers have voluntarily pared down the phosphate content of their formulations so that they contain no more than 8.7 percent phosphorus. Most laundry detergents now deliver no more than 5.5 to 6.5 g of phosphorus per wash load at recommended levels. In addition, soap manfacturers are replacing phosphates with other builders such as soda ash, silicates, citrates, and zeolites.

Sales of liquid laundry detergents, which contain no builders, have increased dramatically recently as environmental regulations have restricted the use of phosphates. Unfortunately, these liquids are not as effective as detergents containing builders, and more liquid soap has to be used to get the same degree of cleaning.

Concern about the environment is leading the industry to develop even more environmentally sensitive products, most of which are being sold in recycled packaging materials. Soap companies have developed new products that perform better in concentrated powdered forms and liquids. Most traditional (not concentrated) laundry detergents contain a filler, such as sodium sulfate Na_2SO_4, that adds bulk to the powder and acts as a drying agent to keep the powder flowing freely. Concentrated detergents have removed the filler and in many cases the phosphate from the product. They also have the advantage that they use less packaging material. Sales of concentrated products, which are called compacts, ultradetergents, or superconcentrates, have captured almost 30 percent of the 6.9 billion pound per year household laundry detergent business.

Questions
1. Superconcentrates come in much smaller packages than do traditional laundry detergents. Does the smaller package make the superconcentrate more environmentally friendly?
2. Should there be a federal law that limits the phosphate content of laundry soap?

SUSPENDED SOLIDS AND SEDIMENTS

As a result of the natural erosion of rock and soil (Chapter 2), all bodies of water contain undissolved particles termed **sediments.** Fine particles, such as clays, remain suspended in water for months; coarser particles such as sand and silt settle out quite rapidly.

Many human activities increase the formation of sediments. For example, bulldozing for housing developments, clear-cutting of timber (removal of every tree), strip mining, overgrazing, and plowing are all practices that remove natural ground cover and accelerate erosion of the land. Soil loss is greatest when the ground slopes steeply. It has been estimated that construction sites contribute 10 times as much sediment per unit area as farmland, 200 times as much as grassland, and 2000 times as much as undisturbed forested land.

Increased loads of sediments in streams, rivers, and lakes can cause many problems. Suspended particles make water turbid, thereby reducing light penetration and thus the rate of photosynthesis. As sediments settle, they can bury bottom-dwelling organisms and fish spawning grounds, generally disrupting aquatic habitats. The effects of suspended and settling sediments can be especially damaging where rivers meet the sea in estuaries and bays—regions that are particularly important as breeding grounds for fish and shellfish and as feeding grounds for birds and many other creatures.

Sediments cause problems by filling irrigation ditches and clogging harbors and lakes. Another concern is that when toxic substances such as metals and pesticides are released into turbid water the toxins adhere to the suspended particles and become concentrated in sediments. A later disturbance in the water, such as an increase in acidity, may then release the toxins. In clear water where there are few suspended particles, toxic substances cause fewer problems because they are more likely to remain in solution and become diluted to insignificant concentrations by the flow of water.

DISSOLVED SOLIDS

Fresh water always contains some dissolved solids. The ions in solution (Table 12-3) are essential for the normal growth of most life forms, but if present at concentrations above certain levels, they become harmful. Water that contains an abnormally high concentration of dissolved ionic salts is termed **saline water**. It is unfit to drink and can destroy freshwater fish and other organisms. The salts in saline water themselves may not be toxic, but if present in concentrations high enough to render the water unsuitable for normal purposes, they are pollutants.

Irrigation is the main cause of increased salinity in natural waters. All over the world, food production has been increased dramatically by irrigating once-arid land. Unfortunately, this practice can lead to serious problems.

Land that receives water through irrigation is located in regions where rainfall is sparse and the climate is generally hot and dry. Consequently, the rate of evaporation of water from the soil is high, and salts tend to accumulate on the soil surface. Runoff from the land and subsurface drainage, when returned to the water supply, carry an increased load of dissolved salts.

Because rainfall is scarce on irrigated land, the only way to remove excess salts is by further irrigation. Frequently, irrigation water is recycled, and with each recycling, the water becomes increasingly saline. It is estimated that 25 to 30 percent of the irrigated cropland in the United States suffers from increased salinity, and that thousands of acres have ceased to be productive. Crops that are particularly susceptible to salinity include beans, carrots, and onions; cabbage, broccoli, and tomatoes are more tolerant.

Other causes of increased salinity of natural waters are discharges from industrial and municipal waste-treatment plants and runoff from fertilized agricultural land and urban areas. Salt spread to de-ice roads can be a factor in regions where winters are severe.

By the natural processes of erosion, solids are continually added to a river as it flows to the sea. Thus, water near the mouth of a river always contains more dissolved solids than water near the source. It has been estimated that in many rivers in the United States, including the Colorado, as much as half the load of dissolved solids present in the water as it approaches the sea is the result of human activities.

TOXIC SUBSTANCES

A substance is said to be **toxic** if, when ingested by an organism, it causes harm by interfering with the organism's normal metabolic processes. In the context of water pollution, only substances that cause harm when present at very low concentrations—parts per trillion to parts per million—are generally described as toxic. Many substances, including common salt, are harmful at high concentrations but would not be termed toxic substances.

Of particular concern in aquatic environments is **bioaccumulation**, the process by which a toxic substance becomes more concentrated as it moves upward through the food chain (Chapter 3). For example, lake water may contain an insignificant concentration of a toxic substance, but as a result of bioaccumulation, the toxin can become increasingly concentrated as it passes from the water to microorganisms, then to small fish, and finally to large fish (Fig. 12-17). The concentration in large fish may be high enough to cause harm to birds of prey or humans who eat the fish. Measuring the concentration of a toxin in fish is, therefore, a useful guide to toxic pollution in a body of water.

Because aquatic ecosystems have from four to six trophic levels (Chapter 3), bioaccumulation is a more serious problem in water habitats than in terrestrial ones, which usually have only two or three levels. Toxic substances that are of particular concern are metals and organic chemicals. A variety of industrial activities and farming practices can result in the release of these substances into water supplies.

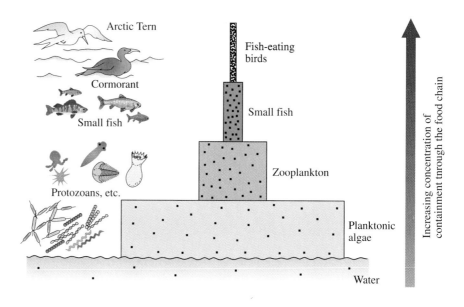

Figure 12-17 Bioaccumulation in an aquatic food chain. As a result of bioaccumulation, the concentration of a toxic substance in a fish-eating bird may be over a million times as great as the concentration of the substance in the water in the fish's habitat.

(Not drawn to scale)

Toxic Metals

Many metals, including sodium, potassium, calcium, magnesium, copper, and zinc, are essential for the normal development and well-being of humans and other creatures. Other metals, even when present in an organism at very low concentrations, are toxic. The EPA has established maximum permissible levels in drinking water for the following toxic metals: antimony, arsenic (which is not a metal but is usually listed with them), barium, beryllium, cadmium, chromium, lead, mercury, nickel, selenium, and thallium (Table 12-6). Municipal water supplies are monitored constantly to check that these limits are not exceeded.

Metals, because they are elements, cannot be broken down into simpler, less toxic forms. They persist unchanged in the environment for many years and bioaccumulate through food chains. Toxic metals can cause brain damage, kidney and liver disorders, and bone damage. Many are carcinogens (Chapter 16). Metals enter waterways from two main sources: industrial waste discharges and particulates in the atmosphere that settle out and enter water in runoff. We will consider three metals—mercury, lead, and cadmium—that because of their widespread use and toxicity, are particularly dangerous.

Mercury Mercury is the only common metal that is a liquid at room temperature. It is a component of many rocks and is released continually, but very slowly, into natural waters by normal chemical weathering processes. Small quantities of mercury are also vaporized into the atmosphere from mercury-containing rocks in the earth's crust. Concentrations of mercury in water from natural sources generally amount to a few parts per billion.

Mercury and mercury compounds in bodies of water tend to settle at the bottom and adhere to sediments. As released, mercury and its compounds are not readily absorbed by aquatic creatures. However, anaerobic bacteria in bottom sediments can convert them to an organic form, methyl mercury (CH_3Hg^+), that is very soluble and readily absorbed. Methyl mercury bioaccumulates as it moves up the food chain, and for this reason, it can be dangerous to eat large quantities of very big ocean fish such as tuna and swordfish.

TABLE 12-6 Maximum Permissible Levels for Metals in Drinking Water

Metal	Maximum Level (ppb)
Antimony	6
Arsenic	50
Barium	2000
Beryllium	4
Cadmium	5
Chromium	100
Lead	50*
Mercury	2
Nickel	100
Selenium	50
Thallium	2

Source: EPA, December 1992.
*EPA is proposing to lower it to 15 ppm.

Mercury and its compounds are used in the production of chemicals, paints, plastics, pharmaceuticals, and photographic and electrical equipment. They are used in the steel industry, as algicides in paper manufacture, and as fungicides on seeds. Mercury and its compounds can reach natural bodies of water from any of these manufacturing or farming sources as well as from sites where mercury-rich zinc and silver ores are mined.

The danger of inhaling mercury vapor has been known for a long time. In the nineteenth century, mercury(II) nitrate was used in the manufacture of felt hats, and workers in the industry developed many neurological symptoms. This is the origin of the expression "mad as a hatter" and explains the name and behavior of the Mad Hatter in Lewis Carroll's *Alice in Wonderland*. Mercury and most inorganic mercury salts, unless swallowed in considerable quantity, cause little harm because they are rapidly excreted from the body. Methyl mercury, however, remains in the body for months and can damage the nervous system, kidneys, liver, and cause birth defects.

In the 1950s, the toxicity of mercury was tragically demonstrated when a chemical plant in Japan released large quantities of mercury compounds into Minimata Bay. Fish and shellfish became contaminated with methyl mercury, and many people who relied heavily on fish from the bay for their diet became sick. Fifty-two people died, and others suffered severe neurological symptoms, including numbness, impaired vision, paralysis, and brain damage.

A worse tragedy occurred in Iraq in 1972 when nearly 500 people died after eating bread made from wheat grain that had been treated with a methyl mercury fungicide. The imported grain was intended for planting, but the warning label was written in a language the Iraqis did not understand.

In the United States, the main concern is the mercury content of fish. Because the conversion of mercury to methyl mercury is accelerated in acidic water, the acidification of lakes (Chapter 8) can increase concentrations of this toxic form of mercury in fish.

Lead   Lead has many useful properties. It has a low melting point and is malleable and dense. It is used as a solder and is a component of many alloys. Lead and lead oxide are used in lead-acid storage batteries (Chapter 8), and, until restricted by law, lead compounds were ingredients in paint (Chapter 11). Now that lead has been virtually eliminated from gasoline (Chapter 14), drinking water is the main source of exposure to lead. The current standard for lead in drinking water is 50 ppb (Table 12-6) but the EPA has proposed lowering it to 15 ppb.

Plumbing is often a source of unacceptably high lead levels in drinking water. Many older buildings still receive water through lead pipes; in newer buildings, where pipes are now made of copper, lead solder is still used to seal the pipes. Water flowing through pipes dissolves small quantities of lead. The amount of lead that dissolves is highest in acidic water that has a low mineral content, particularly if the water has been in contact with lead for a long time.

Lead reaches water from a number of sources besides plumbing fixtures. Lead-containing emissions from automobiles remain a problem in traffic-congested urban areas. Lead in emissions adheres to dust particles in the air, the particles settle out onto streets, and the lead enters water supplies in runoff from the streets. Lead is present in the earth's crust in many ores, primarily lead sulfide (PbS); mining and smelting of these ores add lead compounds to waterways.

Lead is particularly toxic to young children. Even at very low concentrations in the blood, it is known to retard development and cause brain damage. In adults, it causes neurological disorders and complicates pregnancy. Lead is a cumulative poison and tends to concentrate in bones. From there, it can be remobilized and enter the bloodstream. Lead disrupts metabolic processes primarily by inhibiting vital enzyme reactions (Chapters 10 and 16).

Cadmium   Cadmium is used in paints, plastics, nickel-cadmium batteries, and in electroplating. These products are potential sources of cadmium contamination in water, but the main cause for concern is cadmium in zinc products. Most zinc ores contain small amounts of cadmium, and as a result, some cadmium is usually present as an impurity in the zinc metal and zinc compounds produced from the ores. Zinc and its compounds have many commercial uses. Large quantities of zinc are used to galvanize iron—a process in which iron is coated with zinc to protect it from corrosion (Fig. 12-18). Because zinc is an essential element for plants, manufacturers include zinc in most inorganic fertilizers. Leafy vegetables and tobacco leaves absorb zinc—along with any cadmium associated with it—from water in the soil. When cadmium enters the body, it accumulates in the liver. Smokers run a much greater risk of suffering the effects of cadmium poisoning than nonsmokers do.

In the 1950s, numerous cases of cadmium poisoning occurred when effluents from a zinc mine were discharged into the Zintsu River in northern Japan. Water contaminated with the discharge was used to irrigate rice fields. People who ate the rice developed an extremely painful skeletal disorder known as itai-itai byo (or "ouch-ouch" disease). Their bones became brittle from loss of calcium and broke very easily. Many also suffered from abdominal pain, diarrhea, vomiting, liver damage, and kidney failure.

Cadmium is very similar chemically to zinc (both are in Group IIB of the periodic table; Chapter 4), and it is thought that the mechanism of cadmium poisoning may involve the substitution of cadmium for zinc in certain enzymes. Cadmium is a carcinogen; long-term effects include increased blood pressure and liver, kidney, and lung disease.

The word plumbing is derived from *plumbum*, the Latin name for lead.

In 1990, many older lead-lined water coolers had to be replaced when it was discovered that water taken from them had unacceptable levels of lead—in some cases as high as 2000 ppb.

In the past, paints were often colored red with mercury sulfide and cadmium selenide and were colored yellow with cadmium sulfide.

Figure 12-18 A pail made of galvanized steel. The steel is coated with zinc to protect it from corrosion. Cadmium, which is toxic, is often present as an impurity in zinc products.

Synthetic Organic Chemicals

Thousands of different organic chemicals are synthesized each year for use as insecticides, herbicides, detergents, insulating materials, and for many other purposes. If allowed to enter water supplies, some of these chemicals can cause serious health and environmental problems.

Of particular concern as water pollutants are certain organic chlorine compounds, called **polychlorinated hydrocarbons** (Chapter 9). These compounds pose a threat to aquatic life because they are stable and do not readily break down into simpler less toxic forms. They persist in the environment for long periods of time, and, like toxic metals, they bioaccumulate through food chains. Polychlorinated hydrocarbons are insoluble in water but soluble in fats. They become concentrated in the fatty tissues of fish and of birds and humans who eat the fish. We will consider three polychlorinated hydrocarbons that are of particular concern as environmental contaminants: DDT, dioxin, and PCBs.

DDT When **DDT** (dichlorodiphenyltrichloroethane) was introduced in the late 1940s, it appeared to be an ideal insecticide. It was cheap to produce and apparently nontoxic to humans and other mammals. Because it did not break down easily, it continued to kill insects for a long time after it had been applied. Later, however, it was discovered that water contaminated with DDT from aerial spraying and runoff from treated land was having a devastating effect on fish-eating birds. Also, insects became increasingly resistant to the pesticide, and larger and larger amounts had to be applied to achieve the desired results (Chapter 18).

Scientists have observed that DDT concentration may increase as much as 7 million times as it passes through food chains from water to fish-eating birds.

Figure 12-19 As a result of ingestion of DDT, many birds laid eggs that were so fragile that they cracked.

Research showed that DDT interfered with calcium metabolism in birds, and, as a result, eggshells (which are composed primarily of calcium compounds) became thin and broke when parent birds attempted to incubate the eggs (Figure 12-19). By the 1950s and 1960s, populations of bald eagles, peregrine falcons, and brown pelicans had shrunk to disturbingly low levels. In 1973, the use of DDT was banned in the United States, and since that time, these birds have made a dramatic recovery. DDT continues to be produced in the United States for export to developing countries, where its success in controlling the scourge of mosquito-borne malaria outweighs its disadvantages.

Dioxin **Dioxin** is the name of a family of chlorinated hydrocarbons, but it has come to be associated with one particular compound, 2,3,7,8-tetrachlorodibenzo-p-dioxin, usually abbreviated to TCDD. TCDD is a by-product of the manufacture of trichlorophenol, a chemical that is used in the manufacture of a variety of herbicides, including 2,4-D (Silvex). Silvex is widely used on agricultural land and suburban lawns to control broad-leaved weeds.

During the Vietnam War, 2,4-D was mixed in equal quantities with the related herbicide 2,4,5-T to make the defoliant Agent Orange. Present in the latter as an impurity was TCDD. Because TCDD is extraordinarily toxic to guinea pigs and is a known carcinogen for many animals, there was concern that humans could also be affected. As a result of these concerns, the EPA banned the use of 2,4,5-T in 1985.

The effects on humans of exposure to Agent Orange have been debated for many years. In 1993, the National Academy of Sciences issued a report of a comprehensive study of the effects of exposure to the chemical. The report linked Agent Orange to three types of cancer and two skin conditions, but failed to support Vietnam Veterans' claims that the chemical caused birth defects, infertility, and other disorders.

Solution of water-pollution problems associated with the use of pesticides is not easy. Some people would ban all synthetic pesticides, even though pesticides are enormously valuable in the struggle to feed the world's ever-growing population. It is estimated that without their use, crop production would decrease by an average of 20 percent. Pesticides are discussed in more detail in Chapter 18.

PCBs   PCBs (polychlorinated biphenyls) are structurally similar to DDT and, like DDT, bioaccumulate at the upper levels of food webs. PCBs present in water in negligible concentration can become over a million times more concentrated in fish. Since 1977, strict regulation of the use of PCBs in the United States has resulted in a dramatic decrease in PCB contamination of fish. However, because PCBs are stable, they persist in the environment for many years, and fish in many lakes still contain low levels of contaminants.

PCBs are fire-resistant, stable at high temperature, and have a high electrical resistance. They were widely used as insulating materials in transformers, electrical capacitors, and condensers, and are still present in older equipment (Fig. 12-20). PCBs were also used extensively as plasticizers (Chapter 11) in the plastics industry. Plastics tend to be brittle; the addition of PCBs makes them more flexible and resistant to cracking.

PCBs can enter surface waters in leakages of municipal and industrial wastes and in particulates from incinerators. When plastic wastes and other PCB-containing materials are burned, PCB vapors condense on airborne particles that then fall directly onto water or reach water in runoff from the land. PCBs cause eggshell thinning and neurological damage in birds and impair reproduction of aquatic species. In humans, they cause chloracne (a serious form

Figure 12-20   PCBs have to be removed from electrical transformers before disposal. Electrical transformers now contain a paraffin oil derived from petroleum.

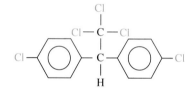

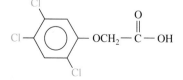

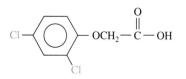

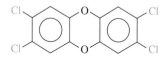

(a) DDT
(Dichloridiphenyltrichloroethane)

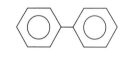

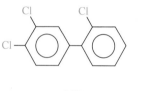

(b)

2,4-Dichlorophenoxyacetic acid
(2,4-D)

2,4,5-Trichlorophenoxyacetic acid
(2,4,5-T)

2,3,7,8-Tetrachlorodibenzo-p-dioxin
(a "dioxin")

(c)   Biphenyl   PCB$_1$ PCB$_2$

Figure 12-21 Chemical formulas of (a) DDT, (b) TCDD ("dioxin") and related compounds, and (c) biphenyl and two PCBs that can be derived from it.

of acne) and liver damage, and, most importantly, they can be transferred from mother to fetus through the placenta and from mother to infant through breast milk. PCBs cause stillbirths and retard growth.

The chemical formulas of DDT, TCDD and related dioxins, and representative PCBs are shown in Figure 12-21.

THERMAL POLLUTION

A different type of pollution, involving heat, is **thermal pollution**. The electric power industry and many other industries draw huge volumes of water from rivers and lakes for cooling purposes. During operation, electric power plants produce enormous quantities of waste heat, which is removed in water that is circulated through the plant. As the water circulates, its temperature may rise by as much as 10 to 20°C. If the warm water is returned directly into a waterway, the temperature of the water at the point of discharge increases.

Higher temperature can adversely affect aquatic life by increasing the body temperature of aquatic organisms. High body temperatures can be fatal to some fish, particularly game fish. Warm water also raises respiration rates, and high respiration rates in turn increase oxygen consumption. At the same time, because the solubility of oxygen in water decreases with rising temperature (Fig. 12-14), the dissolved oxygen content of the water decreases. Fish and other aquatic species may die from lack of oxygen if they are not killed directly by increased temperature. The life cycles of many aquatic species are controlled by temperature. For example, the number of days it takes for trout eggs to hatch depends on temperature, and fish migrate and spawn in response to slight changes in water temperature. Any unusual changes in water temperature can disrupt their normal development.

Thermal pollution can be prevented in several ways. Heated water can be made to flow through a cooling pond, or it can be sprayed into a cooling tower and cooled by evaporation. More efficient, but more costly, is the dry tower, in which heated water transfers heat to the surrounding air as it circulates through pipes.

POLLUTION OF GROUNDWATER

So far in this chapter, we have concentrated on pollution of fresh surface water—lakes, rivers, and streams. Of even greater concern is pollution of groundwater. Normally, groundwater is of such high quality that it meets safe drinking water standards without the need for purification or treatment, and it is usually pumped directly from the ground into homes. Recently, however, in a number of locations, formerly pure groundwater has become contaminated with hazardous substances.

Hazardous is a broader term then *toxic*. It is used to describe any substance that threatens human health or adversely affects the environment.

Not all hazardous substances are toxic, but all toxic substances are hazardous.

The main sources of hazardous materials that threaten groundwater are dumpsites where waste chemicals produced by industry are leaking from corroded metal drums. In the past, many industries disposed of their hazardous wastes carelessly—and often illegally—without considering the consequences. Chemical wastes generated in the production of paints, metals, textiles, fertilizers, pesticides, metals, plastics, and petroleum products were often discarded in open dumps and landfills or buried. The problem of cleaning up these dumpsites will be discussed in Chapter 19.

Other sources of groundwater pollution include pesticides and fertilizers from farmland, sewage from septic tanks and leaking sewer pipes, and gasoline leaking from service station storage tanks.

Septic tanks and cesspools can pollute groundwater if they are not operated properly or if they are located where the soil's natural filtering capacity is inadequate.

As hazardous chemicals from dumpsites and other sources seep down through the ground, some pollutants are filtered out by the soil and travel only short distances. Soluble substances such as nitrates, however, are not filtered out and can percolate down to groundwater directly from septic tanks, feed lots, and fertilized land. Consumption of very dilute solutions of nitrates can cause abortions in cattle and methemoglobinemia (blue baby disease) in infants. In methemoglobinemia, microorganisms in the digestive tract convert the nitrate ions to nitrite ions. Nitrite ions oxidize iron in hemoglobin from Fe(II) to Fe(III), and an abnormal hemoglobin called methemoglobin is formed. Methemoglobin is incapable of combining with oxygen and as a result, the blood becomes deficient in oxygen.

SEWAGE TREATMENT

Pollution problems can result if cities have a single line for both sewage and storm-water drainage. If the volume of water reaching the sewage plant during heavy rains exceeds the plant's capacity, untreated sewage enters into waterways.

Raw sewage is 99.9 percent water. The water comes primarily from flush toilets, kitchens, laundry facilities, car washes, and storm drains. In the United States, sewage is usually purified in two processes: primary and secondary treatment. Presently, tertiary treatment, a process designed to take care of potentially harmful substances not removed in the first two stages, is not fully developed and is used by few municipal treatment plants.

Primary Treatment

As the first step in **primary treatment** (Fig. 12-22), the sewage is passed through a screen to remove large pieces of debris, such as sticks, stones, rags, and plastic bags, that have washed in through storm drains. The sewage then enters a grit chamber, where flow rate is slowed just enough to allow coarse sand and gravel to settle out on the bottom. The grit is collected and disposed of in landfills. As water enters the sedimentation tank, its flow rate is further decreased to permit suspended solids—which account for about 30 percent of the total organic wastes—to settle out as **raw sludge**. Any oily material floats to the surface and is skimmed off. In the past, raw sludge was incinerated, put in landfills, or dumped at sea. Now much of it is composted to produce a nutrient-rich bacteria-free humuslike material that is used as fertilizer.

Calcium hydroxide and aluminum sulfate are often added to speed up the sedimentation process. These two chemicals react to produce a gelatinous precipitate of aluminum hydroxide that settles out slowly, carrying suspended material and bacteria with it.

$$3 \, Ca(OH)_2 \; + \; Al_2(SO_4)_3 \; \longrightarrow \; 2 \, Al(OH)_3 \; + \; 3 \, CaSO_4$$

In older sewage-treatment plants, the water at this stage is often chlorinated to kill pathogens and then discharged into a natural waterway. This practice can cause problems because the discharged water still contains a large amount of oxygen-consuming wastes, which may deplete dissolved oxygen in the waterway and cause eutrophication.

Secondary Treatment

Secondary treatment (Fig. 12-22) is a biological process that relies on aerobic bacteria and other detritus feeders to break down nearly all the remaining oxygen-consuming wastes. In the most frequently used method, a mixture of organisms—termed activated sludge—is added to the sewage effluent. Air or oxygen is vigorously bubbled through pipes into the effluent as it moves slowly through a tank. In this oxygen-rich environment, the organisms digest the organic material and break it down into carbon dioxide and water. After settling out in a sedimentation tank, the organisms, together with any remaining undecomposed material, are returned to the aeration tank and reused.

Most municipal plants chlorinate the water after secondary treatment and then release it into waterways. Although about 90 percent of the original organic matter is removed by primary and secondary treatment, many other substances are much less completely removed. Over 50 percent of the nitrates and

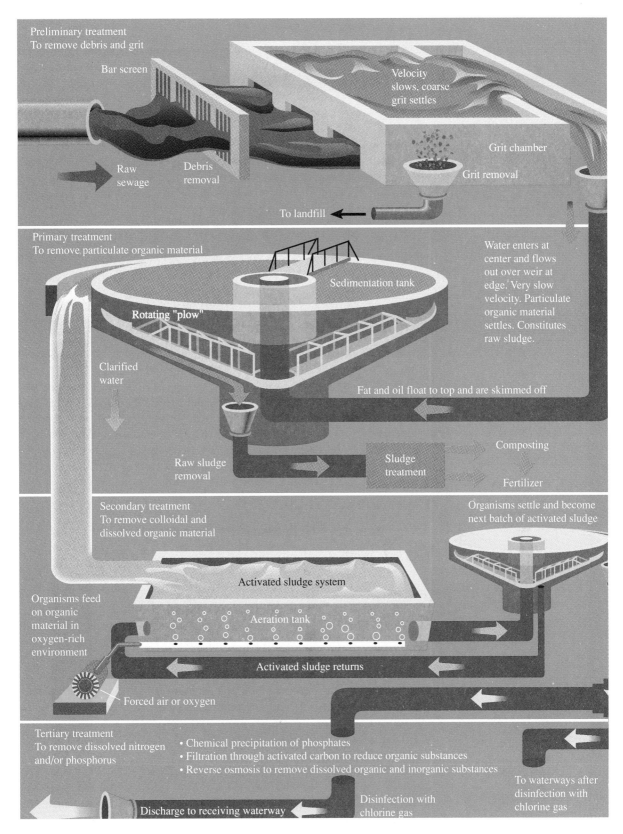

Figure 12-22 Schematic diagram showing the main steps in primary, secondary, and tertiary sewage treatment.

Effluent released from sewage treatment plants after secondary treatment is a major cause of eutrophication because the effluent contains phosphates.

phosphates remain, and metal ions, and many synthetic organic compounds including pesticides, are incompletely removed. Also, the use of chlorine for disinfection may introduce hazardous chemicals.

Tertiary Treatment

Tertiary treatment (Fig. 12-22) is designed to remove harmful contaminants that remain in the water after primary and secondary treatments. Phosphates are removed in one of two ways. One method involves the addition of aluminum sulfate or lime (CaO), which results in the precipitation of phosphate as insoluble aluminum or calcium phosphates:

$$3\ PO_4^{3-}\ +\ Al^{3+}\ +\ 3\ Ca^{2+}\ \longrightarrow\ AlPO_4\ +\ Ca_3(PO_4)_2$$

The second method is similar to the activated-sludge system. Microorganisms that absorb phosphate are added and then removed together with the phosphates. The procedure also removes some remaining organic materials.

Dissolved organic substances are removed primarily by filtration through activated carbon, finely powdered charcoal that has been activated by heat. Organic substances become attached to the surface of the carbon particles and are thus removed. Reverse osmosis can be used to remove remaining dissolved organic and inorganic substances, including toxic metal ions.

In 1972—the year the Clean Water Act was passed—the EPA estimated that one-third of the sewage in the United States was discharged with only primary treatment or no treatment at all. Now, most sewage receives both primary and secondary treatment, and as a result, water quality has improved in many waterways that were once seriously polluted. Tertiary treatment is still experimental and expensive and is not used by many municipal sewage-treatment plants.

Disinfection with Chlorine

After primary and secondary treatment, sewage effluent is usually disinfected with chlorine gas (Cl_2), which reacts with water to form hypochlorous acid (HClO), a powerful oxidizing agent.

$$Cl_2\ +\ H_2O\ \longrightarrow\ HClO\ +\ H^+\ +\ Cl^-$$

Hypochlorous acid kills disease-causing bacteria and some (but not all) viruses and removes color from the water. Chlorine gas is effective and relatively cheap, but it can cause a number of environmental problems. It is very toxic, and there is always the danger of accidental release from tanks during transportation or at the plant. Further, low concentrations of chlorine in water are toxic to fish. Perhaps the most serious problem is the finding that chlorine reacts with residual organic substances to form chlorinated hydrocarbons, including dichloromethane (CH_2Cl_2), chloroform (CH_3Cl), and trichloroethylene (C_2HCl_3), which are all suspected carcinogens.

The dangers associated with chlorinated hydrocarbons are minimized if tertiary treatment is used to remove organic substances. Alternatively, ozone (O_3), a powerful oxidizing agent, can be used. Ozone is very effective in killing bacteria, and in the process it is converted to oxygen, which improves water quality. When it is used, it is manufactured on site by passing oxygen or air through an electric discharge. Ozone disinfection is expensive and still essentially an experimental procedure. It is used quite widely in Europe but infrequently in the United States.

Regulation of Water Quality

In the 1970s, many rivers and lakes in the United States had become so polluted that they were unsuitable for recreational activities, and it was often dangerous to eat fish taken from them. Now, as a result of federal regulations—primarily the Clean Water Acts of 1972 and 1977, and amendments in 1985—water quality has improved dramatically. Water discharged from industries and municipal sewage-treatment plants must meet certain minimum standards before it can be released into natural waterways. Also, all communities are required to monitor their water supplies to be certain that they meet the EPA's standards for drinking water. However, despite these and other measures, water contamination remains a problem in some parts of the nation.

EXPLORATIONS

The Aral Sea: The Great Lake That Is Disappearing

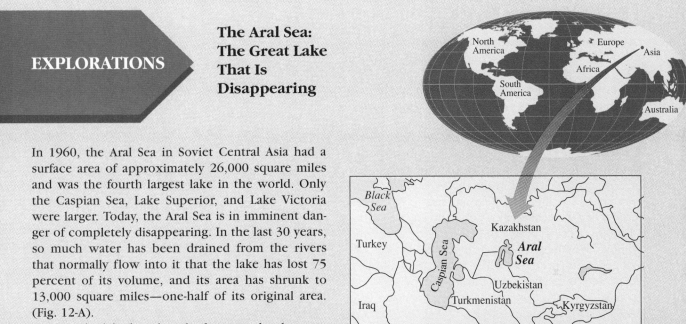

In 1960, the Aral Sea in Soviet Central Asia had a surface area of approximately 26,000 square miles and was the fourth largest lake in the world. Only the Caspian Sea, Lake Superior, and Lake Victoria were larger. Today, the Aral Sea is in imminent danger of completely disappearing. In the last 30 years, so much water has been drained from the rivers that normally flow into it that the lake has lost 75 percent of its volume, and its area has shrunk to 13,000 square miles—one-half of its original area. (Fig. 12-A).

As the lake has shrunk, the water has become increasingly salty. Since 1960, salinity has risen from 10 to 23 percent; fish that once were plentiful have disappeared, and the lake is lifeless. The receding water has left behind an 11,000-square-mile desert of salt and sand. Wetlands in the river deltas that once teemed with deer, boar, muskrats, and egrets have dried up, and the animals and birds have vanished.

For thousands of years, the Aral Sea was replenished by two rivers: the fabled 1578-mile-long Amu-Darya (known to Alexander the Great as the Oxus) and the 1370-mile-long Syr-Darya. These two great rivers, which arise in mountains to the southeast, flowed north to the Aral Sea, bringing life to the hot arid steppe they traversed. Today, the rivers run dry before they reach the Aral Sea, their water diverted into immense canals and drawn off into reservoirs to irrigate the world's largest cotton belt.

The seeds of the Aral Sea's destruction were sown almost as soon as the communists came to power in 1917. In the following year, the Soviet government ordered the cultivation of millions of acres of land south of the Aral Sea for the production of cotton. Water for the project was drawn from the Amu-Darya and the Syr-Darya. The original aim had been to make the Soviet Union self-sufficient in cotton, but it was soon realized that cotton could become a valuable source of revenue. After World War II (1939–1945), cotton became a vital export, and to fulfill quotas set in Moscow, more and more land was brought under cultivation. It became a patriotic duty to increase production, and much of the land previously used for pasture and orchards and to grow vegetables was turned over to cotton.

Cotton requires a great deal of water, and in the 1950s the Kara Kum Canal—the longest canal in the world—was built to meet the government's demand for cotton. Water for the canal, which runs for 850 miles along a course parallel to the borders of Afghanistan and Iran, was siphoned from the Amu-Darya. In the 30 years following the opening of the canal in 1956, cotton production doubled.

With the completion of the immense canal, the continual withdrawal of water from the rivers began to have a significant effect on the water level in the Aral Sea. As the flow of water into the lake slowed and then almost ceased, the delicate balance between inflow and evaporation that had maintained the water at a constant level was upset. Between 1957 and 1984, the water level in the lake fell 6 feet. It is estimated that if depletion continues at the present rate, the Aral Sea will disappear entirely by the year 2010.

The Aral Sea was once the center of a thriving fishing industry. In the early 1960s, 24 native species inhabited the lake, and more than 150 pounds of fish were taken from it daily. Approximately 10,000 fishermen in and around the lakeside town of Muynak made a living catching pike, perch, bream, and other valuable species. Thousands of other men and women had jobs at a nearby cannery. Today, the fish are gone. Muynak is 20 miles inland from the lake shore, and fishing boats that once sailed the waters of the lake lie rusting on a salty desert wasteland (Fig. 12-B). When commercial fishing ended, the cannery

Figure 12-B   Fishing boats left stranded in a desert wasteland.

was kept open to preserve jobs by bringing frozen fish, caught 1700 miles away in the Barent Sea, to Muynak in special refrigerated trains. In the harsh economic times following the breakup of the Soviet Union, this expensive operation may soon have to be discontinued. Already, in 1993, steel for making cans could be obtained only by bartering scrap metal retrieved from beached fishing boats.

The decrease in size of the Aral Sea has had a profound effect on the region's climate. In the past, the huge body of water moderated the temperature, having a cooling effect in summer and a warming effect in winter. As the lake has shrunk, temperatures have become more extreme. The region was always dry, but now summers are hotter and dryer than ever before; winters are longer, more severe, and snowless—and the wind blows almost constantly. Sand and salt from the drying lake are swept up in frequent dust storms that contaminate the land and threaten the health of the people. Much of the land is so encrusted with sodium chloride and sodium sulfate that nothing will grow. When the wind blows, people taste salt on their lips and salt stings their eyes. The number of cases of eye disease, respiratory disease, and throat cancer has risen dramatically in the last 25 years.

By the 1980s, the overworked land had begun to lose productivity, and immense quantities of fertilizers were applied to meet cotton quotas. It has been estimated that each acre of cotton received 48 lb annually of hazardous chemicals from the fertilizers, defoliants, and pesticides that were used. Often the pes-

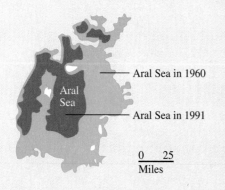

Aral Sea in 1960

Aral Sea in 1991

0   25
Miles

Figure 12-A   Shrinkage of the Aral Sea during the period from 1960 to 1991.

ticides and defoliants were sprayed when the workers were in the fields. Residues of the chemicals are now found in river water and groundwater, and in 1989, over 80 percent of the community's drinking water was declared unsafe for human consumption. Trace quantities of pesticides and heavy metals have been detected in breast milk; infant mortality in the region is higher than in any other state in the former Soviet Union, and there has been a dramatic increase in the number of cases of cancer, hepatitis, typhoid, and internal illnesses.

The Aral Sea is bordered by the republics of Uzbekistan and Kazakhstan. The region most affected by the tragedy is Karakalpakstan, an autonomous republic in Uzbekistan that includes the desiccated Amu-Darya delta. The people live in mud-brick houses, and their only source of water is well water that is now salty and contaminated. The women who pick the cotton are glad to have jobs and have little understanding of the cause of their desperate plight.

Before glasnost, no one dared question the directives from Moscow, but now people are speaking out indignantly and calling for action to prevent further loss of water from the lake. Promises have been made to save water by reducing cotton acreage and lining the irrigation ditches with plastic; there are plans to feed at least five cubic miles of water back to the Aral by the year 2005. But without a central government to dictate policy and provide funding, little is expected to be done. Lack of resources and disagreements among the newly independent republics competing for water continue to jeopardize efforts to save the Aral Sea.

In the span of a single lifetime, the obsession for cotton has devastated practically all of Soviet Central Asia. It has polluted the land, altered the climate, destroyed the wildlife, and impoverished the people and ruined their health. No matter what is done, it is unlikely that the Aral Sea will ever again be the immense and beautiful inland sea, teeming with life, that was extolled by poets and travelers throughout the centuries. Its disappearance must surely rank as one of the worst environmental disasters of all time.

Question:

Prior to the floods of 1993, the Midwest had suffered several years of drought. It had been suggested that water from the Great Lakes (which naturally flows into the Atlantic via the St. Lawrence River) be diverted into the Mississippi. Discuss possible environmental and political consequences of such a diversion.

References: W. S. Ellis, "The Aral: A Soviet Sea Lies Dying," *National Geographic* 177 (1990): 73; J. Rupert, "A Death in the Desert: Soviet Plan Kills a Sea," *Washington Post* (June 20, 1992), p. Al; R. Nasar, "How the Soviets Murdered a Sea," *Washington Post* (June 4, 1989), p. B3.

KEY WORDS AND CONCEPTS

aquifer	hazardous substances	point source
bioaccumulation	heat capacity	polychlorinated hydrocarbons
brackish water	heat of fusion	primary treatment
coliform bacteria	heat of vaporization	saline water
DDT	hydrologic (water) cycle	secondary treatment
desalination	nonpoint source	sludge
dioxin	osmosis	surface water
eutrophication	oxygen-consuming wastes	tertiary treatment
fresh water	pathogen	thermal pollution
groundwater	PCBs	toxic substances

⬛ QUESTIONS AND PROBLEMS

1. Explain why:
   a. water is a liquid at room temperature **b.** ice floats on liquid water **c.** water has such an exceptionally high boiling point

2. Water has a large heat of vaporization.
   a. Define heat of vaporization. **b.** Why is it cooler near the shore than inland? **c.** Is water a more or less effective coolant for our bodies than another liquid with a smaller heat of vaporization?

3. There are many types of natural water.
   a. What percent of the earth's water is seawater? **b.** List four metal ions present in seawater. **c.** What is the approximate concentration of dissolved solids in seawater in parts per million?

4. If the water molecule were linear rather than bent, would ice be less dense than liquid water? Explain.

5. At what temperature does water reach its maximum density? What are the implications of this behavior for life in a pond?

6. During the Persian Gulf War, desalination plants in Saudi Arabia and other Middle Eastern countries were contaminated with crude oil intentionally released by the Iraqis. Describe the following desalination processes. Do you think oil would interfere with the operation of these processes?
   a. distillation **b.** reverse osmosis

7. Describe the natural process that replenishes our freshwater supply.

8. Describe ways in which changes in average global temperature could affect the hydrologic cycle.

9. Some people are suggesting that water from the Great Lakes be diverted to raise the level of the Mississippi River. List possible benefits and problems that need to be considered before such diversion begins.

10. Explain why land subsides when groundwater is depleted. If removal of groundwater is discontinued, will the land necessarily rise up to its previous level?

11. Many urban areas such as Washington, D.C., draw their drinking water from a nearby river. Much of this water is then returned to the river in the form of treated sewage. Does this practice affect the hydrologic cycle of the locality? How does this practice affect water quality?

12. San Diego, California, draws its water supply from the Colorado River, which is several hundred miles away. The water flows in open aqueducts from Colorado to San Diego. Would you expect this method of transporting water to have any effect on the quality of the water?

13. Where does the drinking water used in your community come from? What alternate sources does your community have?

14. Assume that you are a homeowner and you have a limited amount of water available from the following sources: (1) bath water, (2) good well water, and (3) rainwater drained from the roof. You want to cook dinner, give your little dog a bath, wash the car, and water your vegetable garden. Which source of water would be suitable for each task?

15. List five ways you can conserve water on a personal level. In your opinion, should the price of water for all uses be raised to encourage conservation in the United States?

16. List eight ways in which water can become polluted.

17. In the middle of the nineteenth century, what was the main cause of water pollution in the United States and Europe? Explain briefly.

18. Name four waterborne diseases that are caused by human body wastes. Are these diseases common in the United States today? Explain.

19. What are pathogens? Explain the test that is generally used to determine if water is free from these organisms.

20. How is the solubility of oxygen in water affected by:
    a. an increase in water temperature **b.** a decrease in pressure caused by increasing altitude

21. What are oxygen-consuming wastes? Name typical sources.

22. Describe the sequence of events that leads to lowered dissolved oxygen (DO) in a river that is polluted with animal wastes from a meat-packing plant.

23. List the end products of the decomposition of organic materials that contain the elements carbon, sulfur, and nitrogen when decomposition occurs under:
    a. aerobic conditions **b.** anaerobic conditions

24. What is meant by point and nonpoint sources of pollution? Give an example of each.

25. What is eutrophication? How is it caused? Before regulations were introduced to limit its discharge, which consumer product was most responsible for eutrophication?

26. Describe the effects of eutrophication on the aquatic life in an estuary.

27. Describe human activities that cause erosion and the deposition of sediments into waterways. List types of land use in the order in which they cause the most erosion.

28. What are the effects on aquatic life of increasing the amount of sediments in rivers? What are the economic consequences of increased sediments?

29. Water is drawn from a river to irrigate agricultural land. Explain how this practice can lead to increased salinity of the river water.

30. Name the process in which the concentration of a toxic substance increases as the toxin moves up the food chain. Is this process a more serious problem in a lake than in a woodland ecosystem? Explain.

31. Explain how chlorinated hydrocarbons, such as DDT, that are present in very low concentrations in a body of water can become a serious pollution problem.

32. Why are metals a threat to aquatic environments? List sources of mercury, lead, and cadmium.

33. Is the water obtained from the water fountains in your school or office safe to drink? What are the most likely contaminants, and how might they get into the water?

34. Describe how mercury compounds discharged into a lake can be a threat to humans who eat fish taken from the lake.

35. The lethal dose of DDT for a human has been estimated to be 0.5 g/kg of body weight. How much DDT would be fatal to a person weighing 55 kg?

36. List the properties of PCBs that make them valuable insulating materials. How do PCBs discharged into a river threaten wildlife and humans?

37. What was the purpose of spraying Agent Orange over the Vietnamese countryside during the Vietnam War? Discuss the risk Agent Orange may pose to humans.

38. Explain how thermal pollution of natural waters has an adverse effect on fish. What effects does thermal pollution have on the metabolic rate and reproduction of aquatic organisms?

39. Discuss ways in which thermal pollution by industry can be prevented.

40. In arid regions, how should limited water resources be used? Should water be used for agriculture, industry, city water supplies, other uses?

41. With the aid of diagrams to show the flow of material, describe primary, secondary, and tertiary sewage treatments.

42. Chlorine gas is used for disinfection of our drinking water.
a. What is the active oxidizing agent that is formed when chlorine is added to water? **b.** Describe any health risks associated with chlorination.

THE AIR WE BREATHE

13

 *In this view of a sunset in the southern hemisphere taken by astronauts aboard the space shuttle Discovery, the atmosphere is seen as a thin gaseous blanket around the planet*

The atmosphere is a thin blanket of gas that envelops the earth. It provides the carbon dioxide plants need for photosynthesis and the oxygen animals need for respiration, and it is the ultimate source of nitrogen for plant growth. Fresh water reaches the earth from the atmosphere as dew, rain, and snow. The atmosphere shields us from the sun's cancer-causing ultraviolet (UV) radiation, and it also moderates the earth's climate. Without it, the earth would experience the extremes of hot and cold that are found on planets that have little or no atmosphere.

The atmosphere is obviously vital for our existence, and yet we have been polluting it for years. By the end of the nineteenth century, huge quantities of coal were being burned to fuel the Industrial Revolution, and smokestacks belching great brown clouds into the atmosphere became a sign of prosperity. By the middle of the twentieth century, the automobile had become another significant source of air pollution. Today, in the United States, pollution from these sources has been greatly reduced as a result of legislation, but we face other problems. There is growing concern now that the earth's protective ozone layer is being destroyed by certain chemicals produced by industry. Also, our continued dependence on fossil fuels for energy is introducing increasingly large quantities of carbon dioxide into the atmosphere. Many believe this practice will cause the earth to become warmer, a trend that could have disastrous consequences for the world's climate.

In this chapter, we will first study the composition and properties of the atmosphere. We will then study each of the major air pollutants and examine their sources, their effects on health and the environment, and the steps that are being taken to lower emissions. We will review the evidence that strongly suggests that chlorofluorocarbons are the cause of ozone depletion, and we will discuss the way carbon dioxide and certain other gases may be causing a global warming trend. Lastly, we will consider federal legislation that has been passed to control pollutants.

Learning Goals:

In this chapter, you should gain an understanding of:

1. The composition of the atmosphere.
2. The five primary air pollutants (their sources and effects on human health and the environment).
3. Secondary air pollutants.
4. The importance of the ozone layer and how it is being depleted.
5. Ozone as a pollutant.
6. Sources of indoor air pollution.
7. Greenhouse gases and their potential for causing global warming.

THE MAJOR LAYERS IN THE ATMOSPHERE

The gases that make up the atmosphere are held close to the earth by the pull of gravity. With increasing distance from the earth's surface, the temperature, density, and composition of the atmosphere gradually change. On the basis of air temperature, the atmosphere can be divided vertically into four major layers: troposphere, stratosphere, mesosphere, and thermosphere (Fig. 13-1).

Temperature Changes in the Atmosphere

The **troposphere** extends above the earth to a distance of 10–16 km (6–10 mi). The lower part of the troposphere, which interacts directly with the surface of the earth, is the part of the atmosphere we generally call air. Most of our weather occurs in the troposphere. Temperature decreases steadily as the distance from the earth's warm surface increases, until it reaches a minimum of about −56.7°C (−70°F) at the top of the troposphere, in the region called the **tropopause**.

Above the troposphere is the **stratosphere**, which extends to about 50 km (30 mi) and includes the ozone layer. To avoid going through bad weather in the troposphere, jet aircraft fly in the lower stratosphere, where the atmosphere is calm. Temperature remains constant in the lower part of the stratosphere, but with increasing altitude, it begins to rise, reaching a maximum of approximately −1°C (30°F) at the **stratopause**, which is the boundary between the stratosphere and the mesosphere. This rise in temperature is caused by the ozone in the stratosphere, which absorbs ultraviolet (UV) light and converts the radiant energy into heat. It might be expected

The location of the upper limit of the troposphere varies with the temperature at the earth's surface, the nature of the underlying terrain, and other factors. It is generally highest above the equator.

Because there is little exchange of air between the stratosphere and the troposphere, pollutants that enter the stratosphere tend to remain there for long periods of time.

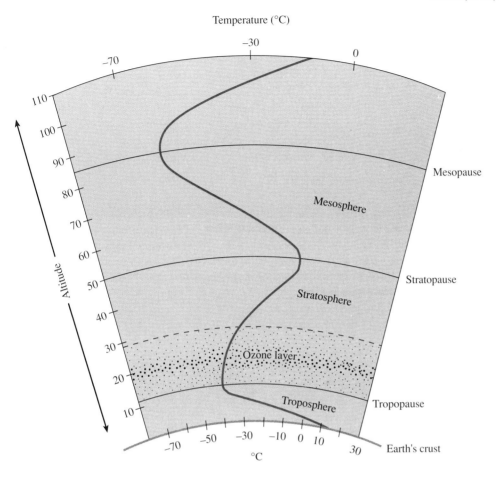

Figure 13-1 The atmosphere is subdivided vertically into four regions based on the air-temperature profile. The ozone layer that protects us from the sun's ultraviolet radiation is found in the stratosphere.

that maximum temperature would be reached in the region where ozone concentration is at a maximum (Fig. 13-1). However, because ozone is such an efficient absorber of UV light, the relatively few ozone molecules in the upper stratosphere are sufficient to absorb most of the radiation before it reaches the region of maximum ozone concentration. The troposphere and stratosphere together are called the **lower atmosphere**. The **upper atmosphere** extends out beyond the stratosphere and is divided into the **mesosphere** and the **thermosphere**.

Continuing outward though the mesosphere, the temperature again falls, and, at an altitude between about 80 and 90 km (50 and 56 mi), the lowest temperature in the atmosphere, approximately −90°C (−130°F), is reached. Above the mesosphere, the temperature rises once more and reaches a maximum of approximately 1200°C (2192°F) in the thermosphere.

There is no precise altitude at which the earth's atmosphere ends and "space" begins.

Pressure and Density Changes

The gases in the atmosphere exert a pressure on the surface of the earth. Although we are not aware of it, we and everything else on the earth's surface are constantly subjected to this pressure. We are adapted to it and are adversely affected by relatively small variations in it. People flying in jet aircraft through the stratosphere, where the air is thin, could not survive if cabin pressure were not adjusted to match the air pressure normally found on the earth's surface.

The average air pressure at sea level is approximately equal to one kg per square centimeter (14.7 lb per square inch).

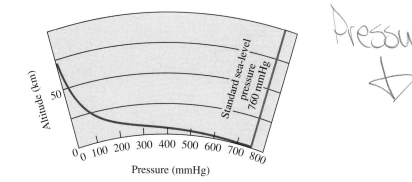

Figure 13-2 Atmospheric pressure decreases rapidly with distance from the earth's surface.

The Italian physicist Evangelista Torricelli (1608-1647) was the first to suggest that the atmosphere exerted a pressure on the earth's surface. He invented an instrument for measuring atmospheric pressure which he called a barometer (from the Greek words *baros*, meaning weight, and *metros*, meaning measure).

With increasing distance from the earth, the pull of gravity becomes less, and air density (mass per unit volume) decreases. The troposphere and stratosphere together account for 99.9 percent of the mass of the atmosphere; almost half of this mass is concentrated within 6 km (3.6 mi) of the earth's surface. As the air becomes thinner with increasing distance from the earth's surface, atmospheric pressure decreases rapidly (Fig. 13-2). At an altitude of 6 km (3.6 mi), atmospheric pressure is reduced to approximately 50 percent of the value at sea level, which normally varies between about 740 and 770 mm mercury (29 to 30 in. Hg).

COMPOSITION OF THE ATMOSPHERE

The major components in the atmosphere are nitrogen (N_2) and oxygen (O_2), which make up approximately 78 percent and 21 percent of the volume of the atmosphere, respectively (Table 13-1). Minor components are the noble gas argon (0.934 percent) and carbon dioxide (0.036 percent). Smaller amounts are contributed by the other noble gases (neon, helium, and krypton) and by methane. The percentage of carbon dioxide in the atmosphere is extremely small, but as the essential raw material for photosynthesis, this very small amount is vital for life on earth. As we shall learn later, it also plays an important role in maintaining the earth's heat balance.

Water vapor was not included in Table 13-1 because its concentration in air is variable. Depending on temperature, precipitation, rate of evaporation, and other factors at a particular location, the percentage of water vapor may be as low as 0.1 percent or as high as 5 percent. It generally lies between 1 and 3 percent; thus, water is the third most abundant constituent of the air.

If water vapor is excluded, the composition of the air is remarkably constant. In the absence of pollution, no matter where we may be on the surface of the earth, the air we breathe will be the same. The homogeneity of air results from the mixing that is brought about by the continuous circulation of the air in the troposphere.

In addition to water vapor and gases, the atmosphere contains many airborne particles. These particles are the nuclei around which ice crystals and water droplets form. Under appropriate conditions, the droplets coalesce to produce clouds and ultimately rain. Airborne particles range in size from those that are visible, such as dust, to others that can be seen only with a high-powered microscope. Minute particles with diameters less than about 10 μm are termed **aerosols**; larger particles are called **particulates**. Both particle types can be in the form of either liquids or solids.

TABLE 13-1   Composition of Pure Dry Air at Ground Level

Gas	Percent by Volume	Parts per Million
Nitrogen (N_2)	78.08	780,840
Oxygen (O_2)	20.95	209,460
Argon (Ar)	0.93	9,340
Carbon dioxide (CO_2)	0.03	350
All other gases	0.01	10

Relative to their size, small particles have very large surface areas that act as sites for chemical interactions. Depending on the nature of the particle and of the impacting molecule (or other species), chemical reactions may occur at the surface of a particle or within it. If impacting molecules become attached to the particle's surface, the process is termed **ad̲sorption**. If, in the case of liquid particles, molecules are drawn inside and dissolved, the process is termed **ab̲sorption**.

Particulates will be studied in more detail when we consider their role as pollutants.

Ozone in the Stratosphere

In the absence of pollution, ozone (O_3) is not present to any appreciable extent in the troposphere, but it occurs naturally in the stratosphere. Its concentration is greatest at a distance of 25 - 30 km (16-19 mi) from the earth's surface (Fig. 13-1). The **ozone layer** is formed when intense UV radiation from the sun causes ordinary molecules of oxygen gas (O_2) in the stratosphere to dissociate into single oxygen atoms (O). Single oxygen atoms are very reactive and combine with O_2 to form O_3:

$$O_2 + UV \text{ radiation} \longrightarrow O + O$$

$$O + O_2 \longrightarrow O_3$$

Because light energy is measured in photons, the dissociation is termed a **photodissociation**. Although ozone formed in this way is continually being converted back to ordinary molecular oxygen, the rates of formation and dissociation are such that a fairly constant concentration of ozone remains in the stratosphere. It varies with season and latitude but averages 10 ppm. Ozone absorbs UV radiation very strongly in the 220–330 nm region of the spectrum (Chapter 1, Fig. 1-13), and, although its concentration in the stratosphere is so low, it is sufficient to screen out 95–99 percent of the sun's dangerous UV radiation.

A beam of light (and other electromagnetic radiation) is considered to consist of particles called photons. Each photon has a characteristic energy that is related to its wavelength.

Because few other molecules in the atmosphere absorb strongly in the UV region, any reduction in the ozone layer increases the amount of UV radiation reaching the earth's surface.

TYPES AND SOURCES OF AIR POLLUTANTS

An air pollutant is defined as a substance that is present in the atmosphere at a concentration sufficient to cause harm to humans, other animals, vegetation, or materials. Each day humans inhale about 20,000 liters (5300 gallons) of air. If harmful gases or fine toxic particles are present in the air, they too will be drawn into the lungs, where they may cause serious respiratory diseases and other health disorders.

Although the emission of carbon dioxide into the atmosphere during the combustion of fossil fuels may be causing global warming, carbon dioxide is not regarded as an air pollutant.

Approximately 56 percent of the electric power generated in the United States is produced by burning coal.

Approximately 90 percent of all air pollution in the United States is caused by five **primary air pollutants**: carbon monoxide, sulfur dioxide, nitrogen oxides, volatile organic compounds (mostly hydrocarbons), and suspended particles. Their major sources and the relative contribution of each to pollution nationwide are shown in Figure 13-3. Emissions of all five are now regulated in the United States.

The transportation industry is responsible for nearly 50 percent of all anthropogenic air pollution (pollution caused by human activities). In addition to carbon monoxide, automobiles emit nitrogen oxides and hydrocarbons. The burning of fossil fuels by stationary sources (power plants and industrial plants) accounts for approximately one-third of air pollutants, mainly in the form of sulfur oxides. Other industries, and a variety of processes including incineration of solid wastes, contribute smaller amounts.

If air pollutants were distributed evenly over the entire country, their harmful effects would be greatly reduced. But because pollutants tend to be concentrated in urban areas, where automobile traffic is congested, large segments of the population are exposed to their harmful effects, particularly during daily rush hours.

In addition to the five primary air pollutants, the atmosphere is contaminated with **secondary air pollutants**, which are harmful substances produced by chemical reactions between primary pollutants and other constituents of the atmosphere. Secondary pollutants include sulfuric acid, nitric acid, sulfates and nitrates (which contribute to acid deposition), and ozone and other photochemical oxidants (which contribute to photochemical smog).

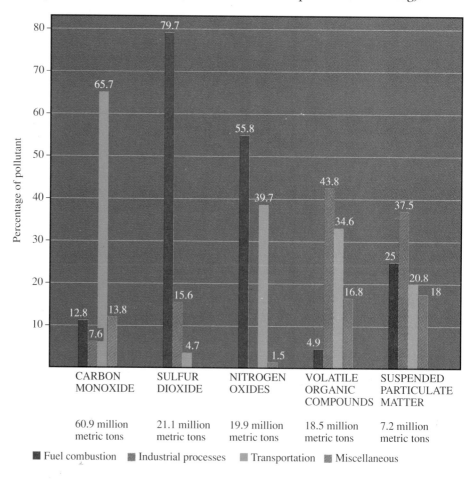

Figure 13-3 1989 nationwide emissions of the five primary air pollutants, according to source. Fuel combustion includes fuel used for generating electrical power and for space heating. Transportation is responsible for approximately half of all pollutants emitted.

Before we study the individual pollutants, we will examine the influence of solar radiation on chemical reactions in the atmosphere. Many reactions, particularly in the upper atmosphere, are quite unlike reactions we have studied so far.

CHEMICAL REACTIONS IN THE ATMOSPHERE

In Chapter 5 we learned that reactions between chemical species (atoms, molecules, and ions) usually involve the loss and gain of electrons in such a way that the product achieves a stable electron configuration, with electrons arranged in pairs. In the atmosphere, species called **free radicals** (or simply, **radicals**) are often formed under the influence of solar radiation. Radicals are uncharged fragments of molecules that, unlike ordinary species, contain an unpaired electron. As a result, radicals are highly reactive and very short-lived. They are responsible for many of the complex, often poorly understood, reactions that occur in the normal and polluted atmosphere.

Central to the chemistry of the troposphere is the **hydroxyl radical**. The hydroxyl radical (OH•) is uncharged and thus quite distinct from the negatively charged hydroxide ion (OH⁻). It is written with a dot (•) beside it to indicate an unpaired electron.

Gamma rays cause harm to humans if they enter the body because they can release hydroxyl radicals from water in tissues. The radicals attack other compounds to produce more radicals that then disrupt the normal functioning of the cells.

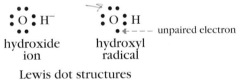

hydroxide ion   hydroxyl radical   — unpaired electron

Lewis dot structures

Hydroxyl radicals are continually formed and consumed in the troposphere. They are produced as the result of a series of complex reactions involving, primarily, ozone, water, and nitrogen dioxide. They play a role in the removal of carbon monoxide from the atmosphere and in the formation of nitric acid, sulfuric acid, and photochemical smog from atmospheric gases.

CARBON MONOXIDE

Sources of Carbon Monoxide

The main anthropogenic source of **carbon monoxide** is the combustion of gasoline in automobile engines. Gasoline is a complex mixture of hydrocarbons (Chapter 9). If it is ignited in an adequate supply of oxygen, the products are carbon dioxide and water, as shown below for octane (C_8H_{18}), a representative gasoline hydrocarbon.

$$2\,C_8H_{18} + 25\,O_2 \longrightarrow 18\,H_2O + 16\,CO_2$$

In the confined space of the internal combustion engine, however, atmospheric oxygen is in limited supply, and combustion is incomplete. Carbon monoxide is formed, and released to the atmosphere in automobile exhaust.

$$2\,C_8H_{18} + 17\,O_2 \longrightarrow 18\,H_2O + 16\,CO$$

Since the introduction of the catalytic converter, which is described later in the chapter, carbon monoxide emissions have been greatly reduced. In addition to the automobile, other sources of carbon monoxide are combustion processes used by the power industry, various industrial processes, and solid-waste disposal.

For every 100 gallons of gasoline burned, 230 pounds of carbon monoxide are produced.

It is perhaps surprising to discover that natural sources release approximately 10 times more carbon monoxide into the atmosphere than all the anthropogenic sources combined. The main natural source is methane gas, which is released during the anaerobic decay of plant materials (Chapter 12, Table 12-5) in swamps, rice paddies, and other wetlands, where vegetation is submerged in oxygen-depleted water (Fig. 13-4). It is also produced in the stomachs of ruminants (cattle and sheep) and the intestines of termites. Flatulence in cattle and consumption of cellulose by termites are significant sources of methane. Oxygen in the atmosphere oxidizes the methane to carbon monoxide.

An estimated 170 million tons of methane are produced annually by termites.

Unlike anthropogenic sources, natural emissions of carbon monoxide are dispersed over the entire surface of the earth. Two mechanisms are thought to maintain the average global level constant at about 0.1 ppm: the conversion of carbon monoxide to carbon dioxide in reactions involving hydroxyl radicals and the removal of carbon monoxide from the atmosphere by soil microorganisms. In cities, where soil has largely been replaced with asphalt and concrete and where emissions are very concentrated, nature's natural defense mechanism is overwhelmed, and carbon monoxide levels increase.

EXAMPLE 13-1 Write balanced equations for the combustion of $C_{10}H_{22}$, a component of gasoline, to yield (a) water and carbon dioxide, and (b) water and carbon monoxide.

Solution:

(a) Write an equation showing the reactants and products.

$$C_{10}H_{22} + O_2 \longrightarrow H_2O + CO_2$$

Balance the C and H atoms.

$$C_{10}H_{22} + O_2 \longrightarrow 11\,H_2O + 10\,CO_2$$

To balance the 31 (11 + 20) O atoms on the product side of the equation we need 31 O atoms on the reactant side. But, because oxygen is diatomic, the number of O atoms must be *even*. Therefore, we need 2 molecules of $C_{10}H_{22}$ and 31 molecules of O_2 (62 O atoms).

$$2\,C_{10}H_{22} + 31\,O_2 \longrightarrow 22\,H_2O + 20\,CO_2$$
$$44\,H + 20\,C + 62\,O = 44\,H + 20\,C + 62\,(22 + 40)\,O$$

(b) Use the same procedure as in (a).

$$C_{10}H_{22} + O_2 \longrightarrow H_2O + CO$$
$$C_{10}H_{22} + O_2 \longrightarrow 11\,H_2O + 10\,CO$$

To balance the 21 (11 + 10) O atoms on the product side we need 21 O atoms on the reactant side. But there must be an even number of O atoms. The equation is as follows.

$$2\,C_{10}H_{22} + 21\,O_2 \longrightarrow 22\,H_2O + 20\,CO$$
$$44\,H + 20\,C + 42\,O = 44\,H + 20\,C + 42\,(22 + 20)\,O$$

PRACTICE EXERCISE: Write balanced equations for the combustion of C_6H_{14} to yield (a) H_2O and CO_2, and (b) H_2O and CO.

Answer: (a) $2\,C_6H_{14} + 19\,O_2 \longrightarrow 14\,H_2O + 12\,CO_2$
(b) $2\,C_6H_{14} + 13\,O_2 \longrightarrow 14\,H_2O + 12\,CO$

Figure 13-4 Anaerobic bacteria which thrive in rice paddies like this one are a major source of methane gas.

Effects of Carbon Monoxide on Human Health

Although carbon monoxide is the most abundant air pollutant (Fig. 13-3), it is not very toxic at the levels usually found in the atmosphere, and it is not usually responsible for the most serious health problems associated with contaminated air.

Carbon monoxide interferes with the oxygen-carrying capacity of blood. Normally, hemoglobin (Hb) in red blood cells combines with oxygen in the lungs to form oxyhemoglobin (HbO$_2$). The oxyhemoglobin is carried in the bloodstream to the various parts of the body, where oxygen is released to the tissues.

Carbon monoxide binds much more strongly to hemoglobin than oxygen does. If carbon monoxide is present in the lungs, it displaces oxygen from hemoglobin and thus reduces the amount of oxygen that can be delivered to the tissues.

Figure 13-5   Levels of carbon monoxide build up if cars are detained in traffic jams such as this one in Cairo, Egypt.

$$HbO_2 \ + \ CO \ \longrightarrow \ HbCO \ + \ O_2$$
$$\text{oxyhemoglobin} \qquad\qquad \text{carboxyhemoglobin}$$

The treatment for carbon monoxide poisoning is inhalation of pure oxygen, which reverses the direction of the above reaction.

The symptoms of carbon monoxide poisoning are those of oxygen deprivation: headache, dizziness, impaired judgment, drowsiness, slowed reflexes, respiratory failure, and eventually loss of consciousness and death. Prolonged exposure to carbon monoxide levels as low as 10 ppm can be harmful. The danger from carbon monoxide is heightened by the fact that the gas is colorless, tasteless, and odorless; people succumb to its effects before they are aware of its presence. On busy city streets, carbon monoxide concentration may reach 50 ppm and may be much higher in underground garages and traffic jams (Fig. 13-5).

Exposure to air containing over 1,000 ppm of CO for about 4 hours, converts approximately 60 percent of the body's hemoglobin to carboxyhemoglobin, and usually results in death.

NITROGEN OXIDES: NO$_x$

Nitrogen dioxide (NO$_2$) is the major **nitrogen oxide** pollutant in the atmosphere. It is formed from nitric oxide (NO). Together these two interconnected oxides are designated NO$_x$.

Sources of NO$_x$

Practically all anthropogenic NO$_x$ enters the atmosphere from the combustion of fossil fuels in automobiles and power plants (Fig. 13-3). At normal atmospheric temperatures, nitrogen and oxygen—the two main components of air—do not react with each other. However, at the very high temperatures that exist in the internal combustion engine and in industrial furnaces, normally unreactive atmospheric nitrogen reacts with oxygen in the air. In a series of complex reactions, the two gases combine to form nitric oxide.

$$N_2 \ + \ O_2 \ \longrightarrow \ 2\,NO$$

When released to the atmosphere, nitric oxide combines with atmospheric oxygen to form nitrogen dioxide.

$$2\,NO \ + \ O_2 \ \longrightarrow \ 2\,NO_2$$

As is the case with carbon monoxide, far more nitrogen oxides are released to the atmosphere by natural processes than by human activities. Recall from Chapter 3 that during electrical storms atmospheric nitrogen and oxygen

A third nitrogen oxide, nitrous oxide (N_2O), is formed naturally in soil during nitrogen fixation. This oxide, which is known as "laughing gas," is used as an anesthetic for minor surgery, particularly in dentistry.

react to form nitric oxide, which then rapidly combines with more atmospheric oxygen to form nitrogen dioxide, as shown in the above equations. Bacterial decomposition of nitrogen-containing organic matter in soil is a further natural source of nitrogen oxides. Because emissions from natural processes are widely dispersed, they do not have an adverse effect on the environment.

The Fate of Atmospheric NO_x

Nitrogen dioxide, regardless of its source, is ultimately removed from the atmosphere as nitric acid and nitrates in dust and rainfall. In a series of complex reactions involving hydroxyl radicals, nitrogen dioxide combines with water vapor to form nitric acid. The overall reaction can be written:

$$4\,NO_2 + 2\,H_2O + O_2 \longrightarrow 4\,HNO_3$$

Much of the nitric acid in the atmosphere is formed within aqueous aerosols. If weather conditions are right, the aerosols coalesce into larger droplets in clouds, and the result is acid rain. Some of the nitric acid formed reacts with ammonia and metallic particles in the atmosphere to form nitrates. Ammonium nitrate is formed as follows:

$$HNO_3 + NH_3 \longrightarrow NH_4NO_3$$

Nitrate salts dissolve in rain and snow or settle out as particles. The combined fallout contributes to **acid deposition**.

Effects of NO_x on Human Health and the Environment

Nitrogen dioxide is a red-brown toxic gas with a very unpleasant acrid odor. It can cause irritation of the eyes, inflammation of lung tissue, and emphysema, but even in badly polluted areas, its concentration in the atmosphere is rarely high enough to produce these symptoms. Nitrogen oxides are a serious health problem because of their role in the formation of secondary pollutants associated with photochemical smog, which is discussed later in the chapter.

Emissions from stationary fuel combustion sources are difficult to control. Lowering the combustion temperature of the furnace decreases NO formation, but it decreases efficiency at the same time. Most research has concentrated on reducing emissions from automobiles by means of the catalytic converter.

VOLATILE ORGANIC COMPOUNDS (VOCS)

A great variety of **volatile organic compounds (VOCs)**, including many hydrocarbons, enter the atmosphere from both natural and anthropogenic sources (Fig. 13-3). Most are not pollutants in themselves, but create a problem when they react with other substances in the atmosphere to form secondary pollutants associated with photochemical smog.

Sources of Hydrocarbons

The petroleum industry is the main anthropogenic source of hydrocarbons in the atmosphere. Gasoline is a complex mixture of many volatile hydrocarbons (Chapters 9 and 14), and, in urban areas, there are numerous situations where gasoline vapors can escape into the atmosphere: at the gas pump, during filling of storage tanks, and from unburned gasoline in automobile exhausts.

In the natural world, the pleasant aroma of pine, eucalyptus, and sandalwood trees is caused by the evaporation of complex volatile hydrocarbons called **terpenes** from their leaves. Natural sources account for about 85 percent of total emissions of hydrocarbons. The remaining 15 percent from anthropogenic sources is of concern because, unlike hydrocarbons from natural sources, it is not evenly distributed but is concentrated in urban areas.

AUTOMOBILE POLLUTANTS AND THE CATALYTIC CONVERTER

Motor vehicles are a major source of the three pollutants we have just discussed: carbon monoxide, nitrogen oxides, and volatile hydrocarbons (Fig. 13-3). Since 1975, when all new cars in the United States were required by law to be equipped with a catalytic converter, emissions of those pollutants have been reduced significantly. Today's cars emit 95 percent less pollutants than pre-1970 vehicles, despite the fact that the number of miles traveled in automobiles has almost doubled in the last 20 years.

In the three-way **catalytic converter** (in use since 1981), two opposing chemical reactions take place: an oxidation and a reduction. Hot exhaust gases from the engine pass through the converter before they enter the muffler. The converter is a very fine ceramic honeycomb structure coated with the precious metals platinum (Pt), palladium (Pd), and rhodium (Rh), which act as catalysts (Fig. 13-6). As the gases enter, rhodium catalyzes the reduction of nitrogen oxides to nitrogen gas. Then air is injected into the exhaust stream to provide oxygen, which, in the presence of the platinum and palladium catalysts, oxidizes carbon monoxide to carbon dioxide and hydrocarbons to water and carbon dioxide.

The overall reaction for the reduction of nitric oxide and the oxidation of carbon monoxide can be written:

$$\text{Pt, Pd, Rh catalysts}$$
$$2\,NO + 2\,CO \longrightarrow N_2 + 2\,CO_2$$

The oxidation of a typical gasoline hydrocarbon occurs as follows:

$$\text{Pt, Pd catalysts}$$
$$2\,C_8H_{18} + 25\,O_2 \longrightarrow 16\,CO_2 + 18\,H_2O$$

The formation of carbon dioxide in these two reactions results in an increased level of carbon dioxide in the atmosphere, which, as we shall discuss later, may be linked to global warming.

Recall from Chapter 6 that a catalyst is a substance that increases the rate of a chemical reaction without itself being used up in the reaction. Platinum and rhodium are very expensive, but since they are not used up, the small amounts needed last a long time. Lead-free gasoline must be used in cars fitted with catalytic converters because lead inactivates the catalysts.

Today's catalytic converters remove 96 percent of carbon monoxide and hydrocarbons and 76 percent of nitrogen oxides from automobile exhausts. By 2002, even better results will have to be achieved to meet the requirements of the new 1990 Clean Air Act. The auto industry is expecting a new electrically heated catalytic converter, which is still in the research and development stage, to achieve these reductions.

Present catalytic converters do not reach their minimum operating temperature until about 1½ minutes after cold-start of an automobile during which time 50 percent of emissions pass unreacted from the tail pipe. Electrically heated catalysts are being designed to reduce the cold-start time.

Figure 13-6 (a) The catalytic converter, which has been standard equipment in automobiles sold in the United States since 1975, reduces engine emissions. (b) A three-way catalytic converter consists of a ceramic honeycomb coated with the precious metals platinum, palladium, and rhodium. (c) The metals catalyze the conversion of nitrogen oxides and carbon monoxide to nitrogen and carbon dioxide, and hydrocarbons (HC) to carbon dioxide and water.

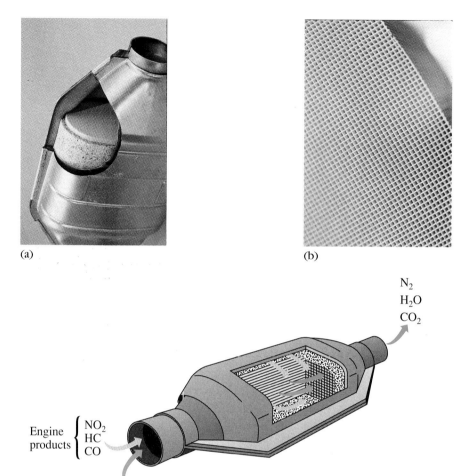

(a)

(b)

(c)

 ### ■ SULFUR DIOXIDE: SO_2

Sources of SO_2

Release of **sulfur dioxide** (SO_2) to the atmosphere is the primary cause of acid rain (Chapter 8). Fuel combustion at electric power generating plants accounts for approximately 80 percent of anthropogenic emissions. Industrial sources contribute approximately 16 percent (Fig. 13-3).

Coal, oil, and all other fossil fuels naturally contain some sulfur because the plant materials from which they were formed included sulfur-containing compounds in their makeup. Coal frequently contains additional sulfur in the form of the mineral pyrite, FeS_2. Coal mined in the United States is, typically, between 1 and 4 percent, by weight, sulfur. The percentage is lower in coal from the western states than in coal mined in the central and eastern states.

When sulfur-containing coal is burned, any sulfur it contains is oxidized to sulfur dioxide:

A 1000-megawatt coal-fired electric power plant burns approximately 700 tons of coal per hour. If the coal is 3% by weight sulfur, 42 tons of SO_2 per hour (nearly 370,000 tons per year), will be produced.

$$S + O_2 \longrightarrow SO_2$$

In addition to direct sources of sulfur dioxide, there are a number of indirect sources. Paper and pulp mills, for example, produce the gas hydrogen sulfide (H_2S), which is rapidly oxidized in the atmosphere to SO_2. The overall reaction is as follows:

$$2\,H_2S + 3\,O_2 \longrightarrow 2\,SO_2 + 2\,H_2O$$

Natural sources account for approximately half of all sulfur dioxide emissions. Hydrogen sulfide produced as an end product of the anaerobic decomposition of sulfur-containing organic matter by microorganisms (Chapter 12, Table 12-5) is the main source. On entering the atmosphere, hydrogen sulfide is oxidized to sulfur dioxide as shown in the above equation. Volcanic eruptions are another, more localized, source of sulfur dioxide. It has been estimated that the eruption of Mount Pinatubo in the Philippines in June 1991 (Fig. 13-7) injected as much as 25 million tons of sulfur dioxide into the stratosphere, where it was converted into sulfuric acid aerosols. — *acid rain*

Figure 13-7   The eruption of Mount Pinatubo in the Philippines in June 1991 ejected huge quantities of sulfur dioxide and ash into the atmosphere.

The Fate of Atmospheric Sulfur Dioxide: Acid Rain

Sulfur dioxide in the atmosphere reacts with oxygen to form sulfur trioxide (SO_3), which then reacts readily with water in liquid aerosols and water droplets to form sulfuric acid. The overall reactions can be written as follows:

$$2\,SO_2 + O_2 \longrightarrow 2\,SO_3$$
$$SO_3 + H_2O \longrightarrow H_2SO_4$$

The actual pathways are more complex and involve hydroxyl radicals.

Sulfuric acid in the atmosphere becomes concentrated near the base of clouds, where pH levels as low as 3 have been recorded. Thus, cloud-enshrouded, high-altitude trees and vegetation may be exposed to unusually high acidity. Since rain is made up of moisture from all cloud levels, it is less acidic than moisture at the lower cloud levels.

Within the atmosphere, H_2SO_4 reacts with ammonia to form ammonium sulfate, $(NH_4)_2SO_4$.

$$2\,NH_3 + H_2SO_4 \longrightarrow (NH_4)_2SO_4$$

Other sulfates are formed in reactions with metal particulates.

Ash particles are usually emitted together with sulfur dioxide from electric power generating plants. Sulfur oxides become adsorbed onto particle surfaces and may be carried many miles from their source before settling out or being washed out by precipitation. Like nitric acid and nitrates formed from nitrogen oxides, sulfuric acid, sulfates, and particulates all contribute to acid deposition.

The formation of SO_3 from SO_2 is influenced by prevailing atmospheric conditions: sunlight, temperature, humidity, and the presence of hydrocarbons, nitrogen oxides, and particulates in the atmosphere. When photochemical oxidants are present during episodes of photochemical smog, the formation of SO_3 is accelerated.

Effects of SO_2 on Human Health and the Environment

Sulfur dioxide is a colorless, toxic gas with a sharp, acrid odor. Exposure to it causes irritation of the eyes and respiratory passages and aggravates symptoms of respiratory disease.

Sulfur dioxide is also harmful to plants. Crops such as barley, alfalfa, cotton, and wheat are particularly susceptible to its damaging effects. The environmental effects of acid deposition associated with sulfur dioxide were considered in Chapter 8.

Plants are even more susceptible to low levels of SO_2 than humans are. Before emissions controls were introduced, land around smelters and coal-burning power plants was often completely devoid of vegetation.

Control of SO₂ Emissions

In response to federal regulations, many industries, including the coal-fired electric power industry, now take steps to reduce sulfur dioxide emissions. Reductions can be achieved in two ways: (1) Sulfur can be removed from coal before combustion, or (2) sulfur dioxide can be removed from the smokestack after combustion—but before it reaches the atmosphere. The second, cheaper, approach is generally chosen.

The most commonly used method is **flue-gas desulfurization (FGD)** in which sulfur-containing compounds are washed (or **scrubbed**) from the chimney (flue) gases by absorption in an alkaline solution. A promising newer method is **fluidized bed combustion (FBC)**, a process in which coal in granular form is burned with finely divided pulverized limestone ($CaCO_3$) or dolomite (Ca-Mg carbonate) or both. Air is introduced to keep the mixture in a semifluid state. Upon heating, the limestone is converted to lime (CaO) and carbon dioxide:

$$CaCO_3 \longrightarrow CaO + CO_2$$

The lime then reacts with SO_2 to form fine particles of calcium sulfite ($CaSO_3$).

$$CaO + SO_2 \longrightarrow CaSO_3$$

Calcium sulfite is removed from the stack by an electrostatic precipitator, a device that will be described when we discuss particulate pollutants. Both FGD and FBC have the problem of disposing of large quantities of waste products.

Another, less desirable approach is the installation of very tall stacks, which reduce pollutants in the immediate neighborhood but disperse them more widely. Communities distant from the source of pollution then suffer most of the adverse effects.

Because prevailing winds carry pollutants across national boundaries, measures to control acid-producing emissions must be agreed upon by many countries to be effective. In 1985, 21 European governments signed an agreement to lower sulfur dioxide emissions by at least 30 percent of 1985 levels by the year 1995. In 1988, the Canadian government undertook a plan to reduce emissions by 50 percent by 1994. In the United States, the 1990 Clean Air Act requires utilities to cut sulfur dioxide emissions by 50 percent of 1990 levels by the year 2000. To meet this requirement, electric utilities must decide whether to install costly efficient scrubbers or switch to low-sulfur western coal, a move that would mean a loss of jobs and great economic hardship for the high-sulfur coal miners of the Ohio Valley and West Virginia.

The world's tallest stack is the 1250 ft stack at the Copper Cliff Nickel Smelter in Sudbury, Ontario, Canada.

SUSPENDED PARTICULATE MATTER

In addition to emissions from natural sources, a significant quantity of harmful particulate matter is emitted by industrial processes. Airborne particles of natural origin include smoke and ash from forest fires and volcanic eruptions (Fig. 13-7), dust, sea salt spray, pollen grains, bacteria, and fungal spores (Fig. 13-8). Liquid aerosols and particulates are generally called mist and include fog and raindrops.

Eventually, all particulate matter is deposited on the earth's surface. Relatively large particles (diameters greater than 10 μm) settle out under the influence of gravity within 1 to 2 days; medium-sized particles (diameters 1–10 μm) remain suspended for several days; fine particles (diameters less than 1 μm) may remain in the troposphere for several weeks and in the stratosphere for up to 5 years. Aerosols, acting as nuclei for the formation of droplets in clouds, reach the ground when the droplets condense and fall as rain and snow. Fine particles can be transported considerable distances by winds before settling to the ground or being washed out by rain or snow.

In cloud seeding, a powdered chemical, usually silver iodide (AgI), is injected into a cloud. The particles of powder act like aerosols and become nuclei, or centers, around which moisture collects to form water droplets that then fall as rain.

Anthropogenic Sources

Suspended pollutant particles are the most visible form of air pollution. A major source is the combustion of coal by electric power generating plants (Fig. 13-3). One product of the incomplete combustion of coal is **soot**, a finely divided impure form of carbon. Other products include metallic and nonmetallic oxides formed from mineral materials present in coal. These materials, called **fly ash**, are swept up smokestacks in drafts from roaring furnaces and emitted into the atmosphere.

Other sources of particulate emissions are automobiles, solid-waste incineration, domestic wood stoves, mining, ore processing, and agricultural activities. Toxic metal particles, cement dust, and pesticide and fertilizer residues that are released in these activities are all potentially hazardous. Examples of particulates from anthropogenic sources are included in Figure 13-8.

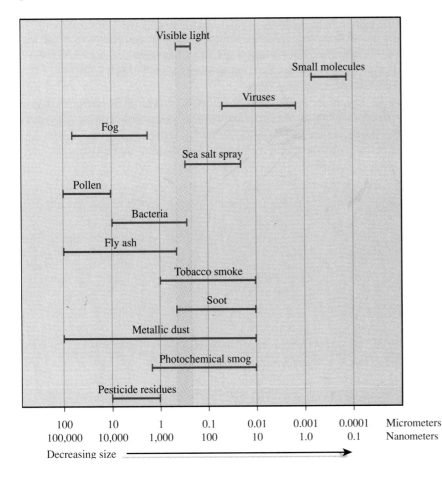

Figure 13-8 Suspended particulate matter of many types and sizes enters the atmosphere from both natural and anthropogenic sources.

Effects of Particulate Matter on Human Health and the Environment

Health and environmental effects of particulates depend on both the size and the nature of the particulates. Very fine particles (diameter less than about 1 μm) are the most hazardous. They are not filtered out by hairs and mucus in the nose but are drawn deep into the lungs, where they can remain indefinitely and cause tissue damage and contribute to the development of emphysema. They take with them any toxic chemicals that are attached to their surfaces or are dissolved within them, such as sulfuric acid.

Particulates and aerosols have a direct effect on climate. Those with diameters greater than 1 μm intercept and scatter incoming light away from the earth's surface and thus exert a cooling effect. Geologists anticipate that fine ash particles emitted into the stratosphere in the eruption of Mount Pinatubo (the largest volcanic eruption since Krakatoa; Chapter 2) will have a temporary global cooling effect, for the next 2–4 years, that will offset any warming trend caused by an increase in atmospheric carbon dioxide.

Control of Particulate Emissions

One method used by industry to control particulate emissions is **electrostatic precipitation** (Fig. 13-9a). In this process, gases and particulate matter are passed through a high-voltage chamber before leaving the chimney stacks. A negatively charged central electrode imparts a negative charge to the particles, which are then attracted to the positively charged walls of the chamber. As their charges are neutralized, the particles clump together and fall to the bottom, where they can be collected.

A second procedure is **bag filtration** in which fabric bags lined up in a bag house function essentially like the bag in a household vacuum cleaner (Fig. 13-9b). Gases can pass through the finely woven bags but particulate matter cannot and is filtered out. The bags are shaken at intervals to dislodge

Figure 13-9 Two methods for controlling particulate emissions. (a) In electrostatic precipitation, the dirty gas is passed through a chamber where particulates acquire a negative charge from a negatively charged central electrode. The charged particles are attracted to the positively charged wall of the chamber where they are deposited. Cleaned gases exit from the top of the chamber. (b) In bag filtration, the dirty gases are forced through fabric bags which trap particulate matter. The bags are shaken at intervals to cause particles to fall into a hopper.

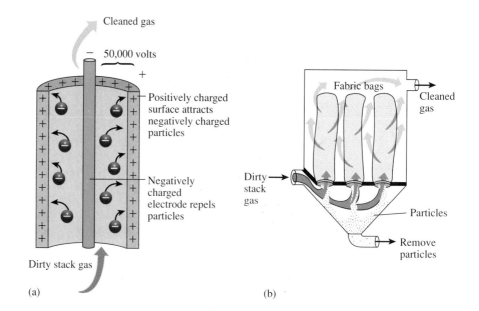

particles, which are collected in a hopper below. Although both methods remove more than 98 percent of particulates, they fail to remove the finest ones, which are the most dangerous.

Another less efficient process, **cyclone separation**, spins the particle-laden gases in an outward motion in a type of centrifuge. The particles hit the walls, settle out, and fall to the bottom, where they are collected. All three methods produce solid wastes, which must be disposed of. Most of the wastes end up in landfills (Chapter 19).

INDUSTRIAL SMOG

Particulate matter and sulfur dioxide can be a deadly combination. Released into the atmosphere together during the burning of coal, they can form **industrial smog**, a mixture of fly ash, soot, sulfur dioxide, and some volatile organic compounds.

In the nineteenth and twentieth centuries, industrial smog (sometimes called London smog) was common in the industrial centers of Europe and the United States. Visibility was often reduced to a few yards, and people in factory towns lived under a pall of black smoke (Fig. 13-10). Smog formed in winter, typically in cities where the weather was cold and wet.

The word smog is thought to be derived from a combination of *smo*ke and f*og*.

PHOTOCHEMICAL SMOG

Photochemical smog is quite different in origin from industrial smog. Typically, it develops as a yellow-brown haze in hot sunny weather in cities, such as Los Angeles, where automobile traffic is congested (Fig. 13-11). The reactions that lead to its formation are initiated by sunlight and involve hydrocarbons and nitrogen oxides emitted in automobile exhausts. Nitrogen dioxide is responsible for the brownish color of the haze.

The reactions that produce photochemical smog are complex and not fully understood. A simplified reaction sequence is shown in the following equations.

$$NO_2 \xrightarrow{UV} NO + O$$
$$O + NO \longrightarrow NO_2$$
$$O + O_2 \longrightarrow O_3$$
$$O + H_2O \longrightarrow 2\,OH\bullet$$

In the initiating step, absorption of sunlight causes nitrogen dioxide to split into nitric oxide and atomic oxygen. Very reactive atomic oxygen reacts in several ways: It reacts with nitric oxide to reform nitrogen dioxide in a continuing cycle; it combines with ordinary molecular oxygen to form ozone; and it reacts with water to form hydroxyl radicals, which then react readily with organic compounds.

Unburned hydrocarbons in automobile exhausts (represented by RH in the following equation) react with hydroxyl radicals to form a number of secondary pollutants, including the hydrocarbon radical $RO_2\bullet$. This radical then re-

Figure 13-10 Industrial smog like that shown in the photograph is derived from the burning of coal and is made up primarily of fly ash, soot, and sulfur dioxide.

Figure 13-11 (a) Los Angeles, CA on a smoggy day. Interactions between nitrogen oxides and hydrocarbons in the presence of sunlight are responsible for the orange-brown haze. (b) Los Angeles on a clear day.

acts with NO to form aldehydes and ketones (Chapter 9) and the hydroperoxide radical $HO_2\bullet$.

$$RH + OH\bullet + O_2 \longrightarrow RO_2\bullet + H_2O$$

$$RO_2\bullet + NO \longrightarrow \text{(aldehydes and ketones)} + NO_2 + HO_2\bullet$$

Further reactions produce peroxyacetyl nitrate (PAN).

$$\underset{\text{acetyl}}{\underline{CH_3-\overset{\overset{\textstyle O}{\|}}{C}-O}}-\underset{\text{nitrate}}{\underline{\overset{\text{peroxy}}{\overbrace{O-O}}-NO_2}}$$

Ozone, aldehydes, and PAN all contribute to the harmful effects of photochemical smog, but ozone—the pollutant produced in greatest quantity—causes the most serious problems.

Ozone: A Pollutant in the Troposphere

Ozone in the stratosphere protects us from damaging UV radiation from the sun, but in the troposphere it becomes a dangerous pollutant. Ozone is a powerful oxidizing agent. It is a colorless, pungent, very reactive gas, and it irritates the eyes and nasal passages. People with asthma or heart disease are particularly susceptible to its harmful effects. Exposure to ozone levels as low as 0.3 ppm for one or two hours can cause fatigue and respiratory difficulties. In Los Angeles, school children are kept indoors if ozone levels reach 0.35 ppm. Fortunately, because of pollution controls, these levels are rarely reached today.

Ozone is also very toxic to plants. In California, crop damage caused by ozone and other photochemical pollutants costs the state millions of dollars a year. Ozone also damages fabrics as well as rubber in tires and windshield wiper blades.

Tomatoes, tobacco, and other leafy plants are very susceptible to damage by ozone. Brief exposure to concentrations as low as 0.1 ppm reduces photosynthesis causing leaves to turn yellow.

TEMPERATURE INVERSIONS AND SMOG

Certain meteorological and geographic conditions favor the formation of both industrial and photochemical smog. Normally, air temperature in the troposphere decreases with increasing altitude (Fig. 13-1). Warm air at the earth's surface expands, becomes less dense, and rises. As it does so, cooler air from above flows in to replace it. In turn, the cooler air is warmed and rises, and in this way, the air is continually renewed and pollutants are dispersed by vertical currents and prevailing winds.

Reversal of the usual temperature pattern, called a **temperature inversion**, sometimes occurs; after an initial decrease, the air temperature, instead of continuing to decrease with increasing altitude, begins to increase. A lid of warm air forms over cooler air near the earth's surface (Figure 13-12). The cooler, denser layer cannot rise through the warm lid of air above it and becomes trapped, sometimes for days. There is no vertical circulation, and pollutants accumulate.

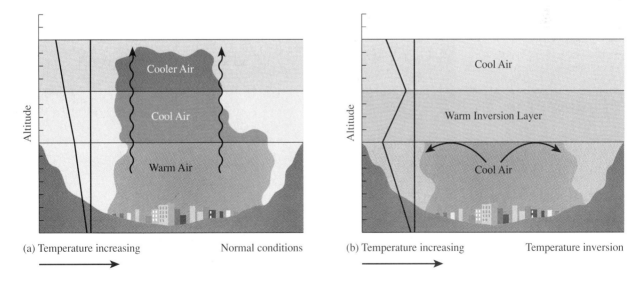

Figure 13-12   Temperature inversion. (a) Under normal conditions air warmed at the earth's surface rises and mixes with cooler air above, and any pollutants are dispersed upward. (b) When a temperature inversion occurs, a layer of warm air settles over the cooler air preventing it from rising. Any pollutants become trapped under the warm lid until atmospheric conditions change.

A particularly serious incident occurred in 1948 in the industrial Donora Valley in Pennsylvania. As a result of a temperature inversion, industrial smog settled over the valley for 5 days. Many people died, and almost half the population suffered from respiratory ailments. In 1952, a similar incident in London resulted in 4000 deaths. In that case, smoke from coal burned for heat in homes and workplaces was the main cause of the pollution.

If a thermal inversion occurs in an area partly surrounded by mountains, as, for example, in Los Angeles, Denver, or Salt Lake City, photochemical smog buildup is particularly serious. Because of the encircling mountains, the pollutants cannot be dispersed horizontally but remain as a blanket over the city until the stalled high-pressure system moves on.

DEPLETION OF THE PROTECTIVE OZONE LAYER IN THE STRATOSPHERE

Since 1983, satellite images have revealed a "hole" in the ozone layer each spring over the South Pole (remember that spring in the Southern Hemisphere is autumn in the Northern Hemisphere) (Fig. 13-13). In September 1992, the **ozone hole** was the largest ever recorded and was almost three times as large as the area of the United States. There is now convincing evidence that a class of synthetic compounds, the **chlorofluorocarbons (CFCs)**, is responsible for this destruction of the ozone layer.

CFCs are very useful inert, nontoxic, nonflammable compounds that have been used for many years as coolants for refrigerators and air conditioners (including those in automobiles) and as spray-can propellants for aerosol forms of hair sprays and deodorants. Also, they are unsurpassed as solvents for cleaning electronic microcircuits. Commercially, the most important CFCs are the halogenated methanes, Freon-11 and Freon-12 (Freon is a DuPont trade name).

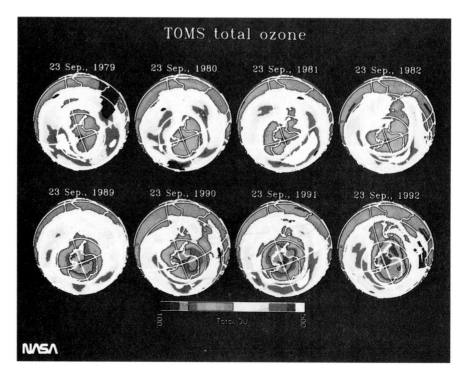

Figure 13-13   Ozone concentrations over the Antarctic from September 1979 to 1982 and from 1989 to 1992, based on information obtained by the NIMBUS–7 satellite. Concentrations are measured in Dobson units as shown on the bottom scale. The pink areas indicate where ozone has been depleted. Since 1989, ozone concentrations have been well-below normal levels.

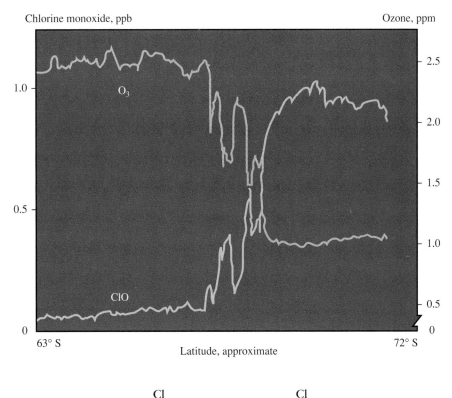

Chlorine monoxide, ppb

Ozone, ppm

O₃

ClO

63° S

72° S

Latitude, approximate

Figure 13-14 In September 1987, instruments aboard a NASA research plane simultaneously measured ozone and chlorine monoxide concentrations as the plane flew from 50°S (Punta Arenas, Chile) to 72°S. As the plane entered the ozone hole, chlorine monoxide concentration rapidly increased to about 500 times its normal level while at the same time ozone concentration decreased dramatically.

$$
\begin{array}{cc}
\text{Cl} & \text{Cl} \\
| & | \\
\text{Cl}-\text{C}-\text{F} & \text{Cl}-\text{C}-\text{F} \\
| & | \\
\text{Cl} & \text{F} \\
\text{Freon 11} & \text{Freon 12}
\end{array}
$$

Because CFCs are so unreactive, they do not break down when released into the air. In time, air currents carry them into the stratosphere, where, under the influence of UV radiation, they form chlorine radicals (Cl•) that initiate the destruction of ozone.

$$
CF_2Cl_2 \xrightarrow{\text{UV}} CF_2Cl + Cl\bullet
$$

Convincing field evidence for the involvement of CFCs in ozone depletion came in 1987. Between August and September of that year, a NASA research plane, equipped with sophisticated analytical instruments, flew 25 missions in the region of the ozone hole in the stratosphere over Antarctica. Data collected showed conclusively that as ozone concentration decreased, the concentration of the chlorine monoxide radical, ClO•, rose (Fig. 13-14). ClO• has been dubbed the "smoking gun" of ozone depletion.

The following chain reaction is now thought to account for about 80 percent of the ozone loss in the stratosphere.

$$
\begin{array}{r}
Cl\bullet + O_3 \xrightarrow{\text{UV}} ClO\bullet + O_2 \\
ClO\bullet + O \longrightarrow Cl\bullet + O_2 \\
\hline
\text{net result:} \quad O_3 + O \longrightarrow 2\,O_2
\end{array}
$$

Notice that the reaction of the chlorine monoxide radical with an oxygen atom

As early as 1974, M. J. Molina and F. S. Rowland at the University of California had demonstrated in the laboratory that CFCs exposed to UV radiation form chlorine radicals. They predicted that these radicals could cause problems by destroying stratospheric ozone.

results in the formation of another chlorine radical—which can then start the chain reaction over again. The net result each time a molecule of ozone reacts with a chlorine radical is the destruction of a molecule of ozone.

Normal meteorological conditions in Antarctica accelerate the destruction of ozone each spring. During winter, when there is no sunshine, frigid temperatures and winds set up a rapidly rotating polar vortex of air in which stratospheric ice clouds form. Active chlorine species condense on the cloud surfaces, and, in the spring, the returning sunlight initiates ozone-destroying reactions. Warmer weather eventually breaks up the vortex, and ozone-poor air masses spread out toward Australia, New Zealand, and the southern tips of Africa and South America, where they may remain for several weeks. During this time, the UV radiation reaching those areas may increase by as much as 20 percent, causing serious problems. Very-short-wavelength UV radiation can kill living organisms, including both beneficial and harmful bacteria. Slightly longer wavelengths cause tanning and sunburn of human skin and also cause several types of skin cancer, including deadly melanoma.

A number of related bromine compounds called Halons (which are used in fire extinguishers; for example, CF_2ClBr and CF_3Br) may also cause ozone destruction. Bromine radicals ($Br\bullet$) are thought to initiate the same type of chain reaction as shown for $Cl\bullet$.

CONSUMER BRIEF

Sunscreens: Boon or Bane?

Many Caucasian Americans seek the perfect suntan. During the Industrial Revolution when workers were forced to spend long daylight hours in factories, a suntan became a status symbol. A suntan showed others that one had the wealth and leisure time to travel to places where there was plenty of sun.

The tanning of skin involves the natural pigment melanin, which is responsible for the color of skin. People with white skin have little melanin, whereas black people have a lot. Melanin reacts to the sun in a two-step process. In the first step, light-colored, unoxidized melanin near the skin's surface is changed by ultraviolet (UV) light to a dark-brown, oxidized form. This process produces a tan, or burn, in a few hours, but the tan fades quickly once the person moves out of the sun. A more lasting tan comes from the second step in which additional melanin is produced from the amino acid tyrosine, which is abundant in skin cells. A tan produced by this process lasts for several days. More

sunbathing produces more melanin and deepens the color of the tan.

Today there is considerable concern about the effect the sun's UV radiation has on skin. UV radiation can damage the proteins that make up the skin's elastic and connective tissue, which can lead to wrinkled, leathery, sagging skin. The term *tanning* comes from a comparison of the exposed skin to leather that has been chemically tanned. Even more alarming is the increasing incidence of skin cancers, especially melanoma which is especially deadly. Studies have indicated a correlation between exposure to sunlight and the incidence of melanoma. Dark-skinned people, who have a higher level of melanin in their skin than do light-skinned people, are more protected and show a lower incidence of melanoma.

Sunscreens have been developed to protect the skin from the harmful UV radiation while allowing some lower-energy sunlight to reach the skin and begin the tanning process. The most ef-

fective sunscreens are zinc and titanium oxide creams. We have all seen lifeguards at the pool with their noses painted white with these creams. Although they may not be attractive, such creams are extremely effective in preventing almost all UV radiation from reaching the skin, and because they are white, it is easy to see where they have been applied. Most other sunscreens are formulated in creams that are invisible on the skin, and it is difficult to see which areas have been covered and which are unprotected.

Most sunscreens contain aromatic organic compounds that absorb UV radiation. The organic compounds are dissolved in a cream base that is similar to hand lotion and provides some moisture to the skin. Some sunscreen preparations are resistant to removal by water. It should be noted that fresh water dissolves sunscreens more effectively than does salt water. Most of the early preparations contained para-aminobenzoic acid (PABA), an aromatic organic compound that is a photosensitizer for a small percentage of the population. A photosensitizer magnifies rather than reduces the intensity of the effects of sunlight. Because of this characteristic, PABA is now considered too dangerous to use. The aromatic organic compounds in today's sunscreens are considered safe. They absorb or reflect UV radiation and they are chemically and photochemically stable. Because they are soluble in the cosmetic cream but insoluble in water or perspiration, they do not leach out of the cream.

Sunscreens are rated according to the protection factor (PF) that they provide. The protection factor is a ratio of the effect sunlight has on sunscreen-protected skin to the effect it has on unprotected skin. If a sunscreen with a protection factor of 10 is used, a person can stay out in the sun 10 times longer than without sunscreen and achieve the same effect. Sunscreens with a low PF, such as 2, have a low concentration of the UV-absorbing compounds, whereas those with a high PF, such as 25, have a high concentration.

One of the problems with sunscreens is that many chemicals can be absorbed through the skin. When sunscreens containing salicylates were applied to the skin in a research project, salicylates could be detected in a tested person's urine within 30 minutes. Few other cosmetic preparations are used as extensively or cover such a large area of skin as sunscreens. Because of this extensive use, the chance of absorbing chemicals contained in the creams is increased. Some claim that the risk of exposure to these chemicals outweighs the benefit to be gained by protecting oneself from the harmful UV radiation of sunlight. Only time will tell whether sunscreen preparations are a boon or a bane.

Questions
1. A recent advertisement tells of a sunscreen that is "Chemical Free." This sunscreen does not contain an aromatic organic molecule as a sunscreen, but rather uses a fine dispersion of sand (in the cream) to block the sun. How does it work? Is it really chemical free?
2. If you are going out in the sun for a period of 2 hours, should you use a sunscreen with a high or a low PF? What are the risks and benefits?

The Arctic Ozone Hole

In 1989, a hole in the ozone layer was detected over the North Pole for the first time. Since three-quarters of the world's population live in the Northern Hemisphere, depletion of the ozone layer over the Arctic could have even more serious consequences than depletion over the Antarctic.

The Montreal Protocol

As early as 1978, CFC aerosol propellants were banned in North America and most of the Scandinavian countries, but not elsewhere in the world. The first international effort to protect the ozone layer came in January 1987 when 24 nations signed the Montreal Protocol on Substances That Deplete the Ozone Layer. The protocol calls for CFC production to be cut back in stages until, by 1998, they are 50 percent of the 1986 level. The protocol was amended in 1990 when 93 countries pledged to halt all production of CFCs and Halons by 2000. Since then, many industrialized nations, including the United States, have agreed to a complete phaseout as early as 1995. As a result of these decisions, scientists now expect that after 2000 the ozone layer will begin to thicken and the hole will gradually disappear.

Alternatives to CFCs include hydrochlorofluorocarbons (HCFCs), such as CF_3CHCl_2, which have a lower percentage of chlorine than CFCs, and hydrofluorocarbons (HFCs), such as CF_3CH_2F, which have no chlorine.

In 1993, the EPA issued regulations requiring technicians to recover and recycle all ozone-depleting chemicals during the servicing and disposal of refrigeration and air conditioning equipment.

INDOOR AIR POLLUTION

One might expect to be safer from pollutants indoors, but in today's well-sealed homes and offices, one often is not. In buildings where there is little or no circulation of fresh air, pollutants may accumulate to dangerous levels. Figure 13-15 shows sources of the major pollutants in the home.

Smoking is a particularly dangerous cause of indoor pollution. In addition to tars and nicotine, tobacco smoke contains high levels of all the primary pol-

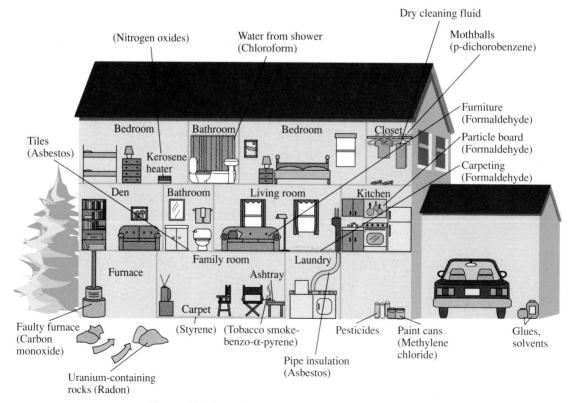

Figure 13-15   Indoor air pollutants and their sources in the home.

lutants associated with combustion. Smoking causes emphysema, lung cancer, and coronary heart disease. Of increasing concern is the risk of smoking to non-smokers. Young children of smokers, for example, suffer more asthma and bronchitis than children of nonsmokers.

Gas stoves, kerosene heaters, wood stoves, and faulty furnaces are potential sources of nitrogen oxides and carbon monoxide. Paint, paint strippers and thinners, gasoline, and pesticides, which many people store in their basements, release harmful vapors and dust particles. Formaldehyde, a toxic, irritating gas, is released from polymers used in the manufacture of certain types of insulation foam and furniture stuffing and from newly installed carpeting and paneling. The clothes brought home from the dry cleaners may also cause a problem. Traces of harmful volatile solvents used in the cleaning process are retained by the garments and later released into the atmosphere. Even taking a hot shower or bath may be harmful if chloroform is released from chlorine-treated water. A particularly insidious indoor air pollutant is radon, which was discussed in Chapter 6.

Apart from preventing pollutants from entering a building in the first place, one way to control indoor air pollution is to install air-to-air heat exchangers that circulate fresh air without adversely upsetting the temperature of the indoor air. Air conditioners, smoke removers, vacuum cleaners, and everyday household water running down the drain (which washes away some pollutants) all help in reducing air pollutants.

Cigarette smokers inhale much higher concentrations of pollutants than would be found in even severely polluted air.

GLOBAL WARMING

There is growing concern that increases in concentrations of certain trace gases in the atmosphere, primarily carbon dioxide, are causing a general warming trend, referred to as **global warming**. Before we can understand how these gases may be causing a rise in temperature, we need to understand how the earth maintains its heat balance.

The Earth's Heat Balance

Solar energy reaches the earth as electromagnetic radiation, which, proceeding from longer to shorter wavelengths, includes radio waves, microwaves, infrared radiation, visible light, UV radiation, and X-rays (Chapter 1, Fig. 1-13). The amount of solar energy that reaches the outer limits of the earth's upper atmosphere is enormous; if all of it penetrated to the earth's surface and were retained, the earth would have burned up long ago. Fortunately for life on earth, as solar energy travels through the atmosphere, interactions with gases and particulates prevent about half of it from penetrating to the earth's surface. The 50 percent of solar energy that eventually reaches the earth is largely absorbed at the surface and then reradiated back to space. If this radiation away from the surface did not occur, the earth would become increasingly warm as solar energy continued to flow in.

Incoming radiation includes the entire visible region of the spectrum together with smaller portions of the adjacent UV and infrared (IR) regions. Outgoing radiation from the earth is in the longer-wavelength IR region.

A considerable amount of the outgoing IR radiation is reabsorbed by gases in the atmosphere called **greenhouse gases** and then reradiated back to earth

A small amount of incoming radiation is scattered when it encounters atmospheric gases and particulates. Scattering of the blue part of visible light by nitrogen and oxygen molecules accounts for the blue color of the sky.

Figure 13-16   The green-house effect. Light from the sun penetrates the atmosphere and warms the earth's surface. Longer-wavelength infrared radiation is radiated away from the earth's surface; some of the outgoing radiation is absorbed by carbon dioxide and other greenhouse gases and then reradiated back to earth. When concentrations of greenhouse gases increase, more infrared radiation is returned toward the earth and the surface temperature rises.

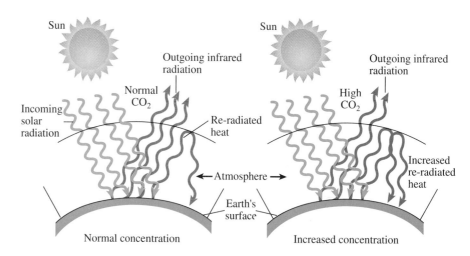

before it can escape into space. As a result of the absorption and reradiation, the atmosphere is warmed (Fig. 13-16). The most important greenhouse gas is carbon dioxide. Others include water vapor, methane, nitrous oxide (N_2O), CFCs, and ozone. Greenhouse gases act somewhat like a pane of glass in a greenhouse: Visible light passes through the glass and is absorbed by the objects inside the greenhouse. The objects are warmed and emit heat energy in the form of IR radiation, which cannot pass through glass. It is trapped and the temperature rises inside the greenhouse. The warming effect caused by the absorption and reradiation of IR radiation by greenhouse gases is commonly called the **greenhouse effect**.

Although carbon dioxide is present in the atmosphere at a much lower concentration than water vapor, it absorbs infrared radiation more strongly than water does. Also, water vapor does not remain for long in the atmosphere but soon falls to earth as precipitation.

The Increase in Atmospheric Carbon Dioxide

Analysis of air trapped in ancient ice samples shows that the average concentration of carbon dioxide in the atmosphere has been increasing for thousands of years. Twenty thousand years ago, the concentration was approximately 200 ppm; before the start of the Industrial Revolution at the end of the nineteenth century, it had risen to about 280 ppm. In 1958, when measurements were first made at Mauna Loa in Hawaii, the concentration was 315 ppm; by 1992, it had reached 360 ppm (Fig. 13-17). Thus, there has been a 26 percent increase in

TABLE 13-2   Concentrations of Greenhouse Gases in the Atmosphere Today and 100 Years Ago

Gas	Average Concentration (ppb)	
	100 Years Ago	1992
Carbon dioxide (CO_2)	290,000	360,000
Methane (CH_4)	900	1,700
Nitrous oxide (N_2O)	285	310
Chlorofluorocarbons (CFCs)	0	3

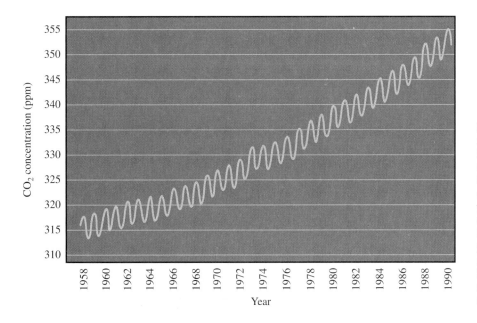

Figure 13-17 Since the first measurements were made at Mauna Loa, Hawaii, in 1958, carbon dioxide concentration in the atmosphere has increased dramatically. The yearly seasonal variations are caused by the removal of carbon dioxide from the atmosphere by growing plants in the summer and the return of carbon dioxide to the atmosphere during winter, when plants decay.

just 100 years. The major cause of the increase in atmospheric carbon dioxide is the burning of fossil fuels by electric utilities, automobiles, and industry.

A secondary cause of the increase is deforestation. As trees grow, they absorb carbon dioxide; when they decay or are burned in forest fires, carbon dioxide is released back into the atmosphere. Any new growth in a cleared area, whether crops or grasses, will absorb much less carbon dioxide than a mature forest. At the present time, the Brazilian rain forest (the largest remaining rain forest on earth) is being cleared at the rate of an area the size of Pennsylvania each year.

Tree-planting is encouraged as a way to combat CO_2 buildup in the atmosphere. It has been estimated that a tree farm the size of Australia would be needed to absorb the excess CO_2 produced each year.

The atmospheric concentrations of the other greenhouse gases have also risen in recent years (Table 13-2). Although their combined total concentration is considerably less than that of carbon dioxide, molecule for molecule, they absorb and reemit more IR radiation than carbon dioxide. If their levels in the atmosphere continue to rise at the present rate, it is estimated that their warming effect will soon equal that of carbon dioxide. The steady rise in methane is attributed primarily to a worldwide increase in the number of rice paddies and the number of cattle, both of which are sources of methane. Nitrous oxide is formed naturally in the soil during microbiological processes associated with nitrogen fixation (Chapter 3). Its increase in the atmosphere is believed to be caused primarily by increased use of artificial fertilizers.

Global Warming and World Climate

Is the increase in atmospheric carbon dioxide causing a global warming trend? Some scientists believe that excess carbon dioxide can be absorbed by the oceans. Others believe that warming is inevitable.

Recent studies of ancient ice cores have provided some convincing evidence that warming is likely to occur. Scientists took mile-long ice cores, dating back 160,000 years, from Antarctic glaciers and analyzed the air trapped in pockets in the ice for carbon dioxide. From other measurements, they were able to estimate the air temperature at the time the air became locked in the ice. As Figure 13-18 shows, there is a remarkable correlation between carbon dioxide concentration and air temperature. The air temperature rises and falls in step with the increases and decreases in the carbon dioxide concentration.

Computer-generated climate models also predict a warming trend. These models are based on numerous mathematical equations representing variables that affect climate throughout the global atmosphere. The information is fed into special supercomputers, which then project temperature changes that are likely to occur if carbon dioxide increases. All models consistently predict that if carbon dioxide concentration doubles by the end of the next century, global temperature will rise by 2–5°C (3–9°F). However, because many factors are not well understood, such as cloud formation and deep ocean currents, and are not included in the models, some scientists are skeptical about the predictions.

The effects of global warming on world climate could be devastating. Warming is likely to be more pronounced at the poles than in equatorial regions. As a result, air circulation and rainfall patterns would change, causing generally violent weather. Some regions of the world, including much of the United States, would experience drought conditions; other regions would become wetter. Melting of polar ice caps and glaciers would raise sea levels in coastal regions around the world, and low-lying countries would be inundated. A rise of just one foot would increase salinity in estuaries and aquifers. Ecosystems worldwide would be disrupted and some species might face extinction.

Scientists in many nations view global warming as the major potential environmental problem of the twenty-first century. A global warming treaty was signed at the United Nations' Conference on the Environment (the Earth Summit) in Rio de Janeiro, Brazil, in June 1992. The treaty set goals for reducing emissions of CO_2 and other greenhouse gases by industrialized nations to 1990 levels by the year 2000. However, because the proposed reductions are not binding there is no assurance that they will be met.

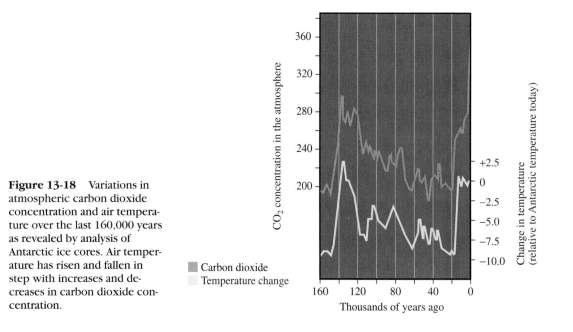

Figure 13-18 Variations in atmospheric carbon dioxide concentration and air temperature over the last 160,000 years as revealed by analysis of Antarctic ice cores. Air temperature has risen and fallen in step with increases and decreases in carbon dioxide concentration.

REGULATING AIR POLLUTION

The Clean Air Act of 1970 mandated air-quality standards for five air pollutants: suspended particles, sulfur dioxide, carbon monoxide, nitrogen dioxide, and ozone. A few years later, lead was added to the list. Until a phaseout began in 1975, lead was added to gasoline to prevent knocking (Chapter 14), and large quantities of lead were released to the atmosphere in engine exhausts, causing serious problems in urban areas. Although many cities have often failed to meet the required standards for air quality, marked improvements in quality have been achieved during the last 20 years (Fig. 13-19). The drop in lead emissions has been particularly dramatic.

A significant problem not addressed by the act, however, was the emission by industry of thousands of tons of unregulated hazardous chemicals, many suspected of being carcinogens. Now, the Clean Air Act of 1990 requires that industries emitting any of 189 specifically named toxic chemicals install control devices capable of reducing emissions by at least 90 percent. All industries must be in compliance by 2003.

The first person to enact legislation to prevent air pollution was probably King Edward I (1239–1307) of England who banned the burning of a high-sulfur coal called "seale" coal, because it produced noxious fumes.

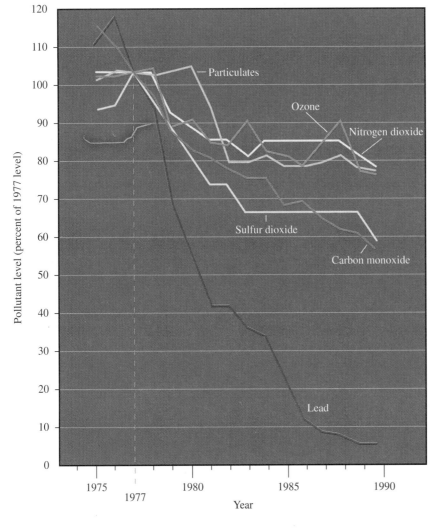

Figure 13-19 Reductions in six air pollutants, using 1977 values as 100 percent. The most dramatic reductions have been made in lead emissions.

As we noted earlier, the 1990 Clean Air Act includes other important provisions. It requires all cities to meet tough emissions standards for ozone by 2010. Coal-burning power plants must reduce annual sulfur dioxide emissions to approximately 50 percent of 1990 levels by 2000, and new cars will have to meet increasingly strict emission standards.

Air pollution is a worldwide problem that is far more serious in many other countries than it is in the United States. For example, in Mexico City, Sao Paulo in Brazil, and many industrial cities in Eastern Europe and the former republics of the Soviet Union, people live under a pall of toxic smog. The sun is obscured, severe respiratory ailments are common, and infant mortality rates are high; often there are few complaints because the smog and fumes mean jobs.

Cleaning up the atmosphere will be costly and will bring economic hardship, particularly to the developing nations of the world. But the alternative of continuing to pour pollutants into the air will have far more serious consequences. As we have seen, pollutants are already destroying the protective ozone shield, and if, as many fear, our activities are warming the earth, the resulting alterations in the world's climate could change our entire way of life.

EXPLORATIONS   **Mars: Setting Up House on the Red Planet**

All the planets in our solar system were born from common cosmic material some 4.6 billion years ago, but only the earth is known to support life. Mercury and Venus, the two planets closer than Earth to the sun, are too hot to sustain life, and Mars, the next planet farther from the sun than Earth, is too cold. But would it not be possible to warm Mars and make it habitable? Then, if at some future date the earth becomes too crowded, or its environment deteriorates beyond repair, there would be another planet to escape to. We have already learned, at a cost, that introducing greenhouse gases into our atmosphere can raise temperature and alter climate on a global scale. Why not use the same approach on Mars?

This notion sounds like the stuff of science fiction, but space scientists believe it could be done. They point out that data from the 1976 *Viking* space program indicate that many of the ingredients necessary for life, including carbon dioxide and water (potential sources of oxygen) and nitrogen, are already present on Mars, frozen into its surface. Mars's gravity (about two-fifths as strong as that of Earth) is adequate for life, and, conveniently, a day on Mars is almost the same length as a day on Earth, although the martian year extends to 687 days.

At present, Mars is very inhospitable. It is dry and barren, its surface frequently swept by wind-blown clouds of the iron-rich dust that give it its characteristic red color (Fig. 13-A). Located 1.5 times as far from the sun as the earth, Mars receives 57 percent less sunlight. Temperature at its equator may reach 20°C (68°F), but average surface temperature is a frigid −60°C (−76°F), very much colder than our own planet's average of 15°C (59°F). The sparse martian atmosphere is 150 times thinner than Earth's and consists primarily of CO_2 (95 percent), with some N_2 (2–3 percent), and smaller amount of other gases.

Early in its history, Mars was very different. Approximately 3.5 billion years ago, when the planet was still very hot, volcanoes were erupting on its sur-

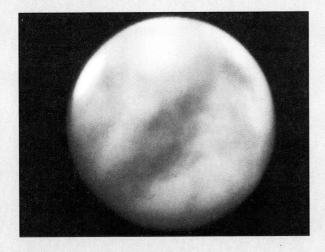

Figure 13-A Mars, the red planet.

face and emitting carbon dioxide, water vapor, and nitrogen into the martian atmosphere. As Mars cooled, liquid water is believed to have flowed over its surface, carving out the riverlike channels that can be seen on the red planet today. In time, most of the atmosphere disappeared. Some was lost to space, but evidence suggests that huge quantities of CO_2 were absorbed into the ground and froze in subsurface reservoirs as the planet continued to cool. Some CO_2 may have reacted with minerals to form carbonates when water was present on Mars, and we know that the extensive polar ice caps visible year-round, today, are composed of CO_2 ice in the south and water ice in the north. Large amounts of water are also probably tied up in permafrost.

The key to making Mars habitable is to warm it up and thus unfreeze these stores of frozen CO_2 and water. Without liquid water on its surface, Mars could not support life. Scientists suggest that the simplest way to warm the planet is to release greenhouse gases into its atmosphere. These gases, by absorbing heat-producing infrared radiation from the sun and reradiating it back to the planet's surface, would raise the temperature. Chlorofluorocarbons (CFCs), the compounds that are of concern on Earth because of their destructive effect on our ozone layer, would be suitable warming agents. Since the mass of CFCs required to cause a significant rise in temperature would far exceed the amount that could be lifted by a space shuttle, the gases would have to be manufactured on Mars. Adequate supplies of chlorine, fluorine, carbon, and other necessary raw materials are believed to be present in the martian soil. All that is needed is the development of technology to manufacture CFCs from these materials in the harsh martian environment.

It is estimated that, allowing for some breakdown of CFCs by the sun's ultraviolet radiation, several million tons of these compounds would be needed to raise the temperature of the martian surface above $-30°C$ (approximately $-20°F$). At that point, CO_2 frozen into the ice cap and absorbed into the martian soil should begin to be released. Since CO_2 is also a greenhouse gas, it would augment the warming effect of the CFCs. More and more CO_2 would be released, and temperature would rise in a continuing cycle until it eventually climbed above the freezing point of water. Ice in the polar cap, in permafrost, and in underground reservoirs would then begin to melt. Water would flow on the martian surface, and water vapor (another greenhouse gas) would enter the atmosphere.

By that stage, Mars should have an atmosphere and a surface environment capable of sustaining simple terrestrial plants. An adequate supply of nitrogen, an essential element for all organisms, is thought to be present in the martian soil in the form of nitrites and nitrates. Other necessary elements, including sulfur, phosphorus, and various trace metals, were detected in the martian rocks and soil by the *Viking* landers. Scientists estimate that, if the frozen stores of CO_2 are close to the surface, where they could be warmed up and degassed relatively rapidly, it would require 100 to 200 years to reach a time where plants could survive. If, however, the stores are buried deeply, where it would take much longer for heat to penetrate, it could require as long as 100,000 years.

A factor that would have to be considered is that once ponds and lakes began to form, the newly discharged CO_2 would dissolve in the water and gradually become tied up in carbonate deposits. This process does not cause a problem on Earth because carbonate deposits are naturally recycled. Plate movements bury the deposits deep underground where they become heated, and CO_2, released from the carbonates, reenters the atmosphere in volcanic eruptions. Mars, however, is a single-plate planet and therefore not subject to this type of tectonic activity. Some means, therefore, might have to be devised to release CO_2 from carbonate rocks to prevent depletion of the atmosphere.

If green plants could be introduced on Mars, they would immediately start to photosynthesize, converting CO_2 and water into carbohydrates and oxygen. If the O_2 content of the atmosphere could be raised sufficiently, animals and humans might be able to take up residence on Mars. Unfortunately, plants are very inefficient producers of oxygen, and, unless genetic engineering could improve the process, it could take up to 100,000 years to reach the required oxygen level.

Assuming it was possible to raise O_2 to a level suitable for humans to breathe, N_2 would have to be added to the atmosphere to act as a buffer gas to prevent oxygen toxicity. The small amount of N_2 known to be present in the martian atmosphere would need to be supplemented by introducing microorganisms capable of releasing nitrogen from compounds in the soil.

Once O_2 entered the martian atmosphere, some of it would be converted into a protective ozone layer by the action of ultraviolet radiation from the sun. To prevent destruction of the ozone by chlorine compounds released from the CFCs, alternatives to these greenhouse gases (probably related compounds containing no chlorine) would have to be manufactured at this point.

At present, of course, this plan for colonizing Mars is no more than scientific speculation. Before any action could even be considered, space scientists would have to determine if, in fact, Mars contains hidden reservoirs of CO_2, H_2O, and N_2. We would also have to decide if we have any right to attempt to alter the natural world on such a vast scale. Certainly, we should not do so until the possibility of life already existing on Mars had been completely ruled out. We must conclude that for the foreseeable future we shall have to accommodate our burgeoning population on our own planet.

Question
Does the idea of colonizing Mars have merit? Discuss the ethical considerations of such a proposal.

References: J. Kluger, "Mars: In Earth's Image," *Discovery* (September 1992), p. 70; C. P. McKay, O. B. Toon, and J. F. Kasting, "Making Mars Habitable," Review article, *Nature* (August 1991), p. 489.

KEY WORDS AND CONCEPTS

absorption
adsorption
aerosols
bag filtration
carbon monoxide
catalytic converter
chlorofluorocarbons (CFCs)
cyclone separation
electrostatic precipitation
flue-gas desulfurization (FGD)
fluidized bed combustion (FBC)
fly ash

free radicals
global warming
greenhouse effect
greenhouse gases
hydroxyl radical
industrial smog
mesosphere
nitrogen oxides
ozone hole
ozone layer
particulates
photochemical smog

photodissociation
primary air pollutants
secondary air pollutants
stratosphere
sulfur dioxide
temperature inversion
terpenes
thermosphere
troposphere
volatile organic
    compounds (VOCs)

QUESTIONS AND PROBLEMS

1. What is the concentration of each of the following gases in clean, dry air? Express the concentration in both percent and parts per million (ppm).
   a. O_2 **b.** CO_2 **c.** N_2 **d.** Ar

2. Draw a diagram showing the four main layers into which the atmosphere can be divided. Label your diagram to show:
   a. the distance the two layers closest to the earth extend out from the earth's surface **b.** the changes in temperature with increasing distance from the surface of the earth **c.** the location of the ozone layer

3. Explain why the stratosphere, which is more than 20 mi thick, has a smaller total mass than the troposphere, which is less than 10 mi thick.

4. Give equations to show how ozone is formed naturally from O_2 in the ozone layer.

5. The percent of water vapor in the atmosphere is variable. It generally lies in the range of:
   a. 0.05–1 percent **b.** 1–3 percent **c.** 5–10 percent **d.** 10–20 percent

6. State briefly why the ozone layer is so important to all living things on earth.

7. Draw a diagram to show how atmospheric pressure varies with altitude.

8. Define air pollutant. What are primary air pollutants? What are secondary air pollutants?

9. List the five types of substances that are responsible for more than 90 percent of air pollution in the United

States. In metric tons per year, which substance accounts for most air pollution in the United States? Does this substance constitute the most serious air pollution health problem? Explain your answer briefly.

10. What are the major air pollutants produced by the following industries?
    a. automobile **b.** electric power generation

11. Give equations to show how carbon monoxide and carbon dioxide are formed during the burning of fossil fuels. Which gas (CO or CO_2) is the main product when:
    a. gasoline is burned in an automobile engine **b.** coal is burned in an open fireplace

12. Draw a diagram of an automobile catalytic converter and describe how it reduces pollutants in automobile exhausts. Catalytic converters reduce automobile emissions by 95 percent. Why are there calls for greater efficiency?

13. Are the following statements true or false?
    a. Anthropogenic sources release more carbon monoxide into the atmosphere than do natural sources. **b.** The electric power industry is a major source of VOCs. **c.** CFCs are very reactive, unstable compounds.

14. What are the health effects of carbon monoxide inhalation?

15. Write the formula for the major nitrogen oxide air pollutant. How is it produced, and how is it ultimately removed from the atmosphere?

16. **a.** What are the main anthropogenic sources of sulfur dioxide? **b.** Why does the burning of coal release sulfur dioxide? **c.** Write equations to show how sulfur dioxide reacts in the atmosphere.

17. What adverse effects does SO_2 have on human health?

18. Describe two methods for removing sulfur dioxide from smokestacks.

19. In your opinion, is the installation of very tall smokestacks a good way to deal with hazardous industrial emissions? Explain.

20. What does VOC stand for? What are the main anthropogenic and natural sources of VOCs?

21. Using equations, explain how photochemical smog is formed.

22. What is a photochemical oxidant? Give equations to show how they are formed. Which photochemical oxidant is produced in greatest quantity?

23. What adverse effects does ozone have on plants and human health?

24. Describe the difference between an aerosol and a particulate.

25. Define the following:
    a. mist **b.** fog **c.** soot **d.** fly ash **e.** smog

26. Give four examples of particulate matter that arises from:
    a. natural sources **b.** anthropogenic sources

27. Explain how the surface areas of particulates and aerosols are a factor in causing them to be health hazards.

28. How is industrial smog formed?

29. In what way does the size of a particle influence its effect on health?

30. Describe two different methods for removing particulate matter from industrial emissions.

31. Use a diagram to show how a temperature inversion occurs. What geographic features promote the formation of an inversion?

32. What does CFC stand for? Write the chemical formula of a representative CFC. List three uses for CFCs.

33. Describe the evidence that implicates CFCs in the destruction of the ozone layer.

34. Describe the reactions that chemists now believe account for the destruction of the ozone layer by CFCs. Include the chemical reactions.

35. Where is the ozone "hole" located? Why is its formation cause for alarm? In your opinion, what steps should be taken to prevent destruction of the ozone layer?

36. What is the fate of outgoing infrared radiation from the earth's surface? How does the fate of infrared radiation affect the temperature of the earth's atmosphere?

37. List the greenhouse gases. Which of these gases occurs in the atmosphere in greatest concentration? How do the effects of other greenhouse gases compare with those of CO_2?

38. With the aid of a diagram, explain how an increase in the level of CO_2 in the atmosphere can produce a warming trend. What are the main causes of the increase in CO_2 in the atmosphere?

39. Make a list of indoor air pollutants. Give the source of each one.

40. Why is tobacco smoke such a dangerous air pollutant? In your opinion, should cigarette smoking be permitted in public buildings?

41. In your opinion, is some level of air pollution an inevitable consequence of living in an industrialized society?

42. What are the major types and sources of air pollution in less developed countries?

FOSSIL FUELS
Our Major Source of Energy

An oil field in California at sunset. Our society is almost entirely dependent on fossil fuels—petroleum, natural gas, and coal—for its energy needs.

Energy exists in many forms, including solar energy, heat energy, mechanical energy, electrical energy, and chemical energy. These different forms of energy are continually being converted from one form to another. For example, when animals run they convert chemical energy stored in carbohydrates in their bodies into mechanical energy to contract muscles. In a power plant, combustion of coal produces heat energy that is used to change water into steam. The steam turns a turbine, and the mechanical energy of the turning turbine produces electricity (electrical energy).

In accordance with the second law of thermodynamics, in these and all other energy conversions, some energy is wasted. For example, when we drive a car, only about 10 percent of the chemical energy in gasoline is converted to mechanical energy of motion. As much as 90 percent is converted to heat, which is lost to the surroundings.

Because energy is wasted, all life forms and all societies require a constant input of energy. The more complex the society is, the greater is its need for energy. Today, the United States and other industrialized nations obtain most of their energy from fossil fuels: coal, petroleum, and natural gas. Fossil fuels are the remains of plants, animals, and microorganisms that lived millions of years ago and, over time, were buried and converted to the three fossil fuels. Thus, we are almost entirely dependent for our energy needs on photosynthesis that occurred billions of years ago.

Figure 14-1 Throughout history, the discovery of new sources of energy has paralleled the development of new technologies and changed the pattern of human life.

Fossil fuels, like all living and once-living materials, are composed of carbon compounds. The majority of these compounds are hydrocarbons, compounds that contain only carbon and hydrogen. As we saw in Chapter 9, these two elements can combine in many proportions and ways to form thousands of different hydrocarbons. The differences in properties of the three fossil fuels are due primarily to differences in hydrocarbon composition. Petroleum, natural gas, and coal are all excellent fuels, because, when they are burned, some of the chemical energy stored in their hydrocarbon bonds is released as high-intensity heat energy, which can perform useful work.

In this chapter, we will review the development of energy sources and learn why fossil fuels are such a good source of energy. We will study the formation, extraction, composition, and uses of each of the three fossil fuels, and we will examine the environmental impact of our reliance on fossil fuels.

Learning Goals:

In this chapter, you should gain an understanding of:

1. Our major sources of energy and how they are used.
2. The difference between power and energy.
3. The production of energy from fossil fuels.
4. The formation, composition, and uses of the three fossil fuels: petroleum, coal, and natural gas.
5. Environmental and health problems associated with the use of fossil fuels.
6. The production of electricity.

ENERGY USE: A HISTORICAL OVERVIEW

Throughout human history, as societies have become more advanced, their energy use has increased (Fig. 14-1). In primitive cultures, people relied on their own physical labor to supply their energy requirements. They obtained the food they needed by hunting animals and gathering wild plants. With the discovery of fire, humans began to use energy in a form other than food. Heat obtained from the burning of wood was used to heat dwellings and cook food.

As civilization advanced, the horse and ox were domesticated and put into service to supply energy needed for farming, transportation, and other activities. Water power and wind power were harnessed to do other useful work. As early as the first century A.D., Egyptians were using waterwheels to grind grain. By the Middle Ages, windmills were common in Europe, furnishing the energy needed to grind grain and pump water.

As late as the first quarter of the nineteenth century, the main sources of energy, apart from human and animal muscle, were the burning of wood, and windmills and waterwheels that could be used to power lumber mills, textile-manufacturing mills, and other small industries. A great surge in the use of energy came with the development of the steam engine in the late 1700s. The steam engine, for the first time, enabled heat energy from fuel to be converted directly into useful work. This development was the beginning of the Industrial Revolution, the period when machines began to replace human and animal labor. In the United States, an essentially agrarian economy changed to one based on manufacturing. Because steam engines could be located almost any-

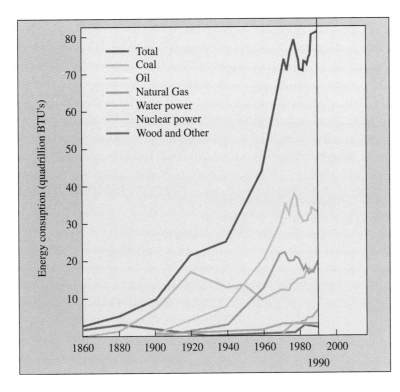

Figure 14-2 The contributions of different energy sources to total energy consumption in the United States from 1860 to 1990. Before 1880, when coal consumption began to increase sharply, wood was the major source of energy. Petroleum consumption increased dramatically as the twentieth century advanced and the number of automobiles on the roads multiplied. At the same time, the use of natural gas for home heating increased. The brief decline in fuel consumption in the 1970s was the result of an oil shortage engineered by the major oil-producing nations (Chapter 15). Nuclear power and water power continue to be minor sources of energy.

where and provided unprecedented amounts of energy, factory sites were no longer dictated by the availability of water, and large-scale industries became possible. Steam engines could be mounted on wheels, and the development of the steam locomotive marked the beginning of mass transportation. Enormous quantities of fuel were required to supply energy to operate the new machinery; in the 200 years between 1725 and 1925, per capita energy consumption in industrialized nations increased tenfold.

The changing pattern of energy use in the United States since 1860 is shown in Figure 14-2. Steam engines—including those in railroad locomotives and riverboats—were initially fueled by wood; in the 1830s, however, wood became scarce, and coal was gradually substituted. By 1910, coal was the source of 70 percent of the energy used in the United States. Petroleum did not become a significant energy source until the end of the nineteenth century. It was first discovered in the United States near Titusville, Pennsylvania, in 1859 (Fig. 14-3) and in the Middle East, in Iran, in 1908. For the first 60 years after its discovery in the United States, petroleum was used principally to produce kerosene for lighting and heating.

The development of petroleum as a major source of energy paralleled the development of the internal combustion engine and the growth of the automobile industry (Fig. 14-4). As a fuel for transportation, it has a tremendous advantage over coal: Gasoline-powered engines are many times smaller and lighter than coal-powered steam engines.

Natural gas, which is often found with petroleum, was not used as a fuel until an economical means for transporting it from the well to the consumer was established. In the early 1900s as much as 90 percent of the natural gas found with petroleum was burned—or "flared"—at the well site as a waste product. By 1945, a network of pipelines had been constructed throughout the United States that

In a steam engine, water is boiled to produce high-pressure steam that thrusts a piston back and forth in a cylinder. The piston is attached to a crankshaft that transmits the power to the drive wheels of the machinery.

From the sixteenth century until the discovery of petroleum in the nineteenth century, whale oil obtained principally from whale blubber, was the main source of fuel for lamps.

Figure 14-3 The first commercial oil well in the United States near Titusville, Pennsylvania, began producing oil in 1859.

Figure 14-4   Motor vehicles clog our highways and consume more than 60 percent of the oil burned in the United States.

formed direct links between the gas fields, located mainly in the Southwest, and the Midwestern and Northeastern states, where gas for heating was in greatest demand.

Current Use of Energy

In the industrialized nations of the world, energy is used for four main purposes: (1) transportation, (2) industrial processes, (3) heating and cooling buildings, and (4) generating electricity. The major sources of energy in the United States and the uses to which they are put are summarized in Figure 14-5.

Transportation, which includes road vehicles, trains, boats, and airplanes, consumes approximately 27 percent of all the energy we use. Energy for transportation comes almost exclusively from gasoline and diesel oil, both of which are derived from crude oil (the word *oil* is frequently used as a synonym for petroleum). Huge amounts of energy are required to manufacture the goods we depend on for our high standard of living; **industries** producing these goods account for about one-quarter of our energy consumption. **Heating** of homes and other buildings accounts for another 13 percent of our energy use. The energy for industry comes almost equally from petroleum, natural gas, and electricity; that for heating and cooling is supplied mainly by natural gas and electricity.

Over one-third of the energy consumed in the United States is used to produce electricity. **Electricity**, unlike fossil fuels, nuclear power, or water power,

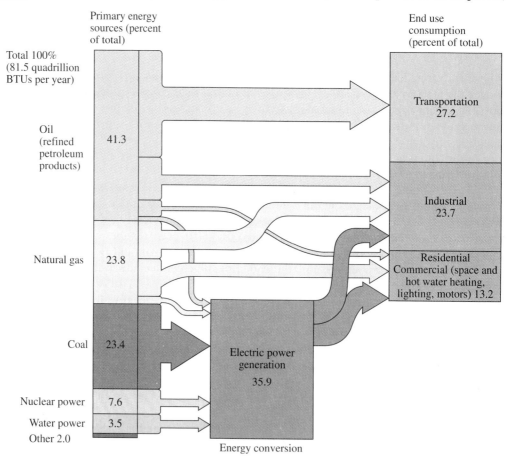

Figure 14-5   How each of the major sources of energy in the United States is used. Transportation depends almost entirely on petroleum products for energy; coal is used almost exclusively for generating electricity; natural gas is used about equally for commercial and residential heating and for industrial purposes.

is not a primary source of energy. It is called a secondary source because it is produced from a primary source.

As Figure 14-5 shows, fossil fuels—petroleum, natural gas, and coal—are used to meet almost all our energy requirements. Nuclear energy, water power, solar energy, and other less conventional sources of energy fulfill a minute fraction of our needs.

ENERGY AND POWER

It is important to understand the distinction between energy and power. **Energy** is the ability to do work, and it is measured in units such as calories, joules, and British thermal units. **Power** is a measure of the *rate* at which energy is used. It is expressed in terms of units of energy used per unit of time, such as calories per second or joules per second (watts). Units in which energy and power are measured are explained in Table 14-1.

ENERGY FROM FUELS

Fuels are defined as substances that burn readily in air and release significant amounts of energy. The process of burning (or combustion) is an oxidation process, and it is exothermic (Chapter 7). Most materials on the earth—minerals in rocks and soil, and water, for example—are already in an oxidized state. They do not burn and cannot be used as a source of heat energy.

What makes petroleum, natural gas, and coal such excellent fuels? The answer lies in their chemical composition. Fossil fuels are composed primarily of hydrocarbons (Chapter 9); when they are burned in an adequate supply of air, they are oxidized, and carbon dioxide and water are formed. Chemical bonds are broken, new bonds are formed, and, as a result, energy is released as heat.

Primitive people consumed about 2,000 kcal per day—the minimum amount of energy needed for survival. Today, the average person in the United States consumes about 230,000 kcal per day.

The value of one horsepower (the power needed to raise 33,000 pounds one foot in one minute) was established in the late eighteenth century, based on results obtained with strong dray horses. It is about 50 percent greater than the rate that can be sustained by the average horse for a working day.

TABLE 14-1   Glossary of Energy Terms

Calorie (cal):	The amount of heat needed to raise the temperature of 1 g of water 1°C
Kilocalorie (kcal):	One thousand calories (1000×1 cal)
Joule (J):	The joule is the standard unit of energy: 4.18 J = 1 cal
British thermal unit (BTU):	The amount of heat needed to raise the temperature of 1 lb of water 1°F
Quad:	One quadrillion (1000 trillion) British thermal units (BTU): one quad = 10^{15} BTU
Horsepower:	One horsepower = 33,000 ft lb of work/min
Watt (W):	An amount of power available from an electric current of 1 ampere (A) at the potential difference of one volt (V): 1W = 1 V $\times$ 1 A; 1W = 1 J/s
Kilowatt (kW):	One thousand watts (1000×1 W)
Megawatt (MW):	One million watts or 1000 kW
Kilowatt-hour (kWh):	Unit of electrical energy equivalent to the energy delivered by the flow of 1 kW of electrical power for 1 hour
Barrel:	Approximately 42 gallons

Equations for the combustion of methane (CH_4, the major component of natural gas) and octane (C_8H_{18}, a component of gasoline) are shown below:

$$CH_4 + 2\,O_2 \longrightarrow CO_2 + 2\,H_2O + \text{heat}$$

$$2\,C_8H_{18} + 25\,O_2 \longrightarrow 16\,CO_2 + 18\,H_2O + \text{heat}$$

When these hydrocarbons are burned in air, C–H bonds and bonds between oxygen atoms are broken. New bonds form between C and O and between H and O, as CO_2 and H_2O are produced. Heat is produced in these reactions because the energy released in making the new bonds in CO_2 and H_2O is greater than the energy required to break the bonds in the hydrocarbons and in O_2. (If the energy released in making the new bonds were *less* than the energy required to break the old bonds, heat would be consumed, and the reaction would be endothermic.)

The amount of energy required to break 1 mole of a particular kind of bond is the **bond energy**. Average values for bond energies are given in Table 14-2. The same amount of energy is *released* if the bond is re-formed from individual atoms. Bond energy is measured in kilocalories (kcal) (see Table 14-1).

We can use the values given in Table 14-2 to calculate the heat energy that will be released when 1 mole of methane (CH_4) is oxidized to yield carbon dioxide and water, as shown in the above equation.

Energy is obtained from a chemical reaction when the energy produced by making bonds to form products is greater than the energy required to break bonds in the reactants.

Bond	Bond Energy (kcal)	Number of Bonds Broken or Formed	Bond Energy/ Mole (kcal)
C–H	99	4	4 × 99 = 396
O–O	118	2	2 × 118 = <u>236</u>
			Energy required to break bonds = 632
C=O	192	2	2 × 192 = 384
O–H	111	4	4 × 111 = <u>444</u>
			Energy released in forming bonds = 828

Thus more energy is released in making new bonds than is consumed in breaking the hydrocarbon and O_2 bonds.

The above calculation explains why fossil fuels are valuable as fuels. When they are burned, numerous C–H bonds are broken, and as CO_2 and H_2O are formed, a relatively large quantity of energy is released. The energy output (fuel value) of various common fuels is given in Table 14-3.

TABLE 14-2
Approximate Bond Energies for Selected Bonds

Type of Bond	Bond Energy (kcal/mole)
C–H	99
C–C	83
C=C	147
C–O	85
C=O	173
C=O in CO_2	192
C–Cl	78
H–H	104
H–Cl	103
O–O	35
O–O in O_2	118
O–H	111
Cl–Cl	58

TABLE 14-3 Fuel Values of Some Common Fuels

Fuel	Fuel Value (kcal/g)
Wood (pine)	4.3
Bituminous coal	7.4
Anthracite	7.6
Crude oil	10.8
Gasoline	11.5
Natural gas	11.7
Hydrogen	33.9

EXAMPLE 14-1   How much energy is released when one mole of propane, C_3H_8, is burned?

Solution:

1. Write the balanced equation for the reaction:

$$C_3H_8 + 5\,O_2 \rightarrow 3\,CO_2 + 4\,H_2O$$

2. Determine the type and number of bonds broken:

Draw the structural formula of C_3H_8:

$$
\begin{array}{c}
\text{H}\ \ \text{H}\ \ \text{H} \\
|\ \ \ |\ \ \ | \\
\text{H}-\text{C}-\text{C}-\text{C}-\text{H} \\
|\ \ \ |\ \ \ | \\
\text{H}\ \ \text{H}\ \ \text{H}
\end{array}
$$

8 C—H and 2 C—C bonds are broken; 5 O—O bonds are broken.

3. Use Table 14-2 to obtain the bond energy (E) of each type of bond. Calculate the energy required to break the bonds.

Bond	Bond E (kcal)	Number Broken	Bond E/mole (kcal)
C—H	99	8	8 × 99 = 792
C—C	83	2	2 × 83 = 166
O—O	118	5	5 × 118 = 590
		Energy required to break bonds	= 1548

4. Similarly, calculate the energy released when the new bonds are formed.

Bond	Bond E (kcal)	Number Formed	Bond E/mole (kcal)
C=O	192	6	6 × 192 = 1152
O—H	111	8	8 × 111 = 888
		Energy released in forming bonds	= 2040

Difference between the energy consumed and the energy released (2040 – 1548) = 492. 492 kcal of energy are released when one mole of propane is burned.

PRACTICE EXERCISE: How much energy is released when one mole of ethane, C_2H_6, is burned?

Answer: 344 kcal/mole.

PETROLEUM

How Oil Fields Are Formed

Petroleum probably originated from microscopic marine organisms that once lived in great numbers in shallow coastal waters. When the organisms died, their remains collected in bottom sediments, where the supply of oxygen was insufficient to oxidize all the organic material. Over millions of years, the organic matter was buried by sediments, and, in some locations, the seas dried up, leaving behind a new landmass. The buried material was subjected to high temperature and pressure; chemical reactions converted some of the organic matter to liquid hydrocarbons (crude oil or petroleum) and gaseous hydrocarbons (natural gas).

Based on their understanding of the earth's history, geologists are able to recognize the major geological features that indicate the presence of large accumulations of oil (Fig. 14-6). They can then assess the probable size of the oil-bearing formation, its location, and other factors, to determine the economic feasibility of extracting the petroleum.

Petroleum and natural gas are generally found together. They accumulate in porous, permeable rock called **reservoir rock**. Rock is described as porous if, like a sponge, it contains open spaces in its structure, where gases and liquids can collect. It is permeable if the spaces are interconnected so that fluids—gases, water, and oil—can migrate through the pores. The porosity of reservoir rock, which is usually sandstone or porous limestone, ranges from 10 to 30 percent; usually, one-half of the pore space is occupied by water.

For petroleum to accumulate in quantity, there must be a layer of impermeable rock (known as **cap rock**) above the reservoir rock to prevent upward escape of the fluids in the oil-bearing rock. There must also be a folding or faulting of the rock strata to create a dome, or **anticline trap**, under which the petroleum and natural gas become trapped. The trap, which may extend over many miles, prevents lateral movement of fluids in the reservoir rock and allows oil and gas to collect. In time, the fluids in the reservoir rock separate out according to density. Petroleum and natural gas, being less dense than water, rise above the water and collect directly under the dome, with the gas above the oil (Fig. 14-6).

Oil cannot be withdrawn from an oil field easily because it does not accumulate in liquid pools. Instead, it is dispersed within the pores of the reservoir rock and must be forced out. The oil and gas are under pressure. Drilling into the reservoir rock releases the pressure, and the crude oil is forced up the bore hole. If the pressure is sufficiently high, the oil may be released as a gusher. As the oil flows out, the pressure gradually decreases, and the rate of flow declines. The average oil field ceases to be productive after only about one-third of the oil in the formation has been recovered. That is, two barrels of oil are left in the ground for every barrel extracted. A further one-third can be obtained by

A drill bit can cut through about 30 m (100 ft) of soft shale per hour and lasts about 2 days before it dulls. If the rock is hard, the bit may dull after cutting through just a few feet.

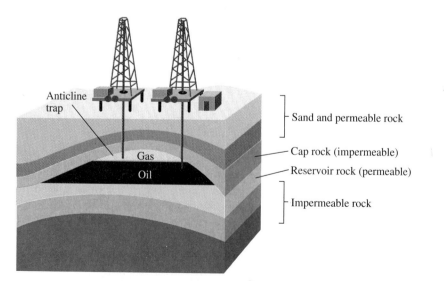

Anticline trap

Gas

Oil

Sand and permeable rock

Cap rock (impermeable)

Reservoir rock (permeable)

Impermeable rock

Figure 14-6 Geologic formations associated with petroleum and natural gas deposits. As a result of folding and faulting of rock layers, oil and gas became trapped under a dome of impermeable cap rock.

repressurizing the oil field. Auxiliary wells are drilled, and water, steam, or gases such as carbon dioxide or nitrogen are injected into the reservoir rock to force more oil out. Repressurizing is costly; whether it is worthwhile depends primarily on the prevailing price of oil.

The Composition of Petroleum

Crude petroleum is a complex mixture of thousands of organic compounds. The majority of these compounds are hydrocarbons, including alkanes, cycloalkanes, alkenes, and aromatic hydrocarbons. Small amounts of sulfur-, nitrogen-, and oxygen-containing compounds are also present. The proportions of different hydrocarbons in a sample of crude petroleum vary with location. For example, the percentage of straight-chain and branched-chain hydrocarbons in crude oil from Pennsylvania is much higher than the percentage in oil from California, which has a high content of aromatic hydrocarbons.

Refining Petroleum

Crude petroleum in the form in which it is recovered has few immediate uses. To be useful, it must be refined (separated into fractions) by **fractional distillation**, a process that separates the hydrocarbons according to their boiling points (Fig. 14-7).

Figure 14-8 shows the essential parts of a fractional distillation tower. The tower may be as tall as 30 m (100 ft). Crude oil at the bottom of the tower is

Figure 14-7 Distillation towers at an oil refinery.

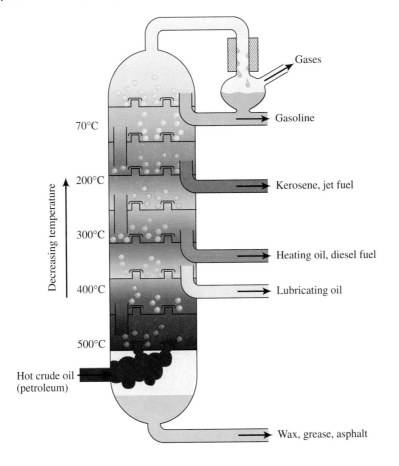

Figure 14-8 Diagram showing the separation of petroleum into different fractions in a distillation tower. Crude petroleum is heated at the bottom of the tower and most of the hydrocarbon constituents vaporize. Temperature in the tower decreases with increasing height. High-boiling components condense in the lower part of the tower; lower-boiling components (including the gasoline fraction) condense near the top. Fractions are removed at various heights in the tower.

heated to about 400°C (750°F), and the majority of the hydrocarbons vaporize. The mixture of hot vapors rises in the fractionating tower; components in the mixture condense at various levels in the tower and are collected. The temperature in the tower decreases with increasing height, and the more volatile fractions with lower boiling points condense near the top of the tower; high-boiling, less volatile fractions condense near the bottom. Any uncondensed gases are drawn off at the top of the tower. Residual hydrocarbons that do not vaporize collect at the bottom of the tower and are transferred to a vacuum distillation tower, where they are vaporized under reduced pressure to yield further fractions. In this way, the full range of hydrocarbons in crude oil becomes available for useful purposes. The properties and uses of typical petroleum fractions are listed in Table 14-4.

> Reducing the external pressure lowers the boiling point of a liquid.

The relative amount of each fraction produced by fractional distillation varies with the composition of the original crude petroleum; it rarely reflects consumer needs. Before the advent of the automobile, the fraction in greatest demand was **kerosene**, which was used for heating and lighting. Today, the greatest need is for the more volatile fraction known as **straight-run gasoline**, which provides the fuel for automobiles (Table 14-4).

To meet the increasing need for gasoline, refineries must convert higher-boiling fractions to gasoline by the process of **cracking**. Those fractions contain large-molecule hydrocarbons, for which there is little demand. When heated under pressure in the absence of air, the long-chain hydrocarbons break (or crack) into shorter-chain hydrocarbons, including both alkanes and alkenes, some of which are in the desired gasoline range (C_6 to C_{10}).

$$C_{14}H_{30} \xrightarrow[\text{of air}]{\overset{\text{heat and pressure}}{\text{absence}}} \underset{\text{alkane}}{C_7H_{16}} + \underset{\text{alkene}}{C_7H_{14}}$$

If a suitable catalyst is added, the reaction proceeds at lower pressure, and the yield of hydrocarbons in the gasoline range is increased.

TABLE 14-4 Typical Petroleum Fractions

Fraction	Boiling Point (°C)	Composition	Uses
Gas	Up to 20	Alkanes from CH_4 to C_4H_{10}	Synthesis of other carbon compounds; fuel
Petroleum ether	20–70	C_5H_{12}, C_6H_{14}	Solvent; gasoline additive for cold weather
Gasoline	70–180	Alkanes from C_6H_{14} to $C_{10}H_{22}$	Fuel for gasoline engines
Kerosene	180–230	$C_{11}H_{24}$, $C_{12}H_{26}$	Fuel for jet engines
Light gas oil	230–305	$C_{13}H_{28}$ to $C_{17}H_{36}$	Fuel for furnaces and diesel engines
Heavy gas oil and light lubricating distillate	305–405	$C_{18}H_{38}$ to $C_{25}H_{52}$	Fuel for generating stations; lubricating oil
Lubricants	405–515	Higher alkanes	Thick oils, greases, and waxy solids; lubricating grease; petroleum jelly
Solid residue			Pitch or asphalt for roofing and road material

To meet the increased demand for heating oil in the winter months, a refinery can switch to the reverse process: **polymerization**. In this catalytic process, small-molecule hydrocarbons are combined to produce larger molecules.

Octane Rating

Straight-run gasoline consists primarily of straight-chain hydrocarbons and is of limited value as an automobile fuel. A high content of straight-chain hydrocarbons causes a fuel to begin burning before it is ignited by the spark plug. This premature ignition produces a knocking sound and leads to loss of engine power and eventual damage to the engine. The tendency of gasoline to cause knocking is rated according to an arbitrary scale known as the **octane rating**.

The octane rating was established in 1927, at which time a large number of straight- and branched-chain hydrocarbons in the gasoline range were tested separately for performance in a standard engine. The branched hydrocarbon 2,2,4-trimethylpentane (isooctane) was found to be a superior fuel that burned without causing knocking; it was assigned an octane rating of 100. The straight-chain hydrocarbon n-heptane, in contrast, caused serious knocking and was given a rating of 0.

2,2,4-trimethylpentane is just one of several isomers of octane; therefore, it is not strictly correct to call it isooctane.

$$CH_3-\underset{\underset{CH_3}{|}}{CH}-CH_2-\underset{\underset{CH_3}{|}}{\overset{\overset{CH_3}{|}}{C}}-CH_3 \qquad CH_3-CH_2-CH_2-CH_2-CH_2-CH_2-CH_3$$

2,2,4-trimethylpentane (isooctane) n-heptane

The octane rating of a particular gasoline is determined by burning the gasoline in a standard engine and comparing its knocking properties with those of standard mixtures of isooctane and n-heptane. Thus, a gasoline that performs in the same way as a mixture containing 90 percent isooctane and 10 percent n-heptane is assigned an octane rating of 90. Table 14-5 lists the octane ratings for a number of the hydrocarbons that are present in gasoline.

TABLE 14-5   Octane Numbers of Selected Hydrocarbons

Name	Formula	Octane Number
n-butane	C_4H_{10}	94
n-pentane	C_5H_{12}	62
2-methylbutane		94
n-hexane	C_6H_{14}	25
2-methylpentane		73
2,2-dimethylbutane		92
n-heptane	C_7H_{16}	0
2-methylhexane		42
2,3-dimethylpentane		90
2-methylheptane	C_8H_{18}	22
2,3-dimethylhexane		71
2,2,4-trimethylpentane (isooctane)		100
Aromatic hydrocarbons		
benzene		106
toluene		118
o-xylene		107
p-xylene		116

Most gas stations offer customers gasoline with a choice of 3 octane ratings: 87, 89, and 92.

The straight-run gasoline fraction from the distillation tower has an octane rating of 50–55, much too low for today's automobile engines, which require gasoline with a rating between 87 and 90. The octane rating of gasoline can be increased in three main ways: (1) cracking, (2) catalytic re-forming, and (3) the addition of octane enhancers.

Cracking increases the octane rating by increasing the percentage of short-chain hydrocarbons, which, in general, have higher octane numbers than longer-chain hydrocarbons (Table 14-5). For example, as the length of the hydrocarbon chain decreases from heptane (C_7H_{16}) to butane (C_4H_{10}), the octane rating increases from 0 to 94.

Branched-chain hydrocarbons have higher octane numbers than the corresponding straight-chain isomers. It can be seen in Table 14-5 that as the degree of branching increases in the three isomeric hexanes (C_6H_{14})—n-hexane, 2-methylpentane, and 2,2-dimethylbutane—the octane numbers increase from 25 to 92. Table 14-5 also shows that aromatic hydrocarbons have higher octane numbers than nonaromatic hydrocarbons do. Benzene, toluene, and the xylenes, for example, have octane numbers above 100. (How octane numbers higher than 100 are obtained is explained below.)

It therefore follows that if the percentages of branched-chain hydrocarbons and aromatic hydrocarbons in gasoline can be increased, the octane rating will be improved. This improvement can be achieved by **catalytic re-forming**, a process in which hydrocarbon vapors in straight-run gasoline are heated in the presence of suitable catalysts such as platinum. By this means, n-hexane, for example, is converted to 2,2-dimethylbutane:

$$CH_3-CH_2-CH_2-CH_2-CH_2-CH_3 \longrightarrow \begin{array}{c} CH_3 \\ | \\ CH_3-C-CH_2-CH_3 \\ | \\ CH_3 \end{array}$$

n-hexane   2,2-dimethylbutane

and n-heptane is converted to toluene:

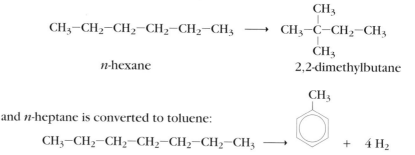

n-heptane   toluene

The octane rating of gasoline can also be increased by adding antiknock agents, or **octane enhancers**. Before 1975, the most widely used octane enhancer was tetraethyl lead (TEL), $(C_2H_5)_4Pb$, which was both cheap and effective.

$$\begin{array}{c} CH_3 \\ | \\ CH_2 \\ | \\ CH_3-CH_2-Pb-CH_2-CH_3 \\ | \\ CH_2 \\ | \\ CH_3 \end{array} \qquad \text{tetraethyl lead (TEL)}$$

Adding as little as 0.1 percent of TEL per gallon of gasoline can increase the octane rating by 10 to 15 points. However, lead is toxic (Chapters 12, 16), and recognition of the health hazards associated with its release into the atmosphere from automobile exhausts led to the mandatory phasing out of TEL in gasoline. The 1975 requirement that all new cars be fitted with a catalytic converter to control pollutants (Chapter 13) was a further factor in reducing the

TABLE 14-6 Octane Number of Gasoline Additives

Additive	Octane Number
Methanol	107
Ethanol	108
Methyl-t-butyl ether (MTBE)	116
Ethyl-t-butyl ether (ETBE)	118

use of TEL. Only unleaded gasoline can be used in automobiles fitted with catalytic converters because lead inactivates catalysts.

Octane enhancers that have been used to replace TEL include methyl-t-butyl ether (MTBE), ethyl-t-butyl ether (ETBE), methanol, and ethanol, all of which have high octane ratings (Table 14-6). Octane ratings of fuels with octane numbers above 100 are obtained by comparing the fuel with reference fuels containing isooctane (octane number, 100) plus known amounts of TEL. A conversion chart is used to obtain the octane number.

Methanol and ethanol, which are both fuels, have been added to gasoline in amounts up to 10 percent to produce gasoline blends known as **gasohol**. The use of alcohol as an alternative fuel will be considered in more detail in the next chapter.

CONSUMER BRIEF

Oxygenated Gasoline: Carbon Monoxide Reductions with Gasohol

In response to the Clean Air Act of 1992, the composition of gasoline is being changed as a means of improving air quality. Beginning in November, 1992, all gasoline sold during the winter months—November through February—in 39 designated areas of the United States, must contain 2.7 percent oxygen. The thirty-nine areas represent the most densely populated regions in the country and include most major metropolitan areas (New York, Washington, Chicago, etc.). These areas have been shown by the EPA to have carbon monoxide (CO) concentrations that exceed the federally mandated ceiling during the winter. In rural areas CO levels average under 10 ppm, but in areas of heavy traffic, CO levels can rise as high as 100 ppm. Poorly ventilated tunnels and parking garages are especially dangerous.

Because the automobile engine is the primary source of carbon monoxide emissions, regulation of tailpipe emissions should improve air quality. Carbon monoxide emissions can be lowered by using a leaner air-fuel mixture, that is, one in which there is much more air than fuel. If the ratio of air to fuel is greater than 16 to 1, CO emissions are greatly re-

duced. This air to fuel ratio, however, is impractical, because if the engine is tuned this "lean" it is difficult to start and may stall while idling. This is especially true in winter when the engine is cold and harder to start and the fuel does not vaporize as easily. Catalytic converters (Chapter 13) also cut down on CO emissions; air pumped into the exhaust system facilitates the oxidation of CO to CO_2. The addition of ethanol to gasoline is another method of oxygenation.

Brazil has been a leader in the use of ethanol as an automotive fuel, producing more than 4 billion liters per year. At one time, Brazil had more than 450,000 cars running on pure ethanol. The carburetor or fuel injection system of an automobile has to be modified to use pure alcohol. As petroleum prices fell in the 1980s, many of these cars were converted back to gasoline. Brazil's interest in ethanol production stems from the fact that the country has few petroleum reserves, but has a huge sugarcane industry producing biomass.

The octane rating of ethanol (108) is higher than that of isooctane (100). Thus, a mixture of gasoline and ethanol has a higher octane rating than regular gasoline. The addition of alcohol also reduces exhaust emissions of CO by as much as 50 percent. Gasoline blended with up to 20 percent ethanol was named gasohol when it was introduced in the 1970s. This mixture, unlike pure alcohol, can be used without any modification of the automobile engine.

Ethanol can be produced by fermentation of carbohydrates from waste agricultural material. The cost of production, however, is higher than that of gasoline, and farmers are lobbying for government support for the creation of an alcohol industry to provide fuel for automobiles. Because it is formed from plant material, alcohol may be considered a renewable resource rather than a depletable fossil fuel, and the fermentation of waste products is a method of recycling .

At present, most U.S. petroleum companies are increasing the oxygen content of their gasoline, not by adding alcohol, but by adding methyl-t-butyl ether (MTBE). The oxygen content of gasoline is increased to 2.7 percent if the concentration of MTBE is set at 15 percent. This use of MTBE necessitates the production of huge quantities of the chemical, and since 1990, MTBE has been on the list of the top 25 industrial chemicals produced in the U.S.

U.S. petroleum companies are choosing MTBE instead of ethanol for two reasons. First, MTBE has an octane rating of 116, which is higher than that of ethanol, and as a result, gasoline formulated with MTBE runs more smoothly and "knocks" less. Second, MTBE does not evaporate as quickly as does ethanol, so it does not vaporize out of the gas tank and add to evaporative emissions.

Although oxygenated gasoline reduces the emission of CO from automobiles, consumers have two complaints about the fuel: gasoline formulated with MTBE costs 5 to 10 cents more than regular gasoline and gas mileage is reduced by 10 percent.

Questions:
1. Do you think the federal government should do more to support an alcohol industry to provide fuel for automobiles?
2. Why do cars that burn oxygenated gasoline get poorer gas mileage?

Oil Shale and Tar Sands

Most oil shales yield approximately 25–50 gallons of oil per ton; a few yield as much as 150 gal/ton.

A largely untapped source of petroleum is **oil shale** (Fig. 14-9a), a common sedimentary rock that contains a solid organic material called kerogen. **Kerogen** consists primarily of heavy hydrocarbons together with small quantities of sulfur-, nitrogen-, and oxygen-containing compounds. Oil shale is found close to the earth's surface and can be mined like coal. When oil shale is crushed and heated in the absence of air at about 48°C (118°F), a thick brown liquid called **shale oil** is produced. After it has been treated to remove sulfur and nitrogen impurities, shale oil can be refined like petroleum. Fractional distillation yields mainly high-molecular-weight hydrocarbons, which can then be converted to more desirable hydrocarbons in the gasoline range by cracking.

The United States has immense deposits of oil shale, mainly in Utah and Wyoming. Although the technology is available to exploit this resource, there are environmental and economic problems that must first be overcome. For example, very large quantities of waste rock are produced when oil shale is mined, and because the wastes are alkaline and poor in nutrients, very little will

It has been estimated that 300 billion barrels of oil are recoverable from tar sands in Canada

grow on it. Furthermore, for every gallon of shale oil produced, two to four gallons of water are required. At the present time, with oil prices relatively low, development of oil shale for fuel is not economically attractive.

Tar sands are another potential source of petroleum. Tar sands (Fig. 14-9b) consist of sandstone mixed with a black, high-sulfur, tarlike oil known as **bitumen**. Large deposits exist in the United States, particularly in Utah, but the most extensive known deposits are found in Northern Alberta, Canada, where they are nearer the surface than in the United States and much easier to mine. Canadian oil producers hope to obtain 40 percent of their oil from tar sands by the year 2000. The extraction process involves injecting steam under pressure into the tar sand to liquefy the bitumen so that it can float to the surface and be collected. After it is purified, the oil can be refined in the usual manner.

Producing oil from tar sands is expensive and requires a great deal of energy. As with oil shale, economic and environmental considerations will determine the future of tar sands as a viable energy source. When conventional oil becomes scarce, interest in both oil shale and tar sands will undoubtedly increase.

The Petrochemical Industry

As we saw in Chapter 9, all organic compounds are either hydrocarbons or derivatives of hydrocarbons. Petroleum is the primary source of the hydrocarbons that are needed as starting materials for the synthesis of organic compounds used in the manufacture of plastics, synthetic fibers, synthetic rubber, pesticides, fertilizers, pharmaceuticals, and numerous other consumer products (Fig. 14-10). Approximately 10 percent of the petroleum refined today is used for this purpose. The organic compounds synthesized from petroleum hydrocarbons are called **petrochemicals**. In the future, as petroleum reserves decline, more organic compounds are expected to be produced using synthesis gas derived from coal (discussed later in this chapter) as the starting material.

(a)

(b)

Figure 14-9   Oil can be extracted from oil shale and tar sands. (a) Oil shale is a sedimentary rock rich in the hydrocarbon material kerogen. (b) Tar sands are a mixture of sandstone and a black, high-sulfur, tarry substance called bitumen.

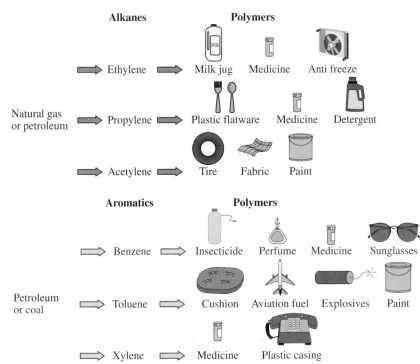

Figure 14-10   Some products made from chemicals derived from petroleum.

Shampoos and Conditioners: Clean and Shining

Hair is formed from long strands of protein. Each protein strand grows from a root in the scalp which is contained in a follicle. The hair shaft is lubricated with sebum, an oily secretion produced by the sebaceous gland which lies near the follicle. The sebum coating prevents hair from drying out, and gives it a gloss. If there is too much sebum, the hair feels greasy and dirty; if there is too little, the hair feels dry and is unmanageable. Washing hair removes sebum and any dirt adhering to it. Although we want to remove dirt and excess oil, we do not want to use a detergent that is strong enough to remove all the sebum.

Modern shampoos for adults contain a synthetic anionic detergent, such as sodium dodecyl (lauryl) sulfate, which forms large negatively charged organic ions in solution. Shampoos for oily hair have a higher concentration of detergent than those formulated for normal hair.

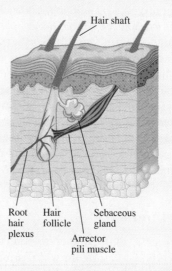

Root hair plexus   Hair follicle   Sebaceous gland

Arrector pili muscle

Hair shaft

$$CH_3-(CH_2)_{10}-CH_2-O-SO_3^- \ Na^+ \text{ Sodium lauryl sulfate}$$
Anionic detergent

"Tear-free" shampoos for children contain an amphoteric type of soap that is less irritating to the eyes. Amphoteric detergents have both a positive and negative charge in the same molecule, and thus can react with both acids and bases. They are not as strongly basic as are the anionic detergents.

$$CH_3-(CH_2)_{10}-\overset{\overset{\displaystyle H}{|}}{\underset{\underset{\displaystyle H}{|}}{N^+}}-CH_2-CH_2-O^-$$
Amphoteric detergent

Hair is composed primarily of the protein keratin, which contains both acidic and basic groups. Hair is slightly acidic. Most shampoos have a pH between 5 and 8. If the pH of a shampoo is too high or too low, the structure of the hair can be damaged. The pH also affects the ability of the shampoo to give clean hair with a good luster. The surface of hair, which is called the cuticle, is scaly. The glossiness of hair depends on the outer cuticle, where stacked transparent plates reflect light. The scales can be held tight to the shaft or ruffled out. If the shampoo is too basic, the cuticle scales fluff out and produce a surface that appears dull. Slightly acidic cuticle, on the other hand, is held tight to the surface and the hair acquires a reflective luster.

Many shampoo formulations contain oily substances that replace some of the oils lost in shampooing. Long-chain organic alcohols, such as oleyl alcohol, and alkyl polyethoxylates, such as polyethylene glycol, are used to lubricate the strands of hair. Many of these constituents are derived from petroleum. Long-chain hydrocarbons containing double bonds that are present in petroleum are treated with various reagents to put functional groups on the ends of the chains.

Prior to World War II most people used ordinary hand soap to wash their hair which often left a dull soap film on the hair after washing. To remove the film, the hair was rinsed with lemon juice or vinegar. Today we use conditioners and conditioning shampoos that contain quaternary ammonium compounds such as tricetylammonium chloride to remove remaining detergent residues. The "quats" in the conditioner form large organic cations that electrostatically attract the lauryl sulfate anions in the shampoo. The quaternary ammonium ion and the lauryl sulfate ion form a water-soluble ion pair which is removed from the surface of the hair as it is rinsed.

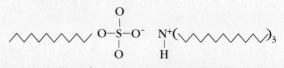

Lauryl sulfate ion
Anionic shampoo

Tricetyl ammonium ion
Quaternary ammonium conditioner

Oil Pollution

The environmental cost of producing, transporting, and using oil is considerable. The worst oil spill in U.S. history occurred in March of 1989, when the supertanker *Exxon Valdez* ran aground 28 miles south of the trans-Alaska pipeline terminal at Valdez and spilled 260,000 barrels (11 million gallons) of crude oil into Prince William Sound. More than 1000 miles of shoreline were contaminated with oil, and tens of thousands of oil-coated seabirds (Fig. 14-11) and at least 1000 sea otters died.

When crude oil is spilled into the ocean, it is dispersed and changed by many physical and chemical processes. Nearly all the components in crude oil are less dense than water and insoluble in it; they float on the surface as a layer that gradually thins as it spreads outward. The most volatile components in the layer (usually about 25 percent of the oil) evaporate into the atmosphere. Less volatile components are broken up by wave action into fine droplets that become dispersed in the water. Near shore, oil droplets tend to adsorb onto any suspended sand and silt particles and sink. At the ocean surface, the oil is whipped by wind and waves into an oil-in-water emulsion (or mousse). Oil that washes ashore and oil near the ocean surface is gradually decomposed by bacteria and photooxidation, processes that occur very slowly in cold arctic waters. As the emulsion breaks up, tar balls form and are washed ashore. The small fraction of crude oil that is heavier than water sinks to the ocean floor, where it can coat and destroy bottom-dwelling organisms.

The long-term environmental effects of oil spills are difficult to assess. Where strong waves pound exposed rocky shores, cleanup is rapid, but on sheltered sandy beaches and in salt marshes and tidal flats, oil may become buried and remain without decomposing for 2 or 3 years. Hydrocarbons are insoluble in water; when ingested by marine creatures, they dissolve in the animals' fatty tissues. Fish metabolize and excrete hydrocarbons rapidly, but shellfish may remain contaminated, smelling and tasting unpleasant for a long time.

Accidental oil spills from supertankers receive the most publicity, but nearly 70 percent of oil pollution in the ocean comes from the normal routine operation of these tankers. Tankers often leak, oil is frequently spilled during unloading and refueling, and most tankers intentionally flush oil wastes directly into the sea.

Offshore oil-drilling operations are a minor cause of oil pollution, except when an occasional well blowout occurs. The main environmental concern is the wastewater that is brought up mixed with oil from below the ocean floor (Fig. 14-12). This ancient fossil water contains barium, zinc, lead, and trace amounts of radioactive materials, and when it is returned to the ocean, it can destroy nearby natural habitats. Another concern is the clay mud that contains lignite, barium, and other chemicals. It is used as a drill lubricant and is usually dumped at sea after use. The clay can bury and destroy bottom-dwelling organisms, which form an essential part of the marine food chain. The most serious

Figure 14-11 When seabirds become coated with oil they are unable to fly. Their feathers no longer insulate them from the cold, and most die of exposure.

Figure 14-12 Offshore drilling for oil deposits beneath the ocean causes environmental problems. Wastewater containing toxic substances is brought up with the oil and then returned to the water. Marine organisms are buried when mud that contains rock chips and lubricants is dumped during drilling operations.

As much as 50 percent of the oil that contaminates the oceans is waste oil generated by industries, cities, gas stations, and individuals, that flows in from sewers, streams, and rivers.

effects of offshore drilling probably occur along the shore, where facilities built to receive the oil destroy wetlands. Because the environmental impact of offshore drilling is not well understood, many people hope that drilling will not be permitted off the coasts of California and southern Florida.

A significant cause of oil pollution is improper disposal of lubricating oil from machines and automobile crankcases. Many car owners who change their own motor oil pour the used oil into storm sewers and other areas that drain into creeks, rivers, and groundwater. Few realize that as little as one quart of oil can contaminate 2 million gallons of drinking water; one gallon can form an oil slick measuring nearly eight acres. Because of this danger, many communities have set up collection centers where used oil can be brought for recycling.

NATURAL GAS

The geological conditions necessary for the formation of natural gas are similar to those required for the formation of oil (Fig. 14-6). As we saw earlier, oil and gas are often found together, but natural gas is also found by itself. In the formation of natural gas, part of the original, buried organic material was converted to light gaseous hydrocarbons; in the formation of oil, heavier liquid hydrocarbons were formed.

Natural gas is composed chiefly of methane and small amounts of ethane, propane, and butane. Typically, it contains 60–80 percent methane, but the exact composition varies with the source. Crude natural gas also contains small quantities of higher alkanes, carbon dioxide, nitrogen, hydrogen sulfide (H_2S), and noble gases. The major natural gas-producing states are Texas, Louisiana, Oklahoma, and New Mexico.

Before crude natural gas can be used as a commercial fuel, it must be treated to remove carbon dioxide, sulfur compounds, water vapor, and most of the hydrocarbons with molecular weights greater than ethane. Carbon dioxide is undesirable because it reduces the heat output of the gas; sulfur compounds cause corrosion and release unpleasant odors. Water vapor can cause problems if it condenses in the pipeline.

Propane and butane are useful by-products. After they have been removed from crude natural gas, these by-products are converted to liquids (**liquefied petroleum gas**, or **LPG**) and marketed as heating fuels. Helium, which is often present in crude natural gas, is another important by-product. It is valuable in providing an inert atmosphere for welding and as a lighter-than-air gas for balloons and other devices.

Natural gas is nontoxic and safe, but many people mistakenly believe that it is dangerous. This misconception dates from the early part of the century when coal gas (discussed in the next section), a toxic mixture of carbon monoxide and hydrogen, was piped into homes.

Natural gas is relatively inexpensive. It is a superior fuel and has many advantages over coal and petroleum. It burns cleanly, leaves no residue, and, weight-for-weight, has a higher heat output when burned than any other common fuel (Table 14-3). It emits less carbon dioxide per unit of energy than other fossil fuels, and, generally, it produces no oxides of sulfur. It is the safest fuel available. The extensive system of pipelines that crisscrosses the United States, taking gas directly to the consumer, reduces the need for expensive storage facilities.

The household gas that has often been used to commit suicide and is mentioned in Agatha Christie's detective novels as a cause of death is coal gas, not natural gas.

Because natural gas is odorless, for commercial use a small quantity of an unpleasant-smelling organic compound (usually a mercaptan) that can be detected in parts per million amounts is added to it. If a leak occurs, the escaping gas can be detected before it can build up to form an explosive atmosphere.

COAL

The Formation of Coal Deposits

Three hundred million years ago, when the earth was much warmer than it is now, plants grew in great abundance in tropical freshwater swamps and bogs, which covered many regions of the earth (Fig. 14-13). When the luxuriant growth died, some of the plant material sank under water before it could be oxidized by atmospheric oxygen in the usual way. In the absence of air, little decomposition occurred. The material accumulated and became buried under sediments, falling leaves, and other vegetation. In time, it was compressed and converted to a porous brown organic material that we know as **peat**. In some locations, because of geologic changes, the peat became more deeply buried; increasing pressure compressed it and changed it to a harder material called **lignite**.

Over thousands of years, deeper burial and the resulting increase in temperature and pressure transformed the lignite into various grades of **bituminous coal** (also known as soft coal). In areas where mountains formed as a result of deformation and uplifting of rock formations (Chapter 2), further changes occurred. The very high pressure and temperature associated with these geologic activities converted bituminous coal to **anthracite** (or hard coal).

With each step in the transformation from peat to anthracite, chemical reactions occur. Volatile compounds containing carbon, oxygen, and hydrogen are released; the water content of the material decreases; and the carbon content increases (Table 14-7). At the same time, the material becomes harder and brighter. The quality of coal as a fuel source increases with its carbon content. Thus, anthracite, which is approximately 90 percent carbon, is the most desirable form of coal. Peat is not a true coal; it is considered a low-grade fuel because of its very high water content (90 percent) and low car-

Figure 14-13 Approximately 300 million years ago huge ferns and other plants grew in great abundance on the earth. Over time, dead plant material became buried under layers of sediments, and some of the material was ultimately converted to coal.

TABLE 14-7   Characteristics of Different Types of Coal

Type of Coal	Carbon (%)	Water (%)	Fuel Value
Peat	5	90	very low
Lignite	30	40	low
Subbituminous coal	40	9	medium
Bituminous coal	65	3	high
Anthracite	90	3	high

Note: Values may vary considerably with the source of coal.

bon content (5 percent). Dried peat, however, is used as a fuel in some parts of the world.

The locations of the principal coal deposits in the United States are shown in Figure 14-14. Most of the anthracite has already been mined and depleted.

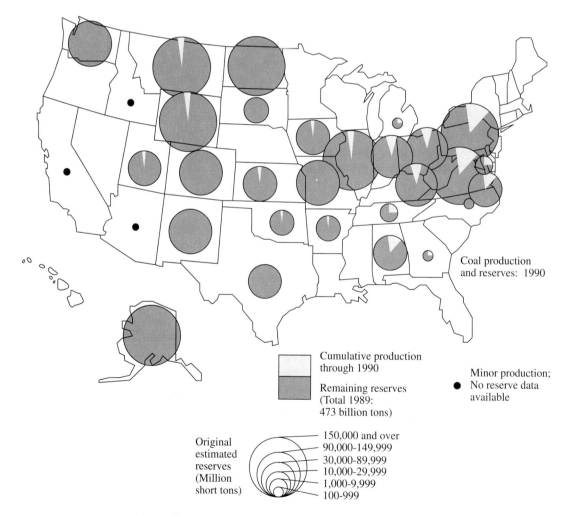

Figure 14-14   Major coal deposits in the United States. The nation has huge reserves of bituminous coal.

The Composition of Coal

Coal is a complex mixture of organic compounds. It is composed primarily of hydrocarbons and small amounts of oxygen-, nitrogen-, and sulfur-containing compounds. Compared with petroleum, coal contains a much higher percentage of aromatic hydrocarbons, including many polycyclic aromatic compounds. The composition varies with the type of coal and the location of the deposit.

(a)

(b)

Figure 14-15 (a) Strip mining for coal can devastate the land, leaving it bare and subject to erosion. (b) Land destroyed by mining can be reclaimed by regrading and planting with trees and grass.

Problems with Coal

Most coal is extracted from the earth by strip-mining, which devastates the land unless adequate steps are taken to reclaim it (Fig. 14-15). Both strip-mining and underground mining produce large quantities of waste rock, and acid leaching from the wastes can contaminate nearby streams and rivers (Chapter 8). Coal is dirty, bulky, and expensive to transport. When coal is burned, the large quantities of ash produced present a disposal problem. All coal contains some sulfur, and sulfur dioxide emitted during the combustion of coal is the primary cause of acid rain (Chapter 13). Also, coal burning produces carbon dioxide, and, as we saw earlier (Chapter 13), the constantly increasing concentration of carbon dioxide in the atmosphere has the potential to change drastically the earth's climate. The environmental effects of coal use are pictured in Figure 14-16.

Apart from the environmental problems associated with its use, solid coal is not a versatile fuel. It is not suitable for today's home and office heating systems or as a fuel for automobiles or airplanes. The main consumer of coal today is the electric power industry. Many of the disadvantages of coal as a fuel can be overcome if it is converted to a gas or liquid before it is burned. In a gaseous or liquid form, coal is easier to transport and it burns more cleanly.

The gaseous and liquid fuels obtained from coal are called syn-fuels.

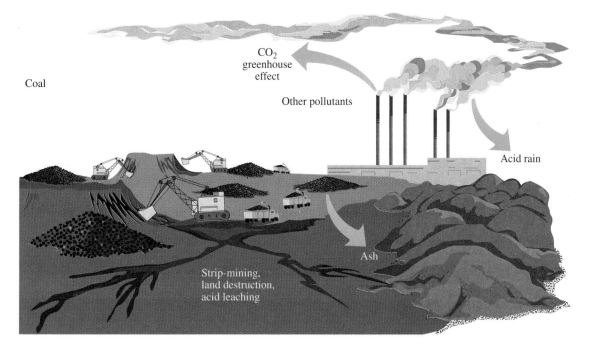

Figure 14-16 Environmental effects of the use of coal as an energy source. Combustion of sulfur-containing coal is the major cause of acid rain. When coal is burned, the greenhouse gas carbon dioxide is added to the atmosphere, increasing the possibility of global warming. Coal mining can result in acid drainage and destruction of land.

Coal Gasification

Coal gasification is not a new idea. As early as 1807, a municipal **coal gas** system was used to light streets in London. Coal was heated in the absence of air—by a process known as pyrolysis—and yielded a mixture of hydrocarbons that could be burned to produce light. As late as 1932, gas supplied to homes in the eastern United States was derived from coal, but, with the construction of the gas pipeline network in the 1940s, less expensive and safer natural gas took over the market. Coal is the United States' largest fossil fuel resource, and as concerns over our dependence on imported oil have grown, interest in both coal gasification and coal liquefaction has been renewed.

Coal gasification processes yield a mixture of carbon monoxide, hydrogen, and methane that is often called **synthesis gas** (or syn gas). This synthetic gaseous fuel can be produced by treating heated, crushed coal with superheated steam under carefully controlled conditions. Carbon monoxide and hydrogen are formed as follows:

$$C + H_2O \longrightarrow CO + H_2$$
$$\text{coal} \quad \text{steam} \qquad\quad \text{coal gas}$$

Carbon monoxide then reacts with hydrogen to produce methane and water, and with steam to produce carbon dioxide and more hydrogen.

$$CO + 3\,H_2 \longrightarrow CH_4 + H_2O$$
$$CO + H_2O \longrightarrow CO_2 + H_2$$

Conditions are adjusted to give the maximum possible yield of methane. Unreacted carbon monoxide and hydrogen are recycled through the system, and sulfur compounds and other impurities are removed. The process, which is relatively inexpensive, can also be used to produce methanol, another useful fuel.

Coal Liquefaction

Petroleum-like liquids can be obtained from coal if the complex molecules in coal are broken down to smaller molecules and if the hydrogen content of the molecules is increased. This result can be achieved by subjecting a mixture of coal and hydrogen to high pressure in the presence of a metal catalyst. A material similar to crude oil is produced, which can be fractionally distilled to yield products similar to those obtained from petroleum: kerosene, diesel fuel, gasoline, and lubricants.

At the present time, the cost of a barrel of this liquid fuel is approximately twice the cost of a barrel of petroleum. The future of coal liquefaction will depend on the world price of oil and our willingness to continue to rely on imported oil. It should be remembered that the conversion of coal to a liquid or a gas in no way reduces the amount of carbon dioxide released to the atmosphere during combustion.

ELECTRICITY

Electricity, our most important secondary source of energy, plays a critical role in modern society. When the first electrical generating plants began operating at the end of the nineteenth century, electricity was used only for lighting. It was not until the turn of the century, when the electrification of industry began, that electrical appliances were introduced to the marketplace. The first clothes washers and vacuum cleaners were produced in 1907. Today, more than one-third of all the energy consumed in the United States is used to generate electricity. Approximately two-thirds of this production powers electric motors, which are vital to the industrial, domestic, and military sectors of our economy. Elevators, air conditioners, communications systems, television sets, microwave ovens, toasters, and numerous other systems and appliances that depend on electricity have become commonplace, taken-for-granted necessities for our way of life.

Electricity is produced whenever a coil of wire is moved through a magnetic field—or whenever a magnetic field is moved within a coil of wire. The movement induces a flow of electrons (electricity) in the wire (Fig. 14-17). Most modern electric generators are operated by rotating a coil of wire within a circular arrangement of magnets. Energy, of course, is needed to rotate the coil. In the United States, this energy is obtained primarily from coal. Heat from the combustion of coal converts water in a boiler into high-pressure steam. The steam strikes the blades of a turbine, which is coupled to a generator. The turbine turns, causing the coil of wire to rotate in a magnetic field, and electricity is generated (Fig. 14-18a). After the steam has passed through the turbine, it is cooled, condensed back to water, and returned to the boiler to be reused.

Generation of electricity is not an efficient process. Inevitably, as a consequence of the second law of thermodynamics, energy is wasted. In the average coal-fired power plant, 60–70 percent of the energy used to produce steam is lost in the form of heat as the steam is cooled. Only 30–40 percent of the energy derived from coal is converted into electricity. The waste heat is removed by circulating large quantities of water, usually taken directly from a nearby river or lake, around the condenser. As we saw in Chapter 12, the increased temperature of the water as it is returned to its natural source can adversely affect fish and other marine organisms.

Steam turbines generate approximately 80 percent of the electricity produced in the United States. Coal is the major heat source used to convert water to steam, but oil and nuclear power are also used. Approximately 10 percent of our electricity is generated with hydroelectric power (Fig. 14-18b). In this process, the kinetic energy released when water held behind a dam or at the top of a waterfall drops to a lower level is used to turn the blades of a turbine. Another 10 percent of our electricity is generated by gas turbines (Fig. 14-18c). High-pressure gases produced in the combustion of natural gas directly turn the turbine blades.

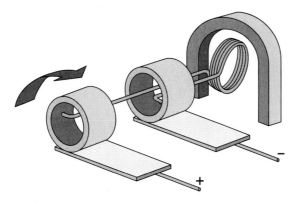

Figure 14-17 The principle of an electric generator. When a coil of wire is rotated in a magnetic field, an electric current flows in the wire.

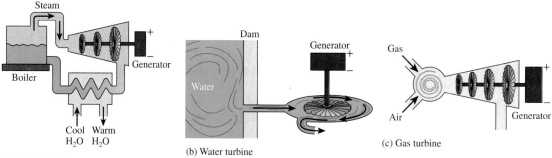

Steam

Generator

Boiler

Cool   Warm
H₂O   H₂O

(a) Steam turbine

Dam

Generator

Water

(b) Water turbine

Gas

Air

Generator

(c) Gas turbine

Figure 14-18   Turbines driven by (a) steam, (b) water, and (c) gas can be used to generate electricity.

An electric bill charges for the kilowatt hours of electrical energy consumed in a stated period of time.

Electricity is a very clean and convenient form of energy as received by the consumer, but its production creates many environmental problems. These problems are related to the primary energy source. In the United States, coal provides 55 percent of the energy used to generate electricity; its combustion by power plants produces the adverse environmental effects that have already been discussed. As our demand for electricity has escalated, we have become increasingly dependent on coal and nuclear energy to meet our needs.

Nuclear energy and other nonfossil-fuel sources of energy will be considered in the next chapter.

EXPLORATIONS

Environmental Terrorism in the Persian Gulf

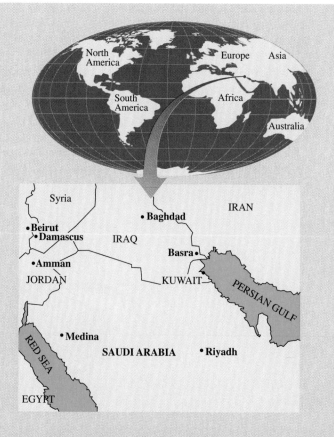

In August 1990, the Iraqi army invaded Kuwait. Six months later, when Saddam Hussein still refused to comply with United Nations demands that he withdraw his forces, troops from the United States and its allies drove the invaders out. In February 1991, during the final days of the war, the retreating Iraqis committed an unprecedented and unforgivable act of environmental terrorism. They fired more than 600 of Kuwait's oil wells, and they purposely released millions of barrels of oil into the Persian Gulf.

The dynamited wells exploded in flaming fountains of burning oil that roared 200 feet into the air (Fig. 14-A). Huge plumes of black smoke from the fires were blown by the wind, creating immense black clouds that blotted out the sun. A steady drizzle of soot and oil droplets fell from the plumes and coated every surface below with a black film. Mixed with sand, the oily soot formed a black crust over the desert. Soot particles, rising as high as 20,000 feet into the air, were carried by the prevailing winds, mostly southward over Saudi Arabia, although some reached India, 1500 miles to the east, before falling to the ground or being washed out in rain or snow. Each day that the fires raged they consumed about 5 million barrels of oil and released approximately half a billion tons of pollutants into the atmosphere.

At some of the dynamited wells, the oil did not catch fire but surged into the air in great dark brown jets. Millions of barrels of crude oil spilled onto the desert, creating huge oil lakes that became a threat to groundwater. Desert plants, insects, lizards, and small mammals drowned in the oil. Thousands of birds became mired in the thick crude when they landed on the oil lakes, mistaking the shimmering surfaces for water.

In January 1991, a few weeks before Saddam Hussein's troops fired the oil wells, they callously poured as much as 6 million barrels (approximately 250 million gallons) of oil into the Persian Gulf, creating the world's largest oil spill. The Iraqis deliberately ruptured pipelines and storage tanks and emptied oil from as many as seven loaded tankers. A 600-square-mile area of the gulf was covered with a layer of oil, and black oil washed up along 300 miles of the Saudi Arabian coast. The spill could have been much worse if employees at the Kuwait Oil Company had not secretly closed a vital valve. Their courageous action prevented 8.5 million barrels of oil in storage tanks from pouring through damaged pipelines to the dynamited offshore loading docks on Sea Island.

The Persian Gulf is a shallow sea, rarely more than 300 feet deep, that opens into the Arabian Sea through the narrow Strait of Hormuz. The environment in the gulf is unique. Although the water is one and a half times saltier than typical sea water, it is rich in nutrients that flow in from the Tigris and Euphrates rivers, and the gulf teems with life. Blue-green algae and sea grasses flourish in the broad tidal shallows that stretch along the Saudi Arabian coast. Minute organisms living in the algae, and plankton in the water, are the first links in the food chain that supports the gulf's wildlife. The seagrass meadows are habitat for juvenile shrimp, fry, and other small creatures. The waters are rich in shrimp and commercially valuable fish, including mackerel, mullet, snapper, and grouper, and have been fished for millennia. The marine life supports large flocks of migratory and resident birds. Ospreys, flamingos, terns, cormorants, wa-

Figure 14-A Oil wells burning in Kuwait during the Persian Gulf War.

terfowl, and shorebirds such as plover and sandpipers are abundant along the intertidal sandflats. Small islands in the gulf, many ringed with fragile coral, provide nesting sites for the endangered hawksbill turtle and for many seabirds. The southern end of the gulf, below Bahrain, is home to the dugong, a relative of the manatee.

As the main shipping route through which Middle East oil reaches the outside world, the Persian Gulf is constantly subject to oil spills. Annually, in the normal course of operations, about a quarter of a million barrels of oil—an amount equal to the 1989 *Exxon Valdez* spill in Prince William Sound in Alaska—spill into the gulf. The Iraqi's act of sabotage in January 1991 released more than 20 times that amount in a matter of days. Prevailing northwesterly winds, combined with the counterclockwise currents in the water, dispersed the oil southward along the Saudi coastline.

Saudi and Kuwaiti authorities were overwhelmed by the enormity of the spill and were completely unprepared to deal with it. Booms were deployed, and cleanup crews engaged to protect refineries, petrochemical facilities, and desalination plants, but little was done to protect natural habitats. Compared with the remedial actions taken after the much smaller *Exxon Valdez* spill, the response to the gulf tragedy was slow and inadequate.

A black ribbon of oil half a mile wide soon covered the tidal zone, and with each incoming tide, more oil seeped into small estuaries and creeks. As the oily tides retreated, they left behind thousands of dead, oil-soaked crabs, fish, and dead and dying birds. A bird rehabilitation center treated hundreds of oil-soaked birds, but it is estimated that

453

at least 20,000 birds died. Unique stands of black mangrove trees were other victims of the oil. Much of the oil in the shallows mixed with sand and sank, smothering numerous bottom-dwelling creatures. Farther out from the shore, the shimmering film of oil on the water gradually thinned. It is estimated that 40 to 50 percent of the slick evaporated during the first week.

The damage to the environment from the oil fires and the spills was enormous, but one year after the disaster, the damage appeared to be less severe than originally feared. Although crabs were still dying in the shallows where oil remained on the bottom, fresh green sea grass was growing vigorously, and small fish were swimming around it. Thousands of terns had bred successfully on Karan Island despite a hard layer of tar that contaminated their rocky nesting sites.

It had been predicted that the fired wells might burn for at least 5 years and have serious effects on climate and human health. In fact, all the damaged wells were capped within 9 months. Even when the fires were at their height, concentrations of sulfur dioxide and other toxic gases in the plumes, although above prewar levels, were within United States air quality limits. High levels of fine particulates in the atmosphere accounted for the increase in respiratory ailments that was observed. After the fires were extinguished, there was still some concern that unidentified carcinogens in the smoke might cause health problems in the future.

Despite cleanup efforts and encouraging signs of healing, it is likely to be decades before the Gulf fully recovers. One year after the catastrophe, oil still sullied the desert and blackened beaches. Oily deposits in the vulnerable tidal flats continued to threaten marine creatures at the bottom of the food chain, and fish and bird populations were expected to decline. For many years, research teams will continue to assess the damage and learn all they can from the tragedy.

Accidental oil spills and other insults to the environment are, unfortunately, likely to occur in the future but we must hope that no nation will ever again intentionally devastate the natural world.

Question
Discuss the short- and long-term health effects of exposure to smoke from burning oil.

References: T. Y. Canby, *National Geographic*, 180, No. 2, August 1991, p. 2. S. A. Early, *National Geographic,* 181, No. 2, February 1992, p. 122.

KEY WORDS AND CONCEPTS

anthracite
bitumen
bituminous coal
bond energy
catalytic reforming
coal gas
cracking
energy

fractional distillation
gasohol
kerogen
kerosene
liquefied petroleum
    gas (LPG)
lignite
octane enhancer

octane rating
oil shale
petrochemicals
polymerization
power
straight-run gasoline
synthesis gas
tar sands

QUESTIONS AND PROBLEMS

1. Describe the difference between energy and power. What units are used to measure each?
2. What is the origin of petroleum?
3. What are the geological features that are required for the formation of petroleum in amounts that can be extracted from the earth?
4. Why is the fuel value (kcal/g) for crude oil higher than the value for coal or wood (see Table 14-3)?
5. Write a chemical equation for the complete combustion of butane.
6. Methane, CH_4, when burned releases 192 kcal/mole. How much energy is released when 10.0 moles of methane are burned?

7. Are all of the following fuels?
    a. CH_4 b. H_2O c. C_4H_{10}
8. How much energy is released when one mole of butane, C_4H_{10}, is burned?
9. Describe what is meant by fractional distillation.
10. What is the difference between gasoline and kerosene?
11. What is straight-run gasoline?
12. Describe what is meant by "catalytic re-forming."
13. Draw all hydrocarbon isomers with the formula C_5H_{12}. Using Table 14-5, list them according to increasing octane number.
14. Which has a higher octane number, toluene (C_7H_8) or heptane (C_7H_{16})? Why?

15. For each of the following pairs of hydrocarbons, indicate which has the higher octane rating? And why?
    a. pentane, octane **b.** 2-methylpentane, 2,2-dimethylbutane **c.** octane or 2-methylheptane

16. Why was tetraethyllead (TEL) used as an octane enhancer? Why was it banned?

17. Describe the octane rating system.

18. Describe three ways for increasing the octane rating of a gasoline.

19. What is the purpose of an antiknock agent? Why does knocking occur in an automobile engine?

20. What is oil shale? What organic material does it contain?

21. Describe the problems associated with oil recovery from oil shale.

22. What are tar sands? What organic material do they contain?

23. What is the difference in chemical composition between commercial natural gas and crude natural gas?

24. What is the difference between bituminous and anthracite coal? Why is anthracite coal a better fuel?

25. What are the two main methods for mining coal, and which one is used to the greatest extent? What is the environmental impact of each of these methods?

26. Write a chemical equation for the complete combustion of coal.

27. Why is coal, which was a major source of residential heating in the 1930s, considered an unsuitable fuel for home heating in the 1990s?

28. Why is coal gasification considered to be a way to convert coal to a more suitable fuel for home use?

29. What is "synthesis gas"? How is it formed in coal gasification? Is it a high energy fuel?

30. Make two lists: one showing the environmental advantages of coal gasification and the other showing the environmental disadvantages.

31. A recently developed coal gasification technique produces methane from coal. List the reactants needed for this process, and write a chemical equation that describes the process.

32. The generation of electricity produces significant quantities of various pollutants. Make a list that shows **a.** the type of fuels used to generate electricity **b.** the percentage that each contributes to total U.S. electrical production and **c.** the pollutants each produces.

33. Describe the relationship between the demand for fossil fuels and economic growth in the following U.S. industries (indicate increasing or decreasing demand of which fossil fuel):
    a. Farming **b.** Food processing **c.** Manufacturing **d.** Transportation

34. Describe the principle of an electric generator.

35. Describe how energy produced by burning coal is used to turn a turbine and generate electricity.

ENERGY SOURCES FOR THE FUTURE

Throughout human history, improvements in living standards have been linked to the availability of energy. Since the end of the eighteenth century, material well-being in the industrialized nations has become almost entirely dependent on fossil fuels as the source of energy. This finite, nonrenewable resource is being rapidly depleted, and, by most estimates, economically recoverable reserves of all fossil fuels are likely to be used up within the next 200 to 300 years. In terms of human history, this period of dependence on fossil fuels for energy will be seen as a very brief episode (Fig. 15-1).

In the United States, our most immediate concern is for the availability of petroleum, which today supplies approximately 40 percent of our energy. If world consumption of petroleum continues at its present rate, there could be shortfalls between world production and world demand by the end of the century. Economically recoverable reserves could be gone by 2025. If we wish to avoid serious economic and social disruptions, we need to begin making a gradual transition to other sources of energy before oil becomes scarce. We also need to use energy more efficiently and less extravagantly. Although the United States has only 5 percent of the world's population, it consumes 30 percent of the world's supply of energy. If other countries used as much energy per capita as the United States does, fossil fuels would be depleted within the next decade.

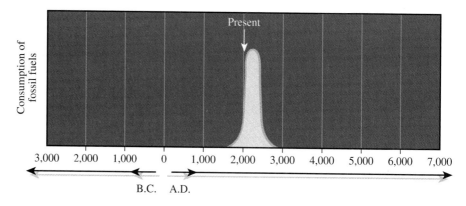

Figure 15-1 The use of fossil fuels during the past 5000 years and projected use for the next 5000 years. Fossil fuel reserves are expected to be used up during the next few hundred years. From a historical perspective, the period of dependence on fossil fuels will be very brief.

In this chapter, we will consider the extent of the world's known reserves of fossil fuels and our present dependence on this energy source. We will examine ways in which we can conserve energy, and we will study the advantages and disadvantages of alternative sources of energy.

Learning Goals:

In this chapter, you should gain an understanding of:

1. Energy conservation's crucial role in future energy policies.

2. Nuclear power generation and the nuclear fuel cycle.

3. Nuclear fusion.

4. Active and passive solar energy.

5. Biomass as a source of energy.

6. Alternative sources of energy: wind, tides, geothermal sources, and fuel cells.

FOSSIL FUEL RESOURCES AND THE ENERGY CRISIS

Eighty percent of the world's energy is supplied by nonrenewable fossil fuels. How long these rapidly dwindling energy resources will actually last depends on several factors: the total amount of oil, natural gas, and coal that exists in the earth; how much of this total is economically recoverable; the probability of discovering new deposits; and the rate at which the resources are consumed. One thing is certain: Sooner or later, the earth's store of fossil fuels will be used up.

Oil Reserves

Oil is very unevenly distributed around the world (Fig. 15-2). Over half the world's proven reserves of oil are found in the Middle East.

As we saw in Chapter 14, oil has replaced coal as the major source of energy in the United States (Fig. 14-2), and today, oil supplies approximately 40 percent of our energy. From 1950 to the mid-1970s, oil was cheap, and consumption steadily increased. During that time, production from the huge oil-fields, discovered in the 1930s, in eastern Texas and in Wilmington, California, and from later discoveries generally kept pace with demand. Since then, however, production has not kept up with demand. The last major discovery of oil in the United States was the giant field found in 1968 on the North Slope in Alaska. In 1982, a large but considerably smaller field was discovered off the coast of California, but because of environmental concerns and the high costs of offshore drilling, exploitation of this field has been limited. Worldwide, the rate of discovery is declining, and petroleum geologists believe that most of the oil that exists has already been found.

In the future, advances in technology and rising oil prices may make it feasible to obtain oil from wells that at present cannot be worked profitably. Also, as oil supplies dwindle, extraction of oil from oil shale and tar sands (Chapter 14) will probably become economically attractive. These measures, however, will serve only to postpone for a short time the inevitable exhaustion of available oil.

The North Slope in Alaska includes the Arctic National Wildlife Refuge. Energy producers want to drill for oil in the refuge's coastal plain; conservationists are opposed to the idea because an oilfield would disrupt the area's abundant and diverse wildlife.

The "Energy Crisis"

By the early 1970s, North America, Western Europe, and Japan were heavily dependent on oil imported from the Middle East and thus were vulnerable to economic and political upheavals there. This vulnerability was demonstrated in 1973 and 1974, after the Arab-Israeli War. At that time, the five members of the Organization of Petroleum Exporting Countries (OPEC) (Saudi Arabia, Iraq, Iran,

OPEC was founded in 1960. The 13 members in the organization today are Algeria, Ecuador, Gabon, Indonesia, Iran, Iraq, Kuwait, Libya, Nigeria, Qatar, Saudi Arabia, United Arab Emirates, and Venezuela.

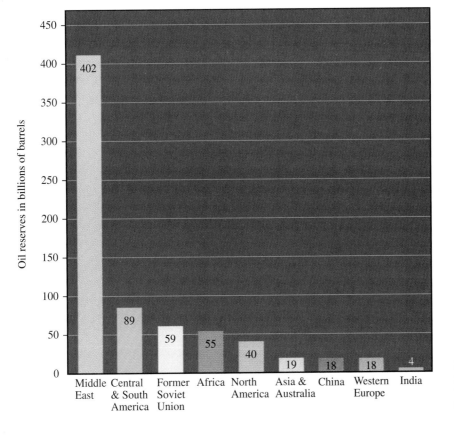

Figure 15-2 The world's oil reserves are very unevenly distributed. More than 50 percent are located in the Middle East.

Kuwait, and Venezuela), which were supplying more than 50 percent of the world's oil, placed an embargo on shipments of oil to all nations that had supported Israel in the war. OPEC quadrupled the price of oil, causing shortages and economic disruption in the industrialized and developing nations of the world. In the United States, gasoline was in short supply, and long lines of automobiles at gas stations became common. Oil prices, which were determined unilaterally by OPEC, continued to rise, reaching an all-time high in 1979.

In 1979, a second oil crisis occurred when, after the fall of the Shah, Iran cut back production and suspended oil exports to the United States. The two energy shocks of the 1970s, coupled with rising oil prices, were a spur to the development of more efficient ways to use energy. Smaller cars became popular, thermostats in homes and offices were lowered in winter and raised in summer, and in 1973 and again in 1979, energy consumption fell (Fig. 14-2). In the 1980s, to reduce its reliance on oil from the countries of the Middle East, the United States began importing more of its oil from other OPEC countries and from non-OPEC countries, primarily Mexico, which now supplies the largest share of our oil imports (Fig. 15-3). In the 1980s, the members of OPEC (which by then numbered 13) could no longer agree on oil production goals, and as a result, overproduction occurred, causing oil prices to fall dramatically. Concern about shortages of oil subsided; energy consumption began to rise again and has continued to do so until the present time. The drop in oil imports that occurred in the early 1980s has been reversed, and we are again becoming increasingly dependent on foreign oil.

The instability of the Middle East was again demonstrated in 1990 when Iraq invaded Kuwait. Although intervention by forces from the United States and other countries soon freed Kuwait, the firing of the oil wells by the retreating Iraqis jeopardized oil supplies (Chapter 14).

The energy crises in the 1970s and again in 1990 were caused by the disruption of oil supplies: An energy crisis in the future is likely to be caused by an irreversible shortage of oil.

The Department of Energy was established by President Jimmy Carter in 1977 to develop long-range energy policy and conduct energy research.

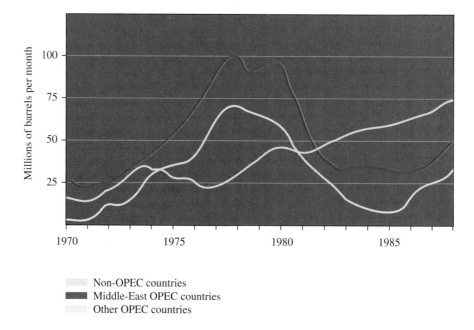

Figure 15-3   United States imports of crude oil from different sources. Beginning in the mid-1970s the United States increased imports of oil from non-OPEC countries and reduced imports from OPEC countries. Since the early 1980s imports from all sources have been increasing.

Non-OPEC countries
Middle-East OPEC countries
Other OPEC countries

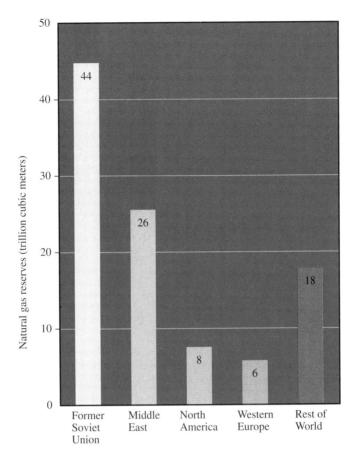

Figure 15-4   The world's reserves of natural gas are unevenly distributed. Most are located in the former Soviet Union.

Natural Gas Reserves

Natural gas, which supplies 17 percent of the world's energy, is, like oil, unevenly distributed around the world (Fig. 15-4). The largest reserves are found in the republics of the former Soviet Union, which produce about 50 percent more natural gas than the United States does. Before the breakup of the Soviet Union, scientists there were developing automobiles that burn compressed natural gas as a means of reducing dependence on oil. Worldwide, natural gas consumption has been increasingly rapidly.

The transportation and storage of natural gas remains a problem. In the United States, natural gas is distributed through pipelines, but in many parts of the world, because of great distances between the source of the gas and potential customers, pipelines are not a practical solution. Consequently, in the Middle East, Mexico, Venezuela, and Nigeria much of the natural gas found with crude oil is burned as a waste product. However, if cooled to −162°C (−259°F) natural gas liquifies, and the volume it occupies decreases by a factor of 600, making it suitable for transportation by tanker. **Liquefied natural gas (LNG)** is expensive to produce, and if leakage occurs, there is always the risk of an explosion. But as oil becomes scarce and demand increases, more LNG will undoubtedly be produced.

The major import market for LNG in the future will be in the Pacific regions where there are few gas pipeline facilities.

In the United States, natural gas production peaked in 1973. If consumption continues at its present rate, remaining reserves are expected to last for about 40

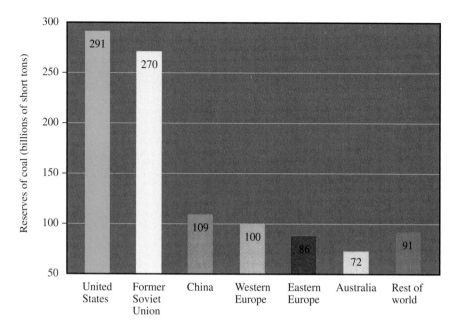

Figure 15-5 The distribution of the world's reserves of coal. The largest reserves are located in the United States.

Today, approximately 95 percent of the natural gas that we use is obtained from sources within the United States. The remaining 5 percent is imported from Canada.

to 60 more years. Supplies can be extended by importing gas by pipeline from Canada and, if we are prepared to face the hazards, by importing LNG from overseas. Synthesis gas can also be produced from coal and oil shale (Chapter 14).

Coal Reserves

Coal is the most abundant fossil fuel in the world. The largest proven recoverable reserves, estimated to be 29 percent of the world's total, are found in the United States (Fig. 15-5). Geologists believe that all the world's coal has been discovered. At the present rate of consumption, reserves are expected to last for approximately 300 years. Currently, coal supplies about 27 percent of the world's energy.

Could coal be a solution to the immediate energy problem we face as oil supplies dwindle? Coal is useful almost exclusively as a fuel for generating electricity, and most power plants are already using it for this purpose. Plants that once used oil changed to coal after the oil shortages in the 1970s. It is probable that coal will gradually replace petroleum as a source of organic chemicals needed by the petrochemical industry, and as oil becomes scarce and prices rise, coal liquefaction (Chapter 14) is likely to become increasingly important. But coal will not be able to fill the energy gap. Apart from practical problems, the environmental costs of coal combustion are very high. Although hazardous emissions can be controlled with new technologies (Chapter 13), the release of carbon dioxide to the atmosphere cannot easily be controlled. Increasing coal combustion would inevitably increase the threat of global warming (Chapter 13).

 ENERGY CONSERVATION

Energy conservation is an important option in planning an energy policy for the future. Measures such as turning off lights, using less hot water, and turning down thermostats are important, but for the long term, conserving energy by improving the efficiency of energy use is far more effective. Improving efficiency means finding a way to decrease the amount of energy that is needed to perform a particular task. Already, great strides have been made by industry. For example, in the 1970s, new cars averaged 13 miles per gallon (mpg); by the 1980s, they averaged 25 mpg. If the automobile industry makes use of available technology, new cars in the late 1990s could average 60 mpg or more. However, unless laws are enacted mandating new standards, very modest improvements in fuel economy are expected in the near future.

Improved insulation in homes and other buildings, greater use of mass transportation, less intensive use of energy for agriculture, the manufacture of more-durable goods, greater use of energy-efficient appliances and fluorescent light bulbs, and more recycling are all measures that can be taken to save energy. The initial costs are often high, but in the long run, savings are considerable.

Many industries have improved efficiency by using waste heat that previously was dissipated to supply energy for a second process. In some power plants, waste heat is being used to heat nearby buildings.

Energy conservation through increased efficiency is not an answer to the energy problem, but it is an essential step if we are to avoid severe shortages while alternative sources of energy are being developed to replace dwindling fossil fuel supplies.

Many utilities now encourage energy conservation. They offer their customers free advice on saving energy by improving insulation because it is more profitable to maintain current levels of energy production than to increase capacity.

CONSUMER BRIEF — The E-Lamp: A Bright Idea for the Future

In June 1992, a new light bulb, the E-lamp (or electronic lamp), was introduced at the annual meeting of the Edison Electric Institute in Columbus, Ohio. Hailed as the most significant advance in lighting technology in 60 years, the E-lamp fits into any ordinary light bulb socket, lasts for at least 14 years, and uses 75 percent less energy than a standard incandescent light bulb.

Unlike conventional incandescent and fluorescent lamps, the E-lamp has no filament to burn out. Instead, high-frequency radio waves are generated inside a bulb containing mercury and argon—the same gases present in fluorescent lamps. The radio waves, which are generated by an electronic oscillator and transmitted by a magnetic coil (or energy coupling antenna), interact with the mercury vapor in the bulb, converting it to a plasma, or ionized vapor. The excited atoms in the vapor emit invisible ultraviolet (UV) light, which strikes a thin phosphor coating on the inside surface of the bulb, causing the phosphor to emit visible light.

The plasma-gas technology on which the E-lamp is based is not new. It was developed in the 1970s by plasma physicist and inventor Donald Hollister. However, it was not until the plasma technology was combined with advanced microelectronics that construction of a practical light bulb became a possibility. The main problem that had to be overcome was preventing the radio waves from leaking out and interfering with TVs, radios, computers, and other electronic devices. According to Intersource Technologies of California, the developers of the E-lamp, the problem has been solved but the company is not divulging their solution. Their silence has led to considerable speculation since substances that block radio waves generally also block light waves.

The E-lamp has many advantages over both incandescent and compact fluorescent light bulbs. Incandescent light bulbs are very inefficient. Only 5 percent of the electricity that passes through the tungsten filament is converted to light; the rest is lost as heat. Typically, a standard 100-watt bulb if switched on for 4 hours a day, will burn out after 6 months to a year (750–1500 hours). An E-lamp with the same lighting intensity uses one-quarter the electric energy, produces much less heat, and

E-lamp

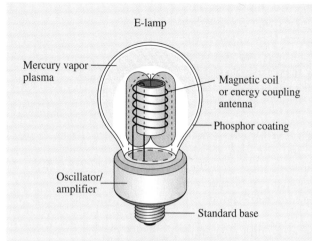

Mercury vapor plasma

Magnetic coil or energy coupling antenna

Phosphor coating

Oscillator/amplifier

Standard base

Standard Incandescent Light Bulb

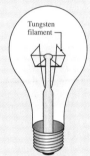

Tungsten filament

lasts for approximately 14 years (20,000 hours). An E-lamp does not go out suddenly, but gradually grows dimmer as the phosphor wears out. The light output at 20,000 hours is estimated to be 30 percent less than that of a new lamp. The compact fluorescent bulb for home use was developed as a means of saving energy. The compact is 75 percent more efficient than an incandescent bulb and lasts about 10 times as long, but because it costs approximately $15 and does not fit all lamps, sales have been limited.

The E-lamp is only slightly more energy-efficient than a compact fluorescent lamp but it lasts twice as long, and, because it is the same size and shape as a regular incandescent bulb, it fits all standard lamps and fixtures. Also, unlike a fluorescent lamp, it does not flicker when turned on, it can be used with a dimmer switch, and it operates at low temperatures. Fluorescent lighting is often described as "cold." E-lamps, by using various phosphor coatings can supply light with the same color quality as an incandescent lamp, or light in many different colors.

Comparison of Incandescent and Compact Fluorescent Light Bulbs with the E-lamp

	Price	Life Span*	Cost per Week†
100-W incandescent bulb	$ 0.75	6 months to 1 year	$0.30
25-W compact fluorescent bulb	$15	5–7 years	$0.11
25-W E-lamp	$15	14 years	$0.09

*If used 4 hours a day.

†Cost of bulb plus electricity averaged over life span of bulb.

Although the initial cost of an E-lamp is expected to be between $12 and $15, in the long term replacing the incandescent bulbs in the home with E-lamps would result in savings. Because of reduced operating costs and less frequent replacement, $50 to $100 would be saved over the life of each bulb. Savings would be even more significant if, as the sponsor's predict, the price of the lamp drops as production volume increases. Industry and businesses would also reap savings. By introducing regularly scheduled maintenance programs they could reduce the labor costs involved in constantly having to replace burnt-out bulbs.

If the E-lamp is to be successful in eliminating incandescent light bulbs, the public will have to be persuaded to regard light bulbs not as disposable items but as durable goods that are worth their high initial cost. Also, the bulb's sponsors will have to demonstrate convincingly that the main technological problem—the containment of the radio signal within the bulb—has been completely solved. If these obstacles are overcome, the E-lamp could revolutionize the lighting industry in the next few years.

Question

If an ordinary incandescent light bulb in your home burned out, would you replace it with another incandescent bulb or with an E-lamp, if one were available? Explain your choice.

Source: John Burgess, "Building a Better Light Bulb," *Washington Post*, Business Section C1, June 2, 1992; Intersource Technologies Inc., Corporate Relations Office, Glendale, CA.

NUCLEAR ENERGY

Few issues have generated as much controversy as the future of nuclear energy. When controlled nuclear fission reactions became a reality in the early 1940s (Chapter 6), nuclear energy appeared to hold great promise as a cheap, clean, and safe source of energy. In the 1950s, it was predicted that by the year 2000, nuclear power plants would supply up to 20 percent of the world's primary energy. Now, however, in response to fears about safety, and with the present relatively low cost of other sources of energy, the predicted figure is 5 percent. In 1993, the United States had 109 operating nuclear power plants that were supplying approximately 10 percent of our energy. Three other plants were under construction (Fig. 15-6). Other industrialized countries with significant numbers of plants are the former Soviet Union (51), France (44), the United Kingdom (37), and Japan (32). In France and Japan, which are not well-endowed with coal, nuclear energy supplies 79 percent and 67 percent of their energy needs.

Since the mid-1970s, nearly all orders for nuclear plants have been cancelled.

The Shoreham nuclear power plant on Long Island, New York, which cost $5.5 billion to build and generated electricity for just 30 hours before it was shut down, is now being demolished.

Nuclear Fission Reactors

Recall from Chapter 6 that when uranium-235 is bombarded with neutrons, it splits into a variety of lighter atoms. At the same time, additional neutrons are formed, and a tremendous amount of heat energy is released. A typical fission reaction is as follows:

$$\,^{1}_{0}n + \,^{235}_{92}U \longrightarrow \,^{141}_{56}Ba + \,^{92}_{36}Kr + 3\,^{1}_{0}n + energy$$

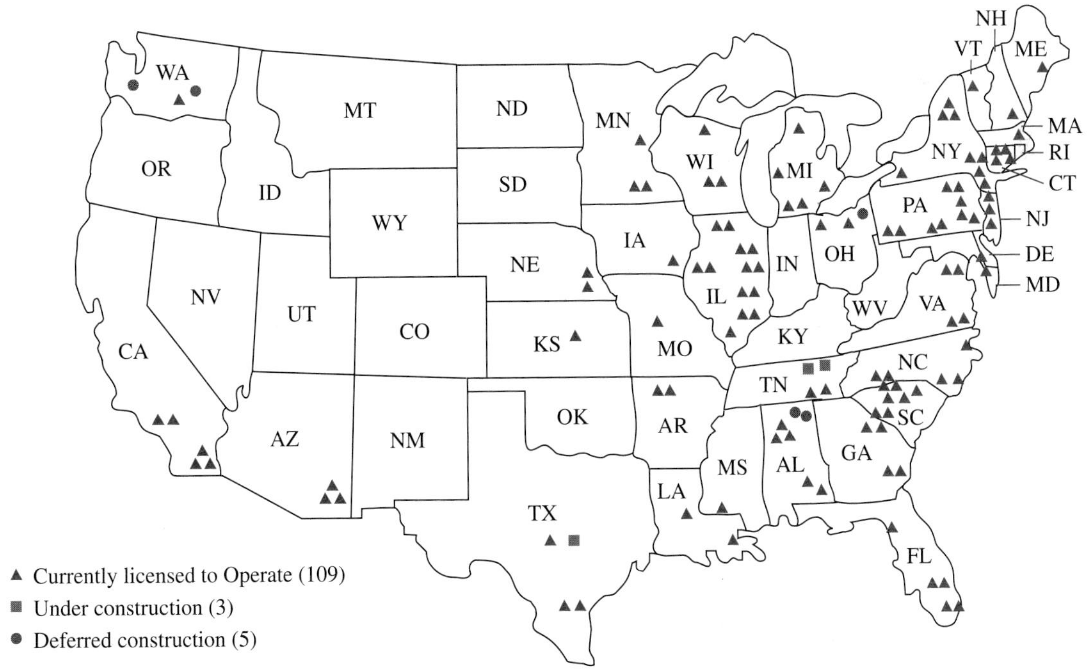

▲ Currently licensed to Operate (109)
■ Under construction (3)
● Deferred construction (5)

Figure 15-6   Nuclear power plants in operation and under construction in the United States in March 1993.

The neutrons formed can initiate further fission reactions, and the resulting chain reaction can lead to a tremendous explosion. In a nuclear power plant, the chain reaction is controlled so that energy is produced at a safe, steady rate.

Nuclear power is used primarily for the production of electricity. A diagram of a pressurized water reactor (PWR), the most common type of reactor in use, is shown in Figure 15-7. The basic design is essentially the same as that for a coal-fired power plant (Fig. 14-18a), except that nuclear fuel is used instead of coal as the source of heat for converting water to steam.

At the **core** of the **nuclear reactor** are several hundred narrow steel **fuel rods** containing the fissionable uranium fuel. Interspersed between the fuel rods are **control rods** made of material—usually cadmium or boron—that absorbs neutrons. When the rods are withdrawn from the core, the rate of fission increases; when the rods are inserted, more neutrons are absorbed, and the rate slows. In the event of an accident, or to make repairs or remove spent fuel, operators can stop the reaction by inserting the rods to their limit. Water circulating around the fuel rods and control rods acts as a **moderator** to slow the neu-

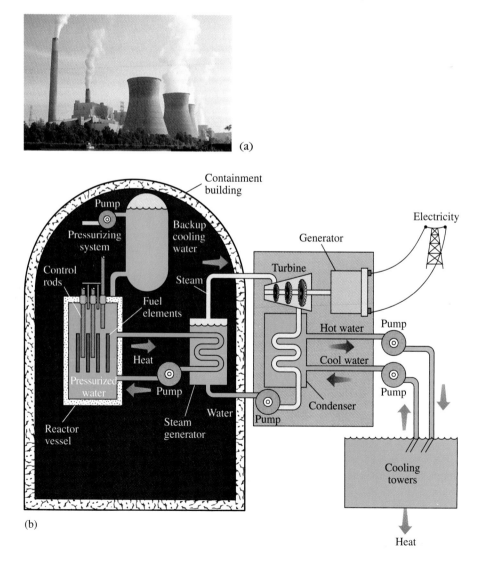

Figure 15-7 (a) A nuclear power plant and (b) a schematic diagram of a pressurized water nuclear reactor.

trons to speeds that are optimal for splitting uranium-235 atoms. The water, which is circulated through a heat exchanger, also serves to keep the fuel rods cool and to prevent the reactor from overheating. The entire reactor is housed in a thick-walled containment building.

The Nuclear Fuel Cycle

Because most uranium ore, as mined, contains no more than 0.2 percent of uranium, producing the fuel suitable for a nuclear power plant is expensive and energy-intensive. In the first steps, milling and chemical treatment convert the ore to a product that is approximately 80 percent uranium oxide (U_3O_8). The uranium in the oxide is 99 percent nonfissionable uranium-238 and only 0.7 percent fissionable uranium-235. It must be **enriched** to about 3 percent uranium-235 by means of a complex and technically difficult process before being fabricated into small pellets, which are packed into the fuel rods.

As fission occurs in the reactor, the uranium-235 concentration decreases, and after about 3 years, fuel rods must be removed and replaced. Spent rods remain radioactive. They can either be permanently stored as wastes or reprocessed so that the approximately 1 percent of uranium-235 and 1 percent of fissionable plutonium-239 in them can be recovered and made into new rods. Because of the high cost of reprocessing, most spent fuel rods are stored.

Each uranium pellet in a fuel rod weighs less than 1 g but has the energy equivalent of 270 kg (595 lb) of coal.

Problems with Nuclear Power

Most people consider the risk of a catastrophic explosion to be the greatest problem with nuclear power, but a power plant could never blow up like a nuclear bomb because the fuel is not sufficiently enriched with uranium-235. But, as was demonstrated in 1979 at Three Mile Island (TMI) in Pennsylvania and in 1986 at Chernobyl in the Soviet Union, human error and mechanical failure can lead to serious accidents. In both incidents, loss of cooling water caused overheating of the reactor and core meltdown. At TMI, the amount of radiation that escaped into the atmosphere was small; in the Chernobyl accident, however, enormous quantities of radioactive gases and particles were released. Huge tracts of contaminated land will remain uninhabitable for hundreds of years. The health hazards associated with exposure to radiation were considered in Chapter 6.

Some risk of radiation exposure exists in every step of the nuclear fuel cycle. The huge quantities of crushed waste rock left from the processing of uranium ores are a source of low-level radioactivity, which may leach into groundwater or be dispersed to the atmosphere in windblown dust. A major problem, which will be studied in Chapter 19, is disposal of the high-level radioactive wastes produced during the operation of a nuclear power plant. These wastes, which include fission products and spent fuel rods, must be stored so that they are completely isolated from the atmosphere.

Nuclear power plants have a limited life expectancy. After about 30 years of operation, continual bombardment of plant components with neutrons makes the metals brittle. The chances of cracking and of radiation leakage are increased, and for safety reasons, the plant must be shut down (decommissioned). Even after spent fuel rods and circulating water have been removed, a plant is still radioactive; the usual procedure is to seal the plant permanently in reinforced concrete. By the end of the century, most of the plants in the United States will be due for decommissioning.

If the cooling system in a nuclear reactor fails, the core and fuel rods can heat to temperatures in excess of 3000°C (twice the temperature required to melt steel) and melt through the floor of the reactor into the ground.

Meltdown has been termed the China Syndrome because if it occurred in the United States, melting would be directed through the earth towards China. In fact, in a meltdown, molten material would penetrate only a few meters into the ground.

Nuclear Breeder Reactors

The world's supply of uranium ores is not abundant; in the 1960s, when rapid expansion of the use of nuclear energy was expected, there were fears that shortages of uranium-235 would develop. The solution appeared to be the development of **nuclear breeder reactors**, which not only produce heat from fission but, at the same time, also produce a new supply of fissionable fuel.

If uranium-238 is bombarded with fast-moving neutrons, the following series of reactions occurs, and fissionable plutonium-239 is formed:

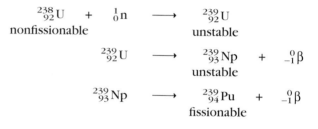

$$\underset{\text{nonfissionable}}{^{238}_{92}\text{U}} + {}^{1}_{0}\text{n} \longrightarrow \underset{\text{unstable}}{^{239}_{92}\text{U}}$$

$$\underset{\text{unstable}}{^{239}_{92}\text{U}} \longrightarrow {}^{239}_{93}\text{Np} + {}^{0}_{-1}\beta$$

$$^{239}_{93}\text{Np} \longrightarrow \underset{\text{fissionable}}{^{239}_{94}\text{Pu}} + {}^{0}_{-1}\beta$$

Fission of uranium-235 provides the neutrons needed to start the reaction sequence. Since two or, in some cases, three neutrons are produced in every

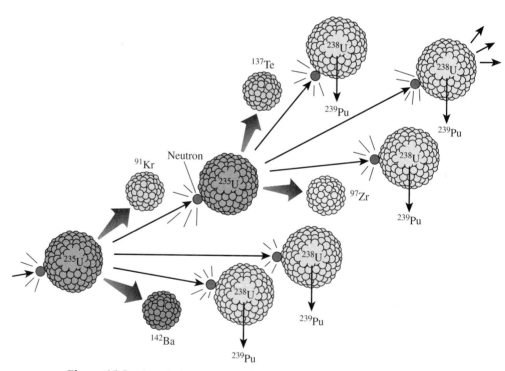

Figure 15-8 A typical reaction in a breeder reactor. When uranium-235 atoms are bombarded with neutrons, fission occurs. In each fission reaction two lighter elements and two or three neutrons are released from the uranium-235 atom. The neutrons bombard further atoms of uranium-235, or they bombard atoms of uranium-238, converting them to atoms of fissionable plutonium-239. Fission of one uranium-235 atom can result in the formation of more than one fissionable plutonium-239 atom. Thus the amount of fuel produced (plutonium-239) exceeds the amount of fuel consumed (uranium-235).

uranium -235 fission reaction (see the equation in "Nuclear Fission Reactors," p. 465), two (or three) plutonium-239 atoms may be formed from each uranium-235 atom (Fig. 15-8). The amount of fuel produced thus exceeds the amount consumed. Water cannot be used as the moderator in a breeder reactor because it slows the neutrons needed to produce the plutonium-239. Instead, liquid sodium is used.

The future of breeder reactors is uncertain. In addition to the problems associated with conventional reactors, there are several other drawbacks. Plutonium-239, which is an alpha emitter, has an extremely long half-life (Chapter 6)—24,000 years—and is one of the most toxic substances known. Inhalation of even minute quantities can cause lung cancer. Another problem is that plutonium-239 can be used more easily than uranium-235 to make nuclear weapons, thus increasing the need for security. The conversion of uranium-238 to fissionable plutonium-239 is difficult to control, and sodium reacts explosively if it comes in contact with water. At the present time, no breeder reactors are in commercial operation in the United States.

The Department of Energy's Clinch River Reactor in Tennessee has never been completed and was closed down in 1983 largely because of cost overruns.

Nuclear Fusion

When two very light atomic nuclei are combined, or fused, a heavier nucleus is formed. There is a loss of mass, and an enormous amount of energy is released. Fusion of hydrogen atoms to form helium is the primary source of the energy emitted by the sun. Theoretically, for each gram of fuel, **nuclear fusion** releases four times as much energy as the fission of uranium-235 does and about one million times as much as the combustion of fossil fuels does. Many scientists believe that if controlled fusion could be achieved on earth, it would solve the world's energy problems. But enormous technical difficulties must first be overcome before this source of energy can be exploited.

A temperature approaching that on the sun is required for hydrogen atoms to combine; even if this temperature is attained, the problem of finding a container that can withstand the heat without vaporizing remains. A fission (or atomic) bomb was used to produce the high temperature needed for the hydrogen bomb, but this is hardly an option for controlled energy production.

Fusion of the two isotopes of hydrogen, deuterium (^2_1H) and tritium (^3_1H) (Chapter 4), to form helium (^4_2He) is receiving most attention because fusion of these two atoms requires a lower temperature (about 100,000,000°C) than do other fusion reactions.

$$\underset{\text{deuterium}}{^2_1H} + \underset{\text{tritium}}{^3_1H} \longrightarrow \underset{\text{helium}}{^4_2He} + {^1_0n} + \text{energy}$$

Deuterium, a naturally occurring isotope, can be obtained in unlimited quantity from seawater. Tritium, however, is an unstable radioactive isotope and must be prepared. It can be produced by bombardment of lithium with neutrons.

$${^1_0n} + {^6_3Li} \longrightarrow {^3_1H} + {^4_2He}$$

Although nuclear fusion produces little radioactive waste, the danger of leakage of tritium (half-life of 12.3 years) and thermal pollution could pose problems. Also, known reserves of lithium ores are very limited.

The word *tokamak* is an abbreviation of the Russian for "torroidal magnetic chamber."

Two approaches to achieving nuclear fusion are being tested. One is the Tokamak reactor pioneered by Soviet physicists. In this device, high temperatures strip electrons from deuterium and tritium atoms, creating a gaslike plasma of energetic nuclei and free electrons. The greater the energy, the more likely the nuclei are to fuse. The hot plasma is contained within a powerful magnetic field (Fig. 15-9). In the second approach, laser beams are focused sharply on a minute pellet containing frozen deuterium and tritium (Fig. 15-10). A rapid increase in pressure and temperature causes the nuclei to fuse. So far, neither method has achieved sustained controlled fusion, and nuclear fusion is unlikely to become a practical source of energy in the near future.

(a)

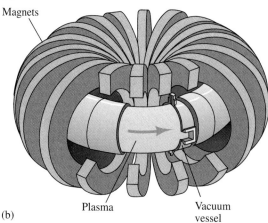

Magnets

Plasma

Vacuum vessel

(b)

Figure 15-9 (a) The Tokamak Fusion Test Reactor at Princeton University. (b) Diagram showing the plasma contained in the tunnel in the doughnut-shaped magnet.

(a)

Figure 15-10 (a) The experimental laser fusion reactor at the Lawrence Livermore Laboratory. (b) A powerful laser beam is focused on a tiny glass pellet containing a mixture of deuterium and tritium; the high pressure and temperature produced cause the hydrogen isotopes to fuse.

SOLAR ENERGY

Probably the most attractive source of energy is the sun. During daylight hours in sunny locations, huge quantities of solar energy reach the earth's surface. This energy comes to us free, is nonpolluting and, for all practical purposes, infinitely renewable. However, it is widely dispersed, and concentrating it and converting it to a usable form are both difficult and costly.

Passive and Active Solar Heating Systems

Solar energy can be used very simply to heat buildings and water. A building, if made of appropriate materials and suitably constructed and oriented, can capture the sun's heat. In a typical **passive solar heating** system for a home, solar energy enters through windows facing the sun, and convection currents passively distribute the heat around the building. Some of the heat is stored in rock below the house for release later when the sun is not shining.

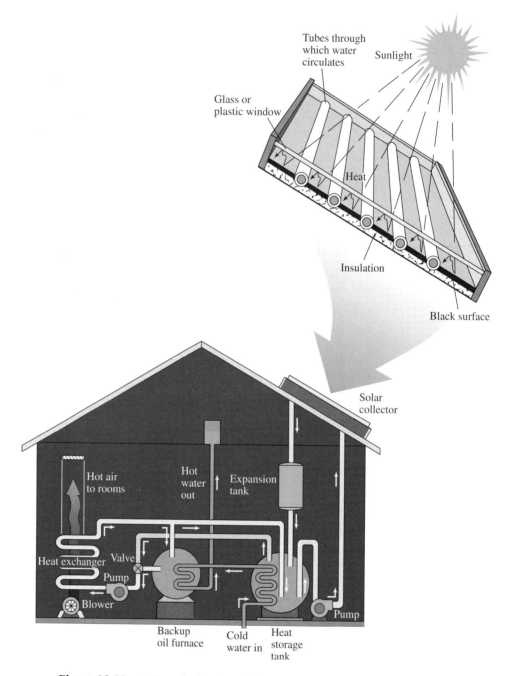

Figure 15-11 Active solar heating of a home. Water circulating through flat-plate solar collectors on the roof is heated by the sun and conveyed to a storage tank; the stored heat is used to provide hot water and hot air for space heating. The system is termed active because energy other than solar energy is required to pump the water through the house.

In an **active solar heating** system, heat gathered in solar collectors located on the roof of a building is circulated by means of pumps (Fig. 15-11). A typical **flat-plate collector** consists of a black surface covered with a glass or plastic plate. As anyone who has walked barefoot on asphalt in summer knows, sunlight is absorbed by a black surface and converted to heat (Chapter 13). In a solar collector, the glass plate prevents the heat from escaping. Circulating water is heated as it passes between the glass and the black surface.

Depending on climate and the availability of sunshine, passive and active solar systems can provide from 50 to 100 percent of home heating requirements. Although initial construction costs are high and a backup system may be needed, savings on energy bills are substantial. As oil becomes scarce and solar technology advances, solar heating should become increasingly attractive and affordable.

Solar Power Plants

Heat from solar collectors can be used to generate electricity. One of the most successful developments is the parabolic trough collector system (Fig. 15-12). The trough is a reflector that focuses sunlight onto a pipe running down its center. Oil or other fluid in the pipe is heated to a temperature up to 400°C (750°F). The heat is used to boil water and produce steam that can be used to turn a turbine and generate electricity. Solar trough facilities in southern California are already producing 350 MW (megawatts). By 1994 they are expected to produce 680 MW, approximately two-thirds the output of a typical nuclear power plant.

The inventor and scientist Charles Abbott introduced the idea of the solar trough collector for generating electricity in 1930.

Photovoltaic Cells

Another approach to converting solar energy directly into electricity is the **photovoltaic cell** (or **solar cell**). Most solar cells consist of two layers of almost pure silicon. The top, very thin layer contains a trace of arsenic; the lower, thicker layer contains a trace of boron. (The addition of a trace of an-

Figure 15-12 The parabolic trough collector system for generating electricity. Sunlight hitting the curved reflector is focused on the pipe running down its center, and oil flowing through the pipe is heated. The heated oil is used to boil water and produce steam to run a generator.

Since the 1950s, U.S. and Soviet satellites and space vehicles have been powered by
solar cells.

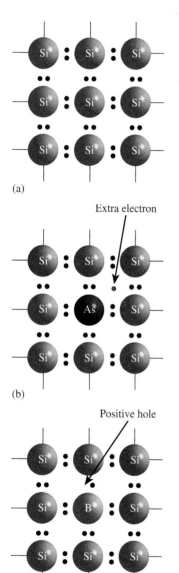

Figure 15-13 (a) A pure silicon crystal. (b) Silicon doped
with arsenic. (c) Silicon doped
with boron.

other element to silicon is known as doping.) Recall that silicon has four valence electrons, arsenic has five, and boron has three (Chapter 5). In a pure
silicon crystal, each silicon atom is covalently bonded to four other silicon
atoms (Fig. 15-13). When arsenic is included in the silicon structure, four of
the arsenic electrons form bonds with silicon atoms. The fifth electron is relatively free to move about. When boron is included, there is a shortage of one
bonding electron around the boron atom, and a positive hole is created. As a
result, there is a tendency for free electrons in the arsenic-doped layer to migrate to the boron-doped layer to fill the holes. Once the holes near the interface between the two layers are filled, the flow of electrons ceases. If the mobile electrons are sufficiently energized by exposure to sunlight and the
layers are connected by an external circuit, however, electrons will flow
through the external circuit, producing an electric current that can do useful
work (Fig. 15-14a).

Solar cells are widely used in calculators, watches, and other small devices. They also provide a vital source of power for space satellites. The efficiency of solar cells increased dramatically during the 1980s from about 10 to
20 percent; as new silicon materials are developed, efficiency is expected to
rise even higher.

Solar cells hold enormous promise for the future. If technology continues
to advance and prices continue to fall as projected, arrays of solar cells could
soon provide electricity for homes. Already, experimental automobiles that run
on solar energy have been built (Fig. 15-14b). An even more significant breakthrough is the development in California of an electric generating plant powered by banks of solar cells that can provide enough electricity to meet the
needs of a city with a population of about 250 thousand. Solar-powered plants
have many advantages over conventional power plants. They are relatively inexpensive to construct, produce no pollution, require little maintenance, and
there are no fuel costs. Many people hope that as the large number of aging nu-
clear-, petroleum-, and coal-fueled power plants in the United States near the
end of their useful lives, serious consideration will be given to replacing them
with solar-powered plants.

ENERGY FROM BIOMASS

Biomass, which is defined as any accumulation of biological materials, can be
used as a source of energy. Like a fossil fuel, biomass can be burned directly to
provide energy in the form of heat. Biomass can also be converted (by **bioconversion**) to methane (natural gas) and the liquid fuels methanol and ethanol.
Burning biomass has the disadvantage that it adds to the burden of carbon dioxide in the atmosphere.

Burning Biomass

Wood, a form of biomass, is the major source of energy for cooking and heating
for about 80 percent of people in the less-developed nations of the world. In
many areas of these countries, the constant search for firewood has led to excessive deforestation, with resulting erosion and degradation of soil. Severe
shortages of firewood are anticipated by the turn of the century.

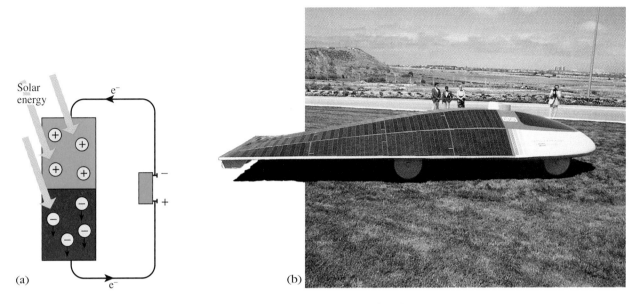

Figure 15-14 (a) A typical photovoltaic cell. When energized by sunlight, electrons flow through an external circuit. (b) The Sunraycer, an experimental automobile built by General Motors, is powered by solar cells. Silver-zinc batteries provide energy on cloudy days or if needed on steep hills.

In response to the relatively high costs of heating oil and electricity, wood-burning stoves have become increasingly popular in the United States. By 1985, about 20 percent of homes were relying partially or entirely on wood for heating. Many sawmills obtain most of their power from burning wood waste, and some cities have burned municipal trash, which is generally about 40 percent wastepaper, as a source of energy.

It is unlikely that burning wood or wood wastes will ever be more than a very minor source of energy in the United States. Wood wastes can supply only a small fraction of our energy needs, and cutting timber on a scale large enough to provide a significant amount of energy would have an adverse effect on the environment. Furthermore, wood stoves, unless they include control devices, are a source of indoor and outdoor air pollution and are often dangerous. For this reason, some cities, including London and cities in South Korea, have banned wood burning.

Crop residues are another source of energy that can be exploited. Already, in Hawaii, the fibrous residue from sugarcane is being burned to produce electricity. Another possibility is raising crops of fast-growing, high-energy-yield plants specifically to be used as fuel. However these "energy plantations" have many disadvantages. Large acres of land are needed, and because of the energy that must be expended to raise, harvest, and dry the crop, net energy yield is low.

Biogas Production

Plants, organic wastes, manure, and other forms of biomass can be used as sources of methane. In the absence of air, digestion of these organic materials by anaerobic bacteria produces **biogas**, which is about 60 to 70 percent methane.

Methane produced in the anaerobic digestion of organic wastes in landfills is now collected from many sites and burned as fuel.

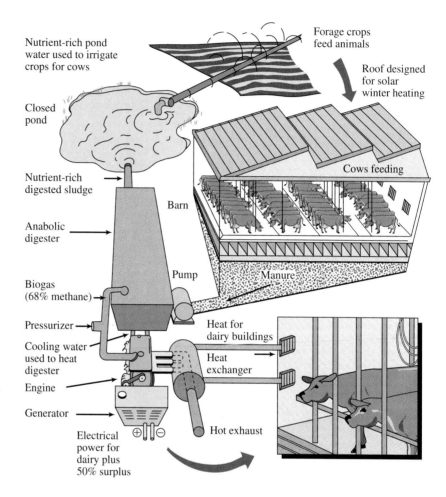

Figure 15-15 Biogas production. The Mason Dixon Dairy located near Gettysburg, Pennsylvania, produces more than enough methane from cow manure to meet its power needs. Surplus power is sold to the local utility. Nutrients in the manure are recycled onto land used to grow crops for the dairy herd.

The most suitable starting materials are sewage sludge or manure. A dairy farm can produce enough methane to power an electrical generator (Fig. 15-15); once the energy needs of the dairy have been met, excess electricity can be sold to the local power company. Production of biogas is economically feasible only in locations where there is a large concentration of the starting materials.

Alcohols from Biomass

The methane in biogas can be converted to methanol as shown in the following equations:

$$CH_4 + \underset{\text{steam}}{H_2O} \longrightarrow CO + 3\,H_2$$

$$CO + 2\,H_2 \longrightarrow \underset{\text{methanol}}{CH_3OH}$$

Natural gas or methane produced by coal gasification (Chapter 14) can be converted to methanol in the same way. Methanol can also be produced by destructive distillation of wood (Chapter 9).

Fermentation of sugars and starches in plants produces ethanol (Chapter 9).

$$C_6H_{12}O_6 \underset{\text{yeast}}{\longrightarrow} 2\,CO_2 + \underset{\text{ethanol}}{2\,C_2H_5OH}$$

Suitable plant materials for alcohol production are sugarcane, sugar beets, cassava, and sorghum, all of which have a high content of sugars and starches. The alcohol can be concentrated by distillation.

Both methanol and ethanol have high octane ratings (see Table 14-6) and can be used directly as automobile fuels. Methanol requires major modifications in conventional engine design, but, with minor changes, ethanol can be used in today's automobiles. Brazil has pioneered the development of alcohol as fuel; a large percentage of its cars now run on ethanol or gasohol, a mixture of gasoline with from 10 to 20 percent (by volume) of ethanol. In the United States, gasohol accounts for about 5 percent of all gasoline sold. Alcohol burns more cleanly than gasoline; however, large-scale use of land to grow suitable fermentable crops could compete with the use of land to produce food crops.

Although it is unlikely that bioconversion will ever become a major source of energy, it is a useful way to supplement other sources. It is particularly valuable as a means of converting plant and animal agricultural wastes and wastepaper in municipal garbage into usable energy.

WIND POWER

Wind power was one of the earliest forms of energy humans harnessed to do useful work. Until the 1930s, rural America relied heavily on windmills for pumping water and generating electricity. Then, as rural electrification schemes brought cheap power to farming communities, most windmills fell into disuse. Now, windmills are making a comeback.

Modern windmills have blades made of fiberglass, metal, or wood and are much stronger and lighter than older, conventional ones. A number of giant windmills with blades measuring 100 m (330 ft) from tip to tip, mounted on 50-m (150-ft) towers are now in operation. These enormous wind machines produce from 1 to 10 MW of electricity, enough to supply the needs of 1000 to 10,000 homes. This amount is much less than the 1000 MW generated by a nuclear power plant, but windmills are not associated with any hazards.

Another development is the construction of "wind farms" consisting of several hundred smaller windmills (Fig. 15-16), each with a blade span of about

The correct name for a windmill that is used to generate electricity is wind turbine.

Figure 15-16   A wind farm at Altamont Pass in California. Arrays of modern wind turbines are an efficient means of producing energy. The land under the windmills can be used for agriculture.

In Mongolia, where winds blow steadily, nomadic people carry small wind driven electric generators to provide electricity for lights and television sets.

In 1991, in the United States, wind power supplied 3,000 million kW of electricity, the equivalent of 5 million barrels of oil.

The first electric generating plant based on steam began operating in Larderello, Italy, in 1904 and has been in continuous use ever since.

In the 1980s, The Geysers was generating enough electricity to supply a city of 2 million people.

Figure 15-17 Old Faithful geyser in Yellowstone National Park has erupted every 40 to 80 minutes for the past 100 years.

6 m (20 ft), and capable of producing 10 to 15 kW of electricity (1 kW = 0.001 MW). Single units can be used for individual homes and farms, but arrays of units are more efficient. The land between windmills can still be cultivated or used for grazing.

For wind power to be cost-effective, winds must blow fairly steadily at about 10 mph (18 km per hour) or more. In California, where many suitable sites exist, 70 wind farms were in operation in 1985. By the year 2000, the state expects to obtain 8 percent of its electricity from wind power. Other states that are expanding the use of wind farms are Hawaii, Oregon, Montana, New Hampshire, and Vermont. Wind is clean, free, and abundant, and rapid growth in the use of wind power is predicted.

GEOTHERMAL ENERGY

Geothermal energy is heat energy that is generated deep within the earth's interior by the decay of radioactive elements (Chapter 2). Geothermal energy is immense, but like solar energy, it is widely dispersed. It becomes accessible only in certain unstable regions of the world, where, as a result of geologic activity, magma rises from great depths to near the earth's surface. The hot magma, at temperatures between 900° and 1000°C (1600° and 1800°F), heats rock and groundwater that comes in contact with the rock. Heated groundwater may emerge as a geyser (Fig. 15-17) or hot spring, or it may remain as a reservoir below the surface, sealed by a layer of impermeable cap rock.

Reservoirs may contain hot water, or steam. Geothermal wells, basically similar to oil wells, can be drilled into the reservoirs to release the hot water or steam. Hot water deposits are the most common and are used primarily to heat buildings. Steam is used to generate electricity. The largest geothermal steam field is The Geysers, located 145 km (90 mi) north of San Francisco. Between 1960 and 1988, this field was producing electricity more cheaply than fossil-fueled or nuclear-powered plants (Fig. 15-18). Recently, however, power output has begun to fall because the underground reservoir is drying up. The largest electric generating plant based on steam is in New Zealand.

Both water and steam deposits are abundant in Iceland. In Reykjavik, the capital, nearly all the buildings—including greenhouses for producing fruits and vegetables—are heated with hot water obtained from geothermal wells. Hot water reservoirs also heat homes in Boise, Idaho, and Klamath Falls, Oregon.

Hot rock can also be exploited to obtain steam. In a procedure that is still in the development stage, cold water is injected into wells drilled into the hot dry rock. The water returns to the surface as steam, which can be used to generate electricity.

The future of geothermal energy is uncertain. Many geothermal fields occur in rugged, inaccessible terrain; many are located in scenic areas, such as Yellowstone National Park, and cannot be developed. Most steam deposits contain hydrogen sulfide, small amounts of which are released during plant operation and pollute the atmosphere. Salts dissolved in the water released with steam from reservoirs can corrode pipes and other fixtures and, if not prevented from reaching streams or rivers, can cause severe ecological damage. Despite these problems, the use of geothermal energy is likely to increase in areas where it can be exploited as oil becomes scarce and more expensive.

Figure 15-18   Part of The Geysers in California. This geothermal system for generating electricity is operated by the Pacific Gas and Electric Company.

WATER POWER

Hydropower is a very efficient (80 percent) means of producing electricity, and it is essentially renewable and nonpolluting. However, the dams that must be built to produce hydroelectric power create many environmental problems (Fig. 15-19). Water impounded behind a dam floods scenic stretches of river, valuable cropland, and, in some cases, places of historic and geological interest. Downstream from a dam, as water flow is adjusted to meet electrical demand, the constantly changing water level alters the natural ecosystem. Silt becomes trapped behind the dam, and the amounts of sediments and nutrients downstream are reduced.

Worldwide, many new dams are planned. In the United States, however, because most of the best dam sites have already been used and because of environmental concerns, few, if any, new large dams are expected to be constructed. But there is renewed interest in increasing hydroelectric power by putting back into service many of the 3000 unused small facilities located at dams built on streams and small rivers. These facilities, designed to generate electricity locally, went out of production when large centralized power stations were built.

Figure 15-19   Hoover Dam.

TIDAL POWER

The twice daily rise and fall of tides—usually from 1 to 10 m (3 to 30 ft)—represents an enormous potential source of energy. This energy can be exploited by building a dam across the mouth of a bay or inlet. The incoming tide generates electricity as it flows through turbines constructed in the dam. The turbine blades are then reversed so that the outgoing tide also produces electricity. To be practical, the difference in level between low and high tide must be 6 m (20 ft) or more. Two tidal power plants are now in operation—one in France (Fig. 15-20) and the other in the former Soviet Union.

Tidal power plants alter the normal flow of water and disrupt and disfigure the natural environment. Because of the adverse environmental effects and lack of suitable sites, tidal power is never likely to be more than a minor source of energy in the United States.

Tides are caused by gravitational attractions among the moon, the sun, and the rotating earth. Their magnitude varies in cycles that depend on the relative positions of the three bodies.

Figure 15-20 Tidal power plant located on the Rance estuary in northern France.

ENERGY FROM HYDROGEN GAS

A possible fuel to replace oil and natural gas is hydrogen gas. Hydrogen could be transported through pipes like natural gas. It could be used for heating homes and for heating water to produce steam for electric power generation. With minor changes to the carburetor, today's cars could run on hydrogen. Hydrogen burns cleanly, combining with oxygen in the air to produce water. One problem is that it is even more flammable than gasoline or natural gas and would have to be handled with great care.

Practically no free hydrogen exists on earth, but hydrogen can be produced by the electrolysis of water (Chapter 1). This process, by which water is separated into hydrogen and oxygen, is not very efficient. Even if efficient solar cells could be used to provide the electricity, the available energy in the hydrogen would still represent only 20 percent of the electrical energy consumed in producing it. However, if solar energy could be used *directly* to decompose water into hydrogen and oxygen, hydrogen could become a useful fuel. Seawater could provide an almost inexhaustible supply of water.

In the initial stages of photosynthesis, water is decomposed to hydrogen and oxygen. This decomposition has been duplicated with some success in the laboratory by exposing blue-green algae to sunlight. The hydrogen and oxygen produced were separated and collected. The process is still in the very early stages of development but has great promise for the future.

FUEL CELLS

Fuel cells are another promising alternative energy source. They are lightweight and have been used successfully to produce electricity in spacecraft.

Fuel cells depend on an oxidation-reduction reaction (Chapter 8) that converts chemical energy directly into electrical energy. The oxidation of hydrogen with oxygen is the basis of the fuel cells used in space.

$$2\,H_2 + O_2 \longrightarrow 2\,H_2O + energy$$

The water produced can be purified to make it suitable for drinking.

The cells consist of an electrolyte solution, usually potassium hydroxide (KOH), and two porous carbon electrodes containing platinum or some other suitable metal. Hydrogen and oxygen are fed continuously to the anode and cathode compartments (Fig. 15-21), where the following reactions occur:

at the anode: $2 H_2 + 4 OH^- \longrightarrow 4 H_2O + 4 e^-$

at the cathode: $O_2 + 2 H_2O + 4 e^- \longrightarrow 4 OH^-$

overall reaction: $2 H_2 + O_2 \longrightarrow 2 H_2O$

The electrodes are connected, and electrons flow from anode to cathode through the external circuit.

Fuel cells are about 60 percent efficient in converting chemical energy into electrical energy. Calculation of overall efficiency, however, must also take into consideration the energy used in producing the hydrogen and oxygen. In the future, it may become possible to obtain both gases by the photodissociation of water as described in the preceding section.

Many other types of fuel cells have been developed. As new technologies become available, it is expected that electricity from fuel cells will be used to power cars and to provide heat for buildings.

Batteries and fuel cells depend on oxidation-reduction reactions. Batteries store energy; fuel cells convert one form of energy (chemical) to another form (electrical).

ENERGY SOURCES FOR THE TWENTY-FIRST CENTURY

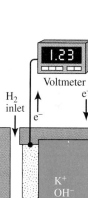

In the next 20 years, as oil supplies in the United States rapidly dwindle, we will become almost entirely dependent on imported oil. In the next century, as recoverable oil and natural gas reserves worldwide are used up, we will be forced to shift to other sources of energy.

Coal is plentiful in the United States, and in the short term, more will be used, primarily to generate electricity. Some oil is expected to be obtained from oil shale, tar sands, and coal liquefaction. But all these fossil fuel sources have a high environmental cost.

Our best immediate option is energy conservation. Since at present we have no suitable substitute for liquid fuels, developing cars that are more fuel-efficient to reduce consumption should be a top priority. Many believe that if another energy crisis is to be avoided, new laws must be passed, mandating that new cars achieve gas mileages substantially higher than the present requirement of 27.5 mpg.

The most promising source of energy for the future is renewable solar energy. Already solar cells show great promise, and as they become cheaper and more efficient, solar-powered electric generating plants, automobiles, and numerous other products are likely to become common. Fuel cells and solar production of hydrogen are other promising long-term options. In suitable areas, wind power, geothermal energy, and hydroelectric power from small generating plants are expected to make an increasingly large contribution locally. The future of nuclear energy is uncertain, but because of fear of accidents and problems of disposing of hazardous wastes, the use of nuclear energy will probably decrease.

Without a strong commitment by the federal government to provide incentives for conservation and to fund the research and development of economically and environmentally acceptable new technologies, the transition

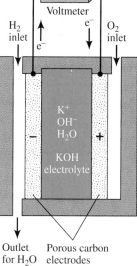

Figure 15-21 Cross section of a hydrogen-oxygen fuel cell. This type of cell was the primary source of electrical energy on the *Apollo* moon flights.

from an economy based on fossil fuels to one based on alternative energy-delivery systems will not be easy. As the following quotation shows, the difficulties inherent in changing the status quo have been understood for centuries:

> There is nothing more difficult to carry out, nor more doubtful of success, nor more dangerous to handle, than to initiate a new order of things. For the reformer has enemies in all who profit by the old order, and only lukewarm defenders in all those who would profit by the new order. This lukewarmness arises partly from fear of their adversaries, who have the law in their favor, and partly from the incredulity of mankind, who do not truly believe in anything new until they have had actual experience of it. (Machiavelli, *The Prince*, 1517)

EXPLORATIONS

The Chernobyl Nuclear Power Plant Disaster

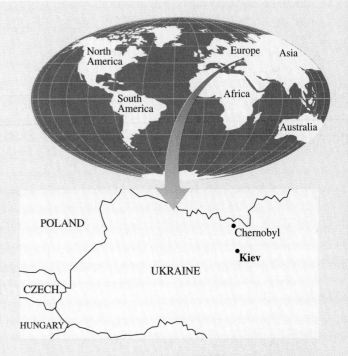

On April 26, 1986, Unit Number 4 at the Chernobyl nuclear power station in the Ukraine, then part of the Soviet Union, exploded, sending more radioactive material into the atmosphere than was released by the bombs that destroyed Hiroshima and Nagasaki in World War II. Ironically, it all started with an experiment designed to increase safety.

The Chernobyl power station was gigantic. Four units were in service at the time of the accident; two more were scheduled to begin operating in 1988. The core of each reactor was made up of more than 1500 fuel tubes filled with enriched uranium fuel and approximately 200 boron-carbide control rods. The fuel tubes and control rods were packed into a stack of graphite blocks that served as the moderator in slowing down fast-moving neutrons to an optimum speed for effective atom splitting.

Nuclear power involves maintaining a delicate balance between keeping the fission reaction going and preventing it from getting out of control. In Unit 4, excess heat in the core was removed by water circulated around the fuel tubes; there was an additional water-cooling system to automatically shut down the reactor in case of an emergency. The rate of fission was regulated by the control rods. When the rods were withdrawn from the core, the reaction rate increased; when the rods were fully inserted, the reactor shut down.

Each unit had backup generators to keep water-cooling pumps running in case of a break in the normal flow of electricity from the turbine generators. However, about 40 seconds were required for the backup system to start up. On April 25, when Unit 4 was scheduled to be shut down for routine maintenance, it was decided to run an experiment to determine whether, when power was cut off, the gradually slowing turbine rotors could be made to supply enough energy to keep the pumps running during this critical time gap.

At the start of the experiment, as the reactor's power was slowly lowered, plant operators turned off the automatic emergency water-cooling system. They did this—in violation of safety rules—to avoid a complete shutdown, which would have prevented a repeat of the experiment if one had become necessary. Then, over the next few hours, because of a failure to set an automatic power control, they had difficulty regulating power, until finally a sudden surge of power triggered an uncontrolled chain reaction. The operators had broken a total of six safety regulations with the result that none of the emergency systems that should have shut down the reactor was operable. The supervisor desperately slammed all the rods into the core, but it was too late. At exactly 1:23 A.M. on April 26, Unit 4 exploded.

The fuel tubes, red hot from lack of coolant water, had broken through into the water-steam system, causing a colossal explosion. The containment walls around the reactor were blasted apart, and a deadly plume of radioactive material shot into the night sky. A second fiery explosion occurred a few seconds later when hydrogen, produced by steam mixing with hot graphite, reacted violently with oxygen in the air. Red-hot graphite and chunks of burning material started numerous fires on adjacent buildings (Fig. 15-A).

It was not until two days later, after high levels of cesium-134—a fission product found only in nuclear reactors—had been detected in Sweden during routine monitoring for radioactivity at one of their own plants, that the Soviets reported the accident. In a brief announcement on April 28, they admitted that a reactor at Chernobyl had been "damaged" and stated that "measures were being taken to liquidate the consequences."

The fire in the reactor core proved very difficult to extinguish. Even after it had been smothered by 5000 tons of material dropped onto it from helicopters, heat continued to build up. Meltdown (complete melting of the rods and core) leading to melting of the concrete containment floor and rock beneath it became a possibility. To prevent this ultimate nuclear nightmare, operators drained the basin of cooling water that lay below the reactor—a very dangerous operation because the radioactivity beneath the core was intense—and then 400 miners tunneled under the reactor to install a heat exchanger mounted on a massive concrete base. By the end of June, the entire reactor was encased in 300,000 tons of concrete.

Immediately after the accident, few people in the rural community around the power station or in the nearby town of Pripyat understood the gravity of the situation. They were told to stay indoors with their windows shut, but schools and businesses stayed open, and marriages continued to be performed. The inhabitants were taken completely by surprise when, 36 hours after the accident—by which time radiation levels had soared to 400,000 times normal—an immediate evacuation of everyone within 6 miles of the power station was ordered. Many were reluctant to leave, but all were loaded into a fleet of buses that drove out over roads specially treated to prevent radioactive dust from being kicked up. Military trucks carrying livestock followed. Six days later, the evacuation zone was widened to 18 miles.

Initially, the radioactive cloud was carried in a northwesterly direction, and fallout was recorded over most of Europe, with levels being highest in areas of rainfall. Then, 9 days after the explosion, the wind veered to the south, and radiation levels almost 100 times the normal level were recorded in Kiev, a city of 2.4 million people 60 miles south of Chernobyl. Buildings and streets were hosed down daily; residents were advised to wash their hair every day, clean the soles of their shoes, and eat only food that had passed inspection. Not surprisingly, thousands of people fled the city despite efforts by the authorities to prevent them from leaving.

In the evacuation zones, the task of cleanup was overwhelming. Tens of thousands of workers—who had to be rotated every few minutes because of high radiation levels—removed topsoil, sprayed trees with decontaminants, demolished contaminated houses, and erected barriers to stop radioactive water from reaching the Dnieper River. Even so, an area greater than 1000 square miles remains uninhabitable for the foreseeable future.

Thirty-one people died in the immediate aftermath of the accident. One man was killed by the explosion, and five died within hours from a combination of severe radiation burns and massive doses of radiation. The remainder died in the next few months

Figure 15-A Unit 4 at the Chernobyl nuclear power station after the accident.

from radiation exposure. Most of those who died were heroic firemen who, well aware of the dangers, gave their lives to prevent fires from spreading to the nearby Unit 3. In the years to come, it will be impossible to estimate how many cases of cancer will be directly linked to the Chernobyl accident. The number is expected to be in the thousands, although recent studies indicate that there may not be as many cases as originally estimated.

After the investigation into the accident, which revealed reckless operating procedures and repeated violations of safety regulations, the director of the station and two engineers were each sentenced to 10 years hard labor. Six others received sentences of between 2 and 5 years.

The accident at Chernobyl was by far the worst that has ever occurred at a nuclear power plant. The radioactivity released—which represented only about 10 percent of the reactor's total inventory—was several million times greater than that released at Three Mile Island (TMI) in 1979. At Chernobyl, long-lived plutonium radioisotopes (which were not released at TMI) will continue to emit radiation for thousands of years. Of particular concern are strontium-90 and cesium-137, both of which have half-lives of approximately 30 years and readily move up through food chains.

Could another Chernobyl occur? The Unit 4 type reactor had several design faults, including its fatal instability at low operating power. Unlike reactors in the United States and Western countries, it used combustible graphite instead of water as the moderator; it lacked a strong concrete dome above the reactor; and the automatic computer-controlled emergency systems could be overridden manually.

Even with the most up-to-date technology and every conceivable safety device, machinery can fail and human beings can make mistakes. As Chernobyl has shown, a nuclear accident, no matter where it occurs, can have disastrous consequences for the entire world.

Question
Discuss the possibility of a nuclear power plant accident on the scale of the Chernobyl disaster occurring in the United States.

References: V. Haynes and M. Bojcun, *The Chernobyl Disaster: The True Story—An Unanswerable Indictment of Nuclear Power* (London: Hogarth Press, 1988); R. P. Gale and T. Hauser, *Final Warning: The Legacy of Chernobyl* (New York: Warner Books, 1988).

KEY WORDS AND CONCEPTS

bioconversion
biogas
biomass
conservation
control rods
energy conservation
fuel cell

fuel rods
geothermal energy
hydropower
liquefied natural gas (LNG)
moderator
nuclear breeder reactor
nuclear fission reactor

nuclear fuel cycle
nuclear fusion
photovoltaic cell
solar energy
solar heating systems
tidal power
wind power

QUESTIONS AND PROBLEMS

1. What percentage of the United States' total energy needs is supplied by petroleum? In approximately what year are petroleum reserves expected to be used up?

2. What is the most abundant fossil fuel in the world?

3. It is estimated that air pollutants from coal-fired electric power plants cause more than 10,000 deaths a year. Carbon dioxide emissions from these plants are measured in millions of tons and may be contributing to global warming. In your opinion, should U.S. energy strategy favor increased use of coal-burning power plants? What are the alternatives?

4. How long are the natural gas supplies of the United States expected to last?

5. What are the disadvantages of liquefied natural gas (LNG)?

6. Which countries of the world place a high reliance on nuclear power?

7. In your opinion, should the United States build a large number of new conventional nuclear fission reactors to meet electrical demand in the future?

8. How enriched in uranium-235 must uranium ore be to be used as a fuel in a nuclear power plant?

9. The licensing time for new nuclear power plants in the United States should be cut in half so that costs would be reduced. Do you agree with this statement?

10. Can nuclear bombs be made from reactor-grade uranium? Explain.

11. Describe the nuclear fuel cycle for the production of reactor-grade uranium.

12. More nuclear power plants should be built in the United States to reduce our dependence on foreign fossil fuel and to slow global warming. Do you agree with this statement?

13. How does the breeder reactor produce more fuel than it uses?

14. The U.S. government should develop the nuclear breeder reactor to conserve uranium resources and to keep the United States from becoming dependent on other countries for uranium. Do you agree? Why?

15. Describe the function of the following in a nuclear power plant:

a. control rods **b.** fuel rods **c.** moderator **d.** containment building

16. What roles do ^2_1H and ^3_1H play in nuclear fusion?

17. In a fusion reaction, a plasma called "the fourth state of matter," is formed. Describe the composition of a plasma.

18. What are some of the problems associated with solar power?

19. What are the differences between active and passive solar heating systems?

20. Draw a schematic diagram of a photovoltaic cell.

21. Describe the chemical "dopants" used to make a silicon photovoltaic cell. Which element creates an excess of electrons? Which element creates an excess of positive charges?

22. Should the U.S. government increase funding for the rapid development of photovoltaic cells? Can the United States generate most of the electricity it needs through direct solar energy?

23. Describe what is meant by biomass.

24. Write chemical equations to show how the following are made from biomass:

a. methyl alcohol **b.** ethyl alcohol **c.** methane

25. List the advantages and disadvantages of using biomass as a source of energy.

26. List two disadvantages of wind power.

27. Will geothermal energy ever be a major source of energy world-wide? Explain.

28. Is the following statement true? Electricity is clean heat. Explain.

29. Rank the following energy sources in the order you think they will contribute to meeting our energy needs in the year 2025. Briefly explain the order you choose.

a. nuclear fusion **b.** solar energy **c.** coal
d. petroleum

30. List two advantages fuels cells have over conventional batteries.

31. A mandatory energy conservation program should be adopted by the U.S. government. Explain briefly why you agree or disagree with this statement.

TOXICOLOGY

16

Insecticides sprayed on crops contain toxic chemicals that are particularly poisonous to insects.

Humans have been familiar with the harmful effects of animal venoms and poisonous plants for thousands of years. Early in our history this knowledge was used in hunting and in warfare (Fig. 16-1). The earliest known document to give information about poisons is the Ebers papyrus (circa 1500 B.C.), which describes early medical practice in Egypt. This document includes nearly 1000 recipes, many containing recognizable poisons. For example, it mentions hemlock, which was the poison used in the execution of the Greek philosopher

Figure 16-1 People in Amazonia still hunt monkeys and other small animals with poisoned darts. A Yagua Indian in Peru fitting a dart into a blowgun.

Socrates (470–399 B.C.); aconite, an arrow poison used by the Chinese; opium, used both as a poison and antidote; and plants containing belladonna alkaloids, which can cause cardiac arrest. During the Roman Empire and continuing through the Middle Ages and into the Renaissance, poisoning was a common practice. The Borgia family in Italy removed many of their enemies in this way.

It was not until the beginning of the nineteenth century that any systematic attempt was made to identify the agents responsible for the toxicity of venoms and poisonous plants. One of the first to identify a poison was the French physiologist Francois Magendie. In 1809, while studying arrow poisons, he isolated the alkaloid strychnine from the plants used to make the poison and showed that it was responsible for the convulsions that victims suffered before they died.

Today, we are concerned with the risk of exposure to toxic substances that may be present in our water supply, in the atmosphere, and in the workplace. There is particular concern about the possible harmful effects of pesticide residues in foods. In this chapter we will examine the sources and effects of exposure to toxic substances. We will focus on what makes certain substances toxic and how the risk associated with each is measured.

Learning Goals:

In this chapter, you should gain an understanding of:

1. How the LD_{50} test is used to determine toxicity.

2. Routes by which toxic substances enter the body.

3. Corrosive poisons and their reactions.

4. The way metabolic poisons cause harm.

5. Poisoning with toxic metals and with neurotoxins.

6. How chemicals are detoxified in the liver.

7. The similarities between carcinogens, teratogens, and mutagens, and how to test for them.

8. The risk associated with exposure to toxic chemicals.

HOW POISONOUS ARE TOXIC CHEMICALS AND BIOLOGICAL TOXINS?

A substance is said to be toxic if it causes harm to a biological system. Only substances that cause harm when present at very low concentration (parts per million or less) are generally described as toxic. Many substances, including common salt, are harmful at high concentration but would not be described as toxic. Toxic substances act in many different ways: Some upset metabolism, others the nervous system. Still others cause genetic changes, cancers, and birth defects.

Although the terms *poison* and *toxin* have different meanings, they are often used interchangeably. **Poison** is a general term for any toxic substance, whether synthetic or of natural origin. The term **toxin** refers to poisons of biological origin. Toxins can enter the body if we eat spoiled foods. Aflatoxins, for example, are an extremely potent group of toxins present in molds that can grow on peanuts and grains while they are in storage. Insect stings and snake bites also can introduce toxins into humans and other animals.

The word *poison* is derived from an English word, *poysoun*, meaning a poisonous drink.

Aspirin, the most widely used drug in the world, is an invaluable medicine, but in large doses it is toxic. For many years before the introduction of child-proof containers, aspirin caused more cases of accidental poisoning in young children than any other substance (Fig. 16-2). Unsupervised children often ingested a toxic quantity of the attractively flavored tablets by eating them like candy.

Testing for Toxicity

Because any chemical can be harmful if consumed in large enough quantity, we need a measure of the level at which a particular substance becomes toxic. The

Figure 16-2 The introduction of child-proof containers has greatly reduced the number of cases of aspirin poisoning in children.

The *dose* is the amount of a chemical, per unit of body mass, to which an animal is exposed. The *response* is the effect this chemical exposure has on the animal.

toxicity of a chemical is usually tested by determining the **dose** (the amount per unit of body mass) that kills a laboratory animal, usually a rat or mouse. Because even inbred animals vary considerably in their response to a specific dose of a chemical, a large group of animals must be tested to obtain a statistically valid response.

LD_{50}

Since 1920, the standard method for determining the toxicity of a chemical has been the **LD_{50}**—lethal dose—test. In this procedure, different doses of the chemical being tested are fed to large groups of laboratory animals (Fig. 16-3). The result is expressed as the LD_{50}: the dose that kills 50 percent of the animals in the group. The LD_{50} values of some common chemicals are listed in Table 16-1. By comparing the LD_{50} values, one can assess the relative toxicity of the chemicals.

The smaller the value of the LD_{50}, the greater the toxicity of the substance. For example, the LD_{50} for aflatoxin-B fed to rats is 0.009 mg/kg, while that for caffeine is 130 mg/kg. This means that aflatoxin-B is about 10,000 times as toxic to rats as caffeine is.

The LD_{50} values in Table 16-1 are a measure of the toxicity of the listed chemicals to rats or mice. Other animal species may vary in their response to these substances. For example, guinea pigs are 10,000 times as sensitive to dioxin (TCDD) as are dogs. Because of these species differences, LD_{50} values for rats, or other animals, cannot be extrapolated to determine the amount of a chemical that would be lethal to 50 percent of a human population. Nevertheless, the values for rats are useful in predicting the probable toxicity in humans. A substance that is extremely toxic to animals is likely to be toxic to humans also.

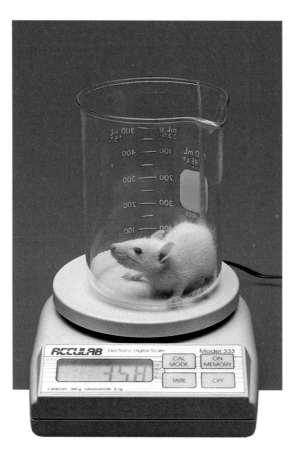

Figure 16-3   Specially bred laboratory rats are used to determine the LD_{50} values of test chemicals.

TABLE 16-1   LD$_{50}$ Values of Selected Chemicals

Chemical	LD$_{50}$(mg/kg)*
Sugar	29,700
Ethyl alcohol	14,000
Vinegar	3,310
Sodium chloride	3,000
Malathion (insecticide)	1,200
Aspirin	1,000
Caffeine	130
DDT (insecticide)	100
Arsenic	48
Strychnine	2
Nicotine	1
Aflatoxin-B	0.009
Dioxin (TCDD)	0.001
Botulinum toxin	0.00001

*For rats or mice.

The method used to introduce a test chemical into an animal's body affects the LD$_{50}$ and must be taken into account. For example, in mice, the LD$_{50}$ for procaine, a local anesthetic, is 800 mg/kg when the drug is injected under the animal's skin, 500 mg/kg when given by mouth, and only 45 mg/kg when injected directly into a vein.

EXAMPLE 16-1   Assuming that humans and rats are equally sensitive to the harmful effects of strychnine, how many grams of this chemical would be required to kill 50% of a group of:
a. 220-lb adult humans?
b. 22-lb children?

Solution:
a. Convert the weight in pounds to kilograms. The conversion factor is: 2.2 lb = 1 kg. Therefore, the man's weight in kilograms is:

$$220 \text{ lb} \times \frac{1 \text{ kg}}{2.2 \text{ lb}} = 100 \text{ kg}$$

Find the LD$_{50}$ of strychnine in Table 16-1. It is 2 mg/kg. Therefore, the lethal dose in 50% of cases is:

$$100 \text{ kg} \times \frac{2 \text{ mg}}{1 \text{ kg}} = 200 \text{ mg} = 0.200 \text{ g}$$

b. Similarly, a 22-lb child weighs 10 kg. The lethal dose in 50% of cases is:

$$10 \text{ kg} \times \frac{2 \text{ mg}}{1 \text{ kg}} = 20 \text{ mg} = 0.020 \text{ g}$$

PRACTICE EXERCISE: The LD$_{50}$ of the anti-inflammatory drug ibuprofen is 495 mg/kg. How many 350-mg ibuprofen tablets taken all at one time would cause death in 50% of a group of 174-lb adults?

Answer: 112 tablets.

Figure 16-4 The sooner a chemical can be washed from the point of contact with the body the less harm it causes.

HUMAN EXPOSURE TO TOXIC SUBSTANCES

Toxic substances in the environment can enter the human body in three main ways: through the lungs (**inhalation**), through the mouth (oral **ingestion**), or through the skin (**dermal contact**). When a toxic substance enters the body it may cause harm locally at the site of entry or it may not exert its effect until it is absorbed into the bloodstream. Once in the bloodstream it is transported to all parts of the body, and, depending on the nature of the toxic substance, it may then disrupt one or more of the body's normal functions.

Toxic gases in the atmosphere, such as carbon monoxide, enter the body primarily by inhalation through the lungs. Fine particulates that are often present in industrial emissions also reach the lungs via this route. Larger particles are filtered out by the hairs and mucus in the nasal passages. Fine particulates, and gases such as sulfur dioxide and ozone, damage lung tissue and cause respiratory problems. Carbon monoxide is absorbed into the bloodstream, where it disrupts the oxygen-carrying capacity of the blood.

Acceptable levels for inhalation exposure of individuals to chemicals in the working environment were determined in response to the Occupational Safety and Health Act (OSHA) of 1970.

Toxic substances ingested in food and water pass from the mouth into the digestive tract. Most cause harm after they have been absorbed into the bloodstream through the walls of the intestines.

Toxic chemicals that cause harm through dermal contact are absorbed through the skin into the tissues below and enter the blood that supplies those tissues. Penetration through the skin is time-dependent. The longer the skin is in contact with the chemical, the greater the amount of the chemical transferred through the skin and into the circulatory system. An individual splashed by a chemical should always try to wash it off immediately with copious amounts of water (Fig. 16-4).

Acute exposure to a chemical is defined as exposure for less than 24 hours. Chronic exposure is defined as exposure for more than 3 months.

CORROSIVE POISONS

Corrosive poisons are substances that destroy tissues on contact. They include strong acids such as hydrochloric acid (HCl) and sulfuric acid (H_2SO_4), strong bases such as sodium hydroxide (NaOH) and potassium hydroxide (KOH) (Chapter 8), and strong oxidizing agents such as ozone (O_3). A list of some common corrosive poisons can be found in Table 16-2.

Acids and Bases

Many strong acids and bases are found in the home. Hydrochloric acid (also called muriatic acid) is often used to clean calcium carbonate deposits from toilet bowls; sulfuric acid is present in automobile batteries; both sulfuric acid and sodium hydroxide are useful as drain cleaners because they destroy hair; sodium hydroxide and potassium hydroxide are ingredients in oven cleaners. There have been many tragic results from children swallowing these products. The strong acids and bases in them destroy the tissues in the mouth and throat, and cause severe pain. Swallowing as little as one-half ounce (15 mL) of concentrated (98 percent) sulfuric acid can be fatal.

Acids and bases destroy tissues by catalyzing the breakdown of peptide bonds in the protein molecules that make up tissues (Chapter 10). In acids, the catalyst is the hydrogen ion; in bases, it is the hydroxide ion.

TABLE 16-2   Corrosive Substances and Their Properties

Name	Effects
Nitric acid (HNO_3)	Reacts with protein in exposed skin to form lesions
Sulfuric acid (H_2SO_4)	Dehydrates protein in skin to form lesions
Hydrochloric acid (HCl)	Splash to the eye causes severe damage, inhalation of HCl gas damages lungs
Hydrofluoric acid (HF)	Produces severe skin burns and causes painful ulcers
Metal hydroxides (NaOH, KOH)	Dissolve tissue and cause severe burns
Halogen gases (Cl_2, Br_2, F_2)	Produce corrosive acids when in contact with moist tissue; inhalation damages lungs
Hydrogen peroxide (H_2O_2)	Oxidizes tissue and causes severe burns

$$R-\overset{\overset{\displaystyle O}{\|}}{C}-\overset{\overset{\displaystyle H}{|}}{N}-R' + H_2O \xrightarrow{\text{H}^+ + \text{OH}^-} R-\overset{\overset{\displaystyle O}{\|}}{C}-OH + H-\overset{\overset{\displaystyle H}{|}}{N}-R'$$

peptide bond

Acids add to the destruction by dehydrating the tissues.

 Sulfur dioxide (SO_2), which is released into the atmosphere when sulfur-containing coal is burned (Chapter 13), is another corrosive poison. In the atmosphere sulfur dioxide is converted to sulfur trioxide (SO_3), which then combines with moisture in the air to form sulfuric acid. If the contaminated air is inhaled, the acid damages lung tissue.

Oxidizing Agents

 Other corrosive poisons destroy tissue by oxidation. For example, ozone, a component of photochemical smog (Chapter 13), causes damage to tissues by deactivating enzymes and oxidizing the lipid coatings on cell walls.

■ METABOLIC POISONS

Metabolic poisons are more specific in their actions than corrosive poisons. They cause harm, and frequently death, by interfering with an essential metabolic process in the body. Examples of metabolic poisons are cyanide and carbon monoxide (Chapter 13).

Cyanide

Many of us have watched a movie in which a captured spy being interrogated suddenly pops something into her mouth and dies instantly. The spy probably bit down on a cyanide capsule. A dose as small as 90 mg (0.003 oz) can be fatal.

Gaseous hydrogen cyanide (HCN) and cyanide salts such as sodium cyanide (NaCN) have many uses. Hydrogen cyanide is used as a fumigant to destroy rodents and insects in warehouses and grain storage bins. Solutions of sodium and potassium cyanide are used in electroplating baths; metal that is dissolved in the bath is plated onto the surface of an object placed into the bath. This process is used to make gold-plated jewelry. The polymer polyacrylonitrile, which is used to make fibers for carpeting and fabrics (Chapter 11), contains cyanide, and deaths in fires are often caused by hydrogen cyanide released from these products when they are burned.

Iron can form two ions: Fe^{3+}, iron(III) ion, and Fe^{2+}, iron(II) ion. When Fe^{3+} gains an electron it is reduced to Fe^{2+}.

If cyanide enters the body, it inhibits an enzyme, cytochrome oxidase, that is essential for a key step in the process that makes oxygen available to the cells in the body. Cyanide binds to the iron(III) in the enzyme and prevents its reduction to iron(II). If this reduction does not occur, oxygen cannot be utilized. Cellular respiration ceases and death results within minutes.

An antidote for cyanide poisoning is sodium thiosulfate ($Na_2S_2O_3$), which reacts with cyanide as shown in the following equation.

$$\underset{\text{cyanide}}{CN^-} + \underset{\text{thiosulfate}}{S_2O_3^{2-}} \longrightarrow \underset{\text{thiocyanate}}{SCN^-} + \underset{\text{sulfite}}{SO_3^{2-}}$$

The transfer of a sulfur atom from the thiosulfate ion to the cyanide ion results in the formation of the relatively harmless thiocyanate ion. Because cyanide acts so rapidly, however, thiosulfate must be administered almost immediately after the poison is ingested to be effective.

TOXIC METALS

Metals are widely used in our society and there are many opportunities for exposure to them. People who live near incinerators and industrial plants are concerned that they may be exposed to fine metal particles released into the atmosphere when materials containing metals are burned. There is also concern that metal-containing wastes produced by industry and released into rivers and lakes may pollute water.

The presence of toxic metals in incinerator ash makes safe disposal of the ash very difficult.

Most metals, including those such as iron that are essential for life, are toxic if consumed in large enough quantities. Some are toxic in very small amounts. Metals that are of particular concern include beryllium, cadmium, chromium, lead, mercury, thallium, and arsenic, which is not a metal but is often included with them because it has many metallic properties.

Heavy metals—those found near the end of the periodic table—derive their toxicity from their ability to react with sulfhydryl (−SH) groups. These groups are present in the active sites of enzyme systems involved in oxygen transport and the production of cellular energy. A heavy metal such as mercury, for example, reacts by replacing the two hydrogen atoms in a sulfhydryl group:

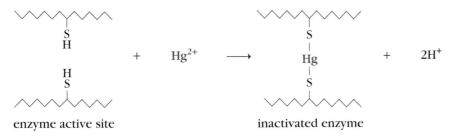

enzyme active site   inactivated enzyme

The free sulfhydryl (−SH) groups are essential for the enzyme's activity; once the bond with the metal is formed, the enzyme becomes inactive.

Arsenic

In the past, arsenic compounds were used extensively as pesticides. Because of their toxicity, inorganic arsenic salts containing arsenate ions (AsO_4^{3-}) or arsenite ions (AsO_3^{3-}) were used to destroy rodents, insects, and fungi. The less toxic organic compound arsphenamine was the first drug able to treat syphilis successfully. In recent years there has been a decline in the use of arsenic compounds and few of today's pesticides contain them.

There are many recorded instances of murderers using arsenic-containing pesticides to dispose of their victims.

Like heavy metals, arsenic compounds inactivate enzymes by reacting with sulfhydryl groups in enzyme systems:

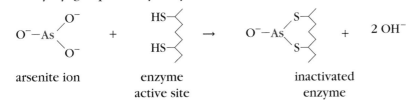

arsenite ion   enzyme   inactivated
                    active site   enzyme

Figure 16-5 During World War I (1914–1918) military personnel were issued gas masks to protect them from toxic gases.

During World War I, the U.S. Army was caught off guard when the German army first used "mustard" gas (Fig. 16-5). The term *mustard* refers to the yellow color of the clouds of chlorine gas (Cl_2) that were the first chemical agents used in modern warfare. Chlorine is a corrosive poison that forms hydrochloric acid in the wet mucous membranes of the nose and lung. Extensive research to find a "more effective" chemical warfare agent led to the production of an arsenic-containing gas, Lewisite, that was named for W. Lee Lewis, who discovered it.

Lewisite was never used in warfare.

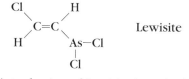

Lewisite

Anticipating the introduction of Lewisite into the battlefield, British scientists began a search for an antidote to it. Knowing that arsenic reacts with sulfhydryl groups, they synthesized a compound containing sulfhydryl groups that would compete with the enzymes for arsenic, and thus prevent the arsenic from disrupting the enzyme system.

The compound, British anti-Lewisite (BAL) also proved effective as an antidote for heavy metal poisoning. BAL is a **chelating agent** (from the Greek *chela*, meaning "claw"). It surrounds and binds a heavy metal atom in the same way that a claw can surround an object.

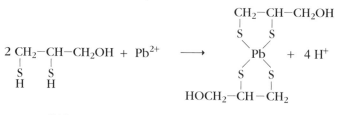

BAL

lead chelated by BAL

Once the metal ions are removed from the enzyme sites, the sulfhydryl groups reform and the enzyme resumes its normal functions.

BAL, in combination with another chelating agent, ethylenediamine-tetraacetic acid (EDTA), is a standard treatment for heavy metal poisoning.

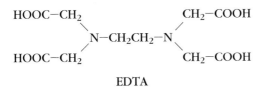

EDTA

The four carboxylic acid (COOH) groups and the two nitrogen atoms in EDTA form bonds with the metal ion.

EDTA is administered as the calcium complex because uncomplexed EDTA could cause problems by combining with calcium ions in the blood. In the body, calcium in the complex is displaced by the heavy metal, which binds more strongly to EDTA than calcium does. The heavy metal-EDTA complex is water-soluble and is excreted in the urine. To be effective, treatment must begin as soon as possible after ingestion of the metal.

Mercury

Mercury (Hg) is the only metal that is a liquid at room temperature. The metal is used in electric discharge tubes (mercury lamps), pressure gauges in laboratories, and in dental amalgams. Mercury salts are used as fungicides; in the past they were used in the felt hat industry (Chapter 12).

Inorganic mercury salts are not very toxic. If swallowed, most of the salts are excreted unchanged from the body. Mercury vapor, however, is very toxic. If inhaled, it passes into the bloodstream and is transported to the brain, where it causes serious damage to the nervous system. Symptoms of mercury poisoning include depression, insomnia, irritability, shaking of the hands, and psychotic behavior.

Organic mercury salts are much more toxic than inorganic salts. The most notorious case of environmental poisoning with mercury occurred in the 1950s in Japan, when a chemical plant discharged mercury-containing wastes into Minimata Bay (Chapter 12). Aquatic organisms became contaminated with mercury, and, in a toxic organic form, the mercury passed up the food chain to small fish and then to larger fish. Through bioaccumulation, the mercury became more concentrated at each level, and more than 100 people who ate fish taken from the bay were poisoned. Forty-three people died and many children were born with birth defects (Fig. 16-6).

Magic properties have been attributed to mercury. For thousands of years the strange, silvery liquid was used to treat almost every imaginable ailment.

Lead

Of all the toxic metals, lead is the most widely distributed in the environment, and the one most likely to be encountered by the average person. In the United States, lead is used primarily to manufacture lead storage batteries for automobiles. It is also used to make plumbing fixtures, pigments, solder, and coverings for cables, and it is present in glazes on some imported ceramic articles. At one time, lead poisoning was common among workers involved in the production of lead or the manufacture of lead products, but now precautions are taken to limit workers' exposure. Today, those at greatest risk are members of the general public, particularly young

Acidic foods can leach lead compounds from lead-containing glazes on ceramic dishes.

Figure 16-6 Villagers who ate mercury-contaminated fish and shellfish taken from Minimata Bay, Japan, suffered from impaired vision, numbness of the limbs, paralysis, and other symptoms; many died.

children, who live in older buildings, or in an area close to a battery-recycling plant, a lead smelter, or an industry that is likely to release lead-containing particulates.

Traces of lead are often present in drinking water (Chapter 12). In older buildings that have lead pipes, small amounts of lead are dissolved from the pipes as the water passes through them, particularly if the water is acidic. Lead may also be dissolved from lead solder on copper pipes. As a precaution, hot water from the faucet should not be used for cooking, and first thing in the morning, cold water should be allowed to run for several minutes.

Another source of lead is old paint. Although the use of lead-based paints was banned in 1978, children continue to be poisoned with it because many older buildings contain lead-based paint under layers of newer paint. If the painted surfaces are sanded to remove the paint, lead-containing dust contaminates the air in the building. In deteriorating buildings in poor neighborhoods where paint is often peeling from the walls (Fig. 16-7), lead poisoning is common. Undernourished, often hungry, children frequently develop an appetite for strange things. This condition, which is termed **pica**, leads them to eat paint.

Lead has been detected in foods in concentrations up to 300 ppb and in municipal water supplies in concentrations up to 100 ppb. These levels do not pose much of a threat because the average adult can excrete 2 mg of lead per day, and most of us do not generally take in this much in the food we eat and the air we breathe. If intake exceeds excretion, however, the excess lead is rapidly transported to the bone marrow and then stored in the bones. Some lead remains in the liver and kidneys but 90 percent eventually ends up in the bones. Lead disrupts hemoglobin metabolism, causing anemia, and, like mercury, it inhibits enzymes containing sulfhydryl groups, which causes damage to the central nervous system.

Children are at greater risk than adults because they cannot excrete as much lead. Consequently they retain a higher percentage of ingested lead than adults do.

Figure 16-7   Paint peeling from an old building.

Also, their growing bones do not absorb lead as rapidly as full-grown bones; the lead remains in the bloodstream longer and has more opportunity to damage developing organs. The concentration of lead in blood is measured to detect lead poisoning. Concentrations as low as 10 µg/dL are considered to be dangerous.

In the 1970s, a concentration of 50 µg/dL of lead in a child's blood was considered acceptable.

A deciliter (dL) is equal to 100 mL.

Exposure to lead can stunt a child's intellectual, behavioral, and physical development. Studies have shown that infants exposed to lead have IQ scores that are 5 percent lower by age 7 than the scores of unexposed children; they are six times more likely to have reading disabilities, and they have a high-school drop-out rate seven times higher than their peers.

NEUROTOXINS

Neurotoxins are metabolic poisons that act by disrupting the normal transmission of nerve impulses in the central nervous system. They include the alkaloid coniine in hemlock, chemical warfare agents (nerve gases), atropine, strychnine, nicotine, organophosphate insecticides, and botulinum toxin, the most deadly poison known.

The Transmission of Nerve Impulses

Nerve impulses travel along nerve fibers by electrical impulses. To pass from the end of one nerve fiber to receptors on the next nerve fiber, the impulse must cross a small gap, called a **synapse** (Fig. 16-8). When an electrical im-

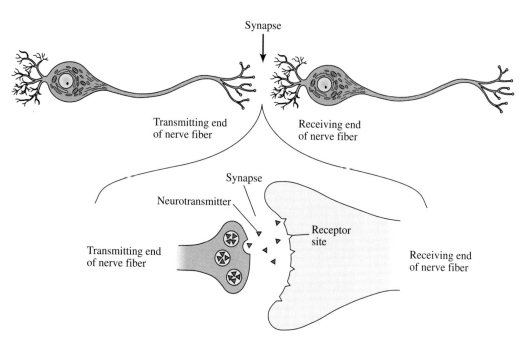

Figure 16-8 The transmission of a nerve impulse. A nerve impulse is transmitted from one nerve fiber to another when an electrical impulse stimulates the release of neurotransmitter molecules from the transmitter end of one nerve; the neurotransmitter molecules cross the synapse and fit into receptor sites on the receiving end of another nerve fiber.

pulse reaches the end of a nerve fiber, chemicals called **neurotransmitters** are released that allow the impulse to cross the gap and travel to receptor cells on the receiving nerve fiber. Each neurotransmitter must fit into a specific receptor to bring about the transfer of the message. Once the nerve impulse has been received, the neurotransmitter is destroyed; the synapse is cleared and is ready to receive the next electrical signal.

An important neurotransmitter is acetylcholine. Once it has mediated the passage of an impulse across the synapse, it is broken down to acetic acid and choline. The reaction is catalyzed by the enzyme cholinesterase.

$$\underset{\text{acetylcholine}}{\overset{\displaystyle\overset{O}{\underset{\|}{}}\qquad\qquad\overset{CH_3}{\underset{|}{}}}{CH_3COCH_2CH_2-\overset{+}{N}-CH_3}} + H_2O \xrightarrow{\text{cholinesterase}} \underset{\text{acetic acid}}{\overset{\displaystyle\overset{O}{\underset{\|}{}}}{CH_3C-OH}} + \underset{\text{choline}}{\overset{\displaystyle\qquad\overset{CH_3}{\underset{|}{}}}{HOCH_2CH_2-\overset{+}{N}-CH_3}}$$

Other enzymes convert acetic acid and choline back to acetylcholine, which can then transmit another impulse across the synapse.

Different neurotoxins disrupt the acetylcholine cycle at different points: They either block the receptor sites, block the synthesis of acetylcholine, or inhibit the enzyme cholinesterase.

Neurotoxins That Block Acetylcholine Receptors

Neurotoxins such as atropine act by occupying the receptor sites on nerve endings that are normally occupied by the neurotransmitter acetylcholine. By occupying those sites, the neurotoxin stops the transmission of impulses across the synapse. The message is sent but cannot be received. Small amounts of atropine can be used as a local anesthetic. Applied to the skin, the chemical relieves pain by deactivating the nerve endings on the skin.

Another compound that blocks receptor sites is nicotine. Nicotine first stimulates and then depresses the central nervous system. Absorption of as little as 50 mg of nicotine can kill an adult in a few minutes. This potent poison is oxidized to less toxic products in air at high temperatures, thus rendering cigarette smoke addictive, not lethal. Nicotine is a good example of the connection between the toxicity of a chemical and the route by which it enters the body. Experiments with mice have shown that nicotine is 1000 times more toxic if it is absorbed into the bloodstream through the skin than if it is administered orally. Heavy cigarette smokers are at risk because nicotine is absorbed very rapidly from the lungs into the bloodstream.

In 1988, the U.S. Surgeon General officially announced that nicotine was an addictive substance.

Neurotoxins That Block the Synthesis of Acetylcholine

Botulinum toxin is produced in improperly canned foods by the bacterium *Clostridium botulinum*, which grows naturally in soil. There is a very small chance of finding the toxin in commercially canned food; most of the known cases of botulinum poisoning are from improperly heated home-canned, nonacidic vegetables. Botulinum toxin, one of the most toxic materials known, binds irreversibly to the nerve endings, preventing the synthesis of acetylcholine. As a result, impulses are not sent to the muscles involved in respiration, and the victim dies from respiratory failure. Fortunately, botulinum poisoning can be avoided by cooking foods at 100°C (212°F) for an adequate time. This treatment kills the bacterium.

Minute quantities of botulinum toxin are used to relieve many muscle disorders, including strabismus (squint eye).

Neurotoxins That Inhibit Cholinesterase

Anticholinesterase poisons prevent the breakdown of acetylcholine by inactivating the enzyme cholinesterase. These poisons bond to the enzyme and thus prevent the breakdown of acetylcholine. As the acetylcholine builds up, nerve impulses are transmitted in quick succession, and nerves, muscles, and other organs are overstimulated. The heart begins to beat erratically, causing convulsions and death.

Anticholinesterase poisons include the organophosphate insecticides malathion and parathion, which will be considered in more detail in the next section. If absorbed into the body, these compounds are converted to cholinesterase inhibitors.

The toxicity of some mushrooms is due to their ability to disrupt the cholinesterase cycle.

Chemical Warfare Agents

At the end of World War I, there was general agreement that poisonous gases should never again be used in warfare. Despite this agreement, many nations continued to develop new and increasingly deadly military poisons.

During World War II, German chemists developed the organophosphate nerve gases Tabun and Sarin. Tabun has a fruity odor while Sarin, which is four times as toxic as Tabun, is odorless, and thus more difficult to detect. After the war, the U.S. army began manufacturing both these poisons (which they named agent GA and agent GB) and other related poisons.

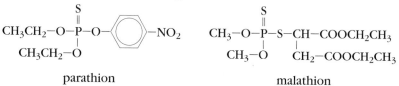

Tabun (GA)   Sarin (GB)

Sarin is one of the most toxic chemicals ever synthesized.

Tabun and Sarin are cholinesterase inhibitors. Both contain an organic group bounded to a P—O group. If troops were exposed to the gases, the gases would be absorbed through the skin as well as through the lungs. Victims would lose muscle control and die in a few minutes from suffocation.

The organophosphate insecticides malathion and parathion (Chapter 18) are structurally similar to the nerve gases.

parathion   malathion

Although parathion and malathion are considerably less toxic than Tabun is, they must be used with extreme caution. Acute poisoning by these phosphate-based insecticides can cause tremors, convulsions, and cardiac or respiratory failure. Parathion is available for application only by licensed pesticide applicators (Fig. 16-9).

Figure 16-9 Agricultural workers who apply anti-cholinesterase insecticides, such as parathion, must wear protective clothing.

DETOXIFICATION OF CHEMICALS

Because the liver has the ability to **detoxify** (render harmless) many chemicals, humans can tolerate moderate amounts of some poisons. The liver can detoxify a chemical by oxidizing it, reducing it, or by coupling it to a natural chemical such as a sugar or an amino acid.

Oxidation-Reduction Reactions

Ethyl alcohol is detoxified by an oxidation reaction. After an alcoholic beverage is consumed, ethyl alcohol is absorbed into the bloodstream from the intestines and transported to the liver, where enzymes catalyze the oxidation of alcohol, first into acetaldehyde, which is further oxidized into acetic acid, and then into carbon dioxide and water.

$$\underset{\text{ethyl alcohol}}{CH_3CH_2OH} \longrightarrow \underset{\text{acetaldehyde}}{CH_3\overset{\overset{\displaystyle O}{\|}}{C}-H} \longrightarrow \underset{\text{acetic acid}}{CH_3\overset{\overset{\displaystyle O}{\|}}{C}-OH} \longrightarrow CO_2 \ + \ H_2O$$

Chronic alcohol consumption causes liver enzymes to build up. The same enzymes that oxidize alcohol also oxidize the male sex hormone, testosterone. Alcoholic impotence, a well-known symptom of the disease of alcoholism, is a direct result of the oxidation of testosterone by the high concentration of liver enzymes.

 The end product of liver oxidation is not always less toxic than the chemical being oxidized. For instance, methyl alcohol (CH_3OH) is oxidized to the more toxic chemical formaldehyde (HCHO). Methyl alcohol poisoning is known to cause blindness, respiratory failure, and death. It is not the methyl alcohol itself that causes the problems, but its oxidation product, formaldehyde.

 Ethyl alcohol is often administered intravenously as an antidote for methyl alcohol poisoning and for ethylene glycol (antifreeze) poisoning. Because ethyl alcohol competes with other alcohols, such as methyl alcohol and ethylene glycol (CH_2OHCH_2OH), for oxidation in the liver, it can be used to slow their conversion to more harmful oxidation products, thus giving the body a chance to excrete the alcohols before damage is done.

Alcohol is the most abused drug in the United States; it is calculated that 10 million of our citizens are alcoholics.

Coupling Reactions

The cytochrome P-450 enzymes in the liver play an important role in detoxifying poisons. These enzymes act on fat-soluble substances to make them water-soluble and thus more easily excreted. For example, benzene (C_6H_6), a component of gasoline, is a toxic aromatic compound that is insoluble in water. If benzene is ingested, it tends to be deposited in fatty tissues in the liver. The P-450 enzymes oxidize it to phenol, which is more soluble in water. The phenol then couples to glucuronic acid, a sugar that is naturally present in the liver, to form phenyl glucuronide, which is even more water-soluble than phenol, and is readily excreted in the urine.

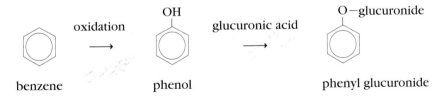

Breathalyzer: A Chemical Test for Intoxication

Over the past several decades, chemical tests for ethyl alcohol intoxication have become commonplace evidence in the courtroom. The value presented for the defendant's blood alcohol concentration (BAC) can result in a jail term or exoneration. Before 1985, most states used a BAC value of 0.15 percent as the concentration of alcohol in the blood that rendered an individual "intoxicated." Since that time, the public has become more concerned about drunk driving, and states have lowered their BAC level for a driving-while-intoxicated (DWI) citation to 0.10 percent. Some states are debating lowering the value even further.

Congress has proposed required alcohol testing for "safety-sensitive" workers as part of the Om-

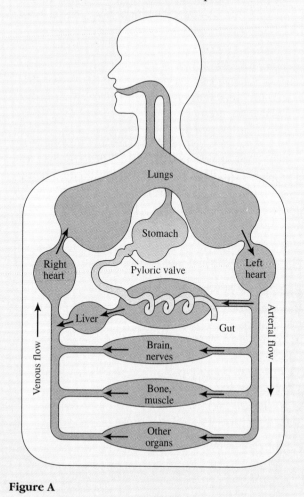

Figure A

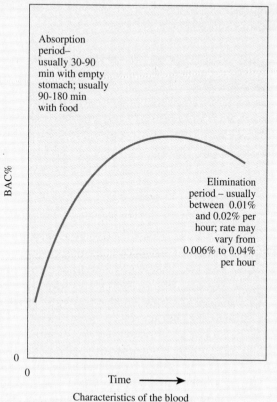

Absorption period– usually 30-90 min with empty stomach; usually 90-180 min with food

Elimination period – usually between 0.01% and 0.02% per hour; rate may vary from 0.006% to 0.04% per hour

BAC%

Time

Characteristics of the blood alcohol concentration (BAC) curve

Figure B

nibus Transportation Employee Training Act. People employed as airline pilots, train engineers, and truck drivers would be required to take a breath test to determine their BAC. If the workers tested above 0.04 percent they would be removed from their jobs for a period of at least 8 hours. To register a 0.04 percent BAC, a 180-pound person would have had to consume two drinks in the preceding 30 to 120 minutes.

A person's BAC varies depending on how much alcohol was consumed, the size of the person, and whether the individual had a full or an empty stomach. Alcohol is absorbed very slowly while in the stomach, but once it passes into the small intestine, it is absorbed rapidly into the bloodstream and carried to all parts of the body (Fig. A). If food is in the stomach, much of the alcohol is retained with it until the food is digested and moves into the small intestine. A stomach filled with a heavy meal with a high fat content may take 4 to 6 hours to empty. Alcohol taken on an empty stomach can be absorbed in 30 to 90 minutes (Fig. B).

Alcohol is removed from the bloodstream in the liver by enzymatic oxidation. The liver works slowly; the rate of elimination of alcohol is only 0.015 percent per hour. Blood also transports alcohol to the lungs, where it is released into the expired breath.

For forensic purposes, a test on expired breath rather than blood, has a lot of advantages. It is a noninvasive technique, it is rapid, and it can be used in the field by police officers rather than requiring laboratory technicians. In the standard test, the individual being tested blows expired air from the lungs into a 52-mL breathalyzer sample chamber. If alcohol is present in the expired air sample, it undergoes an oxidation-reduction reaction with dichromate, which is added to the air sample in the breathalyzer:

$$8\,H^+ + Cr_2O_7^{2-} + 3\,C_2H_5OH$$
$$\text{dichromate ion}$$
$$\text{(orange)} \longrightarrow 2\,Cr^{3+} + 3\,C_2H_4O + 7\,H_2O$$
$$\text{chromium ion}$$
$$\text{(green)}$$

The oxidation reaction converts the orange dichromate ion to the green chromium (III) ion. The light source in the breathalyzer instrument measures the increase in green color and converts this measurement to concentration of alcohol in the sample. The more alcohol present in the expired breath, the more intense the green color.

A conversion factor of 2100:1 is used to convert the breath alcohol to BAC. An assumption is made that 2100 mL of deep lung breath will contain the same weight of alcohol as 1 mL of blood when the body is at equilibrium.

Breath testing is vulnerable to contamination because of the low levels of alcohol involved. For a 0.1 percent BAC sample, the expired air contains only 25 µg of alcohol. If an individual with a zero BAC inhales alcohol vapors from an open bottle of 50 percent alcohol his or her breath can send the breath tester off scale for several minutes afterward. An individual who has consumed some alcohol and who, while being tested, "burps" stomach gas can also cause the breathalyzer to register a high BAC value. In this case, one of the basic assumptions of the test, namely that the alcohol has come from the lungs, is not true and the test is not valid.

Breath testing for BAC is an important safeguard for the public. Public officials have to remember that they are making a scientific measurement that must be done properly to identify intoxication without falsely accusing sober drivers.

Questions

1. Do you think that employees who operate expensive industrial equipment should be required to take a breathalyzer test when asked to do so by their employer?

2. A painter who has worked all day indoors is stopped on his way home from work for speeding. The arresting officer administers a breathalyzer test and obtains a 0.25 percent BAC level for the painter. The painter argues that he hasn't had a drop of alcohol. Can you think of an explanation for this result?

Reference: D. N. Hume and E. Fitzgerald, "Chemical Tests for Intoxication," *Analytical Chemistry* 57 (1985).

TERATOGENS, MUTAGENS, AND CARCINOGENS

The toxic chemicals we have discussed so far act quickly and cause harm almost immediately. The harmful effects of other toxic chemicals are often not apparent until as much as 10 or more years after exposure to them. Toxic substances of this kind include **carcinogens**, chemicals that cause cancer, **teratogens**, chemicals that cause birth defects, and **mutagens**, chemicals that are capable of altering genes and chromosomes sufficiently to cause abnormalities in offspring. Some chemicals are both mutagens and carcinogens, while others are only carcinogens.

Teratogens

During the third to eighth week of pregnancy, the embryo is particularly sensitive to the harmful effects of teratogens. During this critical period the different parts of the body are differentiated, and the limbs, eyes, ears, and the internal organs are developed. If the embryo is exposed to teratogens at this time, abnormalities in development can result. Examples of teratogens are listed in Table 16-3.

Probably the most notorious teratogen is thalidomide, a drug that was prescribed extensively as a tranquilizer and sleeping pill in Europe and Japan in 1960 and 1961. On the basis of animal studies, thalidomide was considered so safe at that time that it was sold in what was then West Germany without a prescription. In 1961, it became shockingly apparent that the drug could cause gross deformities. If it was taken between days 34 and 50 of pregnancy, children were born with no arms or legs or with abnormally short limbs, and with other deformities (Fig. 16-10). Approximately 10,000 children were affected.

The word *teratogen* is derived from the Greek word *teras*, meaning "monster."

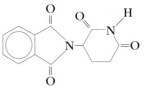

thalidomide

Thalidomide was never approved for distribution in the United States, thanks to the action of Frances Kelsey, an investigator with the Food and Drug Administration (FDA). Dr. Kelsey was not convinced that the animal testing done by the company that was promoting the drug was complete, and, despite pressure from the company, she refused to approve thalidomide for sale. As a result, only 20 "thalidomide babies" were born in the United States.

Thalidomide is now being used in the United States as an experimental drug for the treatment of AIDS.

Another drug that has been shown to be a teratogen is Accutane (retinoic acid), which is used to treat acne. In 1988, the FDA concluded that Accutane could cause heart defects, facial abnormalities, and mental retardation if a developing embryo was exposed to the drug for just a few days. Accutane may have caused birth defects in approximately 1000 children born to women who took the drug between 1982 and 1986.

One of the most dangerous, and most commonly used, teratogens is alcohol. Consumption of even small quantities of alcohol can lead to the development of fetal alcohol syndrome (FAS). Infants with this syndrome are abnormally

TABLE 16-3   Teratogenic and Mutagenic Substances

Teratogens	Mutagens
Arsenic (As)	Aflatoxin
Cadmium (Cd)	Benzo(α)pyrene
Cobalt (Co)	Lysergic acid diethylamide (LSD)
Mercury (Hg)	Nitrous acid (HNO_2)
Diethylstilbestrol (DES)	Ozone (O_3)
Polychlorinated biphenyls (PCBs)	Tris(2,3-dibromopropyl)phosphate
Retinoic acid	
Thalidomide	

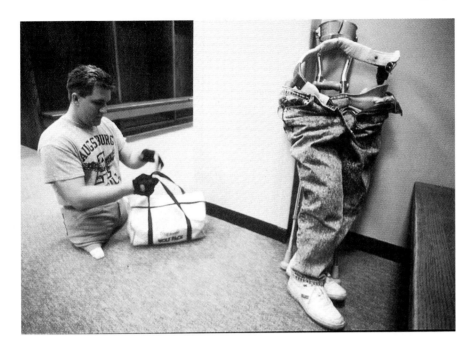

Figure 16-10   David Stevens was born with no legs as a result of his mother's use of thalidomide during her pregnancy. Despite his handicap, Stevens was a back-up defensive lineman for the Augsburg College, Minnesota, team, and he tried out for the Dallas Cowboys.

small, often are mentally retarded, and have facial deformities (Fig. 16-11). Cigarette smoking during pregnancy also affects the fetus. Smoking raises blood levels of carbon monoxide, nicotine, and benzo(α)pyrene (a known carcinogen) in the mother's blood, and these chemicals can then pass into the developing infant's blood. Since any chemical that can pass across the placenta from mother to child is a potential teratogen, pregnant women should not drink alcohol or smoke, and they should limit medications to those prescribed by their doctors.

Mutagens

Mutagens are chemicals that alter the sequence of bases in the nucleic acids that make up DNA, the material that carries genetic traits from one generation to the next (Chapter 10). In addition to chemicals, changes in DNA may be brought about by exposure to radiation, and some occur spontaneously. Changes in DNA are called **mutations**. Examples of mutagens are listed in Table 16-3.

All the cells in the body contain chromosomes, threadlike structures composed of DNA and proteins. Genes are sections of the DNA in the chromosomes. Humans have 46 chromosomes (23 pairs), each of which may contain over 100,000 genes. Each gene contains the information required to produce one specific protein. These proteins determine hereditary traits such as eye and hair color and height. They also determine if an individual has a hereditary disease.

If a chemical substance, or some other agent, causes a mutation, a protein with an incorrect sequence of amino acids may be formed. If the mutation occurs in a body (somatic) cell, very little damage may result, or it may cause uncontrolled cell growth, a cancer. If the mutation occurs in a germ cell (egg or sperm), the alteration in the amino acid sequence is passed on to the offspring and may result in the child's being born with a hereditary disease such as sickle cell anemia, hemophilia, or cystic fibrosis. The abnormal genes in cystic fibrosis, and in several other hereditary diseases, have been identified, but how the faulty genes cause the symptoms associated with the diseases is still not understood.

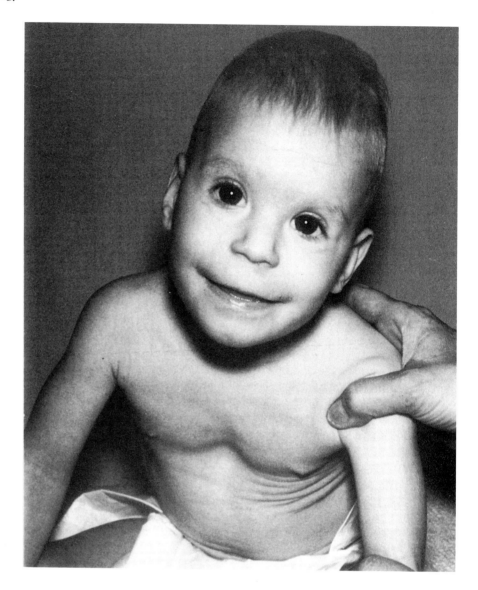

Figure 16-11 Fetal alcohol syndrome. Babies whose mothers drink alcohol during pregnancy are usually underweight; often they are mentally retarded and have slight facial abnormalities.

Many chemicals that alter chromosomes and produce mutations in plants, viruses, bacteria, insects, mice, and other animals have been identified, but there is no conclusive proof that any chemical has caused mutations in human germ cells. There is concern, however, that chemicals that cause mutations in microorganisms and animals may also cause mutations in humans.

A chemical that is a cause for concern is sodium nitrite ($NaNO_2$), which is used to retard spoilage and to give a pink color to processed meats such as hot dogs, bologna, bacon, and smoked ham (Fig. 16-12). When these foods are eaten, sodium nitrite is converted to nitrous acid (HNO_2) by stomach acid.

During curing, nitrites are converted to nitric oxide (NO) which reacts with brown pigments in muscle and blood in meats, converting them to pink compounds.

$$NaNO_2 \ + \ HCl \ \longrightarrow \ HNO_2 \ + \ NaCl$$

sodium nitrite   nitrous acid

Figure 16-12 Processed meats such as bologna, bacon, and hot dogs are colored pink by the addition of sodium nitrite. Without nitrite, the meats would have a gray appearance.

Nitrous acid has been shown to be a bacterial mutagen, and, as a result of this finding, the FDA has recommended limiting the amount of nitrite allowed in foods and is considering a complete ban on the use of the additive.

Another mutagen that has received attention is tris(2,3-dibromopropyl) phosphate, commonly called "tris," that was once used as a flame retardant in children's sleepwear made from synthetic polymer fibers. Tris was an excellent flame retardant, but unfortunately laboratory studies showed that it caused mutations (and also cancer) in animals, and its metabolites were found in the blood of children wearing treated pajamas. The presence of the metabolites indicated that tris is absorbed through the skin, and it was ordered to be removed from the market by the U.S. government. It is still sold in other countries.

Carcinogens

Chemical carcinogens are compounds that cause cells in an organism to reproduce and grow abnormally to produce a **malignant tumor**. Cells in a malignant tumor may divide more rapidly or more slowly than normal cells do. Growth is usually uncontrolled, and cells in the growth migrate through the body to other tissues, which they destroy. Tumors are described as **benign** if they grow slowly and do not spread to other tissues. Benign tumors, unlike malignant tumors, often regress spontaneously.

Chemicals That Cause Cancer Chemicals that are known to cause cancer vary widely in their structures, and scientists still do not know which chemical structures are likely to make a compound a carcinogen. At the present time, about 23 chemicals have been shown conclusively to cause cancer in humans (Table 16-4). Another approximately 300 have been shown to cause cancer in animals. As with mutagens, there is concern that animal carcinogens may also be human carcinogens.

As early as 1775, in England, soot was suspected as the cause of the high incidence of testicular cancer in chimney sweeps. Later it was shown that 3,4-benzo(α)pyrene, a component of soot and coal tar, is a carcinogen.

TABLE 16-4   Substances Known to be Human Carcinogens

Substance	Source or Use
1. Aflatoxins	Produced by fungi on peanuts
2. 4-Aminobiphenyl	Used to retard oxidation in rubber
3. Analgesic mixtures containing phenacetin	Used in over-the-counter pain relievers
4. Arsenic and certain arsenic compounds	Wood preservative, insecticide
5. Asbestos	Insulation
6. Azathioprine	Drug developed to prevent organ rejection in kidney transplants
7. Benzene	Solvent in paint, gasoline additive
8. Benzidine	Intermediate in the production of dyes
9. Bis(chloromethyl)ether	Used in the synthesis of certain plastics
10. 1,4 Butanediol dimethylsulfonate (Myleran)	Chemotherapeutic agent for leukemia
11. Chlorambucil	Chemotherapeutic agent for lymphoma
12. 1-(2-chloroethyl)-3-(4-methylcyclohexyl)-1-nitrosourea (MeCCNU)	Chemotherapeutic agent for malignant melanoma and cancer of the brain, lungs, and digestive tract
13. Chromium	Steel manufacturing
14. Conjugated estrogens	Drug for uterine bleeding
15. Cyclophosphamide	Chemotherapeutic drug for leukemia
16. Diethylstilbestrol	Synthetic hormone, used for menopause, menstrual disorders
17. Erionite	Catalyst used for cracking crude oil
18. Melphalan	Chemotherapeutic agent for ovarian cancer
19. Methoxsalen with ultraviolet A therapy	Used in the treatment of severe psoriasis
20. Mustard gas	Chemical warfare agent
21. 2-Naphthylamine	Intermediate in the production of dyes
22. Thorium dioxide	X-ray imaging medium
23. Vinyl chloride	Monomer in the production of vinyl plastic

Source: National Institute of Environmental Health Sciences, "Sixth Annual Report on Carcinogens," 1991.

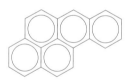

3,4-benzo(α)pyrene

Multi-ringed compounds such as 3,4-benzo(α)pyrene are called polycyclic aromatic hydrocarbons (PAHs). PAHs have been studied extensively and many of them have been shown to be carcinogenic. They are often formed when organic materials are burned. They are present in automobile exhaust, cigarette smoke, charcoal-grilled meat, and stack gases from industrial incinerators. Although not all of the PAHs that have been tested are carcinogens, there is some correlation between the size and shape of the PAH molecule and carcinogenicity.

Other compounds that were studied because they were suspected of causing cancer in humans were the aromatic amines. In 1895, a German physician noted a number of cases of bladder cancer among workers in a dye-making factory. In 1937, it was shown that 2-naphthylamine, one of the amines to which they were exposed, caused bladder cancer in dogs. Since then, other related compounds, including benzidine, which is an intermediate in the manufacture of magenta-colored dyes, have been shown to cause bladder cancers in animals.

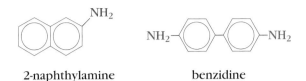

2-naphthylamine   benzidine

Other dyes that have received attention are the azo dyes, several of which have been found to be carcinogens. These compounds include two benzene rings joined by a —N=N— group in their structures. They have been used for many years to color foods (Chapter 17). Recently, a number of them, including FD&C Yellow No. 3, were banned because they can react with stomach acid to produce known carcinogens.

Not all carcinogens are aromatic compounds. The nonaromatic organic compound vinyl chloride (CH_2=CHCl), which is used in huge quantities to make the polymer poly(vinyl chloride) (Chapter 11), is a well known carcinogen. Its carcinogenicity was confirmed after it was suspected of causing a rare form of liver cancer in workers who were exposed to it. Asbestos, a silicate mineral, was a serious problem at one time among shipyard workers who were exposed to large quantities of it in the atmosphere in their workplace. Inhalation of very small, fine asbestos fibers caused lung cancer and various respiratory diseases (Chapter 2). Certain metals and metal compounds are also carcinogenic. Metals that can cause cancer in animals include beryllium, chromium, lead, nickel, and titanium.

Carcinogens are present in the natural environment in many plants and microorganisms. Food items that contain carcinogens include certain mushrooms, pepper, mustard, celery, and citrus oils. Aflatoxin-B, which causes liver cancer in rats, can grow as molds on grains. The possible dangers associated with carcinogens in food are discussed in Chapter 17.

The Development of Cancer   Metabolic poisons and neurotoxins produce effects almost immediately on entering the body, but carcinogens behave differently. Many years may elapse before a person exposed to a carcinogen shows signs of cancer. For example, poly(vinyl chloride) was first manufactured in the 1940s, but it was not until about 25 years later that liver cancer began to appear in workers who had been exposed to it. Similarly, workers exposed to asbestos and to benzidine did not develop cancers until 20 years after exposure.

Some chemicals cause cancer at the point where they make contact with the body; inhaled carcinogens in cigarette smoke and inhaled asbestos fibers cause lung cancers. Other chemicals cause cancer in an organ that is distant from the entry point. The liver, which receives most of the toxic chemicals that enter the body, and the colon, where solid wastes collect, are particularly susceptible to carcinogens.

How cancers develop is still not understood. The pathways leading to the growth of a tumor are complex, and there is evidence that different carcinogens produce cancers in different ways. It is now generally agreed that carcinogenesis is a two-step process. In the first step, which is called **initiation**, a reaction occurs between the carcinogen and the DNA in a cell, and, as a result, an abnormal cell is formed. The abnormal cell has the potential to grow into a malignant tumor. It may start to divide and grow into a tumor immediately after it is formed, but more often it continues its normal functions and does not prolif-

FD&C stands for Food, Drug, and Cosmetic. It indicates that the FDA has approved the chemical identified by an FD&C number for use in food, drugs, or cosmetics.

erate until a second stage, called **promotion**, occurs months or years later. Promotion is thought to occur in response to exposure to a toxic agent (the promoter). The abnormal cell becomes a cancerous cell and develops into a tumor.

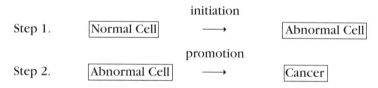

It is expected that cancer will strike almost a million Americans in 1994, and that one-third of the people now living will eventually develop some type of cancer. As life expectancy of the average person continues to rise beyond 70 years, the number of deaths due to cancer will increase.

The World Health Organization has estimated that lifestyle and environmental factors are responsible for the development of up to 90 percent of all cancers. About 10 percent are thought to be caused by inherited genetic traits. Cigarette smoking is blamed for 40 percent of cancers, diet for 25–30 percent, occupational exposure for 10–15 percent, and environmental pollutants for 5–10 percent.

Fortunately, in the United States, about half the people who develop cancer are now cured. Many people could reduce the risk of developing the disease if they would change their lifestyle.

It is estimated that 150,000 deaths per year in the United States are related to cigarette smoking.

EXPLORATIONS

Assessing the Risk of Hazardous Chemicals in the Environment

In the mid-1970s, the Environmental Protection Agency (EPA) began risk analysis to determine the cancer risks associated with toxic chemicals and pesticides. The EPA has a four-step procedure for risk analysis: (1) hazard assessment, (2) dose-response assessment, (3) exposure assessment, and (4) risk characterization. Because the information available to the EPA is often incomplete, the analysis is often imprecise. By constantly reassessing the risks as new information becomes available, the EPA can review and rewrite regulations to protect the public health.

The process of **hazard assessment** involves examining evidence and determining whether a potential hazard has caused harm. Often historical data are available and have been used to calculate risks.

For example, it has been determined that each year 3.6 of every 1000 people who smoke a pack of cigarettes a day will die from smoking-related diseases. Although historical data exist describing exposure of certain individuals to chemicals, it is very difficult to correlate that information with cancer deaths. Because the onset of the cancer occurs long after exposure, many other factors affecting the person's health for the last 20 years may be involved.

In cases where the cause-and-effect relationship is not obvious, epidemiological studies are used. If there is an interest in establishing a link between a particular chemical and a certain cancer, an epidemiological study can be done by comparing a large number of people who were exposed to the chemical and their cancer rates with a large number of people who were not exposed and their cancer rates. The statistical correlation of this type of information has been used by the National Cancer Institute to label a chemical as a "known carcinogen."

A second way to assess a cancer risk where no historical data exist is by testing the chemical of interest on animals. For example, before a new artificial sweetener can be marketed, it must be tested on several thousand mice or rats. Rodents have much shorter lifetimes (2–4 years) than humans, so although they may develop cancers late in their lives, the test takes only 3–5 years. If a significant number of rodents develop tumors after being fed the sweetener, the test indicates that the sweetener may be a carcinogen. The three main criticisms of these studies are (1) large numbers of animals are needed, (2) rodents and humans may respond differently to a particular chemical, and (3) the doses fed to the animals are often very high. Proponents of animal testing point out that all chemicals shown to be human carcinogens by epidemiological testing are also carcinogenic to animals.

A third way to assess the risk of cancer is to conduct a bacterial screening test. The **Ames test** uses *Salmonella typhimurium* bacteria to test for mutations. Although the test is simple to perform and inexpensive, it is the least accurate of the three methods for assessing cancer risk. The Ames test assumes that the first step in the development of cancer is a mutation, and, therefore, a chemical that causes a mutation is likely to be a carcinogen also.

The bacteria that are used in the Ames test are modified biologically so that they are unable to synthesize the amino acid histidine. In the absence of histidine, the bacteria cannot grow. In the test, the modified bacteria are placed in a medium that contains all the ingredients the bacteria need to grow, except histidine. The suspected chemical carcinogen is then added to the medium. If the test chemical is not a mutagen, no bacterial growth will occur. But if it is a mutagen, a mutation will occur, and the bacteria will revert to a form that can synthesize its own histidine, and grow. The growth of bacteria in the test dish identifies the chemical as a mutagen (Fig. 16-A).

Because of the strong correlation between chemicals that cause cancer and those that cause mutations (about 90 percent appear to do both), the Ames test is a useful, and rapid screening test for identifying potential chemical carcinogens. If a chemical is identified as a mutagen in the Ames test, the chemical is then tested on animals to confirm its carcinogenicity.

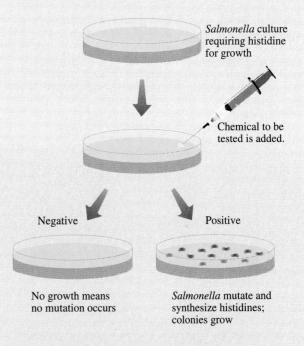

Figure 16-A The Ames test for carcinogens.

The process of **dose-response assessment** begins with establishing a correlation between exposure of animals to a chemical and a harmful effect. Next, the relationship between the dose (concentration of chemical) and the effect is studied. Both the frequency of incidence and severity of the effect are studied. Once this information is available, statisticians use it to predict the effect of the chemical on human health.

If the chemical being studied is shown to be harmful, an **exposure assessment** is undertaken. Information on groups of people who have already been exposed, where and how they came to be exposed, the dose to which they were exposed, and the duration of exposure all must be collected. Only when this process is completed can the risk be characterized.

The information gathered in the first three steps of the risk analysis is compiled, and the risk posed by the chemical in question is calculated. This step is called **risk characterization**, and the risk is expressed as a probability. The 1990 Clean Air Act regulates chemicals that have a cancer risk of greater than one in a million for people who are exposed to high doses. The same standard is used by the FDA for regulating chemicals in food and drugs. Because it is difficult to express this information, risks are often presented as a reduction in life expectancy or the increased probability of dying, as shown in Table 16-A.

TABLE 16-A A Ranking of Hazards According to the Degree of Risk

Hazard	Annual Risk*
Cigarette smoking (1 pk/day)	3.6 per 1000
All cancers	2.8 per 1000
Motor vehicle accident	2.4 per 10,000
Police killed in action	2.2 per 10,000
Air pollution, eastern United States	2.0 per 10,000
Home accidents	1.1 per 10,000
Alcohol, light drinker	2 per 100,000
Radiation, sea level	2 per 100,000
Electrocution	5.3 per 1,000,000
Drinking water containing EPA limit of chloroform	6 per 10,000,000

*Probability of dying.

Question

Should nicotine in cigarettes be classified as a drug? This would make the FDA responsible for regulating the sale and promotion of tobacco products.

Source: R. Wilson and E. A. C. Crouch, "Risk Assessment and Comparisons: An Introduction," *Science*, 236, (1987): 267.

KEY WORDS AND CONCEPTS

Ames test
anticholinesterase poison
benign tumor
carcinogen
chelating agent
corrosive poison
dermal contact
detoxification
dose

ingestion
inhalation
initiation
LD_{50}
malignant tumor
mutagen
mutation
neurotoxin

neurotransmitter
pica
poison
promotion
risk analysis
synapse
teratogen
toxin

QUESTIONS AND PROBLEMS

1. Define:
   a. toxin **b.** poison **c.** LD_{50}
2. Is ethyl alcohol (grain alcohol) poisonous? Explain.
3. Does the toxicity of a chemical depend on the concentration administered to the individual?
4. What are the three main ways a chemical can enter the body? Can you think of another way a chemical can enter the body?
5. A commercial pain relief medication contains 500 mg of acetaminophen per tablet. Assume that the LD_{50} of 338 mg/kg for mice applies to humans as well. How many tablets, taken all at once, would produce a 50 percent chance of a lethal dose of acetaminophen in a 154-lb (70-kg) person?
6. A commercial anti-inflammatory preparation contains 350 mg of ibuprofen per tablet. Assume that the LD_{50} of 1050 mg/kg for rats applies to humans as well. How many tablets, taken all at once, would produce a 50 percent chance of a lethal dose in a 44-lb (20-kg) child?

7. The herbicide Paraquat was once sprayed on marijuana plants by the U.S. government to kill the plants so that they could not be sold. Paraquat has an LD_{50} of 100 mg/kg for rats. Another herbicide, Silvex, has an LD_{50} of 650 mg/kg in rats. Is Paraquat more toxic to humans than Silvex?
8. Dioxin (TCDD) was a toxic by-product formed in the production of Agent Orange during the Vietnam War. Dioxin has an LD_{50} of 0.022 mg/kg in rats. How much dioxin would be lethal to a 154-lb (70-kg) person?
9. Describe how nitric acid can act as a corrosive poison.
10. Write a reaction that shows the effect of sodium hydroxide on protein. Is sodium hydroxide a corrosive poison?
11. How does cyanide act as a poison?
12. Give two sources of mercury poisoning.
13. What is the mechanism of mercury poisoning?
14. How does BAL act as an antidote for mercury poisoning?
15. How does EDTA act as an antidote for lead poisoning?

16. What is pica?

17. Why are children at greater risk of lead poisoning than are adults?

18. Describe what is meant by a neurotoxin.

19. Neurotoxins can interfere with the neurotransmitter acetylcholine. List three different ways that neurotoxins can disrupt the acetylcholine cycle.

20. Nerve gases are acetylcholinesterase poisons. Name two nerve gases.

21. How do nerve gases act on cholinesterase?

22. Why is atropine an antidote for nerve gas poisoning?

23. What do the molecular structures of the nerve gases Sarin and Tabun have in common with the molecular structure of the insecticide malathion?

24. The liver removes toxins from our bodies. How is ethyl alcohol removed from the body?

25. Why is ethyl alcohol used as an antidote for methyl alcohol poisoning?

26. What is the P-450 system? How does it participate in detoxifying water-insoluble poisons?

27. Describe the difference between a teratogen and a mutagen.

28. Describe the difference between a mutagen and a carcinogen.

29. What are the differences between benign and malignant tumors?

30. What are the sources of polycyclic aromatic hydrocarbons (PAHs)?

31. Give two examples of aromatic compounds that are carcinogens.

32. If natural foods contain many natural carcinogens, why don't many young people die from cancer?

33. What is an initiator and what is a promoter of cancer growth?

34. Are all carcinogens also mutagens?

35. The city of New Orleans takes its drinking water from the Mississippi River. It has been shown that the water in the Mississippi contains many carcinogens at low levels (parts per billion). Why isn't the cancer rate in New Orleans much higher than for the rest of the country?

FOOD AND NUTRITION

The bounty of nature. A harvest of fruits and vegetables.

The study of **nutrition** is the study of foods and the ways in which the body utilizes them. It includes the identification of the components in foods that are essential for growth and for the maintenance of health. For humans, a nutritional diet must supply the proper balance of six basic ingredients: carbohydrates, fats, proteins, minerals, vitamins, and water.

517

These essential food components, or **nutrients**, serve three main functions: they provide (1) energy for the performance of essential biological processes, (2) materials for building and replacing body tissues, and (3) materials needed for the regulation of life processes such as the transport of oxygen and digestion. Fats and carbohydrates supply most of the body's energy needs, while proteins supply most of the building blocks for tissue construction. Small, often trace, quantities of nutrients, including vitamins and minerals, act as regulators of vital processes.

In this chapter, we will examine the ways in which the body makes use of the different nutrients. We will discuss nutritional disorders and the effects that deficiencies and overconsumption have on health. Last, we will study food additives, chemicals having little or no nutritional value that are added to foods to prevent spoilage and to flavor, color, or in some other way improve appearance or taste.

Learning Goals:

In this chapter, you should gain an understanding of:

1. The process by which food is digested.

2. Our dietary needs for carbohydrates, fats, and proteins.

3. How energy is obtained from food.

4. Nutritional disorders and diseases.

5. The value and potential danger of food additives.

6. Dietary goals for good health and nutrition.

➡ THE COMPOSITION OF THE BODY

We need food essentially to make and maintain our bodies. Food, therefore, must supply all the elements from which the body is composed. Broken down into its elements, the body is made up as shown in Table 17-1. By weight, four

TABLE 17-1   Approximate Elemental Composition of the Human Body (including water)

Element	Percent by Weight
Oxygen (O)	65
Carbon (C)	18
Hydrogen (H)	10
Nitrogen (N)	3
Calcium (Ca)	1.5
Phosphorus (P)	1
Potassium (K)	0.35
Sulfur (S)	0.25
Chlorine (Cl)	0.15
Sodium (Na)	0.15
Magnesium (Mg)	0.05
Iron (Fe)	0.004
Trace elements	0.546

major elements account for 96 percent of the human body: oxygen, 65 percent; carbon, 18 percent; hydrogen, 10 percent; and nitrogen, 3 percent. Two-thirds of the body's weight consists of water; therefore, much of the oxygen and hydrogen in the body is in water. The remaining oxygen and hydrogen, together with the carbon and nitrogen, are combined to form organic compounds, primarily carbohydrates, fats, and proteins.

Other elements found in the body in significant amounts are calcium, phosphorus, potassium, sulfur, chlorine, sodium, and magnesium. Trace amounts of about 36 other elements are also present. Of these, about 20 are known to be essential for various life processes, but it is not yet known whether the remaining 16 also play essential roles in the body.

The weight contributed by the different atoms in the body is related to the atomic weights of the atoms (O = 16; C = 12; H = 1; N = 14)

THE FATE OF FOOD IN THE BODY

The carbohydrates, fats, and proteins present in food are not in a form in which they can be immediately utilized by the body. They must first be broken down into simpler compounds and then recombined to form the specific compounds that the body needs to build tissues and perform many other functions.

Digestion of Food

Breakdown of food occurs in a series of enzyme-catalyzed reactions as food passes through the digestive tract (Fig. 17-1). Digestion of carbohydrates begins in the mouth when food is chewed and becomes mixed with saliva. **Saliva** contains the enzyme **ptyalin**, or **salivary amylase**, which catalyzes the partial

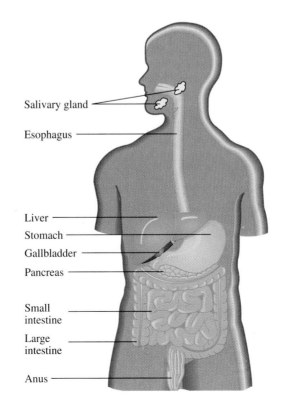

Salivary gland

Esophagus

Liver

Stomach

Gallbladder

Pancreas

Small intestine

Large intestine

Anus

Figure 17-1 The digestive tract.

Starch is a polymer made up of many glucose monomers linked together.

Maltose is composed of two molecules of glucose linked together.

breakdown of starch in food to yield the disaccharide maltose (Chapter 10). When food is swallowed, it enters the stomach and encounters the very acidic gastric juice. In this acid medium (pH of 1–3), the enzyme **pepsin** begins the digestion of proteins by catalyzing the breakdown of approximately 10 percent of the peptide bonds in protein molecules. Polypeptides with molecular weights ranging from 600 to about 3000 are produced. The lining of the stomach is protected from enzyme breakdown by a layer of mucus; but despite this protection, it is slowly digested and must constantly be renewed. Digestion of carbohydrates is halted in the stomach because the acidic conditions inactivate the enzyme ptyalin.

Food passes from the stomach to the small intestine, where secretions from the pancreas neutralize the acid and provide the enzymes needed to complete the digestion of carbohydrates and proteins. Carbohydrates are broken down to the monosaccharides glucose, fructose, and galactose, and polypeptides are broken down to amino acids. The relatively small monosaccharide and amino acid molecules are then absorbed through the intestinal wall into the bloodstream.

Bile, which is manufactured in the liver and stored in the gall bladder, contains bile salts, cholesterol, and bile pigments which give bile its yellow to green color.

Digestion of fats occurs primarily in the small intestine, where triglycerides, which account for more than 90 percent of the lipids in food, are broken down to glycerol and fatty acids. The breakdown is brought about by the combined action of **bile salts** (Chapter 10) and the enzyme **lipase**. Bile salts, which are released into the small intestine from the gallbladder, act rather like soaps in washing (Chapter 10). Like soaps, bile salts have both polar and nonpolar areas in their molecular structures, and they form micelles with fats, thus making it possible for water-soluble lipase to react with water-insoluble triglycerides.

The end products of the digestion of food—monosaccharides, amino acids, glycerol, and fatty acids—are absorbed through the wall of the small intestine into the bloodstream and carried to the liver, where they undergo many further changes. Some glucose is released to the general circulation, but most glucose (and fructose and galactose) is converted to glycogen and stored. Amino acids are converted to enzymes and other proteins, and fatty acids and glycerol are resynthesized to fats (Fig. 17-2).

Cellulose, like starch, is a glucose polymer but the linkages between the glucose monomers in the two polymers are different. Humans do not have enzymes capable of breaking the linkages in cellulose.

Digestion is very efficient; only a small amount of digestible food escapes conversion to small absorbable molecules and is excreted in the feces. Indigestible plant food composed mainly of cellulose—usually referred to as roughage or fiber—accounts for most of the food material in feces.

Building and Maintaining Body Tissues

The end products of digestion, primarily amino acids, are used to build the tissues and organs of the body. These tissues are not static, but are continually being broken down and replaced. Renewal occurs at different rates depending on the type of tissue. Red blood cells, for example, have a life span of 120 days, while the cells lining the intestines are replaced as rapidly as every three or four days. Some tissues last much longer. Collagen, a protein of tendons, has a life span of about 10 years. Many of the amino acids and other compounds formed when tissues break down are reused, but some new material must always be obtained from food.

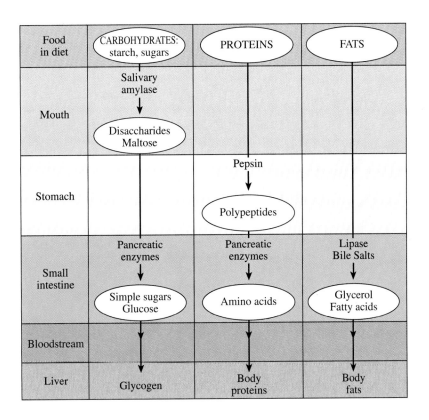

Figure 17-2   Summary of digestion of food, showing breakdown of carbohydrates, proteins, and fats into simple molecules that are absorbed into the bloodstream.

ENERGY FROM FOOD

Food is to the body as fuel is to an engine. Food provides the body with the energy it needs to do work. As we have seen in earlier chapters (Chapters 3, 7, and 10), photosynthesis in plants is the first step in the production of the complex organic molecules—carbohydrates, fats, and proteins—on which all life depends for energy. Most humans obtain energy by eating a combination of plants and animals. This energy is needed for muscular activity, the transmission of messages from the brain to muscles, the maintenance of body temperature, and for other vital functions such as respiration, blood circulation, digestion, and protein synthesis in the liver. When the body is at rest and not performing any external work, the energy liberated from foods is ultimately converted to heat.

The minimum amount of energy required to keep the body functioning normally when completely at rest is termed the basal metabolic rate (BMR).

Energy derived from foods is generally measured in **Calories** (note the capital *C*) rather than in calories (Chapter 14, Table 14-1).

$$1 \text{ Cal} = 1 \text{ kcal} = 1000 \text{ cal} = 4180 \text{ joules}$$

The energy in a sample of food can be determined directly by measuring the heat released when the food is completely oxidized in a **bomb calorimeter** (Fig. 17-3). However, not all the energy in food is available to the body because not all the food we eat is absorbed from the intestines; some is excreted in the feces. Also, some nitrogen-containing compounds derived from proteins in foods are never completely oxidized in the body and are excreted in the urine, mostly as urea ($NH_2-CO-NH_2$). Thus, some of the energy in proteins is not

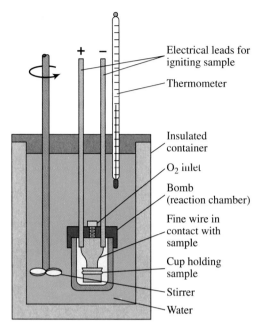

Figure 17-3   Cross section of a bomb calorimeter, which can be used to determine the caloric value of food. A weighed sample of food is placed in a heavy steel, sealed vessel called a bomb. The bomb is filled with oxygen under pressure and placed in a calorimeter, an insulated vessel containing a measured quantity of water. The food is ignited, and the heat produced warms the water. From the increase in temperature and the known heat capacity of the calorimeter, the heat emitted per gram of food can be calculated.

utilized. Table 17-2 shows the approximate amounts of energy provided by carbohydrates, fats, and proteins in foods, after allowance has been made for the energy lost through excretion.

As shown in the following example, if the nutrient content of a food is known, the values given in Table 17-2 can be used to calculate the caloric value of the food.

EXAMPLE 17-1   Whole milk is 3.5 percent protein, 3.5 percent fat, and 4.9 percent carbohydrate. Calculate the number of Calories available from 100 g of milk.

Solution:
1. 100 g of milk contains 3.5 g protein, 3.5 g fat, 4.9 g carbohydrate.
2. Calories from protein   =   3.5 g × 4 Cal/g = 14.0 Cal
   Calories from fat   =   3.5 g × 9 Cal/g = 31.5 Cal
   Calories from carbohydrate   =   4.9 g × 4 Cal/g = 19.6 Cal

3. Therefore, 100 g of milk will provide 65.1 Cal.

PRACTICE EXERCISE: White bread is 50.4 percent carbohydrate, 8.7 percent protein, and 3.2 percent fat. Approximately how many Calories of carbohydrate, protein, and fat will 200 g of bread furnish?

Answer: 403 Cal from carbohydrate, 70 Cal from protein, 58 Cal from fat.

TABLE 17-2   Calories Provided by Carbohydrates, Fats, and Proteins

Type of food	Cal/g
Carbohydrate	4
Fat	9
Protein	4

Fats produce more than twice as much energy per gram as proteins and carbohydrates when they are oxidized to carbon dioxide and water, because, per gram, they contain less oxygen. The following equations show the oxidation of glucose and a typical fatty acid:

$$C_6H_{12}O_6 + 6\,O_2 \longrightarrow 6\,CO_2 + 6\,H_2O + 670 \text{ kcal} \quad (3.7 \text{ kcal/g of glucose})$$
glucose

$$C_{16}H_{32}O_2 + 23\,O_2 \longrightarrow 16\,CO_2 + 16\,H_2O + 2385 \text{ kcal} \quad (9.3 \text{ kcal/g of fatty acid})$$
palmitic acid

The average normal-weight adult has enough body fat to supply the body's energy needs for 30 to 40 days.

Daily Calorie Requirements and Recommended Sources

For normal activity, a young adult male requires about 3000 Cal per day; a young adult female requires about 2100 Cal per day. Some of the foods that are commonly eaten to provide these Calories, together with their Caloric values and nutrient content, are listed in Table 17-3.

In the United States and other developed countries, fats generally provide 40 percent of the energy requirement, with carbohydrates supplying 48 percent and proteins 12 percent. In less developed countries, carbohydrates often provide as much as 80 percent of energy needs, with the remaining 20 percent coming equally from fat and protein.

The distribution of fats, carbohydrates, and proteins in the average American diet is not ideal. As we shall explain in the following sections, it is preferable, for health reasons, to obtain fewer calories from fats and more from carbohydrates.

TABLE 17-3   Calorie Content and Approximate Percentages of Proteins, Fats, and Carbohydrates in Selected Foods

Food	Cal/100 g	Protein %	Fat %	Carbohydrate %
Meat, fish				
Lean beef, broiled	175	31.7	5.3	0
Chicken, whole, boiled	130	23.8	3.8	0
Cod, raw	73	17.6	0.3	0
Dairy products, eggs				
Milk, whole	65	3.5	3.5	4.9
Cheddar cheese	398	25.0	32.2	2.1
Eggs	160	12.9	11.5	0.9
Grains, grain products				
Whole-wheat bread	260	10.5	3.0	47.7
Brown rice, cooked	118	2.5	0.6	25.5
Vegetables				
Carrots, raw	85	1.1	0.2	19.7
Potatoes, cooked	96	2.6	0.1	21.1
Tomatoes, raw	25	1.1	0.2	4.7
Fruits, nuts				
Apples	64	0.2	0.6	14.5
Bananas	95	1.1	0.2	22.2
Oranges	55	1.0	0.2	12.2
Pecans	735	9.2	71.2	14.6

EXAMPLE 17-2 A hamburger at a fast-food restaurant supplies a total of 575 Calories. If fat supplies 275 of these Calories, what percentage of the total Calories are derived from fat?

Solution: To obtain the percentage of fat, divide the number of Calories derived from fat by the total number of Calories, and then multiply by 100.

$$\text{percent of Calories from fat} = \frac{275 \cancel{\text{ Cal}}}{575 \cancel{\text{ Cal}}} \times 100$$

$$= 47.8 \text{ percent}$$

PRACTICE EXERCISE: For lunch, you eat a McDonald's Big Mac that provides a total of 540 kcal. If 280 kcal are derived from fat, do you obtain more or less than half your Calories from fat?

Answer: More than half (51.8 percent).

DIETARY NEEDS FOR CARBOHYDRATES

We need **carbohydrates** in our diet primarily as a source of energy. In 1976, the U.S. Senate Select Committee on Nutrition and Human Health recommended that for good health, 58 percent of dietary Calories should be obtained from carbohydrates, with 38 percent coming from complex carbohydrates, 10 percent from naturally occurring sugars, and no more than 10 percent from sucrose (refined table sugar) (Fig. 17-4).

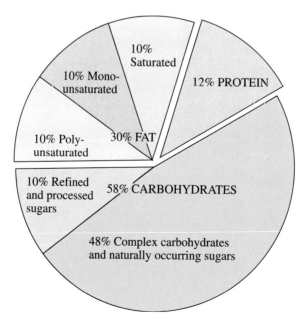

Figure 17-4 Recommended distribution of fat, protein, and carbohydrate in the diet.

Figure 17-5 Many processed foods contain a large quantity of added refined sugar.

Sources of Carbohydrates

Complex carbohydrates in the form of starches are present in potatoes, yams, cereal grains (rice, corn, wheat, barley, oats, millet), and all the products such as pasta and bread that are made from grains. Fructose is present in honey, and both fructose and glucose are found naturally in many fruits and vegetables, with particularly high concentrations in grapes, dates, and figs. Lactose is found in milk and milk products. Refined sugar, sucrose, is obtained from sugarcane and sugar beet. It is used in candy, cookies, preserves, soft drinks, many breakfast cereals, and numerous other processed foods (Fig. 17-5). Many Americans have a "sweet tooth" and add large quantities of sucrose to coffee, tea, tart fruits, and other foods. In the American diet, sucrose often accounts for more than 20 percent of the Calorie intake, more than twice the recommended amount.

In addition to digestible carbohydrates, we need indigestible carbohydrates composed of cellulose to add fiber to the diet. Good sources of fiber are cabbage, lettuce, celery, and whole-grain cereals.

Energy from Glucose

The energy available in carbohydrate foods is derived from the glucose produced in digestion and stored as glycogen. Glucose is released, as needed by the tissues, from stored glycogen and enters the general circulation. In this way, glucose in blood is maintained at a fairly constant level between 70 and 100 mg of glucose per 100 mL of blood. Glucose is carried in the bloodstream to all the body's tissues, where it supplies the energy cells need to perform their functions. In muscle, some of the glucose received from blood is converted back to glycogen and stored.

Energy is obtained from glucose in body cells by means of a complex series of chemical reactions that can be divided into two types: (1) anaerobic reactions (reactions that do not require elemental oxygen), and (2) aerobic reactions (reactions that require elemental oxygen). Each reaction in the two groups is catalyzed by a specific enzyme. We will provide only a general description of these reactions.

The aerobic sequence of reactions involves eight steps known as the Krebs cycle after Hans Krebs who worked it out. Krebs shared the 1953 Nobel Prize for physiology and medicine with Fritz Lipmann who discovered the roles played by ATP and ADP in energy storage.

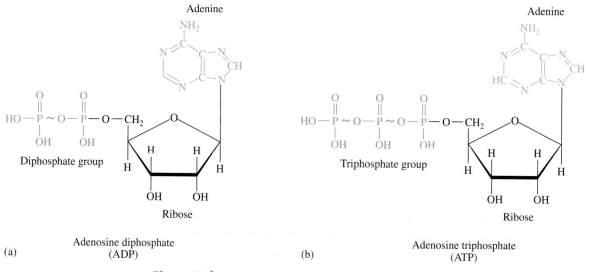

Figure 17-6 (a) Adenosine diphosphate (ADP). (b) Adenosine triphosphate (ATP). The high-energy bonds in the phosphate groups are shown as wavy lines.

The anaerobic reactions involve the formation of a glucose-phosphate intermediate that subsequently breaks down to yield lactic acid. In these reactions, which require the participation of **adenosine diphosphate (ADP)** (Fig. 17-6a), some of the bond energy stored in glucose is transferred to ADP with the formation of **adenosine triphosphate (ATP)** (Fig. 17-6b). The overall reaction can be represented by the following equation:

$$C_6H_{12}O_6 + 2\,ADP + 2\,H_3PO_4 \longrightarrow 2\ \underset{\substack{|\\ OH}}{CH_3-\overset{\substack{H\ \ \ O\\ |\ \ \ \|}}{C-C}}-OH + 2\,ATP + 2\,H_2O$$

glucose   lactic acid

Energy in ADP and ATP is stored in the phosphorus-oxygen bonds. As shown in Figure 17-6a, ADP contains one high-energy P—O bond, while ATP contains two (Fig. 17-6b). In the overall reaction in the anaerobic pathway, a portion of the energy of one molecule of glucose is transferred to two molecules of ATP; the remainder of the energy is retained in the two molecules of lactic acid.

The aerobic sequence of reactions results in the conversion of lactic acid to carbon dioxide and water, and the transfer of energy from lactic acid to further molecules of ATP.

$$C_3H_6O_3 + 18\,ADP + 18\,H_3PO_4 + 3\,O_2 \longrightarrow 3\,CO_2 + 21\,H_2O + 18\,ATP$$
lactic acid

In body tissues, energy is obtained by the conversion of ATP back to ADP.

When any muscle is used—whether in the heart, the intestines, the leg, or any other part of the body—the energy for the contraction is obtained through the reactions just described.

The process by which energy is obtained from glucose is just one example of the extraordinarily complex processes involved in the functioning of the human body.

During prolonged strenuous exercise, lactic acid is often formed more rapidly than it can be converted to carbon dioxide and water. As it accumulates in muscles, it causes fatigue and pain.

DIETARY NEEDS FOR FATS

Fats have two main functions in the body: to supply energy and to provide necessary components for the construction of cell membranes. For good health, it is recommended that fats make up no more than 30 percent of the diet and that these fats be divided equally among saturated fats, monounsaturated fats, and polyunsaturated fats (Fig. 17-4). In the United States, most diets are far from ideal. They have a higher fat content than recommended, and the fats are mainly in the form of saturated and monounsaturated fats. As we saw in Chapter 10, a diet high in saturated fats has been shown to increase the risk of atherosclerosis.

The major sources of dietary fats are butter, suet, and lard from meats, vegetable oils, and margarine prepared by hydrogenating vegetable oils. Although most vegetable oils are low in saturated fats, the tropical oils coconut oil and palm oil are exceptions (Table 17-4); these oils and the products made from them should be avoided by persons seeking to lower their intake of saturated fats. Cholesterol occurs in animal tissues but not in vegetables. Products high in cholesterol include red meat, butter, and cheeses. Particularly high levels are found in egg yolks.

With the exception of linoleic acid (Chapter 10), all the fatty acids that the body needs can be synthesized in the body. This one essential fatty acid is needed for the production of **prostaglandins**, a group of compounds produced in most tissues of the body that play a vital role in regulating hormonal activity. The average diet supplies sufficient linoleic acid to meet the body's needs.

Fats represent the body's most concentrated form of energy. As we have seen, glucose supplies the body's immediate energy needs, but for sustained activity, such as long-distance running, fats are mobilized from the adipose tissues. The metabolic pathways by which energy is obtained from fatty acids will not be considered; like those for glucose, they involve the participation of ADP and ATP.

DIETARY NEEDS FOR PROTEINS

Proteins have numerous functions in the body. Enzymes and antibodies are proteins; the materials from which most tissues and organs in the body are constructed are proteins. Hemoglobin, transferrin, and lipoproteins, which facilitate the transport of oxygen, iron, and lipids, respectively, are all proteins. Children need proteins for tissue building, while adults need them primarily for tissue repair. The daily requirement of protein for an adult is about 0.8 g per kilogram of body weight. Pregnant women and growing children need approximately 1.5 g and 2.0 g per kilogram of body weight, respectively.

To meet the body's needs, the diet must include proteins that, when digested, yield all eight (ten for children) of the essential amino acids (Chapter 10, Table 10-2)). Proteins from animal sources—meat, poultry, fish, eggs, milk, and cheese—are **complete proteins** and provide adequate amounts of all the essential amino acids. Plant proteins, however, are **incomplete proteins**. Cereal grains and legumes (beans, peas, and soy beans) are good sources of protein, but the proteins in them are deficient in one or more of the essential amino acids. Wheat, rice, and corn, for example, all lack lysine. Rice also lacks

TABLE 17-4   Percentage of Saturated Fatty Acids in Selected Vegetable Oils

Oil	%
Coconut	93
Palm	57
Peanut	21
Olive	15
Corn	14
Soybean	14
Safflower	10

A large egg contains about 270 mg of cholesterol.

Synthetic prostaglandins and related compounds have many clinical applications. They are used to induce labor, decrease gastric secretion, raise and lower blood pressure, and treat asthma. The name prostaglandins reflects their original extraction from prostate tissue.

A lipoprotein is a combination of a lipid such as a triglyceride or cholesterol, and a protein.

threonine; corn lacks tryptophan; and beans are deficient in methionine. A vegetarian diet, if it is to supply all the essential amino acids, must therefore contain a suitable, complementary mix of several plant foods.

Many traditional diets that are mainly vegetarian provide all the essential amino acids by combining a cereal with a legume. For example, corn tortillas and refried beans are staples of the Mexican diet; rice and soybean curd (or tofu) form a major part of the diet in Japan. Peanut butter (and jelly) sandwiches are favorites among American children. In these diets, the amino acids missing in the grain are present in the legume, and vice versa.

NUTRITIONAL DISEASES AND DISORDERS

Nutritional diseases and disorders can be caused by (1) inadequate or excessive consumption of food, or of some particular type of food, or (2) an inability to assimilate a particular nutrient in food.

Deficiency Diseases

Except for vitamin-deficiency diseases, which will be considered later in the chapter, most deficiency diseases are caused by a lack of protein in the diet. One of the most widespread and serious is **kwashiorkor**, which was first recognized in Ghana. This disease occurs in children between the ages of one and four who, for some reason—usually the birth of a sibling—are weaned from protein-rich breast milk to a protein-poor starch diet. The diet may supply adequate Calories, but because it is deficient in essential amino acids, synthesis of proteins in the body is disrupted. Fluid collects in the tissues and the limbs, and the face and belly become swollen. The children look plump in the early stages and may not appear to be undernourished. If the condition is untreated, the skin becomes dry and scaly with sores (Fig. 17-7), growth is stunted, and, because of a lack of antibodies, children are very susceptible to infection. If both proteins and Calories are deficient, a more serious condition called **marasmus** develops. The belly becomes very bloated, the children look old, are apathetic and often mentally retarded, and suffer an extreme weight loss. However, if they are treated with a balanced diet in the early stages, the chances for recovery are good (Fig. 17-8).

Overconsumption

In the less developed countries of the world, millions of people have inadequate diets and suffer from nutritional deficiency diseases. In the developed countries, a different nutritional problem exists: obesity. **Obesity** is caused primarily by long-term consumption of more dietary Calories than are required by the body to meet its energy needs. The excess Calories are stored as fat that accumulates in and on the body. In the United States, according to standards set by life insurance companies, 10–25 percent of teenagers and 25–50 percent of adults—particularly women over 40—are obese. Most overweight people take little exercise, and their diets are usually high in saturated fats, processed foods, and sucrose, and low in fruits, vegetables, and fiber. They have a shorter life expectancy than slim people have and suffer disproportionately from diabetes, gallbladder disease (particularly women), high blood pressure, and heart disease.

Kwashiorkor (pronounced kwash-ee-OR-core) was the name used in Ghana for the evil spirit that was believed to cause the disease suffered by the firstborn child after the second child was born.

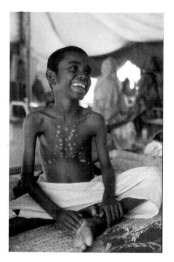

Figure 17-7   A child suffering from kwashiorkor, a disease caused by a lack of protein in the diet. Symptoms include swollen belly, retarded growth, and discolored, scaly skin with sores.

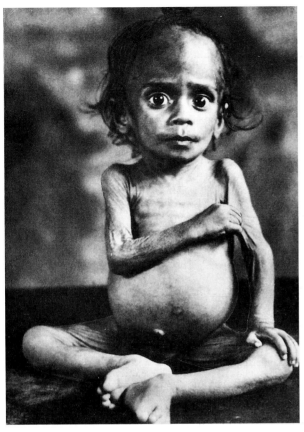
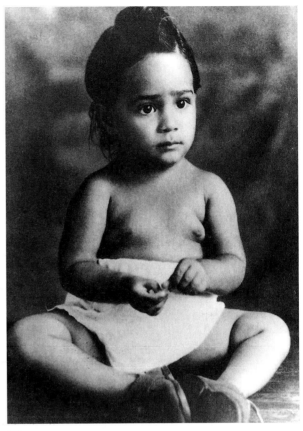

(a)   (b)

Figure 17-8 The wasting disease marasmus is the result of a diet deficient in both protein and Calories. If treated early with a balanced diet, the adverse effects can usually be reversed. (a) Before treatment. (b) After treatment.

Although there is no doubt that overconsumption of food is the main cause of obesity, other factors, including heredity and psychological problems, also play a part. In the United States, obesity has spawned an entire industry devoted to ways to reduce weight. Despite the millions of dollars spent yearly on drugs, surgery, aerobics programs, and numerous fad diets, there is no escaping the fact that the loss of one pound of fat requires the expenditure of 3500 Calories in muscular exercise. There is still only one way to lose weight: Reduce caloric intake and increase the expenditure of energy, while continuing to eat a nutritional diet.

Metabolic Malfunctions and Nutritional Disorders

Diabetes is a condition in which the body is unable to utilize glucose in the normal way; as a result, blood glucose levels become abnormally elevated. Normally, after a meal, the blood glucose level begins to rise as carbohydrates are digested and glucose enters the circulation. In response to this stimulus, the **pancreas** (Fig. 17-1) releases the hormone **insulin** into the bloodstream. Insulin causes glucose to be transferred from the blood into the cells; within about one hour, the blood glucose returns to the fasting level (the level before eating the meal). In diabetes, this transfer process does not operate normally; in severe cases, the blood sugar level may rise to three or four times its normal level. If untreated, a diabetic may lapse into a coma.

Diabetes affects about 15 percent of Americans over the age of 40, particularly those who are overweight and who consume large quantities of sucrose and processed foods. This type of diabetes develops gradually and is usually caused by both a decrease in insulin production and a decrease in the number of insulin receptors in cells. In most cases, the condition can be controlled by diet and exercise. The type of diabetes that develops suddenly in people under the age of 20—juvenile diabetes—is much more serious. It is caused by an almost complete lack of insulin and must be treated with intramuscular insulin injections. If given orally, insulin, which is a protein, would be broken down in the digestive tract.

> In the United States, by law, all infants are tested for PKU at birth.

There are a number of inherited metabolic defects in which the lack of a specific enzyme affects the utilization of food by the body. One is **phenylketonuria (PKU)**, which occurs in 1 out of every 20,000 newborns and is caused by the lack of the enzyme that catalyzes the conversion of the amino acid phenylalanine to the amino acid tyrosine (Chapter 10, Table 10-2). When not converted, phenylalanine and related compounds, which are toxic to brain tissue, accumulate and cause severe mental retardation. If an affected infant is put on a low-phenylalanine diet soon after birth, brain damage is minimized.

> The disaccharide lactose is composed of a molecule of glucose linked to a molecule of galactose.

Another inheritable disease is **galactosemia**, in which there is a deficiency of the enzyme that catalyzes the conversion of galactose to glucose. Infants with this condition are unable to utilize the galactose formed by the breakdown of lactose in milk and fail to thrive. They can be treated successfully with a special lactose-free diet.

CONSUMER BRIEF   Diabetes: Measuring Blood Sugar

One form of diabetes mellitus leaves an individual with an insufficient supply of the hormone insulin. In the absence of insulin, glucose is not transported from the bloodstream into cells, where it is needed as a source of energy. As a result, glucose levels rise in the blood and urine. Measuring glucose in the blood is the first test done to diagnose diabetes. For people afflicted with the disease, the day-to-day monitoring of glucose levels is essential in the management of diabetes.

Most tests for glucose in urine or blood detect a color change that accompanies the oxidation of glucose. Glucose and its oxidation product, gluconic acid, are colorless, so an indicator molecule is added to the sample.

Clinitest tablets are used to test for sugar in urine. A sample of urine is placed in a vial with the

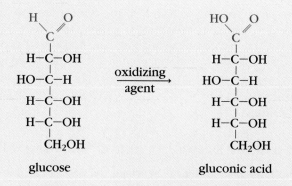

solid tablet. The tablet contains Benedict's reagent, an indicator that is an oxidizing agent. As the glucose undergoes oxidation, the blue copper(II) ion in the Benedict's reagent is reduced to brick-red copper(I) oxide. The color of the solution is compared to a standard color chart and the glucose concentration in the urine estimated. For many years this was the only test available to a diabetic for at-home use. Its major drawback is that the level of sugar in the blood has to be quite high before it shows a positive test in the patient's urine. Also, the test is not specific for glucose. Lactose in the urine of pregnant women will also give a positive test.

Modern methods of glucose detection rely on the action of the enzyme glucose oxidase which is specific for glucose. The enzyme catalyzes the oxidation of glucose to gluconic acid and hydrogen peroxide (H_2O_2) but will not catalyze the oxidation of other sugars. The diabetic pricks his or her finger and mixes the blood with a reagent that contains glucose oxidase, a colorless compound and a second enzyme, peroxidase. In the presence of peroxidase, the hydrogen peroxide produced in the oxidation of glucose oxidizes the colorless compound to a colored form. The intensity of the color is a measure of the glucose concentration in the blood.

$$\text{glucose} + O_2 + H_2O \xrightarrow{\text{glucose oxidase}} \text{gluconic acid} + H_2O_2$$

$$H_2O_2 + \text{colorless compound} \xrightarrow{\text{peroxidase}} H_2O + \text{colored compound}$$

Diabetic individuals can now monitor their blood glucose levels, at home, with a modestly priced instrument called a glucometer. The glucometer reads the color change of the glucose oxidase reaction electronically and displays the glucose level on a built-in meter. The home test is helpful to a diabetic who needs to keep a close watch on blood sugar because it detects rising glucose levels earlier than does a urine test.

Questions
1. A recent clinical study recommends that people with diabetes should measure their blood sugar 3–4 times a day. What advantages does the glucometer have over the urine test? What is the only disadvantage of the glucometer?
2. Do you think that health insurance companies should reimburse diabetic patients for the cost of medical instruments that can be used at home, or pay only for blood sugar tests performed by a clinical laboratory?

The Importance of Fiber in the Diet

In developed countries where the diet is rich in processed foods but low in roughage (**fiber**), appendicitis, diverticulitis, and colon cancer are common. In less developed countries, however, where little Western food is consumed and the diet has a high fiber content, these conditions are uncommon.

There is some evidence that potential carcinogens in the gut are absorbed by fiber and eliminated before they can cause any harm.

As indigestible carbohydrate material passes through the digestive tract, it absorbs water, swells, and becomes soft. This process promotes frequent bowel action, and usually the undigested material passes through the body within 2 days. If the diet lacks the bulk provided by roughage, however, bowel action is less frequent, and material from a meal may remain in the tract for as long as 4 days. This longer retention time gives bacteria—which make up about one-third of the weight of dry feces—more time to react with the material in the tract and produce toxic substances that can cause irritation and also colon cancer. Good sources of dietary fiber are celery, the peel, seeds, and pulp of fruits, and wheat bran.

Problems with Processed Foods

Because many nutrients are lost during processing, most processed foods are less nutritious than unprocessed foods. This deficiency is often remedied by the manufacturer—at least, in part—by adding the removed nutrients to the final product.

In the United States, most bread products are made from milled white flour. The milling process removes the germ and husks of the grain, which are a good source of fiber and which contain proteins, vitamins, and minerals. Enriched flour is milled flour to which vitamins, minerals, and a variety of other additives have been added. Despite these additions, products made from whole grains are better food sources because they provide more fiber.

In a similar way, fiber and nutrients are removed during the production of refined white sugar from sugarcane and sugar beets. Compared with the sugars in fruits and other natural foods, refined sugar has the added disadvantage that it is digested much more rapidly, and the glucose produced quickly builds up in the bloodstream. The body utilizes glucose for energy less efficiently when it enters the bloodstream rapidly, and more glucose is converted to fat for storage.

Because both refined sugar and refined flour provide less fiber than unrefined products, they contribute to the problems associated with a low-fiber diet.

> Brown sugar is more nutritious than sucrose because it contains molasses, a residue of sugarcane that is rich in minerals.

VITAMINS

Vitamins are organic compounds that are required in small amounts for normal metabolism and the maintenance of good health. They are not a source of energy for the body, but they have many vital functions. Most are needed as components of enzyme systems. Vitamins are synthesized by plants but cannot be synthesized in the human body and must, therefore, be obtained from food or food supplements.

> Vitamins were originally called *vitamines* because when first discovered and recognized as essential for life (*vita* means "life" in Latin), they were thought to be amines (Chapter 9). The final *e* was later dropped when it was realized that most are not amines.

Vitamins and Classical Deficiency Diseases

Five major diseases are associated with a lack of a specific vitamin: (1) scurvy (vitamin C), (2) beriberi (vitamin B_1), (3) pellagra (vitamin B_3), (4) blindness (vitamin A), and (5) rickets (vitamin D).

For several hundred years, it has been known that the absence of certain foods in the diet causes specific diseases, but it was not until the early part of the twentieth century that the chemical identity of the missing substances was established. One of the earliest nutrient-deficiency diseases to be recognized was **scurvy**, the scourge of seafarers who went without fresh fruits and vegetables for extended periods of time while on long ocean voyages. The sailors suffered from bleeding gums, loss of teeth, vomiting, weight loss, anemia, and slow-healing wounds, and many died. In 1735, the Scottish naval surgeon James Lind showed that scurvy could be cured and prevented, if citrus fruits were included in the diet. Shortly thereafter, the British Admiralty required all naval ships to carry stores of limes on long voyages—hence, the name "limeys" for British sailors. The factor in citrus fruits, and other fruits and vegetables, that prevents scurvy is **vitamin C**, or **ascorbic acid**. Because vitamin C is destroyed during long storage and by cooking, fruits and vegetables should be eaten when fresh.

> Humans, other primates, and guinea pigs are the only mammals that are unable to synthesize ascorbic acid in their bodies.

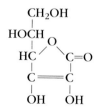

Vitamin C (ascorbic acid)

The disease **beriberi** became common in parts of Asia at the end of the nineteenth century, when milled (or polished) rice began to replace un-processed rice in the diet. Milling improves the keeping qualities of rice, but it removes **vitamin B₁**, or **thiamine**, which is present in the husk and germ. People with the disease suffer from stiffness of the limbs, heart disease, loss of appetite, and, in later stages, paralysis of the limbs and mental disorders.

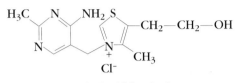

Vitamin B₁ (thiamine)

Another disease associated with overdependence on one type of grain is **pellagra**, which at one time was common in the southern United States among people who subsisted mainly on corn. Corn, as we noted earlier, is deficient in the amino acid tryptophan, which is the precursor of **niacin** (sometimes called vitamin B₃). A lack of niacin causes reddening and drying of the skin and, in severe cases, gastrointestinal and nervous system disorders. The condition can be prevented by including whole wheat in the diet; it can be cured by administering niacin.

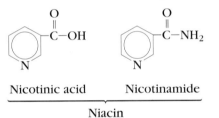

Nicotinic acid   Nicotinamide

Niacin

A major cause of blindness in young children in less developed countries is a **vitamin A**, or **retinol**, deficiency. An early sign of this deficiency is **night blindness**. This condition occurs when there is a lack of retinol, which is required for the regeneration of rhodopsin, the photosensitive material that makes it possible for the eye to adapt to dim light. Retinol is also essential for the maintenance of the body's epithelial cells, which are the cells that cover and line the body and its organs. In severe retinol deficiency, the skin becomes scaly and hard; in the eye, degeneration of the conjunctiva and the cornea can eventually lead to complete blindness. Good sources of vitamin A are milk, cheese, eggs, liver, and some fatty fish. The body can also obtain vitamin A from green, leafy vegetables and yellow and orange vegetables such as carrots, because these products contain β-carotene and other plant pigments that the body can convert to retinol.

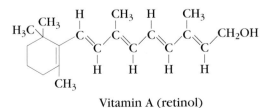

Vitamin A (retinol)

Rickets is practically unknown in the tropics where there is plenty of sunlight, but it was common among poor children in northern industrial cities until the middle of the twentieth century. Smog and latitude combined to limit exposure to sunlight and because of poverty the diet was poor.

Polar bear liver contains such an exceptionally high concentration of vitamin D that it is poisonous to humans.

A deficiency of **vitamin D (calciferol)** causes rickets in children and osteomalacia (soft bones) in adults. The diet need not supply all of the body's needs for vitamin D, because on exposure to sunlight, certain substances in the skin structurally related to cholesterol are converted to vitamin D. Vitamin D is required for absorption of calcium and, to some extent, phosphorus from the intestinal tract, and it regulates the deposition of calcium and phosphorus in bones and teeth. If there is a deficiency, bones do not grow normally, and many skeletal deformities develop, including knock knees and a protruding forehead (Fig. 17-9). Rickets was largely abolished in developed countries when vitamin D was added to infant formula and milk. Foods that are naturally high in vitamin D are dairy products, fatty fish, and fish liver oils.

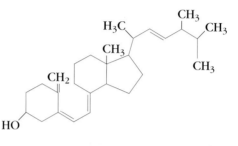

Vitamin D (calciferol)

The B-Complex Vitamins

Since the discovery of the causes of beriberi and pellagra, several other B vitamins, which have the same food sources as vitamins B_1 and B_3, have been rec-

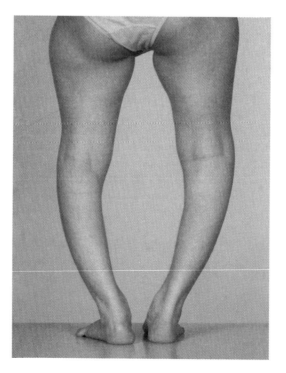

Figure 17-9 A deficiency of vitamin D (calciferol) combined with a lack of sunlight causes rickets in children. As a result of abnormal bone growth caused by a disturbance in calcium metabolism, children develop bow legs and knock knees.

ognized and chemically identified. Collectively, the B vitamins are called the **B-complex vitamins**. They function primarily as coenzymes in metabolic reactions that lead to the release of energy in body cells, including the reactions described earlier for the release of energy from glucose. Vitamins B_{12} (**cobalamin**) and B_9 (**folic acid**) are required for the normal development of red blood cells. A deficiency of either leads to anemia.

Vitamins E and K

Vitamin E (**tocopherol**) is widely distributed throughout foods. An antioxidant, it is essential for the maintenance of cell membranes and is involved in the normal functioning of most organs of the body.

  Vitamin K is needed for the formation of prothrombin, a substance necessary for blood clotting; a deficiency of vitamin K may result in blood loss.

Many people consume large quantities of vitamin E because it has been suggested that it may postpone the aging process.

Classes of Vitamins

As we have seen, vitamins vary greatly in their functions and chemical structures. They are usually divided into two groups according to solubility. **Water-soluble vitamins** include the B-complex vitamins and vitamin C. They dissolve in water because their structures include polar -OH groups. Water-soluble vitamins are not stored, but pass quite rapidly through the body and must be replenished daily.

  Fat-soluble vitamins, which include vitamins A, D, E, and K, have nonpolar hydrocarbon chains in their structures and are soluble in fats and oils. They do not need to be replenished as frequently as water-soluble vitamins because the body stores them in fatty tissues, particularly in the liver. If vitamins A and D are supplied in too great a quantity, they build up to toxic levels and cause serious problems. Too much vitamin D causes bonelike material to be deposited in the kidneys and other parts of the body.

  Although many people take them, vitamin supplements are unnecessary in most cases. Adults who regularly eat a well-balanced diet will obtain all the vitamins they need in their food, much of which in the United States is already fortified with vitamins.

  The recommended daily allowance, sources, and effects of deficiencies of the various vitamins are listed in Table 17-5.

Because water-soluble vitamin C is rapidly excreted, many scientists doubt that large doses of the vitamin, which have been recommended as a preventative for the common cold, do more than add to the amount of the vitamin excreted in the urine.

MINERALS

For nutritional purposes, **minerals** are defined as those elements, other than carbon, oxygen, hydrogen, and nitrogen, that are needed for normal growth and the maintenance of good health. Most minerals are present in the body as ions. Some, including calcium, phosphorus, and magnesium, are required in amounts of 1 g or more per day, while others, such as cobalt, copper, and zinc, are needed in trace quantities only. Since many minerals are excreted in urine, feces, and sweat, they must be resupplied daily in food. The average diet is most likely to be deficient in calcium, iron, and iodine.

> ## CONSUMER BRIEF
>
> ## Multivitamins: A Source of Minerals

Americans are so convinced that they need more nutrients than their diets provide that they spend about $3 billion a year on vitamins and other nutritional products. Despite this expenditure, the public tends to be polarized over vitamin therapy. Some argue that vitamin supplements are unnecessary and megavitamin therapy is dangerous. Others claim that everyone can benefit from vitamin supplements and that megavitamins are the answer to most health problems.

In most cases, average Americans who eat balanced diets do not need multivitamin supplements. On the other hand, the elderly, alcoholics, dieters, and pregnant and lactating women may benefit from supplemental vitamins. Although nutritional deficiencies may lead to disease, more commonly it is a serious illness, such as kidney or liver disease, that causes the deficiency.

The Food and Drug Administration (FDA) publishes a list of recommended daily allowances (RDA) that are used by industry for labeling multivitamins. The amount of each individual vitamin and mineral in a vitamin tablet is listed on the multivitamin bottles as a percentage of the adult U.S. RDA values. Multivitamin tablets should supply close to 100 percent of the RDA for each vitamin listed.

The mineral content of plants varies according to the composition of the soil in which the plants are grown. There have been reports of mineral deficiencies in people who eat meat, fruits, and vegetables raised on soil where the mineral content is low. But the major cause of mineral deficiencies is our increasing use of highly refined foods, which are low in minerals. Processing foods at high temperature and pressure quickly removes minerals.

For good nutrition, multivitamins should contain minerals in addition to vitamin A, D, E, C, and the B vitamins. Almost all multivitamins contain the U.S. RDA for vitamins, but only a few contain minerals as well.

The minerals added to vitamins are simple inorganic salts. The inorganic salt should be very water soluble for rapid uptake and should also be nontoxic. Zinc sulfate, $(ZnSO_4)$, is added to supplement zinc; sodium and potassium phosphate,

$(Na_3PO_4$, and $K_3PO_4)$, are added to supplement phosphorus, and iron sulfate, $(FeSO_4)$, is added to supplement iron. All these compounds are nontoxic, water-soluble, inorganic salts that are cheap and widely available.

The most common sources of calcium are calcium carbonate, $(CaCO_3)$, and calcium phosphate, $(Ca_3(PO_4)_2)$ which are the main constituents of limestone and phosphate rock, respectively. Unlike the other minerals mentioned, calcium salts are insoluble in water and must be digested by stomach acid (hydrochloric acid) before they can be absorbed.

Questions
1. Most people take multivitamins once a day. Why not take a larger vitamin tablet once a week?
2. Pregnant and lactating women may benefit from multivitamins. What inorganic elements should be contained in the multivitamins they choose?

Source: M. Ivey and G. Elmer, "Nutritional Supplement, Mineral and Vitamin Products," *Handbook of Nonprescription Drugs,* 9th Ed., 1991, American Pharmaceutical Association.

TABLE 17-5   Recommended Daily Allowance (RDA), Sources, and Effects of Deficiencies of Vitamins

Vitamin	RDA*	Sources	Effects of Deficiency
Water-soluble			
B_1 (thiamine)	1.5 mg	Whole-grain bread, milk, nuts, legumes	Beriberi: heart disease, mental disorders
B_2 (riboflavin)	1.7 mg	Milk, meat, eggs, whole-grain bread	Dermatitis
B_3 (niacin)	19.0 mg	Red meat, whole-grain bread, leafy vegetables	Pellagra: dry skin, intestinal and mental disorders
B_6 (pyridoxine)	2.0 mg	Eggs, liver, legumes, milk	Dermatitis; susceptible to infection, anemia
B_9 (folic acid)	0.4 mg	Liver, leafy vegetables, whole-grain bread	Anemia, retarded growth
B_{12} (cobalamin)	5.0 μg	Meat, eggs, milk, fish	Pernicious anemia
Biotin	0.3 mg	Meat, peanuts, eggs	Dermatitis, fatigue
C (ascorbic acid)	60.0 mg	Citrus fruits, tomatoes, green peppers	Scurvy: bleeding gums, slow-healing wounds
Pantothenic acid	10.0 mg	Meat, eggs, milk	Retarded growth, mental disorders
Fat-soluble			
A (retinol)	1.0 mg	Milk, cheese, eggs, fatty fish, carrots, leafy vegetables	Night blindness
D (calciferol)	10.0 μg	Fish liver oils, fatty fish, milk	Rickets, skeletal disorders
E (tocopherol)	10.0 mg	Eggs, milk, green vegetables	Anemia, sterility
K	70.0 μg	Green leafy vegetables	Bleeding disorders

*RDA values are for adults and children over 4 years old.

Calcium and **phosphorus** are required for the proper development of bone and teeth. Growing children and women who are pregnant or are breast-feeding need about 1.5 g of each per day compared with the approximately 1.0 g needed by other people. The best source of calcium and phosphorus is milk. In addition to being a component of bone and teeth, calcium is required for the transport of ions in and out of body cells, the coagulation of blood, and the maintenance of the heart's normal rhythm.

Bone, like other structures in the body, is continually being broken down and rebuilt. As people grow older, bone breakdown tends to exceed bone formation, and bones may become thin and brittle and fracture easily. This condition, called **osteoporosis**, is common in women after menopause, when there is a drop in the secretion of the hormone estrogen (Chapter 10), which plays a role in preventing excessive bone dissolution.

Phosphorus, in addition to being present in bone, is a component of nucleic acids (Chapter 10) and ADP and ATP. **Magnesium**, like calcium, is stored in bone. Magnesium ions (Mg^{2+}) are present in all body cells, and they are an essential part of many enzyme systems.

Iron is an essential component of hemoglobin and myoglobin, the predominant protein in muscle. A deficiency of iron in the diet can cause anemia and is most likely to occur in women, who lose blood during menstruation, and in people with bleeding ulcers. Good sources of iron are meat and eggs.

In many less developed countries where meat and eggs are in short supply in the diet, iron-deficiency diseases are common.

Figure 17-10 A woman suffering from goiter. A lack of iodine in the diet causes enlargement of the thyroid gland in the neck.

Goiter is quite common in mountainous areas such as the Alps and the Himalayas where iodine has been leached from the soil.

Iodine is required for the synthesis of the thyroid hormones, secreted in the thyroid gland in the neck, that regulate the consumption of oxygen in the cells of the body. Fish and other seafood are excellent sources of iodine. A serious deficiency of iodine in children's diets results in impaired development and stunted growth, a condition known as cretinism. A deficiency in adults leads to goiter, a disorder characterized by enlargement of the thyroid gland (Fig. 17-10). Iodine deficiencies occur mainly in inland regions, where iodine is lacking in both water and locally grown produce because it has been leached from the soil. Since the introduction of iodized salt (usually 0.1 percent potassium iodide, KI), iodine deficiency diseases have become rare in most parts of the world.

Sodium, **potassium**, **chloride**, and **bicarbonate ions** are present in all body fluids and are essential for proper electrolyte balance. Too much salt (NaCl) in the diet, however, can cause water retention in the tissues, resulting in swelling of the legs and ankles (edema) and high blood pressure. Most American diets include too much salt but may be low in potassium because salt is added to processed foods to replace both the sodium and potassium lost in the manufacturing process.

The **trace minerals**, copper, manganese, molybdenum, and zinc, are essential as coenzymes in various enzyme systems. Cobalt is a component of vitamin B_{12}.

FOOD ADDITIVES

Many people are concerned about the possible health hazards associated with food additives, the chemicals that are added to basic foods during processing and packaging. Of the over 2000 additives used in the United States, very few have any nutritional value. They are added to prevent spoilage, sweeten, enhance flavor, color, emulsify, or in some other way make the product more marketable and attractive to the consumer. Food additives generated approximately $4 billion in sales in 1991. The beverage industry was the largest user (Fig. 17-11).

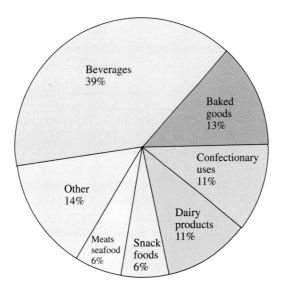

Figure 17-11 The beverage industry is the largest user of food additives.

Food Preservation

Preservatives are the most useful of all food additives. Without some form of treatment most foods soon spoil and become inedible. The main causes of food spoilage are (1) growth of bacteria and fungi (molds) and (2) air oxidation. Any method for improving the keeping qualities of foods must prevent or retard one or both of those processes.

Since prehistoric times, drying has been used to prevent spoilage of meat, fish, plants, and grains. Salting to preserve meat and fish and preservation of fruits in concentrated sugar solution are other ancient techniques. Refrigeration, bottling, and canning have been used for many years. More recent techniques include freezing, freeze drying, irradiation (Chapter 6), and the addition of various chemicals. The chemical preservatives that are routinely used today are of two kinds: (1) antimicrobial and (2) antioxidant.

Drying is an effective technique for food preservation because in the absence of water, microorganisms do not multiply, and air oxidation is slowed.

Antimicrobial Additives The **antimicrobial agents sodium nitrate** ($NaNO_3$) and **sodium nitrite** ($NaNO_2$) have been added to cured meats, bacon, hot dogs, bologna, ham, and smoked fish for at least 30 years. They are responsible for the pink color of the products. Without them, these foods would soon change to an unattractive gray.

Nitrites are particularly valuable in preventing the growth of *Clostridium botulinum*, the bacterium that produces deadly botulism poisoning. Recently, however, the use of both nitrates and nitrites has been questioned because of concerns that they may cause stomach cancer (Chapter 16). In the stomach, nitrates are reduced to nitrites, which react with hydrochloric acid in the stomach to form nitrous acid:

$$NaNO_2 + HCl \longrightarrow HNO_2 + NaCl$$

sodium nitrite, hydrochloric acid, nitrous acid, sodium chloride

Nitrous acid may then react with amines (Chapter 9) in the stomach to form **nitrosamines**, compounds that have been identified as cancer-causing agents.

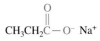
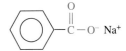

Sodium benzoate

Sodium propionate

Potassium sorbate

Figure 17-12 Antimicrobial agents used as preservatives. Carboxylate groups are in red.

$$\overset{\displaystyle R'}{\underset{\displaystyle}{HNO_2 \; + \; R-\overset{|}{N}-H}} \longrightarrow R-\overset{\displaystyle R'}{\overset{|}{N}}-N=O \; + \; H_2O$$

nitrous   amine   a nitrosamine
acid

The use of nitrate as a food additive has been banned by the FDA because nitrate is converted into nitrite in the body and thus has no value of its own.

In 1986, the FDA banned the spraying of fruits and vegetables with sulfites, and it issued regulations requiring that all food containing more than 10 ppm of sulfite be labeled to indicate the concentration.

There is no conclusive evidence to show that this reaction does occur in the stomach, but the possibility is under investigation by the U.S. Food and Drug Administration (FDA).

Another controversial preservative is sulfite (Na_2SO_3 or K_2SO_3), which has been added to wine and dried fruits for centuries and more recently has been included in jellies and jams. In foods and wine, sulfites appear to be safe, but when sprayed on vegetables and fruits to produce a fresh appearance, sulfites have caused allergic reactions in asthmatics.

In the United States, the most widely used antimicrobial additives are the sodium, potassium, and calcium salts of **benzoic acid, propionic acid,** and **sorbic acid** (Fig. 17-12). Benzoic acid and benzoates are added to carbonated beverages, fruit juices, margarine, pickles, relishes, preserves, and a variety of other products. Propionic acid, sorbic acid, and their salts are added to bread, cakes, chocolate, and cheese. They are very effective in retarding the growth of molds (Fig. 17-13).

Figure 17-13 When sorbates are added to bread, molds grow much more slowly than when they are absent.

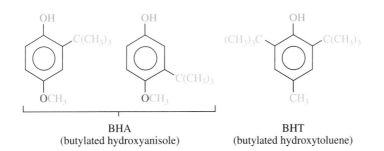

BHA
(butylated hydroxyanisole)

BHT
(butylated hydroxytoluene)

Figure 17-14 Two common antioxidants added to foods to slow the oxidation of fats.

Antioxidants   **Antioxidants** are chemicals added to foods to slow the oxidative process. Foods that deteriorate particularly rapidly because of air oxidation are fats and oils. The products of oxidation include carboxylic acids, aldehydes, and ketones (Chapter 9), which together give spoiled fats their characteristic unpleasant rancid taste and odor. The two antioxidants used most frequently are the phenols (Chapter 9) **butylated hydroxyanisole (BHA)** and **butylated hydroxytoluene (BHT)** (Fig. 17-14). These compounds are added to vegetable oils, shortenings, and products that contain oils and fats, including breakfast cereals, potato chips, bread, and sausages. The oxidation of fats involves the formation of very reactive free radicals (Chapter 13), which set up chain reactions. BHT and BHA interrupt the oxidation process by combining with the free radicals.

The safety of adding BHT and BHA to foods was first questioned in the 1950s, when a study suggested that large doses caused birth defects in rats. Although later studies were unable to confirm this finding, some people remain unconvinced that the additives are safe.

An effective natural antioxidant is vitamin E, which, like BHT and BHA, is a phenol. Vitamin E, however, is more expensive than the synthetic antioxidants and is used less frequently.

A recent study showed that BHT may be beneficial. When rats were fed a diet high in BHT, their average life span increased by an amount equivalent to 20 years for humans.

Flavorings

Many foods are dependent on additives for their flavor. Flavorings, which may be natural or synthetic, represent the largest class of food additives. Of the over 2000 flavors that have been approved by the FDA, approximately 1600 are synthetic.

For centuries, cloves, ginger, cinnamon, nutmeg, pepper, and many other plant materials have been used to flavor foods. Flavors such as vanilla, peppermint, and oil of wintergreen are extracted from the crushed plants with appropriate solvents. Natural flavors are complex mixtures of numerous compounds, many of which can now be synthesized in the laboratory and mixed to produce synthetic flavors. Synthetic and natural flavors are usually slightly different because the synthetic product lacks many of the trace components present in the natural product. The chemical structures of a number of commonly used synthetic flavorings are given in Figure 17-15. Many are esters or aldehydes (Chapter 9).

Unlike the majority of other additives, most natural and synthetic flavors have not been extensively tested to determine if they might cause health problems. They are assumed to be safe in the small quantities added to foods and do not have to be identified on food labels.

A thousand years ago, peppercorns were so valuable and costly that they were often used in trading as a substitute for money.

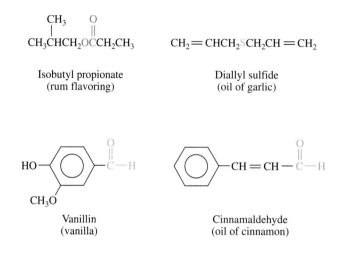

$$CH_3CHCH_2OCCH_2CH_3$$

Isobutyl propionate
(rum flavoring)

$$CH_2\!=\!CHCH_2SCH_2CH\!=\!CH_2$$

Diallyl sulfide
(oil of garlic)

Figure 17-15 Examples of flavorings added to foods. Functional groups are shown in color.

Vanillin
(vanilla)

Cinnamaldehyde
(oil of cinnamon)

Flavor Enhancers

A number of substances that have little or no flavor of their own enhance the flavors of other substances. One of the most widely used flavor enhancers is **monosodium glutamate (MSG)** (Fig. 17-16), the sodium salt of the amino acid glutamic acid (Table 10-2). MSG is added to thousands of processed foods. It occurs naturally in many natural products including tomatoes and mushrooms, and it has been approved as a food additive by the FDA. MSG is currently being reevaluated, however, because recent studies have indicated that large doses may cause brain damage in mice. A few years ago there were several reports of people suffering from unexplained headaches, shortness of breath, chest pains, and sometimes fainting, shortly after eating a meal in a Chinese restaurant—the "Chinese restaurant syndrome." MSG was under suspicion because many Chinese dishes contain relatively large quantities of MSG. Recently, it has been proved that MSG is not the cause of the symptoms.

Food Colors

Ripe Florida oranges often have green skins and although it makes no difference to their flavor, they are sometimes colored orange so that they can better compete with naturally orange California oranges. Similarly, yellow dyes are added to chicken feed to give chicken skin a yellow color.

Food colors increase the asthetic appeal of foods but serve no other function. They are used to make products more attractive to the consumer, who may reject foods if they are not colored as they expect.

About half the food colors used in the United States are extracted from natural products such as carrots, beets, grape skins, and the stamens of crocuses (saffron). The remainder are synthesized from coal tar (Chapters 14 and 15). Most synthetic food dyes consist of one or more aromatic rings joined by a $-N\!=\!N-$ group. FD&C Yellow No. 3 is a typical example (Fig. 17-17).

Figure 17-16 Monosodium glutamate (MSG), the sodium salt of the amino acid glutamic acid, is a commonly used flavor enhancer. Functional groups are shown in color.

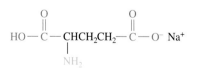

$$HO\!-\!C\!-\!CHCH_2CH_2\!-\!C\!-\!O^-\ Na^+$$
$$NH_2$$

MSG
(monosodium glutamate)

Since 1906, food colorings have been regulated by the FDA, which sets limits on the amounts that can be added and requires that a label show that a color has been added. Several dyes once listed as safe have now been banned. Red Dye No. 2 was removed from the list in 1976 after large doses were shown to cause cancer in laboratory animals. A number of compounds in coal tar that are chemically related to certain food dyes have been shown to cause cancer, and as a result some people contend that all dyes obtained from this source should be banned.

Artificial Sweeteners

In the United States, where so many people are overweight or weight-conscious, there is a thriving market for foods and beverages—particularly "diet" soft drinks—that are sweetened with **artificial sweeteners** instead of with sucrose. These products are particularly useful for diabetics and others who must control their sugar intake. Most artificial sweeteners have no nutritional value and contribute few, if any, Calories.

Saccharin, which for many years was the only artificial sweetener available, was discovered accidentally in 1878. A chemist studying derivatives of toluene (Chapter 9), synthesized a new compound (Fig. 17-18) and noticed that it had a sweet taste. After it was tested for toxicity, the compound was marketed as saccharin. Saccharin is approximately 400 times sweeter than sucrose (Table 17-6) but it has a slightly bitter aftertaste.

In 1937, another sweetener, **cyclamate** (Fig. 17-18), was discovered, again accidentally. Cyclamates are only about 30 times sweeter than sucrose but

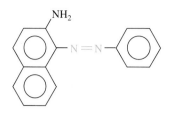

Figure 17-17 The synthetic food color FD&C Yellow No. 3.

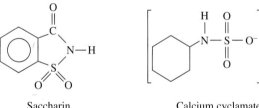

Saccharin   Calcium cyclamate

Aspartame

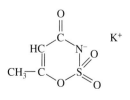

Acesulfame K

Figure 17-18 Artificial sweeteners.

TABLE 17-6 Sweetness of Natural Sugars and Artificial Sweeteners Relative to Sucrose

Compound	Relative Sweetness*
Lactose	16
Glucose	74
Sucrose	100
Honey	145 (average)
Sodium cyclamate	3,000
Aspartame	18,000
Acesulfame K	20,000
Saccharin	40,000

*Sucrose is given an arbitrary value of 100.

it has no bitter aftertaste. Cyclamate was used extensively in beverages and other foods until 1969, when it was banned because a study showed that very large doses could cause bladder cancer in rats. Although later work revealed that the study was flawed and that cyclamate was harmless to laboratory animals, attempts to have the ban lifted have so far failed.

In 1978, the FDA recommended that saccharin, the only artificial sweetener still available at that time, should also be banned because studies indicated that very high doses could cause cancer in mice. However, in response to concerns that a ban would leave diabetics without a sugar substitute, the U.S. Congress decided that saccharin should remain on the market but should be labeled to indicate that it might be a health hazard.

The synthetic compound P-4000 is 400,000 times sweeter than sucrose but because of its toxicity it cannot be used as a sweetener in foods.

Newer artificial sweeteners include aspartame and acesulfame K (Fig. 17-18). **Aspartame**, which is marketed by NutraSweet, is a methyl ester of the dipeptide formed between the amino acids aspartic acid and phenylalanine (Table 10-2), and it is metabolized as a protein. People with phenylketonuria (PKU), who lack the enzyme needed to metabolize phenylalanine, should avoid products containing aspartame. **Acesulfame K** is slightly sweeter than aspartame and has the added advantage that it is not broken down at the temperatures used in cooking.

Chemists still do not know what chemical structures impart a sweet taste. Very minor changes in chemical structures of artificial sweeteners result in loss of sweetness or development of a bitter taste.

The search for artificial sweeteners remains one of trial and error.

Other Food Additives

There are other types of additives besides those that have been described. They include sequestrants, acids, alkalis, emulsifiers, thickeners, and stabilizers.

Sequestrants are substances that are added to tie up trace metals that may get into foods accidentally during harvesting and processing. Sequestrants include citric acid and ethylenediaminetetraacetic acid (EDTA) (Fig. 17-19), which form complexes with metals and thus prevent them from catalyzing the oxidation or decomposition of foods.

Weak organic acids are added to soft drinks and other products to impart a tart taste and, in some cases, to mask undesirable aftertastes. The acids used include citric, lactic, phosphoric, tartaric, and malic acids. Acids also help in preventing oxidation and the growth of microorganisms.

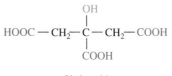

Citric acid

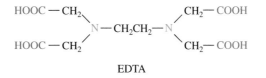

EDTA

Figure 17-19 The sequestering agents citric acid and EDTA (ethylenediaminetetraacetic acid) bond with metal ions, thereby preventing them from catalyzing oxidation reactions. Groups and atoms that bond to metals are shown in color.

Emulsifiers are substances that keep very small droplets of a nonpolar liquid (usually an oil) evenly dispersed in a polar liquid (usually water). They are widely used in diary products and are present in ice cream, mayonnaise, margarine, and peanut butter. Emulsifiers include mono- and diglycerides of fatty acids (Chapter 10) (Fig. 17-20), propylene glycol, polysorbates, and lecithin. They function on the same principle as soaps and detergents: The nonpolar end of the emulsifier molecule dissolves in the oil, while the polar end dissolves in water. In this way, molecules of the two immiscible liquids are held together and do not separate.

How Safe Are Food Additives?

In the United States, food additives are regulated by the FDA. Before 1958, an additive could be used unless the FDA could prove that it was unsafe. But since then, federal law has required that the safety of any new additive must first be tested by the manufacturer and then approved by the FDA before it can be used in foods. Further, the Delaney Amendment, also passed in 1958, requires the FDA to ban any substance that has been shown to cause cancer in animals or humans.

The law did not apply to the several hundred additives already in use before 1958, and, as it was impractical to attempt to test all those substances at that time, the FDA drew up a list of some 600 additives that were **generally regarded as safe** if used in specified foods in specified amounts. The list, which became known as the **GRAS list**, was based on the opinions of experts in toxicology and related fields. Since the list was published in 1959, many substances on the original list have been reevaluated, and, as we noted earlier, some, including cyclamates and several food-coloring agents, have been removed. Today, well over 2000 additives are approved by the FDA.

Many people are concerned that even if individual additives are harmless, several in combination may prove harmful. A single food product on the grocery shelf may contain as many as 100 different additives, and it is obviously impossible to determine how, or if, some of them might interact. However, even though some people would prefer all-natural foods without additives of any kind, such an approach is unrealistic in today's world. Without preservatives, food would spoil, food supplies would be greatly reduced, costs would rise, and our already limited ability to feed the world's ever-growing population would be further jeopardized.

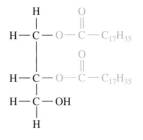

Glycerol distearate

Figure 17-20 Glycerol distearate is used as an emulsifier in dairy products. Emulsifiers keep tiny droplets of oils evenly distributed in aqueous solutions.

▦ DIETARY GOALS FOR GOOD NUTRITION

Good nutrition depends on eating a varied, properly balanced diet. The average American diet, unfortunately, does not meet this requirement. In general, it supplies more Calories than are needed, and, because of its reliance on convenience and processed foods, it is too rich in fats, cholesterol, sugar, and salt, and too low in fiber. This type of diet has been linked to many of the leading causes of death in the United States, including heart disease, colon cancer, and diabetes.

For the sake of our health, we would all be well advised to follow the recommendations listed in the 1988 Surgeon General's Report on Nutrition and Health:

1. Maintain ideal weight. If overweight, reduce energy intake and increase energy expenditure.
2. Eat a varied diet, with energy intake made up as follows:
    a. Carbohydrates: 58 percent

complex carbohydrates	38 percent
naturally occurring sugars	10 percent
refined sugar, no more than	10 percent

    b. Fats: no more than 30 percent

polyunsaturated fats	10 percent
monounsaturated fats	10 percent
saturated fats, no more than	10 percent

    c. Protein: 12 percent
3. Avoid excess consumption of foods high in cholesterol.
4. Avoid excess sodium by keeping salt intake at about 3 g/day.
5. Eat foods that provide adequate fiber.
6. Drink alcohol in moderation.

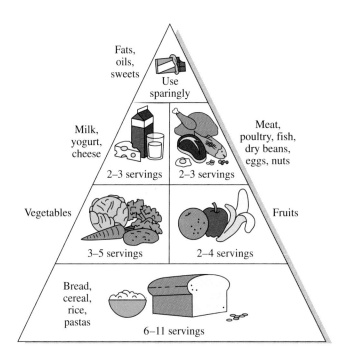

Figure 17-21 The food pyramid proposed by the U.S. Department of Agriculture to show how meals should be planned for good nutrition.

7. Increase calcium intake for adolescent girls and adult women.

8. Increase iron intake for children, adolescents, and women of childbearing age, particularly in low-income families.

To obtain the ideal diet, a person should plan daily meals according to the diagram shown in Figure 17-21. The pyramid emphasizes the importance of increasing consumption of complex carbohydrates and fruits and vegetables and the need to restrict consumption of fats and sweets.

EXPLORATIONS

The Lowly Potato: A Treasure That Grows Beneath the Ground

Potatoes are a favorite food. Baked, boiled, or as French fries and chips, we eat them almost every day. Confined to South America until the sixteenth century, the potato has since traveled the world, improving health, influencing population growth, and changing economies. Today, it joins rice, corn, and wheat as one of the world's four major food crops.

The potato was first cultivated high in the Andean mountains at least 8000 years ago. It belongs to the botanical family Solanacea, and its relatives include the tomato, pepper, eggplant, petunia, tobacco plant, and deadly nightshade. The potato (*Solanum tuberosum*) is a tuber, a swollen underground stem that stores food for the use of the plant above ground (Fig. 17-A). New plants sprout from the "eyes," or buds, on the tuber. The name *potato* is derived from "batata," the Spanish form of the Arawak Indian word for the sweet potato, or yam. (The potato and yam are not related, but the two tubers were often confused by the first Europeans who encountered them, and the name was misapplied.)

In the sixteenth century, when the Spanish conquistadors arrived in the New World in search of gold and souls to convert to Christianity, the potato was the main staple of the great Inca empire that stretched almost 3000 miles down western South America. The Spanish took potatoes home with them, little realizing that they carried an edible treasure of greater lasting value than all the gold they plundered from the Indians. From Spain, potatoes spread to the rest of Europe and, by the seventeenth century, reached Africa, India, China, Japan, and most of Southeast Asia. Potatoes traveled back across the Atlantic Ocean to North America with the early English settlers and were well established in Virginia and Pennsylvania by the end of the seventeenth century.

Figure 17-A
Potato (*Solanum tuberosum*).

Europeans were slow to accept the strange tubers; for many years, they were viewed with deep suspicion. Scots refused to touch them because they were not mentioned in the Bible; when outbreaks of leprosy and tuberculosis occurred in regions where potatoes had been introduced, potatoes were blamed. The innocent potato was even accused of causing aphrodisiac effects deleterious to moral health.

Gradually, the potato was recognized as a very valuable food, and its cultivation during the eighteenth and nineteenth centuries had a profound effect on the history of Europe. Potatoes were easy to grow; they matured more quickly and provided more food per acre than either rye or wheat. With potatoes, a family could, for the first time, grow enough food on its own small plot to meet its basic needs. And potatoes were safer than grains. Although it was not understood at the time, the molds that often grew on stored grains produced potent cancer-causing aflatoxins. By improving nutrition and health, potatoes were undoubtedly a significant contributing factor in the population explosion that provided the workforce needed to power the Industrial Revolution.

In no country has the potato had a greater impact than in Ireland. By the 1800s, the poor were almost entirely dependent on it for their food, and when a deadly blight struck the potato crop in 1845, the results were devastating. The disease spread with terrifying speed through the land. Both harvested potatoes and those still in the ground rotted. Deprived of their only source of food, at least a million Irish died of starvation or sickness, and over a million emigrated to the United States. In time, the potato recovered, but it remained vulnerable to the blight, which we now know was caused by a fungus accidentally introduced from Mexico. Today, the disease is controlled with fungicides.

Andean farmers have always grown thousands of different varieties of potatoes in an astonishing range of shapes, sizes, flavors, and colors, (Fig. 17-B)

but in most countries only a few varieties have ever been cultivated. In the United States, just six varieties of potatoes, all scientifically bred, make up 80 percent of the crop, and one potato—the Russet Burbank—accounts for 40 percent of the market. Lack of diversity can be a prescription for disaster, as was the case in Ireland in 1845. If a crop is genetically uniform, disease, pests, or an unexpected fluctuation in climate can completely wipe it out.

Potatoes are not just an excellent source of carbohydrate, they also contain minerals and vitamins, and they provide twice as much high-quality protein per acre as wheat. Potatoes are 99.9 percent fat-free and an 8-oz baked or boiled potato (if eaten without adding butter or sour cream) supplies approximately 100 Calories. The major potato-growing states in America are Idaho, Washington, and Maine.

More than half the American potato crop is processed into chips, frozen French fries, dehydrated products, and potato starch for use as a low-fat additive or binding agent in soups, ice cream, and bakery goods. Potato processing is an old technique that was first developed hundreds of years ago in the Andes. Potatoes were spread on the ground, and alternate freezing and drying during cold nights and warm days reduced their water content (potatoes are about 80 percent water). Villagers then stomped the potatoes with their bare feet to press out remaining water. Once dry, the dehydrated product, called *chuno*, could be stored for several years.

Potatoes are not just processed into edible products. They are also used in the production of paper, textiles, adhesives, cosmetics, and pills and capsules. In the future, they are likely to find many more applications. Water-soluble, and thus readily disposable, potato starch materials may soon replace Styrofoam for packaging, and in the years to come as petroleum becomes scarce, automobiles may run on potato gasohol.

The most exciting area of potato research is in genetic engineering. By extracting a desirable gene from one potato variety and introducing it into another, scientists are working to produce disease-resistant, high-yielding potatoes that can be grown in many different climates around the world. But genes cannot be invented; they can only be altered. Genetic reserves are kept at the International Potato Center, established in Lima, Peru, in 1971. The center now has a collection of over 13,000 varieties of wild potatoes, which serve as a world bank of potato genes.

Figure 17-B Potatoes of all shapes, sizes, and colors grow in the Peruvian Andes.

The humble potato, at times maligned and often scorned as a poor man's food, is now universally respected. Nutritious, adaptable, hardy, versatile, and delicious, few foods can compare with it. This amazing vegetable, which can be grown almost anywhere in the world, may be our best hope for feeding the millions of hungry people that inhabit our planet.

Question

Discuss the value of the potato as a basic food for feeding the world's ever-increasing population.

References: M. Sayles Hughes, "Potayto, Potahto—either way you say it, they 'a' peel," *Smithsonian* (October 1991), p. 138; R. E. Rhoades, "The Incredible Potato," *National Geographic* (May 1982), p. 668; and "The World's Food Supply at Risk," *National Geographic* (April 1991), p. 74.

KEY WORDS AND CONCEPTS

adenosine diphosphate (ADP)
adenosine triphosphate (ATP)
antimicrobial agent
antioxidant
artificial sweetener
ascorbic acid
beriberi
BHA and BHT
bile salts
calciferol
Calories
carbohydrates
cobalamin
complete protein

diabetes
emulsifier
fats
fiber
folic acid
galactosemia
GRAS list
kwashiorkor
lipase
marasmus
minerals
MSG
niacin
nitrosamine
nutrition
obesity

pellagra
pepsin
phenylketonuria (PKU)
prostaglandins
ptyalin
retinol
rickets
saliva
salivary amylase
scurvy
sequestrant
thiamine
tocopherol
vitamin

QUESTIONS AND PROBLEMS

1. Outline the steps in the digestion of carbohydrates. What are the end products?

2. What is the role of enzymes in digestion? Give examples.

3. How do humans store energy?

4. Briefly describe the role of ATP in the release of energy from glucose.

5. How many kilocalories of energy are released when 180 g of glucose ($C_6H_{12}O_6$) are burned to carbon dioxide and water?

6. Does any food literally contain calories? What does the manufacturer mean when the label on the candy bar says "contains 250 Calories"?

7. Why are carbohydrates considered energy-rich?

8. What is the body's most concentrated form of energy?

9. What compound causes soreness in your muscles after serious exercise? How is it formed?

10. What foods supply fiber? Why do we need fiber in our diet?

11. What is meant by a complete protein? Give examples of complete and incomplete proteins.

12. Are vitamins organic or inorganic compounds? Explain.

13. Identify each vitamin.
    a. tocopherol **b.** calciferol **c.** ascorbic acid **d.** cobalamin **e.** retinol

14. Identify the vitamin deficiency associated with these diseases:
    a. pellagra **b.** night blindness **c.** rickets **d.** beriberi

15. Identify the deficiency disease associated with a diet lacking this vitamin:
    a. thiamine **b.** vitamin B_{12} **c.** niacin

16. Indicate whether the following are water-soluble or fat-soluble:
    a. vitamin C **b.** vitamin K **c.** vitamin B **d.** vitamin A **e.** vitamin B_{12} **f.** riboflavin **g.** niacin **h.** tocopherol **i.** calciferol

17. Is excessive consumption of water-soluble vitamins more dangerous than consumption of fat-soluble vitamins? Explain.

18. What is kwashiorkor? How is it caused?

19. What is marasmus? How is it caused?

20. What causes obesity?

21. Explain why blood sugar rises to abnormally high levels in untreated diabetics after they have eaten a meal.

22. What causes PKU? How is it treated?

23. What causes galactosemia? How is it treated?

24. Describe the problems associated with low-carbohydrate, high-fat diets.

25. Are there any nutritional problems associated with a strict vegetarian diet?

26. Indicate a biological function for each of the following:
a. phosphorus **b.** iodine **c.** calcium **d.** iron

27. Which of the following minerals would you expect to find in a healthy human? Justify each choice.
a. Hg **b.** Na **c.** Zn **d.** Ca **e.** I **f.** P

28. What are the functions of carbohydrates, fats, and proteins in our bodies?

29. Why is it preferable to eat a whole apple rather than one that is peeled?

30. Three fast-food restaurants offer hamburgers and French fries. Each chain's offerings are tested, and the following data are collected:Which restaurant's hamburger with fries offers the lowest Caloric intake? How many Calories will you consume at the highest-calorie restaurant?

	Hamburger	Fries
Restaurant 1	**(g)**	**(g)**
protein	12	2
carbohydrates	30	22
fat	10	20
Restaurant 2		
protein	11	3
carbohydrates	30	26
fat	12	12
Restaurant 3		
protein	14	2
carbohydrates	28	24
fat	11	10

31. Saltine crackers contain 0.2 g of protein and 10 g of carbohydrates per cracker. If your Recommended Daily Allowance (RDA) is 70 g of protein, how many crackers would you need to eat to satisfy your RDA for protein? How many grams of carbohydrates would you also consume? Would you consider crackers to be a sensible source of protein?

32. A glass of milk contains 250 g of milk. Milk is about 3.5 percent protein. How much protein is contained in two glasses of milk? Is milk a sensible source of dietary protein?

33. A cup of corn flakes (28.4 g of flakes) contains 2 g of protein, 24 g of carbohydrates and 0 g of fat. How many Calories are contained in one cup of corn flakes? If 100 g of milk is added to a cup of corn flakes, how many total Calories are available? Refer to Example 17-1 for the composition of milk.

34. What two types of agents are used as preservatives? Give an example of each.

35. What preservative is added to bologna to give it a pink color? Are there any dangers associated with this additive?

36. What do the letters MSG stand for, and why is this substance added to food?

37. Compare the sweetness of the following products: sucrose, honey, saccharin, aspartame.

38. What is a sequestrant? Give an example, and explain why it is added to foods.

39. What is the purpose of an emulsifier? Give an example.

40. How was the GRAS list established?

AGRICULTURAL CHEMICALS
Feeding the Earth's People

Yields of farm produce have been increased by the use of synthetic fertilizers and pesticides, and the introduction of improved crop plants.

How many people can the earth support? Is the population about to overtake food production? In the developed nations, large surpluses of food have been produced, and in many of the developing nations, food production has managed to keep pace with population growth. Food and resources are not evenly distributed, however, and despite the advances, millions of people in the poorer nations of the world suffer from hunger and malnutrition. Drought, war, and other disasters that disrupt food production and distribution continue to cause widespread starvation and disease.

Significant increases in agricultural production began during the nineteenth century as farmers gained a better understanding of the needs of plants. Crop rotation, the use of natural fertilizers, improved irrigation, and the introduction of efficient farm machinery, all helped to increase crop yields. However, the dramatic increases in food production that have been achieved in recent years have been due primarily to (1) the application of synthetic fertilizers to restore lost fertility to soil, (2) the use of pesticides to reduce the amount of food lost to disease and pests, and (3) the introduction of new, improved seeds and crop plants.

In this chapter, we will focus on the agricultural chemicals—fertilizers and pesticides—that help us meet the world's increasing demand for food. We will study the soil conditions and nutrients that plants need to grow and thrive. We will also examine the risks to human health and the environment that the use of synthetic fertilizers and pesticides pose, and we will consider alternative farming techniques that are now available.

Learning Goals:

In this chapter, you should gain an understanding of:

1. How the human population is growing.

2. The composition and properties of soil.

3. The need for primary and secondary nutrients and for micronutrients.

4. The production and use of synthetic fertilizers.

5. The main classes of synthetic pesticides and the benefits and problems associated with their use.

6. Pest control methods that do not rely on synthetic pesticides.

7. Approaches to farming that are less dependent on agrichemicals.

POPULATION GROWTH

As early as 1798, Thomas Robert Malthus, an English clergyman and economist, published an essay in which he argued that populations inevitably tend to grow faster than their food supply. He predicted that, unless birth rate was controlled, population would expand until it reached the limit of subsistence. After that, population would be kept in check by starvation, disease, pestilence, and war. Since Malthus's time, the world's population has increased from less than 1 billion to over 5 billion, but because agricultural productivity has increased faster than Malthus thought possible, so far, his pessimistic predictions have not been realized.

During most of the 3 million years that humans have lived on the earth, the population has grown very slowly, with numerous setbacks caused by famine and epidemic diseases (Fig. 18-1). One of the most devastating killers was the Black Death, or plague, which swept through Europe during the Middle Ages, causing the deaths of one-quarter of the population—approximately 25 million people. In the eighteenth and early nineteenth centuries, childhood diseases took a fearful toll, and many people died from waterborne diseases (Chapter 12). The introduction of vaccinations, the discovery of drugs that could cure disease, and the increased availability of clean water were the major factors responsible for the decrease in infant mortality and the increase in life expectancy that led to the rapid rise in population that continues today.

Today, despite decades of growth in agricultural productivity and steady expansion of cropland at the expense of forest and wilderness, there are more hungry people on the earth than ever before.

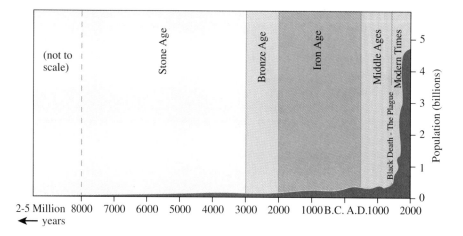

Figure 18-1 The human population grew very slowly until the middle of the seventeenth century, when it began to grow very rapidly.

TABLE 18-1   Growth of World Population

Population	Year	Time Required to Double
1 billion	1850	All of human history
2 billion	1930	80 years
4 billion	1975	45 years
8 billion (projected)	2020	45 years

World population reached 1 billion in 1830. By 1930, just 100 years later, it had reached 2 billion. After another 45 years, it had doubled again to 4 billion (Table 18-1). Today, our population stands at over 5 billion; if the present growth rate continues, it is expected to reach 8 billion by 2020.

Population Arithmetic

Malthus's dire prediction was based on the argument that population can grow geometrically, but food supply can grow only arithmetically. Geometric growth occurs when something increases by a fixed percentage every year. For example, if $1000 is invested at 10 percent interest (compounded daily) at a child's birth, by the time the child is 7 years old, the account will be worth $2000 (Fig. 18-2a).

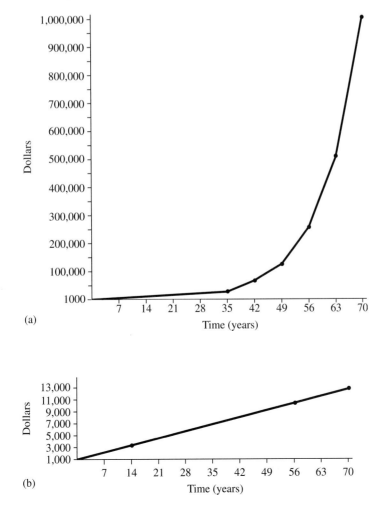

(a)

(b)

Figure 18-2   (a) Geometric growth of a savings account in which $1000 was invested at 10% interest compounded daily. A period of slow growth is followed by increasingly rapid growth. (b) Arithmetic growth of a savings account in which a constant amount ($1000) was added at regular intervals (every 7 years).

Seven years later, it will have doubled to $4000. After 56 years, it will be worth over a quarter of a million dollars; at 70, it will be worth over $1 million.

The most striking characteristic of geometric growth is that it begins slowly, but once it has "rounded the bend," it grows at an ever-increasing rate. This rapidly increasing rate occurs whether the growth being recorded is money in a bank account or population.

Arithmetic growth occurs when a constant amount is added at regular intervals. For example, if a parent distrustful of banks puts aside $1000 at a child's birth and then adds further installments of $1000 at 7-year intervals, the sum will grow slowly and steadily over successive periods (Fig. 18-2b). By the time the child reaches 70 years, the cache, if still intact, will amount to $11,000—far less than the more than one million dollars produced over the same period by the $1000 invested at 10 percent.

EXAMPLE 18-1   You set up a savings account as follows. You deposit 10 cents the first week and then double the amount you deposit each week thereafter (20¢ the second week, 40¢ the third week, and so on). How much money will be in the account at the end of 10 weeks? Is this an example of geometric or arithmetic growth?

Solution:

Week	Amount Deposited ($)	Total in Account ($)	Week	Amount Deposited ($)	Total in Account ($)
1	0.10	0.10	6	3.20	6.30
2	0.20	0.30	7	6.40	12.70
3	0.40	0.70	8	12.80	25.50
4	0.80	1.50	9	25.60	51.10
5	1.60	3.10	10	51.20	102.30

There will be $102.30 in the account at the end of 10 weeks. The growth is geometric.

PRACTICE EXERCISE: A child saves 10 cents every week and puts the money in her piggy bank. How much will she have at the end of 10 weeks? Are her savings growing arithmetically or geometrically?

Answer: $1.00. Arithmetically.

TABLE 18-2   Growth Rates and Doubling Times—1992

Region	Growth Rate (%)	Doubling Time (years)
Africa	3.0	23
Latin America	2.1	33
Asia (excluding China)	2.0	34
China	1.3	52
Oceania	1.2	60
North America	0.8	88
CIS (Former Soviet Union)	0.7	91
Europe	0.2	293

Source: Population Resource Center, Washington, D.C.

An instructive perspective on population growth is revealed by the **doubling time**: the number of years it takes for a population to double if growth continues at its present rate. This number is calculated by using the **Rule of 70**, that is, by dividing the current annual growth rate (as a percentage) into 70. At the present time, world population is growing at the rate of 1.7 percent per year. If this seemingly slow rate continues, world population will double every 41 years (70/1.7 = 41). Doubling times for different regions of the world are given in Table 18-2. It can be seen that the highest birth rates are found in the poorest countries.

EXAMPLE 18-2   The population of a city is 16 million and is growing at an annual rate of 3.5 percent. In how many years will the population double if growth continues at the same rate?

Solution: Use the Rule of 70. Divide the annual rate into 70.

$$\frac{70}{3.5} = 20$$

The population will double in 20 years.

PRACTICE EXERCISE: In 1991, it was estimated that the population of a country was 23.6 million and was growing at a rate of 4.1 percent a year. If the growth rate remains constant, in what year will the population reach 47.2 million?

Answer: 2008.

THE DEVELOPMENT OF MODERN AGRICULTURE

One of the earliest techniques for growing food was **slash-and-burn cultivation** (Fig. 18-3) in which trees and other vegetation were cut down and burned. The ashes returned nutrients to the soil, making the soil sufficiently fertile for crops to grow at the site for 2 to 5 years. When fertility waned, the site was abandoned, and a new plot was cleared. Slash-and-burn agriculture continues today in less developed countries, particularly in the tropical rain forests in the Amazon Basin of South America.

For centuries, farmers in settled communities relied on manure and nutrient-producing legumes (Chapter 3) grown in crop rotations to return nutrients to the soil, but the scientific basis for these practices was not understood until the nineteenth century. The German chemist Justus von Liebig (1803–1873) was a pioneer in the field of agricultural chemistry and was the first to identify many of the chemical substances that plants require for healthy growth. As a result of his work and that of others, farmers began using increased amounts of natural fertilizers to supply plants with the needed nutrients. Crop yields increased dramatically, and, with the help of the new farm machinery introduced with the Industrial Revolution, food production managed to keep pace with the upsurge in population. The need for more and more fertilizer led directly to the introduction of synthetic inorganic fertilizers, upon which modern agriculture today depends so heavily to maintain an adequate food supply.

Legumes enrich the soil with nitrogen compounds because their root nodules contain bacteria that are able to fix atmospheric nitrogen.

Figure 18-3   In slash-and-burn agriculture, nutrients in the ashes make it possible to grow crops for two to five years before the land becomes exhausted and is abandoned.

Although India's population increased by 130 million people during the 1970s, application of synthetic fertilizers increased food production to such an extent that by 1980, the country was able to export grain.

All our food comes ultimately from green plants. We either eat plants or we eat animals that have fed on plants (Chapter 3). To grow and flourish, plants need sunlight, carbon dioxide, suitable temperature, and soil with an appropriate pH that can provide an adequate supply of water, oxygen, and essential nutrients. Before we consider the nutrients that plants need, we will discuss the properties that make a soil productive.

SOIL

In the United States and other developed countries, soil is kept fertile largely by the application of commercial inorganic fertilizers. The productivity of soil is also greatly influenced by its texture and by the amount of organic matter, both living organisms and dead material, that it contains.

Soil Texture: The Inorganic Component

The mineral particles of soil are derived from the weathering of rock (Chapter 2) and are composed primarily of silicates, but, depending on location, soil may also contain phosphates and limestone ($CaCO_3$). The size of the particles determines the soil's **texture**. Relatively coarse particles (0.10–2.0 mm) form **sand**; slightly finer particles form **silt**; and the finest particles (0.002 mm or less) form **clays**. A typical soil is a mixture of all three types. **Loam**, a productive soil that crumbles easily, contains approximately equal proportions of sand and silt.

Plant roots absorb water, nutrients, and oxygen from the soil and release carbon dioxide to it. They need a soil that allows good circulation of air and

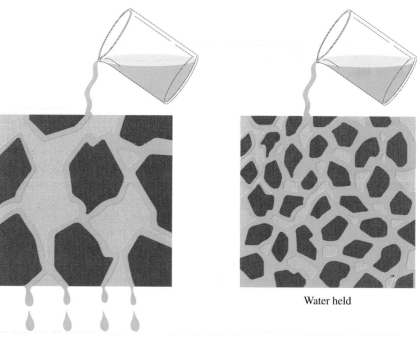

Figure 18-4 The water-holding capacity of soil increases as the size of the soil particles decreases. As particle size decreases, relative surface area increases, and more water can cling to a given quantity of soil.

Water runs through

Water held

(a)   (b)

Figure 18-5   Soil texture.
(a) Soil that lacks humus is
dense and solid. (b) Soil rich in
humus is light and friable.

provides continuous access to water and nutrients. In sandy soils, the relatively large soil particles have large spaces between them. The large spaces promote good air circulation but allow water to percolate rapidly through the soil, taking dissolved nutrients beyond the reach of the roots. As the size of soil particles decreases, relative surface area increases, allowing more water to adhere to a given amount of soil. At the same time, the size of the spaces between particles becomes smaller. Smaller spaces do not allow water to flow through as rapidly and tend to hold it in the spaces by capillary action (Fig. 18-4). The disadvantage is that air circulation is less efficient. In clay soils, the very fine particles are packed so closely together that the soil becomes impermeable to water. Rainfall collects at the surface and in the upper layers, which soon become so waterlogged that gases cannot circulate. The most productive mineral soils (soils that contain no organic matter) are silt and loam.

The water-holding capacity of soil increases as soil particle size decreases.

Humus: The Organic Component

Soil is greatly improved by the addition of decayed organic matter called **humus**. As leaves, twigs, dead plants and animals, and other detritus accumulate at the soil surface, they are attacked by bacteria and other soil organisms and are partially decomposed. The partially decomposed product is humus, a dark-colored mixture of organic matter that is resistant to further decomposition.

Compost, which gardeners produce from vegetable wastes, leaves, grass, and other yard wastes by allowing them to be digested by soil bacteria and worms, is a form of humus.

As humus forms, it becomes mixed with soil particles, and the soil acquires a light spongy structure that retains water well (Fig. 18-5). Humus acts as a reservoir of water and nutrients, which are slowly released to plants in a form they can absorb. In an undisturbed natural ecosystem, nutrients are constantly recycled (Chapter 3), and humus is replenished. If crops are repeatedly grown on the same plot of land, however, nutrients are permanently removed as each crop is harvested, and the soil soon becomes infertile.

A typical soil is made up of three main layers, or horizons (Chapter 2) (Fig. 18-6). The uppermost, or A, horizon, which includes the **topsoil**, is dark in color and contains most of the humus found in soil. The A horizon is made up of decayed plant and animal materials, and it is rich in microorganisms and other creatures including earthworms, insects, and small burrowing animals that keep the soil aerated. Below the topsoil is the lighter-colored, humus-poor, and more-compacted B horizon, or **subsoil**. This layer, which may be several feet thick, is made from inorganic particles of the parent rock and contains minerals and organic materials leached from the A horizon. The C horizon consists of an upper layer of fragmented bedrock mixed with clay that rests on solid bedrock.

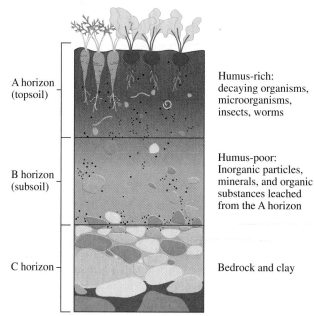

Figure 18-6 A soil profile showing the three main layers (or horizons). The thickness of the layers varies considerably depending on the type of soil.

A horizon (topsoil) — Humus-rich: decaying organisms, microorganisms, insects, worms

B horizon (subsoil) — Humus-poor: Inorganic particles, minerals, and organic substances leached from the A horizon

C horizon — Bedrock and clay

Soil pH

Plants vary considerably as to the pH range they can tolerate. Most prefer a soil with a pH close to neutral but some, such as corn, wheat, and tomatoes, grow best in slightly acidic soil; potatoes and most berries prefer even more acidic conditions.

Acidic soils are often called *sour* soils and slightly basic soils are described as *sweet*.

Soils tend to become acidic if the carbon dioxide released by roots or formed by the oxidation of organic matter accumulates below the soil surface:

$$CO_2 + H_2O \longrightarrow H^+ + HCO_3^-$$

The acid can be neutralized by the application of lime (a process called liming) in the form of quicklime, CaO, or slaked lime, $Ca(OH)_2$, either of which removes hydrogen ions.

$$2H^+ + CaO \longrightarrow Ca^{2+} + H_2O$$

Some soils are alkaline and are neutralized by adding sulfur. Soil bacteria gradually convert the sulfur to sulfuric acid.

PLANT NUTRIENTS

Plants obtain carbon, hydrogen, and oxygen from air and water and, with energy from the sun, convert them to simple carbohydrates and oxygen through the process of photosynthesis:

$$6\,CO_2 + 6\,H_2O \longrightarrow C_6H_{12}O_6 + 6\,O_2$$

The simple carbohydrates are then converted into the complex carbohydrates that form the basic structural material of all plants.

In addition to carbon, hydrogen, and oxygen, plants require at least 13 other nutrients that are called mineral nutrients because they are obtained from the soil. The mineral nutrients are divided into three groups according to the relative amounts of each that plants need to thrive: **primary nutrients** (nitrogen, phosphorus, and potassium), **secondary nutrients** (calcium, magnesium, and sulfur), and **micronutrients** (Table 18-3). Dissolved in water, mineral nutrients are absorbed from the soil through plants' roots.

Carbon, hydrogen, and oxygen are often referred to as the nonmineral nutrients.

Nitrogen

Plants need nitrogen to manufacture amino acids and proteins. After carbon, hydrogen, and oxygen, nitrogen is the nutrient needed in greatest quantity for the formation of leaves and stems. When a bushel (equal to 8 gallons) of corn ripens, it removes approximately 0.5 kg (1 lb) of nitrogen from the soil. If this nitrogen is not replaced, the supply of nitrogen is gradually exhausted. Soil is more likely to be deficient in nitrogen than in any other nutrient.

Although 78 percent of the atmosphere is nitrogen gas (N_2), very few plants can exploit this source directly. For most plants, atmospheric nitrogen must first be changed by **nitrogen fixation** into a water-soluble form that plants can absorb through their roots. As we saw in Chapter 3, nitrogen fixation is accomplished primarily by: (1) soil bacteria, (2) rhizobium bacteria that live in nodules on the roots of leguminous plants, (3) blue-green algae (cyanobacteria) in soil and water, and (4) lightning. Nitrogen-fixing bacteria convert atmospheric nitrogen to ammonia (NH_3), which in the slightly acidic pH of most soils

In aquatic ecosystems, cyanobacteria are the most important nitrogen fixers.

TABLE 18-3 Chemical Nutrients Required by Plants

Name	Chemical Form Absorbed
Nonmineral	
Carbon (C)	CO_2
Hydrogen (H)	H_2O
Oxygen (O)	CO_2, H_2O, O_2
Primary nutrients	
Nitrogen (N)	NH_4^+, NO_3^-
Phosphorus (P)	$H_2PO_4^-$, HPO_4^{2-}
Potassium (K)	K^+
Secondary nutrients	
Calcium (Ca)	Ca^{2+}
Magnesium (Mg)	Mg^{2+}
Sulfur (S)	SO_4^{2-}
Micronutrients	
Boron (B)	$H_2BO_3^-$, $B(OH)_4^-$
Chlorine (Cl)	Cl^-
Copper (Cu)	Cu^{2+}
Iron (Fe)	Fe^{2+}, Fe^{3+}
Manganese (Mn)	Mn^{2+}
Molybdenum (Mo)	MoO_4^{2-}
Zinc (Zn)	Zn^{2+}

It is estimated that about 90 percent of nitrogen fixation is due to biological processes, and 10 percent is due to lightning.

forms ammonium ions (NH_4^+). By the process of **nitrification**, other soil bacteria bring about the oxidation of ammonium ions to nitrate ions (NO_3^-), the preferred form of nitrogen for most plants.

During thunderstorms, electrical discharges cause nitrogen to react with oxygen to produce nitric oxide (NO) and nitrogen dioxide (NO_2):

$$N_2 + O_2 \longrightarrow 2\,NO$$

$$2\,NO + O_2 \longrightarrow 2\,NO_2$$

Nitrogen dioxide dissolves readily in water, forming nitrous acid and nitric acid:

$$2\,NO_2 + H_2O \longrightarrow HNO_2 + HNO_3$$

Nitric acid in rainfall replenishes the supply of nitrates in soil.

By the nineteenth century, in addition to the use of manure, fertility was being restored to soil by crop rotation, the practice of alternating a nitrogen-consuming crop, such as corn or wheat, with a nitrogen-fixing leguminous crop, such as beans, peas, alfalfa, or soybeans. The discovery of huge deposits of sodium nitrate (called Chile saltpeter) in the deserts of Chile provided a valuable additional source of nitrogen, but it was not until 1913 that a seemingly inexhaustible supply of nitrogen became available. The breakthrough came with the development by the German chemist Fritz Haber of a nitrogen-fixing process, which led directly to the production of synthetic nitrogen fertilizers.

In the Haber process (Chapter 7), nitrogen obtained from the atmosphere reacts directly with hydrogen to produce ammonia:

In the United States, ammonia was first produced commercially by the Haber process in 1921.

$$N_2 + 3H_2 \longrightarrow 2\,NH_3$$

During World War I (1914–1918), the Germans used ammonia to make ammonium nitrate for the manufacture of explosives. It was not until the war was over that the process was used as the first step in the manufacture of nitrogen fertilizers. Today, ammonium salts, urea, and ammonia itself are the major compounds used to restore nitrogen to the soil, and all depend on the Haber process as the first step in their production. The pathways by which nitrogen compounds reach the soil are shown in Figure 18-7.

Anhydrous Ammonia   At normal temperatures and pressures, ammonia is a gas. At higher pressure, it is easily compressed into a liquid called anhydrous ammonia that can be stored and transported in tanks. At the farm, the liquid ammonia is injected directly into the soil (Fig. 18-8), where it is converted into ammonium ions.

Ammonium Salts   The first step in the production of ammonium nitrate fertilizer is oxidation of ammonia to nitric acid:

$$2\,NH_3 + 4\,O_2 \longrightarrow 2\,HNO_3 + 2\,H_2O$$

The nitric acid then reacts with more ammonia to produce ammonium nitrate:

$$HNO_3 + NH_3 \longrightarrow NH_4NO_3$$

Ammonium sulfate and ammonium phosphate are prepared by treating ammonia with sulfuric acid and phosphoric acid.

Urea   Under high pressure, ammonia and carbon dioxide react to form urea:

$$2\,NH_3 + CO_2 \longrightarrow NH_2\overset{\overset{\textstyle O}{\|}}{-C}-NH_2 + H_2O$$

urea

Upon contact with water in the soil, urea gradually decomposes, releasing ammonia as it does so. This slow release gives urea an advantage over the inorganic ammonium fertilizers.

Urea is a major constituent of urine.

The use of synthetic nitrogen fertilizers has been the factor most responsible for our ability to produce enough food to keep pace with population growth. But the cost of manufacturing them is high. Although there is an unlimited supply of nitrogen in the atmosphere, the hydrogen that is required to synthesize ammonia is obtained from petroleum (Chapter 14). Small hydrocarbon molecules, such as propane (C_3H_8), are reacted with steam in the presence of suitable catalysts to yield carbon dioxide and hydrogen:

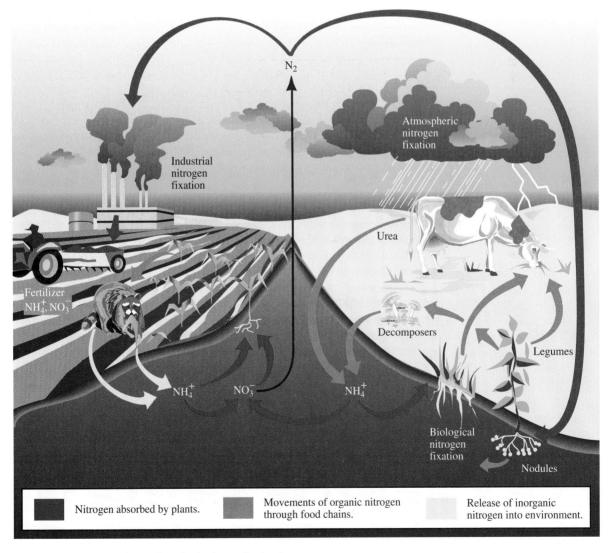

Figure 18-7   The pathways by which plants obtain nitrogen.

Figure 18-8 Application of anhydrous ammonia to a soybean field in Iowa. The ammonia in the tank is under pressure, which keeps it in liquid form. When it is released from the pressure in the tank, it is immediately converted to ammonia gas, which dissolves in moisture in the soil.

$$C_3H_8 + 6\,H_2O \longrightarrow 3\,CO_2 + 10\,H_2$$

Thus, for the present, the cost of fertilizers is tied to the cost and availability of petroleum. In the future it may be possible to use solar energy to obtain hydrogen from the electrolysis of water (Chapter 15).

Scientists are working to understand how the unique enzymes in bacteria in legumes are able to fix atmospheric nitrogen. Using genetic engineering, they hope to create nitrogen-fixing bacteria that will live in the roots of cereal plants in the same way that they live in legumes. If this could be achieved, the impact on agriculture would be enormous. Other approaches include improving the nitrogen-fixing abilities of legumes now being grown and treating soil with nitrogen-fixing blue-green algae (cyanobacteria).

Phosphorus

Plants need phosphorus for the synthesis of DNA and RNA (Chapter 10) and for the synthesis of ATP and ADP (Chapter 17), the compounds involved in energy transfer in many metabolic processes, including photosynthesis. Phosphorus is particularly important for plants such as tomatoes that are cultivated for their fruit.

Plants absorb phosphorus from the soil in the form of phosphate ions. At the near neutral pH of most soils, phosphates exist mainly as HPO_4^{2-} and $H_2PO_4^{-}$ ions, which in natural soils are derived from the weathering of phosphate rock and from ancient deposits of skeletal remains of sea creatures and other animals. Phosphates are common in nearly all soils but often are present at low concentration.

Although it was not until the nineteenth century that a plant's need for phosphorus was established by scientific studies, records show that the practice of improving soil by adding ground bone was used in China as early as 2000 B.C.

The phosphorus in bones and in phosphate rock is tightly bound and not readily available to plants, but it can be converted to a more soluble form, called **superphosphate**, by treatment with sulfuric acid.

Fish meal and guano, the concentrated phosphate-rich droppings of sea birds and bats, have been used to supply phosphorus to the soil.

$$\underset{\substack{\text{phosphate rock}\\\text{or bone}}}{Ca_3(PO_4)_2} + 2\,H_2SO_4 \longrightarrow \underset{\text{superphosphate}}{Ca(H_2PO_4)_2} + 2\,CaSO_4$$

Today, farmers depend primarily on superphosphate obtained by treating the phosphate mineral fluorapatite, $Ca_5(PO_4)_3F$ with sulfuric acid to replenish the phosphorus in soil. An alternative fertilizer is ammonium phosphate, which supplies both phosphorus and nitrogen.

Although deposits of phosphate rock are widespread throughout the world, with major deposits in Morocco, China, and the United States, there is growing concern that unless new deposits are discovered, world reserves of high-grade phosphate will be depleted within the next 40 years. As supplies dwindle, more low-grade phosphate containing significant amounts of cadmium must be used. Cadmium, which is toxic to many life forms (Chapters 12 and 16) and undergoes bioaccumulation (Chapter 3), must be removed before the phosphate can be used; this step adds considerably to the cost of the fertilizer.

Potassium

Potassium, the third primary mineral nutrient, is taken up by plant roots as the ion K^+. Potassium takes part in the formation of starches and cellulose from simpler sugars and is essential for the normal functioning of many plant enzymes. Potassium may also be involved in the movement of carbohydrates within the plant and in the synthesis of proteins.

Potassium is distributed widely in the earth's crust and is abundant in most soils. But if crops are grown repeatedly on the same land without renewing the potassium, the soil's supply of potassium will eventually be depleted. Commercial fertilizers usually contain potassium in the form of potash (K_2CO_3) or potassium chloride (KCl). Large deposits of these salts are found in the United States, Canada, and Germany. The potassium chloride deposits under the Canadian prairies are enormous, but, because they are 1.5 km (nearly one mile) underground, they are difficult and expensive to mine. Although there is no immediate shortage of potassium salts, it should be remembered that the deposits are nonrenewable.

Large deposits of potassium salts are located at Searles Lake, California, and Carlsbad, New Mexico.

Secondary Nutrients: Calcium, Magnesium, and Sulfur

Calcium is abundant in nearly all soils. Not only is it a component of many common minerals (limestone, dolomite, and many silicates), it is frequently spread on soil in the form of lime to neutralize acidity. Also, the application of superphosphate adds calcium ions as well as phosphate ions. Plants absorb calcium as the Ca^{2+} ion.

Magnesium, like calcium, is abundant in most soils, but in some soils (particularly acidic soil), it becomes tightly bound, and there may be too few magnesium ions in solution. The deficiency can be corrected by liming with crushed dolomite ($CaCO_3 \cdot MgCO_3$) or by adding magnesium sulfate ($MgSO_4$). Magnesium, which is absorbed through plant roots as Mg^{2+}, is a constituent of chlorophyll. A magnesium deficiency causes chlorosis, the yellowing of leaves.

Sulfur deficiencies occur most frequently in sandy, well-drained soils. The problem is treated by adding ammonium sulfate, potassium sulfate, or some other sulfate. Plants absorb sulfur as SO_4^{2-} ions. Sulfur is a constituent of some amino acids and thus is essential for the synthesis of proteins.

In the United States, soils most likely to be deficient in sulfur are located in the northwestern and southeastern regions of the country.

Micronutrients

Although required in only trace amounts, micronutrients are as essential for healthy plant growth as the macronutrients are; the effects of deficiencies can be severe. Onions grown in soil lacking zinc, for example, are stunted and develop abnormally (Fig. 18-9).

Figure 18-9 Onions grown in soil that is deficient in zinc do not thrive. Growth is stunted, and shoots are yellow and twisted (right). Onions of the same age and type grown in nutrient-rich soil are shown at left.

It is often difficult to determine which, if any, micronutrient is in short supply in a particular soil, and how much of a nutrient should be applied to remedy a deficiency. Another problem is that the applied nutrient may bond to chemicals already in the soil and become unavailable to the plant. Agriculturists have found that it is preferable to add **iron** as its ethylenediaminetetraacetic acid (EDTA) complex instead of as a simple inorganic salt. In the complexed form, the iron is protected from the soil chemicals that otherwise would bind it, but it is held loosely enough so that it can be released gradually for uptake by plant roots.

Because plants vary considerably in their nutrient requirements, micronutrients must be applied with care. For example, the amount of **boron** that is ideal for sugar beets would be toxic to soybeans and many other plants. Micronutrients are not always applied to the soil. **Molybdenum** compounds, for example, are often sprinkled on seeds before planting; other nutrients are applied to leaves.

MIXED FERTILIZERS

Farmers and home gardeners usually correct soil deficiencies by applying fertilizers that contain a mixture of the three primary nutrients. Bags of **mixed fertilizer** are labeled with a set of three numbers that indicate, in the following order, the percentage of (1) nitrogen (N), (2) phosphorus as P_2O_5, and (3) potassium as K_2O in the fertilizer (Fig. 18-10). If the numbers are 5-10-5, the fertilizer is composed of 5 percent N, 10 percent P_2O_5, and 10 percent K_2O.

The choice of fertilizer depends on the crop being grown. For lawns, the most important nutrient is nitrogen, and a 20-10-10 fertilizer would be suitable. Fruits and vegetables require a fertilizer rich in phosphorus; here, a 10-30-10 fertilizer would be a better choice.

Potassium and phosphorus are not actually present in fertilizers as K_2O and P_2O_5. These designations date back to the late nineteenth century when plant requirements were first studied. Plants were burned and the concentrations of the oxides K_2O and P_2O_5 in the ashes were determined.

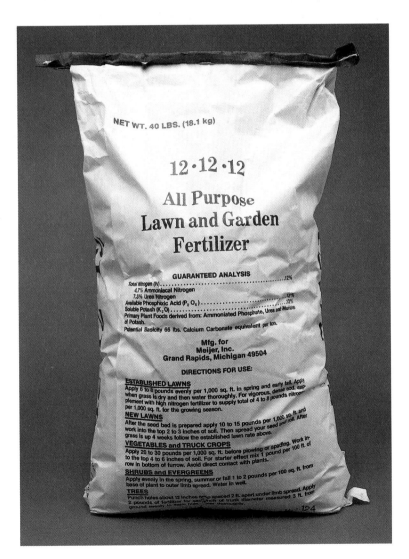

Figure 18-10 The three numbers on a bag of fertilizer refer to the percentage by weight of nitrogen (as N), phosphorus (as P_2O_5), and potassium (as K_2O).

THE USE OF SYNTHETIC FERTILIZERS

Synthetic fertilizers were first manufactured on a large scale in the late 1930s; since then, our dependence on them to return nutrients to the soil has grown dramatically. Between 1955 and 1975, fertilizer use in the United States more than tripled. The application of massive amounts of synthetic fertilizers has undoubtedly been a major factor in the huge increases in food production that have been achieved worldwide during the past 40 years, but, unfortunately, the use of synthetic fertilizers can cause severe environmental problems.

If synthetic inorganic fertilizers are applied repeatedly without the simultaneous addition of sufficient organic materials to sustain the formation of humus, soil becomes compacted and loses its nutrient- and water-holding properties and its ability to fix nitrogen; thus, increasing amounts of fertilizer must be added to maintain crop yields. Eventually, the soil becomes mineralized and increasingly susceptible to erosion. Another problem is that most commonly used fertilizers do not supply the needed micronutrients.

Water pollution is still another problem. As we learned in Chapter 12, if inorganic fertilizers in runoff from agricultural land enter rivers and lakes, they can cause eutrophication, the excessive growth of plants and algae. Also, very soluble ions such as nitrate ions, which are toxic above certain levels, can percolate down through the soil and contaminate groundwater.

CONSUMER BRIEF   Softening Water

Fresh water that contains a relatively high concentration of Ca^{2+}, Mg^{2+}, and Fe^{2+} ions is described as **hard water**. Water that has a low concentration of these hard-water ions is called **soft water**. When hard water is mixed with soap, the metal cations react with anions in soap to form insoluble compounds that precipitate out as "scum," the familiar ring on bathtubs. This problem does not occur with soft water.

Hardness due to calcium or magnesium, present as bicarbonates, is produced when water containing carbon dioxide comes in contact with limestone ($CaCO_3$) and dolomite ($CaCO_3 \cdot MgCO_3$).

$$CaCO_3 + CO_2 + H_2O \longrightarrow Ca^{2+} + 2\,HCO_3^-$$

$$CaCO_3 \cdot MgCO_3 + 2\,CO_2 + 2\,H_2O \longrightarrow$$
$$Ca^{2+} + Mg^{2+} + 4\,HCO_3^-$$

Hardness due to these ions can be removed by boiling the water. On boiling, carbon dioxide and carbonate ions (CO_3^{2-}) are produced.

$$2\,HCO_3^- \longrightarrow CO_2 + CO_3^{2-} + H_2O$$

The CO_3^{2-} ions then react with Ca^{2+} and Mg^{2+} ions to form insoluble carbonates ($MgCO_3$ and $CaCO_3$) that precipitate out of solution and can be removed.

$$M^{2+} + 2\,HCO_3^- \longrightarrow MCO_3 + CO_2 + H_2O$$
$$(M = Ca\ or\ Mg)$$

The scaly deposits that form in tea kettles are formed in this way. Similar deposits in boiler and water pipes can be a serious problem when they build up and clog pipes.

The most frequently used home method for softening water is **ion exchange**, a process in which hard-water ions—Ca^{2+}, Mg^{2+}, and Fe^{2+}—are exchanged for sodium ions. In this method, hard water is passed through a tank containing specially prepared plastic beads (ion exchange resins) covered with sodium ions (see illustration). As the water flows through the tank, sodium ions on the beads exchange with hard-water ions. The hard-water ions are thus removed, and the water leaving the tank contains sodium ions, which do not form a scum with soap or produce scaly deposits.

When all the sites on the plastic beads are occupied with hard-water ions, the material can be

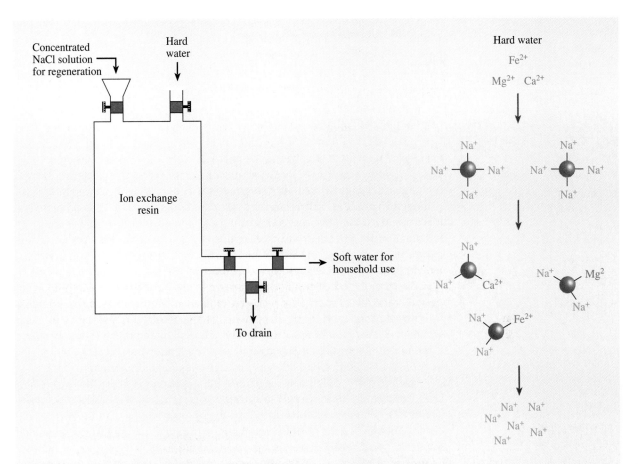

regenerated by flushing with a concentrated solution of sodium chloride. Sodium ions replace the hard water ions that are flushed down the drain.

Water softened in this way contains an increased concentration of sodium ions. Therefore, for health reasons, it may be preferable to drink the hard water and use the softened water for washing and heating purposes only.

Questions
1. Explain how ion-exchange might be used to remove lead from water?

2. Would you expect water that flows over limestone to be harder or softer than water that flows over granite? Explain.

ORGANIC FERTILIZERS

Because of the problems associated with inorganic fertilizers, many people advocate a greater use of **organic fertilizers**, including animal manure, green manure, and possibly sewage sludge. Animal manure can include dung and urine from cattle, horses, and poultry. Green manure refers to the process of plowing a crop of green plants, preferably legumes, into the soil—a practice that has the disadvantage of keeping the land unproductive for a year. Treated sewage sludge (Chapter 12), if free from toxic heavy metals, is a valuable source of nutrients.

Manures return nitrogen and many other nutrients to the soil; at the same time, they improve soil structure by adding valuable humus. They also stimulate the growth of soil bacteria. However, weight for weight, organic fertilizers

In many Asian countries, human wastes, often called night soil, are collected and used as fertilizer.

cannot produce the crop yields achieved with commercial inorganic fertilizers. A typical manure is a 1-2-1 fertilizer. Another problem with using manures is cost. In modern agricultural practice, livestock and crops are seldom raised on the same farm, and packaging and transportation costs generally make organic fertilizers more expensive than inorganic fertilizers.

PESTICIDES

Ever since humans first appeared on the earth, they have had to contend with pests (Chapter 12). The biblical account of the liberation of the Hebrews from bondage in Egypt in the thirteenth century B.C. includes descriptions of plagues of lice, flies, and locusts in the Nile Valley (Exodus 7:20–11:10). The Black Death, which killed millions of people in the Middle Ages, was transmitted to humans by fleas on rats carrying bubonic plague. The failure of the potato crop in Ireland in 1845, which caused widespread starvation and mass emigration to the United States, was caused by the potato blight (Chapter 17). Malaria, spread by the anopheles mosquito, still incapacitates or kills millions of people annually in the less developed countries, and swarms of locusts continue to devastate crops in many areas of the world (Fig. 18-11). Plants on which we depend for food are under attack from insects, fungi, bacteria, viruses, and other microorganisms and must compete with weeds for nutrients. Mice, rodents, rabbits, and other animals also take a share of the crop.

Before 1940, only a few pesticides were available. Among them were several naturally occurring insect poisons extracted from plants, including **pyrethrins**, compounds obtained from the pyrethrum flower (a member of the chrysanthemum family). Other early insecticides were nicotine sulfate obtained from tobacco, rotenone from the tropical derris plant, and garlic oil. A number of inorganic chemicals, primarily compounds of arsenic and lead, were also used. They were less desirable and are rarely used today because they persist in the environment and are toxic to humans and other animals besides insects.

The massive use of pesticides began at the end of World War II with the

Synthetic compounds related to natural pyrethrins are being used in increasing quantities as pesticides. They are biodegradable and nontoxic to birds and mammals.

Figure 18-11 Swarms of locusts like this one in Morocco can completely destroy crops.

introduction of DDT and escalated as other synthetic organic pesticides were developed. Today, in the United States, approximately 500,000 metric tons of pesticides (insecticides and herbicides) are applied annually to control pests.

INSECTICIDES

The three main classes of synthetic organic insecticides in use today are: (1) chlorinated hydrocarbons, (2) organophosphates, and (3) carbamates.

Chlorinated Hydrocarbons

The **chlorinated hydrocarbon** dichlorodiphenyltrichloroethane, which we know as DDT (Fig. 18-12), was first prepared in 1873, but it was not until shortly before the start of World War II that the Swiss chemist Paul Müller discovered that it was an effective insecticide.

After the war, DDT came into widespread agricultural, as well as public-health, use. For almost 20 years, DDT and related chlorinated hydrocarbons, such as aldrin, dieldrin, and chlordan (Fig. 18-12), were major factors in increasing food production and in suppressing insect-borne disease. In 1948, Müller was awarded the Nobel Prize in medicine and physiology for his discovery.

When first introduced, DDT appeared to be an ideal insecticide (Chapter 12). As early as the 1950s, however, it was evident that this was not the case. Insects soon became resistant to it, and, because it did not decompose readily, it persisted in the environment. Microorganisms living in water contaminated with DDT ingested the pesticide, and DDT, being soluble in fats but not in water, accumulated in the organisms' fatty tissues and biomagnified up the food chain. Concentrations of DDT in fatty tissues in fish-eating birds (who are at the top of the food chain) were found to be several million times greater than the concentration of DDT in the water (Fig. 12-17). The DDT disrupted the birds' calcium metabolism, causing eggshells to be so thin and fragile that most of them broke, and only a few young birds hatched. As a result, ospreys, peregrine falcons, and other fish-eating birds nearly became extinct in some parts of the United States.

Because of these adverse effects, DDT was banned in the United States in 1972; since then, the threatened birds have made a dramatic recovery. A few

Combined with talcum powder, DDT was used by Allied forces in World War II to kill body lice and fleas, and it successfully stopped a wartime outbreak of typhus (a disease carried by lice) in Naples, Italy.

DDT is credited with having saved more lives than any other chemical ever synthesized.

DDT

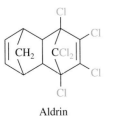

Aldrin

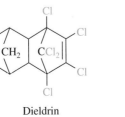

Dieldrin

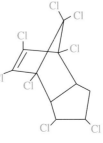

Chlordan

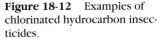

Figure 18-12   Examples of chlorinated hydrocarbon insecticides.

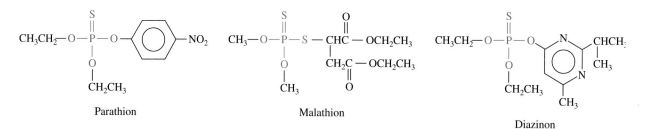

Figure 18-13   Examples of organophosphate insecticides. Organophosphates often contain one of the functional groups that are shown in color.

years later, most of the other chlorinated hydrocarbons were also banned for most uses. But the United States continues to export DDT to developing countries, where its value in controlling malaria outweighs its disadvantages.

Organophosphates

Organophosphate insecticides include **parathion**, **malathion**, **diazinon**, and other related compounds (Fig. 18-13). They are cheap to produce and are very effective against many different insects. They are, however, toxic to humans and mammals; parathion, for example, is more than 20 times as toxic to rats as DDT is (Chapter 16).

Organophosphate pesticides are neurotoxins, nerve poisons that inactivate cholinesterase, the enzyme that plays a vital role in the transmission of nerve impulses between nerve fibers (Chapter 16). Ingestion of organophosphates by humans can result in irregular heartbeat, convulsions, and even death. Because of this, farmers must wear respirators and protective clothing when applying them. Another disadvantage with organophosphates is that, because they break down rapidly in the environment, they must be applied frequently to be effective.

Carbamates

Carbamates, which are derivatives of carbamic acid (Fig. 18-14), include the following functional group in their structures:

$$\begin{matrix} & O \\ & \| \\ N & - C - O \end{matrix}$$

Examples are **carbaryl** (Sevin), **carbofuran** (Furadan), and **aldicarb** (Temik) (Fig. 18-14).

Chlorinated hydrocarbons and most organophosphates kill a wide range of insect types—including many that are beneficial—and are termed **broad-spectrum insecticides**. Most carbamates are **narrow-spectrum insecticides**, which means that they are toxic to only a few types of insects. Unfortunately, one of the insects they kill is the honeybee.

Carbamates, like organophosphates, inactivate cholinesterase. They are rapidly broken down in animal tissues and in the environment, and most are much less toxic to humans and other mammals than organophosphates are. However, granular carbofuran, which for many years was widely used to protect corn and sorghum, was shown to be very toxic to birds. Thousands were poisoned in treated fields, and bald eagles and other birds of prey that ate cont-

Tomatoes and other plants defend against insect attacks by producing compounds in their leaves that interfere with an insect's ability to digest plant tissue.

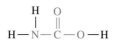

Carbamic acid

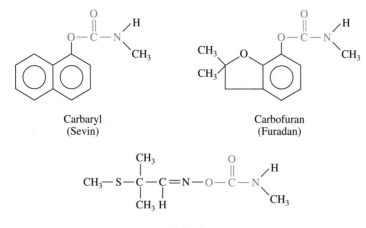

Carbaryl
(Sevin)

Carbofuran
(Furadan)

Aldicarb

Figure 18-14 Carbamic acid and carbamate insecticides derived from it. All contain the functional group that is shown in color.

aminated birds and mammals were threatened. In 1991, the EPA ordered a timed phase-out of carbofuran use.

HERBICIDES

In addition to animal pests, farmers must contend with weeds, which compete with crops for nutrients and water. The traditional method for controlling weeds is **tillage**. Typically, the soil is plowed, which turns over the top 10 to 15 inches of soil and buries and smothers the weeds. The soil is then harrowed to break up large clods before the crop is planted. During the growing season, row crops must be cultivated to destroy weeds that grow up between the rows.

With the introduction of selective herbicides in the 1950s, the need for many of these steps was reduced, and today **no-till agriculture** is practiced extensively in most of the developed countries. Herbicides are applied to a field to kill weeds, and then, in a single operation, a planting machine cuts a furrow, adds fertilizer and seeds, and covers the seeds (Fig. 18-15). The weeds killed by the herbicide remain as mulch and protect the soil from erosion and water loss. The amount of energy and manual labor saved by using herbicides in no-till farming is enormous.

The first chemicals used for weed control were nonselective and included solutions of sodium chlorate ($NaClO_3$), sulfuric acid, copper salts, and various other inorganic chemicals. These herbicides were of limited value in agriculture since they could kill both weeds and crop plants with which they came in contact. The massive use of herbicides began in the 1940s with the introduction of the selective organic herbicides **2,4-D**, **2,4,5-T** (Fig. 18-16a) and other related chlorophenoxy compounds. Today, approximately 50 percent of all pesticides used in the world are herbicides.

Selective herbicides kill only a particular group of plants, such as grasses or broad-leaved plants. Most selective herbicides are **systemic herbicides**; that is, they are absorbed by leaves or roots and then are transferred throughout the plant. They act like growth hormones, causing a plant to grow so fast that it dies. 2,4-D is

No-till farming reduces soil disturbance to a minimum which helps prevent soil erosion.

Figure 18-15 No-till planting of corn in barley stubble. As the planting machine is pulled across the field by a tractor, it does the following in one operation: cuts a furrow, applies fertilizer, plants seeds at regular intervals, and covers the seeds with soil. A machine usually includes more than one planting unit.

very effective against emerging broad-leaved weeds; 2,4,5-T is used to control woody plants. A mixture of these two herbicides, known as Agent Orange, was used extensively as a defoliant during the Vietnam War (Chapter 12). Because of health problems thought to be associated with 2,4,5-T, the EPA banned it in 1985.

Another important herbicide is **atrazine** (Fig. 18-16b), which is used extensively in the production of no-till corn crops. Atrazine acts by interfering with photosynthesis but has no effect on corn and a number of other crop plants because these plants have the ability to change the herbicide into a harmless form. Weeds, however, do not have this ability and quickly die.

Paraquat (Fig. 18-16c) and several related compounds are also widely used as herbicides. These herbicides are toxic to most plants, but because they break down rapidly in the soil, they can be used to kill weeds that emerge before the crop plant does. Paraquat is toxic to humans and must be used with care.

There have been instances in which marijuana sprayed with paraquat to destroy it, has been harvested before the herbicide could take effect. Contaminated marijuana, if smoked, can cause severe lung damage.

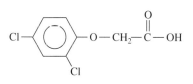

2,4-Dichlorophenoxyacetic acid
(2,4-D)

2,4,5-Trichlorophenoxyacetic acid
(2,4,5-T)

(a)

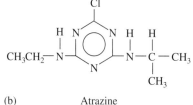

(b)   Atrazine

(c)   Paraquat dichloride

Figure 18-16 Some common herbicides.

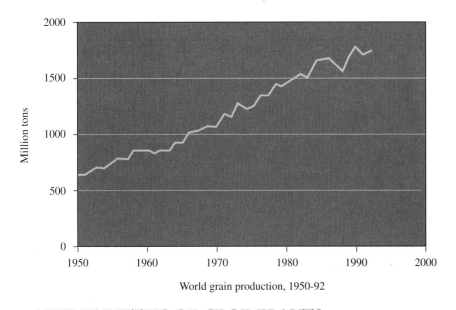

World grain production, 1950-92

Figure 18-17 World grain production from 1950 to 1992.

NEW VARIETIES OF CROP PLANTS

World grain production has increased dramatically during the last 40 years (Fig. 18-17). The third major factor responsible for this increase, in addition to the use of synthetic fertilizers and pesticides, has been the development of new varieties of crop plants. In the 1960s, new high-yielding grains developed by agricultural scientists through selective cross-breeding (Fig. 18-18) were intro-

The dramatic increase in crop yields that occurred during the 1960s was known as the Green Revolution.

Figure 18-18 A new dwarf strain of high-yielding wheat grown in Mexico compared with a native strain.

Some people who would like to become farmers may be discouraged from doing so because they are allergic to pollen, dust, hair, mold, or insect stings. An allergen is a substance that initiates the allergic response. It is usually a protein, but it can also be a polysaccharide.

Pollen, and other particulates in the atmosphere, enter the nose and cling to the mucous membrane. The nasal secretions acting on the pollen release its allergens and other soluble components, which penetrate the outer layer of the mucous membrane and trigger an allergic reaction. By a series of events that are not well understood, the allergen forms a complex with an antibody of a type that is present in unusually high concentrations in allergic persons. The complex causes the release of allergy mediators, one of the most potent being histamine. Histamine is formed by the breakdown of the amino acid histidine, and it causes many of the symptoms of hay fever and other allergies (sneezing, itchy eyes, and runny nose).

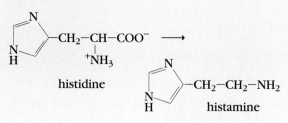

histidine

histamine

Antihistamines are the most widely used medication for treating allergies. Antihistamines are complex organic amines that block the release of histamine. Their structures are similar to the structure of histamine, so they can occupy the same receptor site on the surface of the cells that react to histamine. The antihistamines compete with histamine for these sites. When an antihistamine occupies the receptor site, it remains there passively and prevents the histamine from reaching the site. As long as the antihistamine molecule blocks the receptor site the histamine cannot act, and the symptoms of the allergic response to pollen are reduced.

Like histamines, many antihistamines such as diphenhydramine (Benadryl) contain an ethyl amine group, $-CH_2CH_2N-$. Others, such as oxymetazoline (Dristan) have a five-membered ring with two nitrogens, as does histamine. These structural similarities give the antihistamine molecules a shape very similar to histamine and make them effective competitors for the active sites.

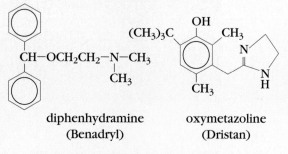

diphenhydramine (Benadryl)

oxymetazoline (Dristan)

These antihistamines not only block the histamine, but they also enter the brain and act on the cells that control sleep. They cause an unpleasant form of sedation with a "blah" or droopy, drowsy feeling. Antihistamines are structurally related to tranquilizers, and they produce a feeling of sleepiness that comes with tranquilizers of the phenothiazine type. The sedation caused by antihistamines is greatly magnified if alcohol or hypnotics are also being taken. People who drink alcohol or take sleeping pills must be very careful when taking antihistamines.

Also available are prescription medications that contain drugs that inhibit the release of histamine but do not enter the brain and therefore do not cause drowsiness. These include terfenadine (Seldane) and astemizole (Histamen). People with serious allergies use those medications to relieve their symptoms without drowsiness.

Questions
1. Some decongestant elixirs contain "hidden" ethyl alcohol that is not disclosed on the label. Is special care required when taking antihistamines and these elixirs?
2. Individuals who are employed in certain occupations should avoid antihistamines while on the job. List five such occupations.

duced into India, South America, and other developing countries. Equally important was the development of crop plants resistant to attack by fungi and insect pests. In the future, scientists hope to produce crops that will be resistant to drought and will grow well in salty soils.

Breeding plants with a new trait is a slow process that may take at least 10 years of painstaking work. Newer genetic engineering techniques can reduce the time required to as little as one year. A genetically engineered tomato that is under review by the FDA is expected to reach the marketplace by the end of 1993. The tomato, called the Flavr Savr, is engineered to postpone the normal softening and decay processes, thus making it possible for farmers to wait until the fruit has ripened naturally on the vine before picking it.

PROBLEMS WITH SYNTHETIC PESTICIDES

Synthetic pesticides have enabled us to control insect-borne diseases and have been responsible for large increases in crop yields, but their use has also created many problems. The major problems are that the pests develop resistance and the pesticides pose threats to the environment and to human health.

The Pesticide Treadmill

Because of genetic variations, some insects in a population have more **resistance** to pesticides than others do. When an area is sprayed with a pesticide for the first time, most insects are killed, but the resistant ones survive to breed the next generation. With repeated spraying of the same pesticide, each generation contains a higher percentage of resistant insects. Control can be maintained only if larger and larger doses of pesticide are applied. Eventually, a new, more potent pesticide may be required. Another problem is that, after a pest has been brought under control, there is often a **resurgence**, with the pest returning in even greater numbers than before.

If beneficial insects are destroyed together with the target pests the natural ecosystem is upset and other problems result. Honeybees that normally pollinate crops are killed, and minor pests that previously were controlled by natural predators multiply, causing serious **secondary pest outbreaks**. Farmers find themselves on a **pesticide treadmill** (Fig. 18-19), continually spending more money for larger quantities and more varieties of pesticides, which become increasingly ineffective.

Since their introduction in the early 1950s, more than 400 different insect pests have become resistant to one or more insecticides.

Health Problems

The people at greatest risk from pesticides are farm workers and workers in pesticide-manufacturing plants. Numerous pesticide-related illnesses and deaths have occurred, particularly in less developed countries, among farm workers who have failed to take adequate precautions when applying pesticides or who have entered sprayed areas before the pesticide has broken down to a harmless form. Organophosphates are especially dangerous in this respect.

There have been several tragedies in pesticide-manufacturing plants. In 1975, employees at a plant near Richmond, Virginia, were exposed to very high levels of **kepone**, a chlorinated hydrocarbon used to kill roaches and ants. Kepone dust contaminated all areas of the workplace, and many workers developed severe neurological disorders and other illnesses. Kepone-containing

According to the EPA, 10,000–20,000 cases of pesticide poisoning are reported annually in the United States. The agency has issued regulations requiring protective clothing for workers applying pesticides and restricting access to newly sprayed areas.

wastes were discharged into the nearby James River, and, by biomagnification through the food chain, fish became so contaminated that commercial fishing on the river had to be banned.

The worst pesticide accident on record occurred in 1984, when highly toxic **methyl isocyanate** (CH_3NCO), a gas used in the manufacture of carbaryl (Sevin) and other pesticides, leaked from a Union Carbide plant in Bhopal, India. More than 3000 people were killed, 14,000 suffered severe injuries including blindness, brain damage, sterility, and liver and kidney damage; many thousands more suffered less severe injuries.

The possibility of water contamination by persistent pesticides that leach through soil into groundwater is a continuing concern for the general public. A disturbing example was the **ethylene dibromide**, **EDB** ($BrCH_2CH_2Br$), contamination of well water that occurred in several areas where the pesticide had been used as a soil fumigant to control root nematodes. EDB was also used to protect stored grain, and unacceptable levels were detected in flour and other foods. Although it had been shown in 1970 that EDB caused cancer in laboratory animals, opposition from users and manufacturers delayed an EPA ban on its use until 1984.

EDB pumped into soil contaminated wells in California, Florida, and Massachusetts.

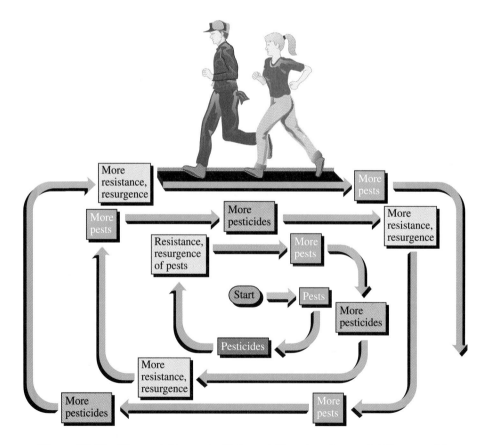

Figure 18-19 The pesticide treadmill. A pesticide, when first used, kills most of the target insects. With repeated spraying, the number of insects resistant to the chemical increases, and a more potent pesticide must be applied. Once controlled, resurgence often occurs, and more and more pesticide with ever greater potency must be applied in an unending cycle.

TABLE 18-4   Comparison of Risk from Exposure to Carcinogens

HERP%*	Daily Average Human Exposure in Diet
0.003	DDT, from remaining residues
0.004	EDB, before it was banned
0.001	Tap water, 32 oz, (chloroform)
0.03	Peanut butter, 1 sandwich, (aflatoxin)
0.03	Comfrey herb tea, 1 cup, (symphytine)
0.07	Brown mustard, 1 teaspoon, (allyl isothiocyanate)
0.1	Mushroom, 1 raw, (hydrazines)
0.1	Basil, 1 g of dried leaf, (estragole)
2.8	Beer, 12 oz, (ethyl alcohol)
4.7	Wine, 8 oz, (ethyl alcohol)

*HERP: Human exposure dose/rodent-potency. Cancer hazard based on a typical person's average daily exposure to the substances over a lifetime. Calculated as a percentage of the LD_{50} for rats (the dose that causes cancer in half the exposed rats).

Probably the greatest concern today is that pesticide residues in foods, particularly in fruits and vegetables, may increase the risk of cancer. Most scientists believe that this fear is unfounded, pointing out that consumption of naturally occurring carcinogens in common foods is many thousand times greater than is consumption of carcinogens in pesticide residues. Biochemist Bruce Ames (Chapter 16) developed a numerical index of the relative cancer hazards of chemicals found in the average person's diet, based on typical daily exposure over a lifetime (Table 18-4). The index is expressed as a percentage of the LD_{50} for rats, and is known as the HERP (human exposure dose/rodent potency) index. It shows, for example, that the natural cancer agents estragole (in basil), allyl isothiocyanate (in mustard), and hydrazines (in mushrooms) are far more potent than either DDT or EDB and that a daily glass of wine or bottle of beer is a far greater cancer hazard than are the trace amounts of synthetic pesticides we consume.

In addition to possible exposure via food and water, people are often exposed to pesticide mists that drift beyond the target areas. Other problems are caused by the enormous quantities of herbicides and other pesticides that are applied annually to private lawns, golf courses, and other nonagricultural land.

In the United States, pesticide use is controlled by the Federal Insecticide, Fungicide, and Rodenticide Act (FIFRA), which was passed by Congress in 1972 (and amended in 1975, 1978, and 1988). This legislation requires that all pesticides be registered with, and approved by, the EPA.

A recent report linking a form of cancer in dogs to regular treatment of lawns with 2,4-D has led to concern that children who play on treated lawns may also be at risk.

ALTERNATIVE METHODS OF INSECT CONTROL

One-third of all the food grown in the world is destroyed by pests each year. In the United States alone, crop losses due to insects cost over $4 billion annually. Obviously, without some form of pest control, there would be a drastic drop in food production, and our ability to feed the world would be jeopardized. Fortunately, some alternatives to the massive use of synthetic pesticides are now available. In this section, we will study biological and nonchemical methods that control insects in ways that do not threaten either the environment or human health.

Pheromones

Pheromones are chemical substances that are released by insects and other animals as a means of communication. They are used to mark trails and territory and to attract mates. A number of species-specific pheromones secreted by female insects to attract males have now been extracted from insects or synthesized in the laboratory and used to control insect pests. The pheromones, which are effective in picogram (1×10^{-12} g) amounts, can be used to lure male pest insects into baited traps, or they can be sprayed over an infested area. Males in a sprayed area smell a female in all directions and become so confused that they fail to mate. Approximately 30 insect pheromones are now available for insect control.

Pheromones represent an environmentally attractive approach to pest control. They are species-specific and thus affect only the target pest; they are biodegradable, nontoxic, and effective at very low concentrations. But they are costly and difficult to produce. Many thousands of insects must be collected to obtain sufficient material for analysis, and identification and synthesis of the desired sex attractant present many problems. Also, sex attractants are effective only against adult insects. They do not control juvenile forms, such as worms and caterpillars, which often do the most harm.

Juvenile Hormones

Another biological approach to insect control is the use of juvenile hormones. Hormones (Chapter 10) are chemicals produced in special plant and animal cells. They are transported to specific target tissues, where they control and mediate many physiological activities. **Juvenile hormones** secreted by immature insects regulate the early stages of development. As the insect matures, production of juvenile hormone ceases, and the insect develops into an adult. If immature insects are sprayed with juvenile hormone, they remain at an immature stage and are unable to mate (Fig. 18-20).

The first commercially available juvenile hormone, called methoprene (Fig. 18-21), was approved by the EPA in 1975. In various formulations, methoprene and related compounds have been used to control mosquitoes, fleas, cockroaches, and a number of other insects that are pests in the adult stage. Ju-

Figure 18-20 Juvenile hormone (JH) sprayed onto immature insects prevents them from maturing and mating. A normal mealworm pupa in center treated with a large dose of JH molts to a second imperfect pupa that dies; treated with a small dose of JH it develops into an intermediate with adult head, thorax, and wings but with pupal abdomen and genitalia.

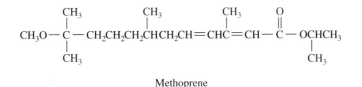

Methoprene

Figure 18-21 Methoprene, the first commercially available juvenile hormone.

venile hormones are not as useful against insects that are agricultural pests in their immature stage. When used against caterpillars, for example, the immediate effect is to cause the caterpillars to grow larger and eat more of the crop for a longer period of time.

Other **insect growth regulators (IGRs)** that disrupt development at various stages are also available, and new ones are constantly being developed. Chitin synthesis inhibitors, which prevent the growth of chitin, the substance that forms an insect's shell, are a promising recent development. Without a protective shell, the insect dehydrates and dies.

Biological Controls

In **biological control** methods, pests are kept in check by natural enemies, including predators, parasites, and disease-causing bacteria or viruses. To be effective, predator insects must be carefully chosen. Praying mantises and ladybugs that are sold commercially to home gardeners are generally of little value in controlling pests. Praying mantises are rapidly reduced in number because they devour each other, and they are as likely to eat useful insects as they are to eat predator insects. Ladybugs, if not at the feeding stage when released, simply fly away.

The bacterium *Bacillus thuringiensis* is toxic to many leaf-eating caterpillars. Marketed as a powder under the name Dipel, it has been used by home gardeners for many years. It is also used as a spray to control gypsy moths. In 1972, a virus was approved for use against the *Heliothus zea* species of bollworms, the moth larvae that attack the tips of corn kernels (corn earworms) and burrow into cotton bolls (cotton bollworms).

Although a great deal of research is needed to identify suitable biological agents, and production is difficult and costly, biological controls have many advantages. In general, they destroy only the target species and are nontoxic to other species. Once established, they are self-perpetuating, and pests do not become resistant to their natural enemies.

The cottony-cushion scale insect that destroyed citrus crops when it was accidentally brought to California from Australia in 1873, was brought under control by importing two of the insect's natural predators from Australia.

Sterilization

Insect sterilization is another option for pest control. In this technique, large numbers of the pest insect are bred and then sterilized by exposure to radiation. The irradiated insects are released in the target area, where they mate with normal insects but produce no offspring. This method is effective only if the number of sterile insects released is sufficient to overwhelm the natural population of nonsterile insects. The procedure is expensive and of limited applicability, but it has been used with great success against the screwworm fly, a deadly pest that lays its eggs in the open wounds of cattle. The developing larvae feed on the wound, and resulting infections often can kill the animal. Sterilization has also been used successfully to control certain tropical fruitflies.

The various alternative methods of insect control techniques just described are not as immediately effective, or as easy to apply, as the commonly used synthetic organic insecticides, but they have many advantages for the environment.

FUTURE FARMING: LESS RELIANCE ON AGRICHEMICALS

Pressure from environmental groups and consumers, and fear that supermarkets may reject products that contain pesticide residues, have encouraged farmers to adopt farming practices that rely less heavily on the massive use of synthetic fertilizers and pesticides. In California, where approximately 50 percent of the vegetables and 40 percent of the fruits and nuts produced in the United States are grown, the number of farmers using no synthetic agrichemicals doubled between 1989 and 1990. To control pests, these farmers rely on **integrated pest management (IPM)**, a technique that was first introduced in the 1960s.

In IPM, several pest control methods are integrated into a carefully timed program. Success depends on knowledge of the metabolism of the crop plant and of the life cycles of the pests to which the crop is susceptible and on an understanding of the interactions of the pests with their whole environment. IPM makes use of sex attractants, male sterilization, natural pesticides such as pyrethrins, and the release of natural predators. Since many pests feed on only one particular crop, crop rotation and crop diversity are introduced to prevent insects, fungi, and microbes from building up in the soil and in crop residues from year to year. Planting and harvesting times are adjusted to maximize the number of insects that die of starvation between successive crops.

The aim of IPM is not to eliminate all pests but to keep them at a level low enough to prevent economic loss. A program does not exclude the use of synthetic pesticides but reserves them for selective use at the lowest possible dose and only where absolutely necessary. IPM programs, which usually must be individualized to suit soil and climate conditions for each crop, are expensive to set up and more labor-intensive than conventional methods, but on a long-term basis, they are less expensive.

The IPM approach has received support from the Board of Agriculture of the National Research Council. In a report entitled "Alternative Agriculture," released in 1989, the board concluded that alternatives to synthetic fertilizers and pesticides are effective and economically viable. They recommended greater emphasis on biologically based methods of farming, including IPM, the use of manure and crop rotation to return fertility to the soil, and mulch and other nonchemical means of controlling weeds. The report was well received by the agricultural establishment but, understandably, was criticized by the agrichemical industry.

Consumers could play an important part in reducing pesticide use if they would tolerate weeds in their lawns (or switch to a ground cover other than grass), and if they would accept fruits and vegetables with some blemishes. At present, many pesticides are applied, not to increase yield, but solely to achieve the perfect appearance that many consumers demand.

EXPLORATIONS

The Neem Tree: A Storehouse of Cures for World Problems

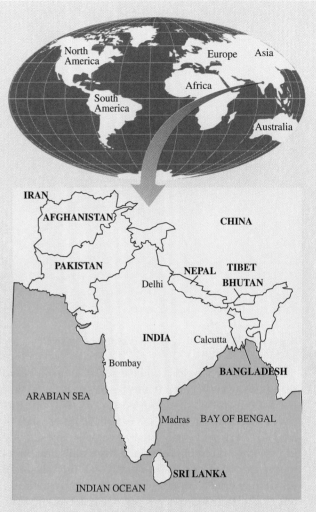

Scientists are calling the neem a wonder tree. According to a report issued by the National Academy of Sciences (NAS) in February 1992, this remarkable tree has the potential to improve pest control, promote health, check population growth, and reduce deforestation.

Western scientists have only recently taken an interest in the neem tree, but in India, the tree has been revered and utilized for generations. Claims for neem

Figure 18-A   The neem tree.

are recorded in the earliest Sanskrit writings, which cite the value of neem seeds and leaves for healing and for pest control. As Noel Vietmeyer, the director of the NAS project, reports: "For centuries, millions have cleaned their teeth with neem twigs, smeared skin disorders with neem leaf juice, taken neem tea as a tonic, and placed neem leaves in their beds, books, grain bins, cupboards, and closets to keep away troublesome bugs. The tree has relieved so many different pains, fevers, infections, and other complaints that it has been called 'the village pharmacy.'"

This amazing multipurpose tree is a member of the mahogany family (Fig. 18-A). It grows widely in many parts of Asia, particularly in India, and was introduced to Africa, the Caribbean, Central and South America, and Saudi Arabia during the last century. The neem tree is a beautiful broad-leaved evergreen that can grow up to 90 feet tall with spreading leafy branches

Figure 18-B The olivelike fruits of the neem tree.

Neem extracts contain four major and many minor active ingredients that act on insects in a variety of ways—although none kill outright. The major active ingredients are limonoids, compounds that are distantly related to steroids. The most potent ingredient is azadirachtin, which is found in greatest concentration in seed kernels and is also present in leaves. It is effective at concentrations of less than 1 ppm. Azadirachtin blocks production of hormones that control normal insect development; as a result, larvae do not molt and cannot develop into pupae. Azadirachtin is also so distasteful to many insects that they will starve to death rather than eat plants that bear traces of it. Two other neem ingredients (melantriol and salanin) also inhibit feeding. The latter appears to be a more effective insect repellent than "DEET" (N,N-diethyl-m-toluamide), the substance present in most commercial mosquito repellents. Still other neem compounds disrupt maturation at various stages and thus prevent breeding. Neem pesticides are promising candidates for inclusion in the integrated pest management programs that are becoming increasingly popular in modern agriculture.

Asians have used neem medicinally for thousands of years and are convinced of its efficacy in treating numerous ailments. However, although neem's value as a general antiseptic is well established and certain extracts have demonstrated antifungal, antiviral, and some antibacterial activity, many health claims for neem have not yet been substantiated. Preliminary evidence suggests that neem has analgesic, antipyretic, anti-inflammatory, hypotensive, and possibly anti-ulcer effects, but these claims have not been tested under controlled conditions.

There is little doubt that neem has value in dentistry. Even the poorest Indians generally have healthy teeth, although their only form of dental hygiene is to scrub their teeth with the frayed end of a neem twig. Research in Germany has shown that compounds in neem bark are effective in preventing tooth decay and healing inflammation of gums, and neem toothpaste is now marketed in both Germany and India.

Neem may also have value for birth control. Oil pressed from neem seeds is a powerful spermicide. Introduced into the vagina before intercourse, it has prevented pregnancy in laboratory animals and a small number of women volunteers. Moreover, neem leaf extracts show promise as a means of male birth control. Reportedly, the extracts reduce fertility in monkeys without inhibiting sperm production, and the effects are temporary.

That is not the end of neem's benefits: Its termite-resistant wood is valued for construction; its

that provide dense, year-round shade. The neem bears masses of sweet-scented white flowers that are attractive to bees. Its fruit resembles olives (Fig. 18-B), and the sweet pulp that surrounds the extremely bitter seeds is food for birds and many other creatures. Neem trees grow in the hottest climates but cannot withstand freezing or extended periods of cold. They thrive in both semiarid and wet tropical regions. Helped by their extensive and unusually deep roots, neem trees are able to extract nutrients even from poor soils.

Scientists in India have studied neem products since the 1920s, but research did not begin elsewhere until the German entomologist Heinrich Schmutterer published a report describing a plague of locusts that he witnessed in the Sudan in 1959. Schmutterer observed that the insects devoured every green thing in their path with the exception of the neem trees. Swarms of locusts landed on the trees but flew off without feeding.

By 1990, research in many parts of the world, including the United States, had shown that extracts of neem seeds and leaves are effective against more than 200 species of insect pests as well as some species of mites and nematodes (microscopic soil-dwelling worms). The extracts appear to be nontoxic to humans, other mammals, birds, and insects such as bees that do not feed on plants. Furthermore, they are biodegradable, and insects do not seem to develop genetic resistance to them.

flowers are a source of high-quality honey; neem oil is used as a lubricant, a fuel for oil-burning lamps, and in making soap; neem cake, the material left after oil is pressed from seeds, improves soil fertility and protects against nematodes; and fast-growing neem trees can be planted to alleviate deforestation and desertification.

The potential for neem seems almost limitless, particularly in less developed countries, where populations are expected to explode and where neem trees generally thrive. Neem oil could provide an acceptable, inexpensive, and easily available contraceptive, and neem preparations could protect crops from destructive pests. Effective pesticides can be made very simply by steeping crushed seeds or leaves in water. The resulting suspension can be used directly in the field or can be filtered and sprayed.

Before neem products can achieve wide acceptance, however, more research is needed. One problem is that extracts vary from region to region because of genetic differences, and no standard of potency has yet been established. It is difficult to evaluate claims made in different studies, particularly medical claims, because harvesting and extracting methods have varied. Also, although neem products appear to be environmentally safe and harmless to humans and other mammals, rigorous toxicity tests have not been conducted, and the persistence of neem residues on food crops is not known. Until more information is available, the one neem product registered in the United States (called Margosan-O after the Spanish name for neem) is authorized for use only against certain insects that do not feed on food crops.

Despite the uncertainties, many scientists are convinced that neem is an extremely valuable resource that could benefit the world. The NAS report stresses the urgent need to provide more funds for research and to make the industrialized nations more aware of neem's extraordinary potential. In the future, the amazing neem tree could play a significant role in helping to solve many of our planet's most intractable problems.

Question
1. Discuss the advantages and disadvantages of replacing synthetic chemicals used in India to destroy pests that attack crops and stored grain with neem tree extracts.

Reference: National Research Council, *Neem: A Tree For Solving Global Problems,* National Academy Press, Washington, D.C., 1992.

KEY WORDS AND CONCEPTS

arithmetic growth
biological control
broad-spectrum insecticide
carbamates
chlorinated hydrocarbon
doubling time
geometric growth
humus
insect growth regulator (IGR)

insect sterilization
integrated pest management (IPM)
juvenile hormone
loam
micronutrient
narrow-spectrum insecticide
nitrification
nitrogen fixation
organic fertilizer

organophosphates
pheromone
primary nutrient
pyrethrins
Rule of 70
secondary nutrient
superphosphate
synthetic fertilizer
topsoil

QUESTIONS AND PROBLEMS

1. A coin collector adds two new coins to his collection every week.
a. How many weeks would it take for him to acquire 100 coins? **b.** Is the collection growing arithmetically or geometrically?

2. You raise carrier pigeons. Starting with two, you have four at the end of the first month and eight at the end of the second month. Assuming that the pigeon population doubles every month:
a. how many pigeons will you have at the end of the first year? **b.** is the pigeon population growing arithmetically or geometrically?

3. The populations of some Latin American countries are growing at a rate of 1.5 percent per year. At this rate, how many years will it take for the population to double?

4. The population of India, now at 1 billion, is growing at an annual rate of 2 percent. At this rate, how many years will it take until the population doubles?

5. What prediction did Thomas Malthus make?

6. What causes a soil to be "sweet" or "sour"?

7. Draw a diagram to show the horizons you would expect to find in a typical soil.

8. If you spread lime on your vegetable garden: **a.** will it raise or lower the pH of the soil? **b.** will it make the soil "sweet"?

9. Which has more air per cubic foot of soil: a soil that has a high clay content or a soil with a high sand content? Explain.

10. Nitrogen is a primary nutrient. In what form is nitrogen used by plants?

11. Describe how ammonia is made commercially. Write balanced equations for its preparation.

12. List three ways that nitrogen can be fixed.

13. Describe how ammonium nitrate fertilizer is made. Start with ammonia and oxygen.

14. Urea is a component of urine and a good source of nitrogen for plants. Why is urea a better fertilizer than other inorganic ammonium fertilizers?

15. What is anhydrous ammonia, and for what is it used?

16. What is the source of phosphorus used in most commercial fertilizers?

17. What is superphosphate, and how is it made?

18. Why must cadmium be removed from phosphate deposits before they can be used for fertilizer?

19. What three elements are obtained from the air and water as nonmineral plant nutrients?

20. Explain the following numbers that are printed on a fertilizer bag: 30-30-10.

21. Which is a bigger problem for a farmer? A soil with a shortage of N, P, and K or a soil with a shortage of Ca, Mg, and S? Explain your answer.

22. In what form is phosphorus readily absorbed by plants?

23. List three secondary plant nutrients. How are they supplied to the soil?

24. List three plant micronutrients. Are they supplied directly to the soil?

25. Why are farmers inclined to use synthetic fertilizers rather than an organic fertilizer such as animal manure?

26. What is the content of N, P, and K in organic fertilizers? How does this compare with the content in commercial fertilizers?

27. Describe how chlorinated hydrocarbons become concentrated in the food chain.

28. What is a narrow-spectrum insecticide? Give an example.

29. What is a pheromone? How is a pheromone used to control insects?

30. What are juvenile hormones, and how are they used to control insect pests?

31. Describe how sterilization can be used as a technique to reduce insect populations.

32. What is a herbicide? Give an example.

33. Are organophosphate insecticides narrow-spectrum insecticides? Describe any disadvantages in the use of organophosphates.

34. Dicamba residues are found in potatoes at a level of 1 mg per kilogram of potato (1 ppm). How much dicamba is contained in one 250-g potato?

35. PCBs are found in a sample of pacific salmon at a level of 5 mg per kilogram of fish (5 ppm). How much PCB is present in a 200-g serving of salmon?

36. Examine the label on a bag of fertilizer used for house plants, lawn, or vegetable garden. List the brand name and the relative amounts of N, P, and K. What do N, P, and K stand for?

THE DISPOSAL OF
DANGEROUS WASTES

In the past, dumpsites like this one in Colorado have leaked hazardous chemicals stored in drums into the environment.

Industrial societies generate enormous quantities of wastes, and the United States produces more waste per person than any other country in the world. We discard products that are worn out and items we no longer want; we throw out food, paper, empty containers, and yard waste. Industrial processes produce chemical wastes; many of these wastes contain toxic substances. The nation's nuclear weapons programs and the nuclear power industry generate

585

wastes that are radioactive. Safe disposal of these wastes is one of the most serious environmental problems facing the United States today.

Until legislation was enacted in the 1960s and 1970s to regulate waste disposal and protect the air and water from contamination, wastes of all kinds were discarded with little concern for human health or the environment. Hazardous wastes were incinerated, contaminating the atmosphere with toxic emissions. They were discarded in sewers, rivers, and streams, leaking landfills, and abandoned buildings and mines, or they were simply dumped on vacant lots and along roadways (Fig. 19-1). Many people were exposed to polluted air and contaminated water; today, thousands of sites in the United States remain severely contaminated with dangerous chemicals.

For purposes of management and disposal, wastes are divided into three categories: (1) solid waste, (2) hazardous waste, and (3) radioactive waste. In this chapter, we will define the different kinds of wastes and discuss the advantages and disadvantages of the various disposal methods that have been recommended for each category. We will examine the legislation that has been enacted (1) to regulate the management and disposal of wastes being generated today and (2) to clean up wastes produced in the past.

Learning Goals:

In this chapter, you should gain an understanding of:

1. The disposal methods used for municipal refuse and the laws that control them.
2. What materials are considered to be hazardous wastes and how they are categorized.
3. The policies that control the disposal of hazardous waste.
4. The sources and classification of radioactive waste.
5. The laws that regulate low-level and high-level radioactive wastes.

 ## CARELESS WASTE DISPOSAL IN THE PAST

Numerous examples of irresponsible, often criminal, methods of waste disposal can be cited. For example, in 1969, oil and other flammable, greasy wastes floating on the surface of the Cuyahoga River, which flows through Cleveland, Ohio, actually caught on fire and destroyed seven bridges. In Toone and Teague, Tennessee, groundwater cannot be used for drinking because it is contaminated with pesticide wastes, which are leaking from 350,000 drums that were dumped in a landfill years ago. Between 1952 and 1972, about 130 million liters (34 million gallons) of chemical wastes, including wastes from DDT manufacture, were dumped at the Stringfellow Acid Pits, a 22-acre site near Riverside, California. Contaminated leachate from this site continues to migrate and threatens groundwater supplies. Many of 17,000 waste drums located at a site in Kentucky known as the "Valley of the Drums" (Fig. 19-2) have leaked, contaminating soil and nearby water. At the huge 560-square-mile Hanford weapons complex in Washington, there are over 1000 separate sites where radioactive and hazardous wastes have been stored or dumped. Contamination at the complex is so extensive that complete cleanup may never be possible.

Figure 19-1 In the past, hazardous wastes were discarded with little regard for health or the environment. Often, wastes were dumped illegally under cover of night to escape detection, a practice known as "midnight dumping."

The most notorious case of hazardous waste dumping in the United States occurred at Love Canal in Niagara Falls, New York, where in the winter of 1978, toxic chemical wastes buried years before began oozing from the ground, forcing the evacuation of many of the town's residents. This tragedy is described at the end of the chapter.

Figure 19-2 At a site in Kentucky known as the "Valley of the Drums," dangerous chemicals are leaking from thousands of drums and contaminating the soil and nearby water supplies.

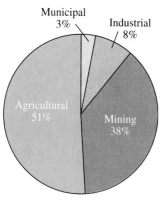

Sources of solid waste
(percent of total)

Figure 19-3 In the United States, agriculture is the major source of solid waste (51 percent). Mining accounts for 38 percent and industry for 8 percent. Municipal waste, which is produced by commercial operations, homes, schools, and other community activities, accounts for 3 percent.

The EPA estimates that, in 1992, the United States produced more than 195 million tons of municipal solid waste.

DEFINING SOLID WASTE

According to the Environmental Protection Agency (EPA), **solid waste** includes garbage, refuse, sludge from waste-treatment plants, wastes from air pollution control facilities and any other discarded material not excluded by regulation. Although termed "solid," these wastes may actually be solid, liquid, semisolid, or contain gaseous material. Agricultural, mining, industrial, and commercial operations and community activities all produce solid waste that must be disposed of (Fig. 19-3). Most agricultural wastes, which consist primarily of manure and crop residues, are plowed back into the land. Mining wastes that are produced when ores are processed are generally left at the mine site (Fig. 19-4). Sludge from sewage-treatment plants, nonhazardous industrial wastes, and municipal refuse generated by community activities are usually disposed of in landfills or incinerated. Increasingly, some industrial and municipal waste is being recycled.

DISPOSAL OF MUNICIPAL REFUSE

Municipal refuse is the solid waste collected locally from homes, institutions, and commercial establishments. It includes garbage (food wastes), paper, glass, cans, plastic containers, yard clippings, and various other discarded household items. The composition varies, depending on whether it is commercial or residential trash and whether it comes from an affluent or a poor neighborhood. Typically, in the United States, municipal trash is made up as shown in Figure 19-5 and amounts to about 2.5 kg (5.5 lb) per person per day. Municipal waste also includes discarded items such as furniture, household appliances, automobiles, and tires.

Figure 19-4 Coppermine tailings. Waste rock produced in mining operations is usually left in large piles at the mine site.

Solid Waste and the Law

Until the 1960s, most municipal waste was trucked to open dumps, where it was discarded and often burned (Fig. 19-6). The dumps became breeding grounds for rats, flies, and other pests, and the burning trash polluted the air with smoke and unpleasant odors. Since the passage of the **Solid Waste Disposal Act** in 1965, management and disposal of solid waste have been subject to increasingly strict controls. Particularly important was the enactment of the **Resource Conservation and Recovery Act (RCRA**, pronounced "rickra") in 1976. The section of the law dealing with solid waste required that all existing open dumps in the United States be closed or upgraded to sanitary landfills by 1983; it banned the creation of new open dumps and set tough standards for landfills. In an effort to reduce the volume of waste generated, the law stressed the need for regeneration and reuse. Federal procurement agencies were directed to make use of recycled materials wherever possible, and the Department of Energy was instructed to conduct research to develop ways of converting solid wastes into energy. Today more than 80 percent of all municipal waste is discarded in regulated sanitary landfills. The remainder is either recycled, incinerated, or composted.

Sanitary Landfills

A **sanitary landfill** site consists of a depression in the ground in which trash is dumped and then compacted by bulldozers. Each day's load of refuse is covered with a 15–30 cm (6–12 in.) layer of soil (Fig. 19-7). Covering the waste minimizes access by vermin and the problem of odors. The landfill site may be a trench excavated for the purpose, but it is often a natural valley or canyon or an abandoned mine or quarry. When completely filled in, landfills are not suitable for building because the ground continues to settle for many

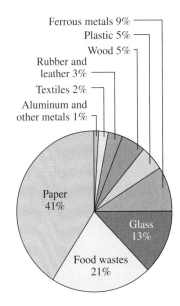

Ferrous metals 9%
Plastic 5%
Wood 5%
Rubber and leather 3%
Textiles 2%
Aluminum and other metals 1%
Paper 41%
Glass 13%
Food wastes 21%

Percent composition of average urban solid waste

Figure 19-5 The composition of typical municipal solid refuse by weight. Percentages vary, depending on the affluence of the neighborhood and on whether the source is primarily commercial or residential. At certain times of the year, yard wastes (branches, leaves, grass clippings) often account for more than 25 percent of the total waste.

Figure 19-6 In the past, municipal waste was often taken to an open dumpsite and burned. The burning trash produced unpleasant odors and polluted the atmosphere near the site.

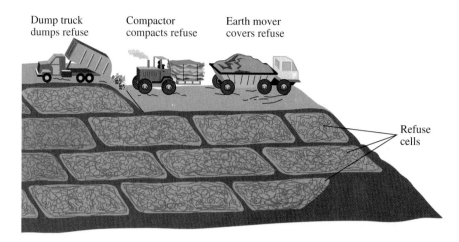

Dump truck
dumps refuse

Compactor
compacts refuse

Earth mover
covers refuse

Refuse
cells

Figure 19-7   In a sanitary landfill, each day's trash is compacted and covered with a layer of dirt. The dirt eliminates odors, keeps rodents away, and prevents trash from blowing.

In 1992, 42 percent of operating landfills in the United States were under 10 acres in size; 51 percent were between 10 and 100 acres, and 6 percent were greater than 100 acres. The largest landfill is Fresh Kills on Staten Island, New York.

Little biodegradation occurs at the center of a landfill. Whole hot dogs and readable newspapers have been found in 40-year-old landfills.

years, but they can be planted with grass and made into attractive recreation areas (Fig. 19-8).

In the early sanitary landfills, there was always the danger that metal salts and other chemicals might leach from the fill and percolate through the underlying soil and contaminate nearby groundwater. Although municipal trash should not contain hazardous materials, many householders often increase the chance of groundwater contamination by discarding unused pesticides, paint, cleaning solvents, and other dangerous wastes together with their nonhazardous refuse (Table 19-1). Since 1983, the EPA has required that all landfills be built with an impermeable clay or plastic liner to contain water seeping through the trash and be fitted with a system for recovering and treating the water. Groundwater in the vicinity must be monitored regularly; in larger landfills, methane produced by the anaerobic decomposition of organic waste must be vented or collected and used for fuel.

Figure 19-8   Part of the attractive Longview golf course in Baltimore County, Maryland, was built over a landfill.

TABLE 19-1 Hazardous Materials Often Found in Household Trash

Products	Hazardous Materials
Pesticides:	
Insect sprays	Insecticides, organic solvents
Other pesticides	Insecticides, herbicides
Paint:	
Oil based	Organic solvents
Latex	Organic polymers
Automotive products:	
Gasoline	Organic solvents
Motor oil	Organic compounds, metals
Antifreeze	Organic solvents, metals
Batteries	Lead, sulfuric acid
Miscellaneous:	
Mercury batteries	Mercury
Nail-polish remover	Organic solvents
Moth balls	Para-dichlorobenzene

Municipal waste disposal has become a major problem for many big cities in the United States. Numerous existing landfills will be filled to capacity within the next few years, and rising land costs and resistance from residents are making it increasingly difficult to find suitable new sites. Many authorities are turning to incineration and recycling as alternatives.

In 1992, despite incentives such as jobs, tax breaks, and economic development, several rural communities in New York state rejected a waste management company's plan to build a regional landfill in their community.

Incineration

Incineration is a considerably more expensive method of dealing with solid waste than is disposal in a nearby landfill, but it has a number of advantages. The volume of the waste is reduced by approximately 85 percent, the high temperature kills disease-causing organisms, there is no risk of groundwater pollution, and the heat produced can be used as a source of energy. At some facilities, valuable metals in the incinerated waste, including iron and aluminum, are recovered before the noncombustible material (assuming it contains no toxic metals or other hazardous materials) is disposed of in a sanitary landfill.

The main disadvantage of incinerating municipal waste is its potential for polluting the atmosphere with toxic chemicals, particularly metals. When discarded items such as cans, jar lids, and batteries are incinerated, many toxic metals, including beryllium, cadmium, chromium, lead, and mercury, are released and carried up the incinerator stack in fly ash. In the past, municipal incinerators were a significant source of air pollution (Chapter 13), but today, in compliance with strict air-quality standards, they are equipped with efficient electrostatic precipitators and other control devices to prevent the escape of fly ash. A modern incinerator in Baltimore, Maryland (Fig. 19-9), which has been in operation since 1984, burns cleanly and consumes up to 2000 tons of trash a day. At the same time, it produces enough electricity to supply the needs of 60,000 homes. Incineration is never likely to contribute more than 1 percent to energy needs in the United States, but it can be a useful local source of energy.

The EPA has estimated that 96 percent of discarded lead-acid automobile batteries were recovered and recycled in 1992.

Nobody wants an incinerator in their neighborhood. Because incinerators often operate intermittently, there is concern that even if yearly average emis-

Figure 19-9 This modern incinerator in Baltimore, Maryland, burns 2000 tons of solid waste per day. The heat produced generates enough electrical power to supply 60,000 homes. Pollution-control devices prevent the escape of hazardous materials into the atmosphere from the smokestack.

Today, about 130 incinerators are in operation in the United States, burning approximately 13 percent of the nation's municipal solid waste each year.

The EPA estimates that, by the year 2000, it should be possible to recover and recycle 30 percent of municipal solid waste.

sions are within allowed limits, emissions at peak periods may be excessive. Despite these difficulties, it is expected that by the turn of the century, as urban areas run out of landfill sites, 30 percent of municipal trash will be incinerated.

Recycling and Resource Recovery

As we saw in earlier chapters (Chapters 11 and 15), more and more waste materials are being recycled in response to the need to conserve mineral resources and energy and reduce environmental pollution. Many communities now require businesses and homeowners to collect discarded items such as newspapers, office paper, aluminum cans, glass containers, and certain plastic items in separate containers so that they can be collected and recycled. Some municipalities have found it more economical to build high-technology resource **recovery plants** that sort the trash after it has been collected.

Everyone is in favor of recycling, but to be successful, recycling must be cost-effective. Dealers and manufacturers will not be willing to accept items for recycling if they cannot count on steady markets for the recycled products.

 ## THE PROBLEM OF HAZARDOUS WASTE

If the public is to be protected from the dangers associated with hazardous waste, two problems must be solved: (1) how to properly manage and dispose of waste generated now and in the future and (2) how to clean up wastes produced in the past. The major pieces of legislation that address these two issues are RCRA (amended in 1984) and the **Comprehensive Environmental Response, Compensation and Liability Act (CERCLA)** of 1980, better known as the **Superfund** Program.

RCRA: REGULATION OF HAZARDOUS WASTE

Prior to passage of RCRA, the two main pieces of legislation enacted to protect the environment were the 1970 Clean Air Act and the 1972 Clean Water Act. RCRA was passed to establish safe methods for the disposal of all wastes, and hazardous wastes in particular. Under RCRA, the EPA was given the responsibility for defining hazardous wastes and for issuing and enforcing regulations to protect the public's health and the environment from improper management and disposal of such wastes.

What Are Hazardous Wastes?

According to RCRA, wastes are described as **hazardous** if they "(a) cause or significantly contribute to an increase in mortality or an increase in serious irreversible, or incapacitating reversible, illness or (b) pose a substantial present or potential hazard to human health or the environment when improperly treated, stored, transported, or disposed of, or otherwise managed." Such wastes may be in solid, liquid, or gaseous form and may result from industrial, commercial, mining, or agricultural operations or from community activities.

The EPA defines hazardous wastes in two different ways. A waste is hazardous if, because of its inherent properties, the EPA has specifically listed it as hazardous. A waste is also hazardous if it exhibits any of the following characteristics: (1) ignitability, (2) corrosivity, (3) reactivity, and (4) toxicity.

If a "characteristic" waste is treated in such a way that it loses its hazardous characteristic, it ceases to be a hazardous waste and can be disposed of in a sanitary landfill. A "listed" waste, however, once generated, retains its hazardous classification regardless of the steps taken to render it nonhazardous. Hazardous wastes are not permitted in a sanitary landfill but must be disposed of according to strict EPA regulations.

The responsibility for deciding if a waste is hazardous lies with the producer of the waste. If the waste has not been listed by the EPA, the producer must determine if the waste possesses any of the four characteristics that define it as hazardous. Once a waste has been identified as hazardous, the proper method for disposing of it must be selected. Any company that generates or transports a hazardous waste is required to keep detailed records of the handling and transferring of the waste from point of origin to point of ultimate disposal or destruction. Chemical plant managers and incinerator operators are required to keep records of all hazardous chemicals that are emitted into the air, land, or water. Landfill operators must maintain groundwater monitoring data. By law, "cradle-to-grave" responsibility for a hazardous waste is assigned to the generator of the waste.

The generator of a hazardous waste must obtain an identification number for the waste from the EPA. The number is used to track the waste from its generation to its disposal.

Listed Hazardous Wastes

The wastes that the EPA, by rule, has specifically listed as hazardous now number nearly 400. The wastes are classified according to source and include nonspecific chemicals—primarily solvents—produced by many industries, sludges and by-products produced by specific industries, discarded chemicals, and chemicals produced as intermediates in manufacturing processes.

Characteristic Hazardous Wastes

The EPA has developed special tests that must be used to determine if a waste exhibits any of the four hazardous characteristics.

Ignitable Wastes   Wastes are classified as **ignitable** if they present a significant fire hazard. Ignitable wastes include (1) liquids such as gasoline, hexane, and other organic solvents that have a flash point (the lowest temperature at which their vapors ignite when tested with a flame) below 60°C (140°F); (2) ignitable compressed gases such as propane; (3) various solid materials that are liable to cause persistent, vigorously burning fires as a result of absorption of moisture, friction, or spontaneous chemical change; and (4) oxidizers. Oxidizers are included because they are likely to intensify an already burning fire. Perchloric acid ($HClO_4$), for example, is a strong oxidizing agent that poses an explosion hazard because it reacts very vigorously with organic matter.

Corrosive Wastes   **Corrosive wastes** include acidic aqueous wastes with a pH equal to or less than 2.0 and basic aqueous wastes with a pH equal to or greater than 12.5. Wastes with these very low or very high pH values are likely to react dangerously with other wastes, and they may leach metals and contaminants from other wastes. If allowed to leak into waterways, corrosives can have a devastating effect on aquatic habitats (Chapter 12).

Also included as corrosive are wastes that can corrode steel at a rate equal to or greater than 0.635 cm (0.25 in.) per year. This criterion was included to prevent the kind of disasters that occurred in the past when acids and other hazardous materials leaked from corroded steel drums in which they had been stored.

Reactive Wastes   **Reactive wastes** include materials that explode readily when subjected to shock or heat; materials that tend to undergo violent, spontaneous chemical change; and materials that react violently with water. Explosions at a dump site in Cheshire, England, were caused by discarded sodium metal, which reacted with water to produce hydrogen gas:

$$2\,Na \;+\; 2\,H_2O \;\longrightarrow\; 2\,NaOH \;+\; H_2$$

The hydrogen ignited and combined explosively with oxygen in the air:

$$2\,H_2 \;+\; O_2 \;\longrightarrow\; 2\,H_2O$$

Also classified as reactive are wastes that generate dangerous quantities of toxic gases when exposed to water, weak acids, or weak bases. For example, wastes that contain cyanides or sulfides can produce toxic hydrogen cyanide (HCN) and hydrogen sulfide (H_2S).

Toxic Wastes   Most characteristic wastes are hazardous because of their toxicity. The terms toxic and hazardous are often used interchangeably, but they are not synonymous. The term **toxic** is applied to substances that cause harm to living organisms by interfering with normal physiological processes. It includes substances that are carcinogenic, mutagenic, teratogenic (Chapter 16), or phytotoxic (poisonous to plants).

Toxic wastes are of particular concern because of their potential as groundwater contaminants. The test used to determine if a waste is toxic was designed to model the leaching that would be likely to occur if the waste were disposed of on land. In the test, a nonliquid waste is treated with an acid solution (pH 5.0 ± 0.2) under specified conditions, and the amount of toxic constituent thus extracted into the acid is determined. If the concentration of toxic substance in the extract (or in the liquid in the case of a liquid waste) exceeds the regulatory level set by the EPA, the waste is considered to be toxic.

Originally, the EPA characterized just 14 chemicals as toxic: 8 metals (arsenic, barium, cadmium, chromium, lead, mercury, selenium, and silver) and 6 pesticides (endrin, lindane, methoxychlor, toxaphene, 2,4-D, and 2,4,5-T [Silvex]). By the end of 1991, a total of 39 chemicals were regulated as toxic including common chemicals such as benzene, carbon tetrachloride, chloroform, nitrobenzene, vinyl chloride—used to make the plastic poly(vinyl chloride)—and several pesticides.

SOURCES OF HAZARDOUS WASTE

Industry is responsible for most of the nearly 300 million tons of hazardous waste that is generated annually in the United States. More than 90 percent of this waste is produced by chemical, petroleum, and metal-related industries. Small business establishments such as dry cleaners, gas stations, and electroplating shops account for a small but significant part of the remaining 10 percent. Household and farm wastes are exempt from hazardous waste regulations, but any business with a monthly output of hazardous waste in excess of 100 kg (220 lb) must dispose of it according to EPA regulations.

TABLE 19-2   Hazardous Wastes Produced by Industry

Industry or Product	Hazardous Wastes
Electrical insulation	Polychlorinated biphenyls (PCBs)
Electroplating	Toxic metals, cyanide
Fertilizers	Sulfuric, nitric, and phosphoric acids; sodium hydroxide, ammonia
Insecticides and herbicides	Polychlorinated hydrocarbons, organophosphates, carbamates, ethylene dibromide
Iron and steel	Acids (including hydrofluoric acid), bases, phenols, benzene, toluene
Leather	Toxic metals, organic solvents
Medicines	Toxic metals, organic solvents
Nonferrous metals	Acids, metals (cadmium, zinc, lead)
Organic chemicals	Many ignitable, corrosive, reactive, and toxic compounds
Paints	Toxic metals, pigments (chrome yellow), organic solvents
Petroleum refining	Oil, phenols, other organic compounds, toxic metals, corrosives (acids, bases)
Plastics	Monomers used in polymer manufacture (ethylene, propylene, styrene, vinyl chloride, phenol, formaldehyde), phthalate plasticizers
Power industry: steam generation (fossil- and nuclear-powered)	Fly ash and bottom ash (organics and heavy metals), sulfur dioxide, wet sludge
Soaps and detergents	Corrosives (sodium hydroxide, sulfuric acid)
Synthetic rubber	Ignitable monomers (styrene, butadiene, isoprene); antioxidants, fillers, etc. in tires
Textiles	Polymer monomers, dyes (azo, nitroso compounds), organic solvents, toxic metals
Wood products, paper	Phenol, formaldehyde, corrosives, sodium sulfide

Thousands of different chemicals and thousands of different processes are used by the petrochemical industry in refining crude oil and producing the synthetic organic chemicals needed for manufacturing polymers, pesticides, medicines, and other consumer products. All these processes inevitably produce wastes, and it has been estimated that 10–15 percent of them are hazardous.

Of greatest concern are (1) toxic metals and (2) synthetic organic chemicals, particularly chlorinated hydrocarbons. As can be seen in Table 19-2, these hazardous substances are generated by a wide variety of industries. Metal wastes are produced primarily by metal-processing industries and by paint, textile, and other manufacturing industries that make or use pigments. Chlorinated hydrocarbons are used in the manufacture of plastics (vinyl chloride), pesticides (chlordane, heptachlor, and others; Chapter 18), electrical insulation (PCBs), and refrigerants (chlorofluorocarbons). They are also used as solvents (chloroform, carbon tetrachloride) in many industrial processes.

 ## POLICY FOR MANAGEMENT AND DISPOSAL OF HAZARDOUS WASTE

In 1983, the National Academy of Sciences issued an influential report that has been a major factor in establishing hazardous waste management policy in the United States. The report described the three basic ways for managing hazardous waste, which, in order of desirability, are (1) minimize the amount produced, (2) convert to a less hazardous or nonhazardous form, and (3) isolate in a secure perpetual-storage site (Fig. 19-10). We will consider each of these options in turn.

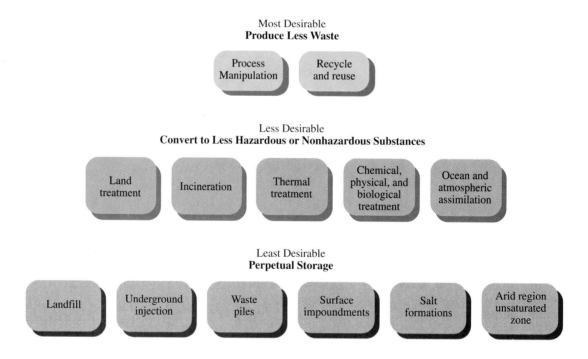

Options for dealing with hazardous waste. (National Academy of Sciences)

Figure 19-10 The top-tier methods are the most desirable. The bottom-tier methods are the least desirable but are often the least expensive.

WASTE MINIMIZATION: PROCESS MANIPULATION, RECYCLING, AND REUSE

The high cost of disposing of hazardous waste in compliance with EPA regulations has made it economically attractive for industries to minimize waste production by modifying their manufacturing processes. For example, some electroplating plants now use ion-exchange resins to selectively remove contaminant metals from electroplating baths. The plating solution, which previously would have been discarded as waste, is then returned to the bath for reuse. Many industries are also purifying wastes to recover chemicals that can be reused, recycled, or sold to another industry for use as raw materials. For example, the pesticide industry has found it can recover hydrochloric acid from its hazardous chlorinated wastes and sell it to other industries. In Europe, several countries have now established waste exchanges, where waste chemicals are successfully traded.

It may be cheaper for a plant to change its process to avoid generating a listed waste than to have to dispose of the waste according to EPA regulations.

CONVERSION OF HAZARDOUS WASTE TO A LESS HAZARDOUS OR NONHAZARDOUS FORM

Since 1990, in compliance with the 1984 amendments to RCRA, nearly all hazardous wastes have had to be treated in some way to make them less hazardous before they can be buried on land. Several options are available (Fig. 19-10); the choice depends on the type of waste to be treated.

Land Treatment

Certain hazardous organic wastes can be destroyed by **land treatment**: spreading them on the land and then mixing them with soil to which nutrients and microorganisms have been added. The microorganisms decompose the wastes and thus detoxify them. Oxidation by sunlight further aids in the degradation. The technique, which is termed land farming, is used primarily to treat petroleum-refinery and paper-mill wastes. It is an inexpensive option, but precautions must be taken to ensure that the wastes do not contain toxic metals or other nonbiodegradable chemicals that could contaminate the soil or groundwater. Microbiologists are working to develop microorganisms that can effectively detoxify polycyclic aromatic hydrocarbons (PAHs), PCBs, trichoroethylene, and other compounds that are resistant to biodegradation.

Land farming, which is a form of composting, can be used to treat sewage sludge.

Incineration and Other Thermal Treatment

If adequately controlled to prevent air pollution, **incineration** has a number of advantages as a means of detoxifying many hazardous wastes. At sufficiently high temperatures, incineration destroys 99.999 percent of toxic organic compounds by decomposing them to carbon dioxide, water, and various gases. At the same time, the volume of the waste is greatly reduced, and the energy released can be used either as a source of heat or to generate electricity. Combustible liquid and solid hazardous wastes such as solvents, pesticides, and many of the organic compounds present in petroleum-refinery wastes can all be destroyed by incineration.

In the Netherlands, Germany, and Denmark, between 50 and 80 percent of all hazardous wastes are incinerated. In the United States, however, only about 5 percent of hazardous wastes are treated in this way. The scrubbers and other devices that must be installed to prevent the escape of toxic gases and particulates make incineration very expensive, but the major factor in preventing construction of more incinerators in the United States is intense community opposition. Many believe that, even with the protection of pollution controls, the potential for air pollution still exists.

During the 1970s and 1980s, hazardous wastes, including Agent Orange (the defoliant used in the Vietnam War; see Chapters 12 and 16), were destroyed at sea on ships equipped with incinerators. This practice was discontinued in 1988, pending adoption by the EPA of measures to regulate incineration at sea. Since 1985, the EPA has operated several mobile incinerators (Fig. 19-11) for the on-site destruction of small quantities of particularly hazardous materials, including dioxin wastes in soil and water.

Certain hazardous organic wastes do not need to be incinerated to render them nonhazardous. They can be decomposed to nonhazardous biodegradable compounds by being heated under pressure at relatively low temperatures (450–600°C [840–1100°F]). In some cases, the products of this thermal treatment can be used to synthesize new compounds.

Chemical and Physical Treatment

Chemical methods that can be used to convert hazardous wastes to nonhazardous materials include neutralization of acids and bases, oxidation-reduction reactions, and removal of metals and other compounds by precipitation or

During the incineration of Agent Orange at sea, the EPA monitored the air to ensure that dangerous quantities of hazardous products were not released to the atmosphere.

Figure 19-11   To avoid the dangers inherent in transporting hazardous materials, the EPA uses mobile incinerators for on-site destruction of small quantities of particularly hazardous materials.

adsorption on carbon. Acidic wastes, which are produced in large quantity by the iron and steel, electroplating, and other metal-related industries, can be neutralized with lime. This process also serves to precipitate out heavy metals that can then be collected and recycled. Toxic chromium(VI) salts in wastes can be chemically reduced to far less dangerous chromium(III) salts. Cyanide, which is used in many industrial processes including the extraction of gold and silver from their ores and electroplating of metals, can be oxidized with sodium hypochlorite (NaOCl) to produce far less toxic cyanate (NCO^-). The cyanate can be oxidized further with chlorine to produce carbon dioxide and nitrogen. In the textile industry, carbon adsorption can be used to remove toxic dyes from liquid wastes.

> Adsorption is the attachment of a gaseous substance, or a substance in solution, onto the surface of a solid such as finely divided carbon.

Certain wastes can be made nonhazardous or less hazardous by immobilizing them in a solid mass. For example, waste material can be mixed with cement or lime to form a concrete block (Chapter 2), which can then be disposed of in a landfill. If the concrete is likely to be exposed to acidic conditions, leaching can be prevented by encapsulating the block in an impermeable plastic container. This technique is useful for certain liquid wastes and for the flue-gas cleaning sludges that are produced in enormous quantity by coal-burning power stations. The sludges, which are formed as a result of treatment to control sulfur oxide emissions (Chapter 13), contain $CaSO_3$ and fly ash that may include toxic metals.

CONSUMER BRIEF   Disposal of Medical Waste

Medical waste is defined by RCRA as waste which is "infectious." Included in the definition are blood, blood products, body fluids, tissues, and sharps (such as needles, razors or scalpels). Different states use different terms for this waste; biomedical, medical, biohazardous, and infectious are all terms used to describe it. Medical waste must be placed in characteristic red bags so that it will not be confused with other types of garbage or hazardous waste. Disposal of medical waste is very expensive. The prices for disposal are set in pounds, not tons as is usual for other types of waste, and it is estimated that by the year 2000, medical waste disposal will cost consumers $6 billion per year.

People who handle medical waste are especially at risk. An increasing pool of information suggests that workers have developed occupational diseases through contact with materials that contain blood or blood products. The spread of hepatitis and human acquired immune deficiency syndrome (AIDS) in the general population has caused concerns over protecting workers whose duties require them to contact potentially infectious materials.

In March 1992, the Occupational Safety and Health Administration (OSHA) issued a new regulation that is referred to as the OSHA bloodborne pathogen rule. This rule is intended to improve the safety of workers and to protect them from exposure to a needle stick, or exposure of eyes or mucous membranes to infectious material from splashes, or wetting. Hospitals and other facilities handling medical waste are required to provide workers with personal protective equipment, such as heavy gloves that a needle cannot pierce, and face shields to protect from splashes. Sharp objects, such as needles, must be placed into puncture-resistant, leakproof containers that are marked with the biohazard sign.

The Center for Disease Control estimates that 600,000 accidental needle sticks occur every year in the United States and each has the potential to transmit disease. Companies are introducing new protective hypodermic needle syringes. The steel

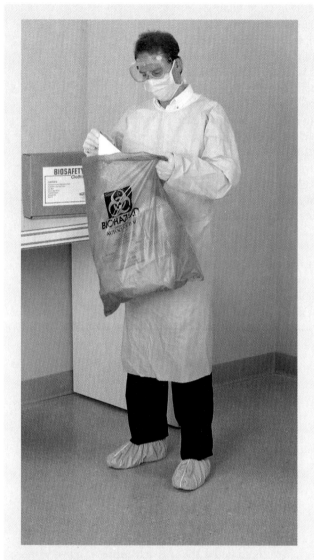

Most of the medical waste that is incinerated is burned in small incinerators located at the hospitals that produce it. On-site incineration has the advantage of minimizing employee accidents. In Pennsylvania, for instance, there are more than 100 on-site incinerators and only 5 commercial ones. In Missouri there are 200 on-site and only 3 commercial incinerators. In compliance with the 1990 Clean Air Amendments, the EPA is developing new emission standards for medical waste incinerators. It is expected that hospitals will be required to install pollution control equipment such as scrubbers in their units which will make disposal of medical waste even more costly.

With the introduction of more stringent state and federal air emission standards, there is increased interest in alternative infectious waste treatment technologies. A microwave treatment system is being used by about twenty hospitals in the United States and Europe. In this system, plastic wastes, such as syringes, tubing, and bags are ground up and sent through a series of microwave ovens that heat the waste to 210° C (410° F). Another technique, which has been on the market since the early 1980s and is approved for use in 30 states, uses chemical treatment with sodium hypochlorite (NaClO), an oxidizing agent similar to laundry bleach, to decontaminate wastes.

State legislatures have been trying to deal with the confusing issues surrounding medical waste. Some states have banned shipments of medical wastes across their borders, while others have placed moratoriums on the construction of new incinerators. States such as California and New York are actively looking at alternatives to incineration.

needle is surrounded by four tough plastic strips that form a protective cap over the needle tip. When the user presses the syringe to the skin, the strips spread and retract the protective cap, allowing the needle to pass through a hole in the cap. When the needle is withdrawn, the strips spread back straight, locking the cap over the needle.

Disposal of medical waste is regulated by individual states. At present more than 75 percent is incinerated. Most of the remaining waste is treated by autoclave, which renders it noninfectious. An autoclave has a high-pressure vessel where the waste is treated with steam at high pressure. Temperature in the autoclave exceeds 178° C (350° F), which will kill all infectious pathogens.

Other states remain cautious about the ability of newer technologies to destroy pathogens completely. Tough new air regulations that took effect in New York in early 1992 have forced several medical incinerators to close.

Hospitals are investigating ways to reduce their waste streams through recycling, reduction, and reuse, where possible. With health care costs as a top national concern, recycling is being carefully considered. But recycling is not always an option with medical waste. The benefit arising from

reuse of medical devices must be weighed against the danger of infecting the public with pathogens.

Questions
1. Should the Federal Government impose standards for the disposal of medical waste, or should states continue to regulate it?
2. Do you think that incineration in hospitals is the best way to handle medical waste?

Source: Waste Age, May, June, August, 1992.

PERPETUAL STORAGE OF HAZARDOUS WASTE

Hazardous wastes that cannot be recycled or converted to nonhazardous forms must be stored. Perpetual storage methods include (1) burial in a secured landfill, (2) deep-well injection, and (3) surface impoundment. None of these methods, however, provides entirely secure storage, and the 1984 amendments to RCRA included restrictions on land disposal of hazardous waste. A 1990 deadline was set for the end of disposal of untreated hazardous materials on land except in specific cases where the EPA has determined that no other option is feasible. Since 1990, wastes not so designated have been automatically banned from land disposal facilities and must be disposed of according to EPA regulations. As a result of legislation, there has been a sharp reduction in improper disposal.

Secured Landfill

A **secured landfill** is a burial site for the long-term storage of hazardous wastes contained in drums (Fig. 19–12). Strict EPA specifications for construction minimize the chance of contaminants migrating in the event of a drum leak. The site must be located in thick, impervious clay soil at least 165 m (500 ft) from a water source in an area that is not subject to floods, earthquakes, or other disturbances. Ideally, the site should be in an arid region, far from any water supplies. The bottom and sides of the trench holding the drums must be lined with two layers of impervious, reinforced plastic, and the drums must rest on a layer of gravel. Below the gravel, there must be a network of pipes that can collect

Soils vary in their ability to retain water. Clays are impervious to water; gravel and sand are very porous.

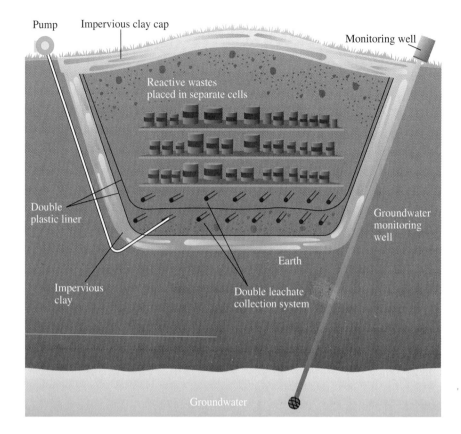

Figure 19-12 A secured landfill for the long-term storage of hazardous wastes contained in drums. The bottom and sides of the fill are lined with plastic; below the gravel on which the drums rest there are pipes that collect any leachate; groundwater below the fill is constantly monitored for contamination. The fill is capped with impervious clay to prevent water from seeping into it.

any leakage from drums and any rainwater seeping through the fill; leachate must be pumped to the surface and treated. When filled, the site must be covered with a layer of impervious plastic material topped, in turn, with layers of sand, gravel, and clay. The cap must be contoured to allow water to drain away from the site. The quality of groundwater in the area of a secure landfill must be monitored for at least 30 years.

Many critics contend that despite the EPA's tough regulations, all landfills will leak eventually, and therefore they should not be used to store materials that retain their hazardous properties almost indefinitely. They maintain that, in time, some drums will corrode and leak, clay layers will be invaded by burrowing animals, and plastic liners will be torn by freezing temperatures or settling, or they will be disintegrated by leachate.

Deep-Well Injection

In the past, liquid hazardous wastes were frequently disposed of by injecting them into deep wells, a technique that was originally developed by the petroleum industry as a means of disposing of brines (salt solutions) that were brought to the surface with crude oil. Liquid waste is pumped into a deep well drilled into a layer of dry porous rock located far below any usable groundwater, and below a layer of impervious rock (Fig. 19-13). The wastes gradually seep into the porous material and are trapped below the impervious layer of rock. The well shaft is surrounded by a sealed casing to prevent contamination as the waste is pumped through the groundwater region.

The RCRA land disposal restrictions apply to **deep-well injection**, but there is a loophole in the law that permits injection of untreated hazardous waste if a company can prove that there will be no migration of the waste from the injection zone during the time the waste remains hazardous. Deep-well injection is one of the least expensive waste-disposal methods, and, with the implementation of the rules restricting disposal in landfills, companies have made use of the loophole to increase the amount of waste they dispose of in deep wells. Critics maintain that, even if properly operated, deep-well injection has the potential for groundwater contamination. In time, the well casing is likely to corrode, and there is always a danger that earth tremors or earthquakes may fracture the impervious layers and lead to the migration of hazardous materials.

Surface Impoundment

Surface impoundment is an inexpensive disposal method used primarily to manage relatively small quantities of hazardous wastes contained in large volumes of wastewater. The wastewater is usually pumped directly from the plant that produced it to a pond lined with impervious clay, a plastic material, or both, to prevent leakage into underlying soil or groundwater (Fig. 19-14). As the water evaporates, the hazardous wastes gradually accumulate on the bottom. Like all methods of disposal on land, surface impoundment is likely to cause environmental contamination sooner or later.

If there is exceptionally heavy rainfall, a surface impoundment pond may overflow, spreading hazardous substances into the surrounding area.

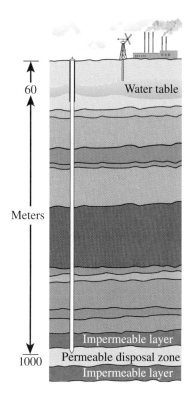

Figure 19-13 Deep-well injection for the disposal of hazardous liquid wastes. Liquid waste is pumped into a deep well drilled into a layer of dry porous rock lying well below any groundwater and below a layer of impervious rock. The well shaft is surrounded by a sealed casing to prevent groundwater contamination.

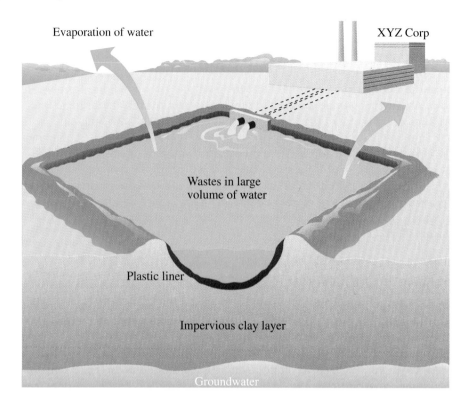

Evaporation of water

XYZ Corp

Wastes in large
volume of water

Plastic liner

Impervious clay layer

Groundwater

Figure 19-14   Surface impoundment for the disposal of small amounts of hazardous waste present in large volumes of water. The water gradually evaporates, and the hazardous materials accumulate at the bottom of the pond, which is lined with clay or plastic to prevent leakage into the ground.

THE UNSOLVED PROBLEM

Society wants the products that generate hazardous wastes—electricity, plastics, gasoline, and pesticides, for example—but no one wants a waste-disposal site in his or her neighborhood. This **NIMBY** (Not In My Back Yard) syndrome has hampered waste disposal but, together with economic considerations, has been an important factor in forcing manufacturers to find ways to minimize waste production. Starting in the 1980s, in-plant process modifications, recycling, and reuse have steadily reduced the production of hazardous waste. But a huge quantity continues to be generated, and we still do not have acceptable ways of dealing with it.

According to the EPA, toxic chemicals released by industry were reduced by 26 percent between 1987 and 1990.

Recently, some oceanographers have suggested testing the ocean as a potential dumping ground for hazardous waste. Although disposal in the oceans is unlikely to be a popular solution, these researchers believe that disposal on flat plains (called abyssal plains) that cover large areas of deep-sea floor may be the only way to save the land and air from additional pollution from landfills and incineration.

In 1979, it was estimated that only 10 percent of hazardous waste was being disposed of in an environmentally safe manner. Today, the situation is very much better, but because of the high cost of proper disposal, large quantities of waste are still dumped illegally. The EPA does not have the resources to keep track of the thousands of operations that generate hazardous wastes, and, consequently, it is often unable to enforce its "cradle-to-grave" provisions. The EPA must also contend with organized crime, which is known to be involved in waste disposal. Mob-controlled trash collectors in New Jersey and New York, for example, have been discovered mixing household refuse with hazardous wastes and illegally disposing of the combined wastes in municipal landfills.

SUPERFUND: CLEANING UP HAZARDOUS WASTE DUMPSITES

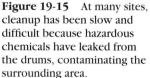

Exposure of the Love Canal disaster in 1978, and increasing evidence that numerous other dangerous abandoned dumpsites existed, made it clear that legislation was urgently needed to deal with problems created in the past. CERCLA, passed in 1980, established the Superfund program, which was aimed primarily at (1) identifying existing hazardous-waste dumpsites and (2) establishing a trust fund to finance the cleanup of the sites.

According to this legislation, the parties responsible for creating a hazardous dumpsite are responsible for cleaning it up. If they fail to do so, the EPA is authorized to undertake the cleanup and then sue the delinquent parties for three times the costs incurred. In cases where responsibility cannot be established, the EPA is authorized to undertake the cleanup using funds provided jointly by the federal government and state governments and by a tax on the chemical and petrochemical industries.

The EPA drew up a **National Priorities List** of sites with the most urgent need of cleanup (Table 19-3), and it soon became evident that the $1.6 billion trust fund provided for cleanup during the first 5 years of the Superfund program was totally inadequate. Many more hazardous sites than expected were discovered, and cleanup was found to be far more difficult than anticipated (Fig. 19-15). By 1990, CERCLA had been amended twice and a trust fund of $13.6 billion had been authorized until 1994. The EPA has located more than 2000 hazardous-

Figure 19-15 At many sites, cleanup has been slow and difficult because hazardous chemicals have leaked from the drums, contaminating the surrounding area.

TABLE 19-3   States and Territories With One or More Sites on the National Priorities List

State or Territory	Number of Sites
New Jersey	109
Pennsylvania	96
California	90
New York	84
Michigan	77
Florida	52
Washington	49
Minnesota	41
Wisconsin	39
Illinois	37
Ohio	33
Indiana	32
Texas	29
Massachusetts	25
South Carolina	23
Missouri	22
North Carolina	22
Delaware	20
Iowa	20
Virginia	20
Kentucky	19
New Hampshire	17
Colorado	16
Connecticut	15
Tennessee	14
Georgia	13
Alabama	12
Rhode Island	12
Utah	12
Kansas	11
Louisiana	11
Arizona	10
Arkansas	10
Maryland	10
New Mexico	10
Oklahoma	10
Idaho	9
Maine	9
Puerto Rico	9
Montana	8
Nebraska	8
Oregon	8
Vermont	8
Alaska	6
West Virginia	5
South Dakota	4
Wyoming	3
Hawaii	2
Mississippi	2
North Dakota	2
Guam	1
Nevada	1
Total	1207

Source: EPA, September, 1991.

waste dumpsites, and 1207 of those sites were included in its National Priorities List in September 1991. The five states with the largest numbers of listed sites are New Jersey (109), Pennsylvania (96), California (90), New York (84), and Michigan (77). The Office of Technology Assessment has estimated that there may be as many as 25,000 hazardous-waste dumpsites in the United States. It is predicted that the cost of cleanup could reach $100 billion.

RADIOACTIVE WASTE

Although radioactive wastes are undoubtedly hazardous to health and the environment, very few are regulated under RCRA. Instead, they are covered by the Atomic Energy Act of 1954 (and later amendments), which is administered by the Department of Energy. The Department of Energy was created in 1977, and, as a legacy from its predecessor, the Atomic Energy Commission, it and not the Department of Defense administers production of the nation's nuclear weapons. The Department of Energy is also responsible for regulating the radioactive wastes generated by nuclear weapons facilities and by commercial nuclear power plants. (Radioactivity and nuclear power were discussed in Chapters 6 and 15.)

Like hazardous waste, radioactive waste presents two problems: (1) waste produced in the past and not disposed of properly and (2) waste currently being generated.

Sources of Radioactive Waste

Nuclear weapons manufacturing facilities and nuclear power plants are the principal sources of the nation's most dangerous radioactive wastes. The nuclear weapons facilities, which are located in 12 states (Table 19-4), generate at least 10 times more waste than the power industry. Relatively small quantities of far less dangerous radioactive waste are generated by hospitals, research laboratories, some industries, and by the mining and processing of the uranium used as nuclear fuel.

Nuclear reactors produce wastes that contain a mixture of radioisotopes with half-lives varying from a few days to many years, including strontium-90 (half-life 28 years), cesium-137 (half-life 30 years), and plutonium-239 (half-life 24,400 years). Since 10 half-lives must elapse before radioactivity is reduced to negligible levels, wastes containing plutonium-239, which emits an alpha particle as it disintegrates (Chapter 6), will remain deadly for tens of thousands of years. Plutonium-239 is used as a nuclear fuel in weapons programs and for many years until the program was discontinued in 1977, it was generated during the reprocessing of uranium. Plutonium-239 is also produced in breeder reactors (Chapter 15), which are used in several European countries to generate electricity. In the United States, however, only a few experimental breeder reactors are in operation.

Classification of Nuclear Waste

Depending primarily on how they are generated, radioactive wastes are classified as high-level wastes or low-level wastes. **High-level wastes**, which all contain plutonium-239, consist of spent fuel-rod assemblies from nuclear reactors from both commercial power plants and weapons plants, and certain other

TABLE 19-4   Nuclear Weapons Facilities Are Located in Twelve States

State	Site
California	Lawrence Livermore National Laboratory
Colorado	Rocky Flats
Florida	Pinellas
Idaho	Idaho Falls
Kansas	Kansas City
Nevada	Nevada Test Site
New Mexico	Los Alamos National Laboratory
	Sandia National Laboratories
	Waste Isolation Pilot Plant (WIPP)
Ohio	Fernald, Mound Plant
South Carolina	Savannah River
Tennessee	Oak Ridge
Texas	Pantex
Washington	Hanford

highly radioactive wastes generated by nuclear weapons facilities. All other radioactive wastes are classified as **low-level wastes**. These wastes are produced primarily by hospitals, research laboratories, and certain industries. Most are dilute and contain radioisotopes with half-lives measured in no more than hundreds of years.

THE LEGACY FROM THE PAST

Radioactive waste, much of it high-level liquid waste, has been accumulating at nuclear weapons facilities (Table 19-4) since World War II. For over 40 years, weapons production took precedence over all other considerations including contamination of the environment and dangers to human health. During this period, most wastes were either dumped directly into the ground, stored in ponds, or deposited in tanks that frequently developed leaks.

As a result of these practices, the environment surrounding all nuclear weapons facilities is extensively contaminated with radioactive material. The Savannah River Weapons Plant near Aiken, South Carolina, is the nation's only facility for producing the radioactive tritium gas used to enhance the explosive power of nuclear weapons. This plant is known to have released tritium gas (half-life of 12.3 years) into the air and water around the plant since the 1950s, and it is also suspected of having deliberately discharged radioactive liquid wastes into the environment for many years. Leaks from old waste storage tanks continue to threaten the Savannah River and nearby groundwater with radioactive contaminants. In 1988, reactors at the plant were shut down for safety reasons. At another weapons plant near Fernald, Ohio, it is estimated that between 1951 and 1985, at least 3 million pounds of uranium dust and other radioactive materials were released into the atmosphere.

The largest, most contaminated, and most dangerous weapons facility in the United States is the 560-square-mile Hanford complex near Richland, Washington, which until operations were halted in 1987, produced plutonium for nuclear weapons. In the past, this facility discharged billions of gallons of ra-

Tritium, $_1^3H$, is an isotope of hydrogen. It decays at a rate of 5.5 percent a year.

dioactive liquid wastes directly into the ground. Wastes that were stored in tanks have leaked, contaminating several cubic miles of soil not only with radioactive material but also with other dangerous by-products of weapons production, including toxic solvents and heavy metals. Some of these wastes continue to migrate through the soil into the Columbia River.

Another severely contaminated weapons facility is the Rocky Flats plant near Denver, Colorado, which is the Department of Energy's only source of the plutonium "triggers" that set off the nuclear chain reactions in warheads. At this plant, which was shut down in 1989 for safety and security reasons, wastes have leaked from corroded storage tanks for many years.

In 1992, Rockwell International Corporation agreed to plead guilty to having illegally disposed of radioactive wastes at the Rocky Flats nuclear weapons plant, and will pay $18.5 million in fines.

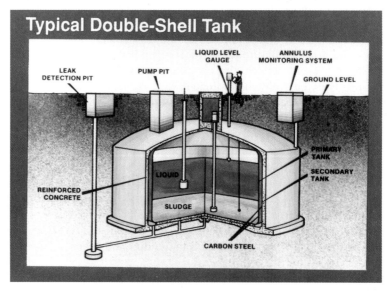

Figure 19-16 High-level liquid radioactive wastes are stored in double-walled steel tanks at the Hanford nuclear weapons facility near Richland, Washington. Each tank can hold 1 million gallons of waste.

Of particular concern are over 70 million gallons of high-level liquid radioactive wastes from weapons production in the past, which are buried underground (some at Hanford and the rest at Savannah River) in single-shell (single-walled) and double-walled steel tanks (Fig. 19-16). Many of the tanks are known to leak, and the wastes in some tanks are potentially explosive. In the 1950s, acidic wastes contained in single-shell tanks were neutralized by adding sodium hydroxide, which resulted in the production of sodium nitrate. Sodium and potassium ferrocyanide and nickel sulfate were also added to the tanks to reduce the volume of the waste by precipitating out the cesium-137. After precipitation, liquid in the tank was pumped out and discarded, leaving behind a radioactive sludge. It is possible that, under the right conditions, the nitrate and ferrocyanide contained in the sludge could react explosively and release radioactive material. In one notorious double-walled tank at Hanford that contains a mixture of radioactive sludge, toxic solvents, and unidentified chemicals, hydrogen gas builds up at intervals and must be vented. Some scientists believe that an electric spark could easily ignite the hydrogen, setting off explosions at nearby tanks and causing the release of large amounts of radioactivity.

In 1993, a tank containing high-level radioactive wastes exploded at a weapons plant in Tomsk, Siberia, contaminating a wide area. The explosion was caused by a rapid rise in temperature that resulted when nitric acid was added to the tank.

One of the biggest problems is finding out exactly what chemicals are in the tanks. Eventually, the Department of Energy plans to remove the liquids from the tanks and solidify them in cement. The sludges in the double-walled tanks are to be vitrified (fused with sand to form a radioactive glass) and then stored in a permanent repository. There is no firm plan for disposal of the wastes in the single-shell tanks.

The Department of Energy had plans to resume operations at its key weapons facilities at Savannah River, Rocky Flats, and Hanford after extensive repairs and improvements had been completed, but early in 1993 it was announced that the facilities would not be restarted.

It has been estimated that the massive task of cleaning up the nation's nuclear weapons plants and other sites contaminated with radioactive wastes will ultimately cost at least $200 billion and take decades to accomplish. Cleanup is hampered by the lack of adequate disposal technology and the difficulty of finding an acceptable site for long-term storage.

REGULATION OF RADIOACTIVE WASTE DISPOSAL

Regulations governing disposal of radioactive wastes vary depending on whether the wastes are high-level or low-level and whether they are generated by weapons programs or by commercial operations.

Low-Level Radioactive Wastes

Until 1970, most low-level radioactive waste generated in the United States was disposed of at sea. When this practice was banned, low-level wastes were buried in shallow, often unlined trenches. The wastes, which in some cases included highly radioactive liquid military waste, were supposed to be contained by impermeable rock, but leaks developed at many locations. Today,

Figure 19-17   Low-level radioactive wastes are permanently stored in steel drums, which are buried in lined trenches.

Since 1993, Washington, Nevada, and South Carolina—the only states with low-level radioactive waste disposal sites— have had authority to refuse wastes from states outside their compacts. No other wastes sites have been built, and waste generators are being forced to store their low-level radioactive wastes on site.

Elements with atomic numbers greater than that of uranium (92) are known as the transuranium elements (or transuranics).

commercially generated low-level radioactive wastes must be buried in steel drums in lined, carefully sited trenches in accordance with rules set by the Nuclear Regulatory Commission (NRC) (Fig. 19-17). During the 1970s, states in which the few NRC-licensed facilities were located objected to receiving wastes generated outside their borders, and there was increasing concern about the dangers inherent in transporting radioactive wastes over long distances. In 1980, in response to these problems, Congress passed the **Low-Level Radioactive Waste Policy Act**, which set 1986 (later extended to 1992) as the deadline for each state to take responsibility for disposal of its own low-level commercial wastes. If states preferred, they had the option of joining in multistate compacts in which one state would provide a disposal site for the other states in the compact. The search for sites touched off furious environmental and political battles, and many states are still not in compliance with the law.

To provide a permanent storage site for low-level radioactive wastes produced by nuclear weapons programs, the Department of Energy in 1983 began building an underground repository in a salt bed under federal land near Carlsbad, New Mexico. The repository, which is known as the **Waste Isolation Pilot Plant (WIPP)**, is also intended for the storage of transuranic wastes produced at weapons facilities. Wastes that contain transuranics are usually dilute but dangerous because they contain elements, including plutonium-239, that have long half-lives. The repository is physically ready to receive wastes,

Figure 19-18 After removal from a nuclear reactor, spent fuel-rod assemblies are stored under water, where they glow as they continue to emit radiation.

but at the present time, safety, environmental, and other problems continue to delay its opening.

High-Level Radioactive Wastes

Pending establishment of a permanent repository for high-level radioactive wastes, spent fuel rods from nuclear reactors are stored under water in tanks at the reactor sites (Fig. 19-18). The water shields the radiation and dissipates the heat that is generated by the rods. Each year, the average 1000-MW nuclear power plant produces approximately 30 metric tons of spent rods; it is estimated that by the year 2000, storage space for 72,000 tons of radioactive waste will be needed. Many commercial plants are already running out of storage space.

To deal with the mounting problem, Congress in 1982 passed the **Nuclear Waste Policy Act**, which set up a timetable for building a repository deep underground in a geologically stable area for permanent storage of high-level waste (Figure 19-19) with the opening date set for 1998. The repository was originally intended for wastes from commercial nuclear reactors only, but under new plans, it will also accept high-level defense weapons wastes after they have been vitrified.

Some utilities have already run out of space for storing used fuel rods and have begun storing them in above-ground facilities that cost $2 million to build and must be guarded and monitored.

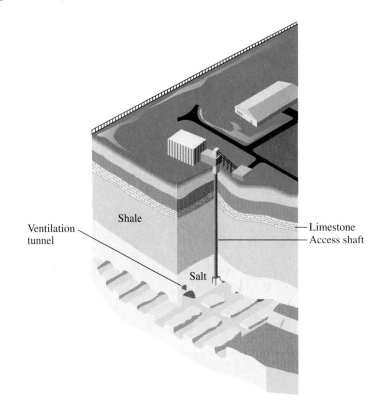

Figure 19-19 A design for the permanent storage of high-level radioactive waste in a deep underground salt mine.

Labels on figure: Ventilation tunnel, Shale, Salt, Limestone, Access shaft

In 1987, after two other sites had been rejected, the Department of Energy chose Yucca Mountain, 100 miles (150 km) northwest of Las Vegas, Nevada, as the site for the repository. The site met two important criteria: It was very dry with an annual rainfall of only 6 inches, and the rock formation was tuff, a compacted volcanic ash that is considered to be one of the best materials for containing radioactivity. After extensive testing at the site, the Department of Energy planned to construct a 1400-acre cavern, located 1000 feet under the mountain and 1500 feet above the water table.

After a few preliminary studies had been completed, however, the state of Nevada, under pressure from citizens' groups, refused to issue drilling permits for further exploratory testing. A further setback was a geological report warning of the probability of volcanic activity at the site during the 10,000-year period that, according to the EPA, is required for spent fuel rods to decay to safe levels. In 1991, a court ruling forced the state to allow resumption of testing but the program is still in deep political and legal trouble. It now seems certain that the new 2010 deadline for opening the repository will not be met. Meanwhile, spent nuclear fuel rods continue to accumulate at power plants.

Disposal of radioactive wastes is one of the most intractable problems that the Department of Energy faces. For environmentalists and the general public alike, opposition to radioactive waste disposal sites is not just a case of NIMBY; it is one of **NOPE** (Not On Planet Earth).

TECHNOLOGIES FOR RADIOACTIVE WASTE DISPOSAL

Currently, the recommended procedure for disposing of high-level radioactive waste is to concentrate it, incorporate it into a glass or ceramic material, seal the glass or ceramic product in a corrosion-resistant steel canister, and then bury the canister in a concrete vault deep underground in a stable geologic formation composed of salt, tuff, or granite.

Although, as we have just seen, an underground repository for high-level waste has yet to be established, a facility for vitrifying high-level liquid waste is nearing completion at Savannah River and is scheduled to begin operation shortly. In a completely automated procedure, the waste will be fused with sand at 2000°F and the molten mixture will be poured into stainless steel canisters and allowed to harden into a black glass. The glass will be radioactive, but it will be very resistant to leaching. Until the canisters can be transferred to a permanent repository, they will be stored in concrete.

Another option for disposing of radioactive waste is burial in deep ocean trenches beneath bottom sediments. Over the long term, however, volcanic activity, earthquakes, and leaking containers could lead to contamination of the ocean. An attractive possibility that is being studied is transmutation of dangerous radioisotopes to harmless or less-harmful isotopes by means of neutron bombardment. But even if such transmutations could be achieved, the process would be extremely costly.

Another idea is to shoot the waste into the sun or into outer space. Although disposal in deep space is technically feasible, it is unlikely to be attempted. Not only would the cost be prohibitive, but if an accident occurred during a launch into space, radioactive material could be dispersed into the atmosphere and contaminate the entire earth's surface.

> If technetium-99 (half-life, 210,000 years)—a radio-isotope often present in nuclear weapons wastes—is bombarded with neutrons, it is transmuted to technetium-100, which decays to stable, non-radioactive ruthenium-100.

THE POST–COLD WAR CHALLENGE

With the end of the Cold War, no additional production of plutonium-239 by the United States is likely to be required, and no new enriched uranium will be needed for weapons. But the difficult problem of disposal of plutonium from dismantled weapons remains to be solved. Some of the plutonium used for weapons may be needed to maintain stockpiles, but ways must be found to dispose of as much as 50 tons in excess of foreseeable needs. In addition to the disposal methods for high-level radioactive wastes that we have just discussed, a number of other procedures for dealing with weapons plutonium have been suggested. These include oxidizing the plutonium to a powder that could be permanently stored in a repository or using the powder to make a nuclear fuel. The option of last resort is controlled detonation.

When mixed with uranium, oxidized plutonium forms a nuclear fuel called MOX that could be used in commercial nuclear power plants. Although today's power plants are not designed to use MOX, several European countries are planning to use it in the future. Since these countries already have plenty of plutonium of their own, however, they are unlikely to need any from the United States. Further, export raises the possibility that the plutonium could get into the hands of unscrupulous nations seeking to make nuclear weapons.

> Japan is committed to using plutonium as an energy source and has already begun importing plutonium from fuel-reprocessing plants in Europe.

EXPLORATIONS

Love Canal: The Nation's Most Infamous Chemical Waste Dumpsite

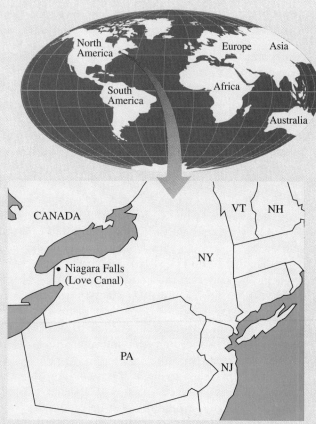

On August 2, 1978, at a public meeting in Albany, New York, Dr. Robert Whalen, the commissioner of health for the state of New York made the following statement: "A review of all the available evidence respecting the Love Canal Chemical Waste Landfill site has convinced me of the existence of a great and imminent peril to the health of the general public residing at or near the site." These ominous words confirmed the worst fears of the people living in Love Canal and introduced the general public to the nation's most infamous toxic chemical dumpsite.

The disaster at Love Canal, a neighborhood of modest homes in the city of Niagara Falls, New York, began many years before Dr. Whalen's announcement. The stage was set in the late 1800s when William Love, attracted to the area by the swiftly flowing waters of the Niagara River and the prospect of cheap power, formed a company to build a model city on the shores of Lake Ontario. He planned to divert water from the river into a 7-mile-long navigation canal and build a hydroelectric generating plant at the site where the water returned to the river. But Mr. Love's ambitious scheme was never completed. After a few homes had been built and a canal—60 feet wide, 10 feet deep, and 3000 feet long—had been dug, his company failed. The canal bed, fed by streams originating in the Niagara River, became the center of a popular recreation area.

In 1905, by which time inexpensive hydroelectric power had become available, the Hooker Electrochemical Company opened a plant near the city of Niagara Falls for the manufacture of chlorine and caustic soda. In 1942, with permission from the city, Hooker began disposing of chemical wastes in the old canal; a few years later, the company purchased the canal and some of the surrounding land. By the 1950s, Hooker was producing pesticides, plastics, and other chemicals and had become one of the area's biggest employers, with annual sales in excess of $75 million.

Chemical wastes, which included caustics, chlorinated hydrocarbons, and solvents, were poured directly into the canal or were poured into metal and fiber drums that were then deposited in it. Twenty-two thousand tons of chemicals had been dumped into the canal by 1952, and it was almost full. The Hooker company, having no further use for the canal, covered it with dirt and clay, and deeded it, along with 16 acres of surrounding land, to the Niagara Falls School Board for the token sum of one dollar. The deed of sale included an interesting disclaimer: The purchaser assumed all risk and liability incident to the use of the land. This meant that no claim could ever be made against Hooker for injury or death caused by the buried wastes.

Despite the disclaimer and the concern of a contractor who encountered chemicals as he dug into the construction site, the school board built an elementary school on land bordering the canal. The school playground and adjoining fields where children played lay directly over the canal dumpsite. Housing developments grew up around the site until eventually nearly 1000 homes were built, some with back-yards bordering the school board's land (Fig. 19-A).

In the ensuing years, many unusual and disturbing occurrences were noted. Black liquid was occasionally observed seeping out of the ground; grass and backyard vegetables failed to grow; and noxious fumes were often a problem. Children running barefoot in the fields near the school complained of irritated sore feet, and many suffered from bronchitis and asthma. Complaints to local officials were disregarded, and the residents, unaware of the existence of the canal and its history, did not appreciate the seriousness of these problems. Then in the winter of 1977–1978, as a result of several years of unusually heavy precipitation, the canal dumpsite became saturated with water. Thick, black, malodorous liquid began oozing to the surface in backyards and the school's playing fields and seeping through cinderblock walls into basements of nearby homes. Corroded drums were brought to the surface, and chemical wastes leaked into storm sewers, contaminating groundwater. Complaints from alarmed citizens could no longer be ignored.

Studies carried out by the Environmental Protection Agency (EPA) in 1978 identified 82 different chemicals, including harmful polychlorinated hydrocarbons, in the area near the dump. Eleven of the chemicals were known or suspected carcinogens. In families living close to the dumpsite, state epidemiologists discovered rates of miscarriages four times greater than average. They found an unusually large number of children with birth defects and a high incidence of cancer and of respiratory, nervous, and kidney diseases.

Residents were confused and frightened. Test results were not properly explained to them, but they knew that their health—particularly that of their children—was threatened and that their property values had plummeted. A fence bearing a "No Trespassing" sign was erected around part of the canal area, and the advice to those close to the dumpsite to stay out of their basements and refrain from eating vegetables grown in their backyards only added to their anxiety. Despite numerous promises of help from elected representatives and other officials, nothing further was done until August of 1978 when public denunciations from outraged residents, and adverse publicity in the media, finally forced the state to take action. A health emergency was declared, the school was closed, and 239 families in the first two rings of houses around the canal (Fig. 19-A) were evacuated. By May of 1980, an additional 710 families in neighboring blocks, who had angrily protested that they too were in danger,

Figure 19-A Photograph of the community of Love Canal that was built around the chemical dumpsite. The Niagara River can be seen at the top; the canal runs from top to bottom between the two rows of houses. After the exposure of leaking chemicals in 1978, residents were evacuated from the first two rings of houses around the canal. Later, families in a wider area were relocated.

had also been relocated. Fifty other eligible families refused to leave their homes. President Carter declared the site a disaster area, and federal money was made available to pay resettlement costs and buy almost 800 contaminated homes. Houses nearest the dumpsite were demolished, and approximately 500 others were boarded up. The surface of the canal was sealed with clay and planted with grass; a drainage system was installed to carry leaking chemicals to a treatment plant. A federal study indicated that chemicals had not migrated away from the dumpsite through underground channels.

In 1990, after more than $275 million had been spent on buying homes from residents, preparing studies, and cleaning up, the New York State Department of Health declared that much of the area was again habitable. Despite the concerns of environmental organizations that health hazards still remained, the EPA supported the state's decision to encourage resettlement. Homes became available at 20 percent below their market value because of the Love Canal stigma, and many people with modest incomes were eager to buy them. Banks, however, were unwilling to hold mortgages on these homes, and in 1991, only a few people had moved back into the area, which had been renamed "Black Creek Village."

The U.S. government sued Occidental Chemical Corporation, the successor to the Hooker company, for $610 million to cover the cost of cleanup and for punitive damages. In 1992, the case was still before the courts. In 1983, former Love Canal residents settled out of court with Occidental, the city of Niagara Falls, and its school board, on claims for health problems arising from their exposure to the Love Canal chemicals. But for the rest of their lives, these people will wonder and worry about what the long-term effects of the years spent living next to the nation's most notorious hazardous waste dumpsite will be on their health and that of their children.

Question
Discuss the precautions that must be taken to protect human health and the environment when drums containing nonradioactive, toxic chemicals are placed in perpetual storage underground.

Reference: A. G. Levine, *Love Canal: Science, Politics, & People* (Lexington Books, Lexington, MA.), 1982.

⬛ KEY WORDS AND CONCEPTS

Comprehensive Environmental
   Response, Compensation and
   Liability Act (CERCLA)
   or Superfund
corrosive waste
deep-well injection
hazardous waste
high-level radioactive wastes
ignitable waste

incineration
land treatment
Low-Level Radioactive Waste
   Policy Act
low-level radioactive wastes
municipal refuse
National Priorities List
Nuclear Waste Policy Act
reactive waste

Resource Conservation and
   Recovery Act (RCRA)
sanitary landfill
secured landfill
solid waste
Solid Waste Disposal Act
surface impoundment
toxic waste
Waste Isolation Pilot Plant (WIPP)

⬛ QUESTIONS AND PROBLEMS

1. Describe how a modern sanitary landfill is constructed. How does this design differ from old "dumps"?
2. List the advantages and disadvantages of refuse incinerators. Would you recommend that the refuse generated by your family be disposed of in a landfill or incinerated? Explain your choice.
3. What is fly ash, and why is it dangerous?
4. Why isn't a larger percentage of our garbage recycled? List three steps that your hometown could take to encourage recycling.
5. Does CERCLA regulate abandoned waste dumps or operation of municipal dumps?
6. Give definitions for each of the following terms:
   a. ignitable waste   **b.** corrosive waste   **c.** reactive waste   **d.** toxic waste
7. Is waste sulfuric acid from automobile batteries classified as ignitable waste or corrosive waste?
8. Should waste sulfuric acid be stored in steel drums prior to proper disposal? What type of container should be used?

9. Picric acid, which was once used by printers to etch copper, is known to explode when heated or subjected to shock. How would picric acid waste be classified?
10. Many printers use hydrocarbon solvents to clean ink from their hands and equipment. These solvents have a flash point below 60°C. How would hydrocarbon wastes be classified?
11. How would waste lead from automobile batteries be classified?
12. How would waste sodium hydroxide be classified?
13. You have to dispose of waste nitric acid. What could you do to make it less corrosive?
14. List three techniques used for "perpetual storage" of hazardous wastes. If you had to make a decision among the three for dioxin-contaminated waste, would you be able to recommend one?
15. Why is deep well injection a problem for communities near the well?
16. What is surface impoundment?

17. What is the Superfund program? What does it regulate?

18. From where does the money for the Superfund trust fund come?

19. What is the National Priorities List?

20. Some people advocate disposing of toxic materials in the ocean far away from human populations. In your opinion is this a viable option?

21. What are the two major sources of radioactive wastes?

22. Are radioactive materials regulated by RCRA?

23. How many half-lives does it take for the radiation in radioactive waste to be considered at negligible levels? How many years for plutonium-239? For strontium-90?

24. How has radioactive waste been disposed of at the Hanford plutonium-producing plant in Richland, Washington?

25. Describe what is meant by vitrifying radioactive wastes.

26. Before 1970, where was the disposal site for most low-level radioactive waste?

27. What government agency now regulates the disposal of low-level radioactive waste?

28. What is the Low-Level Radioactive Waste Policy Act? Should a heavily populated state such as New York be held responsible for the disposal of radioactive waste from its nuclear plants?

29. What is WIPP? Should the state of New Mexico have a say in what goes into WIPP?

30. What happens to spent fuel rods from nuclear power plants? Where will the radioactive material in these rods be permanently stored?

31. What type of radioactive waste does the Nuclear Waste Policy Act regulate?

32. In what state does the Department of Energy plan to permanently store waste from commercial nuclear reactors?

33. Why isn't radioactive material from spent fuel rods recycled?

34. Describe the recommended procedure for the disposal of high-level radioactive waste.

35. What is transmutation, and how could it be used to lessen the amount of radioactive waste needing disposal?

36. Some people have suggested that radioactive waste should be shot into the sun or outer space. List two reasons why this is not an attractive disposal method.

37. Some have suggested that decommissioned nuclear power plants be entombed in concrete rather than being dismantled. List the advantages and disadvantages of entombment relative to dismantling. Consider the exposure of workers and the public to radiation and the cost.

38. What is MOX, and what is its source?

39. With the end of the Cold War, many old nuclear warheads will be decommissioned, and the plutonium in them will need to be disposed of. List three ways to accomplish the disposal. Give two advantages and two disadvantages for each way.

EXPONENTIAL NOTATION

Scientists frequently use numbers that are very large or very small. For example:

A. 12 g of carbon contains 602,000,000,000,000,000,000,000 atoms of carbon

B. a single molecule of water weighs 0.000 000 000 000 000 000 000 03 g

Such numbers are cumbersome to work with and difficult to write without losing track of the zeros. For convenience and to avoid errors, scientists express very large and very small numbers in a compact form called **exponential notation**, or **scientific notation**.

Exponential notation is a system in which numbers are expressed in the form:

$$N \times 10^x$$

N, which is called the **coefficient**, is a number between 1 and 10. *x* is the **exponent**, often referred to as the **power of ten**, and 10^x is the **exponential term**.

The exponent (or power of ten) may be positive or negative. Positive exponents indicate numbers equal to 10 or greater than 10; negative exponents indicate numbers less than 1. (Numbers between 1 and 10 do not need an exponential term.)

In exponential notation, the two numbers in A and B written above in long form become:

A. 6.02×10^{23} (coefficient = 6.02, exponent = 23)
B. 3.0×10^{-24} (coefficient = 3.0, exponent = –24)

Positive Exponents   A positive exponent tells us how many times the number must be *multiplied* by 10 to give the long form of the number. For example,

$$5.43 \times 10^4 = 5.43 \times 10 \times 10 \times 10 \times 10 \ (4 \ \text{times})$$
$$= 54{,}300$$

$$2.1 \times 10^3 = 2.1 \times 10 \times 10 \times 10 \ (3 \ \text{times})$$
$$= 2100$$

In A, above, 6.02 must be multiplied by 10, 23 times to give the long form.

Negative Exponents   A negative exponent tells us how many times the number must be *divided* by 10 to give the long form of the number. For example,

$$1 \times 10^{-1} = \frac{1}{10} \text{ (1 time)}$$
$$= 0.1$$

$$8.06 \times 10^{-5} = \frac{8.06}{10 \times 10 \times 10 \times 10 \times 10} \text{ (5 times)}$$
$$= 0.0000806$$

In B, above, 3.0 must be divided by 10, 24 times to give the long form.

Rules for Writing Numbers in Exponential Notation

1. **For numbers *equal to or greater* than 10:**

   Move the original decimal point to the *left* to obtain a number between 1 and 10. (The decimal point may not be shown, but its position is understood.)

   $$760\ 000\ 000\ 000 \leftarrow \text{original position of decimal}$$

   The original decimal point must be moved 11 places to the left to obtain the number 7.6. The number of decimal places moved (11) gives the exponent (or power of ten); the exponent is *positive*.

   $$760{,}000{,}000{,}000 = 7.6 \times 10^{11}$$

2. **For numbers *less* than 1:**

   Move the original decimal point to the *right* to obtain a number between 1 and 10.

   $$0.00631$$

   The original decimal point must be moved 3 places to the right to obtain the number 6.31. The number of decimal places moved (3) gives the exponent; the exponent is *negative*.

   $$0.00631 = 6.31 \times 10^{-3}$$

EXAMPLE A-1   Express the following numbers in scientific notation:
a. 8400 kg   **b.** 6,000,000 m   **c.** 0.792 mL   **d.** 0.00036 g

Solution:
a. 8400 kg $= 8.4 \times 10^3$ kg
To obtain a number between 1 and 10, the original decimal point must be moved 3 places to the *left*. Therefore, the exponent is +3.
b. 6,000,000 m $= 6.0 \times 10^6$ m
To obtain a number between 1 and 10, the original decimal point must be moved 6 places to the *left*. Therefore, the exponent is +6.
c. 0.792 mL $= 7.92 \times 10^{-1}$ mL
To obtain a number between 1 and 10, the original decimal point must be moved 1 place to the *right*. Therefore, the exponent is –1.
d. 0.00036 g $= 3.6 \times 10^{-4}$ g
To obtain a number between 1 and 10, the original decimal point most be moved 4 places to the *right*. Therefore, the exponent is –4.

To Convert from Exponential Notation to Decimal Notation

The value of the exponent indicates the number of places the decimal point must be moved.

> If the exponent is *positive*, the decimal point must be moved to the *right*.
> If the exponent is *negative*, the decimal point must be moved to the *left*.

Zeros may have to be added as the decimal point is moved.

EXAMPLE A-2

a. Write 2.69×10^5 in decimal notation.
b. Write 4.0×10^{-6} in decimal notation.

Solution:

a. The exponent is +5. Therefore, the decimal point must be moved 5 places to the right. 3 zeros must be added.

added zeros

$$2.69\overline{000} = 269\,000$$

b. The exponent is –6. Therefore the decimal point must be moved 6 places to the left. 5 zeros must be added.

added zeros

$$\overline{000004}.0 = 0.000004$$

Addition and Subtraction

To add or subtract numbers expressed in exponential notation, proceed as follows:

EXAMPLE A-3

coefficient   coefficient
↓   ↓
$$(6.14 \times 10^3) \ + \ (1.73 \times 10^2)$$
↑   ↑
exponential term   exponential term

Solution:

1. Adjust the coefficients so that both exponential terms are the same (i.e., either 10^2 or 10^3 in this example).

$$(61.4 \times 10^2) + (1.73 \times 10^2)$$

2. Add the new coefficients.

$$(61.4 + 1.73) \times 10^2 = 63.1(3) \times 10^2$$

3. Adjust the answer to exponential notation if necessary.

$$6.31 \times 10^3$$

EXAMPLE A-4
$$(2.39 \times 10^{-4}) - (7.81 \times 10^{-6})$$

Solution:
1. Adjust the coefficients so that both exponential terms are the same (i.e., either 10^{-4} or 10^{-6} in this example).

$$(2.39 \times 10^{-4}) - (0.0781 \times 10^{-4})$$

2. Subtract the new coefficients one from the other as indicated.

$$(2.39 - 0.0781) \times 10^{-4} = 2.31(19) \times 10^{-4}$$
$$= 2.31 \times 10^{-4}$$

Multiplication and Division

When numbers are expressed in exponential notation, multiplication and division are greatly simplified. In multiplication, exponents are *added*; in division, they are *subtracted*.

Multiplication

To multiply 2 or more numbers, proceed as follows:

EXAMPLE A-5
$$(1.2 \times 10^3)(3.0 \times 10^2)$$

Solution:
1. Multiply the coefficients in the usual manner.
$$1.2 \times 3.0 = 3.6$$

2. To obtain the new exponent, add the two exponents together.
$$10^3 \times 10^2 = 10^{3+2} = 10^5$$

3. Combine the new coefficient with the new exponent to obtain the answer.
$$3.6 \times 10^5$$

EXAMPLE A-6
$$(5.2 \times 10^9)(7.4 \times 10^{-7})$$

Solution:
1. $5.2 \times 7.4 = 38$
2. $10^9 \times 10^{-7} = 10^{9+(-7)} = 10^2$
3. Combine 1 and 2:
$$38 \times 10^2$$

4. Express in exponential notation to obtain the answer:
$$3.8 \times 10^3$$

Division

To divide two or more numbers, proceed as follows.

EXAMPLE A-7

$$\frac{4.8 \times 10^4}{2.4 \times 10^2}$$

Solution:

1. Divide the coefficients in the usual manner.

$$\frac{4.8}{2.4} = 2.0$$

2. To obtain the new exponent, subtract the exponent of the denominator from the exponent of the numerator.

$$\frac{10^4}{10^2} = 10^{4-2} = 10^2$$

3. Combine 1 and 2 to obtain the answer:

$$2.0 \times 10^2$$

EXAMPLE A-8

$$\frac{7.64 \times 10^{10}}{8.91 \times 10^{-4}}$$

Solution:

1. $\dfrac{7.64}{8.91} = 0.857$

2. $\dfrac{10^{10}}{10^{-4}} = 10^{10-(-4)} = 10^{10+4} = 10^{14}$

3. Combine 1 and 2:

$$0.857 \times 10^{14}$$

4. Express in exponential notation to obtain the answer:

$$8.57 \times 10^{13}$$

COMMON LOGARITHMS

The *common*, or base 10, logarithm (abbreviated log) of a number N is the power to which 10 must be raised to give N. For example, the logarithm of 1000 is 3, because raising 10 to the third power gives 1000.

$$10^3 = 1000$$

$$\log 1000 = 3$$

Thus, a logarithm is an exponent.

When a number N is an integral power of 10, its log is a simple integer; if it is less than one, the integer is negative. For example:

$$N = \quad 1 \quad = 10^0 \qquad \log 10^0 = 0$$
$$N = \quad 10 \quad = 10^1 \qquad \log 10^1 = 1$$
$$N = \quad 100 \quad = 10^2 \qquad \log 10^2 = 2$$
$$N = \quad 0.01 \quad = 10^{-2} \qquad \log 10^{-2} = -2$$
$$N = 0.0001 = 10^{-4} \qquad \log 10^{-4} = -4$$

Logarithms of numbers that are integral powers of 10 are easily determined by inspection. However, logarithms of numbers that are not integral powers of 10, such as 6, 34.2, and 0.00851, must be obtained from a log table or by using a calculator.

Logarithms of nonintegral numbers are not required to solve any of the problems given in this text and will not be considered.

pH

pH, which is a measure of the acidity, or hydrogen ion concentration, of a solution (Chapter 8), is a logarithmic relationship that is used frequently in chemistry.

By definition, the pH of a solution is equal to the negative logarithm of the hydrogen ion concentration, $[H^+]$. Thus

$$pH = -\log [H^+]$$

The pH of a solution is, therefore, equal to the negative power to which 10 must be raised to give the hydrogen ion concentration.

$$\text{If } [H^+] = 10^{-x}$$
$$\text{then pH} = x$$

The pH of a solution whose $[H^+]$ is an exact power of 10 is given by the negative of the exponent.

EXAMPLE A-9   What is the pH of a solution in which $[H^+] = 0.0001$?

$$0.0001 = 10^{-4}$$
$$\log 10^{-4} = -4$$

Therefore

$$-\log 10^{-4} = -(-4) = 4$$
$$pH = 4$$

pH values that are not whole numbers will not be considered in this text.

PROBLEMS

1. Express the following in scientific notation:
   - **a**. 423 million
   - **b**. 903,584
   - **c**. 0.000617
   - **d**. 0.00000006
   - **e**. 0.00000000000000749
   - **f**. 37,693,528
   - **g**. 0.04020
   - **h**. 0.000000001
   - **i**. 6.003
   - **j**. 2,300,000,000

2. What is the pH of a solution that has a hydrogen ion concentration of 0.00001?

3. The pH of a solution is 7. Give the $[H^+]$ in exponential notation.

MEASUREMENT AND THE INTERNATIONAL SYSTEM OF UNITS (SI)

In science and in many other fields it is very important to make accurate measurements. For example, the establishment of the Law of Conservation of Matter was dependent on accurate weight measurements; medical diagnosis relies on accurate measurements of factors such as temperature, blood pressure, and blood glucose concentration; a carpenter building a deck must make careful measurements before cutting the wood.

All measurements are made relative to some **reference standard**. For example, if you measure your height using a ruler marked in meters, you are comparing your height to an internationally recognized reference standard of length called the meter.

Most people in the United States use the English system of measurement and think in terms of English units: feet and inches, pounds and ounces, gallons and quarts, and so forth. If a man is described as being 6 foot 4 inches tall and weighing 300 pounds, we immediately visualize a large individual. Similarly, we know how much to expect if we buy a half gallon of milk or a pound of hamburger at the grocery store. Most of us, however, have a far less clear idea of the meaning of meters, liters, and kilograms, common metric units of measurement that are used by the scientific community and by every other major nation in the world. Although the United States is committed to changing to the metric system, the pace of change, so far, has been extremely slow.

THE INTERNATIONAL SYSTEM OF UNITS (SI)

The International System of Units, or SI (from the French Le Système International d'Unités), was adopted by the International Bureau of Weights and Measures in 1960. The SI is an updated version of the metric system of units that was developed in France in the 1790s, after the French Revolution.

The **standard unit of length** in the SI is the **meter**. Originally, the meter was defined as one ten-millionths (0.0000001) of the distance from the North Pole to the equator measured along a meridian. This distance, however, was difficult to measure accurately, and for many years the meter was defined as the distance between two lines etched on a platinum-iridium bar kept at 0°C (32°F) in the International Bureau of Weights and Measures in Sevres, France. Today, the meter is defined even more precisely as equal to 1,650,763.73 times the wavelength of the orange-red spectrograph line of $^{86}_{36}Kr$.

The **standard unit of mass** is the **kilogram**. It is defined as the mass of a platinum-iridium alloy bar, which, like the original meter standard, is kept at the International Bureau of Weights and Measures. (Note: The difference between mass and weight is explained at the end of this Appendix.)

Base Units

There are seven **base units** of measurement in the SI. They are shown in Table B-1.

TABLE B-1 SI Base Units

Quantity Measured	Name of Unit	SI Symbol
Length	meter	m
Mass	kilogram	kg
Time	second	s
Electric current	ampere	A
Thermodynamic temperature	kelvin	K
Amount of a substance	mole	mol
Luminous intensity	candela	cd

Prefixes

The SI base units are often inconveniently large (or small) for many measurements. Smaller (or larger) units, defined by the use of **prefixes**, are used instead. Multiple and submultiple SI prefixes are given in Table B-2. Those most commonly used in general chemistry are underlined.

TABLE B-2 SI Prefixes

Factor		Prefix	SI Symbol
Exponential Form	Decimal Form		
10^6	1 000 000	mega	M
10^3	1 000	kilo	k
10^2	100	hecto	h
10	10	deka	da
10^{-1}	0.1	deci	d
10^{-2}	0.01	centi	c
10^{-3}	0.001	milli	m
10^{-6}	0.000 000 1	micro	μ
10^{-9}	0.000 000 000 1	nano	n
10^{-12}	0.000 000 000 001	pico	p

Thus, for example:

1 kilogram equals 1000 grams (or 1/000 or 0.001 kg equal 1 g)

1 centimeter equals 1/100 or 0.01 meter (or 100 cm equal 1 m)

Derived SI Units

In addition to the seven SI base units, many other units are needed to represent physical quantities. All are derived from the seven base units. For example, **volume** is measured in cubic meters (m^3), **area** is measured in square meters (m^2), and **density** is measured in mass per unit volume (kg/m^3). Derived units commonly used in general chemistry are listed in Table B-3.

TABLE B-3   SI Derived Units

Physical Quantity	Unit	SI Symbol	Definition
Area	square meter	m^2	$m \cdot m$
Volume	cubic meter	m^3	$m \cdot m \cdot m$
Density	kilogram per cubic meter	kg/m^3	kg/m^3
Force	newton	N	$kg \cdot m/s^2$
Pressure	pascal	Pa	$N/m^2 = kg/m \cdot s^2$
Energy, or quantity of heat	joule	J	$N \cdot m = kg \cdot m^2/s^2$
Quantity of electricity	coulomb	C	$A \cdot s$
Power	watt	W	$J/s = kg \cdot m^2/s^3$
Electric potential difference	volt	V	$J/A \cdot s = W/A$

Note: In 1964, the liter (L) was adopted as a special name for the cubic decimeter (dm^3).

Probably the least familiar of the derived units are the ones used to represent force (force = mass × acceleration), pressure (pressure = force/area), and energy, which is defined as the ability to do work (work = force × distance).

The SI unit of force, the **newton (N)**, is defined as the force that, when applied for 1 second, will give a 1-kilogram mass a speed of 1 meter per second.

The SI unit of pressure, the **pascal (Pa)**, is defined as the pressure exerted by a force of one newton acting on an area of one square meter. Often, it is more convenient to express a pressure in kilopascals (1 kPa = 1000 Pa). For example, atmospheric pressure at sea level is approximately equal to 100 kPa.

The SI unit of energy (or quantity of heat), the **joule (J)**, is defined as the work done by a force of one newton acting through a distance of one meter. Often it is more convenient to express energy in kilojoules (1 kJ = 1000 J). (Note: The relationship between the joule and the more familiar calorie will be discussed later in this appendix.)

Conversions within the SI

Because the SI is based on the decimal system, conversions within it are much easier than conversions within the English system. Subunits and multiple units in the SI always differ by factors of 10, as shown in Table B-2. Thus, conversions from one unit to another are made by moving the decimal point the appropriate number of places. This procedure is best explained by some examples.

EXAMPLE B-1   Convert 0.0583 kilograms to grams.

Solution:
1. Obtain the relationship between kg and g from Table B-2.

$$1 \text{ kg} = 1000 \text{ g}$$

2. From the relationship, determine the factor* by which the given quantity (0.0583 kg) must be multiplied to obtain the answer in the required unit (g).

$$\text{Factor} = \frac{1000 \text{ g}}{1 \text{ kg}}$$

3. Multiply 0.0583 kg by the factor to obtain the answer.

$$0.0583 \text{ kg} \times \frac{1000 \text{ g}}{1 \text{ kg}} = 58.3 \text{ g}$$

*The factor-unit (also called the "dimensional analysis" or "unit-conversion") method of problem solving is explained in more detail in Appendix C.

EXAMPLE B-2 Convert 72,600 grams to kilograms.

Solution:

1.
$$1 \text{ kg} = 1000 \text{ g}$$

2. The answer is required in kg. Therefore, the factor is:

$$\frac{1 \text{ kg}}{1000 \text{ g}}$$

(not 1000 g/1 kg as in Example 1, where the answer was required in g).

3.
$$72600 \text{ g} \times \frac{1 \text{ kg}}{1000 \text{ g}} = 72.6 \text{ kg}$$

EXAMPLE B-3 Change 4236 millimeters to kilometers.

Solution:

1. Table B-2 does not give a direct relationship between mm and km, but it does give relationships between mm and m, and between m and km:

$$1000 \text{ mm} = 1 \text{ m} \qquad 1000 \text{ m} = \text{km}$$

2. The given quantity (4236 mm) must be multiplied by *two* factors to obtain the answer in the required unit (km). In order for the proper terms to cancel out, the two factors must be:

$$\frac{1 \text{ m}}{1000 \text{ mm}} \quad \text{and} \quad \frac{1 \text{ km}}{1000 \text{ m}}$$

3.
$$4236 \text{ mm} \times \frac{1 \text{ m}}{1000 \text{ mm}} \times \frac{1 \text{ km}}{1000 \text{ m}}$$

$$= \frac{4236 \text{ km}}{1\,000\,000} = 0.004236 \text{ km}$$

Conversions from the SI to the English System, and Vice Versa

Units commonly used in the English system of measurement for length, mass, and volume are given in Table B-4.

TABLE B-4   Units of Measurement in the English System

Length	
12 inches (in.)	= 1 foot (ft)
3 feet	= 1 yard (yd)
1760 yards	= 1 mile (mi)
Mass	
16 ounces (oz)	= 1 pound (lb)
2000 pounds	= 1 ton
Volume	
16 fluid ounces (fl oz)	= 1 pint (pt)
2 pints	= 1 quart (qt)
4 quarts	= 1 gallon (gal)

Relationships that must be used to convert from SI units to English units, and vice versa, are given in Table B-5.

TABLE B-5   Conversion Factors:
Common SI and English Units

English Unit	SI Unit
Length	
1 inch (in.)	2.54 centimeters (cm)
1 yard (yd)	0.914 meters (m)
1 mile (mi)	1.61 kilometers (km)
Mass	
1 ounce (oz)	28.4 grams (g)
1 pound (lb)	454 grams (g)
1 pound (lb)	0.454 kilograms (kg)
Volume	
1 fluid ounce (fl oz)	29.6 milliliters (mL)
1 U.S. pint (pt)	0.473 liter (L)
1 U.S. quart (qt)	0.946 liter (L)
1 gallon (gal)	3.78 liters (L)

EXAMPLE B-4   Convert 25 inches to centimeters.

Solution:
1. Table B-5 gives the relationship between inches and centimeters.

$$1 \text{ in.} = 2.54 \text{ cm}$$

2. The answer is required in cm. Therefore, the factor is:

$$\frac{2.54 \text{ cm}}{1 \text{ in.}}$$

3.

$$25 \text{ in.} \times \frac{2.54 \text{ cm}}{1 \text{ in.}} = 63.5 \text{ cm}$$

EXAMPLE B-5   Convert 60 pounds to kilograms.

Solution:
1. From Table B-5:

$$1 \text{ lb} = 0.454 \text{ kg}$$

2. The answer is required in kg. Therefore, the factor is:

$$\frac{0.454 \text{ kg}}{1 \text{ lb}}$$

3.

$$60 \text{ lb} \times \frac{0.454 \text{ kg}}{1 \text{ lb}} = 27.24 \text{ kg}$$

EXAMPLE B-6   Convert 3.5 liters to quarts.

Solution:
1. From Table B-5:

$$1 \text{ qt} = 0.946 \text{ L}$$

2. The answer is required in qt. Therefore, the factor is:

$$\frac{1 \text{ qt}}{0.946 \text{ L}}$$

3.
$$3.5 \, \cancel{L} \times \frac{1 \text{ qt}}{0.946 \, \cancel{L}} = 3.7 \text{ qt}$$

EXAMPLE B-7   Convert 4.83 m to feet.

Solution:
1. From Table B-5:

$$1 \text{ yd} = 0.914 \text{ m} \quad \text{and} \quad 1 \text{ yd} = 3 \text{ ft}$$

2. The answer is required in ft. Therefore, the two factors needed are:

$$\frac{1 \text{ yd}}{0.914 \text{ m}} \quad \text{and} \quad \frac{3 \text{ ft}}{1 \text{ yd}}$$

3.
$$4.83 \, \cancel{m} \times \frac{1 \, \cancel{yd}}{0.914 \, \cancel{m}} \times \frac{3 \text{ ft}}{1 \, \cancel{yd}}$$

$$= 15.8 \text{ ft}$$

Useful Approximations between SI and English Units

To make rough estimates of English units in terms of SI units, and to get a feel for the meaning of SI units, it is useful to memorize the following approximations:

$$1 \text{ kg} \approx 2 \text{ lb} \qquad 1 \text{ m} \approx 1 \text{ yd} \qquad 1 \text{ L} \approx 1 \text{ qt} \qquad 1 \text{ km} \approx 2/3 \text{ mi}$$

OTHER COMMONLY USED UNITS OF MEASUREMENT

For certain measurements, including temperature and energy, scientists continue to use units that are not SI units.

Temperature

The SI unit for temperature is the kelvin (K) (Table B-1), but for many measurements scientists use the **Celsius scale**. On this scale the unit of temperature is the degree (°); 0°C corresponds to the freezing point of water, and 100°C corresponds to its boiling point at atmospheric pressure. The 100 degrees between the two reference points are of equal size.

On the Kelvin scale, zero temperature corresponds to the lowest temperature it is possible to attain, or **absolute zero**. Absolute zero, as determined theoretically and confirmed experimentally, is equal to −273°C (more accurately −273.15°C). The unit of temperature on the Kelvin scale is the same size as a °C. Therefore, to convert from °C to K it is only necessary to add 273.

$$K = °C + 273 \quad \text{or} \quad °C = K - 273$$

Thus, the boiling point of water is 373 K, and its freezing point is 273 K.

In the United States, most temperatures, including those given in weather reports and cooking recipes, are measured on the **Fahrenheit scale**. On this scale, the freezing point of water is 32°F, and its boiling point is 212°F. Thus, there are 180 (212 − 32) degrees between the two reference points, compared with 100 degrees on the Celsius scale.

Conversions between °C and °F

100 divisions in °C equals 180 divisions °F or, dividing both sides of the equation by 20:

$$5 \text{ divisions } °C = 9 \text{ divisions } °F$$

To convert from °C to °F, the following equation is used:

$$°F = \frac{9}{5}(°C) + 32$$

32 is added because the freezing point of water is 32° on the Fahrenheit scale compared with 0° on the Celsius scale.

To convert from °F to °C, the following equation is used:

$$°C = \frac{5}{9}(°F - 32)$$

In this case, 32 must be subtracted from the °F before multiplying by 5/9.

EXAMPLE B-8   Convert 77°F to °C.

$$°C = \frac{5}{9}(77 - 32)$$

$$= \frac{5}{9} \times 45$$

$$= 25$$

$$77°F = 25°C$$

EXAMPLE B-9   Convert 35°C to °F.

$$°F = \frac{9}{5}(35) + 32$$

$$= 63 + 32$$

$$= 95$$

$$37°C = 95°F$$

Energy

The SI unit for measuring heat, or any form of energy, is the joule (J) (Table B-3), but scientists often use the more familiar **calorie (cal)**. For measuring the energy content of food, the kilocalorie (kcal) or **Calorie (Cal)** is used.

A calorie is the amount of heat required to raise the temperature of 1 g of water 1°C.

$$1 \text{ cal} = 4.18 \text{ J}$$

The Difference between Mass and Weight

Although the terms *mass* and *weight* are frequently used interchangeably, they have basically different meanings.

Mass is a measure of the quantity of matter in an object.

Weight is a measure of the force exerted on an object by the pull of gravity.

The difference between mass and weight was dramatically demonstrated when astronauts began to travel in space. An astronaut's mass does not change as he or she is rocketed into space, but once free of the gravitational pull of the earth, the astronaut becomes weightless. On the moon, an astronaut's weight is approximately one-sixth that on earth because of the moon's much weaker gravitational pull.

PROBLEMS

1. Convert 825 mL to L.
2. Convert 153,000 mg to kg.
3. Convert 0.00061 cm to nm.
4. Convert 56,800 mg to oz.
5. Convert 1.7 m to feet and inches.
6. Convert 3 lb 8 oz to kg.
7. A road sign informs you that you are 60 km from Calais. How many minutes will it take you to arrive there if you travel at 60 miles per hour?

8. A doctor orders a one fluid ounce dose of medicine for a patient. How much is this in mL?
9. You need to purchase approximately 2 lb of meat at an Italian grocery store. Should you ask for 1/2, 1, or 2 kg?
10. Normal body temperature is 98.6°F. A patient has a temperature of 40°C. Is this above or below normal? By how many (a) °F and (b) °C?
11. The temperature is 15°C. Should you turn on the air conditioning or the heat?
12. A piece of pie contains 200 Cal. This is equivalent to how many kilojoules?

PROBLEM SOLVING USING DIMENSIONAL ANALYSIS

In solving mathematical problems, including many chemical problems, it is often necessary to convert from one unit of measurement to another: for example, from feet to inches, pounds to kilograms, or grams to moles.

Such problems can often be solved with the aid of the method variously called **dimensional analysis**, **unit conversion**, and the **factor–unit**, or **factor–label**, method.

This method uses the fact that *units* in the denominator and numerator of a fraction cancel in the same way that numbers cancel. For example, if the units in both numerator and denominator are inches, they cancel:

$$\frac{9 \text{ inches}}{5 \text{ inches}} = \frac{9}{5}$$

To convert from one unit to another, it is necessary to develop a **conversion factor** based on the known relationship between the units involved. This process is best explained by an example. Consider the conversion of 6 feet to inches. The relationship between feet and inches is:

$$1 \text{ ft } = 12 \text{ in.}$$

From this relationship we can develop *two* conversion factors, A and B. Since 1 foot equals 12 inches, it follows that:

$$\frac{1 \text{ ft}}{12 \text{ in.}} = 1 \qquad\qquad \text{factor A}$$

Similarly,

$$\frac{12 \text{ in.}}{1 \text{ ft}} = 1 \qquad\qquad \text{factor B}$$

The key to solving problems using dimensional analysis is choosing the correct conversion factor.

To convert 6 feet to inches, multiply by factor A or factor B. Choose the factor so that feet cancel out and the answer is expressed in inches. Use factor B:

$$6 \text{ ft} \times \frac{12 \text{ in.}}{1 \text{ ft}}$$

feet (ft) cancel

$$6 \text{ ft } = 72 \text{ in.}$$

Factor A would have given the wrong numerical answer and the wrong unit:

$$6 \text{ ft} \times \frac{1 \text{ ft}}{12 \text{ in.}} = 0.5 \text{ feet }^2/\text{in}$$

In general terms, the formula for converting from one unit to another is:

(given quantity) × (conversion factor) = (answer in desired unit)

The conversion factor is chosen so that when multiplied by the given unit it gives the desired unit:

$$\cancel{\text{given unit}} \times \frac{\text{desired unit}}{\cancel{\text{given unit}}} = \text{desired unit}$$

EXAMPLE C-1   Convert 100 pounds to kilograms (1 lb = 0.454 kg).

Solution:
1. Given quantity: 100 lb
2. Desired unit: kg
3. Conversion factor: $\dfrac{0.454 \text{ kg}}{1 \text{ lb}}$

4.
$$100 \cancel{\text{ lb}} \times \frac{0.454 \text{ kg}}{1 \cancel{\text{ lb}}} = 45.4 \text{ kg}$$

EXAMPLE C-2   The distance between two cities in Mexico is 80 kilometers. This distance is equal to how many miles? (1 mi = 1.61 km)

Solution:
1. Given quantity: 80 km
2. Desired unit: mi
3. Conversion factor: $\dfrac{1 \text{ mi}}{1.61 \text{ km}}$

4.
$$80 \cancel{\text{ km}} \times \frac{1 \text{ mi}}{1.61 \cancel{\text{ km}}} = 50 \text{ mi}$$

EXAMPLE C-3   Convert 7.0 liters to fluid ounces (1 qt = 0.946 L).

Solution:
1. Given quantity: 7.0 L
2. Desired unit: fl oz
3. Two conversion factors are required: one to convert liters to quarts, and another to convert quarts to fluid ounces (1 qt = 32 fl oz).

4.
$$7.0 \cancel{\text{ L}} \times \frac{1 \cancel{\text{ qt}}}{0.946 \cancel{\text{ L}}} \times \frac{32 \text{ fl oz}}{1 \cancel{\text{ qt}}} = 240 \text{ fl oz}$$

Conversion Factors Based on Equivalence

In the above examples, the numerator of the conversion factor was always *equal* to the denominator. For example, in the factor 12 in./1 ft, 12 inches equal 1 foot; in the factor 1 mi/1.61 km, 1 mile equals 1.61 kilometers.

Conversion factors based on units that are *equivalent* to each other, but not equal, can also be used to solve problems, as shown in Example C-4.

EXAMPLE C-4   A worker is paid at the rate of $10.50 per hour. How many hours will he have to work to earn $84?

Solution:
1. Given quantity: 84 dollars
2. Desired unit: hours
3. Conversion factor: There is an equivalence between time and dollars:

$$1 \text{ hour } \sim \$10$$

The sign, $\approx$, is used to represent "is equivalent to.") The conversion factor is:

4. $$\frac{1 \text{ hour}}{10.50 \text{ dollars}}$$

$$84 \text{ dollars} \times \frac{1 \text{ hour}}{10.50 \text{ dollars}} = 8 \text{ hours}$$

The worker will have to work 8 hours to earn $84.

EXAMPLE C-5   A solution contains 0.040 g of salt per 100 mL. How many grams of salt are contained in 250 mL?

Solution:
1. Given quantity: 250 mL
2. Required quantity: g of salt
3. Conversion factor: In any problem involving concentration (g/L, mg/mL, etc.), the conversion factor is based on the relationship between the mass and the volume. Depending on the quantity that has to be determined, it is either mass/volume or volume/mass.
   In this problem, the conversion factor is $\dfrac{0.040 \text{ g}}{100 \text{ mL}}$

4. $$250 \text{ mL} \times \frac{0.040 \text{ g}}{100 \text{ mL}} = 0.10 \text{ g}$$

0.10 g of salt are contained in 250 mL.

EXAMPLE C-6   When heated, 434 g of mercuric oxide (HgO) yield 32 g of oxygen (O_2). How many grams of HgO are needed to produce 100 g O_2?

Solution:
1. Given quantity: 100 g O_2
2. Required quantity: g HgO
3. To solve this problem we need a conversion factor based on the relationship between two substances (HgO and O_2) measured in the same unit (g).
   Conversion factor: $$\frac{434 \text{ g HgO}}{32 \text{ g O}_2}$$

4. $$100 \text{ g O}_2 \times \frac{434 \text{ g HgO}}{32 \text{ g O}_2} = 1360 \text{ g HgO}$$

1360 g HgO are needed to produce 100 g of O_2.

PROBLEMS

(Conversion tables and more problems are given in Appendix B.)

1. Convert 754.0 centimeters to meters.
2. Convert 0.036 kilograms to milligrams.
3. Convert 6 feet 3 inches to meters.
4. Convert 3 pints to milliliters.
5. It takes a secretary half an hour to type four pages. How long will he take to type 60 pages?
6. The permitted level of lead in drinking water is 50 ppb (50 μg/L). A 12-oz glass of water contains 30 μg of lead. Is this above or below the permitted level? By how much in μg/L?
7. A solution contains 4.0 g of sodium hydroxide per liter. How many grams will be contained in 50 mL?
8. The longest and shortest waves of visible light have wavelengths of 0.000067 cm and 0.000037 cm. Convert these values to nanometers.
9. An athlete runs 100 yards in 10 seconds. What is his speed in miles per hour?
10. A solution contains 6.4 g of barium nitrate per liter. How many milliliters of this solution are needed in order to obtain 1.6 g of barium nitrate?
11. The reaction of 87.0 g of manganese (IV) oxide (MnO_2) with excess hydrochloric acid yields 71.0 g of chlorine (Cl_2). How many grams of MnO_2 are required to obtain 56.8 g of Cl_2?
12. When 114 g of octane (C_8H_{18}), an ingredient of gasoline, are burned, 44 grams of carbon dioxide (CO_2) are produced. Calculate the number of
    a. g,   **b.** kg, and   **c.** tons of CO_2 produced when 20 kg of octane are burned.

SIGNIFICANT FIGURES

A measurement is never exact. Depending on the type of instrument used, there is a limit to the accuracy with which any measurement can be made. For example, if the length of the piece of wood shown in Figure D-1 is measured with the ruler that is shown, it is not possible to determine with certainty whether the length is 3.23, 3.24, or 3.25 cm. There is no uncertainty about the first two digits (3 and 2), but there is a degree of uncertainty about the last digit.

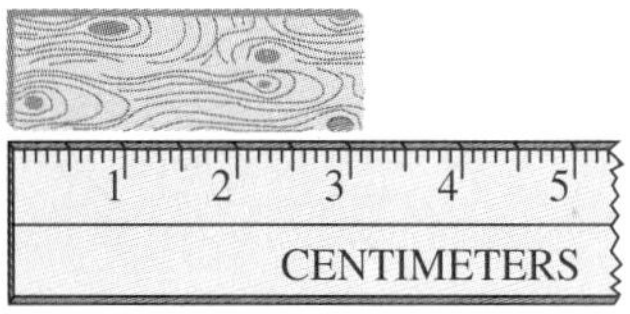

Figure D-1.

The measurements 3.23 cm, 3.24 cm, and 3.25 cm are said to have three significant figures: the first two, which are known with certainty, and the last one, which is an estimate.

The **significant figures** in a number can be defined as *the numbers that express reasonably reliable information*. They include all the numbers known with certainty plus one that is uncertain.

Rules for Determining Significant Figures

The following rules are used to determine how many significant figures there are in a number. It is assumed that the number represents a measurement that has been accurately made and recorded.

1. All nonzero digits (1, 2, 3, 4, 5, 6, 7, 8, and 9) are significant.

    478 has 3 significant figures.

    523.61 has 5 significant figures.

2. Zeros between nonzero digits are significant.

    1.063 has 4 significant figures.

    5002.08 has 6 significant figures.

3. Zeros to the *left* of nonzero digits are not significant. They serve only to indicate the position of the decimal point.

    0.00205 has 3 significant figures.

    0.09 has 1 significant figure.

4. Zeros *following* other digits in a number that contains a decimal point are significant.

    0.5300 has 4 significant figures.

    0.070 has 2 significant figures.

    58.00 has 4 significant figures.

5. Zeros at the end of a number with *no* decimal point may or may not be significant. A number such as 9,400 can usually be assumed to have 2 significant figures unless otherwise indicated. The uncertainty can be removed by expressing the number in exponential notation (Appendix A).

$$9.4 \times 10^3 \text{ denotes 2 significant figures.}$$

$$9.40 \times 10^3 \text{ denotes 3 significant figures.}$$

$$9.400 \times 10^3 \text{ denotes 4 significant figures.}$$

Accuracy and Precision

In science it is important to make measurements that are both accurate and precise. **Accuracy** refers to the extent to which a measurement agrees with the true value of the quantity being measured. **Precision** refers to the closeness of agreement between successive measurements of a quantity. The difference between these two terms is best explained by an example.

Three students each make three measurements of the concentration of sodium chloride in a solution that is known to contain 2.79 g/L, with the following results.

Measurement	Student 1	Student 2	Student 3
1	2.41 g/L	2.24 g/L	2.80 g/L
2	2.43 g/L	3.15 g/L	2.79 g/L
3	2.42 g/L	2.98 g/L	2.78 g/L
Average	2.42 g/L	2.79 g/L	2.79 g/L

Student 1's measurements are very precise—they vary from each other by no more than 0.01 g—but they are not accurate because they differ from the true value (2.79 g/L) by as much as 0.38 g/L. Student 2's measurements are very imprecise, although their average gives an accurate value. Student 3's measurements are both accurate and precise.

Rounding Off Numbers

When measured quantities are added, subtracted, multiplied, or divided, the results must be reported with the correct number of significant figures. This means that any nonsignificant numbers must be **rounded off**. The rules for rounding off are as follows.

1. If the first nonsignificant figure is *5 or more*, increase the last significant figure by one and drop the nonsignificant figures.
Example: Round the number 24.381 to 3 significant figures. The first nonsignificant figure, 8, is greater than 5. The last significant figure, 3, is therefore increased by 1.

$$24.381 = 24.4 \text{ (to 3 significant figures)}$$

Example: Round the number 1273 to 2 significant figures.

$$1273 = 1300 \text{ (to 2 significant figures)}$$

Note that in this case the dropped nonsignificant figures must be replaced by zeros to maintain the correct dimensions.

2. If the first nonsignificant figure is *less than* 5, drop all the nonsignificant figures and do not change the last significant figure.
   Example: Round the number 1.6392 to 2 significant figures.
   The first nonsignificant figure, 3, is less than 5.

$$1.6392 \ = \ 1.6 \text{ (to 2 significant figures)}$$

   Example: Round the number 855,216 to 3 significant figures.

$$855,216 \ = \ 855,000 \text{ (to 3 significant figures)}$$

Addition and Subtraction

In addition and subtraction, the result should be rounded off so that it has only as many digits after the decimal point as the number with the *fewest* number of digits after the decimal point. In these operations, the number of significant figures in the numbers being added or subtracted is irrelevant.
   Example: Add the following numbers: 49.8426, 6.07, 103.5.
The number with the fewest number of digits after the decimal point is 103.5. Therefore, the sum should have only one digit after the decimal point.

$$\begin{array}{r} 49.8426 \\ 6.07 \\ \underline{103.5} \\ 159.4126 \end{array}$$

Answer: = 159.4

Notice that the correct answer (159.4) has 4 significant figures even though one of the numbers (6.07) has only 3 significant figures.
   Example: Add the following numbers: 737, 6.5, 15.316.
The number 737 has no digits after the decimal point (which is understood to be at the right of the last digit). Therefore, the sum should be rounded off so that it has no digits after the decimal point.

$$\begin{array}{r} 737 \\ 6.5 \\ \underline{15.316} \\ 758.816 \end{array}$$

Answer: = 759

   Example: Subtract 1.526 from 5.2308.

$$\begin{array}{r} 5.2308 \\ \underline{-1.526} \\ 3.7048 \end{array}$$

Answer: = 3.705

Multiplication and Division

In multiplication and division, the result should be rounded off to the same number of significant figures as the number with the *least* number of significant figures. In these operations, the number of digits after the decimal point in any of the numbers is not taken into consideration.

Example: Multiply 3.9 by 1.483.

$$3.9 \times 1.483 = 5.7837$$

$$\text{Answer: } = 5.8$$

Example: Divide 27.832 by 1.51.

$$\frac{27.832}{1.51} = 18.431788$$

$$\text{Answer: } = 18.4$$

PROBLEMS

Perform the indicated mathematical operations and give the answers with the proper number of significant figures.

1. 46.105 g + 27.3 g + 208.17 g
2. 0.143 + 123 + 1.51
3. 93.007 mL − 0.024 mL
4. 254.710 mg − 2.71 mg
5. 63.8 L + 5.42 L − 0.095 L
6. 32 m × 53.7 m × 0.41 m

7. 0.293 × 0.028

8. $\dfrac{2.35 \text{ g}}{4600 \text{ g}}$

9. $\dfrac{283.9 \text{ g}}{51.0 \text{ mol}}$

10. 1.236 L × $\dfrac{373 \text{ K}}{263 \text{ K}}$

SOLVING PROBLEMS BASED ON CHEMICAL EQUATIONS

Many problems based on mass and mole relationships in chemical equations can be solved very easily using dimensional analysis (Appendix C).

As we saw in Chapter 7 (Chemical Reactions), a properly balanced chemical equation provides a great deal of information. For example, from the equation for the reaction of hydrogen with oxygen to produce water, we can derive the following information:

$2\ H_2$	+	O_2	$\longrightarrow$	$2\ H_2O$
2 molecules H_2	+	1 molecule O_2	$\longrightarrow$	2 molecules H_2O
2 mol H_2	+	1 mol O_2	$\longrightarrow$	2 mol H_2O
2 formula masses H_2	+	1 formula mass O_2	$\longrightarrow$	2 formula masses H_2O
2×2 amu H_2	+	32 amu O_2	$\longrightarrow$	2×18 amu H_2O
2 molar masses H_2	+	1 molar mass O_2	$\longrightarrow$	2 molar masses H_2O
2×2 g H_2	+	32 g O_2	$\longrightarrow$	2×18 g H_2O
4 g H_2	+	32 g O_2	$\longrightarrow$	36 g H_2O
4 lb H_2	+	32 lb O_2	$\longrightarrow$	36 lb H_2O

Thus, a chemical equation shows the relationships between the various reactants and products taking part in the reaction in terms of molecules, moles, and mass (amu, g, lb, or any other unit of mass). These relationships provide the conversion factors needed for solving problems using dimensional analysis. They enable us, for example, to determine the number of moles or grams of a given reactant that is required to produce a certain number of moles or grams of product.

Mass-to-Mass Calculations

As we have just seen, a balanced chemical equation tells us the **mass relationships** between the substances taking part in a reaction. Knowing these relationships, we can calculate, for example, the number of kilograms of product that can be obtained from a given number of kilograms of a reactant or the number of pounds of reactant needed to produce a required number of pounds of product.

EXAMPLE E-1   How many grams of water can be obtained from the complete reaction of 10 g of hydrogen with oxygen?

Solution:
Step 1. Write the balanced equation for the reaction.

$$2\,H_2 + O_2 \longrightarrow 2\,H_2O$$

Step 2. From the equation, determine the mass relationship between the *given substance* and the *required substance*.
a. Determine the molar masses of the given substance (H_2) and the required substance (H_2O). [Remember that the molar mass of a substance is the mass in grams that is numerically equal to the substance's formula mass. The atomic masses (atomic weights) that are needed to calculate formula masses can be found inside the cover of any chemistry textbook.]

$$\text{Molar mass } H_2 = 2 \times 1 = 2\text{ g}$$

$$\text{Molar mass } H_2O = (2 \times 1) + 16 = 18\text{ g}$$

(The formula mass of O_2 is not required since the question does not ask for the mass of oxygen.)
b. Substitute these values in the equation.

$$2 \times \text{molar mass } H_2 + 1 \times \text{molar mass } O_2 \longrightarrow 2 \times \text{molar mass } H_2O$$

$$2 \times 2\text{ g } H_2 \qquad\qquad\qquad 2 \times 18\text{ g } H_2O$$

$$4\text{ g } H_2 \qquad\qquad\qquad 36\text{ g } H_2O$$

Thus, 4 g H_2 will react with O_2 to yield 36 g H_2O.
c. State the relationship between the given substance and the required substance.

$$4\text{ g } H_2 \text{ are } \textit{equivalent} \text{ to } 36\text{ g } H_2O$$

$$4\text{ g } H_2 \sim 36\text{ g } H_2O$$

Express the relationship as a ratio.

$$\frac{4\text{ g } H_2}{36\text{ g } H_2O} \qquad \text{or} \qquad \frac{36\text{ g } H_2O}{4\text{ g } H_2}$$

Step 3. Solve the problem by dimensional analysis using the following formula.

$$(\text{given quantity}) \times (\text{conversion factor}) = (\text{required quantity})$$

Given quantity: 10 g H_2
Required quantity: g H_2O
Conversion factor: The correct conversion factor is the ratio shown in **c.** that when multiplied by the given quantity will give the required quantity:

$$\frac{36\text{ g } H_2O}{4\text{ g } H_2}$$

$$10\text{ g }H_2 \times \frac{36\text{ g } H_2O}{4\text{ g }H_2} = 90\text{ g } H_2O$$

90 g of water can be obtained from the complete reaction of 10 g hydrogen with oxygen.

EXAMPLE E-2   How many kilograms of iron are needed to prepare 116 kg of the magnetic oxide of iron, Fe_3O_4?

Solution:

Step 1. Balanced equation:

$$3\ Fe\ +\ 2\ O_2\ \longrightarrow\ Fe_3O_4$$

Step 2. From the equation, determine the mass relationship between Fe and Fe_3O_4.

a.   Molar mass Fe = 56 g

Molar mass Fe_3O_4 = $(3 \times 56) + (4 \times 16) = 168 + 64$

= 232 g

b. Substitute these values in the equation.

$3 \times$ molar mass Fe + 2 $\times$ molar mass O_2 $\longrightarrow$ 1 $\times$ molar mass Fe_3O_4

3×56 g   232 g

168 g   232 g

c. State the relationship between the given substance and the required substance.

$$168\ g\ Fe\ \sim\ 232\ g\ Fe_3O_4$$

The answer is required in kg. Convert g to kg.

$$168\ kg\ Fe\ \sim\ 232\ kg\ Fe_3O_4$$

Express the relationship as a ratio.

$$\frac{168\ kg\ Fe}{232\ kg\ Fe_3O_4} \quad \text{or} \quad \frac{232\ kg\ Fe_3O_4}{168\ kg\ Fe}$$

Step 3. Solve by dimensional analysis.

Given quantity: 116 kg Fe_3O_4

Required quantity: kg Fe

Conversion factor:

$$\frac{168\ kg\ Fe}{232\ kg\ Fe_3O_4}$$

$$116\ \cancel{kg\ Fe_3O_4} \times \frac{168\ kg\ Fe}{232\ \cancel{kg\ Fe_3O_4}} = 84\ kg\ Fe$$

84 kg of iron are needed to prepare 116 kg of Fe_3O_4.

EXAMPLE E-3   How many grams of hydrochloric acid are required to convert 80 grams of sodium hydroxide to salt and water?

Solution:

Step 1. Balanced equation:

$$HCl\ +\ NaOH\ \longrightarrow\ NaCl\ +\ H_2O$$

Step 2. From the equation, determine the mass relationship between HCl and NaOH.

Molar mass HCl = 36.5 g

Molar mass NaOH = 40 g

Substitute these values in the equation.

1 × molar mass HCl	+	1 × molar mass NaOH	=	1 molar mass NaCl	+	1 molar mass H_2O

36.5 g HCl   40 g NaOH

36.5 g HCl   40 g NaOH

Express the relationship as a ratio. $\dfrac{36.5 \text{ g HCl}}{40 \text{ g NaOH}}$ or $\dfrac{40 \text{ g NaOH}}{36.5 \text{ g HCl}}$

Step 3. Solve by dimensional analysis.

Given quantity: 80 g NaOH
Required quantity: g HCl
Conversion factor:

$$\dfrac{36.5 \text{ g HCl}}{40 \text{ g NaOH}}$$

$$80 \text{ g NaOH} \times \dfrac{36.5 \text{ g HCl}}{40 \text{ g NaOH}} = 73 \text{ g HCl}$$

73 g HCl are required to convert 80 g NaOH to salt and water.

Mole-to-Mole Calculations

As we saw at the beginning of this appendix, when describing the reaction of hydrogen with oxygen to produce water, a balanced chemical equation shows the **mole relationships** between reactants and products. Knowing these mole relationships, we can easily calculate, for example, the number of moles of product that can be obtained from a given number of moles of a reactant or the number of moles of a reactant that are needed to produce a given number of moles of product.

We will go through the logical steps to be followed in solving this type of problem, although, as the following examples show, most mole-to-mole problems can be solved by inspection.

EXAMPLE E-4   How many moles of water are produced by the complete combustion of 360 moles of methane (CH_4)?

Solution:
Step 1. Balanced equation:

$$CH_4 + 2\,O_2 \longrightarrow CO_2 + 2\,H_2O$$

Step 2. Determine the mole relationship between the given substance (CH_4) and the required substance (H_2O).
a. Enter the number of moles of the substances taking part in the reaction into the equation.

$$1 \text{ mol } CH_4 + 2 \text{ mol } O_2 \qquad 1 \text{ mol } CO_2 + 2 \text{ mol } H_2O$$

b. State the mole relationship between the given substance and the required substance.

$$1 \text{ mol CH}_4 \sim 2 \text{ mol H}_2\text{O}$$

Express the relationship as a ratio.

$$\frac{1 \text{ mol CH}_4}{2 \text{ mol H}_2\text{O}} \quad \text{or} \quad \frac{2 \text{ mol H}_2\text{O}}{1 \text{ mol CH}_4}$$

Step 3. Solve the problem using dimensional analysis.

Given quantity: 360 mol CH_4

Required quantity: mol H_2O

Conversion factor:

$$\frac{2 \text{ mol H}_2\text{O}}{1 \text{ mol CH}_4}$$

$$360 \text{ mol CH}_4 \times \frac{2 \text{ mol H}_2\text{O}}{1 \text{ mol CH}_4} = 720 \text{ mol H}_2\text{O}$$

720 moles of water are produced by the complete combustion of 360 moles of methane.

EXAMPLE E-5   How many moles of hydrogen are needed to react completely with nitrogen to produce 6 moles of ammonia (NH_3)?

Solution:

Step 1. Balanced equation:

$$N_2 + 3 H_2 \longrightarrow 2 NH_3$$

Step 2. Mole relationship between H_2 and NH_3:

$$1 \text{ mol N}_2 + 3 \text{ mol H}_2 \longrightarrow 2 \text{ mol NH}_3$$

$$3 \text{ mol H}_2 \sim 2 \text{ mol NH}_3$$

Step 3. Solve the problem by dimensional analysis.

Given quantity: 6 mol NH_3

Required quantity: mol H_2

Conversion factor:

$$\frac{3 \text{ mol H}_2}{2 \text{ mol NH}_3}$$

$$6 \text{ mol NH}_3 \times \frac{3 \text{ mol H}_2}{2 \text{ mol NH}_3} = 9 \text{ mol H}_2$$

9 moles of hydrogen are needed to react completely with nitrogen to produce 6 moles of ammonia (NH_3).

Mole-to-Mass and Mass-to-Mole Calculations

A balanced chemical equation gives both mass and mole relationships between the substances taking part in the reaction. This information makes it possible to calculate, for example, the number of *grams* of a reactant that will produce a given number of *moles* of product, or the number of *grams* (or kg, or lb) of product that can be obtained from a given number of *moles* of reactant.

EXAMPLE E-6   How many moles of O_2 are produced by heating 49 grams of potassium chlorate ($KClO_3$)?

Solution:
Step 1. Balanced equation:

$$2\ KClO_3 \longrightarrow 2\ KCl + 3\ O_2$$

Step 2. Determine the relationship between the required quantity, *moles* of O_2, and the given quantity, *grams* of $KClO_3$.
a. Enter the number of moles of O_2 and $KClO_3$ taking part in the reaction into the equation.

$$2\ \text{mol}\ KClO_3 \qquad\qquad \textbf{3 mol } \mathbf{O_2}$$

b. Express $KClO_3$ in grams.

$$1\ \text{mol}\ KClO_3 = \text{formula mass}\ KClO_3\ \text{in g} = 122.5\ \text{g}$$

$$2\ \text{mol}\ KClO_3 = 2 \times 122.5\ \text{g}\ KIO_3$$

$$= \textbf{245 g } \mathbf{KClO_3}$$

c. From **a.** and **b.** we can see that the relationship between the required quantity (mol O_2) and the given quantity (g $KClO_3$) is:

$$3\ \text{mol}\ O_2 \sim 245\ \text{g}\ KClO_3$$

Step 3. Solve the problem by dimensional analysis.
Given quantity: 49 g $KClO_3$
Required quantity: mol O_2
Conversion factor:

$$\frac{3\ \text{mol}\ O_2}{245\ \text{g}\ KClO_3}$$

$$49\ \cancel{\text{g}\ KClO_3} \times \frac{3\ \text{mol}\ O_2}{245\ \cancel{\text{g}\ KClO_3}} = 0.60\ \text{mol}\ O_2$$

0.60 moles of O_2 are produced by heating 49 grams of $KClO_3$.

EXAMPLE E-7   In photosynthesis, CO_2 combines with water to produce glucose and oxygen. How many grams of glucose ($C_6H_{12}O_6$) can be produced by the reaction of 10 moles of CO_2 with water?

Solution:
Step 1. Balanced equation:

$$6\ CO_2 + 6\ H_2O \longrightarrow C_6H_{12}O_6 + 6\ O_2$$

Step 2. Determine the relationship between moles and grams.
a. Mole relationships:

$$6\ \text{mol}\ CO_2 \qquad\qquad 1\ \text{mol}\ C_6H_{12}O_6$$

b. Convert mol glucose to g glucose.

$$1\ \text{mol}\ C_6H_{12}O_6 = \text{formula mass}\ C_6H_{12}O_6\ \text{in g}$$

$$= 180\ \text{g}$$

c. Relationship between mol CO_2 and g $C_6H_{12}O_6$.

$$6 \text{ mol } CO_2 \sim 180 \text{ g } C_6H_{12}O_6$$

Step 3. Solve the problem by dimensional analysis.

Given quantity:	10 mol CO_2
Required quantity:	g $C_6H_{12}O_6$
Conversion factor:	$\dfrac{180 \text{ g } C_6H_{12}O_6}{6 \text{ mol } CO_2}$

$$10 \text{ mol } CO_2 \times \frac{180 \text{ g } C_6H_{12}O_6}{6 \text{ mol } CO_2} = 300 \text{ g } C_6H_{12}O_6$$

300 g of glucose can be produced by the reaction of 10 moles of CO_2 with water.

PROBLEMS

Mass-to-Mass

1. How many tons of KOH are needed to completely react with chlorine to yield 49 tons of potassium chlorate ($KClO_3$)?

$$3 \text{ Cl}_2 + 6 \text{ KOH} \longrightarrow 5 \text{ KCl} + \text{KClO}_3 + 3 \text{ H}_2\text{O}$$

2. Calculate the number of kilograms of hydrogen required to reduce 89 pounds of lead oxide (PbO) to lead (454 g = 1 lb).

$$\text{PbO} + \text{H}_2 \longrightarrow \text{Pb} + \text{H}_2\text{O}$$

3. How many milligrams of $BaSO_4$ are produced by the reaction of an excess of Na_2SO_4 with 6.24 g of $Ba(NO_3)_2$?

$$\text{Na}_2\text{SO}_4 + \text{Ba(NO}_3)_2 \longrightarrow \text{BaSO}_4 + 2 \text{ NaNO}_3$$

4. $Mg(OH)_2$, the active ingredient in many common antacids, reacts with stomach acid (HCl) to produce $MgCl_2$ and water. How many grams of $Mg(OH)_2$ are necessary to react with 1.0 g HCl?

$$\text{Mg(OH)}_2 + 2 \text{ HCl} \longrightarrow \text{MgCl}_2 + 2 \text{ H}_2\text{O}$$

Mole-to-Mole

5. How many moles of O_2 are needed to convert 12 moles of ethane (C_2H_6) to CO_2 and water?

$$2 \text{ C}_2\text{H}_6 + 7 \text{ O}_2 \longrightarrow 4 \text{ CO}_2 + 6 \text{ H}_2\text{O}$$

6. Calculate the number of moles of hydrogen produced when 15.9 moles of iron react with water as shown in the following equation.

$$4 \text{ H}_2\text{O} + 3 \text{ Fe} \longrightarrow 4 \text{ H}_2 + \text{Fe}_3\text{O}_4$$

Mole-to-Mass/Mass-to-Mole

7. The catalytic converter, which is mandatory equipment in all automobiles in the United States, converts CO to CO_2 according to the following reaction:

$$2 \text{ CO} + \text{O}_2 \longrightarrow 2 \text{ CO}_2$$

How many pounds of CO_2 are produced from 25 moles of CO?

8. How many moles of O_2 are needed to convert 320 grams of SO_2 to SO_3?

$$2 \text{ SO}_2 + \text{O}_2 \longrightarrow 2 \text{ SO}_3$$

9. How many kilograms of nitric acid can be obtained from the reaction of 40.0 moles of ammonia with oxygen?

$$\text{NH}_3 + \text{O}_2 \longrightarrow \text{HNO}_3 + \text{H}_2$$

10. How many moles of water must be reacted with calcium to produce 296 g of slaked lime, $Ca(OH)_2$?

$$\text{Ca} + 2 \text{ H}_2\text{O} \longrightarrow \text{Ca(OH)}_2 + \text{H}_2$$

ABC VIDEO CASE STUDIES

Each of the following 20 ABC Video Case Studies will bring you face-to-face with a significant environmental issue somewhere in the world, if not in your own backyard. Through the cooperation and courtesy of Prentice Hall and ABC News, the authors of this text have selected video segments from such award-winning ABC news programs as **Nightline, 20/20,** and **World News Tonight.** For each video segment, we have included an overview or abstract, along with several questions that have been designed to encourage you to focus on the issues and controversies within each case study. Each written case study has a chapter reference that serves to link the case study to the text. We hope that each of these case studies will encourage class discussion and debate.

Prentice Hall is making all of these ABC video segments available to your instructor for classroom use on two convenient long-playing video cassettes. We encourage you to explore these cases as you use this text.

ABC CASE STUDY 1

The Sun: The 1991 Solar Eclipse

The relationships between the Earth and the sun have been well documented by geographers, meteorologists, climatologists, and astronomers. One of the most interesting of these relationships is that of the solar eclipse. Solar eclipses are embedded in folklore, considered as omens both positive and negative, but also are very important from a scientific standpoint.

The sun plays an important role in the earth system, such as aiding in the process of photosynthesis and creating wind, rain, and everything else we call the weather. Although extremely rare, solar flares and eclipses have important impacts on the earth and on our day-to-day activities. Flares and eclipses have an unusual effect on the magnetic field that surrounds our planet. This has often led to peculiar distortions in global communications, radio and television reception, navigational equipment, and even electronic garage door openers!

The corresponding video segment demonstrates the fundamentals of solar eclipses, focusing on the solar eclipse that occurred on July 11, 1991. This eclipse lasted 6.5 minutes. Although solar eclipses occur approximately every 11 years, there will not be an eclipse lasting this amount of time for another 141 years.

Buell, Girard reference: Chapter 2. Video title: "The Sun and Today's Solar Eclipse," 20 minutes.

QUESTIONS

1. What are the most serious effects of solar phenomena on our planet? What, if anything, might we do to prevent these?

2. What would happen during a prolonged eclipse of 20 minutes? An hour? A month?

3. What other natural phenomena affect the earth's relationship with the sun?

ABC CASE STUDY 2

Satellites Play an Environmental Role

We use a variety of means to look at the world around us. Satellite imagery is playing an increasingly important role and offers a uniquely powerful view of our planet. They allow us to see more of the planet, and their imagery provides a greater perspective on the earth as a system. They have had immeasurable impact on our understanding of the atmosphere—their benefits to other scientists are multiplying as we become more aware of global processes.

The growing concerns over our environment are distinctly global, and as such, satellites are playing a role in helping understand (and ultimately solve) these problems.

As you watch the corresponding video, you will see how satellite images are used to investigate problems on a global scale and why we should respect satellite technology's ability to monitor the earth. We are all accustomed to seeing satellite images of weather patterns in evening newscasts, and have witnessed meteorologists' use of satellites to track intense weather phenomena like tornadoes and hurricanes. Perhaps one day soon, we'll be just as familiar with satellite imagery concerning ozone depletion, deforestation patterns, and the long range effects of pollution!

Buell, Girard reference: Chapters 2 and 13. Video title: "Lack of Satellites to Observe Earth's Pollution," 5 minutes.

QUESTIONS

1. Besides providing imagery, what are satellite's roles in contemporary North American society? Which of these is more important?

2. Who bears the costs of getting satellites into orbit, and maintaining them once they are there?

3. What other methods do scientists use to collect data about the global environment? Can you think of any other ways satellites might be used?

ABC CASE STUDY 3

Volcanic Eruptions and Climate Change

The massive eruption of Mt. Pinatubo in the Philippines in 1991 cooled Earth's climate for a period of two to four years. The cooling was caused by a fine mist of sulfuric acid droplets that permeated the upper atmosphere. The eruption was the largest since the eruption of Krakatoa in Indonesia in 1883. The largest eruption recorded occurred in 1815 at Tambora, also in Indonesia, which caused pronounced global cooling and the "Winter without Summer" in 1816. Pinatubo put about 25 million tons of sulfur into the upper atmosphere,

Krakatoa 85 million tons and Tambora 300 million tons.

The droplets of sulfuric acid in the upper atmosphere from the volcanic eruptions dim the amount of sunlight reaching Earth's surface by about 2% since they reflect some of the incoming sunlight back into space.

Until Pinatubo's eruption, Earth's surface temperatures have been rising. The 1980s were the warmest decade ever.

The effects of a volcanic eruption are not only cooling. The eruption is also thought to have caused unusually warm winters in North America and Europe. In fact the unusually warm winter of 1991–1992 may have helped prevent civil disorder in Russia after the collapse of the former Soviet Union and resulting disruptions in food and fuel supplies.

Volcanoes are major disturbers of climate. The 1815 eruption of Tambora resulted in the starvation of 90,000 persons after crop failures in the summer of 1816. Snow fell on July 4th in New England that year.

Buell, Girard reference: Chapters 2 and 13. Video title: "Volcano Eruptions Affecting Weather, Ozone Layer," 21 minutes.

QUESTIONS

1. Besides sulfur compounds, what other materials are released in a volcanic eruption?

2. What geological feature is present at a volcanic site?

ABC CASE STUDY 4
Zebra Mussels and the Great Lakes

The natural environment is dynamic. Despite the more dramatic changes we observe in landslides, earthquakes, and severe weather, most changes occur at such a slow pace that it takes thousands (if not millions) of years to see their effects on the landscape. When it comes to the natural distribution of organisms, the effects of human beings, while not as immediately clear as landslides, can often be quite serious— even if they are not done purposefully.

The video showcases an ecological problem occurring in the Great Lakes where the apparently harmless, quarter-sized Zebra Mussel is multiplying out of control. Scientists believe the mussels were accidentally carried to the region within the past 5 years by transport ships from Europe, the Zebra Mussel's native environment. The mussels could have unknowingly been drawn into the ballast of these ships in Europe and then discharged in the Great Lakes as loads were taken off, and ballast was adjusted.

Arriving in a habitat free from predators, and with a bountiful food supply, the mussels have multiplied far faster than ever imagined. Their waterborne young have been transported several hundred miles away, and the problem is growing throughout much of the Great Lakes region. Like the flea and brown rat, the Zebra Mussel has become the latest European stowaway to make an important impact on our continent.

Buell, Girard reference: Chapter 3. Video title: "Abundance of Zebra Mussels Causes Concern," 5 minutes.

QUESTIONS

1. How do you think this problem has affected the local economy in these areas? The economy in other places?

2. Considering the risks to both environment and the economy, what do you think should be done in this situation? Who will pay for the needed action?

ABC CASE STUDY 5
Yellowstone after the Fire

Yellowstone National Park, an extraordinarily beautiful and popular retreat, was devastated by forest fires in 1988. Because they occur naturally from lightning strikes and aid in the forest regeneration process, forest fires present park officials serious questions. When park officials decided to let the fires burn, they did so believing the forest would benefit from the fire, and that the fires would eventually burn themselves out. This did not happen, and controversy soon ensued over the lack of response by Park Service employees. Under public pressure, the Park Service finally intervened and put the fires out.

Naturally occurring fires clean out the old brush that accumulates within a forest, and will produce a greater variety of plants and better animal habitats than those which were in existence before the fires began. In the corresponding video, you will see the effects of naturally occurring forest fires in Yellowstone National Park, and the regeneration that has taken place since. The initial controversy about the Park Service's decision to "let the fires burn," is now illuminated by the fires' effects on the park some years later.

Buell, Girard reference: Chapters 3 and 8. Video title: "Yellowstone Making A Comeback After 1988 Fire," 5 minutes.

QUESTIONS

1. Why are naturally occurring forest fires considered different from those caused by humans? Do you think the Park Service should treat them differently? Why or why not?

2. Ironically, the fires have brought millions of visitors to the park to view the effects, and the new scenery. What effects do human visitors have on the park? Who profits from the visitors? Should access be more limited?

ABC CASE STUDY 6

Economics and Politics of Water in the West

Usable water is a precious commodity. Not only do we need water for consumption, but also for agricultural and industrial activities. More than 70 percent of the Earth's surface is covered by water, but only 2 percent of all the water within the Earth's realm is considered fresh and therefore usable for human activity, and more than half of this is locked in polar ice caps and glaciers.

Securing access to adequate supplies of fresh water is a major problem in many western states. Although large supplies of fresh water exist, they are unevenly distributed. Consequently, certain individuals and groups control large water reserves and therefore also control the access of others to this water. In the West, water rights are bought and sold like any commodity. In addition, no central authority exists to regulate water use. Although this system enables some people to make substantial profits, it places many others in a highly vulnerable position. For farmers, inadequate water supplies can mean loss of valuable cropland. Cities might have to choose between limiting their growth or paying more for water.

Buell, Girard reference: Chapter 12. Video title: "Economics and Politics of Water in the West," 4 minutes.

QUESTIONS

1. As the population continues to grow in the West, and with it the demand for water, what methods might be considered to ensure that water is distributed to all segments of the population?

2. Can you foresee any changes that would affect the demand for water in the West?

ABC CASE STUDY 7

Environmental Concerns under Communism

What would things be like if we simply ignored environmental concerns in the interest of promoting economic production? The answer is becoming clear now that the end of the Cold War has opened up the countries of eastern Europe. This video documents life in Bitterfeld, East Germany, as it was shortly after the Berlin wall came down: air choking with sulfur dioxide and chlorine gas, rivers fouled with industrial waste, land that no longer can grow trees, people dying young with cancer, children with skin diseases. Hans Zimmerman and Rainer Fromann describe how the East German government suppressed information on health and environmental conditions while promoting the "workers' and farmers' paradise" propaganda.

Journalist Bob Brown takes a walk on a Nature Trail that also bears a sign "Keep Out—Dangerous to Life." Zimmerman, now a member of the German parliament, speaks poignantly of shared guilt in response to a question of how people in charge could let things get so bad. The program ends with a sober estimate of what must be done and what it may cost to remedy the environmental problems in the industrial regions of former East Germany.

Buell, Girard reference: Chapters 12, 13, and 19. Video title: "Welcome to Hell," 18 minutes.

QUESTIONS

1. What kind of human health effects were described in the document? What chemicals might be responsible for such effects?

2. What attempt was made by the authorities to lessen the impact of life in Bitterfeld on the children?

3. Compare environmental regulation in the United States and East Germany in the light of grass roots action and government action.

ABC CASE STUDY 8
Hazardous Wastes in the Wrong Place

Hazardous wastes have a way of turning up where they're least expected. The first news segment documents some of our past sins. Massachusetts fishermen have been catching barrels containing dangerous wastes in their nets. Thousands of barrels of radioactive wastes and toxic chemicals were dumped, legally, some 20 miles off the coast from the 1940s to the 1970s. The wastes were dumped in water only 300 feet deep, in areas only poorly identified. EPA is now involved in locating the wastes.

The second news clip presents a disturbing situation: Hazardous waste dumps and incinerators tend to be located in towns and neighborhoods where the residents are nonwhite. The issue is environmental racism, and the video describes some of the possible effects of the toxic wastes. In response to the issue, grass-roots minority groups are organizing and are beginning to be heard across the country.

Buell, Girard reference: Chapters 12, 13, and 19. Video title #1: "Mass. Fishermen Deal with Legally Dumped Toxic Waste," 5 minutes. Video title #2: "Minority Areas Used for Toxic Waste," 5 minutes.

QUESTIONS

1. How does George Perry (the atomic garbageman) figure in the problem of the drums in the ocean?
2. Recount the story told by Amos Favorite from his home in Louisiana. What action has he taken?
3. Why were toxic wastes dumped in the shallow ocean? Why were waste dumps sited near minority neighborhoods?

ABC CASE STUDY 9
Global Warming and the Scientific Community

The Earth Summit meeting of June, 1992, had global warming at the top of its agenda. Citing the scientific uncertainty about global warming, the Bush administration effectively lobbied to water down the global warming treaty. This news segment investigates attempts to document the impacts of early global warming. Biologists examine the phenomena of coral bleaching as a response to warming ocean water and the shift northward of populations of some heat-sensitive plants in the United States.

The complexity of the issue is introduced: What effect will clouds have? Will the oceans absorb the excess heat? Will added smoke from volcanoes and burning forests keep us shaded? Scientists F. Fred Singer and Stephen Schneider give opposing views on the issue, and the continuing uncertainty suggests that the search for evidence of global warming is vital to the future of our civilization.

Buell, Girard reference: Chapter 13. Video title: "Scientists Seek Proof Positive of Global Warming Effects," 5 minutes.

QUESTIONS

1. What evidence for warming comes from the South Pacific?
2. What are the advantages and disadvantages of taking action to prevent global warming before we are absolutely certain it is happening?

ABC CASE STUDY 10
Break in the Ozone Shield

Sometimes, one individual can catalyze action in the environmental arena by careful work and subsequent advocacy. Rachel Carson comes to mind, as does chemist Sherwood Rowland. Rowland, featured in this news clip, has done more than any other single person to bring about a global response to the threat posed by chlorofluorocarbons (CFCs) to the ozone shield. The video briefly documents Rowland's progress from turning down professional basketball to go to graduate school, to initial publication of the atmospheric impacts of CFCs in 1974, to rejection of his claims by industry, to final acceptance by the global community.

The occasion of the broadcast was the directive by President Bush to speed up the phaseout of CFC production because of new evidence of ozone depletion in the Arctic. Rowland has now turned his attention to global warming and closes with a comment on the lesson we should have learned from our experience with the ozone shield and CFCs.

Buell, Girard reference: Chapter 13. Video title #1: "Person of the Week: S. Rowland," 5 minutes. Video title #2: "Southern Exposure," 18 minutes.

QUESTIONS

1. What is the lesson Rowland believes we should learn from this issue?

2. How does Rowland link the ozone depletion problem and global warming?

ABC CASE STUDY 11

Oil and the Arctic National Wildlife Refuge

An ongoing battle in national energy policy pits defenders of the Alaskan wilderness against oil companies. The issue is whether to push ahead with exploration and development of oil suspected to lie under the Arctic National Wildlife Refuge (ANWR). Political and economic pressures to develop the oil are illustrated in this news clip: a new source of oil for the Alaska pipeline when Prudhoe Bay runs out, billions of dollars for the oil companies and the Inupiat Eskimos who live in northern Alaska, and some relief from our dependence on foreign oil. Against these interests is the concern for the fragile tundra wilderness and especially the large herd of caribou that use the land over the oil as crucial breeding grounds. Is more oil money better for the native peoples than continued use of the caribou for food and skins? Can hundreds of miles of new roads and pipelines and scores of drilling sites coexist with wildlife in a fragile wilderness?

The news clip makes it clear that behind this issue is the larger issue of a national energy policy that so far has been much more concerned with using oil than conserving it. The refuge is thought to hold 3.5 billion barrels of oil, about 6 months supply at the current rate of use in the United States. We are left with the question, Shouldn't there be a better way to address our energy needs than to sacrifice one of the remaining wilderness areas of our land (and a refuge at that)?

Buell, Girard reference: Chapter 14. Video title: "Alaska Oil Drilling Debate Renews," 5 minutes.

QUESTIONS

1. What are the reasons given for going ahead with development of the oil under ANWR?

2. What are the potential negative impacts of the exploration and development?

3. How does this conflict fit into the development of a long-term energy strategy for the United States?

ABC CASE STUDY 12

The Case for Energy Conservation

Environmentalists are accustomed to regarding utility companies as part of the problem, but as these two news clips show, this does not have to be true. The first segment documents a Kraft General Foods ice cream plant near Boston that was in serious economic trouble 2 years prior to the report. Boston Edison came in with a complete energy renovation plan that cut energy costs by $400,000 a year and saved the plant from extinction.

Utilities all over New England are aggressively pursuing conservation measures with their commercial and residential customers, providing energy audits, shower heads, and subsidizing use of energy-efficient fluorescent light bulbs. Instead of selling more power, the utilities are selling energy efficiency and making a genuine contribution toward solving our energy problems.

The second news clip crosses the country to California, where the focus is on energy efficiency in order to reduce the threat of global warming. Again, the utilities have realized that it is much less costly to invest in energy savings than to construct new power plants. In promoting energy conservation by helping people to cut their electricity bills, the utilities are reducing their use of fossil fuels and thus holding down global warming. Two power companies are also investing in solar and wind power as the best long-run answer to pollution. The main point: The protests coming from some businesses and the federal government about the economic impacts of cutting our use of fossil fuels appear to be largely unfounded, and the utility industry may well become major proponents of action to curb global warming.

Buell, Girard reference: Chapter 15. Video title #1: "A Look at Energy Conservation Progress for Maine," 5 minutes. Video title #2: "Power Companies Make Profit from Reducing Carbon Dioxide," 5 minutes.

QUESTIONS

1. Why are utility companies, who sell electricity to their customers, trying to get those same customers to use less electricity?

2. What are the main steps being taken by the utility companies to help residents save energy?

ABC CASE STUDY 13
Fuel-Efficient Cars: Detroit versus Japan

In crafting the Clean Air Act of 1990, Congress and the administration refused to increase the fuel-efficiency standard for cars because of intense lobbying by the U.S. automobile industry, and so it remains locked at 27.5 miles per gallon (mpg). This news clip features the 1992 Honda Civic VX, which went from 33 mpg to 51 mpg in 1 year without a major cost increase or downsizing. Detroit carmakers have maintained that you can't have fuel efficiency in anything but the smallest cars; the Geo Metro, shown in the news clip, is their case in point. The Honda Civic has proven otherwise.

Detroit is also losing out in the race to develop useful electric cars. Los Angeles has had to turn to a Swedish firm in a joint venture to develop an electric car to meet the coming requirements of the law in California. The feature emphasizes the difference in basic attitudes of the Japanese and U.S. car manufacturers regarding better mileage and electric cars. The result? U.S. car companies will continue to play catch-up because of their reluctance to change.

Buell, Girard reference: Chapter 15. Video title: "Japanese Prove Us Wrong in Fuel Efficient Car," 5 minutes.

QUESTIONS

1. What did Honda do to get such high mileage in the Civic without sacrificing size and speed?

2. What are the attitudes of the two countries' auto industries, according to Senator Bryan, and why do you suppose there is such a difference?

ABC CASE STUDY 14
Energy Alternatives in Our Future

One-third of all air pollution in the United States comes from burning coal, oil, and gas in electrical power generation. Two important alternatives to the fossil fuels are explored in this newscast: solar and wind power. Not only are they pollution-free, they are also limitless sources of energy. The clip visits California's Altamont Pass, where 4000 wind turbines now generate enough electricity to supply every home in San Francisco. Then we shift to the Mojave Desert, where solar thermal power is producing electricity with solar trough collectors. The costs of these two energy sources are compared with current costs of electricity from coal and gas, and they are getting very close.

The newscast mentions that federal subsidies of clean power are not forthcoming because the Energy Department is unwilling to interfere with the free enterprise system. In the meantime, Japanese and European firms are actively developing wind and solar technologies, as shown in a new California wind farm using Mitsubishi turbines. Sun and wind are seen as the best way to deal effectively with the problems of air pollution and global warming.

Buell, Girard reference: Chapter 15. Video title: "Alternatives to Electricity to Save the Environment," 5 minutes.

QUESTIONS

1. What are costs of fossil fuel-generated electrical power that do not show up on your electric bill?

2. How could wind and solar power be made competitive today?

3. What benefits would come from all-out development of these energy sources?

ABC CASE STUDY 15
The Place of Risk Analysis in Pollution Cleanup

Faced with a host of environmental problems, how does our society make decisions on which ones to address, given the budgetary constraints? The answer is: We have to rank the problems in order of their risk. This segment looks at environmental risks from two perspectives: scientific logic and public concern. The video points out that the experts and the public often do not agree in their ranking of environmental risks. EPA head William Reilly wants to rely more on scientific risk analysis than

we have in the past, and he cites the work of his science advisory board in ranking risks.

Peter Sandman of Rutgers University takes the view that EPA's strategy will backfire if public concerns are not considered. Citizens in Louisiana are shown interacting with their state Department of Environmental Quality to help set the state's priorities in assessing environmental hazards, and the segment closes with the view that both risk analysis and public concern must be considered.

Buell, Girard reference: Chapters 12 and 16. Video title: "Risk Assessment and Public Opinion Combine for Cleanup," 5 minutes.

QUESTIONS

1. What are some risks ranked high by the public and low by the science advisers? Vice versa?
2. How can public concerns be integrated with science-based risk analysis in forming public policy?

ABC CASE STUDY 16
Lead: A Hidden Poison

As we point out in Chapter 16, lead poisoning can be caused by ingestion of lead paint chips or from lead in our drinking water. Many may be surprised to learn that soil that is along interstate highways or near our busiest city roads may be contaminated with as much as 1,000 ppm (parts per million) of lead.

Although leaded gasoline has been phased out, it has been used in automobiles for more than fifty years. Years of heavy traffic have left a residue of lead. The problem is that lead salts are not very soluble in water. Since they don't dissolve and wash away, lead has become the hidden poison in the soil of American neighborhoods.

Children are most at risk for two reasons. Children are more susceptible

to lead than are adults. Lead poisoning attacks a child's brain and nervous system, causing learning disabilities, speech impediments, and hearing loss. Children also play in the dirt and in doing so they accidentally ingest dirt that is contaminated with lead.

In Oakland, California, playgrounds have been closed because of lead levels in the soil that are six times the EPA guidelines for hazardous waste. Although some communities are actively seeking solutions to the problem, EPA does not currently have an urban soil testing and cleanup program. Only three states even have a standard for contaminated dirt.

Once a playground or yard is certified to be lead contaminated, the re-

moval of the contaminated soil will be very expensive. Some are advocating the less expensive alternative of covering the playground soil with concrete, asphalt, or grass to protect children from the contaminated soil.

Buell, Girard reference: Chapter 16. Video title: "Hidden Poison," 18 minutes.

QUESTIONS

1. Is this a problem that should be solved at the federal, state, or local level?
2. If the soil was removed from city playgrounds, where should it be put? Are there ways to clean the lead out of soil?

ABC CASE STUDY 17
Who Wants the Trash?

Some 15 million tons of trash cross state lines each year, and the states are helpless to do anything about it. The problem of where to put the growing heaps of trash is left in the hands of

local towns. This case study explores two very different kinds of responses to this traffic in trash. The first news clip starts in Hinsdale, in western Massachusetts, where a developer

wants permission to put a huge landfill that will take in garbage from New York, New Jersey, and New England. Private landfills in Indiana and Virginia are also featured. It is clear that once town residents approve a private landfill, they basically lose control of how the land is filled; the film shows convoys of trash converging on Centerpoint, Indiana. (As a postscript, in the case of Hinsdale, a referendum was taken, the proposal was turned down in a vote of 830 to 59, and the area was designated as a state Area of Critical Environmental Concern, which precludes landfills.)

Taking an entirely different tack, Riverview, Michigan, has welcomed trash from surrounding municipalities and has built its landfill into a 125-foot-high mound ("Mount Trash-more"), which hosts a skilift in the winter. Profits from the landfill bring in $8 million a year, the town has built a new City Hall, fully funded its school programs, and cut property taxes of its residents. Expensive homes and a 27-hole golf course have been built next to the landfill. Clearly, the town has embraced the traffic in trash and reaped a continuing profit, an example not unnoticed by other towns and cities with their own trash problems.

Buell, Girard reference: Chapter 19. Video title 1: "States Fight the Transport of Garbage Between Cities," 5 minutes. Video title 2: "The Town that Loves Garbage," 12 minutes.

QUESTIONS

1. What happened in Centerpoint, Indiana, and Selma, Virginia, to turn local people against landfills?

2. List the benefits and costs of maintaining the landfill to the people of Riverview.

3. Why do you think the people of Hinsdale were so opposed to the landfill while Riverview people welcomed theirs?

ABC CASE STUDY 18
The Nuclear Waste Dilemma

The focus of this newscast is on the growing stockpile of nuclear wastes from the nuclear power industry. Currently, the wastes are stored in protective pools and above-ground containers at the sites of power plants. Everyone agrees that this is only a temporary answer to the waste storage problem; the permanent answer is geologic disposal—deep burial in the earth. The site selected by the government and the industry is Yucca Mountain in the Nevada desert, shown in the newscast.

Of course, Nevadans, who don't have nuclear plants of their own, are unhappy with this site selection. A million dollars a day is being spent on studies of the Yucca Mountain site, while governmental action clears the way for the eventual disposal by the year 2010. An attempted state veto of the site appears to be doomed, and current work is focusing on the geological properties of the mountain, which may eventually hold 70,000 tons of high-level waste when it is finally developed.

It appears that Yucca Mountain is the only site being considered for waste disposal from nuclear power plants. If this site is blocked by political action or by scientific considerations, the waste will live on at the power plants.

Buell, Girard reference: Chapter 19. Video title: "Debate on Plan to Store Nuclear Waste in Nevada," 6 minutes.

QUESTIONS

1. What are the arguments pro and con regarding the desire of the state of Nevada to block the disposal plan?

2. How is the nuclear power industry involved in this situation?

3. How important is this problem to the future of the nuclear power industry?

ABC CASE STUDY 19
The Disposal of Toxic Waste: Incineration

Wastes of all kinds were discarded with little concern for human health or the environment until legislation was enacted in the 1960s and 1970s to regulate waste disposal and protect the air and water from contamination. Hazardous wastes were incinerated, contaminating the atmosphere with toxic emissions, and they were discarded in sewers, rivers and streams, leaking landfills, abandoned buildings, or simply dumped on vacant lots and along roadways.

Numerous examples of irresponsible, often criminal, methods of waste disposal can be cited. For example, this video describes the conditions at the Caldwell Systems Incinerator, CSI, in Hudson, South Carolina. CSI was shut down and dismantled by the county in 1989 as a danger to public health. Solvents from the local furniture industry, spent torpedo fuel from the Navy, toxins from EPA superfund sites all over the country were sent here, stored, mixed and then burned.

If adequately controlled to prevent

air pollution, incineration has a number of advantages as a means of detoxifying many hazardous wastes. At sufficiently high temperature, incineration destroys 99.999% of toxic organic compounds by decomposing them to carbon dioxide, water, and various gases. At CSI, however, the workers were given little or no safety training and the illegal dumping and mixing of chemicals was a common occurrence. On September 13, 1989, a fire caused by improperly mixed chemicals swept through the plant. Two hundred and fifty residents were evacuated and 54 hospitalized.

Although the state knew of problems at CSI, the incinerator was able to operate through a "loop hole" in the law. Since 1978, an incinerator could apply for an "Interim Status" permit to operate. This permit was designed to give companies time to upgrade their plants to meet new environmental regulatory standards. But the state allowed the interim status to continue for years and CSI was never required to get a final permit.

Buell, Girard reference: Chapter 19. Video Title: "Where There's Smoke," 18 minutes.

QUESTIONS

1. Should incinerators be licensed by federal or state government?
2. Would the workers at CSI be less at risk if they were working at a landfill rather than an incinerator?

ABC CASE STUDY 20

Markets For Recycled Material

More and more waste materials are being recycled in response to the need to conserve mineral resources and energy and reduce environmental pollution. Many communities now require businesses and homeowners to collect discarded items such as newspapers, office paper, aluminum cans, glass containers and certain plastic items, in separate containers so that they can be collected and recycled.

Everyone is in favor of recycling, but to be successful recycling must also be cost effective. Dealers and manufacturers will not be willing to accept items for recycling if they cannot count on steady markets for the recycled products. In 1989 there were 600 curbside collection programs in the United States, by the year 1992, there were more than 4000. This increase has caused a glut of recycled plastic, glass, and newspaper.

It is difficult to recycle paper profitably. Wayne township, New Jersey, used to get paid $25 a ton for recycled newspaper. Now they have to pay $25 a ton to get it hauled away. Because of this, some state governments are trying to boost demand for recycled material. Ten states now require newspapers to use recycled paper, and a dozen states have ordered state agencies to buy recycled office paper.

Philip Bailey of the National Recycling Coalition advocates investment tax credits and low interest-rate loan guarantees to companies who want to get involved in this area. Some of the biggest and best known companies in America are trying to use more recycled materials without government help. Anheuser-Busch is spending more than $2 billion a year for packaging materials with a recycled content. McDonald's restaurant chain is building a prototype for future restaurants, more than half of which will be made from recycled materials. Roof beams, benches, and ceiling tiles will be made from recycled materials for this model located in Seattle.

Buell, Girard reference: Chapter 19. Video title: "Recycling Leads to Glut in Materials Actually Processed," 4 minutes.

QUESTIONS

1. Should the federal government institute a "virgin minerals tax" to encourage recycling?
2. What can we do to create markets for recycled products?

ANSWERS TO SELECTED QUESTIONS AND PROBLEMS

CHAPTER 1

2. Physical change.
4. (a) heterogeneous; (b) heterogeneous;
   (c) homogeneous; (d) homogeneous;
   (e) heterogeneous; (f) homogeneous.
6. (a) element; (b) compound; (c) element;
   (d) element; (e) compound; (f) element;
   (g) compound; (h) compound; (i) element;
   (j) compound; (k) compound; (l) element;
   (m) compound; (n) element.
8. (a) magnesium; (b) copper; (c) cobalt; (d) carbon;
   (e) boron; (f) nickel; (g) phosphorous;
   (h) silicon; (i) sulfur; (j) aluminum; (k) bromine;
   (l) iodine; (m) mercury; (n) oxygen.
10. (a) neutral; (b) anion; (c) cation; (d) anion;
    (e) neutral; (f) cation; (g) neutral; (h) cation;
    (i) neutral; (j) anion.
12. Solids maintain shape and volume; liquids maintain volume but not shape; gases maintain neither shape nor volume.
14. (a) λ = cm; (b) v = cycles/s = hertz;
    (c) c = 3.0×10^{10} cm/s, $c = \lambda v$.
16. (a) Radiant energy the sun transmits through space;
    (b) capacity for doing work; (c) 10^{-9} m;
    (d) distance between adjacent wave crests;
    (e) a plot of wavelength versus intensity;
    (f) no atmospheric pressure.
18. (a) X-rays are ionizing radiation; (b) yes; (c) they are low-energy and are not as great a risk; (d) yes; (e) UV light is higher-energy than visible light, yes there is a risk.
20. Boiling water.

CHAPTER 2

2. (a) Mercury, Venus, Earth, Mars; (b) Jupiter, Saturn, Uranus, Neptune, Pluto; (c) The Terrestrial planets have high density and are composed of metals and minerals. The giant planets have low density and large amounts of gases, such as hydrogen and helium.
4. Zones: core, mantle, and crust; (a) core of the earth is made of iron; (b) see Figure 2–3.

6. Water bound to minerals was released as the interior of the earth heated up. H_2 and O_2 released from the interior combined explosively to form water.
8. Photosynthesis: energy from sun converts carbon dioxide and water into simple carbohydrates and oxygen.
10. Plate tectonics. Pangea was the super continent from which all the existing continents formed. Plates of the rigid lithosphere move slowly over the athenosphere forming mountains and splitting continents.
12. Molten rock rising from the core.
14. Convection currents from earth's interior.
16. (a) Volcano; (b) water and oxygen;
    (c) amino acids.
18. Igneous rock from lava; sedimentary rock from erosion particles or chemical precipitation; metamorphic rock formed by the action of pressure and temperature on sedimentary rock.
20. (a) Sandstone; shale; (b) limestone, dolomite;
    (c) marble, slate.
22. (a) Two-dimensional sheets; (b) chains;
    (c) three-dimensional network.
24. There is a synergism between asbestos and smoking, and a much higher risk of cancer for asbestos workers who smoke.
26. (a) Sand, limestone, sodium carbonate; (b) melts over a wide temperature range and bonds are broken;
    (c) cobalt oxide, CoO.
28. Bauxite, Al_2O_3, Jamaica and Australia.
30. Wire, pipe, and coins.
32. The percentage carbon determines the properties of steel; low-carbon steel is soft, high-carbon steel is hard.
34. (a) Transporting, wearing surface; (b) breaking into fragments.

CHAPTER 3

2. An ecosystem consists of all the different organisms living within a finite geographic region.
4. Herbivores, carnivores, omnivores, decomposers.
6. One or two factors that outweigh all others in determining growth and survival of an organism.
8. Carp can survive in water with low dissolved oxygen and higher concentration of toxic substances.
10. Energy is the ability to do work or bring about change;

potential energy: stored energy; kinetic energy: energy of motion.

12. In every energy transformation some heat is lost in the form of heat energy.

14. (a) Fourth;   (b) first, more energy efficient.

16. 90%.

18. Carbon:

$$6\,CO_2 + 6\,H_2O \longrightarrow C_6H_{12}O_6\text{ [sugar] } + 6\,O_2;$$

see Figure 3-13.

20. Phosphorous cycle; Phosphates are very slowly dissolved from rock, absorbed by plants and eaten by animals; see figure 3-15.

22. Respiration:

$$C_6H_{12}O_6 + 6\,O_2 \longrightarrow 6\,CO_2 + 6\,H_2O + \text{energy}.$$

See Figure 3-13.

24. (a) Dissolve CO_2 in the atmosphere;   (b) calcium reacts with carbon dioxide to form limestone;   (c) when fossil fuel is burned carbon dioxide is released.

26. $N_2 + O_2 \xrightarrow{\text{lightning}} NO_3^-\text{[nitrate]}.$

28. Plant cover, spread manure.

30. Phosphates (PO_4^{3-}) in guano desposits, west coast South America and caves in Arizona and New Mexico.

CHAPTER 4

2. Because of the law of conservation of mass.

4. (a) Reacting substances are made up of atoms and the atoms in different elements are unique. They can be neither created nor destroyed, but only rearranged. The total mass of products must equal the total mass of reacting substances.   (b) All atoms of a certain element are alike. Different samples of any pure compound contain the same elements in the same proportion by mass.   (c) Two elements often combine to form more than one compound. The compounds are distinct from one another, and the masses of one element that can combine chemically with a fixed mass of another element are in the ratio of small whole numbers.

6. See Figure 4-11. Because almost all of the alpha particles pass through the foil without being deflected, the volume occupied by an atom must be mostly empty space. Each alpha particle that was deflected, however, must have encountered a dense, similarly charged particle that repelled it and caused it to change direction.

8. (a) $C_{27}H_{46}O$;   (b) $C_{27}H_{46}O$.

10. The masses of one element that can combine chemically with a fixed mass of another element are in a ratio of small whole numbers.

12. (a) Confirms the law of conservation of mass;
    (b) violates the law of multiple proportions;
    (c) confirms the law of multiple proportions;
    (d) violates the law of multiple proportions.

14. (a) false;   (b) false;   (c) true;   (d) false.

16. (a) Atomic number is the number of protons. Atomic mass is the sum of the number of protons and neutrons.   (b) Mass number is the sum of the number of protons and neutrons for a particular isotope of an element. The atomic mass is the average mass of the individual isotopes that make up the element's natural distribution on earth.   (c) Atomic mass units (amu).

18. (a) 11;   (b) 13;   (c) 11.

20.

Atom	Atomic no.	Mass no.	Protons	Neutrons	Electrons
Zn	30	64	30	34	30
Eu	63	153	63	90	63
U	92	235	92	143	92
Pd	46	106	46	60	46

22.

Atom	Protons	Neutrons	Electrons	Symbol
Neon	10	11	10	Ne
Barium	56	82	56	Ba
Scandium	21	24	21	Sc
Phosphorus	15	16	15	P

24. N, Cl, Zn, Xe, Hg.

26. The seven horizontal rows are called periods. The 18 vertical columns are called groups.

28. a, d, f, h.

30.

	Neutrons	Protons
(a)	8	7
(b)	2	1
(c)	125	82
(d)	88	63
(e)	60	47
(f)	62	47

32. titanium, a metal.

CHAPTER 5

2. The single electron can move to different orbits or energy levels.

4. (a) As: $1s^2\,2s^2\,2p^6\,3s^2\,3p^6\,4s^2\,3d^{10}\,4p^3$;   (b) Na: $1s^2\,2s^2\,2p^6\,3s^1$;   (c) Br: $1s^2\,2s^2\,2p^6\,3s^2\,3p^6\,4s^2\,3d^{10}\,4p^5$.

6. (a) 18;   (b) s, p, d, f;   (c) 1, 2, 3, 4;   (d) 10.

8. (a) Mg;   (b) K;   (c) P.

10. (a) OK;   (b) the second level does not have a d orbital.

12. (a) Li;   (b) F;   (c) O;   (d) B.

14. To gain inert gas electronic structures;   (a) Al^{3+};   (b) S^{2-};   (c) B^{3+};   (d) Cs^+.

16. (a) F;   (b) Na;   (c) P.

18. (a) $\overset{\bullet}{Ba}\bullet$ (b) $\overset{\bullet\bullet}{\underset{\bullet\bullet}{:}}\!Cl\bullet$ (c) $\bullet\,\overset{\bullet}{Al}\,\bullet$ (d) $\overset{\bullet\bullet}{\underset{\bullet\bullet}{:}}\!S\overset{\bullet}{\underset{\bullet}{}}$

20. (a) CsBr; **(b)** SrBr$_2$.

22. (a) MgO; **(b)** Li$_2$O; **(c)** BeO.

24. (a) Na$_2$S, sodium sulfide; **(b)** K$_2$O potassium oxide; **(c)** LiCl lithium chloride.

26. (a) Atoms held together by electrostatic charge; **(b)** sharing of two electrons between atoms.

28. (a) O $::$ O **(b)** $:$ N $::$ N $:$ **(c)** H $:$ H

30. (a) $:$ Br $:$ Br $:$ **(b)** $:$ Br $:$ Cl $:$ **(c)** H $:$ Br $:$

32. Single bond, one e$^-$ pair shared; double bond, two e$^-$ pairs shared; triple bond, three e$^-$ pairs shared.

34. (a) H $:$ C $:$ C $::$ N $:$ **(b)** $:$ N $::$ C $:$ C $::$ N $:$

36. (a) SiO$_2$; **(b)** NaOH; **(c)** SF$_6$; **(d)** CsBr.

38. (a) F; **(b)** N; **(c)** O.

40. (a) H—O, H—Cl, H—F; **(b)** N—O, P—O, Al—O; **(c)** Br—Br, B—N, H—Cl; **(d)** N—P, S—O, Be—F.

42. (a) NH$_4^+$; **(b)** PO$_4^{3-}$; **(c)** CO$_3^{2-}$.

CHAPTER 6

2.

	Protons	Neutrons
(a) ^{37}Cl	17	20
(b) ^{17}O	8	9
(c) ^{99}Mo	42	57
(d) ^{113}Cd	48	65
(e) ^{234}Pu	94	140
(f) ^{115}Ag	47	68
(g) ^{136}Cs	55	81
(h) ^{13}C	6	7
(i) ^{22}Ne	10	12
(j) ^{137}Ba	56	81

4. (a) $^{210}_{82}$Pb $\longrightarrow$ $^{210}_{83}$Bi $+$ $^{0}_{-1}\beta$

6. (a) $^{206}_{81}$Tl
 (b) $^{226}_{88}$Ra
 (c) $^{4}_{2}\alpha$
 (d) $^{25}_{12}$Mg.

8. (a) $^{65}_{28}$Ni $\longrightarrow$ $^{65}_{29}$Cu $+$ $^{0}_{-1}\beta$;
 (b) $^{150}_{60}$Nd $\longrightarrow$ $^{146}_{58}$Ce $+$ $^{4}_{2}\alpha$;
 (c) $^{28}_{12}$Mg $\longrightarrow$ $^{28}_{13}$Al $+$ $^{0}_{-1}\beta$.

10. (a) Helium nucleus, $^{4}_{2}$He; **(b)** the sum of the number of protons and neutrons in the nucleus; **(c)** high-energy waves; **(d)** number of protons in the nucleus; **(e)** identical to an electron: negligible mass and a charge of minus one; **(f)** changing one element into another.

12. 6.25 g, 48 days.

14. Radiation enters Geiger tube and ionizes argon gas causing electric current to flow, see Figure 6-6.

16. Chemical reactions do not cause transmutation of elements.

18. 260 mrem/person.

20. No, ^{238}U has a half-life of 4.5 billion years. The half-life of the isotope should be closer to the age of the object.

22. See Table 6-5.

24. Gamma ray.

26. Ionizing radiation can (1) produce abnormalities in DNA and increase the risk of cancer; (2) lower white blood cell count and increase the risk of infection.

CHAPTER 7

2. (a) 4 Al $+$ 3 O$_2$ $\longrightarrow$ 2 Al$_2$O$_3$;
 (b) 4 Fe $+$ 3 O$_2$ $+$ 6 H$_2$O $\longrightarrow$ 4 Fe(OH)$_3$;
 (c) 2 CH$_3$OH $+$ 3 O$_2$ $+$ 3 O$_2$ $\longrightarrow$ 2 CO$_2$ $+$ 4 H$_2$O;
 (d) CH$_4$ $+$ 2 O$_2$ $\longrightarrow$ CO$_2$ $+$ 2 H$_2$O.

4. (a) 2 Na $+$ Cl$_2$ $\longrightarrow$ 2 NaCl;
 (b) H$_2$ $+$ I$_2$ $\longrightarrow$ 2 HI; **(c)** Si $+$ Br$_2$ $\longrightarrow$ SiBr$_4$.

6. (a) 2 K $+$ Br$_2$ $\longrightarrow$ 2 KBr;
 (b) Ca $+$ Br$_2$ $\longrightarrow$ CaBr$_2$;
 (c) 2 Al $+$ 3 Br$_2$ $\longrightarrow$ 2 Al Br$_3$.

8. (a) 38 moles; **(b)** 28 moles; **(c)** 6 moles.

10. (a) 245.4 g; **(b)** 169.9 g; **(c)** 71.0 g.

12. 46.0 g = 1 mole; 11.5 g = 0.25 mole; 115.0 g = 2.5 mole; 23.0 g = 0.5 mole.

14. (a) 4 moles O; 8 moles N; **(b)** 2 moles Al, 6 moles O, 6 moles H; **(c)** 10 moles Na; 10 moles S; 15 moles O.

16. (a) 2 C$_4$H$_{10}$ $+$ 13 O$_2$ $\longrightarrow$ 8 CO$_2$ $+$ 10 H$_2$O;
 (b) 2.5 moles; **(c)** 5 moles; **(d)** 360 g.

18. (a) Fe$_2$O$_3$ $+$ 3 C $\longrightarrow$ 2 Fe $+$ 3 CO;
 (b) 5 moles; **(c)** 28 g.

20. (a) 157.1 g; **(b)** 28.6 g.

22. (a) Some exothermic reactions must be started (initiated) and are not spontaneous; **(b)** some endothermic reactions must be started and are not spontaneous; **(c)** the equilibrium reaction proceeds more to the right than to the left.

24. Increasing entropy — solid, liquid, gas.

26. As the DDT is dispersed, the entropy rises. Once DDT is dispersed, enormous amounts of energy must be expended to clean it from the environment.

28. Temperature, concentration, catalyst.

30. The lower temperature of the refrigerator will slow the souring reaction.

32. (a) The tobacco is the fuel for combustion in oxygen, the reaction is spontaneous. **(b)** Reactions slow down as the temperature is lowered because the reacting molecules or atoms move more slowly. **(c)** Increasing the concentration of reactants speeds a chemical reaction because the probability that two reactants will hit each other increases with concentration.

34. (a) Flour that is dispersed as small particles has lots of surface and is dispersed in the air that it is reacting with;   **(b)** Ground solids have a larger surface area and can be dissolved by the water more rapidly. Collision frequency between reactants increases.

36. (a) Reaction slows when amount of CH_4 is lowered (lower concentration).   **(b)** "Smothering the fire" is reducing, O_2 concentration, the other reactant.   **(c)** Applying water cools the reaction and slows the rate.   **(d)** CO_2 extinguisher reduces concentration of CH_4 and O_2.

CHAPTER 8

2. (a) K^+, Br^-;   **(b)** Mg^{2+}, CO_3^{2-};   **(c)** Na^+, SO_4^{2-};   **(d)** NH_4^+, Cl^-;   **(e)** Mg^{2+}, NO_3^-.

4. 10 ppb, yes this violates EPA standards.

6. 0.2 ppm, 200 ppb.

8. (a) Ammonia;   **(b)** acetic acid.

10. *B-L Acid*   **(a)** HF,   **(b)** H_2O,   **(c)** H_2CO_3,   **(d)** H_2O
B-L Base   **(a)** H_2O **(b)** S^{2-} **(c)** H_2O **(d)** CH_3NH_2

12. Acetic acid (a).

14. $H_2O \longrightarrow H^+ + OH^-$.

16. (a) Acidic;   **(b)** neutral;   **(c)** basic.

18. a) 4;   **(b)** 8;   **(c)** 1.

20. (a) $1.0 \times 10^{-10} M$;   **(b)** $1.0 \times 10^{-7} M$;   **(c)** $1.0 \times 10^{-4} M$;   **(d)** $1.0 \times 10^{-2} M$.

22. (a) Sour taste, dissolves metals such as tin and magnesium, changes litmus from blue to red, reacts with bases to form salts;   **(b)** bitter taste, feels slippery on skin, changes red litmus to blue, reacts with acids to form salts.

24. $H^+ + H_2O \longrightarrow H_3O^+$; the hydronium ion is a stable ion formed by a proton and a water molecule.

26. $HA \longrightarrow H^+ + A^-$; a strong acid dissociates completely into H^+ and A^-. A weak acid dissociates only to a small extent (acetic acid 5%). A weak acid remains mostly HA.

28. Acid: $B + H_2O \longrightarrow HB^+ + OH^-$;
Base: $HA + H_2O \longrightarrow H_3O^+ + A^-$.

30. Lead-acid car batteries, fertilizer production, manufacturing of dyes and plastics.

32. (a) $CO_2 + H_2O \longrightarrow H_2CO_3$;   **(b)** $H_2CO_3 \longrightarrow H^+ + HCO_3^-$.

34. $CaCO_3 + H^+ \longrightarrow Ca^{2+} + HCO_3^-$.

36. Volcanic eruptions, acid mine drainage.

38. Rainwater with pH less than 4.5:   **(a)** they are made susceptible to damage from disease, insects, and low winter temperatures;   **(b)** pH 4.5.

40. When pyrite-rich coal is mined, the coal is exposed to air and water; $2 FeS_2 + 7 O_2 + 2 H_2O$
$\longrightarrow 2 Fe^{2+} + 4 SO_4^{-2} + 4 H^+$

42. (a) Oxidation;   **(b)** reduction;   **(c)** oxidation;   **(d)** reduction.

44. (c)

46. (a) reduction;   **(b)** oxidation;   **(c)** neither;   **(d)** neither.

CHAPTER 9

2. (a) 2,3-dimethylbutane;   **(b)** 3-ethylpentane;   **(c)** 2-methylbutane;   **(d)** 4-methylheptane.

4. CH_3CH_3, ethane; $CH_3CH_2CH_3$, propane; $CH_3CH_2CH_2CH_3$, butane; $CH_3CH_2CH_2CH_2CH_3$, pentane; $CH_3CH_2CH_2CH_2CH_2CH_3$, hexane.

6. (a)

(b) $CH_2{=}CH{-}CH{-}CH_3$
           |
           CH_3

(c) $CH_3{-}CH{=}C{-}CH_2{-}CH_2{-}CH{-}CH_2{-}CH_3$
                  | |
                  CH_3 CH_3

(d) $CH_3{-}C{=}CH{-}CH_3$
              |
              CH_3

8. (b) Cis, trans-2-butene;

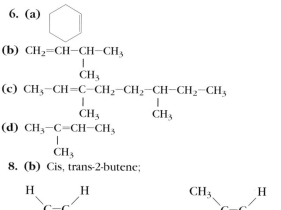

(d) Cis, trans-1-chloropropene.

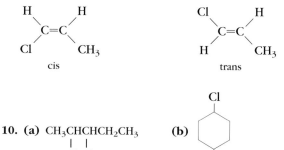

10. (a) $CH_3CHCHCH_2CH_3$ **(b)**
                    | |
                    Cl Cl

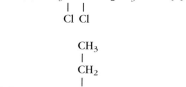

(c) $CH_3{-}CH_2{-}CH{-}CH_2{-}CH_3$ **(d)** $BrCH_2CH_2CH_2CH_3$
                        |
                        CH_2
                        |
                        CH_3

12. (a) carbonyl (ketone);   **(b)** carboxyl (carboxcylic acid;   **(c)** ester.

14. (a) None (hydrocarbon);   **(b)** ketone;   **(c)** alcohol;   **(d)** ketone, alcohol;   **(e)** ester;   **(f)** ether.

16. (a) 2,2-dimethyl-3-hexene;   **(b)** 3-nonene;   **(c)** 3-hexene.

18. (a) RCOOH;   **(b)** RCOR';   **(c)** RCH_2OH;   **(d)** RCOOR'.

20. (a) Isopropyl alcohol;   **(b)** methyl alcohol;   **(c)** ethyl alcohol;   **(d)** ethylene glycol.

22.

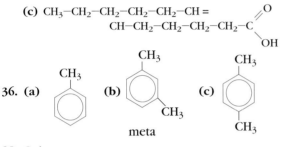

24. Methyl alcohol is poisonous.

26. $CHCOOH \rightleftharpoons CHCOO^- + H^+$

28. **(a)** $HCHO$; **(b)** $CH_3COOCH_2CH_3$;
(c) CH_3CH_2OH; **(d)** CH_3COCH_3.

30. Butyric acid, sharp vomit-like smell; methylbutyrate, sweet smell.

32. **(a)** amine; **(b)** aldehyde (carbonyl); **(c)** carboxyl (carboxylic acid); **(d)** ester.

34. **(a)** $H-O-CH_2-CH=CH-CH_3$;

$$\begin{array}{c} O \\ \parallel \end{array}$$
(b) CH_3-C-CH_3)

(c) $CH_3-CH_2-CH_2-CH_2-CH_2-CH=$
$\qquad CH-CH_2-CH_2-CH_2-CH_2-C \overset{\displaystyle O}{\underset{\displaystyle OH}{}}$

36. **(a)** **(b)** **(c)**

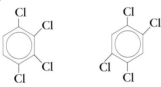

meta

38. Only two.

CHAPTER 10

2. Chiral molecules are optically active. They are able to rotate the plane of polarized light to the right or left. A polarimeter is used to detect the rotation.

4. See Figure 10–10. They differ at carbon 1 and 2; fructose contains a ketone group while glucose contains an aldehyde.

6. Excess starch is converted to glycogen and stored in the liver and in muscles. Large excesses are converted to fats and stored in the adipose tissue. Glycogen forms the storage polysaccharide in animals.

8. See Figure 10-9.

10. **(a)** sucrose: disaccharide; **(b)** glucose: monosaccharide; **(c)** cellulose: polysaccharide; **(d)** lactose: disaccharide.

12. Glycogen: When the starch we eat provides more glucose than the body requires for its immediate needs, some of the excess starch is converted to glycogen and stored in the liver and in muscles.

14. Insulin; glucose and fructose.

16. When carbohydrates are oxidized in the body with molecular oxygen, the reaction yields carbon dioxide, water, and energy.

18. In a saturated fatty acid, all the carbon bonds are single bonds. An example of a saturated fatty acid is stearic acid. An unsaturated fatty acid contains one or more carbon-carbon double bonds. An example of a unsaturated fatty acid is oleic acid.

20. As the polyunsaturated fatty acids are hydrogenated the double bonds are converted to carbon-carbon single bonds and the liquid polyunsaturated fatty acid is converted into a semisolid fat.

22. Triglycerides are formed by the reaction of three fatty acid molecules with a glycerol molecule.

24. Gallbladder.

26. Glycine, see Table 10-2, yes.

28. Nitrogen.

30. The total possible is equal to n! For three amino acids this would be $3 \times 2 \times 1 = 6$:
Leu-Gly-His; His-Gly-Leu; Gly-Leu-His; His-Lue-Gly;
Leu-His-Gly; Gly-His-Leu.

32. Hydrogen bonds.

34. Genetic information and control of protein synthesis

36. DNA, deoxyribose; RNA, ribose.

38. TGACA.

CHAPTER 11

2. **(a)** Monomers are small molecules that can be linked together to form long polymer chains; **(b)** Chemical reaction between rubber and sulfur. Sulfur crosslinks rubber. **(c)** An addition polymer is formed when monomers link together without the loss of any atoms from the monomers.

4. Cellulose nitrate.

6.
$$CH_2 = CHC1 \longrightarrow [-\overset{\overset{\displaystyle H}{\displaystyle |}}{\underset{\underset{\displaystyle H}{\displaystyle |}}{C}}-\overset{\overset{\displaystyle H}{\displaystyle |}}{\underset{\underset{\displaystyle Cl}{\displaystyle |}}{C}}-]_n$$

8. $CF_2 = CF_2$, Addition polymer.

10. Vulcanization is a chemical reaction of rubber and sulfur. The sulfur crosslinks the long polymer chains of rubber and changes the rubber from being soft and elastic to hard, tough, and more resistant to heat and cold.

12. Teflon is polytetrafluoroethylene. It is a derivative of ethylene where all four hydrogen atoms are replaced with fluorine atoms. The presence of the fluorine atoms makes the polymer much more resistant to heat than polyethylene. If polyethylene is heated, it melts and decomposes.

14. **(a)** Polyvinylchloride (PVC); **(b)** Polyacrylonitrile; **(c)** pet; **(d)** PVC.

16.

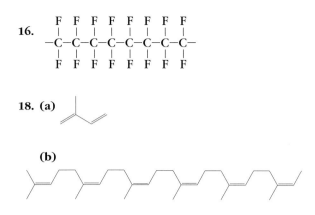

18. (a)

(b)

(c) Yes, the folds in the carbon chain could be flattened out and the molecule stretched.

20. (1) Titanium dioxide, TiO_2, is a white pigment; **(2)** zinc oxide, ZnO, is a white pigment; **(3)** carbon black, C, is a black pigment; **(4)** iron oxide, Fe_2O_3, is a brown or dark red pigment; **(5)** cadmium sulfide, CdS_2, is an orange pigment; **(6)** chromium oxide, Cr_2O_3, is a green pigment.

22. The solvent in the paint, which holds the pigment, binder, and other ingredients in suspension, evaporates as the paint dries. The paint's binder polymerizes and hardens as the paint dries and forms a continuous film that holds the paint to the surface.

24.

$$-\overset{\overset{\displaystyle O}{\|}}{C}-\overset{\overset{\displaystyle H}{|}}{N}-(CH_2)_4-\overset{\overset{\displaystyle H}{|}}{N}-\overset{\overset{\displaystyle O}{\|}}{C}-(CH_2)_6-\overset{\overset{\displaystyle O}{\|}}{C}-\overset{\overset{\displaystyle H}{|}}{N}-$$

26. Dacron is a polyester that is blended with cotton. Together they form a permanent-press fabric. HDPE is tough, but not a good material, for making clothing fibers. It is not very flexible, soft, or comfortable.

28. Silicones are water-repellent, heat-stable, and very resistant to chemical attack.

30. In 1991, only about 2 percent of plastic waste was recycled; 20 percent of paper and 30 percent of aluminum were recycled.

32. Polymers that are brittle and hard can often be made soft and pliable by the addition of plasticizers. PCBs were once used as plasticizers, until they were banned. Now phthalates are the most common plasticizers.

34. Explain your position.

CHAPTER 12

2. (a) The heat of vaporization is the amount of heat required to convert 1 g of a liquid to a vapor at its boiling point. **(b)** The high heat capacity of water allows the oceans to absorb large amounts of heat without there being a corresponding rise in water temperature. **(c)** Because of its high heat of vaporization, a large amount of heat is required to evaporate a small volume of water. Water is much more effective as a coolant than another liquid with a lower heat of vaporization.

4. If the water molecule were linear, then the open lattice arrangement of hydrogen and oxygen atoms could not form when water freezes. The open-lattice arrangement places the molecules of water farther apart in solid water than they are in liquid water and makes ice less dense than liquid water. Solid water forming from linear water molecules would be denser than the liquid.

6. Yes, both processes are fouled by oil.

8. Increased temperature would cause evaporation from the surface of lakes, ponds, and the oceans to increase. More clouds and rainfall would be expected.

10. When massive amounts of groundwater are removed from natural aquifers, the porous rock structure of the aquifers can collapse and cause a sinkhole. The land will not return to its original level if the water removal is stopped.

12. Large amounts of water evaporate as the water is exposed to sunlight. As pure water is removed by evaporation, the concentration of dissolved solids in the remaining water increases, and the water becomes more brackish.

14. (1) Wash your car with the water from your bath; **(2)** water your vegetable garden with rainwater; **(3)** cook your dinner with the well water.

16. Pollutants enter waterways by: **(a)** sewage-treatment plants; **(b)** factories; **(c)** electric power plants; **(d)** mines; **(e)** offshore oil drilling; **(f)** pesticide runoff; **(g)** fertilizer runoff; **(h)** construction runoff; **(i)** logging runoff.

18. Cholera, typhoid, dysentery, polio: None of these is a serious problem in the United States today. As a result of sewage treatment, and disinfection of water supplies, waterborne diseases have been virtually eliminated in developed countries.

20. (a) An increase in water temperature decreases oxygen solubility; **(b)** a decrease in atmospheric pressure decreases oxygen solubility.

22. Organic detritus in aquatic ecosystems is decomposed by aerobic (oxygen-consuming) decomposers, primarily bacteria and fungi. If the water is overloaded with organic wastes, aerobic decomposers proliferate and DO is consumed more rapidly than it can be replaced from the atmosphere.

24. Point sources discharge pollutants into a small area (usually from a pipe). Nonpoint sources discharge pollutants over a wide area (runoff).

26. When eutrophication occurs, dense mats of rooted and floating plants are formed. Blue-green algal blooms that appear on the water's surface release unpleasant-smelling, bad-tasting substances.

28. Sediments fill irrigation ditches and clog harbors. Toxic substances can adsorb on the surface of sediments and become concentrated. Spawning grounds become buried; photosynthesis in turbid waters will decrease.

30. Bioaccumulation. Because aquatic ecosystems have from four to six trophic levels, bioaccumulation is a more serious problem in aqueous habitats than in terrestrial ones, which usually have only two or three levels.

32. Metals, because they are elements, cannot be broken down into simpler less-toxic forms. They persist in the environment for years and bioaccumulate through the food chain. Mercury: used in the production of chemicals, paints, plastics, pharmaceuticals, photographic and electrical equipment; lead: leaded gasoline, batteries, paint, lead pipes, and solder; cadmium: used in the production of paints, plastics, and nickel-cadmium batteries and in electroplating.

34. Fish and shellfish become contaminated with mercury. The mercury bioaccumulates, and humans that eat the fish ingest a higher concentration than that present in the water of the bay.

36. PCBs are fire-resistant, stable at high temperature, and have a high electrical resistance. PCBs bioaccumulate in the food chain because they are fat-soluble.

38. A rise in water temperature increases the body temperature of aquatic organisms, raises respiration rates, and increases oxygen consumption. Oxygen is less soluble in warmer water. Reduced oxygen can be fatal to some fish. The life cycles of aquatic species are controlled by temperature.

40. State your opinion.

42. **(a)** $Cl_2 + H_2O \longrightarrow HClO + H^+ + Cl^-$; **(b)** Cl_2 reacts with organic material in water to form chlorinated hydrocarbons, which are suspected carcinogens.

CHAPTER 13

2. See Figure 13–1.

4. $O_2 \quad + \quad \text{sunlight} \quad \longrightarrow \quad 2O$

$O \quad + \quad O_2 \quad \longrightarrow \quad O_3$

6. The ozone layer forms the earth's main shield against the sun's dangerous UV radiation, and without it life on earth could not exist.

8. An air pollutant is a substance that is present in the atmosphere at a concentration sufficient to cause harm to humans, other animals, vegetation, or materials. Primary air pollutants are emitted directly into the troposphere. Secondary air pollutants are produced by chemical reactions between primary pollutants and other constituents of the atmosphere.

10. **(a)** Oxides of carbon, oxides of nitrogen; **(b)** oxides of carbon, oxides of sulfur.

12. As population continues to increase, more and more automobiles are on the road. More cars cause more total emission of pollutants, see Figure 13–6.

14. Carbon monoxide interferes with the oxygen-carrying capacity of blood by displacing oxygen from hemoglobin:

$$HbO_2 + CO \longrightarrow HbCO + O_2$$

The symptoms of CO poisoning are those of oxygen deprivation: headache, dizziness, impaired judgment, drowsiness, slowed reflexes, respiratory failure, and eventually loss of consciousness and death.

16. **(a)** Burning coal **(b)** Coal contains iron pyrite, FeS_2, the sulfur in the FeS_2 reacts with oxygen when the coal burns:

$$FeS_2 + 2\,O_2 \longrightarrow 2\,SO_2 + Fe$$

$$2\,SO_2 + O_2 \longrightarrow 2\,SO_3$$

(c) Sulfur dioxide (SO_2) reacts with oxygen to form sulfur trioxide (SO_3), which then reacts with water in the air to form sulfuric acid (H_2SO_4).

$$2\,SO_2 + O_2 \longrightarrow 2\,SO_3$$

$$SO_3 + H_2O \longrightarrow H_2SO_4$$

18. **(1)** Flue-gas desulfurization (FGD): The SO_x compounds are washed, or scrubbed from the stack gases by absorption in an alkaline solution. **(2)** Fluidized bed combustion (FBC): Coal in granular form is burned with finely divided pulverized limestone ($CaCO_3$).

20. VOCs are volatile organic compounds. The petroleum industry is the main anthropogenic source of VOCs; The escape of gasoline vapor, and evaporation of organic solvents from paint products. Natural sources account for 85 percent of VOC's; main source is terpene emission from trees.

22. Ozone (O_3) is the photochemical oxidant that is formed in the greatest quantity.

$O_2 \quad + \quad \text{sunlight} \quad \longrightarrow \quad 2O$

$O \quad + \quad O_2 \quad \longrightarrow \quad O_3$

24. Aerosols are very fine particles that remain suspended in the atmosphere and are not heavy enough to be deposited on the earth's surface. Particulates generally have diameters >1μm, and are deposited.

26. **(a)** Natural sources of particulates: volcanic ash, smoke from forest fires, wind-blown soil fragments, pollen, bacteria, sea spray; **(b)** anthropogenic sources of particulates: fly ash, soot, asbestos fibers, cement dust, fertilizer.

28. Industrial smog is formed during the burning of coal. It is a mixture of fly ash, soot, SO_2, and some VOCs.

30. Particulates can be removed from industrial emissions by electrostatic precipitation, bag filtration, cyclone separation.

32. Chlorofluorocarbons (CFCs), are used as coolants in refrigerators and air conditioners, spray can propellants, blowing agents for foam rubber and foam packaging materials, and cleaners for electronic circuits. freon 11: $CFCl_3$; freon 12: CF_2Cl_2.

34. The following is a chain reaction:

$$CFCl_3 + \text{sunlight} \longrightarrow Cl\cdot + CFCl_2\cdot$$

$Cl\cdot$	$+$	O_3	$\longrightarrow$	$ClO\cdot$	$+$	O_2
$ClO\cdot$	$+$	O	$\longrightarrow$	$Cl\cdot$	$+$	O_2
$Cl\cdot$	$+$	O_3	$\longrightarrow$	$ClO\cdot$	$+$	O_2

$$\text{Net:} \quad 2O_3 \longrightarrow 3O_2$$

36. It is absorbed by the greenhouse gases. The temperature of the atmosphere increases.

38. The major cause of the increase in atmospheric CO is the burning of more fossil fuels by electric utilities, automobiles, and industry. A secondary cause is deforestation, see Figure 13-16 and 13-18.

40. Tobacco smoke contains all the primary pollutants associated with combustion (especially CO) in concentrations higher than they are found in polluted air. It also contains tars and nicotine.

42. Carbon monoxide produced from the incomplete combustion of fossil fuels from stoves that are not properly vented.

CHAPTER 14

2. Petroleum probably originated from microscopic marine organisms that once lived in abundance in shallow coastal waters.

4. Petroleum has more C—H bonds per gram that will burn to release energy than does wood.

6. 1920 kcal.

8. $2 C_4H_{10} + 13 O_2 \longrightarrow 8 CO_2 + 10 H_2O$

Bonds Broken

20 C—H	99 kcal/mol	1980 kcal/mol
6 C—C	83	498
13 O—O	118	1534
		Total 4012

Bonds Made

16 C=O	192	3072
20 O—H	111	2220
		Total 5292

$$\frac{(5292 - 4012) \text{ kcal/mol}}{2 \text{ moles}} = 640 \text{ kcal/mol}$$

10. Gasoline is a lower boiling fraction of petroleum than kerosene. The hydrocarbon molecules in gasoline have shorter chains than does kerosene.

12. Catalytic reforming is a process in which hydrocarbon vapors in straight run gasoline are heated in the presence of suitable catalysts, such as platinum or other metals. In this process, straight hydrocarbon molecules are converted to branched hydrocarbon molecules.

14. Toluene has a higher octane number than heptane and is added to gasoline as an octane booster.

16. TEL stopped knocking, but lead was vaporized into the atmosphere.

18. The octane rating can be increased in three ways: cracking, catalytic reforming, and addition of octane enhancers.

20. Oil shale is a common sedimentary rock that contains a solid organic material called kerogen. When this rock is heated in the absence of air, a thick brown liquid called shale oil is produced. Shale oil can be refined like petroleum after sulfur and nitrogen impurities are removed.

22. Tar sands consist of sandstone mixed with a black high-sulfur, tarlike oil known as bitumen.

24. Bituminous coal is soft, and anthracite coal is hard. Anthracite is a better fuel because as it is formed from bituminous coal by the action of heat and pressure, volatile compounds are released, the water content decreases, and the carbon content increases.

26. $C + O_2 \longrightarrow CO_2$

28. Many of the disadvantages of coal as a fuel can be overcome if it is converted into a gas or liquid before it is burned. Coal can produce a gaseous fuel by several different coal gasification reactions. The gas can be easily transported by gas pipeline, burned without leaving an ash and without producing sulfur dioxide.

30. Coal gasification. advantages: gas easy to transport in pipes, no ash formed when gas is burned, gas is not dirty like coal; disadvantages: need water (steam to make gas from coal), large amounts of sulfur-containing solid produced, strip-mining and underground mining of coal are still necessary.

32. (a) Fuels used to generate electricity: coal, oil, nuclear fuel, natural gas, (hydroelectric, geothermal); **(b)** hydroelectric, 14%; coal, 55%; nuclear, 11%; gas, 12%; oil, 8%; **(c)** hydroelectric: none; coal: SO_2, CO_2, CO; nuclear: radioactive waste, thermal pollution; gas: CO, CO_2; oil: CO, CO_2, hydrocarbons.

34. See Figure 14-17. Electricity is produced whenever a coil of wire is moved through a magnetic field. The movement induces a flow of electrons in the wire. Most electric generators operate by rotating a coil of wire within a circular arrangement of magnets.

CHAPTER 15

2. Coal is the most abundant fossil fuel in the world. The largest proven recoverable reserves, estimated to be 29 percent of the world's total, are found in the United States.

4. Gas supplies are expected to last 40 to 60 years.

6. Former Soviet Union, France, United Kingdom, and Japan.

8. The fuel must be enriched to about 3 percent U-235.

10. No. The fuel is not sufficiently enriched to sustain the nuclear chain reaction necessary for detonation.

12. Give your own opinion.

14. Give your own opinion.

16. The fusion of two isotopes of hydrogen, deuterium ($_1^2$H) and tritium ($_1^3$H), seems promising because the temperature required ($100,000,000°C$) is lower than that required for other fusion reactions.

18. Sunlight is not always available (rainy day, night) or intense enough (northern climates) to be a practical heating source. Sunlight is widely dispersed, and concentrating it and converting it to a usable form is difficult and costly.

20. See Figures 15-13 and 15–14.

22 Give your own opinion.

24. (a) The methane in biogas can be converted to methanol as shown in the following equations:

$$CH_4 + H_2O \longrightarrow CO + 3\,H_2$$
$$CO + 2\,H_2 \longrightarrow CH_3OH$$

(b) Fermentation of sugars and starches in plants produces ethanol:

$$C_6H_{12}O_6 \overset{\text{yeast}}{\longrightarrow} 2\,CO_2 + 2\,C_2H_5OH$$
$$\text{sugar} \qquad\qquad\qquad \text{ethanol}$$

(c) In the absence of air, digestion of biomass by anaerobic bacteria produces biogas, which is about 65 percent methane.

26. Wind must blow steadily; costly, unsightly structures.

28. Give your opinion.

30. Lightweight, produces water as reaction product.

CHAPTER 16

2. Ethyl alcohol administered in high amounts can be poisonous. Any chemical, if administered in high enough amounts, will be poisonous.

4. (a) Dermal Contact, Inhalation, and Oral Ingestion; **(b)** Injection.

6. More than 60 tablets.

8. More than 1.54 mg.

10.
$$R{-}\overset{\displaystyle O}{\overset{\|}{C}}{-}N{-}R + H_2O \overset{OH^-}{\longrightarrow} R{-}\overset{\displaystyle O}{\overset{\|}{C}}{-}OH + RNH_2$$

12. Fungicides, industrial waste.

14. BAL chelates the heavy metal and surrounds it as a claw would surround a ball.

16. Undernourished children develop an appetite for strange things. This condition, called pica, leads them to eat paint which used to contain lead.

18. Neurotoxins are metabolic poisons that act by disrupting the normal transmission of nerve impulses in the central nervous system.

20. Tabun, Sarin.

22. Atropine occupies receptor sites on nerve endings that are also occupied by acetylcholine.

24. Ethyl alcohol is removed by oxidation to acetaldehyde, which is then oxidized to acetic acid, which is then oxidized to carbon dioxide and water.

26. The cytochrome P-450 liver enzymes act on fat-soluble substances to make them water-soluble, so they can be excreted.

28. Carcinogens are chemicals that cause cancer. Mutagens are chemicals that can alter genes and chromosomes and cause abnormalities in the offspring.

30. PAHs are formed when organic material is burned; cigarette smoke, automobile exhaust, stack gases from industrial boilers, charcoal-grilled meat.

32. A varied diet, in moderation, provides natural toxins or carcinogens far below levels that might produce an acute illness.

34. About 90 percent of all chemicals that cause mutations also cause cancer.

CHAPTER 17

2. Enzymes are catalysts that facilitate the breakdown of food. The enzyme ptyalin catalyzes the breakdown of carbohydrates, and the enzyme pepsin catalyzes the digestion of proteins. The enzyme lipase digests fats.

4. Glucose reacts with ADP to form lactic acid. Lactic acid reacts with ADP to form CO_2, H_2O, and ATP. In body tissues, energy is obtained by conversion of ATP back to ADP.

6. The energy derived from foods is measured in calories; 250 calories of heat is released when the candy bar is completely oxidized in a bomb calorimeter.

8. Fats.

10. Celery; peel, seed, and pulp of fruit; wheat bran. Fiber increases the frequency of bowel movement and thus minimizes toxin build-up.

12. Organic: fat-soluble, mostly carbon and hydrogen; water-soluble, carbon and hydrogen with more nitrogen and oxygen.

14. (a) Niacin (B_3); **(b)** retinol (A); **(c)** calciferol (D); **(d)** thiamine (B_1).

16. (a) water-soluble; **(b)** fat-soluble; **(c)** water-soluble; **(d)** fat-soluble; **(e)** water-soluble; **(f)** water-soluble; **(g)** water-soluble; **(h)** fat-soluble; **(i)** fat-soluble.

18. A disease caused by protein deficiency.

20. Obesity is caused by long-term consumption of more dietary calories than are required by the body to meet its energy needs.

22. Phenylketonuria (PKU) is an inherited metabolic disease in which the individual lacks a specific enzyme that catalyzes the conversion of the amino acid phenylalanine to tyrosine. Severe mental retardation results.

24. Shorter life expectancy, diabetes, gallbladder problems, high blood pressure, and heart disease.

26. (a) Bone and teeth;   (b) thyroid hormones;   (c) bone and teeth;   (d) hemoglobin.

28. Carbohydrates are a source of energy. Protein is the material from which most tissue and organs are constructed. Fats are a source of energy and provide components for cell membranes.

30. Lowest is number 3 (461 cal), highest is number 1 (534 cal).

32. 17.5 g, yes.

34. Antimicrobial additives; $NaNO_2$; antioxidants: butylated hydroxyanisole (BHA).

36. Monosodium glutamate, flavor enhancer.

38. Sequestrants tie up trace metals that may be in foods. EDTA, ethylenediaminetetraacetic acid forms a complex with metals and prevents them from catalyzing oxidation of food.

40. Established by U.S. Congress in 1958, a list of additives considered safe by the Food and Drug Administration   (FDA).

CHAPTER 18

2. (a) 8192;   (b) geometric.

4. 35 years.

6. Acidic soils are sour, and basic soils are sweet.

8. (a) Raise;   (b) yes.

10. Nitrate, NO_3^-.

12. Soil bacteria, rhizobium bacteria, blue-green algae (cyanobacteria).

14. When wet, urea decomposes slowly, releasing ammonia.

16. Superphosphate from phosphate rock or bone.

18. Cadmium is toxic to humans and plants can bioaccumulate it.

20. 30% nitrogen, 30% phosphorus as P_2O_5, 10% potassium as K_2O.

22. Superphosphate.

24. Boron, copper, iron, manganese, molybdenum, zinc. No, sometimes applied to seeds.

26. Manure: 1:2:1, much lower than synthetic fertilizer.

28. Narrow-spectrum insecticides are toxic to only a few types of insects. Carbamates are narrow-spectrum insecticides.

30. Juvenile hormones, (JH) regulate the development of an insect from juvenile to adult. If juvenile insects are treated with JH, they will remain immature and unable to mate.

32. Herbicides kill a particular group of plants. 2,4-D kills broad-leafed weeds, 2,4,5-T controls woody plants.

34. 0.25 mg.

CHAPTER 19

2. Advantages: volume reduced 85 percent, high temperature kills organisms, no groundwater pollution, heat produced is a source of energy; disadvantages: atmosphere may be polluted with toxic chemicals, ash may contain heavy metals.

4. Recycling must be cost-effective. There must be a market for recycled products.

6. (a) Ignitable wastes. Wastes that present a significant fire hazard. Include (1) liquids such as gasoline, hexane, and other organic solvents that have a flash point (the lowest temperature at which their vapors ignite when tested with a flame) below 60°C (140°F); (2) ignitable compressed gases such as propane; (3) various solid materials that are liable to cause persistent vigorously burning fires as a result of absorption of moisture, friction, or spontaneous chemical change; and (4) oxidizers. Oxidizers are included because they are likely to intensify an already burning fire.   (b) Corrosive wastes. Include acidic aqueous wastes with a pH equal to or less than 2.0 and basic aqueous wastes with a pH equal to or greater than 12.5.   (c) Reactive wastes. Include materials that explode readily when subjected to shock or heat, materials that tend to undergo violent spontaneous chemical change, and materials that react violently with water.   (d) Toxic wastes. Wastes that are hazardous because of toxicity. Substances that cause harm to living organisms by interfering with normal physiological processes. Include substances that are carcinogenic, mutagenic, teratogenic, or phytotoxic (poisonous to plants).

8. No, sulfuric acid will react with steel. Plastic containers should be used.

10. Ignitable waste.

12. Corrosive waste.

14. Hazardous wastes that cannot be recycled or converted to nonhazardous forms must be stored. Perpetual storage methods include (1) burial in a secured landfill, (2) deep-well injection, and (3) surface impoundment.

16. A pool with a nonreactive, plastic liner and/or impervious clay. See figure 19-14.

18. A tax on chemical and petroleum industries; payment from parties responsible for generating the waste and tax money.

20. State your own opinion.

22. Very few radioactive wastes are regulated by RCRA. They are controlled by the Atomic Energy Act of 1954, Department of Energy.

24. Radioactive waste was discharged directly into the ground.

26. At sea.

28. The low-level radioactivity waste policy act set 1992 as the deadline for each state to take responsibility

for disposal of its own low-level commercial wastes. If states prefer, they have the option of joining a multistate compact in which one state would provide a disposal site for the other states in the compact.

30. Spent fuel rods from nuclear reactors are stored under water in tanks at the reactor sites. The water shields the radiation and dissipates the heat that is generated. Many commercial plants are already running out of storage space. Congress in 1982 passed the Nuclear Waste Policy Act, which set up a timetable for building an underground repository for permanent storage of high-level waste.

32. Yucca Mountain, Nevada.

34. Currently, the recommended procedure for disposing of high-level radioactive waste is to concentrate it, incorporate it into a glass or ceramic material, seal the glass or ceramic product in a corrosion-resistant steel canister, and then bury the canister in a concrete vault deep underground in a stable geologic formation composed of salt, tuff, or granite.

36. Deep space disposal is expensive, and an accident occurring during launch could disperse radioactive waste into the atmosphere.

38. MOX is a mixture of oxidized plutonium and uranium. Plutonium is found in nuclear weapons and spent nuclear fuel.

APPENDIX A

1. **(a)** 4.24×10^8 **(b)** 9.03584×10^5 **(c)** 6.17×10^{-4} **(d)** 6.0×10^{-8} **(e)** 7.49×10^{-15} **(f)** 3.7693528×10^7 **(g)** 4.020×10^{-2} **(h)** 1.0×10^{-9} (i) 6.003×10^0 **(j)** 2.3×10^9
2. pH = 5
3. 10^{-7}

APPENDIX B

1. 0.825 L
2. 0.153 kg
3. 61,000 nm
4. 2 oz
5. 5 ft 7 in.
6. 1.6 kg
7. 37 minutes
8. 29.6 mL

9. 1 kg
10. above by **(a)** 5.4° F and **(b)** 3° C
11. 59° F; heat
12. 836 kJ

APPENDIX C

1. 7.540 m
2. 36,000 mg
3. 1.9 m
4. 1420 mL
5. 7 1/2 hours
6. above by 34 ug/L
7. 0.20 g
8. 670 nm and 370 nm
9. 20.5 mi/hr
10. 250 mL
11. 69.6 g
12. 7,700 g, 7.7 kg, 0.008 tons

APPENDIX D

1. 281.6 g
2. 125
3. 92.983 mL
4. 252.10 mg
5. 69.125L
6. 700 m^3
7. 0.0082
8. 0.000508
9. 5.57 g/mol
10. 1.75 L

APPENDIX E

1. 134 tons
2. 3.6 kg
3. 5570 mg
4. 0.80 g
5. 42 moles
6. 21.2 moles
7. 2.4 lb
8. 2.5 moles
9. 2.52 kg
10. 8.00 moles

GLOSSARY

abiotic   Pertaining to factors or things that are separate and independent from living things; nonliving.

acid   A substance that releases hydrogen ions in solution.

acid rain   Precipitation with a pH less than 5.6.

activation energy   The amount of energy reactants must have in order to overcome the energy barrier to reaction.

addition polymers   A polymer in which a small molecule adds to itself, over and over, to form the polymer, without the loss of any atoms from the monomers.

adenosine triphosphate   The triphosphate of adenosine; the energy molecule of metabolism.

adsorption   The process of chemicals (ions or molecules) sticking to the surface of other materials.

aerobic   A process that requires oxygen.

aerosols   Small particles (diameters less than 1 μm) that are dispersed in air.

alcohol   A compound that contains an −OH functional group covalently bonded to a carbon atom; R−OH.

aldehyde   A compound that has a carbonyl group bonded to one carbon atom and one hydrogen atom, RCHO.

alkane   A group of organic hydrocarbons containing only single carbon-carbon bonds.

alkene   Hydrocarbons containing one or more carbon-carbon double bonds.

alkyl group   A group that is formed by removing one or more hydrogens from an alkane.

alkyne   Hydrocarbons containing one or more carbon-carbon triple bonds.

alpha particle   A helium nucleus; $^{4}_{2}$He.

Ames test   A test that uses bacteria to test if a chemical causes mutations; a fast test used to screen chemicals for their potential to cause cancer.

amide   A compound that has a carbonyl group bonded to one carbon atom and one nitrogen atom group, $RCONR'_2$, where R' may be a hydrogen atom.

amine   A compound that has one or more organic groups bonded to nitrogen; RNH_2, R_2NH, or R_3N.

amino acid   Difunctional organic compounds that contain both a carboxyl group and an amino group.

anaerobic   A process that is carried out in the absence of oxygen.

anode   A positively charged electrode.

antacid   Medication used to neutralize excess stomach acid; usually containing hydroxide or carbonate ions.

anthracite coal   Hard coal, which gives much heat and little smoke.

antioxidant   A chemical that slows down or stops oxidation.

aquifer   An underground layer of porous rock, sand, or other material that allows the movement of water between layers of nonporous rock or clay.

aromatic compound   A compound that contains a six-membered ring of carbon atoms with three double bonds.

artificial transmutation   The change of one element into another by bombardment with a high-energy particle.

asthenosphere   The layer of the upper mantle that underlies the lithosphere; its rock is very hot and therefore weak and easily deformed.

atom   The fundamental unit of all elements.

atomic mass unit (amu)   A unit based on the value of exactly 12 amu for the isotope of carbon that has six protons and six neutrons in the nucleus.

atomic nucleus   The very small, dense, positively charged central core of the atom; composed of protons and neutrons.

atomic number The number of protons in the nucleus of an atom of an element.

atomic weight The mass of the atoms of an element in atomic mass units (amu); it is numerically equal to the mass in grams of 1 mol of the element.

Avogadro's number The number of particles (atoms, formula units, ions, or molecules) in one mole; 6.02×10^{23}.

base A substance that releases hydroxide ions in a solution.

beta particle Energetic electron emitted from the nucleus, symbol $_{-1}^{0}\beta$.

BHA and BHT Food additives that prevent spoilage.

bioaccumulation The accumulation of higher and higher concentrations of potentially toxic chemicals in organisms.

biochemical oxygen demand (BOD) The amount of oxygen that will be absorbed as wastes are being digested or oxidized; potential impacts of wastes are measured in terms of the BOD.

bioconversion The use of biomass as fuel; burning materials such as wood, paper, or plant wastes directly to produce energy, or converting such materials into fuels such as alcohol or methane.

biogas The mixture of gases, about two-thirds methane, one-third carbon dioxide, resulting from anaerobic (without air) digestion of organic matter; the methane content enables biogas to be used as a fuel gas.

biological control Control of a pest population by introduction of predatory, parasitic, or disease-causing organisms.

biomass Mass of biological material; usually the total mass of a particular category; for example, biomass of producers.

biome A group of ecosystems that are related by having a similar type of vegetation governed by similar climatic conditions.

biotic Living or derived from living things.

bitumen Heavy black oil, high in sulfur, that is extracted from tar sands.

bituminous coal Soft coal; gives less heat and more smoke than hard (anthracite) coal.

bond energy The energy needed to separate two covalently bonded atoms.

brackish water Water that contains dissolved salt.

breeder nuclear reactor Nuclear fission reactor that produces more fuel than it consumes by converting U-238 to Pu-239.

buffer A substance that will maintain the pH of a solution by reacting with the excess acid.

calorie Fundamental unit of energy. The amount of heat required to raise the temperature of 1 gram of water 1 degree Celsius. All forms of energy can be converted to calories. Calories used in connection with food are kilocalories, or "big" Calories, the amount of heat required to raise the temperature of one liter of water 1 degree Celsius.

carbohydrates Organic compounds that are generally composed of the elements C, H, and O in the ratio 1:2:1.

carbon monoxide A highly poisonous gas, the molecules of which consist of a carbon atom with one oxygen atom attached; not to be confused with the nonpoisonous carbon dioxide, a natural gas in the atmosphere.

carboxyl group The −COOH functional group.

carboxylic acid A compound that has a carbonyl group bonded to one carbon and one −OH group; RCOOH.

carcinogen Having the property of causing cancer, at least in animals and by implication in humans.

carnivore An animal that feeds more-or-less exclusively on other animals.

catalyst A substance that affects the rate at which a chemical reaction occurs.

catalytic converter The device used by U.S. automobile manufacturers to reduce the amount of carbon monoxide, nitrogen oxides, and hydrocarbons in the exhaust; the converter contains catalysts that change these compounds to carbon dioxide, nitrogen, and water as the exhaust passes through.

cathode The electrode at which reduction occurs.

cathode ray Stream of electrons that is produced when high voltage is applied to electrodes in an evacuated tube.

cation A positively charged ion.

cellulose The organic macromolecule that is the prime constituent of plant cell walls and hence the major molecule in wood, wood products, and cotton; it is composed of glucose molecules, but since it cannot be digested by humans its dietary value is only as fiber, bulk, or roughage.

chain reaction   A self-sustaining nuclear reaction.

chemical equation   A written expression in which symbols and formulas are used to represent reactants and products and which is balanced.

chemical formula   Representation with symbols and subscript numbers of the numbers of atoms combined in a chemical compound.

chemical properties   A property that involves a change in the identity of a substance.

chemistry   The study of the nature, properties, and transformations of matter.

chiral carbon atom   A carbon atom bonded to four different groups.

chlorinated hydrocarbons   A substance composed of the elements carbon, hydrogen, and chlorine; for example, the insecticide DDT.

chlorofluorocarbons (CFCs)   Synthetic organic molecules that contain one or more atoms of both chlorine and fluorine.

chlorophyll   A plant pigment that plays a major role in converting solar energy to chemical energy in photosynthesis.

coenzyme   A small, organic molecule that acts as an enzyme cofactor.

complete proteins   Those that contain all the amino acids essential for the maintenance of good health.

compound   The chemical combination of two or more elements.

Comprehensive Environmental Response, Compensation and Liability Act (CERCLA) or Superfund   Cornerstone legislation aimed at protecting groundwater from abandoned chemical waste sites' leakage.

condensation polymer   A polymer formed when two or more monomer units react with each other and split out water.

consumers   In an ecosystem, those organisms that derive their energy from feeding on other organisms or their products.

continuous spectrum   When one color of light emerges into the next; the solar spectrum.

convergent boundary   Area where earth's plates are pushed together.

copolymer   A polymer that is formed by the reaction of two or more monomers.

core (earth's)   Spherical central portion of the earth.

corrosive poison   A poison that acts by destroying tissue; examples are mineral acids, such as sulfuric acid, or strong bases, such as sodium hydroxide.

corrosive waste   Acidic waste with a pH less than 2.0 or basic waste with a pH greater than 12.5.

cosmic rays   Energetic particles, primarily protons, from space.

covalent bond   A bond formed by the sharing of electrons between two atoms.

critical mass   The amount of fissionable (radioactive) material necessary to sustain a chain reaction; amount needed to make a nuclear bomb.

crust (earth's)   The outermost solid layer of the earth.

cyanobacteria   Blue-green algae; photosynthetic plants that live and reproduce entirely immersed in water.

DDT   The first and most widely used of the synthetic organic pesticides belonging to the chlorinated hydrocarbon class.

decomposer   Organism whose feeding action results in decay or rotting of organic material; the primary decomposers are fungi and bacteria.

density   The ratio of an object's mass to its volume.

deoxyribonucleic acid (DNA)   The genetic material found in the nucleus of each cell, which contains all the information for the development of that individual.

detritus   The dead organic matter, such as fallen leaves, twigs, and other plant and animal wastes, that exists in an ecosystem.

diastereomer   Stereoisomers that are not mirror images. Most stereoisomers are either geometric isomers or compounds containing two or more chiral molecules.

dipole   Partial positive and partial negative charges separated by a distance.

dipole-dipole interaction   Attraction between polar molecules.

disaccharide   A carbohydrate containing two sugar groups in each molecule.

divergent boundary   Area where earth's plates are moving apart in opposite directions.

dose   A consideration of the concentration times the length of exposure; for any given material or radiation, effects correspond to the product of these two factors.

double bond   A covalent bond that results from the sharing of two electron pairs between atoms.

ecosystem   A grouping of plants, animals, and other organisms interacting with each other and their environment in such a way as to perpetuate the grouping more-or-less indefinitely.

elastomers   Polymers that have elastic properties; in other words, when stretched they tend to return to their original form.

electromagnetic radiation   A form of energy that has wave characteristics and that propagates through a vacuum at the speed of 3.0×10^{10} cm/s.

electron   A particle with a relative negative charge of one, unit of mass of 0.0005486 amu.

electron configuration   The specific arrangement of an atom's electrons in shells and subshells; represented by notation such as $1s^2 2s^2 2p^6$.

electron-dot symbol   An atomic symbol surrounded by dots that show the number of valence electrons.

electronegativity   The ability of an atom in a molecule to attract electrons; a numerical value on the electronegativity scale.

elements   The basic building blocks of matter.

enantiomers   Two mirror-image molecules of a chiral molecule.

endothermic reaction   A process that absorbs heat from the surroundings.

energy   The ability to do work; energy appears in many forms, for example, heat, chemical, electrical, mechanical, and radiant energy.

energy levels   Various regions, outside the nucleus of an atom, in which the electrons move.

entropy   Describes the randomness of a system; the greater the disorder, the greater the entropy.

environment   The combination of all things and factors external to the individual or population of organisms in question.

enzyme   A group of proteins that are biological catalysts that regulate reactions in our bodies.

equilibrium   A reaction state in which two opposing reactions are occurring at equal rates.

essential amino acids   Amino acids that humans cannot biosynthesize; essential amino acids must be present in the food consumed.

ester   A compound that has a carbonyl group bonded to one carbon atom and one −OR group; RCOOR'.

estrogens   A class of female sex hormones.

ether   A class of organic compounds with the general formula R−O−R'.

eutrophication   The process whereby water becomes nutrient-rich and supports abundant growth of algae and other aquatic life at the water surface.

exothermic reaction   A process that gives off heat to the surroundings.

fat   A mixture of triglycerides that is solid because it contains a high proportion of saturated fatty acids.

fatty acids   Organic compounds composed of a long chain of carbon atoms connected to a carboxyl group.

fiber   Cellulose component of many natural foods.

First Law of Thermodynamics   Matter and energy are neither created nor destroyed, but they can change from one form into another.

flue-gas desulfurization (FGD)   Process to remove sulfur (SO_2) from smokestack gases.

fluorescence   The property of a substance to give off light when acted upon by another source of energy; the coating in a fluorescent light that gives off light when struck by electrons moving through the tube.

fly ash   Solid particulates that are present in smokestack gases from burning coal.

food chain   Transfer of energy and material through a series of organisms as one is fed upon by the next.

food web   The combination of all the feeding relationships that exist in an ecosystem.

formula mass   The sum of the individual atomic masses for all atoms represented in the formula of a chemical compound or polyatomic ion.

fossil fuels   Fuels formed by the breakdown, over millions of years, of plant and animal microorganisms. Coal, oil, and natural gas are the three fossil fuels.

fractional distillation   The separation of materials by boiling followed by condensation into fractions with similar boiling ranges. Low-boiling, small molecules, emerge first followed by higher and higher boiling, larger molecules.

free radicals   Highly reactive uncharged species. They can be atoms or groups of atoms having an odd number of electrons.

fuel cell   A device that directly converts chemical energy into electrical energy through chemical reactions.

fuel rods   The pellets of uranium or other fissionable material that are placed in tubes, which, with the control rods, form the core of a nuclear reactor.

functional group   The reactive part of an organic molecule.

galaxy   A group or cluster of millions or billions of stars.

gamma rays   A high-energy form of electromagnetic radiation given off by many radioactive substances.

gas   One of the states of matter; gases have no definite shape or volume.

Geiger counter   A device used to detect and count nuclear radiation.

geothermal energy   Heat transferred from the earth's interior by water and steam.

global warming   The term given to the possibility that the earth's atmosphere is gradually warming because of the greenhouse effect of carbon dioxide and other gases.

glycogen   A starch formed from branched chains of glucose.

GRAS list   A list of food additives that are generally considered to be safe.

greenhouse effect   An increase in the atmospheric temperature because of increasing amounts of carbon dioxide and certain other gases absorbing and trapping heat radiation that normally escapes from the earth.

greenhouse gases   Gases in the atmosphere that absorb infrared energy and contribute to the air temperature; they include carbon dioxide, water vapor, methane, nitrous oxide, and chlorofluorocarbons and other halocarbons.

groundwater   Water that has accumulated in the ground, completely filling and saturating all the pores and spaces in the rock and/or soil.

half-life   The time required for half of the atoms originally present in a radioactive sample to decay.

hazardous substance   Substance that causes or significantly contributes to an increase in mortality or an increase in serious irreversible illness or poses a substantial present or potential hazard to human health or the environment when improperly treated, stored, transported, or disposed of.

heat capacity   The quantity of heat required to raise the temperature of a fixed quantity of matter by 1 degree Celsius.

heat of fusion   The amount of heat necessary to convert 1 gram (or one mole) of solid to a liquid.

heat of vaporization   The amount of heat necessary to convert 1 gram (or one mole) of liquid into a gas.

herbicide   Any substance used to kill plants, especially weeds, or one that slows down their growth.

herbivore   An animal that feeds on green plants or plant products such as nuts and berries.

heterogeneous mixture   A mixture that is visually nonuniform.

homogeneous mixture   A mixture that is uniform to the naked eye.

hormone   A chemical substance that is formed in one organ of the body and is carried to another organ, where it has a specific function.

humus   A dark brown or black, soft, spongy residue of organic matter that remains after the bulk of dead leaves, wood, or other organic material has decomposed.

hydrocarbon   An organic compound composed of only carbon and hydrogen.

hydrogen bond   Intermolecular force due to attraction between an electronegative atom (O, N, or F) in a receptor molecule and a hydrogen atom bonded to an electronegative atom (usually an O or N) in a donor molecule.

hydronium ion   The ion formed by the reaction of a hydrogen ion with a water molecule; H_3O^+

hydroxyl group   The $-OH$ group.

hypothesis   An educated guess concerning the cause of an observed phenomenon that is then subjected to experimental tests to prove its accuracy or inaccuracy.

igneous rock   Rock formed by solidification of molten magma.

ignitable waste   Waste that poses a serious fire hazard.

incineration   The destruction of material by burning to ashes.

industrial smog   The grayish mixture of moisture, soot, and sulfurous compounds that occurs in local areas where industries are concentrated and coal is the primary energy source.

insecticide   Chemical that kills insects.

integrated pest management   Two or more methods of pest control integrated into a program designed to control pests; an objective is to minimize the use of synthetic pesticides.

intermolecular forces   Forces of attraction and repulsion between partial charges in different molecules.

ion   An atom that has gained or lost electrons and therefore has a negative or positive charge.

ionic bond   A bond formed by the transfer of electrons from one atom to another. The atoms are always different elements.

ionizing radiation   Radiation capable of dislodging an electron from (ionizing) an atom or a chemical compound it strikes.

isomer   Compounds whose molecules have the same overall composition but different structures.

isotopes   Atoms of the same element containing different numbers of neutrons and therefore having different masses.

juvenile hormone   An insect hormone sufficient levels of which preserve the larval state; pupation requires diminished levels, hence artificial applications of the hormone may block development.

kerogen   A hydrocarbon material contained in oil shale that vaporizes when heated and can be recondensed into a material similar to crude oil.

ketone   A compound that has a carbonyl group bonded to two carbon atoms, R_2CO.

kinetic energy   Energy of motion.

lava   Molten magma that is extruded onto the surface of the earth, where it cools and solidifies.

LD$_{50}$ The term used to express the lethal dose for 50% of a population being tested within 30 days.

Lewis dot symbols   A notation that shows the symbol of the element and the number of outer electrons in an atom of the element.

lignite   A soft kind of coal with a woody appearance.

limiting factor   A factor primarily responsible for limiting the growth and/or reproduction of an organism or population. The limiting factor may be physical, such as temperature or light, chemical, such as a particular nutrient, or biological, such as a competing species. The limiting factor may differ at different times and places.

line spectra   Spectra produced by elements that appear as a series of bright lines at specific wavelengths separated by dark bands.

lipids   Biochemical substances found in plant and animal tissues; they include fats and other compounds that resemble fats in physical properties.

liquefied natural gas (LNG)   Natural gas that has been cooled and condensed into liquid form.

liquid   One of the states of matter; liquids have definite volume and indefinite shape.

lithosphere   Outer shell of the earth; composed of the crust and the solid layer above the asthenosphere.

loam   A soil consisting of a mixture of about 40% sand, 40% silt, and 20% clay.

London forces   Intermolecular forces of attraction between short-lived, temporary dipoles.

magma   Molten material in the earth's interior.

malignant tumor   Cancer; a tumor that invades and destroys neighboring tissue.

mantle (earth's)   The portion of the earth beneath the crust and surrounding the core.

matter   The physical material that makes up the universe; anything that has mass and volume.

mesosphere   Atmospheric layer above the stratopause, where temperature decreases with height; also refers to the rigid part of the deep mantle below the asthenosphere.

metal   A malleable element with a lustrous appearance that is a good conductor of heat and electricity.

metamorphic rock   Rock that was originally something else but has been drastically changed by massive forces of heat and/or pressure working on it from within the earth.

micelle   A group of ionic molecules that have an ionic end and an organic end, which cluster together to maximize the contact of the ionic end with water and to minimize contact of the organic end with water; soap and detergent molecules form micelles in water.

micronutrients   A group of elements that have important nutritional value.

mineral   A naturally occurring inorganic substance found in the earth's crust as a solid material.

mixture   Means there is no chemical bonding between the molecules involved; for example, air contains oxygen, nitrogen, and carbon dioxide.

moderator   In a nuclear reactor, the moderator is any material that slows down neutrons from fission reactions so that they are traveling at the right speed to trigger another fission. Water and graphite are two types of moderators.

molar (molarity)   A concentration unit for solutions; moles of a solute per liter of solution.

molar mass   The mass, usually in grams, of 1 mole of a substance.

mole 6.023×10^{23} particles.

molecule The smallest unit of a compound that retains the characteristics of that compound.

monomer A small molecule that can combine with itself or other small molecules to form a polymer.

monosaccharide A carbohydrate containing one sugar group.

mutagen A substance that causes mutations.

mutation Random change in one or more genes in an organism. Mutations may occur spontaneously in nature, but their number and degree are vastly increased by exposure to radiation and/or certain chemicals. Mutations generally result in physical deformity and/or metabolic malfunction.

National Priorities List EPA list of closed hazardous waste dumpsites most in need of cleanup.

neurotransmitter A chemical that carries nerve impulses across the synapse that connects nerve cells.

neutralization The exchange reaction between an acid and a base to yield water and a salt.

neutron An electrically neutral subatomic particle found in the nuclei of atoms.

nitrogen fixation The process of chemically converting nitrogen gas (N_2) from the air into compounds such as nitrates (NO_3^-) or ammonia (NH_3) that can be used by plants in building amino acids and other nitrogen-containing organic molecules.

nonrenewable resources Resources such as ores of various metals, oil, and coal that exist as finite deposits in the earth's crust and that are not replenished by natural processes as they are mined.

nuclear fission The splitting apart of an atom by neutron bombardment to give smaller atoms; a reaction that splits a nucleus into two or more parts.

nuclear fusion The combination of two or more light nuclei to give a heavier nucleus.

nucleic acids The major components of chromosomes.

octane rating A measure of the amount of knocking produced by a gasoline when burned in an automobile.

octet rule When there are eight electrons in the outermost energy level of an atom, the atom tends to be nonreactive.

oil shale Rocks that contain oil. Oil shale deposits represent an enormous storehouse of energy in the United States.

oil A naturally occurring combustible liquid composed of hundreds of hydrocarbons and other organic compounds; also known as petroleum.

omnivore An animal that feeds more or less equally on plant and animal material.

optically active Possessing the ability to rotate the plane of polarized light.

orbitals A region in space in which an electron can be found with a 95 percent probabilty.

ores Minerals rich in a particular element such as iron, copper, or aluminum that can be economically mined and refined to produce a desired metal.

organic compounds compounds containing carbon and hydrogen, and derivatives of these compounds.

osmosis The passage of water through a semipermiable membrane.

oxidation A chemical process that involves the reaction of a compound to add oxygen, or lose hydrogen, or lose electrons.

ozone hole First discovered over the Antarctic, this is a region of stratospheric air that is severely depleted of its normal level of ozone during the Antarctic spring because of CFC's released from anthropogenic sources.

ozone layer The layer of ozone gas in the upper atmosphere that screens out harmful ultraviolet radiation from the sun.

pathogen An organism, usually a microbe, capable of causing disease.

PCBs (polychlorinated biphenyls) A group of widely used industrial chemicals of the chlorinated hydrocarbon class. They have become serious and widespread pollutants, contaminating most food chains on earth. Because they are extremely resistant to breakdown, they are subject to bioaccumulation. They are known to be carcinogenic.

peptide bond The bond that holds amino acids together in proteins.

periodic table The arrangement of elements in order of increasing atomic number, with elements having similar properties placed in vertical columns.

pesticide Any chemical designed to inhibit the growth of or kill pests (insects, rodents, plants).

pH The negative logarithm of the hydrogen ion concentration. A measure of acidity.

pheromones Chemical sex attractants used by many insects for mating purposes; some insect pheromones have been synthesized in the laboratory and are used for pest control.

photochemical smog The result of reactions of chemicals in the atmosphere; in these reactions, sunlight is a catalyst.

photosynthesis   A chemical reaction involving carbon dioxide and water in the presence of sunlight to produce a simple sugar and oxygen.

physical change   Change that occurs with no change in chemical composition.

physical properties   Properties that can be measured without changing the composition of a substance.

plane-polarized light   Light whose electromagnetic oscillations are arranged in one plane.

plastic   A polymeric substance that can be molded into various shapes and forms.

plasticizers   Substances that can be added to plastics to make them more pliable.

plate tectonics   A theory of crustal rearrangement based on the movement of continent-sized crustal plates.

point source   Pollutants coming from specific points such as discharges from factory drains or outlets from sewage-treatment plants.

polar covalent bond   A covalent bond in which there is unequal sharing of electrons

polyatomic ion   An electrically charged group of two or more atoms.

polymer   A large organic molecule made up of repeating units of a smaller molecule.

polymerization   A reaction in which monomers combine to form a polymer.

polysaccharide   Carbohydrate containing many sugar groups per molecule.

power   The rate at which work is done.

primary air pollutants   The major air pollutants that are the direct products of combustion and evaporation: suspended particulate matter, carbon monoxide, nitrogen oxides, sulfur oxides, and lead and other heavy metals.

primary plant nutrients   Nitrogen, phosphorus, and potassium.

primary sewage treatment   Consists of passing water very slowly through a large tank, which permits the solid organic material to settle out.

producer   In an ecosystem, those organisms, mostly green plants, that use light energy to construct their organic constituents from inorganic compounds.

products of a chemical reaction   The result of a chemical reaction.

prostaglandins   Compounds that are involved in the contractions of muscle, increases in blood pressure, and other physiological processes. Derived from arachidonic acid.

proteins   Important biological molecules composed of amino acids linked together.

proton   A particle with a relative charge of plus one unit and a mass of 1.0072766 amu.

quantum   The smallest increment of radiant energy that may be absorbed or emitted.

RAD   The amount of radiation absorbed by living tissue, regardless of type of radiation.

radioactive decay   The reduction of radioactivity that occurs as an unstable isotope gives off radiation and becomes stable.

radioactivity   The energy released by the nucleus of an atom as it undergoes nuclear decay; for example, alpha particles, beta particles, and gamma rays.

radioisotope   An isotope of an element that is unstable and may tend to gain stability by giving off radiation.

radon   A radioactive gas produced by natural processes in the earth that is known to seep into buildings; can be a major hazard within homes and is a known carcinogen.

range of tolerance   The range of conditions within which an organism or population can survive and reproduce.

reactants   The starting materials in a chemical reaction.

reactive waste   Waste that explodes when heated or shocked, or materials that react violently with water.

rearrangement polymers   Polymers formed by rearrangement of atoms in the monomers, without the loss of any atoms.

reduction   The gain of electrons by a substance undergoing a chemical reaction.

REM (sievert)   A weighted unit of radiation.

renewable resources   Biological resources such as trees that may be renewed by reproduction and regrowth.

resin   A solid or semisolid polymeric material.

respiration   The process by which living organisms take in oxygen and exhale carbon dioxide.

ribonucleic acid (RNA)   One of the two major nucleic acids.

risk assessment The process of evaluating the risks associated with a particular hazard before taking some action.

Rule of 70 Method for calculating the doubling time for a process that is increasing exponentially; 70% growth rate = doubling time in years.

sanitary landfill A depression in the ground in which trash is dumped and then compacted; a dump.

saturated fatty acid An organic carboxylic acid that contains only carbon and hydrogen bonded by single bonds.

saturated hydrocarbon An organic compound that contains only carbon and hydrogen atoms, bonded only by single bonds.

scientific method The methodology by which scientific information is gathered; involves observing, formulating specific questions and a hypothesis, then testing the hypothesis through experimentation.

Second Law of Thermodynamics There is no way of getting 100% usable energy from any energy source, since some energy is always lost as waste heat.

secondary air pollutants Air pollutants resulting from reactions of primary air pollutants while resident in the atmosphere.

secondary sewage treatment Also called biological treatment; organic material is removed by enabling organisms to feed on it and oxidize it through respiration.

secured landfill Burial site for long-term storage of hazardous wastes contained in drums.

sedimentary rock Rock formed of sediment that has been consolidated through the combination of pressure and temperature.

sequestrants Chemicals that are added to water, or products containing water, to bind metal ions in solution.

silicates A category of minerals composed of silicon and oxygen combined with other elements.

silicones Polymers with a silicon backbone; formed from reaction of silicon with organic molecules.

sludge The mudlike material that forms from the solids removed from wastewater during wastewater treatment.

soil erosion The loss of soil caused when particles are carried away by wind or water.

solar energy Energy derived from the sun.

solid One of the states of matter; solids have definable volume and shape.

solid waste The total of materials that are discarded as "trash" and handled as solids, as opposed to those that are flushed down sewers and handled as liquids.

solute The substance being dissolved in a solution.

solution A mixture of substances that has uniform composition; a homogeneous mixture.

solvent The substance that causes another substance to dissolve in a solution.

stereoisomers Isomers that have the same molecular and structural formulas but different arrangements of their atoms in space.

steroid A lipid whose structure is based on a tetracyclic (four-ring) carbon skeleton.

stratosphere The atmospheric layer directly above the troposphere.

structural isomers Compounds that have the same overall elemental composition, but different structures.

subduction zone Descent of the edge of a crustal plate under the edge of an adjoining plate.

suborbitals Subdivisions within energy levels of atoms occupied by electrons of different energies; designated by s, p, d, and f.

subsoil In a natural situation the soil beneath the topsoil; in contrast to topsoil, subsoil is compacted and has little or no humus or other organic material living or dead.

substrate The reactant in an enzyme-catalyzed reaction.

synapse Narrow gap between nerve cells across which a signal is carried.

tar sands Sedimentary material containing bitumen that can be "melted out" using heat and then refined in the same way as crude oil.

technology The application of scientific information to solve practical problems or reach a desired goal.

temperature inversion The weather phenomenon in which a layer of warm air overlies cooler air near the ground and prevents the rising and dispersion of air pollutants.

teratogen Chemical that causes birth defects in the fetus, when taken by the mother.

tertiary treatment   Third stage of sewage treatment, designed to remove nitrogen and/or phosphorus from the wastewater; removes plant nutrients to reduce eutrophication.

testosterone   A male hormone secreted by the testes; responsible for the development of secondary male sex characteristics.

theory   A conceptual formulation that provides a rational explanation or framework for numerous related observations.

thermal pollution   The addition of abnormal amounts of heat to air or water; most significant with respect to the discharge of waste heat from nuclear electric generating plants.

thermoplastic polymer   Polymer that can be heated and formed into a different shape.

thermosetting polymer   Polymer that decomposes when reheated; once formed, thermosetting polymers cannot be melted and changed to a different shape.

thermosphere   The highest thermal layer in the atmosphere, above the mesopause.

topsoil   The surface layer of soil, which is rich in humus and other organic material, both living and dead.

toxic waste   Waste that causes harm to living organisms by interfering with normal physiological processes.

toxin   A natural substance, from a biological source, that causes harm.

transform fault   A fault produced by shearing, with adjacent plates being displaced laterally with respect to one another.

transmutation   Conversion of one kind of nucleus to another.

triglycerides   Fats in which all three glycerol groups have undergone esterification.

triple bond   A covalent bond in which each atom donates three electrons to form the bond.

trophic level   Feeding level with respect to the primary source of energy.

troposphere   The layer of the earth's atmosphere from the surface to about 10 miles (16 km) in altitude.

unsaturated fatty acids   Organic acids that contain one or more carbon-carbon double bonds.

unsaturated hydrocarbon   An organic compound that contains only carbon and hydrogen and that has some carbon-carbon double bonds or carbon-carbon triple bonds.

valence electrons   The outermost electrons of an atom; these electrons are used for bonding.

vitamins   Biologically active compounds that are required by living organisms for them to function properly.

volatile organic compounds (VOCS)   A category of major air pollutants present in the air in vapor state, including fragments of hydrocarbon fuels from incomplete combustion and evaporated organic materials such as paint solvents, gasoline, and cleaning solutions. They are major factors in the formation of photochemical smog.

vulcanization   The process by which rubber is heated with sulfur in order to form a polymer with cross-linking to give the rubber better elastic properties.

wavelength   The distance between identical points on successive waves.

weathering   The gradual breakdown of rock into smaller and smaller particles, caused by natural chemical, physical, and biological factors.

work   Any change in motion or state of matter. Any such change requires the expenditure of energy.

x-rays   Electromagnetic radiation of extremely short wavelength.

CREDITS

PHOTOGRAPHS

Chapter 1 **CO** Biophoto Associates/Photo Researchers **1–2** Granger Collection **1–3a** Dr. E. R. Degginger **1–3b** W. Eastep/Stock Market **1–5** Stephen P. Parker/Photo Researchers **1–6a** Paolo Koch/Photo Researchers **1–6b** Tom McHugh/Photo Researchers **1–6c** Francis G. Mayer/Photo Researchers **1–8a** Photo Researchers **1–8b** Connie Hansen/Stock Market **1–8c** Diane Schiuso/Fundamental Photographs **1–10** Richard Megna/Fundamental Photographs

Chapter 2 **CO** E. R. Degginger/Photo Researchers **2–1** NASA **2–6** John Reader/Science Photo Library/Photo Researchers **2–15** Photo Researchers **2–16a** Rod Planck/Photo Researchers **2–16b** Bill Bachman/Photo Researchers **2–18** Roberto De Gugliemo/Science Photo Library/Photo Researchers **2–21** Paul Silverman/Fundamental Photographs **2–23a** G. Schiff/Photo Researchers **2–23b** Musee des Arts Decoratifs/Giraudon/Art Resource **2–24** Dr. E. R. Degginger **2–25** Ken Straiton/Stock Market

Chapter 3 **CO** G.C. Kelley/Photo Researchers **3–2** Harold Hoffman/Photo Researchers **3–3** Dr. Morley Read/Science Photo Library/Photo Researchers **3–7a** Bruce Roberts/Photo Researchers **3–7b** Calvin Larsen/Photo Researchers **3–8a** Dr. E. R. Degginger **3–8b** John Gillmoure/Stock Market **CB1** Andrew Andrew McClanaghan/Photo Researchers

Chapter 4 **CO** Andrew Syred/Science Photo Library/Photo Researchers **4–1** National Library of Medicine/Mark Marten/Photo Researchers **4–2** North Wind Picture Archives **4–3** Jean-Loup Charnet/Science Photo Library/Photo Researchers **4–6** Science Photo Library/Photo Researchers **4–10** Photo Researchers **4–14** Dr. Mitsuo Ohtsuki/Science Photo Library/Photo Researchers **4–15** North Wind Picture Archives **4–16** Lawrence Migdale/Photo Researchers **4–17** Richard Megna/Fundamental Photographs **CB2** Drug Enforcement Administration **Box** North Wind Picture Archives

Chapter 5 **CO** Craig Tuttle/Stock Market **5–1** Science Photo Library/Photo Researchers **5–3a** Bausch & Lomb Incorporated **5–3b** Bausch & Lomb Incorporated **5–3c** Bausch & Lomb Incorporated **5–7** Science Photo Library/Photo Researchers **Box** Oregon State University

Chapter 6 **CO** Tom Bean/Stock Market **6–5** Argonne National Lab **6–8** Granger Collection **CBa** Frank LaBua **CBb** Frank LaBua **6–9a** Visuals Unlimited **6–9b** Visuals Unlimited **6–9c** Visuals Unlimited **6–10a** R.J. Erwin/Photo Researchers **6–10b** E.R. Degginger/Photo Researchers **6–14** Official U.S. Navy Photograph **Box** Mary Evans Picture Library/Photo Researchers

Chapter 7 **CO** James H. Carmichael Jr./The Image Bank **7–1** Kaz Mori/The Image Bank **7–2** Aluminum Company of America **7–3b** Gabe Palmer/Stock Market **7–4** Science Photo Library/Photo Researchers **7–5** Grant Heilman Photography **7–8** Richard Megna/Fundamental Photographs **7–9** Richard Megna/Fundamental Photographs **7–11a** Mathias Oppersdorff/Photo Researchers **7–11b** Tobi Zausner **7–14** Visuals Unlimited **Box** Burton Silverman

Chapter 8 **CO** Grant Heilman Photography **8–5a** Richard Megna/Fundamental Photographs **8–5b** Richard Megna/Fundamental Photographs **8–6a** Richard Megna/Fundamental Photographs **8–6b** Richard Megna/Fundamental Photographs **8–11** Ned Haines/Photo Researchers **8–13** J.H. Robinson/Photo Researchers **8–14** Harvey Lloyd/Peter Arnold **8–15a** Richard Megna/Fundamental Photographs **8–15b** Richard Megna/Fundamental Photographs **8–16** Richard Megna/Fundamental Photographs **8–20** Richard Megna/Fundamental Photographs **CB1** Frank LaBua **CB2** Biophoto Associates/Photo Researchers **Box** Bill Longcore/Science Source/Photo Researchers

Chapter 9 **CO** Dennis Purse/Photo Researchers **9–2a** Sinclair Stammers/Science Photo Library/Photo Researchers **9–2b** M. Claye/Jacana/Photo Researchers **9–8** Philip Bailey/Stock Market **9–17** North Wind Picture Archives **CB1** Roy Morsch/Stock Market **Box A** North Wind Picture Archives **Box B** Courtesy Field Museum **Box C** North Wind Picture Archives

Chapter 10 **CO** David Pollack/Stock Market **10–11b** Chris Sorensen/Stock Market **10–13b** Biophoto Associates/Science Source/Photo Researchers **10–19** Biophoto Associates/Photo Researchers **CB1a** Focus on Sports **CB1b** Focus on Sports **10–21a** CNRI/Science Photo Library/Photo Researchers **10–21b** Jackie Lewin, Royal Free Hospital/Science Photo Library/Photo Researchers **CB2** Gunther Explorer/

Photo Researchers **10–28** Omikron/Photo Researchers **10–29a** Dr. Richard J. Feldman/National Institute of Health **Box** Jan Luyken/Bettmann

Chapter 11 **CO** Jimmy Rudnick/Stock Market **11–2b** Chris Jones/Stock Market **11–3c** Ed Bohon/Stock Market **11–4** Frank LaBua **11–5** Leonard Lessin/Peter Arnold **11–6** Leonard Lessin/Peter Arnold **11–7a** Frank LaBua **11–7b** Frank LaBua **11–9** Tom Bochsler **11–13a** David Leah/Science Photo Library/Photo Researchers **11–13b** James Stevenson/Science Photo Library/Photo Researchers **11–14b** Frank LaBua **11–14c** Thomas Braise/Stock Market **CB1** Frank LaBua **Box A** North Wind Picture Archives **Box B** Gamma-Liaison

Chapter 12 **CO** Kunio Owaki/Stock Market **CB1** Frank LaBua **12–4b** Scott Camazine/Photo Researchers **12–7** Richard Megna/Fundamental Photographs **12–9** Gamma-Liaison **12–16** Michael P. Gadowski/Photo Researchers **CB2** Frank LaBua **12–18** Rich Treptow/Visuals Unlimited **12–19** L. Kiff/Visuals Unlimited **12–20** Tim Davis/Photo Researchers **Box** Novosti/Gamma-Liaison

Chapter 13 **CO** NASA **13–4** Robert Semeniuk/Stock Market **13–5** Louis Goldman/Photo Researchers **13–6a** AC Rochester **13–b** AC Rochester **13–7** Sygma **13–11a** Pete Saloutos/Stock Market **13–11b** Pete Saloutos/Stock Market **13–13** NASA **CB1** Frank LaBua **Box** Hale Observatories/Photo Researchers

Chapter 14 **CO** Tom Hollyman/Photo Researchers **14–1(1)** Patrick Bordes/Explorer/Photo Researchers **14–1(2)** David Ball/Stock Market **14–1(3)** Brownie Harris/Stock Market **14–1(4)** Gabe Palmer/Stock Market **14–1(5)** Harvey Lloyd/Stock Market **14–3** A. Copley/Visuals Unlimited **14–4** Thomas Braise/Stock Market **14–7** Bryan Peterson/Stock Market **14–9a** John D. Cunningham/Visuals Unlimited **14–9b** Science VU/Visuals Unlimited **14–11** Photri/Stock Market **14–13** Ludek Pesek/Science Photo Library/Photo Researchers **14–15a** T. Havill/Visuals Unlimited **14–15b** C.P. Hickman/Visuals Unlimited **CB1** Michael J. Balick/Peter Arnold **CB2** Frank LaBua **Box** Wesley Bocxe/Photo Researchers

Chapter 15 **CO** Gabe Palmer/Stock Market **15–7a** Mark Burnett/Photo Researchers **15–9a** Princeton University **15–10a** Alexander Tsiaras/Science Source/Photo Researchers **15–12** Energy Technology Visual Collection **15–14b** Energy Technology Visual Collection **15–16** Alex Bartel/Science Photo Library/Photo Researchers **15–17** David Halpern/Photo Researchers **15–18** Lowell Georgia/Photo Researchers **15–19** Lowell Georgia/Photo Researchers **15–20** Science VU-API/Visuals Unlimited **Box** Sygma

Chapter 16 **CO** Christopher Springman/Stock Market **16–1** Jack Fields/Photo Researchers

16–2 Chris Priest and Mark Clarke/Science Photo Library/Photo Researchers **16–3** Runk/Schoenberger/Grant Heilman Photography **16–4** David Sprague/Los Angeles Daily News/Gamma-Liaison **16–5** Bettmann **16–6** W. Eugene Smith/Magnum Photos **16–7** John D. Cunningham/Visuals Unlimited **16–9** Alan Pitcairn/Grant Heilman Photography **16–10** Jones/Gamma-Liaison **16–11** From the Journal of the American Medical Association, 1976, Vol. 235, 1458-1460. Courtesy James W. Hanson, M.D. **16–12** Frank LaBua

Chapter 17 **CO** Anne Heimann/Stock Market **17–7** Carole Spencer/Gamma-Liaison **17–8a** WHO Photo **17–8b** WHO Photo **17–9** Biophoto Associates/Science Source/Photo Researchers **17–10** John Paul Kay/Peter Arnold **17–13a** Mark C. Burnett/Photo Researchers **17–13b** Dr. E.R. Degginger **17–5** Frank LaBua **CB1** Bonnie Rauch/Photo Researchers **Box** Steven R. King/Peter Arnold

Chapter 18 **CO** Gabe Palmer/Stock Market **18–3** Asa C. Thoresen/Photo Researchers **18–5a** John D. Cunningham/Visuals Unlimited **18–5b** Matt Meadows/Peter Arnold **18–8** Thomas Hovland/Grant Heilman Photography **18–9** USDA Photo **18–10** Matt Meadows/Peter Arnold **18–11** Janvie/Gamma-Liaison **18–15** Larry Lefever/Grant Heilman Photography **18–18** Dr. Nigel Smith/Animals Animals/Earth Scenes **Box** U.S.D.A.

Chapter 19 **CO** Kent & Donna Dannen/Photo Researchers **19–1** James Barnett/Stock Market **19–2** Science VU/Visuals Unlimited **19–4** Charles Rushing/Visuals Unlimited **19–6** D. Newman/Visuals Unlimited **19–8** Howard Gaskill **19–9** Dennis A. Waters **19–11** Tim Lynch/Gamma-Liaison **19–15** Nancy J. Pierce/Photo Researchers **19–16a** Courtesy, Howard Gaskell, Baltimore County, MD **19–16b** Courtesy, Howard Gaskell, Baltimore County, MD **19–17** Holt Confer/The Image Works **19–18** Department of Energy, Washington, DC **CB1** Photo, courtesy of Lab Safety Supply, Inc., Jamesville, WI **CB2** Photo courtesy of Lab Safety Supply, Inc., Jamesville, WI **CB3** Photo courtesy of Lab Safety Supply, Inc., Jamesville, WI **Box** Andy Levin/Photo Researchers

LINE DRAWINGS

Figure 3A. The Far Side cartoon by Gary Larson is reprinted by permission of Chronicle Features, San Francisco, CA. All rights reserved. Figure 13-14. *Source:* James G. Anderson, Harvard University. The following illustrations were adapted from other texts published by Prentice Hall. Theodore L. Brown, Eugene LeMay, and Bruce Bursten, *Chemistry: The Central Science,* Fifth Edition, 1991: Figures 1-11, 1-13, 1-14, 4-9, 4-11, 5-2, 5-10, 5-16, 5-17, 6-1, 6-3, 6-6, 6-11, 6-12, 6-13, 7-3a, 7-12, 8-17, 8-18, 8-19, 9-2, 10-1, 10-22, 10-24, 10-26, 10-29, 15-21, and 17-3. Ronald Gillespie and David Humphreys, *Chemistry,* Second Edition, 1989: Figures 7-7, 7-14, 11-8, 15-14. Sheldon Judson and Marvin E. Kauffman, *Physical Geology,* Eighth Edition, 1990: Figures 2-9 and 2-17. Frederic Martini, *Fundamentals of Anatomy and Physiology,* Second Edition, 1992: Figures 14-CB2, 17-1, and 17-21. Tom L. McKnight, *Essentials of Physical Geography,* 1992: Figures 2-3, 2-7, 2-8, 2-13, 2-14, 12-1, 12-8, and 18-6. J. N. Miller, *Modern Analytical Chemistry,* 1992: Figures 1-7, 5-4, 6-13, 9-1, 9-4, 9-5, 9-6, 9-7, 13-16, 13-18, and 19-10. Robert T. Morrison and Robert K. Boyd, *Organic Chemistry,* Sixth Edition, 1992: Figures 10-5 and 10-16. Bernard J. Nebel, *Environmental Science: The Way the World Works,* Third Edition, 1990: Figures 3-1, 3-5, 3-7b, 3-11, 3-12, 3-13, 3-14, 3-15, 8-9, 8-10, 8-12, 12-2, 12-10, 12-17, 13-19, 14-5, 14-14, 15-10b, 15-11, 18-7, 18-19, and 19-7. Bernard J. Nebel, *Environmental Science: The Way the World Works,* Second Edition, Figures 18-1 and 18-4.

CASE STUDIES

Case Studies 1, 2, 4, 5, 6, adapted from Tom L. McKnight, *Physical Geography: A Landscape Appreciation,* 4th ed., Prentice-Hall, Englewood Cliffs, New Jersey, 1993, Used by permission of the publisher. Case Study 3, adapted from M. Grant Gross, *Oceanography: A View of Earth,* 6th Ed., Prentice-Hall, Englewood Cliffs, New Jersey, 1993. Used by permission of the publisher. Case Studies 7, 8, 9, 11, 12, 13, 14, 15, 17, 18, adapted from Bernard J. Nebel and Richard T. Wright, *Environmental Science: The Way the World Works,* 4th Ed., Prentice-Hall, Englewood Cliffs, New Jersey, 1993. Used by permission of the publisher. Case Study 10, adapted from Bernard J. Nebel and Richard T. Wright, *Environmental Science: The Way the World Works,* 4th Ed., Prentice-Hall, Englewood Cliffs, New Jersey, 1993.

INDEX

PERIODIC TABLE OF THE ELEMENTS

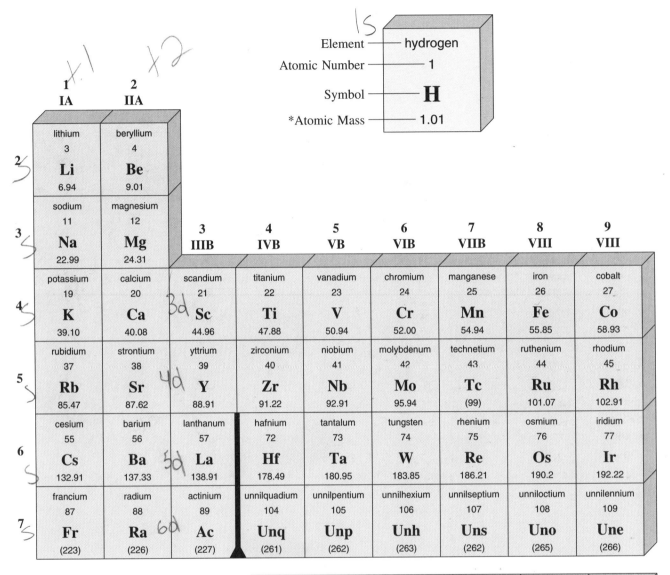

1 IA	**2** IIA							

Element — hydrogen
Atomic Number — 1
Symbol — **H**
*Atomic Mass — 1.01

1 IA	2 IIA	3 IIIB	4 IVB	5 VB	6 VIB	7 VIIB	8 VIII	9 VIII
lithium 3 **Li** 6.94	beryllium 4 **Be** 9.01							
sodium 11 **Na** 22.99	magnesium 12 **Mg** 24.31							
potassium 19 **K** 39.10	calcium 20 **Ca** 40.08	scandium 21 **Sc** 44.96	titanium 22 **Ti** 47.88	vanadium 23 **V** 50.94	chromium 24 **Cr** 52.00	manganese 25 **Mn** 54.94	iron 26 **Fe** 55.85	cobalt 27 **Co** 58.93
rubidium 37 **Rb** 85.47	strontium 38 **Sr** 87.62	yttrium 39 **Y** 88.91	zirconium 40 **Zr** 91.22	niobium 41 **Nb** 92.91	molybdenum 42 **Mo** 95.94	technetium 43 **Tc** (99)	ruthenium 44 **Ru** 101.07	rhodium 45 **Rh** 102.91
cesium 55 **Cs** 132.91	barium 56 **Ba** 137.33	lanthanum 57 **La** 138.91	hafnium 72 **Hf** 178.49	tantalum 73 **Ta** 180.95	tungsten 74 **W** 183.85	rhenium 75 **Re** 186.21	osmium 76 **Os** 190.2	iridium 77 **Ir** 192.22
francium 87 **Fr** (223)	radium 88 **Ra** (226)	actinium 89 **Ac** (227)	unnilquadium 104 **Unq** (261)	unnilpentium 105 **Unp** (262)	unnilhexium 106 **Unh** (263)	unnilseptium 107 **Uns** (262)	unniloctium 108 **Uno** (265)	unnilennium 109 **Une** (266)

cerium 58 **Ce** 140.12	praseodymium 59 **Pr** 140.91	neodymium 60 **Nd** 144.24	promethium 61 **Pm** (147)	samarium 62 **Sm** 150.36	europium 63 **Eu** 151.97
thorium 90 **Th** 232.04	protactinium 91 **Pa** (231)	uranium 92 **U** 238.03	neptunium 93 **Np** (237)	plutonium 94 **Pu** (244)	americium 95 **Am** (243)

Lanthanide Series

Actinide Series

*Note: For radioactive elements, the mass number of an important isotope is shown in parenthesis; for thorium and uranium, the atomic mass of the naturally occurring radioisotopes is given.